Handbook of
Aviation
Human Factors

SECOND EDITION

HUMAN FACTORS IN TRANSPORTATION

Series Editor
Barry H. Kantowitz
Industrial and Operations Engineering
University of Michigan

Handbook of
Aviation
Human Factors

SECOND EDITION

Edited by
John A. Wise
V. David Hopkin
Daniel J. Garland

CRC Press
Taylor & Francis Group
Boca Raton   London   New York

CRC Press is an imprint of the
Taylor & Francis Group, an **informa** business

CRC Press
Taylor & Francis Group
6000 Broken Sound Parkway NW, Suite 300
Boca Raton, FL 33487-2742

Library of Congress Cataloging-in-Publication Data

Handbook of aviation human factors / edited by Daniel J. Garland, John A. Wise, and V. David Hopkin.
-- 2nd ed.
       p. cm. -- (Human factors in transportation)
    Includes bibliographical references and index.
    ISBN 978-0-8058-5906-5 (alk. paper)
    1. Aeronautics Human factors--Handbooks, manuals, etc. 2. Aeronautics--Safety
measures--Handbooks, manuals, etc. I. Garland, Daniel J. II. Wise, John A., 1944- III. Hopkin, V.
David.

    TL553.6.H35 2010
    629.13--dc21   2009024331

Visit the Taylor & Francis Web site at
http://www.taylorandfrancis.com

and the CRC Press Web site at
http://www.crcpress.com

Dedication

To my family

John A. Wise

To Betty

V. David Hopkin

To Danny, Brianna, and Cody

Daniel J. Garland

* * *

*Dedicated to those pioneers of aviation human factors
who made this book possible.*

*In particular to: Lloyd Hitchcock and David Meister, our colleagues
in the first edition who died before the second edition was
completed. Their participation was very much missed.*

Contents

PART III   Aircraft

PART IV   Air-Traffic Control

PART V Aviation Operations and Design

Preface

Nearly a decade ago, the authors of the first edition of this book were writing their contributions. In the interim, much development and progress has taken place in aviation human factors, but they have been far from uniform. Therefore, although the original authors, or their collaborators, and the new authors were all asked to update their chapters and references for this second edition, the actual work entailed in responding to this request differed markedly between chapters, depending on the pertinent developments that had occurred in the meantime. At one extreme, represented by the continued application of human factors evidence to a topic with few major changes, this steady progress could be covered by short additions and amendments to the relevant chapter, and this applies to a few chapters. At the other extreme, major changes or developments have resulted in completely recast and rewritten chapters, or, in a few cases, even in completely new chapters. Many chapter revisions, though substantial, lie between these two extremes.

Human factors as a discipline applied to aviation has come of age and is thriving. Its influence has spread to other applications beyond aviation. Less effort now has to be expended on the advocacy of human factors contributions or on marketing them because the roles of human factors in aviation activities are accepted more willingly and more widely. Both the range of human factors techniques and the nature of human factors explanations have broadened. The relationships between the humans employed in aviation and their jobs are changing in accordance with evolving automation and technological advances.

The demand for aviation continues to expand, and aviation must respond to that demand. The safety culture of aviation imposes a need, in advance of changes, for sound evidence that the expected benefits of changes will accrue, without hidden hazards to safety and without new and unexpected sources of human error. The human factors contributions to aviation must share its safety culture and be equally cautious. Safety ultimately remains a human responsibility, dependent on human cognitive capabilities exercised directly through aviation operations and indirectly through the constructs, planning, design, procurement, and maintenance of aviation systems. Human factors applied to aviation remains primarily a practical discipline, seeking real solutions and benefits and driven by requirements rather than theories. Theory is not ignored, but theory building is seldom an end product. Theories tend, rather, to be tools that can guide the interpretation and generalization of findings and can influence the choice of measures and experimental methods.

Much of this book recounts human factors achievements, but some prospective kinds of expansion of human factors may be deduced from current discernible trends. Teams and training can furnish examples. The study of teams is extending the concept of crew resource management to encompass the organization of the broader aviation system and the cabin, though considerations of cockpit security may restrict the latter development. Team concepts relate to automation in several ways: machines may be treated as virtual team members in certain roles; functions may be fulfilled by virtual teams that share the work but not the workspace; established hierarchical authority structures may wither and devolve into teams or multi-teams; close identification with teams will continue to influence the formation of attitudes and professional norms; and interpersonal skills within teams will gain in interest. Training is evolving toward training in teams, measuring team functioning, and judging success by measuring team achievements. Learning at work is becoming more formalized, with less reliance on incidental on-the-job learning and more emphasis on continuous lifelong planned learning and career development. Associated with this is a closer study of the implicit knowledge, which is an integral part of the individual's professional expertise and skill.

Further future trends are emerging. Aviation human factors may benefit from recent developments in the study of empowerment, since many jobs in aviation rely heavily on the self-confidence of their personnel in the capability to perform consistently to a high standard. The introduction of human factors certification as a tool for evaluating designs in aviation may become more common. The recently increased interest in qualitative measures in human factors seems likely to spread to aviation, and to lead to more studies of such human attributes with no direct machine equivalent as aesthetic considerations and the effects of emotion on task performance. This seems part of a more general trend to move away from direct human–machine comparisons when considering functionality. While studies are expected to continue on such familiar human factors themes as the effects of stress, fatigue, sleep patterns, and various substances on performance and well-being, their focus may change to provide better factual evidence about the consequences of raising the retirement age for aviation personnel, which is becoming a topic of widespread concern. There have been remarkably few cross-cultural studies in aviation despite its international nature. This neglect will have to be remedied sooner or later, because no design or system in aviation is culture free.

I

Introduction

1

A Historical Overview of Human Factors in Aviation

Jefferson M. Koonce
University of Central Florida

Anthony Debons
University of Pittsburgh

Human factors in aviation are involved in the study of human's capabilities, limitations, and behaviors, as well as the integration of that knowledge into the systems that we design for them to enhance safety, performance, and general well-being of the operators of the systems (Koonce, 1979).

1.1 The Early Days: Pre-World War I (Cutting Their Teeth)

The role of human factors in aviation has its roots in the earliest days of aviation. Pioneers in aviation were concerned about the welfare of those who flew their aircraft (particularly themselves), and as the capabilities of the vehicles expanded, the aircraft rapidly exceeded the human capability of directly sensing and responding to the vehicle and the environment, to effectively exert sufficient control to ensure optimum outcome and safety of the flight. The first flight in which Orville Wright flew at 540 ft was on Thursday, December 17, 1903, for a duration of only 12 s. The fourth and final flight of that day was made by Wilbur for 59 s, which traversed 825 ft!

The purposes of aviation were principally adventure and discovery. To see an airplane fly was indeed unique, and to actually fly an airplane was a daring feat! Early pioneers in aviation did not take this issue lightly, as venturing into this field without proper precautions may mean flirting with death in the fragile unstable crafts. Thus, the earliest aviation was restricted to relatively straight and level flight and fairly level turns. The flights were operated under visual conditions in places carefully selected for elevation, clear surroundings, and certain breeze advantages, to get the craft into the air sooner and land at the slowest possible ground speed.

The major problems with early flights were the reliability of the propulsion system and the strength and stability of the airframe. Many accidents and some fatalities occurred because of the structural failure of an airplane component or the failure of the engine to continue to produce power.

Although human factors were not identified as a scientific discipline at that time, there were serious problems related to human factors in the early stages of flight. The protection of the pilot from the elements, as he sat out in his chair facing them head-on, was merely a transfer of technology from bicycles and automobiles. The pilots wore goggles, topcoats, and gloves similar to those used when driving the automobiles of that period.

The improvements in the human–machine interface were largely an undertaking of the designers, builders, and fliers of the machines (the pilots themselves). They needed some critical information to ensure proper control of their craft and some feedback about the power plant. Initially, the aircraft did not have instrumentation. The operators directly sensed the attitude, altitude, and velocity of the vehicle and made their inputs to the control system to achieve certain desired goals. However, 2 years after the first flight, the Wright brothers made considerable effort trying to provide the pilot with information that would aid in keeping the airplane coordinated, especially in turning the flight where the lack of coordinated flight was most hazardous. Soon, these early crafts had a piece of yarn or other string, which trailed from one of the struts of the airplane, providing yaw information as an aid to avoid the turn-spin threat, and the Wright brothers came up with the incidence meter, a rudimentary angle of attack, or flight-path angle indicator.

Nevertheless, as the altitude capabilities and range of operational velocities increased, the ability of the humans to accurately sense the critical differences did not commensurately increase. Thus, early instrumentation was devised to aid the operator in determining the velocity of the vehicle and the altitude above the ground. The magnetic compass and barometric altimeter, pioneered by balloonists, soon found their way into the airplanes. Additionally, the highly unreliable engines of early aviation seemed to be the reason for the death of many aviators. The mechanical failure of the engine or propeller, or the interruption of the flow of fuel to the engine owing to contaminants or mechanical problems, is presumed to have led to the introduction of tachometer and gauges, which show the engine speed to the pilot and critical temperatures and pressures of the engine's oil and coolant, respectively.

1.2 World War I (Daring Knights in Their Aerial Steeds)

The advantages of an aerial view and the ability to drop bombs on ground troops from the above gave the airplane a unique role in World War I. Although still in its infancy, the airplane made a significant contribution to the war on both the sides, and became an object of wonder, aspiring thousands of our nation's youth to become aviators. The roles of the airplane were principally those of observation, attack of ground installations and troops, and air-to-air aerial combat. The aircraft themselves were strengthened to take the increased G-loads imposed by combat maneuvering and the increased weight of ordinance payloads.

As a result, pilots had to possess special abilities to sustain themselves in this arena. Thus, problems related to human factors in the selection of pilot candidates emerged. Originally, family background, character traits, athletic prowess, and recommendations from significant persons secured an individual applicant a position in pilot training. Being a good hunter indicated an ability to lead and shoot at other moving targets, and strong physique and endurance signified the ability to endure the rigors of altitude, heat and cold, as well as the forces of aerial combat. Additionally, the applicant was expected to be brave and show courage.

Later, psychologists began to follow a more systematic and scientific approach for the classification of individuals and assignment to various military specialties. The aviation medics became concerned about the pilots' abilities to perform under extreme climatic conditions (the airplanes were open cockpits without heaters), as well as the effects of altitude on performance. During this period, industrial engineers began to utilize the knowledge about human abilities and performance to improve factory productivity in the face of significant changes in the composition of the work force. Women began to play a major role in this area. Frank Gilbreath, an industrial engineer, and his wife Lillian, a psychologist, teamed up to solve many questions about the improvement of human performance in the workplace, and the knowledge gained was useful to the industry as well as the armed forces.

Early in the war, it became apparent that the allied forces were losing far more pilots to accidents than to combat. In fact, two-thirds of the total aviation casualties were not due to engagement in combat. The failure of the airframes or engines, midair collisions, and weather-related accidents (geographical or spatial disorientation) took greater toll. However, the performance of individuals also contributed significantly to the number of accidents. Fortunately, with the slower airspeeds of the airplanes at that time and owing to the light, crushable structure of the airframe itself, many aviators during initial flight training who crashed and totaled an airplane or two, still walked away from the crash(es) and later earned their wings. Certainly, with the cost of today's airplanes, this would hardly be the case.

The major problems of the World War I era related to human factors were the selection and classification of personnel, the physiological stresses on the pilots, and the design of the equipment to ensure mission effectiveness and safety. The higher-altitude operations of these airplanes, especially the bombers, resulted in the development of liquid oxygen converters, regulators, and breathing masks. However, owing to the size and weight of these oxygen systems, they were not utilized in the fighter aircraft. Cold-weather flying gear, flight goggles, and rudimentary instruments were just as important as improving the reliability of the engines and the strength and crash-worthiness of the airframes. To protect the pilots from the cold, leather flight jackets or large heavy flying coats, leather gloves, and leather boots with some fur-lining, were used. In spite of wearing all these heavy clothing, the thoughts of wearing a parachute were out. In fact, many pilots thought that it was not sporting to wear a parachute, and such technologies were not well developed.

The experience of the British was somewhat different from other reported statistics of World War I: "The British found that of every 100 aviation deaths, 2 were by enemy action, 8 by defective airplanes, and 90 for individual defects, 60 of which were a combination of physical defects and improper training" (Engle & Lott, 1979, p. 151). One explanation offered is that, of these 60, many had been disabled in France or Flanders before going to England and joining the Royal Air Corps.

1.3 Barnstorming Era (The Thrill of It All)

After the war, these aerial cavalrymen came home in the midst of public admiration. Stories of great heroism and aerial combat skills preceded them, such that their homecoming was eagerly awaited by the public, anticipating for an opportunity to talk to these aviators and see demonstrations of their aerial daring. This was the beginning of the post-World War I barnstorming era.

The airplanes were also remodeled such that they had enclosed cabins for passengers, and often the pilot's cockpit was enclosed. Instead of the variations on the box-kite theme of the earliest airplanes, those after World War I were more aerodynamic, more rounded in design than the boxlike model. Radial engines became more popular means of propulsion, and they were air-cooled, as opposed to the earlier heavy water-cooled engines. With greater power-to-weight ratios, these airplanes were more maneuverable and could fly higher, faster, and farther than their predecessors.

Flying became an exhibitionist activity, a novelty, and a source of entertainment. Others had visions of it as a serious means of transportation. The concept of transportation of persons and mails via air was in its infancy, and this brought many new challenges to the aviators. The commercial goals of aviation came along when the airplanes became more reliable and capable of staying aloft for longer durations, connecting distant places easily, but with relatively uncomfortable reach. The major challenges were the weather and navigation under unfavorable conditions of marginal visibility.

Navigation over great distances over unfamiliar terrain became a real problem. Much of the western United States and some parts of the central and southern states were not well charted. In older days, where one flew around one's own barnyard or local town, getting lost was not a big concern. However, to fly hundreds of miles away from home, pilots used very rudimentary maps or hand-sketched instructions and attempted to follow roads, rivers, and railway tracks. Thus, getting lost was indeed a problem. The IFR flying in those days probably meant I Follow Roadways, instead of Instrument Flight Rules!

Writing on water towers, the roofs of barns, municipal buildings, hospitals, or airport hangars was used to identify the cities. As pilots tried to navigate at night, natural landmarks and writing on buildings became less useful, and tower beacons came into being to "light the way" for the aviator. The federal government had an extensive program for the development of lighted airways for the mail and passenger carriers. The color of the lights and the flashing of codes on the beacons were used to identify a particular airway that one was following. In the higher, drier southwestern United States, some of the lighted airway beacons were used even in the 1950s. However, runway lighting replaced the use of automobile headlights or brush fires to indicate the limits of a runway at night. Nevertheless, under low visibility of fog, haze, and clouds, even these lighted airways and runways became less useful, and new means of navigation had to be provided to guide the aviators to the airfields.

Of course, weather was still a severe limitation to safe flight. Protection from icing conditions, thunderstorms, and low ceilings and fog were still major problems. However, owing to the developments resulting from the war effort, there were improved meteorological measurement, plotting, forecasting, and dissemination of weather information. In the 1920s, many expected that "real pilots" could fly at night and into the clouds without the aid of any special instruments. But, there were too many instances of pilots flying into clouds or at night without visual reference to the horizon, which resulted in them entering a spiraling dive (graveyard spiral) or spinning out of the clouds too late to recover before impacting the ever-waiting earth. In 1929, Lt. James Doolittle managed to take off, maneuver, and land his airplane solely referring to the instruments inside the airplane's cockpit. This demonstrated the importance of basic attitude, altitude, and turn information, to maintain the airplane right-side-up when inside the clouds or in other situations where a distinct external-world reference to the horizon is not available.

Many researches had been carried out on the effects of high altitude on humans (Engle & Lott, 1979), as early as the 1790s, when the English surgeon Dr. John Sheldon studied the effects of altitude on himself in balloon ascents. In the 1860s, the French physician, Dr. Paul Bert, later known as the "father of aviation medicine," performed altitude research on a variety of animals as well as on himself in altitude chambers that he designed. During this post-World War I era, airplanes were capable of flying well over 150 miles/h and at altitudes of nearly 20,000 ft, but only few protective gears, other than oxygen-breathing bags and warm clothing, were provided to ensure safety at high altitudes. Respiratory physiologists and engineers worked hard to develop a pressurized suit that would enable pilots to maintain flight at very high altitudes. These technologies were "spinoffs" from the deep sea-diving industry. On August 28, 1934, in his supercharged Lockheed Vega Winnie Mae, Wiley Post became the first person to fly an airplane while wearing a pressure suit. He made at least 10 subsequent flights and attained an unofficial altitude of approximately 50,000 ft. In September 1936, Squadron Leader F. D. R. Swain set an altitude record of 49,967 ft. Later, in June 1937, Flight Lt. M. J. Adam set a new record of 53,937 ft.

Long endurance and speed records were attempted one after the other, and problems regarding how to perform air-to-air refueling and the stress that long-duration flight imposed on the engines and the operators were addressed. In the late 1920s, airplanes managed to fly over the North and South Poles and across both the Atlantic and Pacific Oceans. From the endurance flights, the development of the early autopilots took place in the 1930s. Obviously, these required electrical systems on the aircraft and imposed certain weight increases that were generally manageable on the larger multiengine airplanes. This is considered as the first automation in airplanes, which continues even till today.

1.4 The World War II Era (Serious Business)

Despite the hay day of the barnstorming era, military aviation shrunk after the United States had won "the war to end all wars." The wars in Europe in the late 1930s stimulated the American aircraft designers to plan ahead, advancing the engine and airframe technologies for the development of airplanes with capabilities far superior to those that were left over from World War I.

The "necessities" of World War II resulted in airplanes capable of reaching airspeeds four times faster than those of World War I, and with the shifted impellers and turbochargers altitude capabilities that exceeded 30,000 ft. With the newer engines and airframes, the payload and range capabilities became much greater. The environmental extremes of high altitude, heat, and cold became major challenges to the designers for the safety and performance of aircrew members. Furthermore, land-based radio transmitters greatly improved cross-country navigation and instrument-landing capabilities, as well as communications between the airplanes and between the airplane and persons on the ground responsible for aircraft control. Ground-based radar was developed to alert the Allied forces regarding the incoming enemy aircraft and was used as an aid to guide the aircraft to their airfields. Also, radar was installed in the aircraft to navigate them to their targets when the weather prevented visual "acquisition" of the targets.

The rapid expansion of technologies brought many more problems than ever imagined. Although the equipments were advanced, humans who were selected and trained to operate them did not significantly change. Individuals who had not moved faster than 30 miles/h in their lifetime were soon trained to operate vehicles capable of reaching speeds 10 times faster and which were far more complex than anything they had experienced. Therefore, the art and science of selection and classification of individuals from the general population to meet the responsibilities of maintaining and piloting the new aircraft had to undergo significant changes. To screen hundreds of thousands of individuals, the selection and classification centers became a source of great amounts of data about human skills, capabilities, and limitations. Much of these data have been documented in a series of 17 "blue books" of the U.S. Army Air Force Aviation Psychology Program (Flanagan, 1947). Another broader source of information on the selection of aviators is the North and Griffin (1977) *Aviator Selection 1917–1977*.

A great deal of effort was put forth in the gathering of data about the capabilities and limitations of humans, and the development of guidelines for the design of displays and controls, environmental systems, equipment, and communication systems. Following the war, *Lectures on Men and Machines: An Introduction to Human Engineering* by Chapanis, Garner, Morgan, and Sanford (1947), Paul Fitts' "blue book" on *Psychological Research on Equipment Design* (1947), and *the Handbook of Human Engineering Data for Design Engineers* prepared by the Tufts College Institute for Applied Experimental Psychology and published by the Naval Special Devices Center (1949) helped to disseminate the vast knowledge regarding human performance and equipment design that had been developed by the early human-factors psychologists and engineers (Moroney, 1995).

Stevens (1946), in his article "Machines Cannot Fight Alone," wrote about the development of radar during the war. "With radar it was a continuous frantic race to throw a better and better radio beam farther and farther out, and to get back a reflection which could be displayed as meaningful pattern before the eyes of an operator" (p. 391). However, as soon as the technology makes a step forward, a human limitation may be encountered or the enemy might devise some means of degrading the reflecting signal, so that it would be virtually useless. Often weather conditions may result in reflections from the moisture in the air, which could reduce the likelihood of detecting a target. Furthermore, in addition to the psychophysical problems of detecting signals in the presence of "noise," there was the well-known problem that humans are not very good at vigilance tasks.

Without pressurization, the airplanes of World War II were very noisy, and speech communications were most difficult in the early stages. At the beginning of the war, the oxygen masks did not have microphones built in them, and hence, throat microphones were utilized, making speech virtually unintelligible. The headphones that provided information to the pilots were "leftovers" from the World War I era and did little to shield out the ambient noise of the airplane cockpit.

In addition to the noise problem, as one might expect, there was a great deal of vibration that contributed to apparent pilot fatigue. Stevens (1946) mentioned that a seat was suspended such that it "floated in rubber" to dampen the transmission of vibrations from the aircraft to the pilot. Although technically successful, the seat was not preferred by the pilots because it isolated them from a sense of feel of the airplane.

Protecting the human operator while still allowing maximum degree of flexibility to move about and perform tasks was also a major problem (Benford, 1979). The necessity to protect aviators from antiaircraft fire from below was initially met with the installation of seat protectors—plates of steel built under the pilot's seat to deflect rounds coming up from below. For protection from fire other than the one below, B. Gen. Malcolm C. Grow, surgeon of the 8th Air Force, got the Wilkinson Sword Company, designer of early suits of armor, to make body armor for B-17 aircrew members. By 1944, there was a 60% reduction in men wounded among the B-17 crews with body armor.

Dr. W. R. Franks developed a rubber suit with a nonstretchable outer layer to counter the effects of high G-forces on the pilot. The Franks flying suit was worn over the pilot's underwear and was filled with water. As the G-forces increased, they would also pull the water down around the lower extremities of the pilot's body, exerting pressure to help prevent pooling of blood. In November 1942, this was the first G-suit worn in actual air operations. Owing to the discomfort and thermal buildup in wearing the Franks suit, pneumatic anti-G suits were developed. One manufacturer of the pneumatic G-suits, David Clark Co. of Worcester, Massachusetts, later became involved in the production of microphones and headsets. The Gradient Pressure Flying suit, Type NS-9 or G-1 suit, was used by the Air Force in the European theater in 1944.

Training of aviators to fly airplanes soon included flight simulators in the program. Although flight simulation began as early as 1916, the electromechanical modern flight simulator was invented by E. A. Link in 1929 (Valverde, 1968). The Link Trainer, affectionately known as the "Blue Box," was used extensively during World War II, particularly in the training of pilots to fly under instrument conditions.

Although the developments in aviation were principally focused on military applications during this period, civilian aviation was slowly advancing in parallel to the military initiatives. Some of the cargo and bomber aircraft proposed and built for the military applications were also modified for civilian air transportation. The DC03, one of the most popular civil air-transport aircraft prior to the war, was the "workhorse" of World War II, used for the transportation of cargo and troops around the world. After the war, commercial airlines found that they had a large experienced population from which they could select airline pilots. However, there were few standards to guide them in the selection of the more appropriate pilots for the tasks of commercial airline piloting: passenger comfort, safety, and service. McFarland (1953), in *Human Factors in Air Transportation*, provided a good review on the status of the commercial airline pilots selection, training, and performance evaluation, as well as aviation medicine, physiology, and human engineering design. Gordon (1949) noted the lack of selection criteria to discriminate between airline pilots who were successful (currently employed) and those who were released from the airlines for lack of flying proficiency.

The problems of air-traffic control in the civilian sector were not unlike those in the operational theater. Though radar was developed and used for military purposes, it later became integrated into the civilian air-traffic control structure. There were the customary problems of ground clutter, precipitation attenuating the radar signals, and the detection of targets. Advances in the communications between the ground controllers and the airplanes, as well as communications between the ground control sites greatly facilitated the development of the airways infrastructure and procedures, till date. Hopkin (1995) provided an interesting and rather complete review on the history of human factors in air-traffic control.

Following the war, universities got into the act with the institution of aviation psychology research programs sponsored by the government (Koonce, 1984). In 1945, the National Research Council's Committee on Selection and Training of Aircraft Pilots awarded a grant to the Ohio State University to establish the Midwest Institute of Aviation. In 1946, Alexander C. Williams founded the Aviation Psychology Research Laboratory at the University of Illinois, and Paul M. Fitts opened the Ohio State University's Aviation Psychology Laboratory in 1949. These as well as other university research programs in aviation psychology and human engineering attracted veterans from the war to use the G.I. Bill to go to college, advance their education, and work in the area of human-factors psychology and engineering.

Although developed under the blanket of secrecy, toward the end of World War II, jet aircraft made their debut in actual combat. These jet airplanes gave a glimpse to our imaginations on what was to come in terms of aircraft altitude and airspeed capabilities of military and civilian aircraft in the near future.

1.5 Cold Weather Operations (Debons)

In the vast wastelands of Alaska, climatic levels and day–night seasonal extremes can define human performance and survival in the region. An understanding of the human–technological–climatic interface that prevails both in civil and military aviation activity thus became an important issue. The exploratory character of that effort was well documented and has been archived at the University of Alaska-Fairbanks. Only a few of the many programs of the Arctic Aeromedical Laboratory (AAL) are described here. A close relationship was maintained between the Aeromedical Laboratory located at Right Patterson Air Force Base, Dayton, Ohio (Grether & Baker, 1968), and the AAL located at Ladd Air Force Base, Fairbanks, Alaska. The AAL also collaborated with the ergonomic research activities of Paul M. Fitts, Human Engineering Laboratory, Ohio State University (Fitts, 1949).

The studies undertaken by the AAL included the following:

1. The impact that short–long, day–night variations have on personnel work efficiency
2. Difficulties encountered by military personnel in their ability to engage and sustain work performance import to ground flight maintenance
3. Significant human factors faced by military personnel during arctic operations
4. Study of the human factors and ergonomic issues associated with nutrition and exposure to temperature extremes
5. Optimal clothing to engage and sustain work efficiency during survival operations

1.6 The Jet Era (New Horizons)

The military airplanes developed after World War II were principally jet fighters and bombers. The inventory was "mixed" with many of the leftover piston engine airplanes, but as the United States approached the Korean War, the jet aircraft became the prominent factor in military aviation. Just before World War II, Igor Sikorsky developed a successful helicopter. During the Korean War, the helicopters found widespread service. These unique flying machines were successful, but tended to have a rather high incidence of mechanical problems, which were attributed to the reciprocating engines that powered them. The refinement of the jet engine and its use in the helicopters made them much more reliable and in more demand, both within the armed forces as well as in the civilian sector.

Selection and classification of individuals in the military hardly changed even after the advances made during the pressure of World War II. Furthermore, the jet era of aviation also did not produce a significant effect on the selection and classification procedures, until the advent of personal computers. Commercial air carriers typically sought their pilots from those who had been selected and trained by the armed forces. These pilots had been through rigorous selection and training criteria, were very standardized, had good leadership skills, and generally possessed a large number of flight house.

Boyne (1987) described the early entry of the jet airplanes into commercial air travel. In the United States, aircraft manufacturers were trying to develop the replacement for the fabled DC-3 in the form of various two- and four-radial-engine propeller airplanes. There were advances made such that the airplanes could fly without refueling, the speed was increased, and most of the airplanes soon had pressurization for passenger safety and comfort. In the meantime, Great Britain's Vicker-Armstrong came out with the Vicount in 1950, a four-engine turboprop airplane that provided much faster, quieter, and smoother flight. Soon thereafter, in 1952, the deHavilland Comet 1A entered commercial service. The Comet was an innovative full jet airliner capable of carrying 36 passengers at 500 miles/h between London and Johannesburg. These advances in the jet era had a significant impact on America's long-standing prominence in airline manufacturing. After two in-flight breakups of comets in 1954, deHavilland had difficulty in promoting any airplane with the name Comet. Thus, the focus of interest in airliner production shifted back to the United States, where Boeing, which had experience in developing and building the B-47 and B-52 jet bombers, made its entry into the commercial jet airplane market. In 1954, the Boeing 367–80 prototype of the resulting Boeing 707 made its debut. The Boeing 707 could economically fly close to Mach 1 and was very reliable but expensive. Later, Convair came out with its model 880 and Douglas made its DC-9, both closely resembling Boeing 707 (Josephy, 1962).

The introduction of jet airplanes brought varied responses from the pilots. A number of pilots who had served many years flying airplanes with reciprocating engines and propellers exhibited some "difficulties" in transitioning to the jet airplanes. The jet airplanes had few engine instruments for the pilots to monitor, few controls for the setting and management of the jet engines, and with the advancement of technology, more simplistic systems to control. However, the feedback to the pilot was different between piston propeller and jet airplanes. The time to accelerate (spool-up time) with the advance of power was significantly slower in the jet airplanes, and the time with which the airplane transited the distances was significantly decreased. Commercial airlines became concerned about the human problems in transition training from propeller to jet airplanes. Today, that "problem" seems to be no longer an issue. With the advent of high sophisticated flight simulators and other training systems and jet engines that build up their thrust more rapidly, there have been very few reports on the difficulties of transition training from propeller to jet airplanes.

Eventually, the jet era resulted in reductions in the size of the flight crews required to manage the airplanes. In the "old days," the transoceanic airliners required a pilot, a copilot, a flight engineer, a radio operator, and a navigator. On the other hand, the jet airliners require only a pilot, copilot, and in some instances, a flight engineer. With the aid of computers and improved systems engineering, many of the jet airplanes that previously had three flight crew members eliminated the need for a flight engineer and now require only two pilots.

The earlier aircraft with many crew members, who were sometimes dispersed and out of visual contact with each other, required good communication and coordination skills among the crew and were "trained" during crew coordination training (CCD). However, with the reduction in the number of crew members and placing them all within hand's reach of each other, lack of "good" crew coordination, communication, and utilization of available resources became a real problem in the jet airline industry. The tasks of interfacing with the on-board computer systems through the flight management system (FMS), changed the manner in which the flight crewmembers interact. Reviews on accident data and reports on the Aviation Safety Reporting Systems (ASRS) (Foushee, 1984; Foushee & Manos, 1981) revealed crew coordination as a "new" problem. Since the mid-1980s, much has been written about crew resource management (CRM; Weiner, Kanki, & Helmreich, 1993), and the Federal Aviation Administration (FAA) has issued an Advisory Circular 120-51B (FAA, 1995) for commercial air carriers to develop CRM training. Despite over 10 years of research, programs, and monies, there still seems to be a significant problem with respect to the lack of good CRM behaviors in the cockpits.

The jet engines have proven to be much more reliable than the piston engines of the past. This has resulted in the reliance on their safety, and sometimes a level of complacency and disbelief when things go wrong. With highly automated systems and reliable equipment, the flight crew's physical workload has been significantly reduced; however, as a result, there seems to be an increase in the cognitive workload.

1.7 The Cold War: Arctic Research

1.7.1 The New Technology Era (The Computer in the Cockpit)

In the 1990s, and although many things have changed in aviation, many other things have not. The selection of pilots for the armed forces is still as accurate as it has been for the past 40 years. However, there have been new opportunities and challenges in selection and classification, as women are now permitted to be pilots in the military, and they are not restricted from combat aircraft. The selection and classification tests developed and refined over the past 40 years on males might not be suitable for the females with the greatest likelihood of successfully performing as pilots (McCloy & Koonce, 1982). Therefore, human-factors engineers should reconsider the design of aircraft cockpits based on a wider range of anthropometric dimensions, and the development of personal protective and life-support equipment with regard to females is a pressing need.

With the advent of the microcomputers and flat-panel display technologies, the aircraft cockpits of the modern airplanes have become vastly different from those of the past. The navigational systems are extremely precise, and they are integrated with the autopilot systems resulting in fully automated flight, from just after the takeoff to after the airplane's systems, while the automation does the flying. Thus, a challenge for the designers is regarding what to do with the pilot during the highly automated flight (Mouloua & Koonce, 1997).

Recently, a great amount of attention has been paid to the concept of situation awareness in the advanced airplanes (Garland & Endsley, 1995). Accidents have occurred in which the flight crew members were not aware of their location with respect to dangerous terrains or were unaware of the current status of the airplane's systems, when that knowledge was essential for correct decision-making. Numerous basic researches have been initiated to understand more about the individual differences in situation awareness, the potential for selection of individuals with that capability, and the techniques for improving one's situation awareness. However, much of the studies have been reminiscent of the earlier research on attention and decision-making.

Thus, in future, human-factors practitioners will have numerous challenges, from the effects of advanced display technologies and automation at all levels of aviation, right down to the general aviation recreational pilot. The effectors to invigorate general aviation to make it more affordable, thus attracting a larger part of the public may include issues of selection and training down to the private pilot level, where, historically, a basic physical flight and a source of funds were all that were necessary to get into pilot training.

Economics is restructuring the way in which the airspace system works (Garland & Wise, 1993; Hopkin, 1995). Concepts such as data links between controlling agencies and the aircraft that they control, free flight to optimize flight efficiency, comfort and safety, automation of weather observation and dissemination, and modernization of the air-traffic controllers' workstations will all require significant inputs from aviation human-factors practitioners in the near future.

The future supersonic aircraft, to reduce drag and weight costs, might not provide windows for forward visibility, but might provide an enhanced or synthetic visual environment that the pilots can "see" to maneuver and land their airplanes. Other challenges might include the handling of passenger loads of 500–600 persons in one airplane, the design of the terminal facility to handle such airplanes, waiting and loading facilities for the passengers, and the systems for handling the great quantity of luggage and associated cargo. In addition, planners and design teams including human-factors practitioners may also have to face the future problems in airport security.

References

Benford, R. J. (1979). *The heritage of aviation medicine.* Washington, DC: The Aerospace Medical Association.

Boyne, W. J. (1987). *The Smithsonian book of flight.* Washington, DC: Smithsonian Books.

Cattle, W. & Carlson, L. D. (1954). Adaptive changes in rats exposed to cold, coloric exchanges, *American Journal of Physiology, 178,* 305–308.

Chapanis, A., Chardner, W. R., Morgan, C. T., & Sanford, F. H. (1947). *Lectures on men and machines: An introduction to human engineering.* Baltimore, MD: Systems Research Laboratory.

Debons, A. (1951, March) *Psychological inquiry into field cases of frostbite during operation "Sweetbriar."* Ladd AFB, AL: AAL.

Debons, A. (1950) *Personality predispositions of Infantry men as related to their motivation to endure tour in Alaska: A comparative evaluation*: (Technical Report). Fairbanks, AL: Arctic Aeromedical Laboratory, Ladd Airforce Base.

Debons, A. (1950a, April) *Human engineering research* (Project no.22-01-022. Part 11, Progress E).

Debons, A. (1950b, February) *Gloves as factor in reduce dexterity. Individual reactions to cold* (Project 212-01-018. Phase 1. Program A., ATT 72-11).

Deese, J. A. & Larzavus, R. (1952, June). *The effects of psychological stress upon perceptual motor response.* San Antonio TX: Lackland AFB, Air Traibning Command. Human Resource Center.

Engle, E. & Lott, A. S. (1979). *Man in flight: Biomedical achievements in aerospace.* Annapolis, MD: Leeward.

Federal Aviation Administration. (1995, March). *Crew resource management training* (Advisory Circular AC 120-51B). Washington, DC: Author.

Fitts, P. M. (1947). *Psychological research on equipment design* (Research Rep. No. 17). Washington, DC: Army Air Forces Aviation Psychology Program.

Fitts, P. M. & K. Rodahl (1954). Modification by Light of 24 hour activity of white rats. *Proceedings of Iowa Academy Science, 66,* 399–406

Flanagan, J. C. (1947). *The aviation psychology program in the army air force* (Research Rep. No. 1). Washington, DC: Army Air Forces Aviation Psychology Program.

Foushee, C. J. (1984). Dyads and triads at 36,000 feet. *American Psychologist, 39,* 885–893.

Foushee, C. J. & Manos, K. L. (1981). Information transfer within the cockpit: Problems in intracockpit communications. In C. E. Billings, & E. S. Cheaney (Eds.), *Information transfer problems in the aviation system* (NASA Rep. No. TP-1875, pp. 63–71). Moffet Field, CA: NASA-Ames Research Center.

Garland, D. J. & Endsley, M. R. (Eds.). (1995). *Experimental analysis and measurement of situation awareness.* Daytona Beach, FL: Embry-Riddle Aeronautical University Press.

Gordon, T. (1949). The airline pilot: A survey of the critical requirements of his job and of pilot Evaluation and selection procedures. *Journal of Applied Psychology, 33,* 122–131.

Grether, W. F. (1968). Engineering psychology in the United States. *American Psychologist, 23*(10), 743–751.

Hopkin, V. D. (1995). *Human factors in air traffic control.* London U.K.: Taylor & Francis.

Jospehy, A. M., Jr. (Ed.). (1962). *The American heritage history of flight.* New York: American Heritage.

Koonce, J. M. (1979, September). Aviation psychology in the U.S.A.: Present and future. In F. Fehler (Ed.), *Aviation psychology research.* Brussels, Belgium: Western European Association for Aviation Psychology.

Koonce, J. M. (1984). A brief history of aviation psychology. *Human Factors, 26*(5), 499–506.

McCollum, E. L. (1957). *Psychological aspects of arctic and subarctic living.* Science in Alaska, Selected Papers of the Arctic Institute of America Fairbanks, AK.

McCloy, T. M. & Koonce, J. M. (1982). Sex as a moderator variable in the selection and training of persons for a skilled task. *Journal of Aviation, Space and Environmental Medicine, 53*(12), 1170–1173.

McFarland, R. A. (1953). *Human factors in air transportation.* New York: McGraw-Hill.

Moroney, W. F. (1995). The evolution of human engineering: A selected review. In J. Weimer (Ed.), *Research techniques in human engineering* (pp. 1–19). Englewood Cliffs, NJ: Prentice-Hall.

Mouloua, M. & Koonce, J. M. (Eds.). (1997). *Human-automation interaction: Research and practice.* Mahwah, NJ: Lawrence Erlbaum Associates.

Naval Special Devices Center. (1949, December). *Handbook of human engineering data for design engineers.* Tufts College Institute for applied Experimental Psychology (NavExos P-643, Human Engineering Rep. No. SDC 199-1-2a). Port Washington, NY: Author.

North, R. A. & Griffin, G. R. (1977). *Aviator selection 1917–1977.* (Technical Rep. No. SP-77-2). Pensacola, FL: Naval Aerospace Medical Research Laboratory.

Pecora, L. F. (1962). Physiological measurement of metabolic functions in man. *Ergonomics,* 5.7(1-1962).

Rohdah, K. & Horvath, S. M. (1961, January). Effects of dietary protein on performance on man in cold environment (Rep. No.). Philadelphia, PA: Lankenau Hospital Research Institute.

Stevens, S. S. (1946). Machines cannot fight alone. *American Scientist, 334,* 389–400.

Valverde, H. H. (1968, July). *Flight simulators: A review of the search and development* (Technical Rep. No. AMRL-TR-68-97). Wright-Patterson Air Force Base, OH: Aerospace Medical Research Laboratory.

Weiner, E., Kanki, B. G., & Helmreich, R. (Eds.). (1993). *Cockpit resource management.* New York: Academic Press.

2

Aviation Research and Development: A Framework for the Effective Practice of Human Factors, or "What Your Mentor Never Told You about a Career in Human Factors..."

John E. Deaton
CHI Systems, Inc.

Jeffrey G. Morrison
Space Warfare Systems Center

2.1 The Role of Human-Factors Research in Aviation

Since its humble beginning in the chaos of World War II, human factors have played a substantial role in aviation. In fact, it is arguably in this domain that human factors have received their greatest acceptance as an essential part of the research, development, test, and evaluation cycle. This acceptance has come from the critical role that humans, notably pilots, play in these human–machine systems, the unique problems and challenges that these systems pose on human perception, physiology, and cognition, and the dire consequences of human error in these systems. As a result, there have been numerous opportunities for the development of the science of human factors that have contributed significantly to the safety and growth of aviation.

Times keep changing, and with the end of the Cold War, funding for human-factors research and development started shrinking along with military spending. Being a successful practitioner in the field of human factors requires considerable skills that are beyond those traditionally taught as a part of a graduate curriculum in human factors. New challenges are being presented, which require a closer strategic attention to what we do, how we do it, and what benefits accrue as a result of our efforts. This chapter offers snippets of the authors' experience in the practice of human factors. It describes the questions and issues that the successful practitioner of human factors must bear in mind to conduct research, development, testing, and engineering (RDT&E) in any domain. A large part of the authors' experience is with the Department of Defense (DoD), and this is the basis of our discussion. Nonetheless, the lessons learned and advices made should be applicable across other endeavors related to the science of human factors.

2.1.1 Focus Levels of RDT&E

An important part in succeeding as a human-factors practitioner is recognizing the type of research being funded, and the expectancies that a sponsor is likely to have for the work being performed. The DoD identifies four general categories of RDT&E, and has specific categories of funding for each of these categories.* These categories of research are identified as 6.1–6.4, where the first digit refers to the research dollars and the second digit refers to the type of work being done (Table 2.1). The DoD sponsors are typically very concerned with the work being performed, as Congress mandates what needs to be done with the different categories of funding, and has mechanisms in place for the different categories of funding and to audit how it is spent. This issue is also relevant to the non-DoD practitioner as well, because regardless of the source of RDT&E funding, understanding the expectations that are attached to it is critical to successfully conclude a project. Therefore, the successful practitioner should understand how their projects are funded and the types of products expected for that funding.

Basic research is the one typically thought of as being performed in an academic setting. Characteristically, a researcher may have an idea that he or she feels would be of some utility to a sponsor, and obtains funding to try to explore the idea further. Alternatively, the work performed may be derived from the existing theory, but may represent a novel implication of that theory. Human-factors work at the 6.1 level will typically be carried out with artificial tasks and naïve subjects, such as a university laboratory with undergraduate students as subjects. Products of such work may be theoretical development, a unique model, or theory, and the work typically may entail empirical research to validate the theory. This work is generally not focused on a particular application or problem, although it may be inspired by a real-world problem and may utilize a problem domain to facilitate the research. However, this research is not generally driven by a specific operational need; its utility for a specific application may only be speculated. This type of research might have to address questions such as

- How do we model strategic decision-making?
- How is the human visual-perception process affected by the presence of artificial lighting?
- What impact do shared mental models have on team performance?

Applied research is still very much at the research end of the research–development spectrum; however, it is typically where an operational need or requirement first comes into the picture in a significant way. This research can be characterized as the one considering established theories or models shown to have

---

* In fact, these categories are being redefined as a part of the downsizing and redefinition of the DoD procurement process. For instance, there was until the early 1990s, a distinction in the 6.3 funding between core-funded prototype demonstrations (6.3a) and the actual field demonstrations (6.3b) that received specific funding from the Congressional budget. However, this distinction has been eliminated. The authors were unable to locate a specific set of recent definitions that have been employed when this chapter was written. Therefore, these definitions are based on the authors' current understanding of the DoD procurement system, based on the current practice rather than an official set of definitions.

TABLE 2.1 Types and Characteristics of DoD Research and Development

Number	Type	Definition	Research Questions	Products
6.1	Basic research	Research done to develop a novel theory or model, or to extend the existing theory into new domains. The work may be funded to solve a specific problem; however, there is typically, no single application of the research that drives the work	Can we take an idea and turn it into a testable theory? Can we assess the utility of a theory in understanding a problem?	Theoretical papers, describing empirical studies, mathematical models, recommendations for continued research, and discussion of potential applications
6.2	Applied research	Research done to take an existing theory, model, or approach, and apply it to a specific problem	Can we take this theory/ model and apply it to this problem to come up with a useful solution?	Rudimentary demonstrations, theoretical papers describing empirical studies, recommendations for further development
6.3	Advanced development	Move from research to development of a prototype system to solve a specific problem	Can we demonstrate the utility of technology in solving a real-world need? What are the implications of a proposed technology? Is the technology operationally viable?	Working demonstrations in operationally relevant environments Assessment with intended users of the system Technical papers assessing the operational requirements for the proposed system/technology
6.4	Engineering development	Take a mature technology and develop a fieldable system	Can we integrate and validate the new technology into existing systems? What will it cost? How will it be maintained?	The products of this stage of development would be a matured, tested system, ready for procurement— notably, detailed specifications and performance criteria, life-cycle cost estimates, etc.
6.5	System procurement	Go out and support the actual buying, installation, and maintenance of the system	Does it work as per the specification? How do we fix the problems?	Deficiency reports and recommended fixes

some scientific validity, and exploring their use to solve a specific problem. Owing to its applied flavor, it is common and advisable to have some degree of subject expertise involved with the project, and to utilize the tasks that have at least a theoretical relationship with those of the envisaged application being developed. Questions with regard to this type of human-factors research might include

- How is command-level decision-making in tactical commanders affected by time stress and ambiguous information?
- How should we use advanced automation in a tactical cockpit?
- How do we improve command-level decision-making of Navy command and control staff?
- How can synthetic three-dimensional (3D) audio be used to enhance operator detection of sonar targets?

Advanced development is the point when the work starts moving away from the research and toward development. Although demonstrations are often done as a part of 6.2 and even 6.1 research, there is an implicit understanding that these demonstrations are not of fieldable systems to be used by specific operators. However, a major product of 6.3 R&D is typically a demonstration of a fairly well-reflected system in an operationally relevant test environment with the intended users of the proposed system. As a result, this type of research is typically more expensive than that which takes place at 6.1 or 6.2, and often involves contractors with experience, and requires the involvement of subjects and subject experts with operational experience related to the development that is going to take place. Research questions in advanced development are typically more concerned with the demonstration of meaningful performance gains and the feasibility of transferring the underlying technology to fielded systems. Representative questions in 6.3 human-factors research might include

- Is the use of a decision-support system feasible and empirically validated for tactical engagements?
- What are the technical requirements for deploying the proposed system in terms of training, integration with existing systems, and so on?
- What are the expected performance gains from the proposed technology and what are the implications for manning requirements based on those gains?

As the procurements process for a technology or system moves beyond 6.3, the human factors may typically play lesser dominant role. However, this does not mean that it is not necessary for the human factors to have continued involvement in the RDT&E process. It is just that at the 6.4 level, most of the critical human-factors issues are typically solved, and the mechanics of constructing and implementing technology tend to be the dominant issue. It becomes more difficult (as well as more and more expensive) to implement changes as the system matures. As a result, only critical shortcomings may be addressed by the program managers in the later stages of technology development. If we, as human-factors practitioners, have been contributing appropriately through the procurement process, our relative involvement at this stage may not be problematic and may naturally be less prominent, than it was earlier in the RDT&E process. Human-factors issues still need to be addressed to ensure that the persons in the human–machine systems are not neglected. Typically, at this stage of the procurement progress, we are concerned with testing issues such as compliance and verification. The questions become more related to testing and evaluation of the developed human–machine interfaces, documenting the final system, and the development of the required training curriculum. Thus, although it is imperative that human-factors professionals continue to have a role, there are in fact few dedicated research and development funds for them at the 6.4 and 6.5 stages. The funding received for human factors at this stage typically comes from the project itself, and is at the discretion of the project management. Research done at these levels might comprise questions related to the following:

- Can the persons in the system read the displays?
- What training curriculum is required for the people in the system to ensure adequate performance?
- What criteria should be used in selecting the individuals to work in this system?

2.2 Development of an Effective R&D Program

The R&D process is similar irrespective of the application domain. Unfortunately, R&D managers often lose track of the real purpose of behavioral research: solving a problem. In particular, sponsors may want and deserve to have products that make their investment worthwhile. They (and you) need to know where you are, where you are heading to, and have a pretty good sense of how you are going to get there. Keeping these issues in the forefront of your mind as a program manager or principal investigator may result in further support in the near future. Having said that, what makes an R&D program successful? One can quickly realize that successful programs require the necessary resources, and that there is a "critical mass" of personnel, facilities, and equipment resources that must be available to be effective. It is also intuitively obvious that proper program management, including a realistic funding base, is crucial if research is to be conducted in an effective manner. However, what are the factors that we often neglect to attend to, which may play a deciding role in defining the eventual outcome of the research program? What does one do when the resources do not match the magnitude of the task required to get to the end goal?

You must understand your customers and their requirements. Often, particularly in the DoD domain, there are multiple customers with different, sometimes competing, and sometimes directly conflicting agendas. You must understand these customers and their needs, and find a way to give them not only what they ask for or expect, but what they need. The successful practitioner should understand what they need, and sometimes may have to understand their needs better than they do if the project is to succeed. Needless to say, this can be something of an art rather than a science, and often requires significant diplomatic skills. For example, in the DoD model, there are typically two customers: the sponsors or the people responsible for the money being spent in support of RDT&E, and the users or those who will make use of the products of this effort. In the Navy, the former is typically the Office of Naval Research (ONR) and the latter is the Fleet. The ONR may typically be interested in the theory and science underlying the RDT&E process, and may be interested in an audit trail whereby it can show: (a) that quality science is being performed as measured by meaningful research studies and theoretical papers, and (b) the successful transition of the science through the various levels of the RDT&E process. The Fleet may also be interested in transition, but may be more interested in the applicability of the developed technology in solving its real-world needs in the near future. Thus, the users may be interested in getting the useful products out to the ships (or airplanes or whatever), and may be less interested in the underlying science. The competing needs of these two requirements are often one of the most challenging aspects of managing a human-factors project, and failure to manage them effectively is often a significant factor in the project's failure. One must understand the level of one's technology/research in the RDT&E process, and where it needs to go to be successful, and do whatever one can to facilitate its shift to the next stage in the procurement process. Understanding this process and knowing what questions to ask from a management perspective are vital to meet one's own objectives as a researcher/practitioner, as well as those of the sponsors/customers. However, how can this be accomplished?

First, we suggest that the successful human-factors practitioner should emphasize on providing information that best fits the nature of the problem and the environment in which it is to be applied. In other words, providing a theoretical treatment of an issue when the real problem involves an operational solution may not be met with overwhelming support. There has to be a correlation between theory and application. However, this does not indicate that the theory does not have an important role to play in aviation human factors. The problems arise when researchers (usually more comfortable in describing issues conceptually) are faced with sponsors who want the "bottom line" and they want it now, and not tomorrow. Those in academics may not be comfortable with this mindset. The solution is to become familiar with the operational issues involved, and know the best way to translate the input to the sponsor so that the sponsor can, in turn, communicate such information into something that can be meaningful to the user group in question.

Second, the most common reason for the research programs to get into trouble is that they propose to do more than that which is feasible with the available resources. Initially, one might get approving gestures from the sponsors; however, what might happen a year or two down the road when it becomes evident that the initial goals were far too ambitious? Successful R&D efforts are underscored by their ability to meet project goals on time and within specified funding levels. Promising and not delivering is not a strategy that can be repeated twice. Therefore, it is critical that the program manager keeps track of where the program is, where it is committed to going, and the available resources and those required to reach the goal. When there is a mismatch between the available and required resources, the program manager must be proactive in redefining objectives, rescoping the project, and/or obtaining additional resources. It is far better to meet the most critical of your research objectives and have a few fall to the wayside (for good reason), than to have the entire project be seen as a failure. In recent years, many programs have been jeopardized less by reductions in the funding than by the inability or unwillingness of the program management to realistically deal with the effects of those cuts.

Third, and perhaps the most important (certainly to the sponsor), is how you measure the effectiveness of a new system or technology that you have developed. This issue is often referred to as "exit criteria" and deals with the question: How do you know when you are done? This is by no means a trivial task, and can be critical to the success of obtaining and maintaining funding. Many projects are perceived as failure by the sponsors, not because they are not doing good work, but because there is no clear sense as to when it will pay off. Measures of effectiveness (MOEs) to assess these exit criteria are often elusive and problematic. However, they do provide a method for assessing the efficacy of a new system. Determining the criteria that will be used to evaluate the usefulness of a system is a process that needs to be upfront during the developmental stage. In this way, there are no "surprises" at the end of the road, where the system (theory) does wonderful things, but the customer does not understand why he or she should want it. A researcher once stated that the best he could imagine was a situation where there were no surprises at the end of a research project. It is interesting to note that such a statement runs against the grain of what one is taught in doing the academic research. In academic research, we prize the unexpected discovery and are taught to focus on the identification of additional research. This is often the last thing that a user wants to hear; users want answers—not additional questions. One of the most important things learned by novice practitioners is how to reconcile the needs of the customer with their research training.

Fourth, it is advantageous to make personal contact (i.e., face to face) with the sponsor and supporting individuals. The people whose money you are spending will almost universally appreciate getting "warm fuzzies" that can only come from one-to-one contacts. New developments in the areas of communications (i.e., teleconferencing, e-mail, etc.) are not a substitute to close contact with individuals supporting your efforts. As you become a proficient practitioner of human factors, you may learn that there is no better way to sense what aspects of a project are of greatest interest to your customers and what are problematic, than to engage in an informal discussion with them. Further, your value to the customer will be significantly increased if you are aware of the hidden agendas and their priorities. Although often these may not be directly relevant to you or your project, your sensitivity to them may make you much more effective as a practitioner. This may become painfully obvious when things go wrong. Your credibility is, in part, established through initial contact.

Fifth, do you have external endorsements for the kind of work you are attempting? In other words, who really cares what you are doing? Generating high-level support from the intended users of your effort is indispensable in convincing the sponsors that there is a need for such work. In the military environment, this process is de facto mandatory. Few projects receive continued funding unless they have the support of specific program office within the DoD. Operational relevancy and need must be demonstrated if funding is to be secured, and defended in the face of funding cuts.

Sixth, the interagency coordination and cooperation will undoubtedly enhance the probability of a successful research program. Your credibility as a qualified and responsible researcher depends on being aware of the ongoing related work elsewhere, and its relevance to the issues going on in your project. Generally, efforts made to leverage off this ongoing work to avoid duplication of the effort have become increasingly critical in this era of limited research and development resources. The lack of senior-level support and ineffective coordination among external research organization may in fact be a significant impediment to execute the program goals. However, through the use of coordinating and advisory committees, working groups, cooperative research agreements, and widespread dissemination of plans and products, duplication of effort can be minimized.

Finally, you must be prepared to discuss where your research will go after the conclusion of the project: What transition opportunities are available in both the civilian and military sectors? or describe the applicability of your work to other domains including civilian and military sectors, and particularly, those of interest to your sponsors and customers. This is critical to develop any success achieved in a particular research project, and maintain your credibility. Will there be additional follow-up work required? What other sponsors/customers would be interested in your findings/products? Who could most benefit from the results of your work? Extracting the critical information from your project and demonstrating how this will assist other works is often neglected once a project has been finished. The successful practitioner may not entirely walk away from an area once a particular project is finished, but will track its transitions, both planned and unplanned. An excellent way to build credibility and develop new contracts and funding opportunities is to contact those people whose work you are building on to (a) advise them about their work and (b) make them aware of your expertise and capability. Not only are these people generally flattered by the interest, but they may advocate you as a resource when they meet colleagues with similar interest.

2.3 Some Words of Wisdom Regarding Dealing with the Sponsor, Management, and User

Be honest. Do not tell them what you think and want to hear—unless that bears some resemblance to realty. Be honest to yourself as well. There is nothing more dangerous to a project or an organization than someone who does not know what he or she is talking about. Trying to bluff your way through a discussion will only damage your credibility, and that of your cause, particularly if you are with people who do know what they are talking about. Colleagues and sponsor generally will not confront you with your ignorance, but they will be impressed by it—negatively. If you are not sure of something, the best bet is to ask an intelligent, appropriate question to an appropriate person, at the appropriate time and appropriate place. You can use this strategy to turn a potentially negative situation into a positive one by displaying your sensitivity, judgment, and wisdom, despite your possible lack of technical knowledge.

Management really does not want to hear about your problems. If you must present a problem, then the management expects you to identify the prospective solutions and present the recommended solution with underlying rationale and implications for the decision. It is advisable to deal with problems at the possible lowest level of management. Do not jump the chain in doubt, and try to document everything. It is in everyone's best interests in the midst of turbulence to document discussions, alternatives, and recommended solutions. In this way, if the problem becomes terminal to your efforts, you have the ammunition to fend off accusations and blame, and to potentially demonstrate your wisdom and foresight.

If the problem being discussed is threatening one's project or career, document this situation in the form of memos distributed to an appropriate group of individuals. Generally, this may be given to all the affected parties, with copies to supervisory personnel, if necessary (note that this is almost never appropriate for the first memorandum). Memos of this nature must be well-written and self-explanatory. Assume the reader knows nothing, particularly if you are going to use one of the most powerful features of a memo—the courtesy copies (cc) routing. This is one of the best tools available to ensure that you have covered your backside, and that management recognizes that you appreciate the significance of problems in your project, your skills in dealing with them at an appropriate level, and the consequences of not dealing with the problems effectively. The tone of such memoranda is critical with regard to their effectiveness. Never be vindictive, accusatory, or in any way judgmental in a memorandum. State the facts (as you see them) and be objective. Describe in a clear, concise manner about what has been done and when, as well as what needs to be done by when, and, if appropriate, by whom. One of the most effective techniques in writing such a memorandum is to demonstrate the awareness of the constraints and factors creating your problem, and limiting yourself and the other relevant parties from getting the problem solved. Again, such a strategy will demonstrate your appreciation of conflicting agendas and convey the message that you wish to work around them by building bridges to the other parties involved.

2.4 Developing a Long-Term Research Strategy

It has been the authors' experience that the most successful and interesting research is in fact not only a single program, but related programs operating at several levels of the RDT&E process in parallel. This is an effective strategy for a variety of reasons. First, it offers built-in transition from basic through

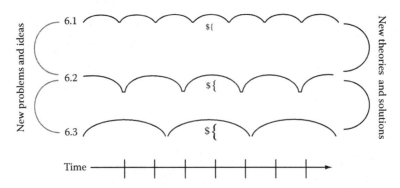

FIGURE 2.1   Representation of ideal R&D investment strategy.

applied research as well as advanced development. Second, it provides a vehicle to address interesting, important, and often unexpected problems that may appear in more advanced R&D at more basic levels of R&D, when appropriate resources might not be available to explore the problem at the higher level of research. Third, it provides a basis for leveraging of resources (people, laboratory development, and maintenance costs, etc.) across a variety of projects. This will make you more effective, efficient, and particularly, cost-effective in this era of down-sizing. Further, such efforts go a long way toward establishing the critical mass of talent necessary to carry out quality research on a regular basis. Finally, a multi-thrust strategy provides the necessary buffer when one or another line of funding comes to an end.

Figure 2.1 shows how such a strategy could be laid out over time. Note that the lower levels of research tend to cycle more rapidly than the projects performing advanced development. In addition, further shift along the project in the R&D process tends to become more expensive and resource-intensive. New problems and ideas for additional research are observed to be inspired by the needs of ongoing applied research. The products of each level of research are found to be feeding down into the next available cycle of more developmental research. It must also be noted that the products of one level of research need not necessarily flow to the next level of research. They may jump across the levels of research or even spawn entirely new research efforts within the same line of funding.

2.5 Critical Technology Challenges in Aviation Research

Several excellent sources are available, which may assist in developing a realistic perspective regarding the future opportunities in aviation research. For example, the recent National Plan for Civil Aviation Human Factors developed by the Federal Aviation Administration (FAA, March 1995) supports several critical areas within aviation. This initiative describes the goals, objectives, progress, and challenges for both the long- and short-term future of human factors.

Research and application in civil aviation, more specifically, the FAA plan, identifies the following five research thrusts: (a) human-centered automation, (b) selection and training, (c) human performance assessment, (d) information management and display, and (e) bioaeronautics. The primary issues in each of the first four thrust areas are summarized in Tables 2.2 through 2.5. These issues certainly exemplify the challenges that the human-factors specialists may face in the upcoming years. These are the areas that will most certainly receive sponsorship support, as they have been deemed to be impacting the rate of human error-related incidents and accidents.

Researchers are expected to be aware of several changes within the R&D environment in the last few years, which may have significant influence on new initiatives. These changes will substantially change the role of human-factors researchers conducting aviation research. First, there has been an increased awareness and sensitivity to the critical importance of the human element in safety. With this

TABLE 2.2   Issues in Human-Centered Automation

Workload	1. Too little workload in some phases of flight and parts of air-traffic control (ATC) operations to maintain adequate vigilance and awareness of systems status 2. Too much workload associated with reprogramming when flight plans or clearances change 3. Transitioning between different levels of workload, automation-induced complacency, lack of vigilance, and boredom on flight deck, ATC, and monitoring of system and service performance
Operational situation awareness and system-mode awareness	1. The ability of operators to revert to manual control when the advanced automation equipment fails 2. An inadequate "cognitive map," or "situational awareness" of what the system is doing 3. Problematic recovery from automation failures 4. The potential for substantially increased head-down time 5. Difficulty and errors in managing complex modes
Automation dependencies and skill retention	1. The potential for controllers, pilots, and others to over-rely on computer-generated solutions (e.g., in air-traffic management and flight decisions) 2. Hesitancy of humans to take over from an automated air-traffic and flight deck system 3. Difficulty in maintaining infrequently used basic and critical skills 4. Capitalizing on automation-generated alternatives and solutions 5. Monitoring and evaluating pilot and controller skills where computer-formulated solutions disguise skill weaknesses 6. Supporting diagnostic skills with the advent of systems that are more reliable and feature built-in self-diagnostics (e.g., those in "glass cockpit" systems and fully automated monitoring systems)
Interface alternatives	1. Major system-design issues that bridge all the aviation operations including selecting and presenting information for effective human–computer interface 2. Devising optimal human–machine interfaces for advanced ATC systems and for flight deck avionics 3. Devising strategies for transitioning to new automation technologies without degrading individual or contemporary system performance

TABLE 2.3   Issues in Selection and Training

New equipment training strategies	1. Training pilots, controllers, security personnel, and systems management specialists to transition to new technologies and the associated tasks for new equipment 2. New training concepts for flight crews, controller teams, security staffs, and system management teams 3. Measuring and training for the performance of new tasks associated with equipment predictive capabilities (vs. reactive-type tasks) for pilots and air-traffic controllers 4. Methods to train personnel in the use of computer decision-aiding systems for air and ground operations 5. Improved strategies for providing the required student throughput within training resource constraints on centralized training facilities, training devices, and simulation
Selection criteria and methods	1. Evaluation of individual and aggregate impacts on personnel selection policies of changing requirements in knowledge, abilities, skills, and other characteristics for flight crew, controller, and airway facilities operations associated with planned and potential changes in the national airspace system (NAS) 2. Expanded selection criteria for pilots, controllers, technicians, and inspectors from general abilities to include both more complex problem-solving, diagnostic, and metacognitive abilities, as well as the social attributes, personality traits, cultural orientation, and background biographical factors that significantly influence the operational performance in a highly automated NAS 3. Development of measures to evaluate these more complex individual and team-related abilities in relation to job/task performance

TABLE 2.4 Issues in Human Performance Assessment

Human capabilities and limitations	Determining the measures and impacts of (a) cognitive factors underlying successful performance in planning, task/workload management, communication, and leadership; (b) the ways in which skilled individuals and teams prevent and counteract errors; (c) ways to reduce the effects of fatigue and circadian dysrhythmia on controllers, mechanics, and flight deck and cabin crews; (d) baseline performance characteristics of controllers to assess the impact of automation; and (e) qualifying the relationship between age and skilled performance
Environmental impacts (external and internal)	1. Assessing the influence of "culture" on human performance, including the impact of different organizational and ethnic cultures, management philosophies and structures, and procedural styles 2. Determining methods to accommodate mixed corporate, regional, and national views of authority, communication, and discipline 3. Addressing variations in aviation equipment-design philosophies and training approaches 4. Understanding the population's stereotypical responses in aviation operations
Methods for measurement	Devising effective aviation-system monitoring capabilities with emphasis upon: (a) expansion of the collection, usage, and utility of human performance data and databases; (b) standardization and improved awareness of critical human-factors variables for improved collection, classification, and use of reliable human performance data; (c) standardization of classification schemes for describing human-factors problems in human–machine systems; (d) better methods and parameters to assess team (vs. individual) performance parameters for flight and maintenance crews, air-traffic controllers, security and aviation operations personnel; and (e) improved understanding of relationship between actual performance and digital data measurement methodologies for the flight deck to predict future air crew performance based on trend data

increased understanding, we can observe a renewed interest on safety, even if that results in less funding for nonsafety-related research. Second, programmatic changes within the organizations, such as increased National Aeronautics and Space Administration (NASA) emphasis on aeronautics and DoD technology transfer programs, are very likely to generate cooperative agreements between the agencies that heretofore had not considered sharing technological advances. Moreover, the emphasis away from strictly military applications is obviously one of the "dividends" resulting from the end of the Cold War and the draw-down of the military complex. Finally, technological changes in the design and development of aviation systems continue at an increasing level of effort. Systems are becoming more complex, requiring modifications to training regimens. Advances in the development of aircraft structures have surpassed the capabilities of the operator to withstand the environmental forces impinging upon him or her. These new developments will certainly stimulate innovative efforts to investigate how to enhance the capabilities of the human operator, given the operator's physiological limitations. These indicate that those in the human-factors field must be aware of what these changes are, and, more importantly, of how we can be more responsive to the needs of both civilian and military research agencies.

With regard to these ongoing and future challenges, there are several driving factors that contribute to the role that aviation human factors will play in the near future. Some of these drivers are: (a) technology, (b) demographics, (c) cultural, and (d) economic. Each of these drivers is subsequently discussed in the light of its impact on the direction of future aviation research efforts.

Technology. With the advent of new aircraft and future changes in the air-traffic control systems, we may see even higher levels of automation and complexity. However, how these changes impact the operator performance and how the system design should be modified to accommodate and minimize human error need to be determined. A blend of the best of computer and human capabilities should result in some type of human–computer interaction designed to minimize errors.

Demographics. With the military draw-down becoming a reality, there will be fewer pilots trained by military sources. Changing the skill levels and other work-related demographics will probably affect

TABLE 2.5   Issues in Information Management and Display

Information exchange between people	1. Identify requirements for access to critical NAS communications for analysis purposes 2. Determine the effects of pilot response delays in controller situation awareness and controller/pilot coordination (particularly with regard to delayed "unable" responses) 3. Set standards for flight crew response to messages 4. Assess the changes in pilot/controller roles 5. Enhance the communication training for pilots and controllers 6. Identify sources, types, and consequences of error as a result of cultural differences 7. Develop system design and procedural solutions for error avoidance, detection, and recovery
Information exchange between people and systems	1. Assess and resolve the effects of data communications on pilots/controllers situational awareness 2. Determine the best display surfaces, types, and locations for supporting communication functions in the cockpit, at the ATC workstation, and at monitoring and system maintenance control centers 3. Identify sources, types, and consequences of error, as well as error avoidance, detection, and recovery strategies 4. Establish requirements and set standards for alerting crew, controller, and system management personnel to messages of varying importance
Information displays	1. Establish policies for operationally suitable communication protocols and procedures 2. Set standards for display content, format, menu design, message displacement, control and interaction of functions, and sharing 3. Assess the reliability and validity of information-coding procedures 4. Provide design guidelines for message composition, delivery, and recall 5. Prescribe the most effective documentation and display of maintenance information 6. Prototype technical information management concepts and automated demonstration hardware to address and improve the content, usability, and availability of information in flight deck, controller, aircraft, maintenance, security, AF system management, and aviation operations
Communication processes	1. Devise methods of reconstructing the situational context needed to aid the analysis of communications 2. Analyze relationships between workload factors and errors in communication 3. Evaluate changes in information-transfer practices 4. Set standards and procedures for negotiations and modifications to clearances 5. Establish procedures for message prioritization and response facilitation 6. Set standards for allocation of functions and responsibilities between pilots, controllers, and automated systems 7. Provide guidelines on the distribution of data to and integration with other cockpit systems 8. Prescribe communication policies related to flight phases and airspace, such as use in terminal area and at low altitudes 9. Determine the impact of data communications on crew and controller voice-communication proficiency

personnel selection and training of pilots as well as ancillary personnel, that is, controllers, maintenance, and operations. However, how these changes drive the development of new standards and regulations remains to be seen. We have already seen a change from strict adherence to military specifications in DoD system-acquisition requirements, to industrial standards. Not only is the "learner, meaner" workforce the hallmark of the new military, but it also gives justification to support future developments in the area of personnel training. The acquisition of additional weapon systems will most probably decrease, resulting in a redoubling of our efforts to train the existing personnel to operate the current generation of weapon systems to a more optimal and efficient level.

Cultural. Opportunities to collaborate with our foreign counterparts will increase, as organizations become increasingly international. The development of aviation standards and practices will take into account the incompatible cultural expectations that could lead to increased human errors and unsafe conditions. We have already observed these developments in the area of air-traffic control, and we will certainly see analogous efforts in other areas in the near future.

Economic. Economic factors have vastly affected the aerospace industry. Available funding to continue R&D efforts has steadily decreased. Under this kind of austere environment, competition for limited research funds is fierce. Many agencies, especially the military, are cutting back on the development of new systems and are now refocusing on improving the training programs to assure a high-level skill base, owing to the reduction in available personnel.

The role that the human-factors field plays in aviation research is not different from the role it plays in any research endeavor. The methods, for the most part, remain the same. The difference lies in the impact it has on our everyday lives. In its infancy, human factors focused on the "knobs and dials" issues surrounding the aircraft and aircraft design. Today, we are faced with more complex issues, compounded by an environment that is driving scarce resources into areas that go beyond theoretical pursuits to that of practical, applied areas of concentration. However, this does not indicate that this area is not vital, progressive, or increasing in scope and value. It merely means that we, as professionals working in the field of aviation human factors, have to be aware of the technology gaps and know the best way to satisfy the needs of our customers. This can be accomplished, but it requires a certain kind of flexibility and visionary research acumen to anticipate what these problems are and the best ways to solve them.

2.6 Major Funding Sources for Aviation Research

In the past, many educational institutions manually searched a selection of sources, from the *Commerce Business Daily* and the *Federal Register*, to periodicals and agency program directories and indexes that were updated on a regular basis. Today, much of this search can be done online, electronically. An array of available technologies can significantly improve the ease of retrieval of information in areas, such as funding opportunities, announcements, forms, and sponsor guidelines. If you have an Internet connection of some type, you can find federal opportunities through Federal Information Exchange Database (FEDIX), an online database retrieval service about government information for college, universities, and other organization. The following agencies are included in the FEDIX database:

1. Department of Energy
2. ONR
3. NASA
4. FAA
5. Department of Commerce
6. Department of Education
7. National Science Foundation
8. National Security Agency
9. Department of Housing and Urban Development
10. Agency of International Development
11. Air Force Office of Scientific Research

A user's guide is available from FEDIX that includes complete information on getting started, including an appendix of program titles and a list of keywords by the agency.

All the government agencies can also be accessed through the Internet. Most colleges and universities provide Internet access. Individuals who require their own service need to subscribe to an Internet provider, such as America Online or CompuServe. Generally, a subscription service fee is paid which may include a specified number of free minutes per month.

In addition to online searches, you may wish to make direct contact with one of the many federal sources for research support. The DoD has typically funded many human-factors programs. Behavioral and social science research and development are referred to as manpower, personnel, training, and human-factors R&D in the DoD.

Although it is beyond the scope of this chapter to review each and every government funding source, the following sources would be of particular interest to those conducting aviation human-factors research. These agencies can be contacted directly for further information.

U.S. Air Force

Air Force Office of Scientific Research
Life Sciences Directorate
Building 410
Bolling Air Force Base
Washington, DC 20332

Armstrong Laboratory
Human Resources Directorate (AL/HR)
7909 Lindbergh Drive
Brooks AFB, TX 78235–5340

Armstrong Laboratory
Crew Systems Directorate (AL/CF)
2610 7th Street
Wright-Patterson, AFB, OH 45433–7901

ASAF School of Aerospace Medicine
ASAFSAM/EDB
Aerospace Physiology Branch
Education Division
USAF School of Aerospace Medicine
Brooks AFB, TX 78235–5301

U.S. Army

Army Research Institute for the Behavioral and Social sciences
5001 Eisenhower Avenue
Alexandria, VA 22233

U.S. Army Research Laboratory
Human Research & Engineering
Directorate ATTN: AMSRL-HR
Aberdeen Proving Ground, MD 21005–5001

U.S. Army Research Institute of Environmental Medicine
Commander
U.S. Army Natick RD&E Center
Building 42
Natick, MA 01760

Walter Reed Army Institute of Research
ATTN: Information Office
Washington, DC 20307–5100

U.S. Army Aeronautical Research Laboratory
P.O. Box 577
Fort Rucker, AL 36362–5000

U.S. Navy

Office of Naval Research
800 North Quincy Street
Arlington, VA 22217–5000

Space Warfare Systems Center
Code D44
53560 Hull Street
San Diego, CA 92152–5001

Naval Air Warfare Center, Aircraft Division
Crew Systems
NAS Patuxent River, MD 20670–5304

Naval Air Warfare Center, Training Systems Division
Human Systems Integration
12350 Research Parkway
Orlando, FL 32826–3224

Naval Air Warfare Center, Weapons Division
Crew Interface Systems
NAS China Lake, CA 93555–6000

Naval Health Research Center
Chief Scientist
P.O. Box 85122
San Diego, CA 92138–9174

Naval Aerospace Medical Research Laboratory
NAS Pensacola, FL 32508–5700

Naval Biodynamics Laboratory
Commanding Officer
P.O. Box 29047
New Orleans, LA 70189–0407

Miscellaneous

National Science Foundation
4201 Wilson Boulevard
Arlington, VA 22230

Federal Aviation Administration Technical Center
Office of Research and Technology Application
Building 270, Room B115
Atlantic City International Airport, NJ 08405

3

Measurement in Aviation Systems

David Meister*
Human Factors Consultant

Valerie Gawron
MITRE

3.1 A Little History†

One cannot understand the measurement in aviation human factors (HF) without knowing a little about its history, which goes back to World War I and even earlier. In that period, new aircraft were tested at flight shows and selected partly on the basis of the pilot's opinion. The test pilots were the great fighter aces, men like Guynemer and von Richtoffen. Such tests were not the tests of the pilot's performance as such, but examined the pilot and his reactions to the aircraft.

Between the wars, HF participation in aviation system research continued (Dempsey, 1985), and the emphasis on the Army Air Force was primarily medical/physiological. For example, researchers using both animals and men studied the effects of altitude and acceleration on human performance. "Angular accelerations were produced by a 20 ft-diameter centrifuge, while a swing was used to produce linear acceleration" (Moroney, 1995). Work on anthropometry in relation to aircraft design began in 1935. As early as in 1937, a primitive G-suit was developed. This was also the period when Edwin Link marketed his flight simulator (which became the grandfather of all later flight simulators) as a coin-operated amusement device.

During World War II, efforts in aircrew personnel selection led to the Air-Crew Classification Test Battery to predict the success in training and combat (Taylor & Alluisi, 1993). The HF specialists were also involved in a wide variety of activities, including determining human tolerance limits for high-altitude bailout, automatic parachute-opening devices, cabin pressurization schedules, pressure-breathing equipment, protective clothing for use at high altitudes, airborne medical evacuation facilities, and ejection seats (Van Patten, 1994). Probably, the best-known researcher during World War II was Paul Fitts, who worked with his collaborators on aircraft controls and displays (Fitts & Jones, 1947).

During the 1950s and 1960s, HF personnel contributed to the accommodation of men in jet and rocket-propelled aircraft. Under the prodding of the new U.S. Air Force, all the engineering companies

---

* It should be noted that our friend and colleague, David Meister, died during the preparation of the second edition and his input was sincerely missed. The chapter was updated by the second author.

† The senior author is indebted to Moroney (1995) for parts of this historical review.

that bided on the development of military aircraft had to increase their staffs to include HF specialists, and major research projects like the Air Force Personnel and Training Research Center were initiated. Although the range of HF investigations in these early days was considered to be limited, Section 3.1.4 of this chapter shows that it has expanded widely.

3.1.1 The Distinctiveness of Aviation HF Measurement

Despite this relatively long history, the following question may arise: Is there anything that specifically differentiates aviation HF measurement from that of other types of systems, such as surface ships, submarines, railroads, tanks, or automobiles? The answer to this question is: Except for a very small number of specific environment-related topics, no, there is not. Except for the physiological areas, such as the topics mentioned in the previous historical section, every topic addressed in aviation HF research is also addressed in connection with other systems.

For example, questions on workload, stress, and fatigue are raised with regard to other transportation and even with nontransportation systems. Questions dealing with such present-day "hot" topics in aviation research as situational awareness (addressed in Chapter 11) and those dealing with the effects of increasing automation (see Chapter 7) are also raised in connection with widely different systems, such as nuclear power plants.

Hence, what is the need for a chapter on measurement in a text on aviation HF? Although the questions and methods are much the same as in other fields, the aircraft is a distinctive system functioning in a very special environment. It is this environment that makes aviation HF measurement important. Owing to this environment, general behavioral principles and knowledge cannot automatically be generalized to the aircraft. Aviation HF measurement emphasizes the *context* in which its methods are employed.

Therefore, this chapter is not based on general psychological measurement, and only sufficient description about the methods employed is provided to enable the reader to understand the way in which the methods are used. We have mentioned statistics and experimental design, but not in detail. Even with such constraints, the scope of aviation HF measurement is very wide; almost every type of method and measure that one finds in the general behavioral literature has been used in investigating aviation issues. These measurements are largely research-oriented, because, although there are nonresearch measurements in aircraft development and testing, they are rarely reported in the literature.

3.1.2 Major Measurement Topics

One of the first questions about measurement is: What topics does this measurement encompass? Given the broad range of aviation HF research, the list that follows cannot be all-inclusive, but it includes the major questions addressed. Owing to space constraints, a detailed description of what is included in each category is not provided, although many of these topics are subjects for subsequent chapters. They are not listed in any particular order of importance, and the references to illustrative research are appended. Of course, each individual study may investigate more than one topic.

1. Accident analysis
   a. Amount of and reasons for pilot error (Pawlik, Simon, & Dunn, 1991)
   b. Factors involved in aircraft accidents and accident investigation (Schwirzke & Bennett, 1991)
2. Controls and displays
   a. The effect of automation on crew proficiency (e.g., the "glass cockpit"; McClumpha, James, Green, & Belyavin, 1991)
   b. Perceptual cues used by flight personnel (Battiste & Delzell, 1991)
   c. Checklists and map formats; manuals (Degani & Wiener, 1993)
   d. Cockpit display and control relationships (Seidler & Wickens, 1992)

e. Air-traffic control (ATC) (Guidi & Merkle, 1993)

f. Unmanned aerial vehicles (Gawron & Draper, 2001)

3. Crew issues

a. Factors leading to more effective crew coordination and communication (Conley, Cano, & Bryant, 1991)

b. Crew health factors, age, experience, and sex differences (Guide & Gibson, 1991)

4. Measurement

a. Effects and methods of predicting pilot workload, stress, and fatigue (Selcon, Taylor, & Koritsas, 1991)

b. Measurement in system development, for example, selection among alternative designs and evaluation of system adequacy (Barthelemy, Reising, & Hartsock, 1991)

c. Situational awareness (see Chapter 11)

d. Methods of measuring pilot performance (Bowers, Salas, Prince, & Brannick, 1992)

5. Selection and training

a. Training, training devices, training-effectiveness evaluation, transfer of training to operational flight (Goetti, 1993)

b. Design and use of simulators (Kleiss, 1993)

c. Aircrew selection, such as determination of factors predicting pilot performance (Fassbender, 1991)

d. Pilot's personality characteristics (Orasanu, 1991)

e. Pilot's decision-making and information processing: flight planning; pilot's mental model (Orasanu, Dismukes, & Fischer, 1993)

f. Evaluation of hand dominance on manual control of aircraft (Gawron & Priest, 1996)

g. Airplane upset training (Gawron, Berman, Dismukes, & Peer, 2003)

6. Stressors

a. Effects of environmental factors (e.g., noise, vibration, acceleration, lighting) on crew performance (Reynolds & Drury, 1993)

b. Effects of drugs and alcohol on pilot performance (Gawron, Schiflett, Miller, Slater, & Ball, 1990)

c. Methods to minimize air sickness (Gawron & Baker, 1994)

d. High *g* environments and the pilot (Gawron, 1997)

e. Psychological factors (Gawron, 2004)

7. Test and evaluation

a. Evaluation of crew proficiency (McDaniel & Rankin, 1991)

b. Evaluation of the human-engineering characteristics of aircraft equipment, such as varying displays and helmets (Aretz, 1991)

c. Lessons learned in applying simulators in crew-station evaluation (Gawron, Bailey, & Lehman, 1995)

3.1.3 Performance Measures and Methods

Aviation HF measurement can be categorized under four method/measure headings: flight performance, nonflight performance, physiological, and subjective. Before describing each category, it may be useful to mention about how to select them. For convenience, we refer to all the methods and measures as *metrics*, although there is a sharp distinction between them. Any individual method can be used with many different measures.

Numerous metric-selection criteria exist, and the most prominent ones are *validity* (how well does the metric measure and predict operational performance) and *reliability* (the degree to which a metric reproduces the same performance under the same measurement conditions consistently). Others include *detail* (does it reflect performance with sufficient detail to permit meaningful analysis?), *sensitivity* (does it reflect significant variations in performance caused by task demands or environment?), *diagnosticity* (does it discriminate among different operator capacities?), *intrusiveness* (does it cause degradation in task performance?), *requirements* (what does it require in system resources to use it?), and personnel *acceptance* (will the test personnel tolerate it?). Obviously, one would prefer a metric that, with all the other things being equal, is *objective* (is not mediated by a human observer) and *quantitative* (capable of being recorded in numerical format). *Cost* is always a significant factor.

It is not possible to make unequivocal judgments of any metric outside the measurement context in which it will be used. However, certain generalizations can be made. With all the other things being equal, one would prefer objective to subjective, and nonphysiological to physiological metrics (because the latter often require expensive and intrusive instrumentation, and in most cases, have only an indirect relationship to performance), although if one is concerned with physiological variables, they cannot be avoided. Any metric that can be embedded in the operator's task and does not degrade the task performance is preferable. The cheaper metric is (less time to collect and analyze data) considered better. Again, with all other factors being equal, data gathered in operational flight or operational environment are preferred than those collected nonoperationally.

3.1.3.1 Flight Performance Metrics

The following paragraph is partly based on the study by Hubbard, Rockway, and Waag (1989). As pilot and aircraft are very closely interrelated as a system, the aircraft state can be used as an indirect measure to determine how the pilot performs in controlling the aircraft. In state-of-the-art simulators and, to a slightly lesser extent, in modern aircraft, it is possible to automatically obtain the measures of aircraft state, such as altitude, deviation from glide slope, pitch roll and yaw rates, airspeed, bank angle, and so forth. In a simulator, it is possible to sample these parameters at designated intervals, such as fractions of a second. The resultant time-series plot is extremely useful in presenting a total picture of what happens to the pilot/aircraft system. This is not a direct measurement of the pilot's arm or hand actions, or the perceptual performance, but is mediated through the aircraft's instrumentation. However, measurement of arm and hand motions or the pilot's visual glances would be perhaps a little too molecular and probably would not be measured, except under highly controlled laboratory conditions. The reader can refer to Chapter 14 that discusses the capabilities of the simulator in measurement of aircrew performance. Measurement within the operational aircraft has been much expanded, as aircraft such as the F-16, have become highly computer-controlled.

As the pilot controls the aircraft directly, it is assumed that deviations from specified flight performance requirements (e.g., a given altitude, a required glide slope) represent errors directly attributable to the pilot, although one does not obviously measure the pilot's behavior (e.g., hand tremor) directly. This assumes that the aircraft has no physical malfunctions that would impact the pilot's performance.

In the case where the pilot is supposed to react to a stimulus (e.g., a topographic landmark) appearing during the flight scenario, the length of the time that the pilot takes to respond to that stimulus is also indicative of the pilot's skill. Reaction time and response duration measures are also valuable in measuring the pilot's performance.

The time-series plot may resemble a curve with time represented horizontally and aircraft state shown vertically. Such a plot is useful in determining when and for how long a particular parameter is out of bounds. Such plots can be very useful in a simulator when a stimulus condition like wind gust or aircraft malfunction is presented; the plot indicates how the pilot has responded. In pilot training, these plots can be used as feedback for debriefing the students.

In the study on flight performance, researchers usually compute summary measures based on data that have been sampled in the course of the flight. This is necessary, because large amounts of data must be reduced to a number that can be more readily handled. Similarly, the flight course is characteristically broken up into segments based on the tasks to be performed, such as straight and level portions, ridge crossings, turns, and so on. Subsequently, one can summarize the pilot's performance within the designated segment of the course.

One of the most common summary metrics is root mean square error (RMSE), which is computed by taking the square root of the average of the squared error or deviation scores. A limitation of RMSE is that the position information is lost. However, this metric is often used. Two other summary metrics are the mean of the error scores (ME) and the standard deviation of those scores (SDE). The RMSE is completely defined by ME and SDE, and according to Hubbard et al. (1989), the latter are preferred because RMSE is less sensitive to differences between the conditions and more sensitive to measurement bias.

There are many ways to summarize the pilot's performance, depending on the individual mission goals and pilot's tasks. In air-combat maneuvering, for example, the number of hits and misses of the target and miss distance may be based on the nature of the mission. The method and measure selected are determined by the questions that the investigator asks. However, it is possible, as determined by Stein (1984), to develop a general-purpose pilot performance index. This is based on the subject experts and is revised to eliminate those measures that failed to differentiate experienced from novice pilots. Another example is from a study evaluating airplane upset recovery training methods (Gawron, 2002) (see Table 3.1). One can refer to Berger (1977) and Brictson (1969) for examples of studies in which flight parameters were used as measures to differentiate different conditions.

TABLE 3.1 Measures to Evaluate Airplane Upset Training Methods

Number	Data	Definition
1	Time to first rudder input	Time from start-event marker to change in the rudder position
2	Time to first throttle input	Time from start-event marker to change in the throttle
3	Time to first wheel column input	Time from start-event marker to change in the wheel column position
4	Time to first autopilot input	Time from start-event marker to change in the autopilot disengagement
5	Time to first input	Shortest of measures 1–4
6	Time to first correct rudder input	Time from start-event marker to change in the rudder position
7	Time to first correct throttle input	Time from start-event marker to change in the throttle
8	Time to first correct wheel column input	Time from start-event marker to change in the wheel column position
9	Time to recover	Time from start-event marker to end-event marker
10	Altitude loss	Altitude at start time minus altitude at wings level
11	Procedure used to recover the aircraft	Video of evaluation pilot's actions from start-event marker to end-event marker
12	Number of correct actions in recovery	Sum of the number of correct actions executed in the correct sequence
13	Number of safety trips tripped (per flight)	Number of the safety trips tripped summed across each evaluation pilot (including safety pilot trips)
14	Number of correct first inputs	Number of correct first inputs summed across each of the five groups
15	Number of first correct pitch inputs	Number of first correct pitch inputs summed across each of the five groups
16	Number of first correct roll inputs	Number of first correct roll inputs summed across each of the five groups
17	Number of first correct throttle inputs	Number of first correct throttle inputs summed across each of the five groups

Source: Gawron, V.J., Airplane upset training evaluation report (NASA/CR-2002-211405). National Aeronautics and Space Administration, Moffett Field, CA, May 2002.

The crew-station evaluation process is not standardized, with a variety of metrics and procedures being used (Cohen, Gawron, Mummaw, & Turner, 1993). As a result, data from one flight test are often not comparable with those of another. A computer aided engineering (CAE) system was developed to provide both standardized metrics and procedures. This system, the Test Planning, Analysis and Evaluation System, or Test PAES, provides various computerized tools to guide the evaluation personnel, who, in many cases, are not measurement specialists. The tools available include a measures database, sample test plans and reports, questionnaire development and administration tools, data-analysis tools, multimedia data analysis and annotation tools, graphics, and statistics as well as a model to predict system performance in the field based on simulation and test data.

3.1.3.2 Nonflight Performance Metrics

Certain performances are not reflected in aircraft state. For example, the aircrew may be required to communicate on takeoff or landing with ATC, to use a radar display or direct visualization to detect possible obstacles, or to perform contingency planning in the event of an emergency. Each such nonflight task generates its own metric. Examples include content analysis of communications or speed of the target detection/acquisition or number of correct target identifications.

All flight performance metrics must be collected during an operational or a simulator flight; nonflight metrics can be used at any time during an operational or simulated flight following that flight (on the ground), or can be used in a nonflight environment, such as a laboratory. Some nonflight metrics are related to flight, but do not measure a specific flight. An example is a summary measure of effectiveness, such as the number of flights or other actions performed by the pilot to achieve some sort of criterion (mostly in training). In the study of map displays or performance of map-of-the earth helicopter flight, the pilot may be asked to draw a map or make time or velocity estimates. Researchers have developed extensive lists of measures (Gawron, 2002; Meister, 1985) from which one can select those that appear appropriate for the task to be measured. Review of the papers in the literature of aviation psychology (see the references at the end of this chapter) may suggest others.

The metrics referred to so far are an integral part of the flight task, but there are also those that are not, which are used purely for research purposes, and therefore, are somewhat artificial. The emphasis on pilot workload studies during the 1980s, for example, created a great number of subjective workload metrics (see Chapter 7). Besides the well-known scales such as subjective workload assessment technique (SWAT) or task load index (TLX) (Vidulich & Tsang, 1985), which require the pilot to rate his or her own performance, there are other scales that demand the pilots to perform a second task (in addition to those required for flight), such as sort cards, solve problems, make a choice reaction, or detect a specific stimulus event. The problem that one faces with secondary tasks is that in the actual flight situation, they may cause deterioration of performance in the primary flight task, which could be dangerous. This objection may not be pertinent in a flight simulator. In general, any secondary task that distracts the pilot from flight performance is undesirable in actual flight.

Performance measures taken after the flight is completed, or where a copilot takes the controls while the pilot performs a research task, are safer. Measurement of flight performance variables is usually accomplished by sensors linked to a computerized data collection system. Such instrumentation is not available for measurement of nonflight performance variables. The following is a description of the instrumentation that could be particularly useful for aviation HF variables.

Although there are many instruments that can measure human performance variables and the measurement environment (e.g., photometer, thermometer, sound-level meter, vibration meter, and analyzer; American Institute for Aerospace and Aeronautics, 1992, describes these in more detail), two are of particular interest for us. The accelerometer, such as a strain gauge or piezoelectric-force transducer, is a device that measures the acceleration along one or more axes. Obviously, such a device would be necessary for any study of G-forces. However, more commonly used device is the video recorder, which is becoming increasingly popular for providing records of visual and audio-operator performance for posttest analysis. A complete system includes a camera, video recorder, and monitor (Crites, 1980).

3.1.3.3 Physiological Measures

Only a relatively small percentage of aviation HF studies use physiological instrumentation and measures, because such measures are useful only when the variables being studied involve a physiological component. In particular, studies involve acceleration (McCloskey, Tripp, Chelette, & Popper, 1992), hypoxia, noise level, fatigue (Krueger, Armstrong, & Cisco, 1985), alcohol, drugs, and workload. One of the most complete reviews of physiological measures is a North Atlantic Treaty Organization (NATO) report edited by Caldwell, Wilson, Centiguc, Gaillard, Gundel, Legarde, Makeig, Myhre, and Wright (1994).

Table 3.2 from the work by Meister (1985) lists the physiological measures associated with the major bodily systems. Heart rate and heart-rate variability have been the most commonly used physiological assessment methods, primarily because they are relatively nonintrusive and portable devices for recording these data are available. These metrics have been employed in a number of in-flight studies involving workload (Hart & Hauser, 1987; Hughes, Hassoun, Ward, & Rueb, 1990; Wilson & Fullenkamp, 1991; Wilson, Purvis, Skelly, Fullenkamp, & Davis, 1987). Itoh, Hayashi, Tsukui, and Saito (1989) and Shively Battiste, Matsumoto, Pepitone, Bortolussi, and Hart (1987) have demonstrated that heart-rate variability can discriminate differences in the workload imposed by flight tasks.

Nevertheless, all these metrics have certain disadvantages. Many of them require intrusive instrumentation, which may not be acceptable in an actual flight environment. However, they are more supportable in a simulator. For example, in a simulator or study of helicopter crew performance, stress, and fatigue over a week-long flight schedule, Krueger et al. (1985) had three electrocardiogram chest electrodes wired to a monitoring system to assess the heart rate and heart-rate variability as indicators of alertness. Oral temperatures were taken at approximately 4 h intervals, and urine specimens (for catecholamines) were provided at 2 h intervals between the flights. Illustrative descriptions of physiological studies in the flight simulator have also been provided by Morris (1985), Armstrong (1985), and Lindholm and Sisson (1985).

Unfortunately, the evidence for the relationship between physiological and performance indices is at best, ambiguous. Often, the meaning of such a relationship, even when it is documented, is unclear. Moreover, the sensitivity of these metrics to possible contaminating conditions, for example, ambient temperature, is very high.

TABLE 3.2 Physiological Measures of Workload

System	Measure
Cardiovascular system	* Heart rate
	* Heart-rate variability (sinus arrhythmia)
	* Blood pressure
	Peripheral blood flow
	* Electrical changes in skin
Respiratory system	* Respiration rate
	Ventilation
	Oxygen consumption
	Carbon dioxide estimation
Nervous system	* Brain activity
	* Muscle tension
	* Pupil size
	Finger tremor
	Voice changes
	Blink rate
Biochemistry	* Catecholamines

Note: Those measures most commonly used have been indicated by an asterisk.

3.1.3.4 Subjective Measures

Subjective measures (whatever one may think about their validity and reliability) have always been and still are integral parts of aviation HF measurement. As mentioned previously, during World War I, ace fighter pilots like Guynemer and von Richtoffen were employed to evaluate the handling qualities of prototype aircraft. Ever since the first aviation school was established, expert pilots have been used not only to train, but also to evaluate the performance of their students. Even with the availability of sophisticated, computerized instrumentation in the test aircraft, the pilot is routinely asked to evaluate handling qualities. Automated performance measurement methods, although highly desirable, cannot entirely replace subjective techniques (Vreuls & Obermayer, 1985).

Muckler (1977) pointed out that all measurement is subjective at some point in test development; the objective/subjective distinction is a false issue. Therefore, the problem is to find ways to enhance the adequacy of the subjective techniques. There is need for more research to develop more adequate methods, to train and calibrate expert observers.

The subjective techniques described in the research literature include interviews, questionnaire surveys, ratings and rankings, categorization, and communications analyses. Subjective data, particularly ratings, are characteristically used to indicate pilot preference, performance evaluations, task difficulty, estimates of distance traveled or velocity, and, in particular, workload, which is one of the "hot" topics in aviation HF research.

Owing to the variability in these subjective techniques, efforts have been made to systematize them quantitatively in scales of various sorts (for a discussion of scales, see Meister, 1985 or Gawron, 2000). The Likert 5-point scale (e.g., none, some, much, very much, all) is a very common scale that can be created in moments, even by someone who is not a psychometrician. However, the validity of such self-created scales may be susceptible. Development of valid and reliable scales requires prior research on the dimensions of the scale, and empirical testing and analysis of the test results. Most complex phenomena cannot be scaled solely on the basis of a single dimension, because most behavior of any complexity is multidimensional. The interest in measurement of workload, for example, has created a number of multidimensional scales: SWAT, which has been used extensively in simulated and actual flight (see American Institute of Aeronautics and Astronautics, 1992, pp. 86–87), has three scalar dimensions: time load, mental effort load, and psychological stress. The scales, either individually or as a part of the questionnaire surveys, have probably been used more frequently as a subjective measurement device than any other technique, as it is difficult to quantize interviews, except as part of formal surveys, in which case they turn into rating/ranking scales.

3.1.4 Characteristics of Aviation HF Research

What has been described so far is somewhat abstract and only illustrative. One may wonder how can one describe the aviation HF measurement literature as a whole?

One way to answer this question is to review the recent literature in this area. The first author examined the *Proceedings of the Human Factors and Ergonomics Society* (HFES) in 1990, 1991, 1992, and 1993, and the journal that the society publishes, *Human Factors*, for the same period, for all the studies of aviation HF variables. To check on the representativeness of these two sources, the 1991 *Proceedings of the International Symposium on Aviation Psychology*, sponsored by Ohio State University (OSU), were examined. One hundred and forty-four relevant papers were found in the HFES *Proceedings* and the journal, and 87 papers were found in the OSU *Proceedings*. Only papers that described specific measurement were included in the sample. Those that were reviews of the previous measurement research or described the prospective research were excluded. Those papers selected as relevant were content-analyzed by applying seven taxonomies:

1. General topic, such as flight, navigation, design, workload
2. Specific topic, such as situational awareness

3. Measures employed, such as tracking error, reaction time
4. Measurement venue, such as laboratory, simulator, operational flight
5. Type of subject, such as pilot, air-traffic controllers, nonflying personnel
6. Methodology, such as experiment, questionnaire, observation, incident reports
7. Statistical analysis employed, such as analysis of significance of differences, correlation, factor analysis, etc.

Owing to space constraints, the listing of all the taxonomic categories employed is not provided, because of their large number. The categories were developed on the basis of the individual papers themselves. The numbers by category are: general topic (47); specific topic (71); measures (44); measurement venue (8); subject type (12); methodology (16); and statistical analysis (16). The categories were not mutually exclusive. Every category that could describe a particular paper was counted. For example, if a paper dealt with instrument scanning and in the process, described the visual factors involved in the scanning, both the categories were counted. Thus, categories overlapped, but the procedure employed resulted in a more detailed measurement picture, than would otherwise be the case. Only those categories that described 5% or more of the total number of papers are listed in the following tables. As the number of these categories is small when compared with the total number of categories reported, it is apparent that although aviation HF measurement is extensive in its subject and its tools, it is not very intensive, except in relatively few areas. These presumably are the areas that most excite the funding agencies and individual researchers.

An analysis was performed to ensure that the two data sources (HFES and OSU) were not so different such that they could not be combined. Roughly, the same data patterns could be discerned (broad but not intensive), although there were some differences of note. For example, the OSU sample dealt much more with flight-related topics than HFES (OSU 72%, HFES 35%). Such differences could be expected, because the two sources were drawn from different venues (e.g., OSU is international, HFES almost exclusively American; OSU preselects its topic areas, HFES does not). Therefore, the differences were not considered sufficient to make combination impossible.

Of the 47 categories under "general topic," 13 met the 5% criterion. These are listed in Table 3.3, which indicates that most of the researches were basic. This means that the researches dealt with general principles rather than specific applications. Applied researches (see Table 3.4) were only 11% of the total number of researches. Both basic and applied researches totaled to 91%. The fact that the figures do not add to 100% simply indicates that a small number of papers, although dealing with measurement, did not involve empirical research. The second point is that only half the papers presented dealt directly with flight-related topics; the others involved activities incident to or supportive of the flight, but not directly the flight. For example, 10% of the papers dealt with ATC, which is of course necessary for aviation, but which has its own problems.

TABLE 3.3   General Topic Categories

1.	Military or commercial flight	50%	113 papers
2.	Design	10%	23 papers
3.	Workload/stress	8%	17 papers
4.	Air-traffic control	10%	23 papers
5.	Training	14%	32 papers
6.	Automation	8%	18 papers
7.	Basic research	80%	189 papers
8.	Instrument scanning	7%	16 papers
9.	Visual factors	9%	20 papers
10.	Evaluation	6%	13 papers
11.	Accidents	6%	14 papers
12.	Applied research	11%	25 papers
13.	Pilot personality	5%	12 papers

TABLE 3.4   Specific Topic Categories

1.	Display design/differences	21%	50 papers
2.	Transfer of training	5%	11 papers
3.	Personnel error	6%	14 papers
4.	Personnel demographics	5%	12 papers
5.	Perceptual cues	16%	36 papers
6.	Decision-making	6%	13 papers
7.	Workload	14%	33 papers
8.	Communications	6%	14 papers
9.	Coding	5%	11 papers
10.	Tracking	9%	21 papers
11.	Crew coordination	5%	12 papers
12.	Incidents	6%	14 papers
13.	Head-up displays (HUD)/ helmet-mounted displays (HMD)	5%	12 papers
14.	Mental model	8%	17 papers
15.	Dual tasks	6%	13 papers
16.	Cognition	6%	13 papers

Table 3.4 lists the 16 specific topics that were most descriptive of the papers reviewed. As one can see, only 16 categories out of the 71 met the 5% criterion. Although the table reveals a wide assortment of research interests, only three, namely, display design/differences, perceptual cues (related to display design), and workload, are described in a relatively large number of papers.

Table 3.5 describes the measures employed by researchers. Of the 44 measures found, only 10 satisfied the 5% criterion. Of course, many studies included more than one type of measure. Obviously, error and time are the most common measures. The frequency and percentage of measures was the most common statistical treatment of these measures. The relatively large number of ratings of, for example, attributes, performance, preferences, similarity, difficulty, and so on, attest to the importance of subjective measures, particularly when these are used in a workload measurement context (e.g., SWAT, TLX).

Table 3.6 describes about where the measurements took place. Of the nine categories, five met the 5% criterion. This is because, a laboratory does not simulate any of the characteristics of the flight; however, a full-scale simulator with at least two degrees of motion may achieve this. Furthermore, a part-task simulator or simulated display reproduces some part of the cockpit environment. In addition, some measures were taken in-flight. In the case where the measurement venue is unimportant, the situation was usually one in which questionnaire surveys were administered by mail or elsewhere.

There is great reliance on flight simulators, both full-scale and part-task, but in many cases, there exists no flight relationship at all (e.g., the laboratory). The fact that only 26 of the 231 papers dealt with the actual

TABLE 3.5   Measures Employed

1.	Reaction time	13%	31 papers
2.	Response duration	16%	48 papers
3.	Response error	33%	76 papers
4.	Tracking error	12%	29 papers
5.	Frequency, percentage	33%	80 papers
6.	Ratings	30%	66 papers
7.	Interview data	5%	11 papers
8.	Workload measure	8%	18 papers
9.	Flight performance variables	10%	22 papers
10.	Categorization	8%	17 papers

TABLE 3.6 Measurement Venue

1.	Laboratory (not simulator)	16%	36 papers
2.	Full-scale simulator	23%	52 papers
3.	Part-task simulator or simulated displays	27%	63 papers
4.	Operational flight	11%	26 papers
5.	Irrelevant	16%	46 papers

flight environment in the air is somewhat surprising, because measurements taken outside that environment are inevitably artificial to a greater or lesser extent. Of the 12 categories describing the type of subject used in these studies, only three were significant: 60% of the subjects were pilots (140 papers), 33% (75 papers) of the subjects were nonflying personnel (college students, government workers, the general public), and 9% (20 papers) were air-traffic controllers. The fact that the largest proportion of the subjects is pilots is not at all surprising, but the relatively large number of nonflying personnel is somewhat daunting.

Nine of the 16 categories under the heading of methodology (Table 3.7) met the 5% criterion. As one would expect, more than half the number of papers published were experimental in nature. What was somewhat less expected was the large number of studies that were not experimental, although there was some overlap, because some of the experimental studies did make use of nonexperimental methodology in addition to the experiment. There was heavy reliance on subjective techniques, observation, questionnaires, interviews, and self-report scales. Pilot opinion was, as it has always been, extremely important in aviation.

Of the 16 statistical analysis categories, 4 were most frequently employed (Table 3.8). Again, as one would expect, the tests of the significance of differences between the conditions or groups were observed in most of the analyses. The percentage might have even been greater if one included such tests as multiple regression, discriminant analysis, or factor analysis in this category. Although the categories in this content area tend to overlap, the relatively large number of studies in which the analysis stopped at frequency and percentage should be noted.

What does this review tell us about the nature of aviation HF research? The large number of topics, both general and specific, ranging from information processing to geographical orientation, electroencephalography, and pilot attitudes (note: only a few topics taken at random), indicates that many

TABLE 3.7 Methodology

1.	Experiment	54%	126 papers
2.	Observation	12%	29 papers
3.	Questionnaire survey	16%	48 papers
4.	Rating/ranking scale	30%	65 papers
5.	Performance measurement (general)	21%	50 papers
6.	Interviews	10%	22 papers
7.	Physical/physiological data recording	8%	17 papers
8.	Analysis of incident reports	8%	17 papers
9.	Verbal protocol analysis	5%	11 papers

TABLE 3.8 Statistical Analysis

1.	Tests of significance of differences	67%	155 papers
2.	Correlation	70%	22 papers
3.	Frequency, percentage	24%	56 papers
4.	None	5%	12 papers

areas have been examined, but very few have been studied intensively. The major concerns are the basic research, as it relates to flight and displays. In spite of the fact that presumably automation (the "glass cockpit"), situational awareness, and workload are all "hot" topics in the aviation research community, they received only a modest degree of attention. If one adds up all the topics that deal with sophisticated mental processes (e.g., decision-making, mental models, and cognition) along with crew coordination, it can be observed that a fair bit of attention is being paid to the higher-order behavioral functions. This represents some change from the earlier research areas.

Most of the behavioral research in aviation is conducted on the ground, for obvious reasons: non-availability of aircraft and cost of flights. Another reason is perhaps that much of the research deals with cockpit or display variables, which may not require actual flight. Reliance on opinion expressed in questionnaires, incident/accident reports, and full-scale simulators diminishes the need to measure in the actual flight. It may also reflect the fact that behavioral research, in general (not only in aviation), rarely takes place in the operational environment, which is not conducive to sophisticated experimental designs and instrumentation. However, this leaves us with the question on whether results achieved on the ground (even with a high degree of simulation) are actually valid with respect to flight conditions. Case studies comparing the ground and in-flight evaluations have been carried out by Gawron and Reynolds (1995). The issue of generalizability to flight is compounded by the fact that one-third of all the subjects employed in these studies were not flying personnel.

The HF research in aviation is not completely devoted to an experimental format; only half the studies reported were of this type. It is remarkable that with a system whose technology is so advanced, there is so much reliance on nonexperimental techniques and subjective data.

3.1.5 Summary Appraisal

This review of the aviation HF literature suggests that future research should endeavor to concentrate on key issues to a greater extent than in the past. "Broad but shallow" is not a phrase one would wish to describe that research in general. One of the key issues in aviation HF research (as it should be in general behavioral research as well) is that of the effects of automation on human performance. It seems inevitable that technological sophistication will increase in the coming century and that some of that sophistication will be represented on the flight deck. Its effects are not uniformly positive, and hence, the match between human and the computer in the air must be explored more intensively.

Another recommendation based on the literature review is that the results achieved in the simulator should be validated in the air. Simulators have become highly realistic, but they may lack certain features that can be found only in-flight. The frequency with which part-task simulators and laboratories are used in aviation HF research makes one wonder whether the same effects will be precisely found in actual flight. It is true that in behavioral research as a whole, there is little validation in the operational context of effects found in the laboratory, but flight represents a critically distinct environment in which most aviation behavioral studies are conducted, as shown in the case studies by Gawron and Reynolds (1995).

A similar recommendation refers to test subjects. Although it is true that the majority of the subjects in the studies reviewed were pilots, it is somewhat disturbing to see the large number of nonflying personnel who were also used for this purpose. It is true that almost all nonpilots were employed as subjects in nonflight studies, such as those of displays, but if one believes that the experience of piloting is a distinctive one, it is possible that such experience generalizes to and subtly modifies the nonpiloting activities. In any event, this issue must be addressed in empirical research.

Finally, we noted that the highest percentage of studies dealt with flight variables, and this is quite appropriate. However, the comparative indifference to other aviation aspects is somewhat disturbing. In recent years, increasing attention is being given to ground maintenance in the aviation research, but proportionately, this area, although critical to flight safety, is underrepresented. However, ATC has been observed to receive more attention, probably because of the immediacy of the relationships between ATC personnel and pilots. We would recommend a more intensive examination of how well the ground maintainers function and the factors that affect their efficiency, and a good start can be made from the Aviation Maintenance Human Factors Program at the Federal Aviation Administration (Krebs, 2004). Furthermore, a little more attention to flight attendants and passengers too, may also be necessary. Though the role of the passenger in flight is a very passive one, on long-distance flights, particularly, the constraints involved in being a passenger are very evident.

References

American Institute of Aeronautics and Astronautics. (1992). *Guide to human performance measurement* (Rep. No. BSR/AIAA, G-035-1992). New York: Author.

Aretz, A. J. (1991). The design of electronic map displays. *Human Factors, 33,* 85–101.

Armstrong, G. C. (1985). Computer-aided analysis of in-flight physiological measurement. *Behavior Research Methods, Instruments, & Computers, 17,* 183–185.

Barthelemy, K. K., Reising, J. M., & Hartsock, D. C. (1991, September). Target designation in a perspective view, 3-D map using a joystick, hand tracker, or voice. *Proceedings of the Human Factors and Engineering Society* (pp. 97–101). San Francisco, CA.

Battiste, V., & Delzell, S. (1991, June). Visual cues to geographical orientation during low-level flight. *Proceedings of the Symposium on Aviation Psychology* (pp. 566–571). Columbus: Ohio State University.

Berger, R. (1977, March). *Flight performance and pilot workload in helicopter flight under simulated IMC employing a forward looking sensor* (Rep. No. AGARD-CP-240). *Proceedings of the Guidance and Control Design Considerations for Low-Altitude and Terminal-Area Flight.* Neuilly-sur-Seine, France: AGARD.

Bowers, C., Salas, E., Prince, C., & Brannick, M. (1992). Games teams play: A method for investigating team coordination and performance. *Behavior Research Methods, Instruments, & Computers, 24,* 503–506.

Brictson, C. A. (1969, November). *Operational measures of pilot performance during final approach to carrier landing* (Rep. No. AGARD-CP-56). *Proceedings of the Measurement of Aircrew Performance-The Flight Deck Workload and its Relation to Pilot Performance.* Neuilly-sur-Seine, France: AGARD.

Caldwell et al. (Eds.) (1994). *Psychophysiological assessment methods* (Report No. AGARD-AR-324). Neuilly-sur-Seine, France: NATO Advisory Group for Aerospace Research and Development.

Cohen, J. B., Gawron, V. J., Mummaw, D. A., & Turner, A. D. (1993, June). Test planning, analysis and evaluation system (Test PAES), a process and tool to evaluate cockpit design during flight test. *Proceedings of the Symposium on Aviation Psychology* (pp. 871–876). Columbus: Ohio State University.

Conley, S., Cano, Y., & Bryant, D. (1991, June). Coordination strategies of crew management. *Proceedings of the Symposium on Aviation Psychology* (pp. 260–265). Columbus: Ohio State University.

Crites, D. C. (1980). Using the videotape method. In *Air force systems command design handbook DH-1-3,* Part 2, Series 1-0, General Human Factors Engineering, Chapter 7, Section DN 7E3 (pp. 1–6). Washington, DC: U.S. Government Printing Office.

Degani, U., & Wiener, E. L. (1993). Cockpit checklists: Concept, design, and use. *Human Factors, 35,* 345–359.

Dempsey, C. A. (1985). *50 years of research on man in flight.* Dayton, OH: Wright-Patterson AFB, U.S. Air Force.

Fassbender, C. (1991, June). Culture-fairness of test methods: Problems in the selection of aviation personnel. *Proceedings of the Symposium on Aviation Psychology* (pp. 1160–1168). Columbus: Ohio State University.

Fitts, P. M., & Jones, R. E. (1947). *Psychological aspects of instrument display. I. Analysis of 270 "pilot-error" experiences in reading and interpreting aircraft instruments* (Rep. No. TSEAA-694-12A). Dayton, OH: Aeromedical Laboratory, Air Materiel Command.

Gawron, V. J. (1997, April). High-g environments and the pilot. *Ergonomics in Design: The Quarterly of Human Factors Applications, 6,* 18–23.

Gawron, V. J. (2000). *Human performance measures handbook*. Mahwah, NJ: Lawrence Erlbaum.

Gawron, V. J. (2002, May). *Airplane upset training evaluation report* (NASA/CR-2002-211405). Moffett Field, CA: National Aeronautics and Space Administration.

Gawron, V. J. (2004). Psychological factors. In F. H. Previc, & W. R. Ercoline (Eds.), *Spatial disorientation in aviation* (pp. 145–195). Reston, VA: American Institute of Aeronautics and Astronautics, Inc.

Gawron, V. J., Bailey, R., & Lehman, E. (1995). Lessons learned in applying simulators to crewstation evaluation. *The International Journal of Aviation Psychology, 5*(2), 277–290.

Gawron, V. J., & Baker, J. C. (1994, March). A procedure to minimize airsickness. *Association of Aviation Psychologists Newsletter, 19*(1), 7–8.

Gawron, V. J., Berman, B. A., Dismukes, R. K., & Peer, J. H. (2003, July/August). New airline pilots may not receive sufficient training to cope with airplane upsets. *Flight Safety Digest*, pp. 19–32.

Gawron, V. J., & Draper, M. (2001, December). Human dimension of operating manned and unmanned air vehicles. *Research and Technology Organisation Meeting Proceedings 82 Architectures for the Integration of Manned and Unmanned Aerial Vehicles (RTO-MP-082)*, Annex F. Neuilly-sur-Seine, France: North Atlantic Treaty Organization.

Gawron, V. J., & Priest, J. E. (1996). Evaluation of hand-dominance on manual control of aircraft, *Proceedings of the 40th Annual Meeting of the Human Factors and Ergonomics Society* (pp. 72–76). Philadelphia.

Gawron, V. J., & Reynolds, P. A. (1995). When in-flight simulation is necessary. *Journal of Aircraft, 32*(2), 411–415.

Gawron, V. J., Schiflett, S. G., Miller, J. C., Slater, T., & Ball, J. E. (1990). Effects of pyridostigmine bromide on in-flight aircrew performance. *Human Factors, 32*, 79–94.

Goetti, B. P. (1993, October). Analysis of skill on a flight simulator: Implications for training. *Proceedings of the Human Factors Society* (pp. 1257–1261). Seattle, WA.

Guide, P. C., & Gibson, R. S. (1991, September). An analytical study of the effects of age and experience on flight safety. *Proceedings of the Human Factors Society* (pp. 180–183). San Francisco, CA.

Guidi, M. A., & Merkle, M. (1993, October). Comparison of test methodologies for air traffic control systems. *Proceedings of the Human Factors Society* (pp. 1196–1200). Seattle, WA.

Hart, S. G., & Hauser, J. R. (1987). Inflight applications of three pilot workload measurement techniques. *Aviation, Space and Environmental Medicine, 58*, 402–410.

Hubbard, D. C., Rockway, M. R., & Waag, W. L. (1989). Aircrew performance assessment. In R. S. Jensen (Ed.), *Aviation psychology* (pp. 342–377). Brookfield, IL: Gower Technical.

Hughes, R. R., Hassoun, J. A., Ward, G. F., & Rueb, J. D. (1990). *An assessment of selected workload and situation awareness metrics in a part-mission simulation* (Rep. No. ASD-TR-90-5009). Dayton, OH: Wright-Patterson AFB, Aeronautical Systems Division, Air Force Systems Command.

Itoh, Y., Hayashi, Y., Tsukui, I., & Saito, S. (1989). Heart rate variability and subjective mental workload in flight task validity of mental workload measurement using H.R.V. method. In M. J. Smith, & G. Salvendy (Eds.), *Work with computers: Organizational, management stress and health aspects* (pp. 209–216). Amsterdam, the Netherlands: Elsevier.

Kleiss, J. A. (1993, October). Properties of computer-generated scenes important for simulating low-altitude flight. *Proceedings of the Human Factors Society* (pp. 98–102). Seattle, WA.

Krebs, W. K. (2004). *Aviation maintenance human factors program review*. Washington, DC: Federal Aviation Administration, www.hf.faa.gov/docs/508/docs/AvMaint04.pdf

Krueger, G. P., Armstrong, R. N., & Cisco, R. R. (1985). Aviator performance in week-long extended flight operations in a helicopter simulator. *Behavior Research Methods, Instruments, & Computers, 17*, 68–74.

Lindholm, E., & Sisson, N. (1985). Physiological assessment of pilot workload in simulated and actual flight environments. *Behavior Research Methods, Instruments, & Computers, 17*, 191–194.

McCloskey, K. A., Tripp, L. D., Chelette, T. L., & Popper, S. E. (1992). Test and evaluation metrics for use in sustained acceleration research. *Human Factors, 34*, 409–428.

McClumpha, A. J., James, M., Green, R. C., & Belyavin, A. J. (1991, September). Pilots's attitudes to cockpit automation. *Proceedings of the Human Factors Society* (pp. 107–111). San Francisco, CA.

McDaniel, W. C., & Rankin, W. C. (1991). Determining flight task proficiency of students: A mathematical decision aid. *Human Factors, 33,* 293–308.

Meister, D. (1985). *Behavioral analysis and measurement methods.* New York: Wiley.

Moroney, W. R. (1995). Evolution of human engineering: A selected review. In J. Weimer (Ed.), *Research techniques in human factors.* Englewood Cliffs, NJ: Prentice-Hall.

Morris, T. L. (1985). Electroocculographic indices of changes in simulated flying performance. *Behavior Research Methods, Instruments, & Computers, 17,* 176–182.

Muckler, F. A. (1977). Selecting performance measures: "Objective" versus "subjective" measurement. In L. T. Pope, & D. Meister (Eds.), *Productivity enhancement: Personnel performance assessment in navy systems* (pp. 169–178). San Diego, CA: Naval Personnel Research and Development Center.

Orasanu, J. (1991, September). Individual differences in airline captains' personalities, communication strategies, and crew performance. *Proceedings of the Human Factors Society* (pp. 991–995). San Francisco, CA.

Orasanu, J., Dismukes, R. K., & Fischer, U. (1993, October). Decision errors in the cockpit. *Proceedings of the Human Factors Society* (pp. 363–367). Seattle, WA.

Pawlik, E. A., Sr., Simon, R., & Dunn, D. J. (1991, June). Aircrew coordination for Army helicopters: Improved procedures for accident investigation. *Proceedings of the Symposium on Aviation Psychology* (pp. 320–325). Columbus: Ohio State University.

Reynolds, J. L., & Drury, C. G. (1993, October). An evaluation of the visual environment in aircraft inspection. *Proceedings of the Human Factors Society* (pp. 34–38). Seattle, WA.

Schwirzke, M. F. J., & Bennett, C. T. (1991, June). A re-analysis of the causes of Boeing 727 "black hole landing" crashes. *Proceedings of the Symposium on Aviation Psychology* (pp. 572–576). Columbus: Ohio State University.

Seidler, K. S., & Wickens, C. D. (1992). Distance and organization in multifunction displays. *Human Factors, 34,* 555–569.

Selcon, S. J., Taylor, R. M., & Koritsas, E. (1991, September). Workload or situational awareness?: TLX vs. SART for aerospace systems design evaluation. *Proceedings of the Human Factors Society* (pp. 62–66). San Francisco, CA.

Shively, R. et al. (1987, June). Inflight evaluation of pilot workload measures for rotorcraft research. *Proceedings of the Symposium on Aviation Psychology* (pp. 637–643). Columbus: Ohio State University.

Stein, E. S. (1984). *The measurement of pilot performance: A master-journeyman approach* (Rep. No. DOT/FAA/CT-83/15). Atlantic City, NJ: Federal Aviation Administration Technical Center.

Taylor, H. L., & Alluisi, E. A. (1993). Military psychology. In V. S. Ramachandran (Ed.), *Encyclopedia of human behavior* (pp. 503–542). San Diego, CA: Academic Press.

Van Patten, R. E. (1994). *A history of developments in aircrew life support equipment, 1910–1994.* Dayton, OH: SAFE-Wright Brothers Chapter.

Vidulich, M. A., & Tsang, P. S. (1985, September). Assessing subjective workload assessment: A comparison of SWAT and the NASA-bipolar methods. *Proceedings of the Human Factors Society* (pp. 71–75). Baltimore, MD.

Vreuls, D., & Obermayer, R. W. (1985). Human-system performance measurement in training simulators. *Human Factors, 27,* 241–250.

Wilson, G. F., & Fullenkamp, F. T. (1991). A comparison of pilot and WSO workload during training missions using psychophysical data. *Proceedings of the Western European Association for Aviation Psychology, R* (pp. 27–34). Nice, France.

Wilson, G. F., Purvis, B., Skelly, J., Fullenkamp, F. T., & Davis, L. (1987, October). Physiological data used to measure pilot workload in actual and simulator conditions. *Proceedings of the Human Factors Society* (pp. 779–783). New York.

4

Underpinnings of System Evaluation

Mark A. Wise
IBM Corporation

David W. Abbott
University of Central Florida (Retd.)

John A. Wise
The Wise Group, LLC

Suzanne A. Wise
The Wise Group, LLC

4.1 Background

Rapid advances in software and hardware have provided the capability to develop very complex systems that have highly interrelated components. Although this has permitted significant efficiency and has allowed the development and operation of systems that were previously impossible (e.g., negative stability aircraft), it has also brought the danger of system-induced catastrophes. Perrow (1984) argued that highly coupled complex systems (i.e., having highly interdependent components) are inherently unstable with a disposition toward massive failure. This potential instability has made the human factors-based evaluation more important than it has been in the past; while the component coupling had made the traditional modular evaluation methods obsolete.

Systems that are highly coupled can create new types of failures. The coupling of components that were previously independent can result in unpredicted failures (Wise & Wise, 1995). With more systems being coupled, the interdisciplinary issues have become more critical. For example, there is a possibility that new problems could reside in the human–machine interface where disciplines meet and interact. It is in these intellectual intersections that new compromises and cross-discipline trade-offs will be made. Furthermore, new and unanticipated human factors-based failures may be manifested in these areas.

As systems grow in both complexity and component interdependence, the cost of performing adequate testing is rapidly approaching a critical level. The cost of certification in aviation has been a significant cost driver. The popular aviation press is continually publishing articles on an aviation part (e.g., an alternator) that is exactly the same as an automobile part (i.e., comes off exactly the same assembly line), but costs two to three times more owing to the aviation certification costs. Therefore, human factors-based verification, validation, and certification methods must not only be effective, but also be cost-effective.

"Technically adequate" human factors testing may not even be sufficient or even relevant for a system to become safely operational. The political and emotional issues associated with the acceptance of some technically adequate systems (e.g., nuclear power, totally automatic public transportation systems) must also be considered. For many systems, the human factors evaluation must answer questions beyond safety and reliability, such as "What type of evaluation will be acceptable to the users and the public?," "How much will the public be willing to spend to test the system?," and "What level of security and reliability will they demand from the system?" In the wake of the September 11, 2001 terror attacks, public scrutiny of aviation systems and security procedures has increased. The threat of aircraft-related terror acts has added a new dimension to the evaluation of passenger safety, with the introduction of intentional system incidents or accidents.

In spite of the fact that the importance of human factors-based evaluation of the complex systems is increasing, the processes by which it is accomplished may be the most overlooked aspect of system development. Although a considerable number of studies have been carried out on the design and development process, very little organized information is available on how to verify and validate highly complex and highly coupled dynamic systems. In fact, the inability to adequately evaluate such systems may become the limiting factor in society's ability to employ systems that our technology and knowledge will allow us to design.

This chapter is intended to address issues related to human factors underpinnings of system evaluation. To accomplish this goal, two general areas have been addressed. The first section addresses the basic philosophical underpinnings of verification, validation, and certification. The second is a simple description of the basic behavioral-science statistical methods. The purpose of this section is to provide the statistically naïve reader with a very basic understanding of the interpretation of results using those tools.

4.2 Definitions

Verification and validation are very basic concepts in science, design, and evaluation, and form the foundation of success or failure of each. Both verification and validation should be considered as processes. In scientific inquiry, *verification* is the process of testing the truth or correctness of a hypothesis. With regard to system design, Carroll and Campbell (1989) argued that verification should also include determination of the accuracy of conclusions, recommendations, practices, and procedures. Furthermore, Hopkin (1994) suggested that one may need to extend the definition of verification to explore major system artifacts, such as software, hardware, and interfaces.

Validation has been defined broadly by Reber (1985) as the process of determining the formal logical correctness of some proposition or conclusion. In hypothesis testing, there are several threats to the validity of the results (Campbell & Stanley, 1963). In the human factors context, it may be seen as the process of assessing the degree to which a system or component does what it purports to do.

With regard to the human factors in aviation, an example of verification and validation is illustrated by the following (fictitious) evaluation of an interface for a flight management system (FMS). As a type of in-cockpit computer, the FMS provides ways for the pilot to enter data into it and to read information from it. The design guidelines for a particular FMS might call for the input of information to be carried out through a variety of commands and several different modes. If these requirements are implemented as documented, then we have a system that is verifiable. However, if the system proves to be unusable because of the difficult nature of the commands, poor legibility of the display output, or difficultly in navigating the system modes, then it may not be an operationally valid implementation (assuming that one of the design goals was to be usable).

Hopkin (1994) suggested that:

- Verification and validation tend to be serial rather than parallel processes.
- Verification normally precedes validation.
- Usually both verification and validation occur.
- Each should be planned considering the other.
- The two should be treated as complementary and mutually supportive.

4.3 Certification

Certification can be considered as the legal aspect of verification and validation: that is, it is verification and validation carried out such that a regulatory body agrees with the conclusion and provides some "certificate" to that effect. The concept of the certification of aircraft and their pilots is not new. For many years, the engineering and mechanical aspects of aviation systems have had to meet certain criteria of strength, durability, and reliability before they could be certified as airworthy. Additionally, pilots of the aircraft have to be certificated (a certification process) on their flight skills and must meet certain medical criteria. However, these components (the machine and the human) are the tangible aspects of the flying system, and there remains one more, less-readily quantifiable variable—the interface between human and machine (Birmingham & Taylor, 1954).

4.3.1 Why Human Factors Certification?

Why do we conduct human factors certification of aviation systems? On the surface, this may seem like a fairly easy question to answer. Society demands safety. There is an underlying expectation that transportation systems are safe. Western society has traditionally depended on the government to ensure safety by establishing laws and taking actions against culpable individuals or companies when they are negligent. It is therefore not a surprise that there is a collective societal requirement for the certification of the human factors of an aviation system. It is not enough to independently certify the skills of the operator and the mechanical integrity of the machine. To assure system safety, the intersection between these two factors must also receive focus to guarantee that a "safe" pilot can effectively operate the engineered aircraft "safely."

If the intended goal of human factors certification is to insure the safety and efficiency of the systems, then one might consider the following questions about certification: Would the process of human factors certification improve system safety by itself?, Would the threat of a human factors audit merely provide the impetus for human factors considerations in system development?, Would the fact that a design that passed a human factors certification process inhibit further research and development for the system?, Would the fact that something was not explicitly included in the process, cause it to be neglected?, or Would it inhibit the development of new approaches and technologies so as to decrease the cost of certification? (one can observe the effects of the last question in the area of general aviation where 30- to 50-year-old designs predominate).

As mentioned earlier, the nature of the relationship between a human factors certification process and a resultant safe system may not be a causal one. Another way to view the effectiveness of a certification program is to assume that the relationship is a "Machiavellian certification." In his political treatise, *The Prince*, Niccolò Machiavelli described the methods for a young prince to gain power, or for an existing prince to maintain his throne. To maintain and perpetuate power, it is often necessary that decisions are made based on the anticipated outcome, while the means to achieving that outcome are not bound by ethical or moral considerations. In other words, the ends justify the means. Could a similar view be applied to human factors certification? While there needs to be an ethical imperative, is it possible to restate the idea such that a process of undetermined causal impact (certification) results in a desirable end (a safer and more efficient air transport system)?

Similarly, Endsley (1994) suggested that the certification process may be not unlike a university examination. Most exams do not claim to be valid reflections of a student's knowledge of the course material; however, by merely imposing an exam on the students, they are forced to study the material, thus learning it. System certification can be viewed similarly—that is, certification, in and of itself, may not *cause* good human factors design. However, the threat of a product or system failing to meet the certification requirements (resulting in product delays and monetary loss) for poor human factors may encourage system designers to consider the user from the beginning.

Another view suggests that a formal, effective human factors certification process may not be a feasible reality. It is possible that an institutionalized certification process may not improve the system safety or efficiency by any significant amount, but instead may merely be "a palliative and an anodyne to society" (Hancock, 1994).

It is not the purpose of this chapter to address the legal issues associated with human factors certification of aviation (or any other type of system). Rather, this chapter addresses the technical and philosophical issues that may underpin the potential technical evaluation. However, for simplicity, the word *evaluation* is used to imply verification, validation, and certification processes.

4.4 Underpinnings

Effective evaluation of large human–machine systems may always be difficult. The complexity and integration of such systems require techniques that seek consistent or describable relationships among several independent variables, with covariation among the dependent variables according to some pattern that can be described quantitatively. It cannot rely on tools that identify simple relationships between an independent variable and a single dependent measure, which one normally uses in classical experimental psychology research. However, Hopkin (1994) warned that although more complex multivariate procedures can be devised in principle, caution is required because the sheer complexity can ultimately defeat meaningful interpretation of the findings, even where the methodology is orthodox.

Hopkin (1994) even went further to suggest that the following data sources can contribute to the evaluation process of new systems:

- Theories and constructs that provide a basis and rationale for generalization
- Data representative of the original data, but which may be at a different level (e.g., theories vs. laboratory studies)
- Similar data from another application, context, or discipline
- Operational experience relevant to expectations and predictions
- Expert opinion compared with the preceding items
- Users' comments based on their knowledge and experience
- Case histories, incidents, and experience with the operational system

This list is not intended to be all-inclusive, but rather is a model of the types of data that should be considered.

A fundamental decision that needs to be made early in the evaluation process relates to the identifying measures and data that may be relevant and meaningful in the evaluation of the target system. Experience has shown that data are often collected based on the intuition, rather than how the data are related and how they contribute to the evaluation process.

4.4.1 When Should Human Factors Evaluation Be Conducted?

The timing of the human factors evaluation within the project timeline will affect the type of evaluation that can be applied. There are three different types or times of system evaluation: a priori, ad hoc, and post hoc.

A priori evaluation includes the consideration of human factors requirements during the initial conceptual design formation. This would require human factors input at the time when the design specifications are being initially defined and documented. Ad hoc evaluation takes place concurrent to the production of the system. This may involve iterative reviews and feedback concurrent to early development. Post hoc evaluation involves an evaluation of the completed system. This would include the hardware, software, and human, and most importantly, their intersection.

"You can use an eraser on the drafting table or a sledge hammer on the construction site" (Frank Lloyd Wright). The cost of implementing a change to a system tends to increase geometrically as the project moves from conceptual designs to completed development. Cost considerations alone may require a priori or ad hoc approaches, where a human factors evaluation process is carried out in a manner that allows the needed changes to be made when the cost impact is low.

Ideally, evaluation of complex aviation systems would require human factors consultation throughout the conceptual (predesign), design, and implementation process. The involvement of a human factors practitioner during the process would guarantee consideration of the users' needs and insure an optimal degree of usability.

4.4.2 How Should Human Factors Evaluation Be Conducted?

Current standards and guidelines, such as the various military standards, provide a basis for the evaluation of products. These standards can be useful for checking workspace design; however, the conclusions gained from "passing" these guidelines should be interpreted with a critical eye.

Evaluation should not only be based on traditional design standards (e.g., Mil-Specs). Hopkin (1994) used the design of the three-pointer altimeter to illustrate this point. If the task was to ensure that a three-pointer altimeter followed good human factors standards (good pointer design, proper contrast, text readability, etc.), then it could be concluded that the altimeter was in fact certifiable. However, research has shown that the three-pointer altimeter is poor in presenting this type of information. In fact, errors of up to 10,000 ft are not uncommon (Hawkins, 1987). Hence, by approving the three-pointer altimeter based on the basic design standards, a poorly designed instrument might be certified. On the other hand, principle-based evaluation may have noted that a three-pointer altimeter is inappropriate even if it does meet the most stringent human factors standards. Therefore, principle-based evaluation may recommend a different type of altimeter altogether.

Wise and Wise (1994) argued that there are two general approaches to the human factors evaluation of systems: (a) the top-down or systems approach and (b) the bottom-up or monadical approach. The top-down approach is developed on the assumption that evaluation can be best served by examining the systems as a whole (its goals, objectives, operating environment, etc.), followed by the examination of the individual subsystems or components.

In an aircraft cockpit, this would be accomplished by first examining what the aircraft is supposed to do (e.g., fighter, general aviation, commercial carrier), identify its operating environment (IFR, VFR, IMC, VMC, combat, etc.), and looking at the entire working system that includes the hardware, software, liveware (operators), and their interactions; subsequently, evaluative measures can be applied to the subsystems (e.g., individual instruments, CRT displays, controls) (Wise & Wise, 1994).

Top-down or the systems approach to evaluation is valuable, as it requires an examination of the systems as a whole. This includes the relationship between the human and the machine—the interface.

On the other hand, the bottom-up approaches look at the system as a series of individual parts, monads that can be examined and certified individually. Using this method, individual instruments and equipments are tested against human factors guidelines. Subsequently, the certified components are integrated into the system. The bottom-up approach is very molar; that is, it tries to break down the whole into its component parts. The benefit of this method is that the smaller parts are more manageable and lend themselves to controlled testing and evaluation. For example, it is obviously much easier to certify that a bolt holding a tier in place is sound, than to certify the entire mechanical system.

However, the simplicity and apparent thoroughness of this approach are somewhat counteracted by the tendency to lose sight of the big picture, such as what the thing is supposed to do. For a given purpose, a weak bolt in a given location maybe acceptable; in another case, it may not be. Unless the purpose is known, one may end up with a grossly overengineered (i.e., overpriced) system.

Additionally, the sum of the parts does not always add up to the whole. A set of well-designed and well-engineered parts may all do their individual jobs well (verification), but may not work together to perform the overall task that they are expected to perform (validation). A good example of this drawback, outside the world of aviation, can be found in the art of music. Molecularly, a melody is simply made up of a string of individual notes; however, the ability to recognize and play the notes individually does not give sufficient cause for believing that the melody will in fact be produced. Thus, individual subcomponents may individually function as designed, but may not be capable of supporting an integrated performance in actual operational settings.

Human factors evaluation of an aviation system's interface may be difficult, to say the least. However, it has been argued that the top-down evaluation produces the most operationally valid conclusions about the overall workability of a system (Wise & Wise, 1994), and perhaps, only full systems evaluation within high-fidelity operational-relevant simulation settings should be utilized.

4.5 Human Factors Evaluation and Statistical Tools

The traditional method of evaluating the "truth" of a hypothesis (the most basic function in the evaluation process) in behavioral science and human factors has been the experimental paradigm. The basic guarantor of this paradigm is the statistical methods that support the experimental designs and establish whether the results are meaningful or "truthful." Thus, an understanding of the basic concepts of statistics is necessary for anyone who even reviews one of the processes. To examine the results of an evaluation process without understanding the capabilities and limits of statistics would be like reviewing a book written in an unfamiliar language.

Unfortunately, there are a number of common misunderstandings about the nature of statistics and the real meaning or value of the various classes of statistical tools. Although it is impossible to provide the readers with adequate tools in a part of a chapter, a chapter itself, or probably even a complete book, the goal of the following section is to provide:

- Awareness of the basic types of statistical tools
- Basic description of their assumptions and uses
- Simple understanding of their interpretations and limits

Anyone who is serious about this topic should prepare to undertake a reasonable period of study. A good place to start would be from the book by Shavelson (1996).

4.5.1 Introduction to Traditional Statistical Methods

Reaching valid conclusions about complex human–machine performance can be difficult. However, research approaches and statistical techniques have been developed specifically to aid the researchers in the acquisition of such knowledge. Familiarity with the logical necessity for various research designs, the need for statistical analysis, and the associated language used are helpful in understanding the research reports in the behavioral science and human factors areas.

This section may help the statistics-naïve reader to better understand and interpret the basic statistics used in behavioral science and human factors research. It addresses the following issues:

- Estimates of population values
- Relationships between factors
- Differences between groups

However, this chapter is not intended to be a "how to" chapter, as that is far beyond the scope of this work. Rather, it may help the statistics-naïve reader to better understand and evaluate the human factors and behavioral science research that utilizes the basic techniques covered in this text.

4.5.2 Estimates of Population Values

To understand or evaluate the studies on human performance, one can begin with the most basic research question: What is typical of this population? This describes a situation where a researcher is interested in understanding the behavior or characteristics that are typical of a large defined group of people (the population), but is able to study only a smaller subgroup (a sample) to make judgments. What is the problem here? A researcher who wants to discover the typical number of legs that human beings have, can pick a few and note that there is no person-to-person variability in the number of legs; all people have two legs. As people do not vary in their number of legs, the number of people a researcher selects for his/her sample, the type of people selected, how they are selected, etc., may make a very little difference. The problem for researchers using human behavior and many human characteristics as the object of study is that virtually all nontrivial human behaviors vary widely from person to person. Consider a researcher who wants some demographic and skill-level information regarding operators of FMS-equipped aircraft. The research may involve selecting a subset (sample) of people from the entire defined group (population), and measuring the demographic and performance items of interest. How does a researcher select the sample? A researcher who seeks findings that may be applicable to the entire population may have to select the people in such a way that they do not give an unrepresentative, biased sample, but a sample that is typical of the whole group that will allow the researcher to state to what extent the sample findings might differ from the entire group.

The correct selection techniques involve some methods of random sampling. This simply means that all members of the population have an equal chance of being included in the sample. Not only does this technique avoid having a biased nonrepresentative sample, but researchers are able to calculate the range of probable margin of error that the sample findings might have from actual population. For example, it might be possible to state that the sample mean age is 40.5 years, and that there is a 95% chance that this value is within 1.0 year of the actual population value. If the researcher gathered this type of information without using a random sample—for example, by measuring only those pilots who fly for the researcher's friend, Joe—then the researcher might get a "sample" mean of 25 if Joe has a new, under-funded flight department, or of 54, if Joe has an older, stable flight department. In either case, the researcher may not know how much representative these group means are of the population of interest and would not know how much error might be present in the calculation. In this example, there would have been an unrepresentative sample resulting in data of dubious value.

Random sampling provides an approximate representation of the population, without any systematic bias, and allows one to determine how large an error may be present in the sample findings. This sort of research design is called a survey or a sample survey. It can take the form of a mailed questionnaire sent to the sample, personal interviews with the selected sample, or obtaining archival data of the selected sample. In all the cases, the degree of likely error between the sample findings and the population values is determined by the person-to-person variability in the population and the size of the sample. If the population members have little individual difference on a particular characteristic, then the "luck of the draw" in selecting the random sample may not produce a sample that differs from the population. For example, in assessing the number of arms that our pilot population have, as all have the same amount (i.e., "0" variability in the population), the sample mean may be identical to the population mean (i.e., both will be "2"), irrespective of how the researcher selects the sample, with no error in the sample value. For the characteristics on which the pilots differ, the greater variability in the individuals in the population indicates greater probable difference between any random sample mean and the actual population. This difference is called *sampling error* and is also influenced by the size of the sample selected. The larger the sample, the smaller is the sampling error. Consider a sample of 999 pilots from the entire population of 1000 pilots. Obviously, this sample will have a mean on any characteristic that is very close to the actual population value. As only one score is omitted from any selected sample, the sample may not be much influenced by the "luck" of who is included. The other extreme in the sample size is to take a sample of only one pilot. Obviously, here, the sample-to-sample fluctuation of "mean" would be equal to the individual variability in the measured characteristic that exists in the population. Very large sampling error may exist, because our sample mean could literally take on any value from the lowest to the highest individual population score value.

Thus, the design considerations for sample surveys must be certain to obtain a random (thus, unbiased) sample as well as to have a large enough sample size for the inherent variability in the population being studied, so that the sample value will be close to the actual population.

There are two additional research questions that are frequently asked in behavioral research. One is, within a group of people, do scores on two variables change with each other in some systematic way? That is, do people with increasing amounts of one variable (e.g., age) also have increasing (or decreasing) amounts of some other variable (e.g., time to react to a warning display)? The second type of research question that is asked is, for two or more groups that differ in some way (e.g., type of altimeter display use), do they also have different average performance (e.g., accuracy in maintaining assigned altitude) on some other dimension? Let us get deeper into these two questions and their research design and statistical analysis issues.

4.5.3 Questions of Relationships

In questions of relationships, researchers are interested in describing the degree to which increases (or decreases) in one variable go along with increased or decreased scores of a second variable. For example, is visual acuity related to flying skill? Is the number of aircraft previously flown related to the time required to train to become proficient in a new type? Is time since last meal related to reaction time or visual perception? These example questions can be studied as relationships between variables within a single group of research participants.

The statistical index used to describe such relationships is Pearson correlation coefficient, *r*. This statistic describes the degree and direction of a straight-line relationship between the values of the two variables or scores. The absolute size of the statistic varies from 0 to 1.0, where 0 indicates that there is no systematic variation in one score dimension related to the increase or decrease in the other score dimension. A value of 1.0 indicates that as one variable increases, there is an exact and constant amount of change in the other score, so that a plot of the data points for the two variables may all fall perfectly along a straight line. The direction of the relationship is indicated by the algebraic sign of the coefficient, with a minus sign indicating that as values on one dimension increase, those on the other decrease, forming a negative relationship. A plus sign indicates a positive relationship, with increases in one dimension going along with the increases on the other.

To study such questions of relationship, one must have a representative sample from the population of interest and two scores for each member of the sample, one on each variable.

Once the degree and direction of linear relationship have been calculated with the Pearson *r*, it is then necessary to consider whether the described relationship in our sample came about owing to the actual existence of such a relationship in the population, or owing to some nonrepresentative members in our sample who demonstrate such a relationship even though the true population situation indicates that no such relationship exists. Unfortunately, it is possible to have a relationship in a sample when none exists in the general population.

Was the result obtained because of this relationship in the population, or was the observed sample relationship a result of a sampling error when the population has no such relationship? Fortunately, this apparent dilemma is easy to solve with statistical knowledge of sampling variability involved in random selection of correlational relationships, just as the calculation of random sampling variability for sample means. A typical method for deciding whether the observed correlation is real (exists in the population) or is simply owing to the nonrepresentative sampling error, is to calculate the probability of the sampling error that provides the observed size of the sample correlation from a population where there is zero correlation. Thus, if a researcher found an observed $r = 0.34$ ($n = 50$), $p = 0.02$, then the p value (probability) of 0.02 indicates that the chance of having sampling error producing a sample r of 0.34 when the population r is 0.0 is only 2 times in 100. As a general rule in the behavioral sciences, when sampling error has a probability as small as 5 in 100, or less, to produce our observed r, we can conclude that our observed r is from a population that really has such a relationship, rather than having come about by this sampling error from a population with zero correlation. We may reach this conclusion by stating that we have a statistically significant, or simply a significant, correlation. We may actually conclude that our sample correlation is too big to have come just from the sampling luck, and thus, there exists a real relationship in the population.

A random sample of corporate pilots showed a significant degree of relationship between total flying hours and the time required to learn the new FMS, $r(98) = -0.40$, $p = 0.01$.

The interpretation of these standard results is that the more flying hours that corporate pilots have, the less time it takes for them to learn a new FMS. The relationship within the sample of pilots is substantial enough that the researcher can conclude that the relationship also exists among corporate pilots in general, because the chance of a nonrepresentative sample with this relationship being selected from a population not having this relationship is less than 1 in 100.

The researcher who finds a significant degree of relationship between the two variables may subsequently want to calculate an index of the *effect size*, which will give an interpretable meaning to the question of how much relationship exists. This can be easily accomplished with the correlation relationship by squaring the r value to obtain the coefficient of determination, r^2. The coefficient of determination indicates the proportion of variability in one variable, which is related to the variation in the other variable. For example, an $r = 0.60$ between the years of experience and flying skill may lead to an r^2 of 0.36. Thus, it could be said that 36% of the variability in pilot skill is related to the individual differences in pilot experience. Obviously, 64% of variation in pilot skill is related to something(s) other than experience. It is this effect-size index, r^2, and not the size of the observed p value, which gives us the information on the size or importance of the relationship. Although the size of the relationship does have some influence on the p value, it is only one of the several factors. The p value is also influenced by sample size and variability in the population, such that no direct conclusion of the effect size can be obtained with respect to the p value. Therefore, the coefficient of determination, r^2, is needed.

However, what interpretation can be made about the relationship between two variables when a significant r is found (i.e., $p \leq 0.05$)? Is it possible to conclude that one variable influences the other, or is the researcher limited only to the conclusion that performance on one variable is related to (goes along with) the other variable without knowing why? The distinction between these two types of valid conclusion of significant research findings may appear negligible, but actually, it is a major and important distinction. This is particularly true for any application of our results. However, what can be concluded from this significant ($r = 0.60$, $p = 0.012$) correlation between pilot experience (hours flown) and pilot skill (total simulation proficiency score)? There are essentially two options.

The decision on what is a legitimate interpretation is based on the way in which the research study has been conducted. One possibility is to select a representative random sample of pilots from our population of interest and obtain scores on the two variables from all the pilots in our sample. The second possibility may be to start again with a random sample, but the sample must be obtained from initial pilots who need a certain amount of experience, and after obtaining the experience, the skill measurements may be taken.

What is the difference in the legitimate interpretation of the two studies? In the first approach, by simply measuring the experience and skill, it is not possible to know why the more experienced pilots have good skills. It could be possible that experience develops skills, or pilots who have good skills get the opportunity to acquire flight-time experience. Furthermore, it could also be possible that highly motivated pilots work hard to acquire both skills and experience. In short, the data show that experience and skills go together, but it cannot show whether experience develops skills, or skills lead to experience, or both follow from some other unmeasured factor.

For pilot-selection applications of this study, this may be all that is needed. If a company selects more experienced pilots, then they may on an average be more skillful, even though they may not know the reason for it. However, for training applications, sufficient information is not available from this study; that is, this study could not propose that obtaining experience will lead to improved skill. This type of research design is called a *post facto* study.

Researchers simply selected people who have already been exposed to or selected to be exposed to some amount of one variable, and evaluated the relationship of scores on that variable to another aspect of behavior. Such designs only permit relatedness interpretations. However, no cause-and-effect interpretation or the conclusion that the first variable actually influences the behavior has been justified. A casual influence may or may not exist—one simply cannot decide from this type of design. If it does exist, then its direction (which is the cause and which is the effect, or are both variables "effects" or some other cause) is unknown. The researcher observes a relationship after the research participants are exposed to different amounts of the variable of interest. Thus, if a statistically significant post facto relationship between the two variables is found, then it will show that the relationship does exist in the population, but it will be impossible to determine its reason.

4.5.4 Questions of Group Difference

This approach to design involves creating groups of research participants that differ on one variable, and then statistically evaluating them to observe if these groups also differ significantly on the behavior of interest. The goal of the research may be either to find out if one variable is simply related to another (post facto study), or to establish if one variable actually influences another (true experiment). With either goal, the question being asked using this method is whether or not the groups differ, as opposed to the previous correlational design that questioned on whether the scores were related to a single group.

If the groups are formed based on the amount of one variable that the participants currently possess (e.g., age, sex, height) and assigning them to the appropriate group, then it is a post facto design. If there is a significant group difference on the behavior performance, then the interpretation may still be that the group difference variable and behavior are related without knowing the reason for it. Furthermore, the information obtained from a post facto group-difference study is similar to that obtained from the correlational relationship post facto study described earlier.

The statistical evaluation for "significance" may not be based on a correlation coefficient, but may use procedures like t-test or analysis of variance (ANOVA). These two techniques allow a researcher to calculate the probability of obtaining the observed differences in the mean values (assuming random sampling), if the populations are not different. In other words, it is possible that the samples have different means when their populations do not have different means. Sampling variability can certainly lead to this situation. Random samples may not necessarily match the population accurately, and hence, two samples can easily differ when their populations do not. However, if the observed groups have different mean values that have a very low probability (≤ 0.05) of coming from equal populations, that is, differing owing to sampling error only, then it is possible to conclude that the group variable being studied and the behavior are truly related in the population, not just for the sample studied.

This is similar to the result from a post facto relationship question evaluated with a correlation coefficient described in the previous section. The legitimate interpretation of a post facto study may be the same, irrespective of whether the researcher evaluates the result as a relationship question with a correlation coefficient, or as a group difference question with a test for significant differences between the means. If the more powerful interpretation that a variable actually influences the behavior is required, then the researcher may need to conduct a true experiment.*

To obtain the cause-and-effect information, a research design where only the group difference variable could lead to the observed difference in the group performance is required. This research would begin by creating two or more groups that do not initially differ on the group difference variable, or anything else that might influence the performance on the behavior variable; for example, research participants do not decide which group to join, the top or lowest performers are not placed in "groups," and existing intact groups are not used. Instead, equal groups are actively formed by the researcher, and controls are imposed to keep unwanted factors from influencing the behavior performance. Experimental controls are then imposed to make sure that the groups are treated equally throughout the experiments. The only factor that is allowed to differ between the groups is the amount of the group difference variable that the participants experience. Thus, the true experiment starts with equal groups and imposes differences on the groups to observe whether a second set of differences is obtained.

In this way, it is possible to determine whether the imposed group difference actually influences the performance, because all the alternate logical possibilities for why the groups differ on the behavior of interest are eliminated. In practice, the equal groups are formed either by randomly assigning an existing pool of research participants into equal groups, or by selecting several equal random samples from a large population of research participants. In either procedure, the groups are formed so that the groups are equal on all factors, known and unknown, which have any relationship or potential influence on the behavior performance. The researcher then imposes the research variable difference on the groups, and later measures the individuals and compares the group means on the behavior performance.

As discussed earlier, random sampling or random assignment might have assigned people to groups in such a way that it failed to produce exact equality. Thus, the researcher needs to know if the resulting group differences are greater than the initial inequality that the random chance might have produced. This is easily evaluated using a test for statistical significance. If the statistic value of the test has a probability of 0.05, then the sampling variability only may have a 5/100 chance of producing the group-mean difference as large as the one found. Again, for any observed result that has a probability of being produced by sampling luck alone, which is as small as or smaller than 5/100, one may conclude that the difference may be from something other than this unlikely source and is "statistically significant." In this case, the researcher may conclude that the reason for the groups to have different behavior performance means is that the imposed group difference variable created these performance differences, and, if these performance differences are imposed on other groups, then one may expect to reliably find similar performance differences.

4.5.5 Examples

As an example of a group difference of true experiment versus a group difference of post facto study, consider an investigation to determine whether unusual attitude training influences the pilot performance in recovering from an uncommanded 135 degree roll. Researcher A investigates this by locating 30 pilots in his company, who have had unusual attitude training within the past 6 months and who volunteered for such a study. He compares their simulator performance with that of a group of 30 pilots from the company, who have never had such training and have expressed no interest in participating in the study. A statistical comparison of the performance of the two groups in recovering from the

---

* Although it is possible to conduct a true experiment as a relationship question evaluated with a correlation coefficient, this is very rare in practice. True experiments producing information on one variable and actually influencing the performance on another, are almost always conducted as a question of group differences and evaluated for statistical significance with some factors other than correlation coefficient.

uncommanded 135 degree roll indicated the mean performances for pilots who were or were not trained in unusual attitude recover, which were 69.6 and 52.8, respectively. These means do differ respectively with $t(38) = 3.45, p = 0.009$.

With such a design, one can conclude that the performance means for the populations of trained and untrained pilots do differ in the indicated direction. The chance of obtaining nonrepresentative samples with such different means (from populations without mean differences) is less than 1 in 100. However, as this is a post facto study, it is impossible to know whether the training or other pilot characteristics are responsible for the difference in the means.

As Researcher A used a post facto study—that is, did not start with equal groups and did not impose the group difference variable (i.e., having or not having unusual attitude training) on the groups—there are many possible reasons that trained group performed better. For example, the more skilled pilots sought out such training and thus, could perform any flight test better because of their inherent skill, not because of the training. Allowing the pilots to self-select the training created groups that differ in ways other than the training variable under study.

It is, of course, also possible that the attitude training is the real active ingredient leading to the roll-recovery performance, but this cannot be investigated using Researcher A's study. It is only possible to know that seeking and obtaining attitude training *is related* to better roll recovery. Is it because better pilots seek such training, or because such training produces increased skill? It is impossible to know. Is this difference in interpretations relevant? If one is selecting pilots to hire, perhaps not. One cannot simply hire those who have obtained such training, and think that they will (based on group averages) be more skilled. If one is trying to decide whether to provide unusual attitude training for a company's pilots and the cost of such training is expensive, then one would want to know if such training actually leads to (causes) improved skill in pilots in general. If the relationship between attitude training and performance is owing to the fact that only highly skilled pilots have historically sought out such training, then providing such training to all may be a waste of time and money.

On the other hand, Researcher B has a better design for this research. Sixty pilots are identified in the company, who have not had unusual attitude training. They are randomly assigned to one of the two equal groups, either to a group that is given such training or to a group that gets an equal amount of additional standard training. Again, the mean performance of the two groups are observed to differ significantly with $p = 0.003$.

This research provides much better information from the significant difference. It is now possible to conclude that the training produced the performance difference and would reliably produce improved performance if imposed on all of the company's pilots. The pilot's average performance on unusual attitude recovery would be better because of the training. The extent of improvement could be indicated by looking at our effect-size index. If eta squared equaled to 0.15, then we can conclude that the training leads to 15% of the variability among pilots on the performance being measured.

Often, these questions on group difference are addressed with a research design involving more than two groups in the same study. For example, a researcher might randomly assign research participants to one of the three groups and then impose a different amount of training or a different type of training on each group. One could then use a statistical analysis called ANOVA to observe whether the three amounts or types differ in their influence on performance. This is a very typical design and analysis in behavioral science studies. Such research can be either a true experiment (as described earlier) or a post facto study. The question of significance is answered with an F statistic, rather than the t in a two-group study, but eta squared is still used to indicate the amount or size of the treatment effect.

For example, unusual attitude recovery was evaluated with three random samples of pilots using a normal attitude indicator, a two-dimensional outside-in heads-up display (HUD), or a three-dimensional HUD. The mean times to recovery were 16.3, 12.4, and 9.8 s, respectively. The means did differ significantly with a one-way ANOVA, $F(2, 27) = 4.54, p < 0.01$. An eta squared value of 0.37 indicated that 37% of the pilot variability in attitude recovery is owing to the type of display used. One can conclude that the three methods would produce differences among the pilots in general, because the probability of finding such large sample differences just from random assignment effects, rather than training effects, is less than 1 in 100. Further, the display effects produced 37% of the individual pilot variability in time to recover. The ANOVA established that the variance among the means was from the display effects, and not from the random assignment differences regarding who was assigned to which group. This ANOVA statistical procedure is very typical for the analysis of data from research designs involving multiple groups.

4.5.6 Surveys as an Evaluation Tool

In addition to the experimental design discussed earlier, there are numerous evaluation tools that are utilized by human factors professionals. As many people consider surveys as an easy way of answering evaluation questions, it seemed appropriate to include a small section on surveys to caution potential users of potential design and interpretation issues. While human factors scientists normally rely on the cold hard data of experimental design, surveys can be used in many areas of investigation to collect data. While only post facto or quasi-experimental data can be obtained from the use of surveys, the vast amounts of data that can be collected by surveys make them an attractive option. Additionally, one can use surveys to triangulate data and further validate results found by experimental or observational methods. In the process of human factors evaluation, surveys can be used to gauge the political and emotional issues associated with the acceptance of systems, and determine the type of evaluation that will be acceptable to users and the public.

Surveys are cost-effective and relatively quick for data collection. One can reach thousands of people all over the world in seconds spending mere pennies via internet surveys. A well-designed and researched survey can provide a multitude of valuable evaluation information from the data sources, as mentioned by Hopkin (1994). Surveys can efficiently gather information that can contribute to the evaluation process of new systems, such as information on operational experience related to expectations and predictions, expert opinions, and users' comments based on knowledge and experience.

With tools like "Survey Monkey™" even a novice can put together a professional looking survey within hours. However, developing a survey that will produce meaningful and valid results is not very simple. Developing multiple choice and even short-answer questions that truly elicit the desired concepts of inquiry requires careful planning and consideration of the multiple interpretations that a question may elicit. As surveys are a language-based measurement, a researcher must consider the readers' comprehension and context when designing survey questions (Sudan, Bradburn, & Schwartz, 1996). Even something as simple as the ordering of questions can impact the survey results. Without properly and carefully designed questions, the results of the survey may become meaningless and potentially misleading.

Researchers who focus on survey development acknowledge that there is no great or even good theory behind good survey design (Sudan et al., 1996). There are many poorly designed surveys in circulation. Interpretation of data derived from poorly designed surveys must be done with extreme caution. A few key issues to consider to avoid making some common survey mistakes and to help recognize a quality survey that may yield useful data are as follows:

- Questions should be simple and relatively short to avoid respondent confusion. Use of simple phrases and terminology may avoid potential errors in comprehension (Dillman, 2000). If needed, break longer questions into two questions. This is especially important if your question is addressing two unrelated questions on the same topic. The question "Is the use of HUDs necessary and efficient?" should be broken into two questions, as the areas of interest (necessity and efficiency) could elicit different responses (Judd, Smith, & Kidder, 1991).
- Survey questions often ask respondents to estimate time or frequency of an event. For example, how much time do you spend reviewing checklists on a "typical" mission? Terms like "typical" and "average" have been shown to be confusing to respondents (Sudan et al., 1996). More accurate

results can be obtained by asking specific questions asking for recollection in a fairly recent time frame. For example, "How long did you review the manual for the 'x-brand simulator'?" or "How long did you review the manual for your last three projects?"

- Clearly defining key terms in a survey question is imperative for valid and useful results. For example, in the question, "How many breaks do you take a day?," breaks could be defined as pauses in work, the amount of time a person spends away from their work station, or simply the time that they spend by not directly engaging in work. Such a broadly defined term could result in very different responses depending on the interpretation (Sudan et al., 1996). Clearly defining what you are looking for and what you are not looking for will help to increase the accuracy of the response.

- Question wording should be very specific, especially when you are trying to measure an attitude (Judd et al., 1991). If you are trying to determine a person's attitude about automation, you may get very different results by asking "How do you feel about automation" vs. "How do you feel about automation in X flight deck design." It is important to remember that attitudes do not always lead to behavior. If a behavior is the matter of interest, it must be the subject of the question, not an attitude related to the behavior. For example, "Do you believe more safety precautions should be designed?" does not indicate whether the person would actually use the precautions, but rather may show that they consider it as a generally good idea. A better option might be, "Would you use additional safety precautions if they were designed?"

- Question order is also fairly important. Grouping of similar questions is a generally recommend practice. It adds continuity and aids in respondents' ability to recall the events related to the questions (Dillman, 2000; Judd et al., 1991). However, the sequence of questions may also form a bias for responses to subsequent responses (Dillman, 2000). Having a series of questions related to accidents followed by a question on readiness training may cause a bias in responses owing to the framing of the question. It also advisable to put objectionable, sensitive, and difficult questions at the end of the survey, as the respondents may feel more committed to respond once they have reached the end (Dillman, 2000; Judd et al., 1991).

- Apart from the question design, one must also consider the response options, especially when using close-ended or multiple choice responses. One must maintain a careful balance between overly specific or vague response choices. Terms such a "regularly" or "frequently" are vague and open to individual interpretation, whereas options such as "1 h a day" or "2 h a day" are so specific that respondents may feel torn over how to respond. Whenever possible, number values or ranges should be assigned (e.g., 4–5 days a week or 2–3 h a day). When using ranges, one needs to be careful not to provide overlap in responses (Dillman, 2000). Assigning negative or zero number value to qualitative labels (e.g., 0 = very unsatisfied vs. 1 = very unsatisfied) may reduce the likelihood of respondents choosing the lower response, and should therefore, be avoided (Sudan et al., 1996).

Owing to the complexity of the survey design, hiring an expert in survey design may help to ensure the validity of the measure. A well-crafted survey may require significant effort and research on the part of the responsible party. Pretesting the questions to ensure that they are concise and direct is a vital step in survey design. Additional information on survey design can be found in Dillman (2000) and Sudan et al. (1996).

4.5.7 Statistical Methods Summary

These are the basics of design and statistical procedures used in human factors research. This foundation can be expanded to several dimensions, but the basics remain intact. Questions are asked about what is typical of a group, about relationships between variables for a group, and about how groups that differ on one variable differ on some behavior. More than one group difference can be introduced in a single study, and more than one behavior can be evaluated. Questions can be asked about group frequencies of some behavior, such as pass/fail rather than average scores. Furthermore, rank order of the performance rather than actual score can be evaluated. Statistical options are numerous, but all answer the same questions, that is, is the observed relationship or difference real or simply sampling variability?

Throughout all simple or elaborate designs and statistical approaches, the basics are the same. The question being answered may be either of relationships between the variables or differences between the groups. The design may be either only post facto-yielding relatedness information or a true experiment with information on the influence that a variable has on behavior. If one considers the group differences as they are found and observes whether they differ in other behaviors, then it is a post facto design and it determines if the two differences are related, but not its reason. If the design starts with equal groups and then imposes a difference, then it is a true experiment, and such a design can determine if the imposed difference creates a behavior difference.

In reviewing or conducting research on the effects of design evaluation on system operational safety, the research "evidence" needs to be interpreted in light of these statistical guidelines. Has an adequate sample size been used to assure representative information for the effect studied? Did the research design allow a legitimate cause-and-effect interpretation (true experiment), or was it only post facto information about relatedness? Were the sample results evaluated for statistical significance?

4.6 How Would We Know Whether the Evaluation Was Successful?

One of the arguments against all types of evaluation is that evaluation drives up cost dramatically, whereas it adds little increase in safety. This is especially true for aviation systems, which have fewer accidents and incidents than any other type of transportation system (Endsley, 1994; Hancock, 1994). However, if society tolerates fewer accidents in aviation than it accepts in other modes of transportation, designers working in aviation must acknowledge and accept this judgment and work toward improved safety. Fewer operator errors in a simulator for certified systems than for poorly designed systems may be a better design evaluator, than waiting for infrequent fatal accidents in actual operation.

A second problem inherent within this issue is on deciding when the evaluation process should stop. In a test of system (interface) reliability, there will always be some occurrences of mistakes. What is the minimum number of mistakes that the evaluation should strive for? The problem is that the answer goes on and on and is never completely done. The challenge is to find how "reliable" a system needs to be, before the cost of additional evaluation overcomes its benefits. Rather than slipping into this philosophical morass, perhaps the evaluation questions should be: Does this certified system produce significantly fewer operational errors than other currently available systems? From a purely economic basis, insurance costs for aviation accidents are probably always cheaper than good aviation human factors evaluation design. This should not be an acceptable reason to settle for a first "best guess" with respect to design. Rather, the best possible evaluation with human factors consultation and evaluation at the predesign, design, and implementation stages should be utilized.

References

Birmingham, H. P., & Taylor, F. V. (1954). A design philosophy for man-machine control systems. *Proceedings of the IRE, 42*, 1748–1758.
Campbell, D. T., & Stanley, J. C. (1963). *Experimental and quasi-experimental designs for research.* Chicago: Rand McNally.
Carroll, J. M., & Campbell, R. L. (1989). Artifacts as psychological theories: The case of human-interaction. *Behaviour and Information Technology, 8*, 247–256.
Dillman, D. A. (2000). *Mail and internet surveys: The tailored design method* (2nd ed.). New York: John Wiley & Sons, Inc.

Endsley, M. R. (1994). Aviation system certification: Challenges and opportunities. In J. A. Wise, V. D. Hopkin, & D. J. Garland (Eds.), *Human factors certification of advanced aviation technologies* (pp. 9–12). Daytona Beach, FL: Embry-Riddle Aeronautical University Press.

Hancock, P. A. (1994). Certification and legislation. In J. A. Wise, V. D. Hopkin, & D. J. Garland (Eds.), *Human factors certification of advanced aviation technologies* (pp. 35–38). Daytona Beach, FL: Embry-Riddle Aeronautical University Press.

Hawkins, F. H. (1987). *Human factors in flight*. Hampshire, U.K.: Gower.

Hopkin, V. D. (1994). Optimizing human factors contributions. In J. A. Wise, V. D. Hopkin, & D. J. Garland (Eds.), *Human factors certification of advanced aviation technologies* (pp. 3–8). Daytona Beach, FL: Embry-Riddle Aeronautical University Press.

Judd, M. C., Smith, E. R., & Kidder, L.H. (1991). *Research methods in social relations* (6th ed.). Fort Worth, TX: Hartcourt Brace Jovanich College Publishers.

Perrow, C. (1984). *Normal accidents: Living with high-risk technologies*. New York: Basic Books.

Reber, A. S. (1985). *The penguin dictionary of psychology*. London, U.K.: Penguin Books.

Shavelson, R. J. (1996). *Statistical reasoning for the behavioral sciences* (3rd ed.). Needham Heights, MA: Allyn & Bacon.

Sudan, S., Bradburn, N. M., & Schwartz, N. (1996). *Thinking about answers: The application of cognitive processes to survey methodology*. San Francisco, CA: Jossey-Bass Publishers.

Wise, J. A., & Wise, M. A. (1994). On the use of the systems approach to certify advance aviation technologies. In J. A. Wise, V. D. Hopkin, & D. J. Garland (Eds.), *Human factors certification of advanced aviation technologies* (pp. 15–23). Daytona Beach, FL: Embry-Riddle Aeronautical University Press.

Wise, J. A., & Wise, M. A. (1995, June 21–23). In search of answers without questions: Or, how many monkeys at typewriters will it take… *Proceedings of the Workshop on Flight Crew Accident and Incident Human Factors*. McLean, VA.

5

Organizational Factors Associated with Safety and Mission Success in Aviation Environments

Ron Westrum
Eastern Michigan University

Anthony J. Adamski
Eastern Michigan University

This chapter examines the organization factors in aviation safety and mission success. The organizations involved comprise the entire range of aviation organizations, from airline operations departments to airports, manufacturing organizations, air-traffic control, and corporate flight departments. Organizational factors include organizational structure, management, corporate culture, training, and recruitment. Although the greater part of this chapter is focused on civil aviation, we have also devoted some attention to space and military issues. We have also used examples from other high-tech systems for the illustration of key points. Obviously, a full description of such a broad field could result in a publication of the size of this book. Hence, we have concentrated on key organizational processes involved in recent studies and major accidents, which may open general issues.

The authors have tried to integrate empirical studies within a broader framework, a model of effective operation. We believe that failures occur when various features of this model are not present. In choosing any model, we risk leaving out some critical factors. This is known as calculated risk. We believe that further discussion will progress best with such an integrative framework.

5.1 High Integrity

The underlying basis for this chapter is a model of *high integrity* for the development and operation of equipment and people. The model is guided by the principle stated by Arthur Squires. Squires was concerned about the integrity of the engineering design process in large systems. Considering several major failures, Squires (1986) proposed the following criterion: "An applied scientist or engineer shall display utter probity toward the engineered object, from the moment of its conception through its commissioning for use" (p. 10). Following Squires' idea, we propose to state the principle as follows:

> The organization shall display utter probity toward the design, operation, and maintenance of the aviation and aerospace systems.

Thus, organizations with "utter probity" will get the best equipment for the job, use it with intelligence, and maintain it carefully (Figure 5.1). In addition, they will display honesty and a sense of responsibility appropriate to a profession with a high public calling. Organizations that embody this principle are "high-integrity" organizations. These organizations can be expected to do the best job they can with the resources available. The concept unites two related emphases, both common in the organization literature: high reliability and high performance.

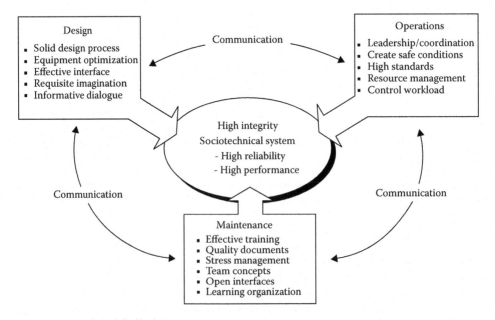

FIGURE 5.1   Central model of high integrity.

High reliability. The high-reliability organization concentrates on having few incidents and accidents. Organizations of this kind typically have systems in which the consequences of errors are particularly grave.

For example, operations on the decks of aircraft carriers involve one of the most tightly coupled systems in aviation. During the Vietnam war, for instance, two serious carrier fires, each with high loss of life, war materiel, and efficiency, were caused when minor errors led to chains of fire and explosion (Gillchrist, 1995, pp. 24–26). Today, aircraft-carrier landings are one of the archetypical "high-reliability" systems (Roberts & Weick, 1993).

High performance. The high-performance organization concentrates on high effectiveness. Here, instead of the multifaceted approach of the high-reliability organization, there is often a single measure that is critical. "Winning" may be more important than flawless operation, and the emphasis is on getting the job done (e.g., beating an adversary) rather than on error-free operation.

For example, during the Korean conflict, the Naval Ordnance Test Station at China Lake designed and produced an anti-tank rocket, the RAM, in 29 days. The need for this weapon was so critical that safety measures usually observed were suspended. The Station's Michelson Laboratory was turned into a factory at night, and the production line ran down the main corridor of the laboratory. Wives came into the laboratory to work alongside their husbands to produce the weapon. The RAM was an outstanding success, but its production was a calculated risk.

A suggestive hypothesis is that in high-performance situations, there is a more masculine emphasis on winning, on being an "ace," and individual achievement, whereas high-reliability situations put the emphasis on balanced objectives and team effort. The context will determine which of these two emphases is more critical to the situation at hand. Usually, in civilian operations, high reliability is given stronger emphasis, whereas in a military context, high performance would be more important than error-free operation. As organizations may face situations with differing performance requirements, effective leadership may shift emphasis from one of these orientations to the other. However, we believe that high-integrity operation implies protection of critical information flows. Maintaining utter probity is possible only when information is freely shared and accurately targeted. Thus, high-integrity organizations may have certain common features involving information including the following:

1. All decisions are taken on the best information available.
2. The processes that lead to or underlie decisions are open and available for scrutiny.
3. Personnel are placed in an environment that promotes good decision-making and encourages critical thought.
4. Every effort is made to train and develop personnel who can and will carry out the mission as intended.
5. Only those persons who are in a fit state to carry out the mission are made responsible to do so.
6. Ingenuity and imagination are encouraged in finding ways to fulfill the organization's objectives.

The rest of this chapter is concerned with the development of organizations that exhibit these performance characteristics. We believe that these features allow high-integrity systems to operate with safety and effectiveness. Conversely, organizations where incidents or accidents are likely to occur are those where one or more of these principles are compromised. The authors believe that every movement away from these principles is a movement away from high integrity and toward failure of the system (cf. Maurino, Reason, Johnston, & Lee, 1995).

5.2 Building a High-Integrity Human Envelope

Around every complex operation, there is a human envelope that develops, operates, maintains, interfaces, and evaluates the functioning of the sociotechnical systems (STS). The system depends on the integrity of this envelope, its thickness, and strength. Compromises to its strength and integrity uncover the system's weakness and make it vulnerable. Accordingly, an aviation organization that

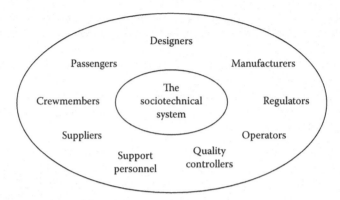

FIGURE 5.2 Members of the human envelope.

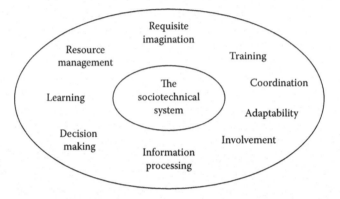

FIGURE 5.3 Essential activities of the human envelope.

nurtures this envelope will be strong. On the other hand, one that weakens it is heading for trouble (Figures 5.2 and 5.3).

"Concorde mafia." It is worthwhile to ponder the reflections of an accomplished chief engineer, Thomas J. Kelly, whose responsibility was the development of the Lunar Lander, and who built a strong human envelope to develop that system.

> The legacy of Apollo has played a major role in raising America to leadership in a global economy. I saw this on a personal level and watched it diffuse through the general practice of management. Apollo showed the value of (1) quality in all endeavors; (2) meticulous attention to details; (3) rigorous, well-documented systems and procedures; (4) the astonishing power of teamwork. I applied these precepts directly to Grumman's aircraft programs when I was vice president of engineering. They have since become the main thrust of modern management practices, developing into widely used techniques, such as total quality management, computer-aided design and manufacturing, employee empowerment, design and product teams, to name but a few (Kelly, 2001 p. 263).

A powerful human envelope, by the same token, may sustain an otherwise fragile and vulnerable system. According to knowledgeable sources, the Anglo-French Concorde airliner was kept aloft only by a kind of "Concorde Mafia." Each Concorde was basically a flying prototype, and only modest standardization existed between the various planes that bore the name. The aircraft's human envelope included many brilliant and strenuous engineers, designers, and maintenance technicians. This "mafia" worked very hard to keep the planes flying, and without it the fleet would have come rapidly to a standstill.

In the following sections, we have examined the activities that provide the high-integrity human envelope including

1. Getting the right equipment
2. Operating the equipment
3. Growing a high-integrity culture
4. Maintaining human assets
5. Managing the interfaces
6. Evaluation and learning

5.3 The Right Stuff: Getting Proper Equipment

5.3.1 Design: Using Requisite Imagination

The focus of this section is on the design process and the subsequent interactions over design, rather than the technical aspects of the designs themselves. It may seem strange to begin with the design of the equipment, because in many cases, aviation organizations take this aspect for granted. However, getting proper equipment is essential to high-integrity functioning. The organization that uses bad equipment will have to work harder to achieve success than the one that starts out with the proper equipment. The equipment that the organization uses should be adequate to insure a reasonable level of safety as well as the best available for the job—within the constraints of cost. The principle suggests that no aviation organization can afford to be indifferent to the equipment that it uses to its development, manufacture, and current state of functioning. It should systematically search out for the best equipment that it can afford to match the mission requirements, test it carefully, and endeavor to use it with close attention to its strengths and weaknesses.

An example of a conspicuous success was the Apollo space program, with its "lunar-orbit rendez-vous" concept. A careful study of the concept's genesis will show how important the openness of the design organization was to the success of Apollo. John C. Houbolt, associate chief of dynamic loads at the Langley Space Center, was not the first to conceive of the lunar-orbit rendezvous, but his studies and advocacy clinched this alternative as the solution. Starting in about 1960, Houbolt began to argue the advantages of a lunar-orbit rendezvous over the other alternatives: earth-orbit and a giant single two-way rocket called Nova. Other, more powerful, experts in NASA were unconvinced. Houbolt's first briefings encountered stiff resistance, but he kept coming back with more data and more arguments. The loose nonmilitary structure of NASA encouraged diverse strands of thinking, and eventually Houbolt won over the doubters. The key support of Wernher von Braun eventually closed the issue, at a time when even von Braun's engineers still favored the big rocket over the lunar rendezvous. (Hansen 1995).

Design should serve human purpose in an economical and safe way. However, system design, par-ticularly on a large scale, often fails owing to lack of foresight. In designing big systems, mistakes in conception can lead to large and costly foul-ups, or even system failure (Collingridge, 1992). This seems to be particularly true regarding software problems. About 75% of the major software projects actually get put into operation; the other 25% are canceled (Gibbs, 1994). Furthermore, many large systems may need considerable local adjustment, as has happened with the ARTS III software used by the Federal Aviation Administration (FAA) to manage major airport-traffic control (Westrum, 1994).

Recent years have provided many examples of compromised designs that affected safety. The destruc-tion of the Challenger, the Hyatt Regency disaster, and the B-1 and B-2 bombers are some major exam-ples. In each case, the designers did not think through the design or executed it badly.

Another major example of design failure is the Hubble Space Telescope. Hubble failed because nei-ther the National Aeronautics and Space Administration (NASA) nor the contractor insisted on carry-ing out all the tests necessary to determine if the system was functioning correctly. Instead, overreliance on a single line of testing, failure to use outside critical resources, and rationalization of anomalies ruled the day. When the telescope was launched, there was already ample evidence that the system had problems; however, this evidence was ignored. In spite of the many indications showing that the telescope was flawed, none were pursued. Critical cross-checks were omitted, inquiry was stifled, and in the end, a flawed system was launched, at a great public cost (Caspars & Lipton, 1991). The failure of the Hubble Space Telescope was a failure of the design process, and repairs were expensive. Another failure of cross-checks took place when an engineer inserted a last-minute correction into the software of the Mars Polar Lander, without checking through all the implications. The result was that the lander's motor cut off 40 feet above the Martian surface, causing loss of the lander and the mission. (Squyres, 2005, pp. 56–71.)

An equally flagrant example was the Denver Airport automated baggage-handling system. Here, an unproven system for moving the passengers' luggage was a key interface between parts of the airport. The concept demanded a careful scale-up, but none was carried out. When the airport opened, the automated baggage system did not work, and instead, a manual backup was used, at a great cost (Hughes, 1994).

The Hubble telescope and Denver Airport cases were mechanical failures. In other cases, the equipment may work mechanically, but may not interface well with people. This can happen through poor interface design (such as error-encouraging features), or because of unusual or costly operations that are necessary to maintain the equipment (cf. Bureau of Safety, 1967). A Turkish DC-10 crashed shortly after takeoff at Orly Airport, in France on March 3, 1974. "The serviceman who closed the door that day was Algerian and could not read the door instructions placard. As a result he failed to check that the latches were closed—as the printed instructions advised he should do. A glance through the door latch-viewing window would have shown that the latches were not fully stowed." (Adamski & Westrum, 2003, p. 194)

Some-years ago, a group of French researchers carried out a major study on French pilots' attitudes about automation (Gras, Morocco, Poirot-Delpech, & Scardigli, 1994). One of the most striking findings of this study was the pilots' concern about lack of dialogue with the engineers who designed their equipment. Not only did the pilots feel that there was insufficient attention to their needs, but they also felt that designers and even test pilots had a poor grasp of the realities that the pilots faced. Although attitudes toward automation were varied, pilots expressed very strong sentiments that more effort was needed to get designers in dialogue with the pilots before the equipment features were finalized.

One of the key skills of a project manager is the ability to anticipate what might go wrong, and test for that when the system is developed. Westrum (1991) called this as "requisite imagination" (cf. Petroski, 1994). Requisite imagination often indicates the direction from which trouble is likely to arrive. Understanding the ways in which things can go wrong often allows one to test to make sure that there are no problems. As demonstrated by Petroski (1994), great designers are more likely to ask deeper and more probing questions, and consider a wider range of potential problems.

Although foresight is valuable, it cannot be perfect. Even the best systems-design strategy (Petroski, 1994; Rechtin, 1992) cannot foresee everything. Hence, once the system is designed and produced, monitoring must be continued, even if nothing appears to be wrong. If things begin to go wrong, a vigilant system will catch the problems sooner. The Comet and Electra airliners, for instance, needed this high level of vigilance, because each had built-in problems that were unanticipated (Schlager, 1994, pp. 26–32, 39–45). Such examples show that, even today, engineering is not advanced to such an extent that all the problems can be anticipated beforehand. Even maestros (discussed later) do not anticipate everything. Joseph Shea, a fine systems engineer, blamed himself for the fire that killed three of the Apollo astronauts. Yet, Shea had done far more than most managers in anticipating and correcting problems (Murray & Cox, 1989).

5.3.2 Getting the Knowledge as Well as the Hardware

No equipment comes without an intellectual toolkit. This toolkit includes, but is not limited to, the written manuals. Kmetz (1984), for instance, noted that the written documentation for the F-14 Tomcat fighter comprised 300,000 pages. However, these abundant materials often are deficient in both clarity and usability. We have observed that the creators of many operational documents—that is, checklist, operational manuals, training manuals, and so on—assume that their message is transparent and crystal clear. Often, the message is anything but transparent and clear. Its faults can include documents that are difficult to use, and therefore, are not used; complex procedures that encourage procedural bypasses and workarounds; and difficult-to-understand documents composed by writers who have not considered the needs of the end users. The writers of such documents can unwittingly set up future failures.

Manuals always leave things out. All equipment is surrounded by a body of tacit knowledge regarding the fine points of its operation, and getting this tacit knowledge along with the formal communication may be vital. Tacit knowledge may include matters that are difficult to put into words or unusual modes of the equipment that are included for liability for reasons. Organizational politics has been known to lead to the inclusion or deletion of material. (e.g., Gillchrist, 1995, pp. 124–125). What goes into the manuals may involve erroneous assumptions about what people would "naturally" do. For instance, during an investigation on the two Boeing 737 accidents, an FAA team discovered that the designers assumed that pilots would respond to certain malfunctions by taking actions that were not in the written manual for the 737. Among other assumptions, the designers believed that if one hydraulic system was jammed, then the pilots would turn off both the hydraulic systems and crank the landing gear down by hand. Of course, if the plane was on landing approach, then there might not be time to do this. Although the hydraulic-device failure is rare in the landing situation, the key point is that the expected pilot actions were not communicated in the manual (Wald, 1995). The Boeing 737 is one of the safest jets in current use, yet, this example illustrates that not all information regarding the equipment is expressed in the manual, and some that is expressed, may not be necessary, because there are lots of things that one need not know. However, sometimes, critical things can get left out. In accepting a new airliner, a used airliner, or any other piece of machinery, care needs to be taken to discover this tacit knowledge.

The designers may not be the only holders of this tacit knowledge. Sometimes, other pilots, operators of air-traffic control equipment, or mechanics may hold this not-written-down knowledge. A study on Xerox-copier repair people, for instance, showed that much of the key information about the machines was transmitted orally through scenario exchange between repair people (Brown & Dugid, 1991). Similarly, process operators in paper pulp plants often solved problems through such scenario exchange (Zuboff, 1984). Kmetz (1984) found that unofficial procedures ("workarounds") were committed only to the notebooks of expert technicians working on avionics repair. Sensitivity to such off-the-record information, stories, and tacit knowledge is important. It is often such knowledge that gets lost in layoffs, personnel transfers, and reshuffling (cf. Franzen, 1994).

The use of automation particularly requires intensive training in the operation and the quirks of the automated system. However, training requires constant updates. Some key problems may be pinpointed only with field experience of the hardware. Failure of the organization to collect and transmit information about quirks in a timely and effective way could well lead to failure of the equipment, death, and injury.

For instance, on December 12, 1991, an Evergreen Air Lines 747 over Thunder Bay in Canada ran into trouble with its autopilot. The autopilot, without notifying the pilots, began to tip the plane over to the right, at first slowly, then more rapidly. The pilots did not notice the motion because it was slow. Finally, with the right wing dipping radically, the plane lost lift, and began plummeting downward. After much struggle, the pilots succeeded in regaining control, and landed in Duluth, Minnesota. An FAA investigation revealed that over the years similar problems had occurred with 747 autopilots used by other airlines. However, particularly intriguing was the discovery that the Evergreen plane's roll computer had previously been installed in two other planes in which it also had caused uncommanded rolls. Nevertheless, the exact cause of the problem in the roll computer remains unknown (Carley, 1993).

As automation problems are more fully covered elsewhere in this book (see Chapters 6, 7, and 20), we have not discussed them in detail. However, it is worth noting that hardware and software testing can, in principle, never be exhaustive (Littlewood & Stringini, 1992) and therefore, the price of safety is constant vigilance and rapid diffusion of knowledge about the equipment problems.

The issue of constant vigilance recalls the dramatic repair of the Citicorp Building. The design and construction (1977) of the Citicorp building in New York City was an important architectural milestone. With an unusual "footprint," the Citicorp building rose 59 stories into the skyline. However, unknown to its designer, William J. LeMessurier, the structure had a built-in vulnerability to high quartering winds. LeMessurier had specified welds holding together the vertical girders of the building. The structure LeMessurier had designed would handle the high winds that struck the building from the diagonal. However, it had not been built strictly to plan. The contractor had substituted rivets for the welds that had been specified. Ordinarily, this would have been fine, but not on this building. The riveted structure might fall to winds expected only once every 16 years. All this was unknown to LeMessurier when he received a call from an engineering student doing a research project. The architect reassured the student that all was fine, but the call got LeMessurier thinking and finally he checked with the contractor. The variance was discovered. The architect met with the contractor, the police, and Citicorp, and they decided that the problem needed to be fixed without raising alarm. Every night after the secretaries left the building, welders came in and did their work. The building was gradually welded into a safe configuration, and then the repair was finally announced to the public (Morgenstern, 1995).

5.3.3 Sustaining Dialogues about Key Equipment

For aviation organizations, we should think about information in terms of a constant dialogue rather than a single transmission. Once a system is turned over to the users, the design process does not stop, it simply scales down. Furthermore, around each piece of key equipment in the aviation organization, a small or large dialogue may be needed. This dialogue includes manufacturers, operators, and regulators as the most obvious participants. Obviously, aircraft and its engines are particularly important subjects of such dialogue, but other items of equipment also require consideration. When there is a lack of dialogue, unpleasant things can happen.

Consider, for instance, the disastrous fire on a Boeing 737 at Ringway Airport near Manchester in the United Kingdom, on August 22, 1985. The fire involved an engine "combustion can" that fractured, puncturing a fuel tank. The can had been repaired by a welding method that had met British CAA standards, but was not what the manufacturer called for in the manual issued to the British Airways. This accident was the most dramatic of a series of problems with the cans. Earlier problems had been written off as improper repairs, but this masked a key breakdown. One sentence in the accident report highlighted this key breakdown in communication between the operators (British Airways) and the engine makers (Pratt & Whitney):

> It has become evident from the complete absence of dialogue between British Airways and Pratt & Whitney on the subject of combustion-can potential failures that, on the one hand, the manufacturer believed that his messages were being understood and acted upon, and on the other, that the airline interpreted these messages as largely inapplicable to them at the time (cited in Prince, 1990, p. 140).

It was the management's responsibility to notice and eliminate the discrepancy between what the manual called for and what was expected from the maintenance technicians. Obviously, the bad practices continued only through the management's willingness to allow variance from the recommended practice.

The November 2001 crash of an American Airlines plane in Belle Harbor, Queens (New York) was the second worst accident in U.S. airlines history. The crash of flight 587 came even though the manufacturer, Airbus, had anticipated that the maneuver causing the accident—rapid back-and forth movement of the tail—could be fatal. Airbus had not shared a memo that discussed an incident near West Palm Beach, Florida in 1997, when rapid tail maneuvering nearly caused a similar fatal crash. The internal Airbus memorandum was not communicated to American Airlines. Thus, it was not incorporated into the pilots' training. Flight 587 was taking off from Kennedy International Airport. When the aircraft was caught in the turbulence following another aircraft, the pilots reacted by moving the tail rapidly back and forth. After 8 s of this rapid movement, the tail broke off. The crash caused the death of 285 people, including 5 on the ground (Wald, 2004).

Therefore, it should be obvious that the security of an airplane is shaped—in part—by the quality of dialogue between the maker and the user. The combustion-can problems were evidently a case of the "encapsulation" response (explained later), in which the system did not pay attention to the fact that it was having a problem.

A particularly important study was conducted by Mouden (1992, p. 141) for the Aviation Research and Education Foundation to determine the most significant factors in preventing airline accidents. Mouden's study included personal interviews with senior airline executives, middle management personnel, and airline safety officers to determine the actions by the management, which they considered the most effective for accident prevention. Several of those interviewed indicated that they thought complete safety was probably an unattainable goal. Many also indicated that risk-management managers may have a strong influence on the safety through effective communication, training, and standard operating procedures.

Mouden's study demonstrated the need for sensitivity to the communication channels in the organization. He noted that sometimes the designated communication channels in the organization are less effective than that believed, but their failure is discovered only after the occurrence of some unpleasant event. Thus, latent failures may accumulate but remain unseen (cf. Reason, 1990). Mouden presented a series of case studies that showed these problems with communication. While the organization chart emphasized vertical communication, Mouden discovered that managers at virtually all levels considered lateral communication as more effective than vertical.

5.3.4 Customizing the Equipment

Equipment in constant use does not stay unchanged for long. Through use, repair, and on-the-spot redesign, its form mutates. Customizing equipment can lead to two situations, each of which is worth consideration:

1. *Enhancements may improve safety.* Changes may provide substantial advantages by improving the ease, efficiency of operations, or aesthetic qualities for the local users.

> Eric Von Hippel, in the studies on "lead users," found that lead users are more likely to customize their equipment (Peters, 1992, pp. 83–85). Often, in the changes that lead users make, there exist the secrets for improving equipment, which, if carefully studied, will provide better manufactured products in the future. This certainly appeared to be true with regard to the ARTS-III traffic control software, developed by the FAA. A considerable number of "patches" had to be made to the software to allow local conditions. These patches, furthermore, were more likely to be spotted and transmitted face-to-face, rather than through any official channels. Many of the patches were tested late at night, when traffic was light, before being officially submitted for approval. The FAA, however, seemed slow to pick up on these changes (Westrum, 1994).

There has been intense interest in the "high-performance team" ever since Peter Vaill wrote his 1978 article. We can define a high-performance team as the one operating beyond ordinary expectations under the situation in which the group finds itself. Just as the ace or the virtuoso embodies unusual individual performance, the "crack" team shows a group performing at virtuoso level. This does not simply mean a group of virtuosos, but rather a group whose interactions allow performance of the task at a high effectiveness level. Although the literature on high reliability seems to have ignored Vaill's work, it is evident that high reliability shares many of the same characteristics as high performance. In any case, high-integrity teams get more out of their equipment. It is a common observation that such teams can get the same equipment that may turn out a lackluster performance for others, to perform "like a Stradivarius" for them. There are two reasons for this.

First, these teams know their equipment better. High-integrity teams or organizations take little for granted and make few assumptions. The equipment is carefully studied, and its strengths and limitations are recognized (Wetterhahn, 1997, p. 64). The team checks out and understands what it has been given, and subsequently "tunes it up" for optimal performance. High-performance teams will often go beyond the usual boundaries to discover useful or dangerous features. When the "Top Gun" air-combat maneuvering school was formed, the characteristics of the F-4 Phantom were carefully studied, and so, the team was able to optimize its use in combat (Wilcox, 1990). Similarly, in the Falklands war, one of the two British Harrier squadrons, the 901, carefully studied and learnt how to use its Blue Fox radar, whereas, the companion 800 squadron considered the Blue Fox unreliable and of limited value. The combat performance of the two groups strongly reflected this difference, with the 801 outperforming the other. Captain Sharkey Ward, Officer in Charge of the 801, summed up what he learnt from the conflict: "I have no hesitation in presenting the following as the most important lessons of the Falklands air war. The two main lessons must be: Know your weapons platforms, their systems, and operational capabilities; then employ them accordingly and to best effect" (Ward, 1992, p. 355). Thus, it is not just discovering the "edge of the envelope" that is important for high-performance teams, but also training to exactly exploit the features discovered.

High-integrity teams may sometimes even reject the equipment that they have been given. If what they have been given is not good enough, they may go outside the channels to obtain the equipment that they need. They are also natural "tinkerers." In a study about nuclear power plants and their incident rates, Marcus and Fox (1988) noted that the teams that carefully worked over their equipment were likely to have lower incident rates. Peters (1988, p. 166) also remarked that high-performance R&D teams customize their equipment more.

Often, the procedures of high-integrity teams skirt or violate official policy. Sometimes, this can affect safety. High-level policies are sometimes shaped by forces that have little to do with either the mission success or safety. Hence, when high performance is the principle criterion for the front line, policy may get violated. In Vietnam, when Air Force Falcon missiles did not work, they were replaced by Sidewinder missiles (Wetterhahn, 1997, p. 69). In a study on the use of the VAST avionics, check-outs were not the official policy, but were used to get the job done (Metz, 1984). Similarly, in Vietnam, American technicians often used "hangar queens," contrary to the official policy untouched, which is the essence of managerial judgment.

2. *Safety-degrading changes.* Wherever there is choice, there is danger as well as opportunity. Failure to think through actions with equipment may lead to human-factors glitches. One example was the United Airlines' new color scheme, dark gray above and dark blue below, which some employees called a "stealth" look. The poor visibility created for both planes and airport vehicles owing to matching colors evidently was not considered. It apparently led to a number of airport "fender benders" (Quintanilla, 1994). Similarly, methods for saving time, money, or hassles with equipment can often lead to the danger zone. Some airliners, for instance, may "fly better" with certain circuit breakers pulled. Although it is good to know such things, overuse of this inside knowledge can encourage carelessness and cause incidents.

Bad maintenance or repairs may cause equipment failures almost as dramatic as the use of substandard parts. In the Manchester fire case, there would have been no problem if the manufacture's instructions for maintenance had been followed.

Yet, it may be almost as bad to accept the equipment "as delivered," and "hope for the best" along with manuals and supportive documentation. Cultural barriers that impede or impair information search or active questioning may be one reason for this issue. Unwillingness to question may be particularly strong when the providers of the hardware are a powerful technical culture (e.g., the United States) and the recipients do not have a strong indigenous technical culture of their own. Airliners delivered to some developing countries may thus arrive with inadequate dialogue.

The organization receiving the equipment may cause further problems by dividing up the information involved and using it in adversarial ways. In fact, for groups with low team skills or internal conflicts, equipment may become a center for organization struggle. Different subgroups may assert their prerogatives, hiding knowledge from the groups using computer tomography (CT) scanners, and it has been found that cooperation between doctors and technicians may be difficult to achieve (Barley, 1986). When such knowledge is divided between the groups that do not communicate well, the best use of the equipment is not possible.

5.4 Managing Operations: Coordination of High-Tech Operations

5.4.1 Creating Optimal Conditions

One of the key functions for all levels of management in an aviation system is creating optimum human-factors situations in which others will operate. This means making sure that all the human-factors environments in the aviation organization provide contexts and personnel, resulting in a safe accomplishment of the job. In high-integrity organization, pilots, flight attendants, maintenance personnel, and dispatchers are more likely to find themselves in situations where they can operate successfully, when they have received the appropriate training for the activity, and where they get an adequate flow of information to do the job correctly.

Thus, environmental design is a management responsibility. At the root of many accidents is the failure to manage the working environment. For instance, on March 1, 1994, the crew of a Boeing 747–251B in a landing rollout at Narita Airport found one of its engines dragging (National Transportation Safety Board, 1994). The reason, it seemed, was that pin retainers for a diagonal engine brace lug had not been reinstalled during the "C" check in St. Paul, Minnesota. In looking into the accident, the National Transportation Safety Board (NTSB) found that the conditions in the Northwest Airlines Service Facility in St. Paul constituted an error-prone environment. Mechanics' understanding of the procedures was inconsistent, training was not systematically carried out, and the layout of the inspection operations was inefficient, causing stress to the inspectors. Clearly, these were the conditions that the management had to identify and improve.

James Reason, in introducing his well-known theory of accidents, noted that errors and mistakes by the operators at "the sharp end" are often promoted as the "cause" of accidents, when actions by management have actually created unsafe conditions in the first place. These management actions create situations that Reason termed as *latent pathogens*—accident-prone or damage-intensifying conditions (Reason, 1990). Therefore, it is important to be aware of the potential of putting personnel in situations where they should never be in the first place. A reluctance to create hazardous situations needs to go hand-in-hand, but with a willingness to deal with them when they appear.

For instance, both British airlines and the British pilots union, BALPA, were reluctant to admit that pilot fatigue was a problem. Fatigue is a proven killer, yet a good many senior managers used a "public relations" strategy (discussed later) to overcome the problem (Prince, 1990, pp. 111–129). A latent pathogen existed, but the organization steadfastly hid it from the sight. Unfortunately, the problem did not go away, but just its visibility was curtailed.

Similarly, when a fire broke out on a grounded Saudi Arabian Airlines flight in Riyadh on August 19, 1980, the three Saudi Arabian Airlines pilots involved failed to take crucial actions in a timely way. Their casualness and inaction apparently caused the entire people onboard flight SV 163—301 persons—to die needlessly. All the three pilots had records that indicated severe problems (Prince, 1990, p. 130). Thus, who placed these pilots at the controls? It would appear a serious failure for management at any airline to place such men at the controls of a Lockheed L-1011.

5.4.2 Planning and Teamwork

Emphasis on planning is a strong indicator of high integrity. High-integrity organizations do not just "let it happen." More of their activities and decisions are under conscious and positive control. A popular bumper sticker in the United States states that "Shit Happens." The implication is that bad things happen in ways that are difficult to predict or control. This expresses a common working-class attitude about the level of control of the person over his or her life—that is to say, very little. The "shit happens" philosophy of life is at the opposite pole from that of the high-reliability team. Very little "shit" is allowed to happen in a high-integrity organization, and what it does is carefully noted, and, if possible, designed out of the next operation.

High-integrity organizations often appear to have mastered the disciplines that others have not, and thus, are able to do things that other organizations consider outside their realm of control. In civilian operations, this has meant a higher degree of safety; for the military, it has meant higher mission-success rates.

A remarkable picture of a high-reliability team is given in Aviel's article (1994) on the tire repair shop at United Airlines' San Francisco maintenance facility. High integrity is evident in the small team's self-recruitment, self-organization, high morale, excellent skills, customized layout, and obvious comprehensive planning. We would all like to know how to build such teams in the first place. However, to refrain from interfering with them is something that every management group can learn. Aviel pointed out that United Airlines was willing to give up some apparent economies to keep the team together.

Some high-integrity teams require extensive practice. But what should be done when the crew—such as an airliner flight deck team—needs to be a team temporarily? It appears that high-integrity characteristics may form even in a short span of time with the right leadership, right standard operating procedures, and proper training. The captain, in the preflight briefing, shapes the crew atmosphere, and this in turn, shapes the interactions during the flight (Ginnett, 1993). Thus, a cockpit with a crew resources management (CRM) atmosphere can be created (or destroyed) rapidly.

One instance of excellent CRM skills took place on United Airlines flight 811, flying from New York to New Zealand. Flight 811 was a Boeing 7474. The front cargo door blew out, killing several passengers, and a 50% power loss was experienced. The company policy in such a situation was to lower the landing gear. However, after considerable discussion, the crew decided not to lower the gear because they did not really know the state of the equipment. This decision was later revealed to have saved their lives. United Airlines' Captain Ken Thomas associates this deliberative behavior with the intense CRM training rendered by United Airlines (K. Thomas, personal communication, October 20, 1994).

5.4.3 Intellectual Resource Management

High-integrity organizations are marked by intelligent use of intellectual resources. As CRM is covered in detail in Chapter 9 by Captain Daniel Maurine, we have concentrated only on the more general application of the same principles. The wise use of intellectual resources is critical to all aviation operations inside, outside, and beyond the aircraft. There are basically three principles.

1. *Use the full brainpower of the organization.* Coordinate leadership is vital for this principle. Coordinate leadership is to allow a person who is the best to make a particular decision to take control—temporarily. Coordinate leadership is basic to aviation. In flying the plane, for instance, control on the flight deck will shift back and forth between the left- and right-hand seats, even though the pilot retains ultimate authority. However, we would like to suggest that coordination has wider implications that need to be examined.

For instance, General Chuck Yeager, in command of a Tactical Air Command squadron of F-100 Supersabres, managed to cross the Atlantic and deploy his planes to Europe without any failures. His perfect deployment was widely considered as exemplary. Yet, one of the keys to this accomplishment was Gen. Yeager's insistence on allowing his maintenance staff to decide whether the airplanes were fit to fly. Yeager had been in maintenance himself, but his basic attitude was that the maintenance people knew the best whether the equipment was ready to fly.

> I never applied pressure to keep all of our airplanes in the air; if two or three were being serviced, we just lived with an inconvenience, rather than risking our lives with aircraft slapdashed onto the flight line. I wouldn't allow an officer-pilot to countermand a crew chief-sergeant's decision about grounding an unsafe airplane. A pilot faced with not flying was always the best judge about the risks he was willing to take to get his wheels off the ground. And it paid off. My pilots flew confident, knowing that their equipment was safe (Yeager & Janos, 1985, p. 315).

Yeager's examples show that great leadership may include emphasis on high reliability as well as winning. This might seem surprising in the view of Yeager's overall "ace" qualities. When coordinate leadership does not take place, problems occur. In the BAC One-Eleven windscreen accident on June 10, 1990 (Birmingham, United Kingdom), a windscreen detached at 17,300 ft because it had been badly attached, nearly ejecting the pilot with it. A maintenance supervisor had done the job himself, owing to the shortage of personnel. As the supervisor did the job in a hurry, he installed the wrong bolts. No one else was present. He needed to have someone else to check his work, but instead, he became lost in the task (Maurino et al., 1995, pp. 86–101). Thus, failure to coordinate leadership can overload the person in charge.

2. *Get the information to the person who needs it.* The information based on which decisions are made should be the best available, and the information possessed by one member of the organization has to be available in principle to anybody who needs it. Probably, no better example of intellectual resource management can be cited than the Apollo moon flights. The organization was able to concentrate the needed intellectual resources to design systems and solve problems. Apollo 13's emergency and recovery took place at the apogee of NASA's high-integrity culture (Murray & Cox, 1989, pp. 387–449). By contrast, a conspicuous example of failure to notify occurred in U.S. air force operations in northern Iraq on April 14, 1994. Two F-15 fighters shot down two U.S. army Blackhawk helicopters, killing all 26 peacekeepers on board. The accident took place through a series of mistaken perceptions, including Identification Friend or Foe, AWACS mistakes, and failure to secure a good visual identification. The army helicopters were also not supposed to be in that place at that time. A disturbing feature was that a similar misidentification had taken place a year and a half before, but without a fatal result. In September 1992 two air force F-111's nearly annihilated two army Blackhawks on the ground, realizing only at the last minute that they were American. A chance meeting at a bar revealed how close the air force had been to wiping out the army helicopters. But when this original "near miss" had taken place, no one had notified the higher command about it, so no organizational learning occurred. Someone should have had the presence of mind to anticipate that another such incident would happen, and pick up the phone. (Snook, 2000, p. 215) In fact, one might use this criterion for cognitive efficiency of the organization: "The organization is able to make use of information, observations or ideas, wherever they exist within the system, without regard for the location or status of the person or group originating such information, observations or ideas" (Westrum, 1991). We will see later in this chapter that an organization's cognitive adequacy can be assessed by just noting how closely it observes this principle.

3. *Keep track of what is happening, who is doing what, and who knows what.* The ability to secure appropriate vigilance and attention for all the organization's tasks, so that someone is watching everything that needs to be watched, is critical to safety. We are all familiar with the concept of mental workload from the studies of pilots and other operators of complex machinery. Yet, often the most important workload is that shouldered by top management. If "situational awareness" is important for the pilot or flight deck crew, "having the bubble" is what top management needs

(Roberts & Rousseau, 1989). The importance of management keeping track cannot be underestimated. Managements having "too much on their minds" was implicated in the Clapham Junction railroad accident (Hidden, 1989), but it is a common problem in aviation as well. John H. Enders, vice chairman and past president of the Flight Safety Foundation, stated that the distribution of contributing cases for the last decade's fatal accidents included "perhaps 60%–80% management or supervisory inattention at all levels" (Enders, 1992).

5.4.4 Maestros

A key feature promoting high integrity in any aviation organization is the standards set by the leaders. The most powerful standards are likely to be those set by the *maestros*, who believe that the organization should operate in a manner consistent with their own high expectations (Vaill, 1982). In these organizations, persons of high technical virtuosity, with broad attention spans, high energy levels, a nd an ability to ask key questions, shape the culture. The maestro's high standards, coupled with the other personal features, force awareness and compliance with these standards on the rest of the organization. Arthur Squires, in his book on failed engineering projects, noted that major technical projects without a maestro often founder (Squires, 1986).

The absence of a maestro may cause the standards to slip or non-performance of critical functions. Such failures can be devastating to aerospace projects. An excellent example of such a project is the Hubble Space Telescope. Although the telescope's primary mirror design and adjustment were critical for the mission, the mirror had no maestro. No single person was charged with the responsibility of making the system work (Caspars & Lipton, 1991). Likewise, historical analysis might well show that safety in the American space program was associated with the presence or absence of maestros. During the balmy days of Apollo, NASA fairly bristled with maestros (see Murray & Cox, 1989). Michael Collins, an astronaut, made this comment about NASA Flight Directors:

> I never knew a "Flight" who could be considered typical, but they did have some unifying characteristics. They were all strong, quick, and certain. [For instance] Eugene Kranz, as fine a specimen of the species as any, and the leader of the team during the first lunar land. A former flight pilot... he looked like a drill sergeant in some especially bloodthirsty branch of the armed forces. Mr. Kranz and the other Flight—Christopher C. Kraft, Jr., John Hodge, Glynn Lunney, Clifford Charlesworth, Peter Frank, deserve a great deal of the praise usually reserved for the astronauts, although their methods might not have passed muster at the Harvard Business School. For example, during practice sessions not only were mistakes not tolerated, but miscreants were immediately called to task. As one participant recalls, "If you were sitting down in Australia, and you screwed up, Mr. Kraft, or Mr. Kranz, or Mr. Hodge would get on the line and commence to tell you how stupid you were, and you knew that every switching center... ships at sea, everybody and his mother, everybody in the world was listening. And you sat there and took it. There was no mercy in those days." (Collins, 1989, p. 29)

And they could hardly afford to have any mistakes. Space travel is even less forgiving than air travel when it comes to mistakes. This maestro-driven environment defined the atmosphere for Project Apollo. In the days of the Space Shuttle, maestros were much harder to find. When NASA standards weakened, safety also decreased (Cooper, 1986; McCurdy, 1993).

Maestros shape climates by setting high standards for aviation organizations. Consider Gen. Yeager's description of Colonel Albert G. Boyd in 1946. Colonel Boyd was then head of the Flight Test Division at Wright Field:

> Think of the toughest person you've ever known, then multiply by ten, and you're close to the kind of guy that the old man was. His bark was never worse than his bite: he'd tear your ass off if you screwed up. Everyone respected him, but was scared to death of him. He looked mean, and he was.

And he was one helluva pilot. He flew practically everything that was being tested at Wright, all the bombers, cargo planes, and fighters. If a test pilot had a problem you would bet Colonel Boyd would get in that cockpit and see for himself what was wrong. He held the three-kilometer low altitude world speed record of 624 mph, in a specialty built Shooting Star. So, he knew all about piloting, and all about us, and if we got out of line, you had the feeling that the old man would be more than happy to take you behind the hangar and straighten you out (Yeager & Janos, 1985, p. 113).

However, standards are not only strong because of the penalties attached. They must be intelligently designed, clear, well understood, and consistently applied. Not all maestros are commanding personalities. Some maintain standards through more subtle means. Leighton I. Davis, Commanding Officer of Holloman Air Force Missile Development Center in the 1950s, managed to elicit a fierce loyalty from his officers to such an extent that many of them worked 50 or 60 h a week so as to not to let him down. He got this loyalty by providing a highly supportive environment for research and testing (Lt. Col. Thomas McElmurry, personal communication, August 15, 1993).

Maestros protect the integrity through insistence on honest and free-flowing communications. Maestro systems exhibit a high degree of openness. Decisions must be open and available, as opposed to a secretive or political one. Maestros may also be critical for organizational change. A maestro at United Airlines, Edward Carroll, a vice president, acted as the champion who sponsored United's original program "Command, Leadership, and Resource Management," which was the organization's version of CRM. Carroll responded to the Portland, Oregon, crash of 1978 by promoting understanding of the root causes and devising a comprehensive solution (K. Thomas, personal communication, October 20, 1994).

5.4.5 Communities of Good Judgment

We speculate that a high-integrity organization must constitute a "community of good judgment." Good judgment is different from technical competence. Although technical knowledge is objective and universal, judgment pertains to the immediate present. Judgment is the ability to make sound decision in real situations, which often involve ambiguity, uncertainty, and risk. Good judgment includes knowledge of how to get things done, who can be counted on to do what, and usually reflects deep experience. Maestros exemplify good judgment.

High integrity demands a culture of respect. When good judgment is compromised, respect is impossible. In communities of good judgment, the individual's position in the system is proportional to the recognized mastery. Each higher level in the system fosters an environment below it which encourages sound decisions. Individual capabilities are carefully tracked, and often, knowledge of individuals' abilities will not be confined to the next higher level, but will go two levels higher in the system, thus, providing the higher-ups with the knowledge of the organizational tasks which run parallel to the knowledge of people. In other words, there exists awareness not only of what people can do, but also of what they are supposed to do. Though this knowledge allows a high degree of empowerment, it is also demanding.

By the way, a good example of the formation of a high-integrity culture on board a destroyer of the United States Pacific Fleet is described by its initiator, Captain Michael Abrashoff (2002). When Abrashoff assumed command of the USS *Benfold* in 1978, he found a culture of distrust and disrespect. Determined to change the situation, Abrashoff systematically built teamwork and cross-training in all departments of the ship. As he interviewed the entire crew, he found a stagnant flow of ideas, so he opened up the channels, built trust and respect, and used the crew's ideas to improve operations. He strongly improved the crew's quality of life as it improved its operational capability. The result was a model ship, high capable, and highly integrated. Its crew solved problems not only for the ship itself, but for the Fleet as a whole. The ship was awarded the Spokane Trophy as the most combat-ready ship in the Pacific Fleet. This remains the best description of the formation of a generative culture (see below) in the armed services of which we are aware.

If this speculation is accurate, then the most critical feature may be that respect is given to the practice of good judgment, wherever it occurs in the organization, rather than to hierarchical position. This observation leads to an interesting puzzle: If the organization is to operate on the best judgment, how does it know what the best judgment is?

5.5 Organizational Culture

5.5.1 Corporate Cultural Features That Promote or Degrade High Integrity

Organizational culture. Organizations move to a common rhythm. The organization's microculture ties together the diverse stands of people, decisions, and orientations. This organizational culture is an ensemble of patterns of thought, feeling, and behavior that guide the actions of the organization's members. The closest analogy one can make is to the personality or character of an individual. The ensemble of patterns is a historical product, and it may reflect the organization's experiences over a surprisingly long span of time (Trice & Beyer, 1993). It is also strongly shaped by external forces, such as national cultures and regional differences. Finally, it is shaped by conscious decisions about structure, strategy, and policy taken by top management (cf. Schein, 1992).

Organizational culture has powerful effects on the individual, but it influences rather than determining the individual actions. An organization's norms, for instance, constrain action by rewarding or punishing certain kinds of acts. However, individuals can violate both informal norms and explicit policy. Furthermore, some organizational cultures are stronger than others, and have a greater influence on the organization's members. For the individual, the norms constrain only to the extent that the organization is aware of what the individual is doing, and the individual in turn may decide to "buy into" or may remain aloof from the norms.

The relative development success of two models of the Sidewinder missile, the AIM-9B and the AIM-9R, was shaped by these buy-in issues. Test pilots are very influential in shaping the perception of Navy top brass about novel weapon systems. Whereas careful efforts were made to put test pilots psychologically "on the team" by the test personnel of the AIM-9B (1950s), such efforts stalled on the AIM-9R (1980s). The test pilots failed to "buy in" to the new digital missile. The result was that the AIM-9R, in spite of technical successes, got a bad reputation in the Pentagon, and was eventually cancelled. (Westrum, 1999, pp. 100, 202)

Organizational culture is an organic, growing concept, which changes over time—and of course, sometimes it changes more rapidly than at other times. Different parts of the organization may reflect variations of the culture, sometimes showing very substantial variations owing to different backgrounds, varying experiences, local conditions, and different leaders.

Aspects of culture. Anthropologists, sociologists, and psychologists (including human-factors specialists) have addressed organizational culture from the perspectives of their respective disciplines. As culture has several facets, some researchers have emphasized on one, some another, or some a combination of these facets. Three of the facets are cognitive systems, values, and behavior.

Culture exists as a share *cognitive system* of ideas, symbols, and meanings. This view was emphasized by Trice and Beyer (1993), who saw ideologies as the substance of organizational culture. Similarly, Schein, in his discussion (1992) on culture, described about organizational assumptions. An organization's assumptions are the tacit beliefs that members hold about themselves and others, shaping what is seen as real, reasonable, and possible. Schein saw assumptions as "the essence of a culture" (Schein, 1992, p. 26), and maintained that a culture is (in part) "a pattern of shared basic assumptions that the group learned as it solved its problems of external adaptation and internal integration, that has worked well enough to be considered valid and, therefore, to be taught to new members as the correct way to perceive, think, and feel in relation to those problems" (p. 12).

Assumptions are also similar to what others addressed as "theories-in-use." Argyris, Putnam, and Smith (1985) and Schon (1983) distinguished between espoused theory and theory-in-use. The former is what the group presents itself as the one that they believe, and the latter is what it really believes. Espoused theory is easy to discuss, but changing it will not change the behavior. On the other hand, theory-in-use may be hard to bring to the surface.

Values reflect judgments about what is right and wrong in an organization. They may be translated into specific norms, but norms may not always be consistent with the values, especially those openly espoused. For instance, Denison (1990, p. 32) defined perspectives as "the socially shared rules and norms applicable to a given context." The rules and norms may be viewed as the solutions to problems encountered by the organizational members; they influence how the members interpret situations and prescribe the bounds of acceptable behavior. However, the values held by an organization may be very difficult to decipher, as what is openly proclaimed may in fact not be the one enforced (Schein, 1992, p. 17). Espoused values (Argyris et al., 1985) may reflect what people may say in a variety of situations, but not what they do. Many participants in unsuccessful "quality" programs were too late to find out that quality is a concept supported by management only as an espoused value, and not as a value-in-use. This separation is parallel to the differences in "theory" mentioned earlier. In any case, values may be different for different subgroups, regions, and levels of responsibility. Sometimes, constellations of values are described as an organization's climate. Dodd (1991), for instance, defined organizational culture as the communication climate rooted in a common set of norms and interpretive schemes about phenomena that occur as people work toward a predetermined goal. The climate shapes how organizations think about what they do, and thus, how they get things done. Some aviation organizations may have a strong common vision and we-feeling (e.g., Southwest Airlines), while others may represent an amalgam of competing values, loyalties, and visions. Lautman and Gallimore (1987) found that management pilots in 12 major carriers thought that standards were set at the top of the organization, but so far, there has been a lack of in-depth studies to confirm this assertion.

Finally, culture is a pattern of observable behavior. This view is dominant in Allport's theory (1955) of social structure. Allport argued that social events involve observable patterns that coalesce into structures. He explored patterns that defined the social structures and implied them by examining the ongoing structure of interacting events. Although Allport did not define the structures as cultures, his research provides a basis for the study of organizational culture. Similarly, Linebarry and Carleton (1992) cited Burke and Litwin regarding organizational culture as "the way we do things around here" (p. 234). Emphasizing behavior suggests that cultures can be discovered by watching what people do.

These definitions and orientations constitute only a handful of those available. While they are intellectually stimulating, none has been compelling enough to gain general acceptance. Even the outstanding survey of the literature by Trice and Beyer (1993) is short of a synthesis. Thus, no one has yet developed a complete and intellectually satisfying approach to organizational culture. However, while this basic task is being accomplished, incidents and accidents occur, and lives and money are being lost. Hence, some researchers have tried to focus on specific cultural forms that affect safety. For instance,

- Pidgeon and O'Leary (1994) defined safety culture "as the set of beliefs, norms, attitudes, roles, and social and technical practices within an organization which are concerned with minimizing the exposure of individuals, both within and outside an organization, to conditions considered to be dangerous" (p. 32).
- Lauder (1993) maintained that safe corporate culture requires clear and concise orders, discipline, attention to all matters affecting safety, effective communications, and a clear and firm management and command structure.
- Wood (1993) stated that culture, taken literally, is what we grow things in. He stated that:

  The culture itself is analogous to the soil and water and heat and light needed to grow anything. If we establish the culture first, the safety committee, the audit program, and the safety newsletter will grow. If we try to grow things, such as safety programs, without the proper culture—they will die (p. 26).

- Westrum suggested that the critical feature of organizational culture for safety is information flow. He defined three types of climates for information flow: the pathological, the bureaucratic, and the generative (Westrum, 1991). As these types bear directly on the concept of high integrity, we have elaborated them in the following section.

5.5.2 Communications Flow and the Human Envelope

Using his well-known model, Reason (1990) suggested that accidents occur when latent pathogens (undetected failures) are associated with active failures and failed defenses by operators at "the sharp end" (Figure 5.4). Ordinarily, this is represented by a "Swiss cheese model" in which accidents occur when enough "holes" in the Swiss cheese slices overlap. However, this can also be represented by the "human envelope" model proposed earlier. Each of Westrum's organization types, because of its communication patterns, represents a different situation vis-à-vis, the buildup of latent pathogens in the human envelope. Effective communication is vital for identifying and removing these latent pathogens. We can represent each one in terms of both a diagram (Figure 5.5) and typical behaviors.

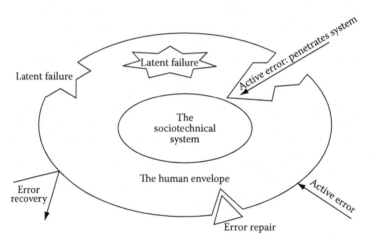

FIGURE 5.4   Active and latent failures in the human envelope.

Pathological	Bureaucratic	Generative
• Information is hidden	• Information may be ignored	• Information is actively sought
• Messengers are shot	• Messengers are tolerated	• Messengers are trained
• Responsibilities are shirked	• Responsibilities are compartmentalized	• Responsibilities are shared
• Bridging is discouraged	• Bridging allowed but not encouraged	• Bridging rewarded
• Failure is covered up	• Organization is just and merciful	• Failure causes inquiry
• New ideas are crushed	• New ideas create problems	• New ideas are welcomed

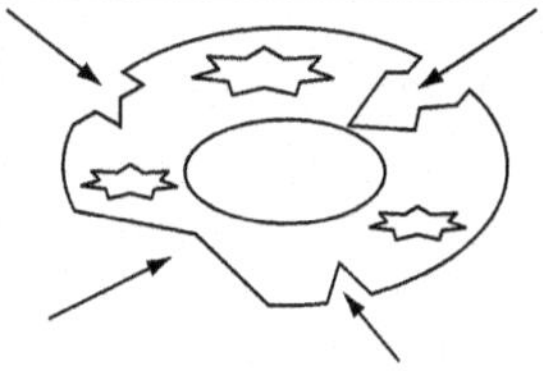

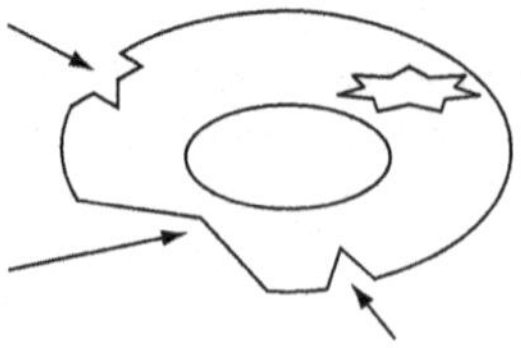

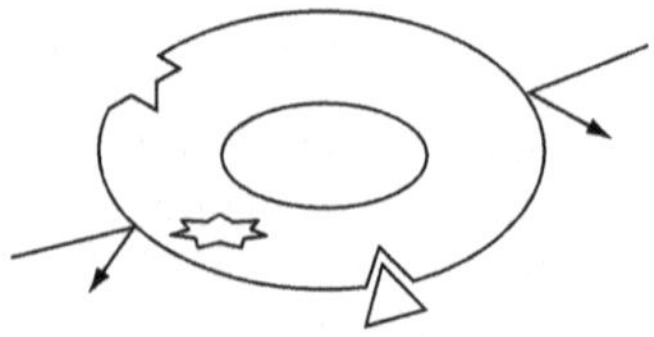

FIGURE 5.5   How organizational cultures treat information.

1. The *pathological* organization typically chooses to handle anomalies by using suppression or encapsulation. The person who spots a problem is silenced or driven into a corner. This does not make the problem go away, but just the message about it. Such organizations constantly generate "latent pathogens," as internal political forces act without concern for integrity. Pathogens are also likely to remain undetected, which may be dangerous for the place where it exists.
2. *Bureaucratic* organizations tend to be good at routine or predictable problems. They do not actively create pathogens at the rate of pathological organizations, but they are not very good at spotting or fixing them. They sometimes make light of the problems or only address those immediately presenting themselves, and the underlying causes may be left untouched. When an emergency occurs, they find themselves unable to react in an adaptive way.
3. The last type of organization is the generative organization, which encourages communication as well as self-organization. There exists a culture of conscious inquiry that tends to root out and solve problems that are not immediately apparent. The depth protects the STS. When the system occasionally generates a latent pathogen, the problem is likely to be quickly spotted and fixed.

Although Westrum's schema is intuitive and is well known in the aviation community, it is yet to be shown through quantitative studies that "generativity" correlates with safety.

Subcultures. In addition to coping with organization cultures, the problem is compounded by the existence of subcultures within the aviation organization. Over a period of time, any social unit that produces subunits will produce subcultures. Hence, as organizations grow and mature, subcultures arise (Schein, 1992). In most cases, the subcultures are shaped by the tasks each performs. Differing tasks and backgrounds lead to different assumptions. Within aviation organizations, subcultures have been identified primarily by job positions. For example, distinctive subcultures may exist among corporate management, pilots, mechanics, flight attendants, dispatch, and ground handling. Furthermore, these subcultures may show further internal differentiation, such as maintenance technicians versus avionics technicians, male flight attendants versus female flight attendants, sales versus marketing personnel, day-shift versus night-shift dispatch, and baggage versus fuel handlers.

Subcultural differences can become important through varying assumptions. Dunn (1995) reported on five factors identified at the NASA Ames Research Center that led to differences between the cabin crew and cockpit crew. Four of the five factors were rooted in assumptions that each group held about the other. Dunn reported that

- The historical background of each group influences the attitudes that they hold about each other.
- The physical separation of the groups' crew stations leads to a serious lack of awareness of each group's duties and responsibilities.
- Psychological isolation of each group from the other leads to personality differences, misunderstanding of motivations, pilot skepticism, and flight attendant ambivalence regarding the chain of command.
- Organizational factors such as administrative segregation and differences in training and scheduling create group differences.
- Regulatory factors lead to confusion over sterile cockpit procedures and licensing requirements.

Dunn argued that often the subcultures, evolving from shared assumptions, are not in harmony with each other—nor do they always resonate with the overall organizational culture. These groups are very clearly separated in most companies. The groups work for different branches of the company, have different workplace conditions, power, and perspectives. This lack of harmony can erode the integrity of the human envelope. Dunn provided a number of examples to depict the hazardous situations that can result from differences between the cockpit crew and the flight attendant crew. She noted that a Human-Factor Team that investigated the 1989 Dryden accident found that such separation was a contributing factor to the accident. These problems were further confirmed in an important study by Chute and Wiener (1995). Chute and Wiener documented the safety problems caused by lack of common training, physical separation, and ambiguous directives—such as the sterile cockpit rule. When emergencies arise, the resulting lack of coordination can have lethal consequences (Chute & Wiener, 1996).

Schein (1992) proposed that in some cases, the communication barriers between subcultures are so strong that organizations have to invent new boundary-spanning functions or processes. One example of such efforts is the recent initiative by the FAA and some industry groups calling for joint training programs between pilots and flight attendants. Such joint training can be very effective. Some years ago, one of us (Adamski) spoke about pilot and flight attendant relationships with a close friend, a captain with a major U.S. airline. The captain said that he had just attended his first joint training session between pilots and flight attendants, since his employment with the airline. With some amazement, he said that previously he never had any idea about the problems or procedures faced by the cabin crew. This joint training was the airline's first attempt to provide a bridge between the two subcultures. Joint training efforts have often produced positive results (Chute & Wiener, 1996).

Major Empirical Studies. Much research has been conducted to explore the many facets of organizational culture in the aviation community and the related high-tech industries. In most of these researches, improving safety and reliability has been the primary purpose. Although the findings are valuable, generally, they have been advanced without a previously articulated theory.

One of the earliest and most interesting examples of subtle creation of a safety culture in an aviation operation was provided by Patterson (1955), who managed to shift attitudes about accidents at a remote airbase in World War II as well as accomplish cross-functional cooperation, at the same time. Patterson's approach later became well known as "sociotechnical systems theory" and under the leadership of Eric Trist and others, it accumulated an imposing body of knowledge (e.g., Pasmore, 1988). The CRM concepts and sociotechnical idea have many factors in common. Nevertheless, although STS theory may be enormously helpful in aviation, it is yet to move out of the industrial environment that spawned it. Instead, current aviation research has focused on the organizational antecedents of "systems accidents" and CRM-related attitude and behavior studies.

The work on "systems accidents" was initiated by Turner (1978) and Perrow (1984), with major contributions from Reason (1984, 1990) and others. Turner and Perrow showed that accidents were "man-made disasters" and that the dynamics of the organizations routinely generated the conditions for these unhappy events. Reason traced the psychological and managerial lapses leading to these accidents in more detail. Reason noted that in accident investigations, blame was often placed on the operators at the "sharp end," whereas the conditions leading up to the accident (the "soft end") are given less emphasis. However, in fact, more probing has demonstrated that management actions are strongly implicated in accidents. For instance, the Dryden, Ontario, accident (1989), was initially dismissed as pilot error; however, investigation showed that it was rooted to problems far beyond the cockpit (Maurino et al., 1995, pp. 57–85). Similarly, in the controlled-flight-into-terrain accident on Mt. Saint-Odile, near Strasbourg, on January 20, 1992, a critical deficiency was the lack of a ground proximity warning system (Paries, 1994). The reasons for the lack of such systems reached far beyond the pilots, to management and national regulation.

5.5.3 Climates for Cooperation

In a parallel development, there was some outstanding ethnographic work by the "high-reliability" group at the University of California, Berkeley. In contrast to Perrow, the Berkeley group decided to find out why some organizations could routinely and safely carry out hazardous operations. Gene Rochlin, Todd LaPorte, Karlene Roberts, and other members of the "high-reliability group" carried out detailed ethnographic studies of aircraft carriers, nuclear power plants, and air-traffic control to determine why the accident rates for some of these operations were as low as they were found to be. These studies suggested some of the underlying principles for safe operation of large, complex systems, including

1. "Heedful interaction" and other forms of complex cooperation.
2. Emphasis on cooperation instead of hierarchy for task accomplishment. Higher levels monitor lower ones, instead of direct supervision at times of crisis.
3. Emphasis on accountability and responsibility and avoidance of immature or risky business.
4. High awareness about hazards and events leading to them.
5. Forms of informal learning and self-organization embedded in organizational culture.

The richness of the Berkeley studies is impressive, yet they remain to be synthesized. A book by Sagan (1993) sought to compare and test the Perrow and Berkeley approaches, but after much discussion (*Journal of Contingencies and Crisis Management*, 1994) by the parties involved, many issues remain unresolved.

Meanwhile, another approach developed from the work on CRM (see Maurino, Chapter 9, this volume). Robert Helmreich and his colleagues developed and tested materials for scoring actions and attitudes indicative of effective CRM. Originally, these materials grew out of the practical task of evaluating pilots' CRM attitudes, but have since been developed and extended to be used as measures of organizational attributes as well—for example, the presence of safety-supportive cultures in organizations. The more recent work has been strongly influenced by scales developed by Hofstede (1980) for studying differences in the work cultures of nations (discussed later). Using the Flight Management Attitudes Questionnaire, Merritt, and Helmreich (1995) made some interesting observations about safety-supportive attitudes in airlines. For instance, they observed that national cultures differed on some attitudes relevant to safety (see Figures 5.6 through 5.8).

The data in Figures 5.6 through 5.8 require some discussion. It is evident, for instance, that there are differences among nations as well as within a nation. In terms of differences between nations, one might expect "Anglo" (U.S./northern European) cultures to have features that support better information flow. Hence, it is not surprising to find that pilots in Anglo cultures seem more willing to support a flattened command structure (Figure 5.6). However, pilots from more authoritarian cultures apparently support a higher degree of information sharing than their Anglo counterparts (Figure 5.7)! According to Merritt and Helmreich, in authoritarian cultures, because of the large status differences in command, information-sharing needs to be particularly emphasized. However, the most-interesting features are the dramatic differences between the airlines from the same nation and the positive organizational culture (Figure 5.8). Positive organizational culture reflects questions about positive attitudes toward one's job and one's company. The airline designated USA 1 has a culture in the doldrums, when compared with the remarkable showing for USA 5, especially, considering that these are averaged scores for the

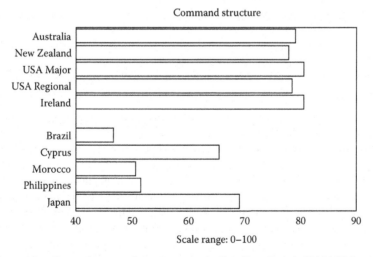

FIGURE 5.6   Support for a flattened command structure among pilots. (Data from the NASA/University of Texas/FAA Crew Resource Project.)

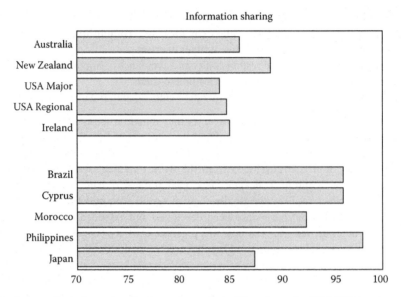

FIGURE 5.7 Support for information sharing among pilots. (Data from the NASA/University of Texas/FAA Crew Resource Project.)

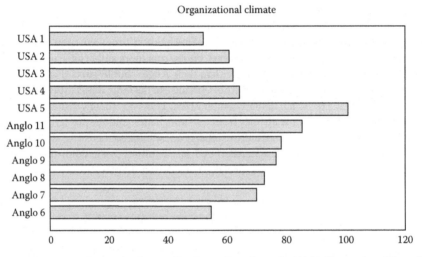

FIGURE 5.8 Positive organizational culture will airlines. (Data from the NASA/University of Texas/FAA Crew Resource Project.)

organization's members. One can only ponder on the impacts that these organizational attitudes have on safety, because the airlines in the study are anonymous.

In a related paper, Law and Willhelm (1995) showed that there are equally remarkable behavioral differences between the airlines. Using the Line/LOS Checklist developed by the NASA/University of Texas/FAA Aerospace Crew Project, raters observed and scored 1300 pilots. Figure 5.9 shows the results for two airlines identified only as "1" and "2." These assessments of behavioral markers show even greater variations in safety-related behavior than the attitudes studied by Merritt and Helmreich. In addition, Law and Willhelm (1995) showed that there are differences in CRM among the fleets of the same airline (Figure 5.10). However, the underlying features (history, recruitment, leadership, etc.) that account for these differences are unknown. However, both sets of data provide very strong evidence that organizational culture is related to safety.

Overall crew effectiveness rating in two airlines

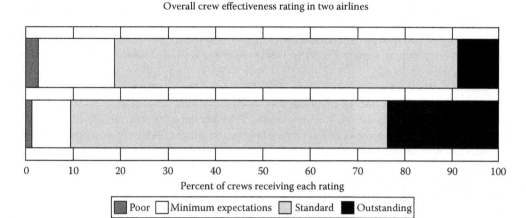

FIGURE 5.9   Overall crew effectiveness in two airlines. (Data from the NASA/University of Texas/FAA Crew Resource Project.)

Ratings of crew effectiveness by fleet

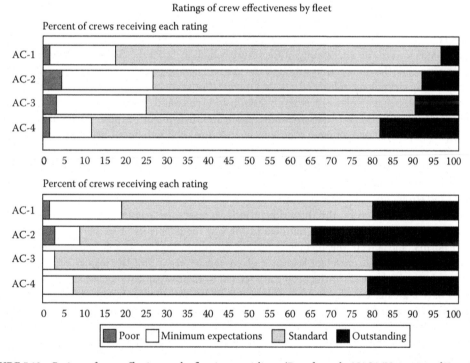

FIGURE 5.10   Ratings of crew effectiveness by fleet in two airlines. (Data from the NASA/University of Texas/FAA Crew Resource Project.)

5.5.4 National Differences in Work Cultures

Aviation operates in a global community. Some aviation organizations are monocultural: They operate within a specific area of the world and employ people largely from that same national culture. These aviation organizations manifest many of the features of the national cultures from which they developed. Others are multicultural: They have facilities throughout the world and employ people from a variety of national cultures. Multicultural crews represent a particular challenge. Recently, a physical struggle over the controls of an Airbus 300 broke out on a Korean Airlines flight deck as a Canadian captain and a Korean first officer struggled over how the landing should be managed. The first officer's command of English was insufficient to express his concerns, and hence, he simply grabbed the wheel. Finally, the plane crash-landed and then burned; fortunately, there were no casualties (Glain, 1994). Obviously, getting multicultural groups to work well together will be one of the key tasks that the aviations community has to face in the next decade.

As pointed out by anthropologists such as Hall (1959) for many years, each society is observed to provide its members with a "mental program" that specifies not only the general orientations, but also minute details of action, expression, and use of space. Travelers are often taken aback when foreigners act in ways that seem incomprehensible at home. However, on a flight deck or in a control tower, these differences can have serious consequences.

One useful framework for sorting out the differences in organization-relevant values between cultures was developed by Hofstede (1980). He identified four dimensions of national culture: power distance, uncertainty avoidance, individualism/collectivism, and masculinity.

Power distance is the degree to which members of a culture will accept differences in power between the superiors and subordinates. An unequal distribution of power over action is common in aviation organizations. It provides a way through which organizations can focus control and responsibility. However, the power distance varies considerably. In some cultures, the "gradient" is far steeper than others. As we have seen in the data provided by Helmreich and Merrit, discussed earlier, this trait shows strong variations, especially between Anglo and non-Anglo cultures.

The second dimension that Hofstede identified is the uncertainty avoidance. This is the tolerance that a culture holds toward the uncertainty of the future, which includes the elements of time and anxiety. Cultures cope with this uncertainty through the use of technology, law, and religion, while organizations cope using technology, rules, and rituals. Organizations reduce the internal uncertainty caused by the unpredictable behavior of the members by establishing rules and regulations. According to Hofstede (1980, p. 116), organizational rituals are nonrational, and their major purpose is to avoid uncertainty. Training and employee development programs may also be used to reduce uncertainty. As technology creates short-term predictability, it can also be used to prevent uncertainty. One way in which this takes place is through over-reliance on flight management systems (FMS) as opposed to CRM. Sherman and Helmreich (1995) found a stronger reliance on automation, for instance, in cultures with high power distance and strong uncertainty avoidance.

Individualism/collectivism, the third dimension, expresses the relationship between a member of a culture and his or her group. It is reflected in the way the people live together and are linked with societal norms, and affects the members' mental programming, structure, and functioning of the organizations. The norm prevalent within a given society regarding the loyalty expected from its members obviously shapes how the people are related to their organizations. Members of collectivist societies have a greater emotional dependence on their organizations. Organizations may emphasize individual achievement or the welfare of the group. The level of collectivism affects the willingness of an organization's members to comply with the organizational requirements. Willingness to "go one's own way" is at one pole of the continuum. At the other pole is the willingness to keep silent and go along with the group—often a fatal response in an emergency.

How different societies cope with masculinity/femininity is the fourth dimension identified by Hofstede (1980, p. 1976). Although masculine and feminine roles are associated with the roles for males and females, respectively, in many societies, how polarized the sexes are on this dimension varies to a greater extent. This dimension is obviously important for aviation. The "macho" attitude so often complained about in CRM seminars reflects a high masculinity orientation, and "task leadership" versus "socioemotional leadership" is also associated with this dimension (Bales, 1965). Similarly, some cultures may value masculine roles more highly than feminine ones. Recently, it was reported by the *Chicago Sun Times* that 20 Indian Airline flights were canceled because the pilots were upset that some senior flight attendants were getting more paid than themselves. The article stated that the pilots sat at the their seats with arms folded and refused to fly if certain flight attendants were onboard. The flight attendants retaliated by refusing to serve tea to the pilots.

Helmreich (1994) made a convincing argument that three of Hofstede's four variables were important in the crash of Avianca 052, which ran out of fuel in a holding pattern over Long Island on January 25, 1990. The pilots failed to communicate successfully with each other and with the ground, allowing a worsening situation to go unrecognized by the air-traffic control. Many of the CRM failures that Helmreich identified as being present during the flight seem to be associated with the high power distance, collectivist, and uncertainty-avoiding features of the pilots' Colombian culture.

Johnston (1995) speculated that differences in the cultural orientations might affect the response to and acceptance of CRM. The CRM itself is a value system, and may or may not collate with the local value systems. However, it is dangerous, as pointed out Johnston, to assume that regional differences in accident rates reflect the CRM orientations. He cited a paper by Weener (1990) that showed that although small aircraft accident rates vary strongly based on the different regions, accident rates for intercontinental aircraft are similar between developed and developing nations. The reason, as suggested by Johnston, is that international airports are more likely to operate on a world standard, while differences in the infrastructure show up more strongly in general accident rates. Hence, economic differences may be similar to that of culture in understanding accident rates. Thus, culture may be an important explanatory variable, but other differences between the nations need to be taken into account.

5.6 Maintaining Human Assets

5.6.1 Training, Experience, and Work Stress

Maintaining the human assets of an organization is critical to high integrity. Yet human assets are often neglected. Accident and incident reports are filled with descriptions of inadequate training, inappropriate tasking, fatigue, job-related stress, boredom, and burnout.

Huge differences can be found in the approaches that organizations take with regard to their members. Although high-integrity organizations are careful with their people, obviously many others are not. High-performance teams, for instance, are anything but passive in their attitude toward the people who are members. They show special care in hiring, making sure their people get trained correctly, giving personnel appropriate tasks, and monitoring how they are doing. New members are carefully vetted and "checked out" to observe their capabilities. Previous training is not taken for granted, and rather, new recruits are given a variety of formal and informal tests to assess their abilities.

Evaluating new member is not enough. Once skills have been certified, personnel have to join the team psychologically as well as legally. Aviation systems are often tightly coupled (Perrow, 1984). This means that all personnel need to be considered as a part of the system, because a failure by any one of them may cause grave problems. Yet, often higher managers fail to secure "buy in" by the organization's less visible members, and hence, the resulting disaffection by the "invisibles" can be costly. For example, maintenance personnel often have important roles in protecting safety, but seldom receive anything like the attention lavished on the flight deck crew by the management, public, and academics (Shepherd, 1994). Securing "buy in" by this group will be difficult, because while their failure receives swift attention, their successes are seldom so visible.

In a high-integrity organization, human assets are carefully maintained and assigned, and the experience of the operators is matched with the requirements of the task. If inexperienced or stressed workers are present, then they are put under careful supervision. In the study by Mouden (1992), mentioned earlier, frequent high-quality training was presumed to be the most important means of preventing accidents within the aviation organizations. However, training, especially high-quality training, is expensive. Organizations on the economic margin or in the process of rapid change or expansion, often neither do not have the money nor the time to engage in the training needed. In these organizations, integrity is often compromised by economic pressures.

One reason for lower integrity is the higher managers who allow the standards to slip. This appears to have been the case at Continental Express prior to the stabilizer detachment accident (discussed later). The NTSB Board Member John Lauber, in a minority opinion, noted that:

> The multitude of lapses and failure committed by many employees of Continental Express discovered in this investigation is not consistent with the notion that the accident originated from isolated, as opposed to systematic, factors. It is clear based on this [accident] record alone, that the series of failures that led to the accident were not the result of an aberration, but rather resulted from the normal, accepted way of doing business at Continental Express (NTSB, 1992, p. 53).

In an Addendum to this report, Brenner further explored the probability that two managers, in particular, the subsidiary's president and its senior director of maintenance and engineering, allowed the airline's maintenance standards to deteriorate (NTSB, 1992, Addendum). Continental's president had been an executive for Eastern Airlines and during this period, had made positive statements about the quality of maintenance during his watch which did not accord with the Eastern practices, as discovered by investigators. The maintenance director had earlier been director of quality control at Aloha Airlines when one of its planes suffered a preventable structural failure, resulting in the detachment of a fuselage upper lobe. Placing such people in critical positions in an airline suggests that higher management at Continental did not put high integrity in the foremost place.

Another way to create hazardous conditions is to turn operations over to undertrained or temporary personnel. It is well known that training flights, for instance, have unusually high accident rates. Furthermore, the accident literature describes many major blunders, sometimes fatal, which have taken place owing to inexperienced people at the controls of the airplane, the bridge of the ship, the chemical or nuclear reactor, and so on (cf. Schneider, 1991). Having such people in control often causes incidents or accidents because:

1. They make decisions based on lack of knowledge, incorrect mental models, or fragmentary information. For instance, they may not have an adequate idea on what a lapse on their part may mean for another part of the operation.
2. Newcomers or temporaries may not be part of the constant dialogue and may intentionally be excluded from participation in informal briefings, story-swapping, and so on.
3. Those who need surveillance by the supervisor increase the latter's mental workloads and thus, distract him or her.
4. Newcomers and temporary workers may have little commitment to the organization's standards, values, and welfare.
5. If they make errors or get into trouble, they are less likely to get the problem fixed rapidly, for fear of getting into trouble.

Even trained people can become risks if they are overstressed or tired. Moreover, often, economic pressures during highly competitive times or periods of expansion will encourage dubious use of human assets. This can happen even in the best firms. For instance, in 1988, users of Boeing 737s and 767s found that some of the fire extinguishers on these planes had crossed connections—that is, when one side was called for, the other side's sprinklers came on. Although the crossed connections were not implicated in an accident, the possibility was present. An even more serious problem with engine overheat wiring was discovered on a Boeing 747 of Japan Airlines. Investigation showed that hoses as well as wires were misconnected, and that the problem was widespread. Ninety-eight instances of plumbing or wiring errors were found on Boeing aircraft in 1988 alone. The FAA inspections in the Boeing plant at Everett, Washington, showed that quality control had slipped. Even the maintenance manual for the 757 was found to be incorrect, showing that the connections were reversed. A possible explanation for these various problems was the sudden brisk demand for Boeing products. Boeing's response may have been to use its assets outside the envelope of safe operation. According to one engineer:

> ...a too ambitious schedule for the new 747-400 aircraft has caused wiring errors so extensive that a prototype had to be completely rewired last year, a $1 million job... The Boeing employee also said the long hours some employees were working last year [1988] on the 747-400 production line—12 hour days for seven days a week, including Thanksgiving, Christmas, and New Year's Day—had turned them into zombies (Fitzgerald, 1989, p. 34).

Such high-stress situations are likely to result in errors that are easier to commit and harder to spot, thus, creating latent pathogens.

Layoffs of experienced people, whether owing to strikes, downsizing, or retirement policies, are likely to endanger integrity in aviation organizations and elsewhere. When the Chicago Post Office retired large numbers of its senior, experienced personnel, it shortly encountered severe problems: mails piled up, were put in trash baskets, or even were burned. The senior managers were badly needed to keep the system running, and the effects of their retirement were both unexpected and damaging to the integrity of the Post Office operations (Franzen, 1994). Similarly, when the PATCO strike led to large numbers of experienced air-traffic controllers being fired, extreme measures were needed to keep the system running. In fact, the air-traffic control system experienced many anxious moments. Although the feared increase in accidents did not take place, the stress experienced by many managers and others who took the place of the fired controllers in control towers was evident.

Major changes of any kind are likely to cause stress. Such changes include mergers, expansions, downsizing, or moving to new facilities. One of the most severe impacts on safety was the deregulation of U.S. airlines in 1978. Deregulation imposed additional pressures on many marginal operators, and led to mergers that brought together incompatible cultures. A study of one unstable and two stable airlines by Little, Gaffney, Rosen, and Bender (1990) showed that pilots in the unstable airline showed significantly more stress than those in the stable airline. This supports what the common sense suggests: A pilot's workload will increase with worries about the company. The Dryden, Ontario, accident also took place in the wake of a merger between Air Ontario and Austin Airways Limited. Investigation showed that the merger resulted in unresolved problems, such as unfilled or overburdened management roles, minimal flight following, and incompatible operations manuals (Maurino et al., 1995, pp. 57–85).

Pilots' worries about the companies in trouble may be well founded. A company in economic trouble may encourage pilots to engage in hazardous behavior, may confront the pilot with irritable supervisors, or may skimp on maintenance or training. It may be tempting to operate on the edge of the "safe region." An investigation of the airline U.S. Air by the *New York Times* showed that a climate existed in which fuel levels might not be carefully checked, resulting in some cases when the planes leave the airport with less fuel than they should have had (Frantz & Blumenthal, 1994).

Furthermore, government organizations are also not immune from the economic pressures. The American Federal Aviation Administration often uses undertrained inspectors to carry out its critical role of monitoring the safety of air carriers. It has a huge workload and a relatively a small number of staff to do the job. Thus, it may not be surprising to note that inspections are often perfunctory and sometimes overlook serious problems (Bryant, 1995b).

These examples suggest that while human assets may be expensive to maintain, failure to maintain them may well prove to be more expensive.

5.7 Managing the Interfaces

5.7.1 Working at the Interface

One of the biggest problems faced by aviation organizations is handling transactions across the boundaries of organizational units. This includes subsystems of the organization as well as the organization's relations with external bodies, ranging from unions to regulators. It is in these interfaces that things frequently go wrong.

One interface problem is hand-offs. When there is a failure to communicate across interfaces, the breakdown can set up some of the most dangerous situations in aviation. As an airplane is handed off from one set of controllers to another by the air-traffic control, as a plane is turned over from one maintenance crew to another, and as initiative on the flight deck is passed back and forth, loss of information and situational awareness can occur. It is essential that the two spheres of consciousness, that of the relinquisher and that of the accepter, intersect long enough to transfer all the essential facts.

The loss of a commuter aircraft, Embraer-120RT on September 11, 1991, belonging to Continental Express (Flight 2574), took place when the leading edge of the left horizontal stabilizer detached during the flight. The aircraft crashed, killing all onboard. Investigation showed that the deicer boot bolts had been removed by one maintenance shift, but were not replaced by the succeeding one, owing to faulty communications. The accident report (NTSB, 1992) commented that management was a contributing factor in setting up the conditions that led to confusion at the interface.

Another common problem is the failure to put together disparate pieces of information to get a picture of the whole situation. This apparently was one of the problems that led to the shoot-down of two U.S. Navy helicopters by Air Force fighters in Iraq. Inside the AWACS aircraft monitoring the airspace, each radarmen at different positions each had a piece of the puzzle; however, they failed to compare the notes. Thus, the failure in crew coordination led to the helicopters being identified as unfriendly, and they were shot down (Morrocco, 1994; see also Snook, 2000).

When two organizations are jointly responsible for action at an interface, neither may assume responsibility. We have already noted the breakdown of an interface in the Manchester fire of 1985. The following is the comment by John Nance on the source of the deicing failure that led to the Air Florida (Potomac) Crash in 1982:

There were rules to be followed, inspections to be made, and guidelines to be met, and someone was supposed to be supervising to make certain it was all accomplished according to plan. But neither Air Florida's maintenance representative nor American's personnel had any idea whose responsibility it was to know which rules applied and who should supervise them. So no rules were applied at all and no one supervised anything. They just more or less played it by ear (Nance, 1985, p. 255).

In contrast to this catch-as-catch-can approach, high-integrity organizations carefully control what comes into the organization and what goes out. An excellent example of such management of an interface is Boeing's use of customer information to provide better design criteria for the 777. Airlines were actively involved in the design process, providing input not only about layout, but also about factors that affected inspection and repair (O'Lone, 1992). By contrast, the Airbus 320 development seems to have made many French pilots, at least, feel that dialogue between them and the designers was unsatisfactory (Gras et al., 1994).

The best interfaces include overlapping spheres of consciousness. We can think of the individual "bubble," or field of attention, as a circle or sphere (in reality, an octopus or a star might be a better model). The worst situation would be if such spheres do not overlap at all; in this case, there would be isolation, and the various parties would not communicate. The best situation would be when the overlap is substantial, so that each would have some degree of awareness of the other's activities. However, sometimes the spheres, only touch at a single tangent point. In this case, there is a "single-thread" design, a fragile communication system. Single-thread designs are vulnerable to disruption, because the single link is likely to fail. Therefore, redundant channels of communication and cross-checking characterize the high-integrity teams. Unfortunately, some individuals do not want to share information, as it would entail sharing power. This is one of the reasons for the pathological organizations to become very much vulnerable to accidents: In such organizations there are few overlapping information pathways.

5.7.2 External Pressures

Another problem for the aviation community is with regard to coping with external forces. Aviation organizations are located in the interorganizational "fields of force," and are affected by social pressures. These fields of force often interfere with integrity. The actions of organizations are often shaped by political, social, and economic forces. These forces include airlines, airports, regulators and the public. One air charter organization, B & L Aviation, experienced a crash in a snowstorm in South Dakota. The crash was blamed on pilot error. However, after the crash, questions were raised about the regulatory agencies' oversight of B & L's safety policies. One agency, the FAA, had previously given the flying organization a clean bill of health, but the Forest Service, which also carries out aviation inspections, described it as having chronic safety problems. Further investigations disclosed that a U.S. Senator and his wife (an FAA official) had tried to limit the Forest Service's power and even eliminate it from inspecting B & L (Gerth & Lewis, 1994). The FAA, in general, is caught in such fields of local political and economic forces, and some have questioned its ability to function as a regulator owing to conflicting pressures and goals (e.g., Adamski & Doyle, 1994; Hedges, Newman, & Carey, 1995). Similarly, groups monitoring the safety of space shuttles (Vaughn, 1990) and the Hubble Space Telescope (Lerner, 1991) were subtly disempowered, leading to major failures.

Other individuals and groups formally "outside" the aviation organization may have a powerful impact on its functioning. Terrorists are an obvious example, but there are many others. Airport maintenance and construction crews, for instance, can cause enormous damage when they are careless. In May 1994, a worker in Islip, New York, knocked over a ladder and smashed a glass box, turning on an emergency power button; and the aircraft in three states were grounded for half an hour (Pearl, 1994). In September 1994, a worker caused a short circuit that snarled the air traffic throughout the Chicago region (Pearl, 1994). On January 9, 1995, power to Newark International Airport was shut down when a construction crew drove pilings through both the main and auxiliary power cables for the airport (Hanley, 1995).

5.8 Evaluation and Learning

5.8.1 Organizational Learning

All aviation organizations learn from experience, but how well they learn is another issue. In the aviation community, learning from mistakes is critical because failure of even a subsystem can be fatal. As aircraft parts are mass-produced, what is wrong with one plane may be wrong with others. Therefore, systematic error must be detected soon and rooted out quickly. When compared with other transport systems, aviation seems to have a good system for making such errors known and get corrected quickly (Perrow, 1984). For instance, when two rudders on Boeing 737s malfunctioned, all the units that had been modified by the procedure and thought to have caused the problem were checked (Bryant, 1995a). Similarly, when some propellers manufactured by Hamilton Standards proved defective, the FAA insisted that some 400 commuter planes be checked and defective propellers be replaced (Karr, 1995). This form of "global fix" is typical of, and somewhat unique to, the aviation industry. However, many other problems are not dealt with so readily.

It may be useful to classify the cognitive responses of aviation organizations to anomalies into a rough spectrum, such as the one presented in Figure 5.11 (based on Westrum 1986).

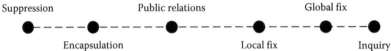

FIGURE 5.11   Organizational response to anomaly.

5.8.2 Suppression and Encapsulation

These two responses are likely to take place when political pressures or resistance to change is intense. In suppression, the person raising questions is punished or eliminated. *Encapsulation* happens when the individuals or group raising the questions are isolated by the management. For instance, an Air Force lieutenant colonel at Fairchild Air Force Base, in Washington state, showed a long-term pattern of risky flying behavior that climaxed in the spectacular crash of a B-52. Although similar risky behavior continued over a period of years, and must have been evident to a series of commanding officers, none prevented the officer from flying, and in fact, and he was put in charge of evaluating all B-52 pilots at the base (Kern, 1995). When this case and others were highlighted in a report by Allan Diehl, the Air Force's top safety official, Diehl was transferred from the Air Force Safety Agency in Albuquerque, New Mexico, to a nearby Air Force testing job (Thompson, 1995). The attempts to get photos of the shuttle *Columbia* during its last flight suffered encapsulation. When questions about the foam strike arose while the *Columbia* was orbiting in space, several individuals wanted photos of the potential damage. For instance, a group of NASA engineers, whose chosen champion was structural engineer Rodney Rocha, felt that without further data, they could not determine if the shuttle had been damaged seriously by the foam strike. Rocha made several attempts to get permission to have the Air Force take photos. The Air Force was willing to get the photos. But it was told by the Mission Management Team and by other NASA officials that it did not want further photographs. Rocha's requests were rebuffed by the Mission Management Team, the Flight Director for Landing, and NASA's shuttle tile expert, Calvin Schomburg. Whether such photos would have affected the shuttle's ultimate fate is unknown, but in retrospect NASA seems reckless not to have gotten them. (See Cabbage & Harwood, 2004, p. 134 and elsewhere). "Fixing the messengers......." Fixing the messengers instead of the problems is typical of pathological organizations. Cover-ups and isolation of whistle-blowers are obviously not a monopoly of the U.S. Air Force.

5.8.3 Public Relations and Local Fixes

Organizational inertia often interferes with learning. It makes many organizations respond to failure primarily as a political problem. Failure to learn from the individual event can often take place when failures are explained through public relations, or when the problem solved is seen as a personal defect or a random glitch in the system. For instance, even though the Falklands air war was largely won by the Royal Navy, public relations presented the victory as a triumph for the Royal Air Force (Ward, 1992, pp. 337–351). The public relations campaign obscured many RAF failures, some of which should have forced a reexamination of doctrine. Similarly, it has been argued that problems with Boeing 737-200s' pitching-up needed more attention than the situation, even after the Potomac crash of an Air Florida jet (Nance, 1986, pp. 265–279). Previously, Boeing had responded to the problem with local fixes, but without the global reach that Boeing could easily have brought to bear. When Mr. Justice Moshansky was investigating the Dryden, Ontario accident, legal counsel for both the carrier and the regulatory body sought to limit the scope of the inquiry and its access to evidence. Fortunately, both these attempts were resisted, and the inquiry had far-reaching effects (Maurino et al., 1995, Foreword).

5.8.4 Global Fix and Reflective Inquiry

In a high-integrity organization, failures are considered as occasions for inquiry, not blame and punishment (cf. Johnston, 1993). Aviation organizations frequently use global fixes (e.g., airworthiness directives) to solve common problems. However, the aviation community also has a large amount of "reflective inquiry" (Schon, 193), in which particular events trigger more general investigations, leading to far-reaching action. A comprehensive system of inquiry is typical of a community of good judgment, and it is this system that spots and removes the "latent pathogens." This system gives each person in the system a "license to think" and thus, empowers anyone anywhere in it to identify the problems and suggest solutions. Such a system actively cultivates maestros, idea champions, and internal critics. The Dryden, Ontario, accident inquiry and the United Airlines Portland, Oregon (1978), accident were both used as occasions for "system learning" far beyond the scope of the individual accident.

One can see in this spectrum, an obvious relationship among the three types of organizational cultures discussed earlier. Pathological organizations are more likely to choose responses from the left side of the spectrum, and generative organizations from the right side. We also expect that organizations with strong CRM skills would favor responses toward the right. We believe that studying this aspect may show that higher mission success and lower accident rates are more typical of organizations choosing responses toward the right of this distribution. Although anecdotal evidence supports the relationship, such a study remains to be done.

5.8.5 Pop-Out Programs

One of the features of reflective inquiry is the willingness to bring the otherwise hidden problems into view. These problems may be "hidden events" to management, suppressed because of unwritten rules or political influence (cf. Wilson & Carlson, 1996). Nonetheless, in high-integrity organizations, considerable effort may be exerted to make such invisible events visible, so that action can be taken on them. A "pop-out program" brings those aspects into the organization's consciousness which may otherwise have remained unknown. For instance, a factor in United Airlines developing its Command, Leadership, and Resources (CLR) program was a survey among United's pilots, which brought to the surface a number of serious unreported incidents. With this expanded database, management became ready to take stronger actions than it might otherwise have done (Sams, 1987, p. 30).

Similarly, the use of anonymous reporting from third parties was critical in the development of the Aviation Safety Reporting System (ASRS) in the United States. Through ASRS, information on a wide variety of incidents is obtained through confidential communications from pilots and others (Reynard, Billings, Cheaney, & Hardy, 1986). The ability to get information that would otherwise be withheld allows decision-making from a broader base of information, and also allows hidden events to become evident. However, the ASRS does not confer complete immunity on those who report to it, and some critics have noted that key information can be withheld (Nance, 1986).

Putting the right information together is sometimes the key to get hazards to stand out. Information not considered as relevant for cultural or other reasons is sometimes ignored. Disaster may follow such a lapse. Information relevant to icing problems on a small commuter plane called the ATR-72 was ignored by the FAA (Engelberg & Bryant, 1995a). Failure to collate the external evidence—in part, owing to political pressures—about the design's hazards meant that the FAA did not arrange the information such that the failure pattern stood out (Frederick, 1996). Similarly, failure of the Space Shuttle Challenger occurred partly because the statistics that pointed clearly to a problem with low temperatures were not assembled in such a way that the pattern linking temperature and blow-by was evident (Bell & Esch, 1989; Tufte, 1997, pp. 38–53).

A famous example of the encouragement for pop-out is Wernher von Braun's reaction to the loss of a Redstone missile prototype. After a prototype went off-course for no obvious reason, von Braun's group at Huntsville tried to analyze what might have gone wrong. When this analysis was fruitless, the group faced an expensive redesign to solve the still unknown problem. At this point, an engineer came forward and told von Braun that he might inadvertently have caused the problem through creating a short circuit. He had been testing a circuit before launch, and his screwdriver had caused a spark. Although the circuit seemed fine, obviously, the launch had not gone well. Investigation showed that the engineer's action was indeed at fault. Rather than punishing the engineer, von Braun sent him a bottle of champagne (von Braun, 1956).

5.8.6 Cognition and Action

Recognizing problems, of course, is not enough, and organizations have to do something about them. It must be remarked that although high-performance teams often have error-tolerant systems, the teams themselves are not tolerant of error, do not accept error as "the cost of doing business," and constantly try to eliminate it. High-performance teams spend a lot of time going over the past successes and failures, trying to understand its reasons, and subsequently, they fix the problems.

However, many organizations do not always follow this after the recognition of problems. Politically influenced systems may respond with glacial slowness while key problems remain, as with the systems used to carry out air-traffic control in the United States (Wald, 1996). Many of the computers used to direct traffic at U.S. airports can otherwise be found only in computer museums. At other times, aviation organizations are caught up in political pressures that influence them to act prematurely. New equipment may be installed (as in the case of the new Denver Airport) before it has been thoroughly tested or put through an intelligent development process (Paul, 1979).

Sometimes, aviation organizations seem to need disaster as a spur to action. Old habits provide a climate for complacency, while problems go untreated (Janis, 1972). In other cases, the political community simply will not provide the resources or the mandate for change unless the electorate demands it and is willing to pay the price. Often, it can require a horrendous event to unleash the will to act. For instance, the collision of two planes over the Grand Canyon in 1956 was a major stimulus to providing more en route traffic control in the United States (Adamski & Doyle, 1994, pp. 4–6; Nance, 1986, pp. 89–107). When FAA chief scientist, Robert Machol, warned about the danger of Boeing 757-generated vortices for small following aircraft, the FAA did not budge until two accidents with small planes occurred killing 13 people (Anonymous, 1994). After the accident, the following distance was changed from 3 to 4 miles. It is possible to trace the progress of the aviation system in the United States, for instance, through the accidents that brought specific problems to public attention. Learning from mistakes is a costly strategy, no matter how efficient the subsequent action is after the catastrophe. The organization that waits for a disaster to act is inviting one to happen.

5.9 Conclusion

"Human factors" has moved beyond the individual and even group level. Human factors are now observed to include the nature of the organizations that design, manufacture, operate, and evaluate aviation systems. Yet, although recent accident reports acknowledge the key roles that organizations play in shaping human factors, this area is usually brought in only as an afterthought. It needs to be placed on an equal footing with other human-factors concerns. We have recognized that "organizational factors" is a field at its infancy. Nonetheless, we hope to have raised some questions that further investigations can now proceed to answer.

However, we are sure about one point: high integrity is difficult to attain, as suggested by its rarity in the literature. Nonetheless, it is important to study those instances where it exists, and understand what makes it operate successfully. In this chapter, we have attempted to show that "high-integrity" attitudes and behaviors form a coherent pattern. Those airlines, airports, corporate and commuter operations, government agencies, and manufacturers that have open communication systems, high standards, and climates supporting inquiry may know things that the rest of the industry could learn. Furthermore, civilians could learn from the military and vice versa. From such inquiries and exchanges, we may learn to design sociotechnical systems that are more likely to get us safely to our destinations.

Acknowledgments

The authors acknowledge the kind assistance rendered by Timothy J. Doyle and Asheigh Merritt in writing this chapter.

References

Abrashoff, M. (2002). *It's your ship: Management lessons from the best damn ship in the navy.* New York: Warner.

Adamski, A. J., & Doyle, T. J. (1994). *Introduction to the aviation regulatory process* (2nd ed.). Westland, MI: Hayden McNeil.

Adamski, A. J., & Westrum, R. (2003). Requisite imagination: The fine art of anticipating what might go wrong. In E. Hollnagel, (Ed.), *Handbook of cognitive task design.* Mahwah, NJ: Erlbaum.

Allport, F. H. (1955). *Theories of perception and the concept of structure: A review and critical analysis with an introduction to a dynamic-structural theory of behavior.* New York: John Wiley & Sons.

Argyris, C., Putnam, R., & Smith, D. M. (1985). *Action science.* San Francisco, CA: Jossey-Bass.

Aviel, D. (1994, November 14). Flying high on auto-pilot. *Wall Street Journal,* p. A10.

Bales, R. F. (1965). The equilibrium problem in small groups. In A. P. Hare, E. F. Borgatta, & R. F. Bales (Eds.), *Small groups: Studies in social interaction* (pp. 444–476). New York: Alfred A. Knopf.

Barley, S. (1986). Technology as an occasion for structuring: Evidence from the observation of CT scanners and the social order of radiology departments. *Administrative Science Quarterly, 31*(1), 78–108.

Bell, T., & Esch, K. (1989). The space shuttle: A case of subjective engineering. *IEEE Spectrum, 26,* 42–46.

Brown, J. S., & Dugid, P. (1991). Organizational learning and communities of practice: Toward a unified view of working, learning, and innovation. *Organization Science, 2*(1), 40–57.

Bryant, A., (1995a, March 15). FAA orders rudder checks on all 737s. *New York Times.*

Bryant, A. (1995b, October 15). Poor training and discipline at FAA linked to six crashes. *New York Times,* pp. 1, 16.

Bureau of Safety. (1967, July). *Aircraft design-induced pilot error.* Washington, DC: Civil Aeronautics Board, Department of Transportation.

Cabbage, M., & Harwood, W. (2004). *Comm-check: The final flight of shuttle columbia.* New York: Free Press.

Carley, W. M. (1993, April 26). Mystery in the sky: Jet's near-crash shows 747s may be at risk of Autopilot failure. *Wall Street Journal,* pp. A1, A6.

Caspars, R. S., & Lipton, E. (1991, March 31–April 3). Hubble error: Time, money, and millionths of an inch. *Hartford Courant.*

Chute, R. D., & Wiener, E. L. (1995). Cockpit-cabin communications I: A tale of two cultures. *International Journal of Aviation Psychology, 5*(3), 257–276.

Chute, R. D., & Wiener, E. L. (1996). Cock-pit communications II: Shall we tell the pilots? *International Journal of Aviation Psychology, 6*(3), 211–231.

Collingridge, D. (1992). *The management of scale: Big organizations, big decisions, big mistakes.* London: Routledge.

Collins, M. (1989, July 16). Review of Murray and Cox, *Apollo: The race to the moon. New York Times Book Review* (pp. 28–29). New York: Simon and Schuster.

Cooper, H. S. F. (1986, November 10). Letter from the space center. *The New Yorker,* pp. 83–114.

Denison, D. R. (1990). *Corporate culture and organizational effectiveness.* New York: John Wiley & Sons.

Dodd, C. H. (1991). *Dynamics of intercultural communications.* Dubuque, IA: Wm. C. Brown.

Dunn, B. (1995). Communication: Fact or fiction. In N. Johnston, R. Fuller, & N. McDonald (Eds.), *Aviation psychology: Training and selection* (Vol. 2, pp. 67–74). Aldershot, England: Avebury Aviation.

Edmondson, A. C. (1996, March). Learning from mistakes is easier said than done: Group and organizational influences on the detection of human error. *Journal of Applied Behavioral Science, 32*(1), 5–28.

Enders, J. (1992, February). Management inattention greatest aircraft accident cause, not pilots, says enders. *Flight Safety Foundation News, 33*(2), 1–15.

Engelberg, S., & Bryant, A. (1995a, February 26). Lost chances in making a commuter plane safer. *New York Times*, pp. 1, 14, 15.

Engleberg, S., & Bryant, A. (1995b, March 12). Since 1981, federal experts warned of problems with rules for icy weather flying. *New York Times*, pp. 1, 12.

Fitzgerald, K. (1989, May). Probing Boeing's crossed connections. *IEEE Spectrum, 26*(5), 30–35.

Frantz, D., & Blumenthal, R. (1994, November 13). Troubles at USAir: Coincidence or more? *New York Times*, pp. 1, 18, 19.

Franzen, J. (1994, October 24). Lost in the mail. *The New Yorker*, pp. 62–77.

Frederick, S. A. (1996). *Unheeded warning: The inside story of American Eagle flight 4184*. New York: McGraw-Hill.

Gerth, J., & Lewis, N. A. (1994, October 16). Senator's bill to consolidate air inspection is questions. *New York Times*, pp. 1, 14.

Gibbs, W. W. (1994, September). Software's chronic crisis. *Scientific American*, 86–95.

Gillchrist, P. T. (1995). *Crusader! Las of the gunfighters*. Atglen, PA: Shiffer.

Ginnett, R. C. (1993). Crews as groups: Their formation and their leadership. In E. L. Wiener, B. G. Kanki, & R. L. Helmreich (Eds.), *Cockpit resource management*. New York: Academic Press.

Glain, S. (1994, October 4). Language barrier proves dangerous in Korea's skies. *Wall Street Journal*, pp. B1, B4.

Gras, A. C., Morocco, S. Poirot-Delpech, L., & Scardigli, V. (1994). *Faced with automation: The pilot, the controller, and the engineer*. Paris: Publications de la Sorbonne.

Hall, E. (1959). *The silent language*. New York: Doubleday.

Hanley, R. (1995, January 12). Blackout at Newark Airport leads to study of cable rules. *New York Times*.

Hedges, S. J., Newman, R. J., & Cary, P. (1995, June 26). What's wrong with the FAA? *U.S. News and World Report*, pp. 29–37.

Helmreich, R. (1994). Anatomy of a system accident: The crash of Avianca 052. *International Journal of Aviation Psychology, 4*(3), 265–284.

Heppenheimer, T.A. (1997). Antique machines your life depends on. *American Heritage of Invention and Technology, 13*(#1, Summer), 42–51.

Hidden, A. (1989). *Investigation into the Clapham Junction Railway accident*. London: HMSO.

Hofstede, G. (1980). *Culture's consequences: International differences in work-related values*. Beverly Hills, CA: Sage.

Hughes, D. (1994, August 8). Denver Airport still months from opening. *Aviation Week & Space Technology*, pp. 30–31.

James, R., & Hansen, J. R. (1995). Enchanted rendezvous: John C. Houbolt and the genesis of the lunar-orbit rendezvous concept. Monographs in aerospace history #4. Washington, DC: NASA History Office.

Janis, I. L. (1972). *Victims of groupthink: A psychological study of foreign-policy decisions and fiascoes*. Boston: Houghton Mifflin.

Johnston, N. (1993, October). Managing risk and apportioning blame. *IATA 22nd Technical Conference*, Montreal, Quebec, Canada.

Johnston, N. (1995). CRM: Cross-cultural perspectives. In E. Wiener, B. G. Kanki, & R. L. Helmreich (Eds.), *Cockpit resource management* (pp. 367–398). San Diego, CA: Academic Press.

(1994). [Special issue]. *Journal of Contingencies and Crisis Management.* 2(4).

Karr, A. R. (1995, August 28). Propeller-blade inspection set on small planes. *Wall Street Journal*, p. A34.

Kelly, J. T. (2001). *Moon lander: How we developed the Apollo lunar module*. Washington, DC: Smithsonian Institution Press.

Kern, T. (1995). *Darker shades of blue: A case of study of failed leadership*. Colorado Springs: United States Air Force Academy.

Kmetz, J. L. (1984). An information-processing study of a complex workflow in aircraft electronics repair. *Administrative Science Quarterly, 29*(2), 255–280.

Lauder, J. K. (1993, April). A safety culture perspective. *Proceedings of the Flight Safety Foundation 38th Annual Corporate Aviation Safety Seminar* (pp. 11–17). Arlington, VA.

Lautman, L. G., & Gallimore, P. L. (1987, June). The crew-caused accident. *Flight Safety Foundation Flight Safety Digest*, 1–8.

Law, J. R., & Willhelm, J. A. (1995, April). Ratings of CRM skill markers in domestic and international operations: A first look. *Symposium Conducted at the 8th International Symposium on Aviation Psychology*. Columbus, OH.

Lerner, E. (1991, February). What happened to Hubble? *Aerospace America*, pp. 18–23.

Lineberry, C., & Carleton, J. R. (1992). Culture change. In H. D. Stolovitch, & E. J. Keeps (Eds.), *Handbook of human performance technology* (pp. 233–246). San Francisco, CA: Jossey-Bass.

Little, L., Gaffney, I. C., Rosen, K. H., & Bender, M. (1990, November). Corporate instability is related to airline pilots' stress symptoms. *Aviation, Space, and Environmental Medicine, 61*(11), 977–982.

Littlewood, B., & Stringini, L. (1992, November). The risks of software. *Scientific American*, pp. 62–75.

Marcus, A., & Fox, I. (1988, December). *Lessons learned about communicating safety. Related concerns to industry: The nuclear regulatory commission after three mile island.* Paper presented at the Symposium on Science Communication: Environmental and Health Research, University of Southern California, Los Angeles.

Maurino, D. E., Reason, J., Johnson, N., & Lee, R. (1995). *Beyond aviation human factors*. Aldershot, England: Avebury Aviation.

McCurdy, H. E. (1993). *Inside NASA: High technology and organizational change in the U.S. space program*. Baltimore, MD: Johns Hopkins Press.

Merritt, A. C., & Helmreich, R. L. (1995). *Culture in the cockpit: A multi-airline study of pilot attitudes and values*. Paper presented at the 1995 papers: The NASA/University of Texas/FAA aerospace crew research project: VIIIth International Symposium on Aviation Psychology, Ohio State University, Columbus.

Morgenstern, J. (1995, May 29). The 59 story crisis. *The New Yorker, 71*, 45–53.

Morrocco, J. D. (1994, July 18). Fratricide investigation spurs U.S. training review. *Aviation Week & Space Technology*, pp. 23–24.

Mouden, L. H. (1992, April). Management's influence on accident prevention. *Paper presented at The Flight Safety Foundation 37th Corporate Aviation Safety Seminar: The Management of Safety*. Baltimore, MD.

Murray, C., & Cox, C. B. (1989). *Apollo: The race to the moon*. New York: Simon & Schuster.

Nance, J. J. (1986). *Blind trust*. New York: William Morrow.

National Transportation Safety Board (NTSB). (1992). *Aircraft accident report: Britt Airways, Inc. d/b/a Continental Express Flight 2574 in-flight structural breakup, EMB-120RT, N33701, Eagle Lake, Texas, September 11, 1991*. Washington, DC: Author.

National Transportation Safety Board. (1994). *Special investigation report of maintenance anomaly resulting in dragged engine during landing rollout of Northwest Airlines Flight 18, Boeing 747-251B, N637US, New Tokyo International Airport, Narita, Japan, March 1, 1994*. Washington, DC: Author.

O'Lone, R. G. (1992, October 12.) 777 design shows benefits of early input from airlines. *Aviation Week and Space Technology*.

Paries, J. (1994, July/August). Investigation probed root causes of CFIT accident involving a new generation transport. *ICAO Journal, 49*(6), 37–41.

Pasmore, W. A. (1988). *Designing efficient organizations: The sociotechnical systems perspective*. New York: John Wiley.

Patterson, T. T. (1955). *Morale in war and work*. London: Max Parrish.

Paul, L. (1979, October). How can we learn from our mistakes if we never make any? *Paper presented at 24th Annual Air Traffic Control Association Fall Conference.* Atlantic City, NJ.

Pearl, D. (1994, September 15). A power outage snarls air traffic in Chicago region. *Wall Street Journal*, p. 5.

Perrow, C. (1984). *Normal Accidents: Living with high-risk technologies.* New York: Basic Books.

Peters, T. (1988). *Thriving on chaos: Handbook for a management revolution.* New York: Alfred A. Knopf.

Peters, T. (1992). *Liberation management: Necessary disorganization for the nanosecond nineties.* New York: Alfred A. Knopf.

Petroski, H. (1994). *Design paradigms: Case histories of error and judgment in engineering.* New York: Cambridge University Press.

Pidgeon, N., & O'Leary, M. (1994). Organizational safety culture: Implications for aviation practice. In N. Johnston, N. McDonald, & R. Fuller (Eds.), *Aviation psychology in practice* (pp. 21–43). Aldershot, England: Avebury Technical.

Prince, M. (1990). *Crash course: The world of air safety.* London: Collins.

Quintanilla, C. (1994, November 21). United Airlines goes for the stealth look in coloring its planes. *Wall Street Journal*, pp. A1, A4.

Reason, J. (1984). Little slips and big accidents. *Interdisciplinary Sciences Reviews, 11*(2), 179–189.

Reason, J. (1990). *Human error.* New York: Cambridge University Press.

Rechtin, E. (1992, October). The art of system architecting. *IEEE Spectrum, 29*(10), 66–69.

Reynard, W. D., Billings, C. E., Cheaney, E. S., & Hardy, R. (1986). *The development of the NASA aviation safety reporting system* (NASA Reference Publication 1114). Moffett Field, CA: Ames Research Center.

Roberts, K. H. (Ed.). (1993). *New challenges to understanding organizations.* New York: Macmillan.

Roberts, K. H., & Rousseau, D. M. (1989, May). Research in nearly failure-free, high-reliability organizations: Having the bubble. *IEEE Transactions on Engineering Management, 36*(2), 132–139.

Roberts, K. H., & Weick, K. (1993, September). Group mind: Heedful interaction on aircraft carrier decks. *Administrative Science Quarterly, 38*(3), 357–381.

Sagan, S. D. (1993). *The limits of safety: Organizations, accidents, and nuclear weapons.* Princeton, NJ: Princeton University Press.

Sams, T. L. (1987, December). *Cockpit resource management concepts and training strategies.* Unpublished doctoral dissertation, East Texas State University, Commerce.

Schein, E. H. (1992). *Organizational culture and leadership* (2nd ed.). San Francisco, CA: Jossey-Bass.

Schlager, N. (Ed.). (1994). *When technology fails: Significant technological disasters, accidents, and failures of the twentieth century.* Detroit, MI: Gale Research.

Schon, D. A. (1983). *The reflective practitioner: How professionals think in action.* New York: Basic Books.

Schneider, K. (1991, July 30). Study finds link between chemical plant accidents and contract workers. *New York Times*, p. A10.

Shepherd, W. T. (1994, February 1). Aircraft maintenance human factors. *Presentation at International Maintenance Symposium.* San Diego, CA.

Sherman, P. J., & Helmreich, R. L. (1995). Attitudes toward automation: The effect of national culture. *Paper presented at the 1995 Papers: The NASA/University of Texas/FAA Aerospace Crew Research Project. VIII International Symposium on Aviation Psychology.* Ohio State University, Columbus.

Snook, S. (2000). *Friendly fire: Shootdown of U.S. blackhawks over northern Iraq.* Princeton, NJ: Princeton University Press.

Squires, A. (1986). *The tender ship: Government management of technological change.* Boston, MA: Birkhauser.

Squyres, S. (2005). *Roving Mars: Spirit, Opportunity, and the Exploration of the Red Planet.* New York: Hyperion.

Thompson, M. (1995, May 29). Way out in the wild blue yonder. *Time*, pp. 32–33.

Trice, H., & Beyer, J. M. (1993). *The cultures of work organizations.* Englewood Cliffs, NJ: Prentice-Hall.

Trotti, J. (1984). *Phantom in Vietnam.* Novato, CA: Presidio.

Tufte, E. R. (1997). *Visual explanations.* Cheshire, CT: Graphics Press.

Turner, B. A. (1978). *Man made disasters.* London: Wykeham.

Vaill, P. B. (1978). Toward a behavioral description of high-performing systems. In M. McCall, & M. Lombardo (Eds.) *Leadership: Where else can we go?* Durham, NC: Duke University Press.

Vaill, P. B. (1982). The purposing of high-performing systems. *Organizational Dynamics, 11*(2) 23–39.

Vaughn, D. (1990, June). Autonomy, interdependence, and social control: NASA and the space shuttle challenger, *Administrative Science Quarterly, 35*(2), 225–257.

von Braun, W. (1956, October). Teamwork: Key to success in guided missiles. *Missiles and Rockets,* pp. 38–43.

Waddell, S. (2002). *The right thing.* Brentwood, TN: Integrity Publishers.

Wald, M. L. (2004, October, 12). 1997 memo cited hazard of maneuver in air crash. *New York Times,* A28.

Wald, M. (1995, May 7). A new look at pilots' role in emergency. *New York Times,* p. 12.

Wald, M. (1996, January 29). Ambitious update of air navigation becomes a fiasco. *New York Times,* pp. 1, 11.

Ward, S. (1992). *Sea harrier over the falklands.* London: Orion.

Weener, E. F. (1990). *Control of crew-caused accidents: The sequal* (Boeing flight operations regional seminar: New Orleans). Seattle, WA: Boeing Commercial Aircraft Company.

Westrum, R. (1986, October). Organizational and inter-organizational thought. *Paper presented at the World Bank Conference on Safety and Risk Management.*

Westrum, R. (1991). *Technologies and society: The shaping of people and things.* Belmont, CA: Wadsworth.

Westrum, R. (1994). Is there a role for the "test controller" in the development of new ATC equipment? In J. Wise, V. D. Hopkin, & D. Garland (Eds.), *Human factors certification of new aviation technologies.* New York: Springer.

Westrum, R. (1999). *Sidewinder: Creative missile development at China lake.* Annapolis, MD: Naval Institute Press.

Wetterhahn, R. F. (1997, August). Change of command. *Air and Space,* pp. 62–69.

Wilcox, R. K. (1990). *Scream of eagles: Top gun and the American aerial victory in Vietnam.* New York: John Wiley.

Wilson, G. C., & Carlson, P. (1996, January 1–7). The ultimate stealth plane. *Washington Post National Weekly Edition,* pp. 4–9.

Wood, R. C. (1993). Elements of a safety culture. *Proceedings of the Flight Safety Foundation 38th Annual Corporate Aviation Safety Seminar* (pp. 26–29).

Yeager, G. C., & Janos, L. (1985). *Yeager: An autobiography.* New York: Bantam.

Zuboff, S. (1984). *In the age of the smart machine: The future of work and power.* New York: Basic Books.

II

Human Capabilities and Performance

6

Engineering Safe Aviation Systems: Balancing Resilience and Stability

Björn Johansson
Saab Security

Jonas Lundberg
Linköping University

6.1 Introduction

A recent development in safety management that has caught attention is "resilience engineering" (Hollnagel & Rigaud, 2006; Hollnagel, Woods, & Leveson, 2006; Woods & Wreathall, 2003). What "resilience engineering" exactly means is still a subject of discussion, but it is clear from the response of the scientific community that the concept appeals to many. According to Hollnagel et al. (2006), "resilience engineering" is "a paradigm for safety management that focuses on how to help people cope with the complexity under pressure to achieve success," and one should focus on developing the practice of resilience engineering in socio-technical systems. The term "socio-technical system" here refers to the constellation of both humans and the technology that they use, as in the case of a nuclear power plant or an air-traffic control center. Systems like those mentioned earlier share the characteristic that the tolerance toward failure is low. The costs of failure in such systems are so high that considerable effort is spent on maintaining an "acceptable" level of safety in them. Indeed, most of such systems can present an impressive record of stable perfor mance over long time-spans. However, the few cases of failure have led to catastrophic accidents where costs have been high, both in terms of material damage as well as the lives lost. Such accidents often lead to large revisions of safety procedures and systems, reenforcing the original system with altered or completely new parts aimed at improving safety. This process normally reoccurs in a cyclic fashion, moving the current level of performance and safety from one point of stability to another (McDonald, 2006). This kind of hindsight driven safety development is a common practice. The process continues until the system is considered as "safe" or the resources for creating new safety systems are depleted. Entirely new systems may be designed, encapsulating the original system with the purpose of making it safer. This is referred to as the "Matryoschka problem," using the metaphor of the Russian dolls, which states that it is impossible to build completely fail-safe systems as there will always be a need for yet another safety-doll to maintain the safety of its subordinate dolls. According to this metaphor, failure cannot be avoided completely; it may only become very improbable according to our current knowledge about it. Thus, we must accept that any system can fail (Lundberg & Johansson, 2006). In resilience engineering, it is proposed that the focus should lay on the ability to adapt to changing circumstances. A system should thus be designed in such a way that it can cope with great variations in its environment. In this chapter, we argue that the focus on such "resilience" is not sufficient in itself. Instead, we propose that systems should be designed in such a way that resilient properties are balanced with the properties aimed at coping with common disturbances.

6.2 What Is Resilience?

Originally, the term "resilience" comes from ecology and refers to the ability of a population (of any living organism) to survive under various conditions (Holling, 1973). Resilience has also been used to analyze individuals and their ability to adapt to changing conditions (e.g., Coutu, 2002). A common approach in the field of ecology is the assumption of "stability," indicating that systems that could recover to a state of equilibrium after a disturbance in their environment would survive in the long run. Holling (1973) presented the idea of resilience, stating that the variability of most actual environments is high, and that stable systems in many cases actually are more vulnerable than the unstable ones.

> Resilience determines the persistence of relationships within a system and is a measure of the ability of these systems to absorb changes of state variables, driving variables, and parameters, and still persist. In this definition resilience is the property of the system and persistence or probability of extinction the result. Stability, on the other hand, is the ability of a system to return to an equilibrium state after a temporary disturbance. The more rapidly it returns, and with the least fluctuation, the more stable it is (Holling, 1973, p. 17).

Some researchers interested in the field of safety/resilience engineering seem to confuse the notion of resilience and stability, actually discussing what Holling referred to as stability rather than resilience, as Holling stated that "With this definition in mind a system can be very resilient and still fluctuate greatly, i.e., have low stability" (Holling, 1973, p. 17). From Holling's perspective, the history of a system is an important determinant regarding how resilient it can be. He exemplified this by showing that species that exist in stable climates with little interaction with other species tend to become very stable, but may have low resilience. On the other hand, species acting in uncertain, dynamic environments are often subjected to great instability in terms of population, but they may as such be resilient and survive over very long time periods. This is in line with a later description of resilience provided by McDonald (2006), in which resilience in socio-technical systems is discussed:

> If resilience is a system property, then it probably needs to be seen as an aspect of the relationship between a particular socio-technical system and the environment of that system. Resilience appears to convey the properties of being adapted to the requirements of the environment, or otherwise being able to manage the variability or challenging circumstances the environment throws up. An essential characteristic is to maintain stability and integrity of core processes despite perturbation. The focus is on medium to long-term survival rather than short-term adjustment per se. However, the organisation's capacity to adapt and hence survive becomes one of the central questions about resilience—because the stability of the environment cannot be taken for granted (McDonald, 2006, p. 156).

McDonald's description of resilience is similar to that of Holling, distinguishing between stability and resilience. However, safety in a socio-technical system can be increased by improving both stability and resilience. In the following section, we discuss about the importance of a balanced perspective between these two aspects.

6.3 Balancing Resilience and Stability

A lesson learned from Holling's original ideas is that systems not only should be designed for stability, even if this is often desired, especially in production systems, but should also have a sole focus on resilience, which is hardly appropriate either. Instead, we need to have a balance between resilience and stability. Stability is needed to cope with expected disturbances, while resilience is needed to survive unexpected events. Westrum (2006) described the unwanted events according to three different categories: the *regular event*, the *irregular event*, and the *unexampled event*. The regular event obviously describes the events that often occur with some predictability. We know, for example, that machines malfunction, fires occur, and cars collide in traffic. We have procedures, barriers, and entire organizations designed to cope with these kinds of disturbances. Irregular events are foreseeable, but not expected. Earthquakes, Tsunamis, nuclear accidents, etc., are all examples of things we know might happen, but we do not expect them to. If they happen, society sometimes has prepared resources to handle them, or at least the possibility to gather such resources. If severe events happen, measures sometimes are taken to increase the preparedness, like earthquake warning systems. Irregular events represent the unimaginable. Westrum used the 9/11 attacks on the World Trade Centre in New York as an example. To these kinds of events, there is no prior preparation and, in some cases, no known countermeasure. In such cases, it is mostly only possible to deal with the event post facto, with whatever resources available.

This leads us to the fundamental problem of designing "safe" systems. It is impossible to prevent some events like Tsunamis, or prevent all the events of some kinds like forest fires or car accidents. Instead, the focus should be on the reactions to these kinds of events, and on the general ability to handle the consequences of such harmful events. The most blatant error that can be made is to assume that a system is completely safe or "immortal" and thus, ignore the need for coping with the unthinkable (Foster, 1993). Even if we cannot imagine a situation where a system loses control, we need to consider what to do if it ever should happen. There are examples, such as the Titanic, where the designers of the ship were so convinced that it could not sink, that they neglected to supply it with a sufficient amount of lifeboats. When reviewing the three kinds of threats described by Westrum (2006), these also seem to match the division between resilience and stability. For regular events, the recommendation might not be to alter or improve resilience in the system, but rather to fine-tune the system to reattain stability. Thus, when moving from regular to irregular and unexampled events, the demand for resilience increases (see Figure 6.1).

According to Lundberg and Johansson (2006), a balanced approach should be encouraged so that both everyday disturbances and unanticipated events can be managed. A simple example of an unbalanced approach is the way automation is often used. In many cases, automation is introduced to improve performance and safety in a system, simply by reducing the human involvement in a process. On the surface, it may look as if the automation has increased safety, as performance and accuracy of the man–machine system is higher than that without the automation. This often leads to an increased usage of automation to increase capacity, gradually reducing the human operator to a supervisor who only monitors the automation. As far as everything works as intended, this is unproblematic, but in case of major disturbances, for example, a breakdown in the automation, performance may degrade dramatically. In the worst case, the man–machine system may cease to function completely, as the human counterpart is suddenly left in a situation that is far beyond his/her performance boundaries (see Figure 6.2).

Thus, simply increasing the "stability" of a system, as in the case of automation, is only acceptable in situations where a loss of such an increase is tolerable. In many instances, this is not the case, and there is an apparent need for resilience so that a system can survive when its stable equilibrium is lost.

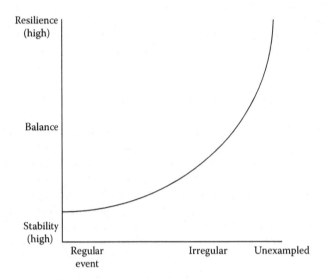

FIGURE 6.1 An outline of the relation between the need for resilience or stability in the face of different types of unwanted events. (From Lundberg, J. and Johansson, B., Resilience, stability and requisite interpretation in accident investigations, in Hollnagel, E. and Rigaud, E. (Eds.), *Proceedings of the Second Resilience Engineering Symposium*, Ecole des Mines de Paris, Paris, November, 8–10, 2006, pp. 191–198.)

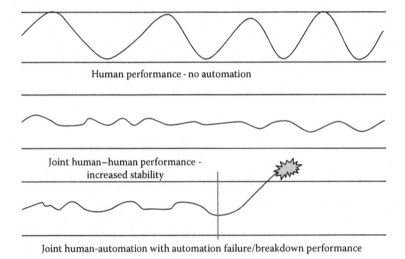

FIGURE 6.2 Effects of automation—increasing speed and accuracy increases stability, but introduces new risk.

Thus, there is a demand for a back-up plan that can be taken into action when stability is lost. Instead of trying to maintain stability in the face of irregular or unexampled events, the system must respond by adapting itself to the new circumstances. In an irregular event, a different use of the existing resources than the normal use might suffice. In such a case, to improve resilience, the resilience engineer might enhance the ability to adapt (before the event), for instance, by training personnel. During the event, the trained personnel might use the human abilities of improvisation and innovation, based on their experience from training. During training, they would have gained skills and got experience regarding the situations, with which they can draw parallels to the new situation and know how to react in similar circumstances as the current one (Woltjer, Trnka, Lundberg, & Johansson, 2006). They may know also how their coworkers act. This is in contrast to the stability-enhancing strategy of trying to predict the event in advance, and prescribe rules for action. After the occurrence of the event, if the new circumstances seem likely to recur, it might also be useful to make the system more stable, perhaps by making the temporary process resulting from the adaptation of a permanent part of the system. Thus, we should understand that there is no alternative situation, we have to accept the fact that rules cannot cover every possible situation, and the prescribed procedures which are seldom executed, with people previously unknown, set a rather fragile frame for actions. At the same time, we have to learn from previous events, and rules and checklists can be useful in the face of a recurring situation.

For the unexampled event, there might be a need to reconfigure the system more drastically, by hiring new staff, reorganizing work, creating new tools, physically moving the entire system, and so forth (Foster, 1993). In that case, resilience comes in the form of accepting the need for a total reconfiguration, and thus, may not indicate adaptation from the current system but a complete change with the purpose of surviving rather than maintaining. If changes are carried out at the cost of consuming the ability to make new changes in the face of a new unexampled event, then the changes can be made to achieve stability in the face of a specific threat, and not to achieve resilience against threats in general. If we also consider the costs of being resilient in this sense, then we can understand the risk that using resources to be resilient in the face of one crisis might use them up, making the system vulnerable to the subsequent different crisis, rather than increasing the safety in the system. This is in line with the way in which the problem is described by Westrum: "A resilient organization under Situation I will not necessarily be resilient under Situation III" (2006, p. 65).

6.4 Structural versus Functional Resilience

As stated earlier, resilience is the ability of a system to survive under extreme circumstances. However, it is important to define what "survive" indicates. In our case, we refer to it as the functional survival, in contrast to the structural survival, even though these two often are inseparable. In many cases, the function of a system depends on its structure, but it is not always so. For example, the personnel of a company may move to another building and keep on doing their work even if the original building in which the employees worked is destroyed, thus, keeping their function or performance "alive." In other cases, a part of a system may be replaced completely, allowing a system to survive, although the individual part is destroyed. Thus, modularity may be a way of achieving resilience (Foster, 1993), as long as there are "spare parts" available.

6.5 Resilience against What?

Resilience can refer to different properties of a system, which might be in conflict with each other. One, often conflicting, issue is whether a system should be resilient in terms of being competitive or being safe. These aspects are both important for the survival of a system. Glaser (1994, quoted in Sanne, 1999) stated that air-traffic control is signified by a continued quest for further scientification and automation. Although the purpose of such work may be based on a wish to improve safety and efficiency in the air-traffic domain, these two desirable ends are often not possible to pursue to their fullest at the same time. Instead of increasing both safety and efficiency, there might be a temptation to use all the new capacity to increase efficiency, and none of it to increase safety margins. The basic idea in increasing the level of automation in a system is to move the current point of both stable performance and safety to a higher level. The problem is that a driving variable in most socio-technical systems is efficiency in terms of money, meaning that the preferred way is to improve performance and reduce costs. Thus, the end result will often be a system that is safe in terms of stability, as described earlier, but not necessarily a resilient system from a safety perspective. This points to the importance of discussing resilience in relation to specific variables: being resilient as a company (surviving on the market), is in many cases, not the same thing as being resilient in terms of safety (maintaining functionality under various conditions).

As stated earlier, these two ends may actually contradict each other. Changing a system completely may also be fundamentally difficult; even in the midst of severe problems, many organizations fail to change simply because they refuse to see the need for it:

> From our evidence, for many organisations, inability to change may be the norm. We have described 'cycles of stability' in quality and safety, where much organisational effort is expended but little fundamental change is achieved. Professional and organisational culture, by many, if not most, definitions of culture, reinforces stasis (McDonald, 2006, p. 174).

Thus, an organization can often present a form of resilience—resistance—against "disturbances" that they should be responsive to. In other cases, individuals may refuse to accept that they need to act upon a disturbance, simply because they cannot or do not want to interpret the consequences of the disturbance even when the facts are clear. Lundberg and Johansson (2006) coined the expression "requisite interpretation" to describe this phenomenon, stating that to be resilient, a system must have "requisite interpretation" so that it actually acts upon changes in the environment, instead of adopting an ostrich-tactic of ignoring potentially dangerous situations. The response from the Swedish foreign ministry during the Asian Tsunami, where the foreign minister did not want to be disturbed as she was on a theatre play and no one dared to contact her, or the fact that New Orleans was not evacuated although it was known that a hurricane was about to hit the city, are both examples of a lack of requisite interpretation.

6.6 The Matryoschka Problem of Designing Safe Systems

When designing safe systems, one strategy, called defense-in-depth, is to encapsulate systems in successive layers of protective gear and hierarchical control levels of the organization at large. Leveson (2004) described a general form of a model of socio-technical control. In this model, "all" factors influencing control and safety on a system is described, from the top level with congress and legislation down to the operating process. The model not only presents the system operations, but also describes the system development and how these two stages interact with each other. It is quite clear that the actual operating process is encapsulated by a number of other systems, both physical and social, that are intended to ensure safe operation.

Similarly, in his 1997 book, Reason described that one could, in theory, go back as far as to the Big Bang in search for causes, and that one has to find the point of diminishing returns to get to a reasonable point of analysis.

> Where do you draw the line? At the organizational boundaries? At the manufacturer? At the regulator? With the societal factors that shaped these various contributions? [...] In theory, one could trace the various causal chains back to the Big Bang. What are the stop rules for the analysis of organizational accidents? (Reason, 1997, p. 15)

Thus, adding a control layer to impose safety in a system, adds the problem of protecting the control layer. Furthermore, adding a control layer to protect the protective layer means that we now have to worry about the protection of that control layer. The situation soon starts to resemble a Russian Matryoschka doll, with larger dolls added to encapsulate the smaller dolls. You can always reach the innermost doll by starting to dismantle the outermost doll.

When engineering a safe system, the problem is even worse. The outermost dolls might stay in place, but start to get large holes. They might be stable or even resilient as organizational entities, but at the same time, lose their protective function, which might be neither stable nor resilient. At that time, the protective system only provides an illusion of safety, making people think that they are safer than they really are, and might also block the way for new, safer systems. As we have emphasized earlier, it is impossible to design in advance for all possible events, and for all future changes of the environment, and that the system has to adapt to maintain its structural and functional integrity. Thus, unlike in the doll metaphor, holes continuously appear and disappear in the protective layers, going all the way from the outermost doll to the system that we aim at protecting. Therefore, the innermost system must, despite or thanks to its encapsulating layers, be able to adapt to new events that it perceives as upcoming, and quickly grasp events that do happen despite their unlikelihood. At the same time, it would be foolish not to increase the stability against known hazards.

Adding protective layers can never assure safety. The protective layers may fail, or may contribute to a bad situation by maximizing resilience in terms of being more competitive, and overemphasize stability concerning the stability-resilience trade-off for safety. Moreover, some of the layers, such as society, might be beyond the control of the resilience engineer. Also, the resilience engineer is a part of a protective layer, tuning the system to assure stability, and looking for new strategies for resilience. Thus, we can aim at engineering resilience into the layers within our control, making them more resilient against changing circumstances. By being a part of the protective system, the resilience engineering effort is also subjected to the Matryoschka problem, just like the other protective systems. This problem was also noted by Rochlin (1999) in his discussion about what distinguishes high-reliability organizations from other organizations. Organizations with high reliability, despite complexity, can be described in terms of properties, such as agency, learning, duality, communication, and locus of responsibility, rather than merely in terms of structure. However, even organizations that do have high reliability are sometimes disturbed by external events, such as the introduction of new technical systems. This might disrupt their ability to judge whether they are in a safe state or not, and hence, Rochlin was concerned about the resilience of that ability.

> Some organizations possess interactive social characteristics that enable them to manage such complex systems remarkably well, and the further observation that we do not know enough about either the construction or the maintenance of such behaviour to be confident about its resilience in the face of externally imposed changes to task design or environment (Rochlin, 1999, pp. 1556–1557).

6.7 Future Directions

The development in most complex socio-technical systems is toward further technical dependency. Human operators are to a large extent being pushed further and further away from the actual processes that they are to control, and this introduces a new kind of hidden brittleness based on the fact that demands for safe and reliable technology increase at the same time as the number of interconnected processes increases. This signifies that the consequence of failure in any component has a potential to cause dramatic resonance through the entire system in which it is part. A system that presents a stable performance may be pushed into a state of uncontrollable instability if its components fail to work. The paradox is that there is an (seemingly) ever increasing demand for increased capacity in systems like aviation; the possibilities given by new technical solutions to cram the air space with more traffic are willingly taken on by companies, as long as the manufacturers can "promise" safe operations. In this way, the safety margins taken to ensure safe operation have been decreasing. By introducing more efficient air-traffic management systems, we can have more aircraft in the same sector. This is based on the assumption that the technology used to monitor and handle air traffic is fail-safe. Thus, the way that we choose to design these systems is of uttermost importance from a safety perspective, as a system where stability and resilience are unbalanced may become very vulnerable.

Resilience and stability are like efficiency and safety—they cannot be pursued to their greatest extent at the same time, and how they are valued depends ultimately on the value judgments. Increased stability indicates that the system can withstand more, while maintaining its performance level. Increased resilience signifies that if the system goes unstable despite the efforts to keep it stable, it can reach a new stable performance equilibrium under the new circumstances. Therefore, resources must be spent on preparing for the change between states, rather than on maintaining the current state.

When considering the balancing of stability and resilience, there are some issues that need to be addressed. In accident investigations, for instance, different kinds of recommendations give rise to increased resilience (e.g., train personnel taking on new roles) than on increased stability (e.g., train personnel more in their current role). However, the balancing does not have to be carried out in hindsight. When designing and implementing new systems, the old ones might stay in place, unused for a while, representing a lower-performance stable equilibrium. This was the case at ATCC at Arlanda airport in Stockholm, Sweden. When a new system (EuroCat) was introduced, it was decided to retain the old system "alive" in the background. Under normal conditions, the air-traffic controller does not even see the old system. However, in the case of a complete breakdown of the new system, it may be possible to step back to the old system, allowing the air-traffic controllers to make a "graceful degradation" into a nonoperating mode. Thus, it is possible to use the old system to reroute the incoming flights to other sectors and to land the flights that are close to landing. As long as personnel who know how to operate the older system are still in place, this gives an opportunity for resilient behavior in the case of a breakdown of the new system. Since the introduction of the new system, this has happened at least once. However, if the know-how wanes, the resilience becomes eroded.

The challenge for the resilience engineer is how to design transitions between states of stability, design and maintain alternative structural configurations for irregular events, and design for the innovation and rapid adaptation needed in the face of unexampled events. This effort has to be balanced against the need for stable performance during normal operations with regular disturbances.

References

Coutu, D. L. (2002, May). How resilience works. *Harvard Business Review, 80*(5), 46–50.

Foster, H. D. (1993). Resilience theory and system evaluation. In J. A. Wise, V. D. Hopkin, & P. Stager (Eds.), *Verification and validation of complex systems: Human factors issues* (pp. 35–60). Berlin: Springer Verlag.

Holling, C. S. (1973). Resilience and stability of ecological systems. *Annual Review of Ecology and Systematics, 4*, 1–23.

Hollnagel, E., & Rigaud, E. (2006). *Proceedings of the Second Resilience Engineering Symposium.* Paris: Ecole des Mines de Paris.

Hollnagel, E., Woods, D. D., & Leveson, N. (2006). *Resilience Engineering: Concepts and Precepts.* Aldershot, U.K.: Ashgate.

Leveson, N. (2004). A new accident model for engineering safer systems. *Safety Science, 42*, 237–270.

Lundberg, J., & Johansson, B. (2006). Resilience, stability and requisite interpretation in accident investigations. In E. Hollnagel, & E. Rigaud (Eds.), *Proceedings of the Second Resilience Engineering Symposium* (pp. 191–198), November, 8–10, 2006. Paris: Ecole des Mines de Paris.

McDonald, N. (2006). Organizational resilience and industrial risk. In E. Hollnagel, D. D. Woods, & N. Leveson (Eds.), *Resilience engineering: Concepts and precepts* (pp. 155–179). Aldershot, U.K.: Ashgate.

Reason, J. T. (1997). *Managing the risks of organizational accidents.* Burlington, VT: Ashgate.

Rochlin, G. (1999). Safe operations as a social construct. *Ergonomics, 42*(11), 1549–1560.

Sanne, J. M. (1999). *Creating safety in air traffic control.* Lund, Sweden: Arkiv Förlag.

Westrum, R. (2006). A typology of resilience situations. In E. Hollnagel, D. D. Woods, & N. Leveson (Eds.), *Resilience engineering: Concepts and precepts* (pp. 55–65). Aldershot, U.K.: Ashgate.

Woltjer, R., Trnka, J., Lundberg, J., & Johansson, B. (2006). Role-playing exercises to strengthen the resilience of command and control systems. In G. Grote, H. Günter, & A. Totter (Eds.), *Proceedings of the 13th European Conference on Cognitive Ergonomics—Trust and Control in Complex Socio-Technical Systems* (pp. 71–78). Zurich, Switzerland.

Woods, D. D., & Wreathall, J. (2003). *Managing risk proactively: The emergence of resilience engineering.* Columbus: Ohio University, Available: http://csel.eng.ohiostate.edu/woods/error/About%20 Resilience%20Engineer.pdf

7

Processes Underlying Human Performance

Lisanne Bainbridge
University College London

Michael C. Dorneich
Honeywell Laboratories

Two decades ago, a chapter on aviation with this title might have focused on the physical aspects of human performance, representing the control processes involved in flying. However, today there has been such a fundamental change in our knowledge and techniques that this chapter focuses almost exclusively on cognitive processes. The main aims are to show that relatively few general principles underlie the huge amount of information relevant to interface design, and that context is a key concept in understanding human behavior.

Classical interface human factors/ergonomics (HF/E) consists of a collection of useful but mainly disparate facts and a simple model of the cognitive processes underlying the behavior—these processes consist of independent information, decision, action, or units. (the combined term HF/E is used, because these terms have different meanings in different countries. Cognitive processing is the unobservable processing between the arrival of the stimuli at the senses and initiation of an action.) Classic HF/E tools are powerful aids for interface design, but they make an inadequate basis for designing to support complex tasks. Pilots and air-traffic controllers are highly trained and able people. Their behavior is organized and goal-directed, and they add knowledge to the information given on an interface in two main cognitive activities: understanding what is happening, and working out what to do about it.

As the simple models of cognitive processes used in classic HF/E do not contain reminders about all the cognitive aspects of complex tasks, they do not provide a sufficient basis for supporting HF/E for these tasks. The aim of this chapter is to present simple concepts that could account for behavior in complex dynamic tasks, and provide the basis for designing to support people doing these tasks. As the range of topics and data that could be covered is huge, the strategy is to indicate the key principles by giving typical examples, rather than attempting completeness. This chapter does not present a detailed model of the cognitive processes suggested or survey HF/E techniques, and does not discuss the collective work. The chapter offers four main sections on simple use of interfaces; understanding, planning, and multi-tasking; learning, workload, and errors; and joint cognitive systems. The conclusion outlines how the fundamental nature of human cognitive processes underlies the difficulties met by HF/E practitioners.

7.1 Using the Interface, Classic HF/E

This chapter distinguishes between the cognitive functions or goals, that is, what is to be done, and the cognitive processes, that is, how these are done. This section starts with simple cognitive functions and processes underlying the use of displays and controls, on the interface between a person and the device that the person is using. More complex functions of understanding and planning are discussed in the following main section.

Simple operations are affected by the context in which they are carried out. Someone does not press a button in isolation. For example, a pilot keys in a radio frequency for contacting the air-traffic control as well as for navigation, which is multitasked with checking for aircraft safety, and so on. From this point of view, an account of cognitive processes should start with complex tasks. However, this may be too difficult. In this section, the simple tasks involved in using an interface are described first, and how even simple processes are affected by a wider context is subsequently presented. The next main section is developed from this topic and describes more complex tasks.

Five main cognitive functions are involved in using an interface:

- Discriminating a stimulus from a background or from the other possible stimuli. The process usually used for this is decision making.
- Perceiving "wholes." The main process here is the integration of parts of the sensory input.
- Naming.
- Choosing an action. The cognitive process by which the functions of naming and choosing an action are carried out (in simple tasks) is recoding, that is, translating from one representation to another, such as (shape → name) or (display → related control).
- Comparison, which may be done by a range of processes from simple to complex.

As discriminating and integrating stimuli are usually done as the basis for naming or choosing an action, it is often assumed that the processes for carrying out these functions are independent, input driven, and done in sequence. However, these processes are not necessarily distinct or carried out in sequence, and they all involve the use of context and knowledge.

This section does not discuss displays and controls separately, as both involve all the functions and processing types. Getting information may involve making a movement, such as visual search or accessing a computer display format, whereas making a movement involves getting information about it. The four subsections present detecting and discriminating; visual integration; naming and simple action choices; and action execution.

7.1.1 Detecting and Discriminating

As the sense organs are separate from the brain, it may be assumed that at least the basic sensory effectiveness, the initial reception of signals by the sense organs, would be a simple starting point, before considering the complexities that the brain can introduce, such as naming a stimulus or choosing an action. However, sensing processes may not to be simple: there can be a large contribution of prior knowledge and present context.

This part of the chapter is divided into four subsections on detecting, discriminating one signal from the others that are present, or that are absent (absolute judgment), and the sensory decisions. It is artificial to distinguish between sensory detection and discrimination, although they are discussed separately here, because they both involve (unconscious) decision making about what a stimulus is. In many real tasks, other factors have more effect on the performance than any basic limits to sensory abilities. Nevertheless, it is useful to understand these sensory and perceptual processes, because they raise points that are general to all cognitive processing.

Detecting. Detection is one of those words that may be used to refer to different things. In this chapter, detection indicates sensing the presence of a stimulus against a blank background, for example, detecting the presence of light. A human eye has the ultimate sensitivity to detect one photon of electromagnetic energy in the visible wavelength. However, we can only detect at this level of sensitivity if we have been in complete darkness for about half an hour (Figure 7.1). The eyes adapt 50 and are sensitive to a range of light intensities around the average (Figure 7.2); however, this adaptation takes time. Adaptation allows the eyes to deal efficiently with a wide range of stimulus conditions, but it indicates that sensing is relative rather than absolute.

The two curves on the dark adaptation graph (Figure 7.1) indicate that the eyes have two different sensing systems, one primarily for use at high light intensities, and the other for the use at low light intensities. These two systems have different properties. At higher levels of illumination, the sensing cells are sensitive to color. There is one small area of the retina (the sensory surface inside the eye)

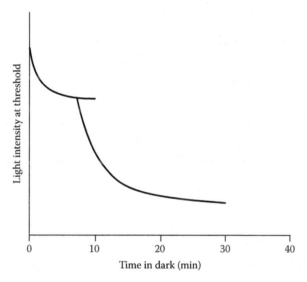

FIGURE 7.1 Increasing sensitivity to light after time in darkness (dark adaptation). (From Lundberg, J. and Johansson, B., Resilience, stability and requisite interpretation in accident investigations. In Hollnagel, E. and Rigaud, E. (Eds.), *Proceedings of the Second Resilience Engineering Symposium*, Ecole des Mines de Paris, Paris, November 8–10, 2006, pp .191–198.)

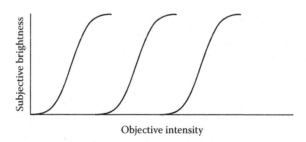

FIGURE 7.2 The sensitivity of the eye when adapted to three different levels of average illumination. At each adaptation level, the eye is good at discriminating between the intensities around that level.

that is best able to discriminate between spatial positions and detect stationary objects. The rest of the sensory surface (the periphery) is better at detecting moving than stationary objects. At lower levels of illumination intensity, the eyes mainly see in black and white, and peripheral vision is more sensitive for detecting position.

Therefore, it is not possible to make a simple statement like "the sensitivity of the eyes is ...." The sensitivity of the eyes depends on the environment (e.g., the average level of illumination) and the stimulus (e.g., its movement, relative position, or color). The sensitivity of sense organs adapts to the environment and the task, and hence, does not have an absolute value independent of these influences. This means that it is difficult to make numerical predictions about sensory performance in particular circumstances, without testing directly.

However, it is possible to draw practical implications from the general trends in sensitivity. For example, it is important to design to support both visual sensing systems in tasks that may be carried out in both high and low levels of illumination, such as flying. It is also sensible to design in such a way that the most easily detected stimuli (the most "salient") are used for the most important signals. Visual salience depends not only on the intensity, but also on the color, movement, and position of the stimulus. Very salient stimuli attract attention; they override the usual mechanism for directing the attention (see the next main section). This indicates that very salient signals can be either useful as warning signals or a nuisance, owing to irrelevant distractions that interrupt the main task.

7.1.1.1 Discriminating between Stimuli

In this section, the word discrimination refers to distinguishing between two (or more) stimuli. As with detection, the limits to our ability to discriminate between the stimulus intensities are relative rather than absolute. The merely noticeable difference between two stimuli is a ratio of the stimulus intensities (there is a sophisticated modem debate about this, but it is not important for most practical applications). This ratio is called the Weber fraction. Again, the size of this ratio depends on the environmental and task context. For example, in visual-intensity discriminations, the amount of contrast needed to distinguish between two stimuli depends on the size of the object (more contrast is needed to see smaller objects) and the level of background illumination (more contrast is needed to see objects in lower levels of background illumination).

The Weber fraction describes the difference between the stimuli that can merely be discriminated. When stimuli differ by larger amounts, the time needed to make the discrimination is affected by the same factors: Finer discriminations take longer, and visual discriminations can be made more quickly in higher levels of background illumination.

Touch and feel (muscle and joint receptor) discriminations are made when using a control. For example, a person using a knob with tapered sides may make three times more positioning errors than when using a knob with parallel sides (Hunt & Warrick, 1957). As neither of the sides of a tapered knob actually points in the direction of the knob, the touch information from the sides is ambiguous.

Resistance in a control affects the effortless discrimination by feel between positions of the control. Performance in a tracking task, using controls with various types of resistance, shows that inertia makes performance worse, whereas elastic resistance can give the best results. This is because inertia is the same irrespective of the extent of the movement made, and hence, it does not help in discriminating between the movements. Elastic resistance, in contrast, varies with the extent of the movement, and thus, gives additional information about the movements being made (Howland & Noble, 1955).

7.1.1.2 Absolute Judgment

The Weber fraction describes the limit to our abilities to discriminate between two stimuli when they are both present. When two stimuli are next to each other we can, at least visually, make very fine discriminations in the right circumstances. However, our ability to distinguish between the stimuli when only one of them is present is much more limited. This process is called absolute judgment. The judgment limits to our sensory abilities are known, in general, for many senses and dimensions (Miller, 1956).

These limits can be affected by several aspects of the task situation, such as the range of possible stimuli that may occur (Helson, 1964).

When only one stimulus is present, distinguishing it from the others must be done by comparing it with mental representations of the other possible stimuli. Hence, absolute judgment must involve knowledge and/or working memory. This is an example of a sensory discrimination process that has some processing characteristics in common with those that are usually considered much more complex cognitive functions. There may not always be a clear distinction between simple and complex tasks with regard to the processing involved.

Although our ability to make absolute judgments is limited, it can be useful. For example, we can discriminate among eight different positions within a linear interval. This means that visual clutter on scale-and-pointer displays can be reduced; it is only necessary to place a scale marker at every five units that need to be distinguished. However, our ability is not good enough to distinguish between 10 scale units without the help of an explicit marker.

In other cases, the limitations need to be taken into account in design. For example, we can only distinguish among 11 different color hues by absolute judgment. As we are very good at distinguishing between colors when they are next to each other, it can be easy to forget that color discrimination is limited when one color is seen alone. For example, a color display might use green-blue to represent one meaning (e.g., main water supply) and purple-blue with another meaning (e.g., emergency water supply). It might be possible to discriminate between these colors and use them as a basis for identifying the meaning, when the colors are seen together, but not when they are seen alone (a discussion on meaning is presented later).

Again, discrimination is a process in which the task context, in this case, whether or not the stimuli occur together for comparison, has a strong effect on the cognitive processes involved and on our ability to make the discriminations.

7.1.1.3 Sensory Decision Making

Detections and discriminations involve decisions about whether the evidence reaching the brain is sufficient to justify in deciding that a stimulus (difference) is present. For example, detection on a raw radar screen involves deciding whether a particular radar trace is a "blip" representing an aircraft, or something else that reflects radar waves. A particular trace may only be more or less likely to indicate an aircraft, and hence, a decision has to be made in conditions of uncertainty. This sort of decision can be modeled by signal detection or statistical decision theory. Different techniques are now used in psychology, but this approach is convenient here, because it distinguishes between the quality of the evidence and the observer's prior biases about the decision outcomes.

Consider that the radar decisions are based on intensity. The frequencies with which the different intensities appear on the radar screen when there was no aircraft, are shown in Figure 7.3a at the top, while the intensities that appear when an aircraft was present are shown in Figure 7.3a at the bottom. There is a range of intensities that occur only when an aircraft is absent or only when an aircraft is present, and an intermediate range of intensities that occur both when an aircraft is present and absent (Figure 7.3b). How can someone make a decision when one of the intermediate intensities occurs? Generally, the decision is made on the basis of signal likelihood. The height of the curve above a particular intensity indicates the probability of the intensity to occur when an aircraft is present or absent. At the midpoint between the two frequency distributions, both the possibilities are equally probable. Thus, intensities less than this midpoint are more likely not to come from an aircraft, and intensities greater than this midpoint are more likely to come from an aircraft.

It must be noted that when a stimulus is in this intermediate range, it is not always possible to be right about a decision. A person can decide a trace is not an aircraft when it actually is (a "miss"), or can decide it is an aircraft when it is not (a "false alarm"). These mistakes are not called *errors*, because it is not always mathematically possible to be right when making uncertain decisions. The number of wrong decisions and the time to make the decision increase when signals are more similar (overlap more).

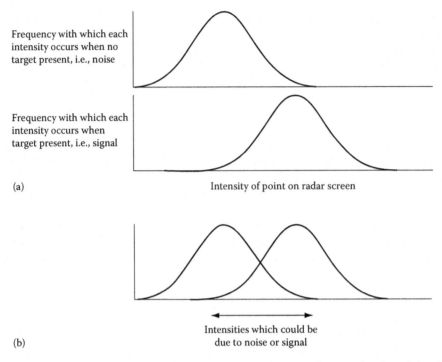

Frequency with which each
intensity occurs when no
target present, i.e., noise

Frequency with which each
intensity occurs when
target present, i.e., signal

(a)   Intensity of point on radar screen

Intensities which could be
(b)   due to noise or signal

FIGURE 7.3 Knowledge about the occurrence of intensities. Decision making employs knowledge about the alternatives, based on previous experience.

It must be noted that when the radar operator is making the decision, there is only one stimulus actually present with one intensity. The two frequency distributions, against which this intensity is compared with to make the decision, must be obtained from the operator's previous experience of radar signals, stored in the operator's knowledge base. Decisions are made by comparing the input stimulus (bottom-up) with the stored knowledge about the possibilities (top-down).

In addition to the uncertainty owing to similarity between the possible interpretations of a stimulus, the second major factor in this type of decision making is the importance or costs of the alternative outcomes. In the example given earlier, the person's decision criterion, the intensity at which the person changes from deciding "yes" to deciding "no," is the point at which both possibilities are equally probable. However, it is very important not to miss a signal—for instance, when keeping radar watch in an early warning system. In this case, it might be sensible to use the decision criterion presented in Figure 7.4. This would increase the number of hits and would also increase the number of false alarms, but this might be considered a small price to pay when compared with the price of missing a detection. Alternatively, imagine people working to detect a signal, for which they have to do a lot of work, and they feel lazy and not committed to their job. In this case, they might move their decision criterion to the other direction, to minimize the number of hits.

This shift in decision criterion is called bias. Decision bias can be affected by probabilities and costs. The person's knowledge of the situation provides the task and personal expectations/probabilities as well as the costs that are used in setting the biases, and thus, top-down processing again can influence the sensory decisions. There are limits to human ability to assess biases (Kahneman, Slovic, & Tversky, 1982). At extreme probabilities, we tend to substitute determinacy for probability. We may think something is sure to happen, when it is just highly probable. Some accidents happen because people see

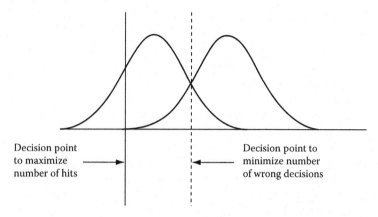

FIGURE 7.4 An example of change in the bias used in decision making. If rewarded for "hits," the bias changes to maximize payoff ("false alarms" also increase).

what they expect to see, rather than what is actually there (e.g., Davis, 1966). Inversely, we may think something will never happen, when it is objectively of very low probability. For example, when signals are very unlikely, then it is difficult for a human being to continue to direct attention to watch for them (the "vigilance" effect).

7.1.2 Visual Integration

The effects of knowledge and context are even more evident in multidimensional aspects of visual perception, such as color, shape, size, and movement, in which what is seen, is an inference from combined evidence. These are discussed in the subsections on movement, size, and color; grouping processes; and shape (there are also interesting auditory integrations, more involved in music perception, but these are not discussed here).

7.1.2.1 Movement, Size, and Color Constancies

It is actually quite odd that we perceive a stable external world, given that we and other objects move, and the wavelength of the environmental light that we see changes. Thus, the size, position, shape, and wavelength of light reflected from the objects onto the retina all change. As we do perceive a stable world, this suggests that our perception is relative rather than absolute: We do not see what is projected on the retina, but a construction based on this projection, made by combining evidence from different aspects of our sensory experience. The processes by which a wide variety of stimuli falling on the retina are perceived as the same are called *constancies*.

When we turn our heads, the stimulation on the retina also moves. However, we do not see the world as moving, because the information from the turning receptors in the ear is used to counteract the evidence of movement from the retina. The changes on the retina are perceived in the context of changes in the head-rotation receptors. When the turning receptors are diseased, or when the turning movements are too extreme for the receptors to be able to interpret quickly, then the person may perceive the movement that is not actually occurring, as in some flying illusions.

There is also constancy in size perception. As someone walks away from us, we do not see them becoming smaller and smaller, although there are large changes in the size of the image of that person that falls on the retina. In interpreting the size of objects, we take into account all the objects that are at the same distance from the eye, and then perceive them according to their relative size. Size constancy is more difficult to account for than movement constancy, as it involves distance perception, which is a complex process (Gibson, 1950). Distance is perceived by combining evidence about texture, perspective, changes in color of light with distance, and overlapping (a construct, discussed later). Information from the whole visual field is used in developing a percept that makes best overall sense of the combination of inputs. Cognitive psychology uses the concept that different aspects of the stimulus processing are carried out simultaneously, unless an aspect is difficult and slows the processing down. Each aspect of processing communicates its "results so far" to the other aspects via a "blackboard," and all the aspects work together to produce a conclusion (Rumelhart, 1977).

Color perception is also an integrative process that shows constancy. Research on the color-receptive cells in the retina suggests that there are only three types of cells that respond to red, green, and blue light wavelengths. The other colors we "see" are constructed by the brain, based on the combinations of stimulus intensities at these three receptors. The eyes are more sensitive to some colors, and hence, if a person looks at two lights of the same physical intensity but different wavelengths, the lights may be of different experienced intensity (brightness). The effectiveness of the color-construction process is such that there have been some visual demonstrations in which were observed to people see a range of colors, even though the display consists only of black and white along with one color. This constructive process also deals with color constancy. The wavelength of ambient lighting can change quite considerably; thus, the light reflected from the objects also changes its wavelength, but the objects are perceived as having a stable color. The wavelengths of light from all the objects change in the same way, and the color is perceived from the relative combinations of wavelengths, and not the actual wavelength. This constancy process is useful for perceiving a stable world despite transient and irrelevant changes in the stimuli, but it does make designing of color displays more difficult. Similar to our response to the stimulus intensity, our perception of color is not a fixed quantity that can easily be defined and predicted. Instead, it depends on the interaction of several factors in the environment and task contexts, and hence, it may be necessary to make color-perception tests for a particular situation.

7.1.2.2 Grouping Processes

Another type of perceptual integration occurs when several constituents of a display are grouped together and perceived as a "whole." The Gestalt psychologists in the 1920s first described these grouping processes that can be at several levels of complexity.

1. Separate elements can be seen as linked into a line or lines. There are four ways in which this can happen: when the elements are close together, are similar, lie on a line, or define a contour. The grouping processes of proximity and similarity can be used in the layout of displays and controls on a conventional interface, to show which items go together.
2. When separate elements move together, they are seen as making a whole. This grouping process is more effective if the elements are also similar. This is used in the design of head-up displays and predictor displays, as shown in Figure 7.5.
3. Something that has uniform color or a connected contour is seen as a "whole"—for example, the four sides of a square are seen as a single square, not as four separate element.
4. The strongest grouping process occurs when the connected contour has a "good" form, that is, a simple shape. For example, a pull-down menu on a computer screen is seen as a distinct unit in front of other material, because it is a simple shape, and the elements within the shape are similar and (usually) different from those on the rest of the screen. When the visual projections of two objects touch each other, then the one with the simplest shape is usually seen as in the front of (overlapping) the other.

The visual processes by which shapes and unities are formed suggest recommendations for the design of symbols and icons that are easy to see (Easterby, 1970).

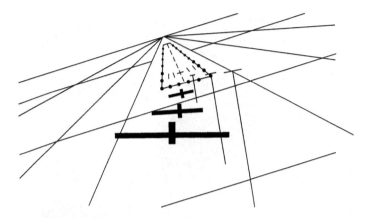

FIGURE 7.5 Gestalt grouping processes relate together the elements of a predictor landing display. (Reprinted from Gallaher, P.D., et al., *Hum. Factors*, 19(6), 549, 1977.)

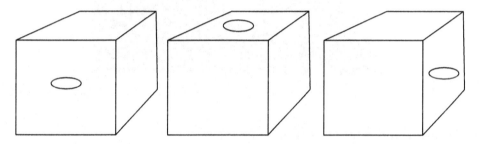

FIGURE 7.6 Shape and size "constancy": the same cube with the same ellipse in three different positions. The ellipses are computer-generated duplicates.

7.1.2.3 Shape Constancy

Visual integrative processes ensure that we see a unity when there is an area of same color or a continuous contour. The shape we see depends on the angles of the contour lines (there are retinal cells that sense angle of line). Again, there are constancy processes. The shape perceived is a construction, taking into account the various aspects of the context, rather than a simple mapping of what is projected from the object onto the retina. Figure 7.6 shows a perspective drawing of a cube, with the same ellipse placed on each side. The ellipse on the front appears as an ellipse on a vertical surface; the ellipse on the top appears to be wider and sloping at the same angle as the top; and the ellipse on the side is ambiguous—is it rotated or not a part of the cube at all? The ellipse on the top illustrates shape "constancy," and is perceived according to the knowledge about how shapes look narrower when they are parallel to the line of sight; thus, a flat narrow shape is inferred to be wider. Again, the constancy process shows that the surrounding context (in this case, the upper quadrilateral) affects the way in which particular stimuli are seen.

The Gestalt psychologists provided dramatic examples of the effects of these inference processes in their reversible figures, as shown in Figure 7.7. The overall interpretation given to this drawing affects how the particular elements of it are grouped together and named—for example, whether they are seen as parts of the body or pieces of clothing. It is not possible to see both interpretations at the same time, but it is possible to quickly change from one to the other. As the interpretation given to an object affects the way in which parts of it are perceived, this can cause difficulty in the interpretation of low-quality visual displays, for example, from infrared cameras or on-board radar.

FIGURE 7.7 Ambiguous "wife/mother-in-law" figure. The same stimulus can be given different interpretations.

7.1.3 Naming and Simple Action Choices

The subsequent functions to consider are the identification of name, status, or size, and choosing the nature and size of actions. These cognitive functions may be met by a process of recoding (association) from one form of representation to another, such as

Shape → name
Color → level of danger
Spatial position of display → name of variable displayed
Name of variable → spatial position of its control
Length of line → size of variable
Display → related control
Size of distance from target → size of action needed

Identifications and action choices that involve more complex processing than this recoding are discussed in the section on complex tasks, including the interdependence of the processes and functions; identifying name and status—shape, color, and location (codes; size → size codes; and recoding/reaction times). Furthermore, computer displays have led to the increased use of alphanumeric codes, which are not discussed here (see Bailey, 1989).

7.1.3.1 Interdependence of the Functions

Perceiving a stimulus, naming it, and choosing an action are not necessarily independent. Figure 7.7 shows that identification can affect perception. This section gives three examples that illustrate other HF/E issues.

Naming difficulties can be based on discrimination difficulties. Figure 7.8 shows the signal/noise ratio needed to hear a word against background noise. The person listening not only has to detect a word against the noisy background, but also has to discriminate it from other possible words. The more

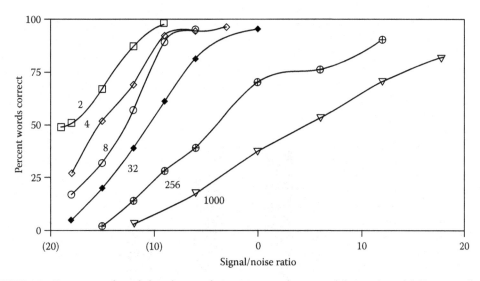

FIGURE 7.8 Percentage of words heard correctly in noise, as a function of the number of different words that might occur. (From Miller, G.A. et al., *J. Exp. Psychol.*, 41, 329, 1951.)

alternatives there are to distinguish, the better must be the signal/noise ratio. This is the reason for using a minimum number of standard messages in speech communication systems, and for designing these messages to maximize the differences between them, as in the International Phonetic Alphabet and standard air-traffic control language (Bailey, 1989).

An important aspect of maximizing the differences between the signals can be illustrated using a visual example. Figure 7.9 shows some data on reading errors with different digit designs. Errors can be up to twice as high with design A than with design C. A quick glance may indicate that these digit designs do not look very different, but each digit in C has been designed to maximize its difference from the others. Digit reading is a naming task based on a discrimination task, and the discriminations are based on differences between the straight and curved elements of the digits. It is not possible to design an 8 that can be read easily, without considering the need to discriminate it from 3, 5, 6, and 9, which have elements in common. As a general principle, design for discrimination depends on knowing the ensemble of alternatives to be discriminated, and maximizing the differences between them.

However, ease of detection/discrimination does not necessarily make naming easy. Figure 7.10 shows an iconic display. Each axis displays a different variable, and when all the eight variables are on target, the shape is symmetrical. It is easy to detect a distortion in the shape, to detect that a variable is off the target. However, studies show that people have difficulty in discriminating one distorted pattern from another by memory, and in identifying which pattern is associated with which problem. This display supports detection, but not discrimination or naming. It is important in task analysis to note which of the cognitive functions are needed, and observe whether the display design supports them.

7.1.3.2 Shape, Color, and Location Codes for Name and Status

Conventional interfaces often consist of numerous displays or controls that are identical both to sight and touch. The only way of discriminating and identifying them is to read the label or learn the position. Even if labels have well-designed typeface, abbreviations, and position, they are not ideal. Hence, an easy-to-see "code" is needed for the name or status, which is easy to recode into its meaning. The codes used most frequently are shape, color, and location (felt texture can be an important code in the design of controls). The codes need to be designed for ease of discrimination as well as translation from code to meaning.

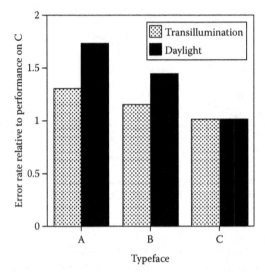

FIGURE 7.9 Reading errors with three different digit designs. Errors are fewest with the design that minimizes the number of elements that the alternatives have in common. (From Atkinson, W.H. et al., A study of the requirements for letters, numbers and markings to be used on trans-illuminated aircraft control; panels. Part 5: the comparative legibility of three fonts for numerals (Report No. TED NAM EL-609, part 5), Naval Air Material Center, Aeronautical Medical Equipment Laboratory, 1952.)

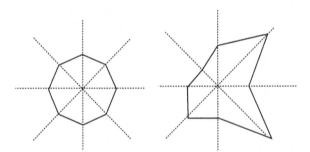

FIGURE 7.10 "Iconic" display: Eight variables are displayed, measured outward from the center. When all the eight variables are on target, the display has an octagon shape.

7.1.3.2.1 Shape Codes

Good shape codes are "good" figures in the Gestalt sense, and also have features that make the alternatives easy to discriminate. However, ease of discrimination is not the primary criterion in good shape-code design. Figure 7.11 shows the materials used in discrimination tests between sets of colors, military look-alike shapes, geometric forms, and aircraft look-alike shapes. Color discrimination is easiest, and military symbols are easier to distinguish than aircraft symbols because they have more different

	C-54	C-47	F-100	F-102	B-52
Aircraft shapes	✈	✈	✈	✈	✈
Geometric forms	Triangle ▲	Diamond ◆	Semicircle ◗	Circle ●	Star ★
Military symbols	Radar	Gun	Aircraft ✈	Missile	Ship
Colors (Munsell notation)	Green (2.5 G 5/8)	Blue (5BG 4/5)	White (5Y 8/4)	Red (5R 4/9)	Yellow (10YR 6/10)

FIGURE 7.11 Symbols used in discrimination tests. (From Smith, S.L. and Thomas, D.W., *J. Appl. Psychol.*, 48, 137, 1964.)

features, and the geometric forms can be discriminated more easily than aircraft shapes (however, geometric forms are not necessarily easier to discriminate. For example, the results would be different if the shapes included an octagon as well as a circle). The results from naming tests rather than discrimination tests would be different if geometric shapes or colors had to be given a military or aircraft name. Naming tests favor look-alike shapes, as look-alike shapes can be more obvious in meaning.

Nevertheless, using a look-alike shape (symbol or icon) does not guarantee obviousness of meaning. The way in which people make the correct link from shape to meaning needs to be tested carefully. For each possible shape, people can be asked regarding (1) what picture they think it represents; (2) what further meaning, such as an action, they think it represents; and (3) to choose the meaning of the shape from the given list of possible meanings. To minimize confusions when using shape codes, it is important not to include any shape that is assigned several meanings, or several shapes that could all be assigned the same meaning in the coding vocabulary. Otherwise, there could be high error rates in learning and using the shape codes. It is also important to test these meanings on the appropriate users, naive or expert people, or an international population. For example, in Britain, a favored symbol for "delete" would be a picture of a space villain from a children's TV series, but this is not understood by people from other European countries!

Besides the potential obviousness of their meaning, the look-alike shapes have other advantages over geometric shapes. They can act as a cue to a whole range of remembered knowledge about this type of object (see later discussion on knowledge). Look-alike shapes can also vary widely, whereas the number of alternative geometric shapes that are easy to discriminate is small. An interface designer using geometric shape as a code runs out of different shapes quite quickly, and may have to use the same shape with several meanings. As a result, a person interpreting these shapes must notice when the context has changed to a different shape → meaning translation, and then should remember this different translation before the person can work out what a given shape means. This multistage process can be error prone, particularly under stress. Some computer-based displays have the same shape used with different meanings in different areas of the same display. A person using such a display has to remember to change the coding translation used every time when the person makes an eye movement.

7.1.3.2.2 Color Codes

Using color as a code poses similar problems as using geometric shape. Except for certain culture-based meanings such as red → danger; the meanings of colors have to be learned specifically rather than being obvious. Furthermore, only a limited number of colors can be discriminated by absolute judgment.

Thus, a designer who thinks color is easy to see, rapidly runs out of different colors, and has to use the same color with several meanings. There are computer-based displays on which color is used simultaneously with many different types of meaning, such as

Color → substance (steam, oil, etc.)
Color → status of item (kg, on/off)
Color → function of item
Color → subsystem item belongs to
Color → level of danger
Color → attend to this item
Color → click here for more information
Color → click here to make an action

A user has to remember which of these coding translations is relevant to a particular point on the screen, with a high possibility of confusion errors.

7.1.3.2.3 Location Codes

The location of an item can be used as a basis both for identifying an item and for indicating its links with the other items.

People can learn where a given item is located on an interface, and then look or reach to it automatically, without searching. This increases the efficiency of the behavior. But, this learning is effective only if the location → identity mapping remains constant; otherwise, there can be a high error rate. For example, Fitts and Jones (1961a), in their study about pilot errors, found that 50% of the errors in operating aircraft controls were with respect to choosing the wrong control. The layout of controls on three of the aircraft used at that time showed why it was easy to get confused (Table 7.1).

Consider that *n* pilot had flown a B-25 very frequently such that he is able to reach to the correct control without thinking or looking. If he is transferred to a C-17, then two-thirds of his automatic reaches would be wrong, and if to a C-82, then all of them would be wrong. As with other types of coding, location → identity translations need to be consistent and unambiguous. Locations will be easier to learn if related items are grouped together, such as items from the same part of the device, with the same function or the same urgency of meaning.

Locations can sometimes have a realistic meaning, rather than an arbitrary learned one. Items on one side in the real world should be on the same side when represented on an interface (ambiguity about the location of left/right displays could have contributed to the Kegworth air crash; Green, 1990). Another approach is to put items in meaningful relative positions. For example, in a mimic/schematic diagram or an electrical wiring diagram, the links between items represent the actual flows from one part of the device to another. On a cause–effect diagram, the links between the nodes of the diagram represent the causal links in the device. On such diagrams, the relative position is meaningful and the inferences can be drawn from the links portrayed (see later discussion on knowledge).

Relative location can also be used to indicate which control goes with which display. When there is a one-to-one relation between displays and controls, the choice of control is a recoding that can be made more or less obvious, consistent, and unambiguous by the use of spatial layout. Gestalt proximity processes the link items together if they are next to each other. However, the link to make can be ambiguous, such as in the layout: O O O O X X X X. In this case, which X goes with which O? People bring expectations about the code meanings to their use of an interface. If these expectations are consistent among a particular group of people, then the expectations are called *population stereotypes*. If an interface uses codings that are not compatible with a person's expectations, then the person is likely to make errors.

TABLE 7.1

Aircraft	Position of Control		
	Left	Center	Right
B-25	Throttle	Prop	Mixture
C-47	Prop	Throttle	Mixture
C-82	Mixture	Throttle	Prop

If two layouts to be linked together are not the same, then it has been observed that reversed but regular links are easier to deal with than random links (Figure 7.12). This suggests that recoding may be done, not by learning individual pairings, but by having a general rule from which one can work out the linkage.

In multiplexed computer-based display systems, in which several alternative display formats may appear on the same screen, there are at least two problems with location coding. One is that each format may have a different layout of items. We do not know whether people can learn locations on more than one screen format sufficiently well, to be able to find items on each format by automatic eye movements rather than by visual search. If people have to search a format for the item that they need, then it is suggested that this could take at least 25 s. This means that every time the display format is changed, the performance will be slowed down while this search process interrupts the thinking about the main task (see later discussion on short-term memory). It may not be possible to put the items in the same absolute position on each display format, but one way of reducing the problems caused by inconsistent locations is to locate items in the same relative positions on different formats.

The second location problem in multiplexed display systems is that people need to know the search "space" of alternative formats available, their current location, and how to get to other formats. It takes ingenuity to design so that the user of a computer-based interface can use the same sort of "automatic" search skills to obtain information that are possible with a conventional interface.

In fact, there can be problems in maximizing the consistency and reducing the ambiguity of all types of coding used on multiple display formats (Bainbridge, 1991). Several of the coding vocabularies and coding translations used may change between and within each format (watch out for the codes used in figures in this chapter). The cues that a person uses to recognize which coding translations are relevant must be learned, and are also often not consistent. A display format may have been designed such that the codes are obvious in meaning for a particular subtask, when the display format and the subtask are tested in isolation. However, when this display is used in the real task, before and after other formats used for other subtasks, each of which uses different coding translations, then a task-specific display may not reduce either the cognitive processing required or the error rates.

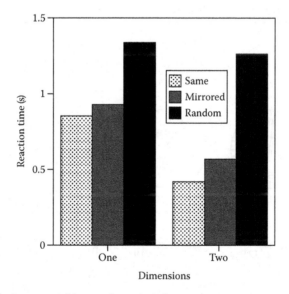

FIGURE 7.12 Effect of relative spatial layout of signals and responses on response time. (From Fitts, P.M. and Deininger, R.L., *J. Exp. Psychol.*, 48, 483, 1954.)

7.1.3.3 Size → Size Codes

On an analogue interface, the length of the line is usually used to represent the size of a variable. The following arguments apply both to display scales and the way in which the control settings are shown. There are three aspects: the ratio of the size on the interface to the size of the actual variable; the way comparisons between sizes are made; and the meaning of the direction of a change in size.

7.1.3.3.1 Interface Size: Actual Size Ratio

An example of the interface size to actual size ratio is that, when using an analogue control (such as a throttle), a given size of action has a given size of effect. Once people know this ratio, they can make actions without having to check their effect, which gives increased efficiency (see later discussion).

The size ratio and direction of movement are again codes used with meanings that need to be consistent. Size ratios can cause display-reading confusions if many displays are used, which may all look the same but differ in the scaling ratio used. If many controls that are similar in appearance and feel are used with different control ratios, then it may be difficult to learn automatic skills in using them to make actions of the correct size. This confusion could be increased by using one multipurpose control, such as a mouse or tracker ball, for several different actions each with a different ratio.

A comparison of alternative altimeter designs is an example that also raises some general HF/E points. The designs were tested for reading the speed and accuracy (Figure 7.13). The digital display gives

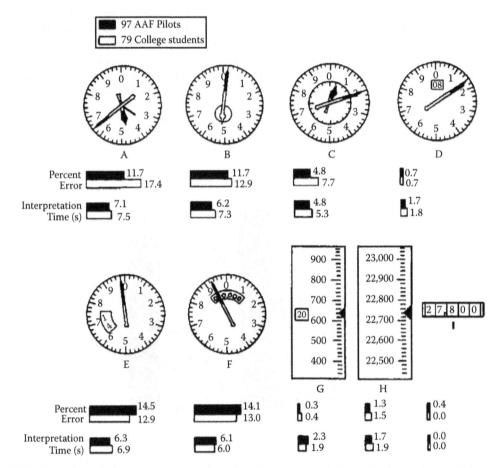

FIGURE 7.13   Speed and accuracy of reading different altimeter designs. (From Grether, W.F., *J. Appl. Psychol.*, 33, 363, 1949.)

the best performance, and the three-pointer design (A) is one of the worst. The three-pointer altimeter poses several coding problems for someone reading it. The three pointers are not clearly discriminable. Each pointer is read against the same scale using a different scale ratio, and the size of the pointer and the scale ratio are inversely related (the smallest pointer indicates the largest scale, 10,000 s, the largest pointer, 100 s).

Despite these results, a digital display is currently not used. A static reading test is not a good reflection of the real flying task. In the real task, altitude changes rapidly, and hence, a digital display would be unreadable. Furthermore, the user also needs to identify the rate of change, for which the angle of line is an effective display. Nowadays, unambiguous combination altimeter displays are used, with a pointer for rapidly changing small numbers, and a digital display for slowly changing the large numbers (D). Before this change, many hundreds of deaths were attributed to misreadings of the three-pointer altimeter, yet, the display design was not changed until these comparative tests were repeated two decades later. This delay occurred for two reasons, which illustrates that HF/E decisions are made in several wider contexts. First was the technology: In the 1940s, digital instrument design was very much more unreliable than the unreliability of the pilot's instrument readings. Second, cultural factors influence the attribution of responsibility for error. There is a recurring swing in attitudes between the statement that a user can read the instrument correctly, so the user is responsible for incorrect readings, and the statement that if a designer gives the users an instrument that it is humanly impossible to read reliably, then the responsibility for misreading errors lies with the designer.

7.1.3.3.2 *Making Comparisons between Sizes*

There are two important comparisons in control tasks: Is the variable value acceptable/within tolerance (a check reading), and if not, how big is the error? These comparisons can both usually be done more easily on an analogue display. Check readings can be made automatically (i.e., without processing that uses cognitive capacity) if the pointer on a scale is in an easily recognizable position when the value is correct. Furthermore, linking the size of the error to the size of action needed to correct it can be done easily if both are coded by the length of the line.

An example shows why it is useful to distinguish cognitive functions from the cognitive processes used to meet them. Comparison is a cognitive function that may be done either by simple recoding or by a great deal of cognitive processing, depending on the display design. Consider the horizontal bars in Figure 7.13 as a display from which an HF/E designer must get information about the relative effectiveness of the altimeter designs. The cognitive processes needed involve searching for the shortest performance bar by comparing each of the performance bar lines, probably using iconic (visual) memory, and storing the result in the working memory, then repeating to find the next smallest, and so on. Visual and working memory are used as temporary working spaces while making the comparisons; working memory is also used to maintain the list of decision results. This figure is not the most effective way of conveying a message about alternative designs, because most people do not bother to do all this mental work. The same results are presented in Figure 7.14. For a person who is familiar with graphs, the comparisons are inherent in this representation. A person looking at this does not have to do cognitive processing that uses processing capacity, which is unrelated to and interrupts the main task of thinking about choice of displays (see later discussion for more on memory interruption and processing capacity). This point applies in general to analogue and digital displays. For many comparison tasks, digital displays require more use of cognitive processing and working memory.

7.1.3.3.3 *Direction of Movement → Meaning*

The second aspect to learn about interface sizes is the meaning of the direction of a change in the size. Here, cultural learning is involved, and can be quite context-specific. For example, people in technological cultures know that clockwise movement on a display indicates increase, but on a tap or valve control indicates closure, and therefore, decrease. Again, there can be population stereotypes in the

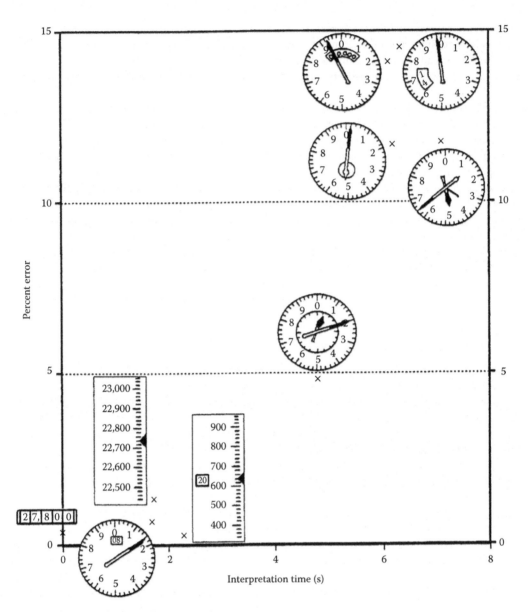

FIGURE 7.14   Graph of pilot data presented in Figure 7.13.

expectations that people bring to a situation, and if linkages are not compatible with these assumptions, error rates may be at least doubled.

Directions of movements are often paired. For example, making a control action to correct a displayed error involves two directions of movement, on the display and on the control. It can be straightforward to make the two movements compatible in direction if both are linear, or both are circular.

It is in combining three or more movements that it is easy to get into difficulties with compatibility. One classic example is the aircraft attitude indicator. In the Fitts and Jones (1961b) study on pilots' instrument reading errors, 22% of the errors were either reversed spatial interpretations or attitude

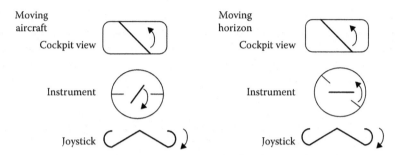

FIGURE 7.15 Two designs for the attitude indicator, showing incompatible movements.

illusions. In the design of the attitude indicator, four movements are observed to be involved: of the external world, the display, the control, and the pilot's turning receptors (see Figure 7.15). The attitude instrument can show a moving aircraft, in which case, the display movement is the same as the joystick control movement, but opposite to the movement of the external world. Else, the instrument can show a moving horizon, which is compatible with the view of the external world but not with the movement of the joystick. There is no solution in which all the three movements are the same, and hence, some performance errors or delays are inevitable. Similar problems arise in the design of moving scales and remote-control manipulation devices.

7.1.3.4 Reaction Times

The evidence quoted so far about recoding has focused on error rates. The time taken to translate from one code representation to another also gives interesting information. Teichner and Krebs (1974) reviewed the results of reaction time studies. Figure 7.16 shows the effect of the number of alternative

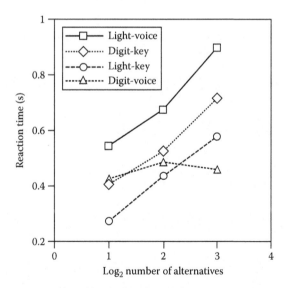

FIGURE 7.16 Response times are affected by the number of alternatives to be responded to, the nature of the "code" linking the signal and response, and the amount of practice. (From Teichner, W.H. and Krebs, M.J., *Psychol. Rev.*, 81, 75, 1974.)

items and the nature of the recoding. The effect of spatial layout is illustrated in Figure 7.12. Teichner and Krebs also reviewed the evidence that, although unpracticed reaction times are affected by the number of alternatives to choose between, after large amounts of practice, this effect disappears and all the choices are made equally quickly. This suggests that response choice has become automatic; it no longer requires processing capacity.

The results show the effect of different code translations—using spatial locations of signals and responses (light, key) or symbolic ones (visually presented digit, spoken digit, i.e., voice). The time taken to make a digit → voice translation is constant, but this is already a highly practiced response for the people tested. Otherwise, making a spatial link (light → key) is quickest. Making a link that involves a change of code type, between spatial and symbolic (digit → key, or light → voice), takes longer time (hence, these data show that it can be quicker to locate than to name). This coding time difference may arise because spatial and symbolic processes are handled by different areas of the brain, and it takes time to transmit information from one part of the brain to another. The brain does a large number of different types of coding translation (e.g., Barnard, 1987).

The findings presented so far are from the studies of reactions to signals that are independent and occur one at a time. Giving advance information about the responses that will be required, allows people to anticipate and prepare their responses, and reduces response times. There are two ways of doing this, as illustrated in Figure 7.17. One is to give a preview, allowing people to see in advance, the responses needed. This can reduce the reaction time to more than half. The second method is to have sequential relations in the material to be responded to. Figure 7.16 shows that the reaction time is affected by the number of alternatives; the general effect underlying this is that reaction time depends on the probabilities of the alternatives. Sequential effects change the probabilities of items. One way of introducing sequential relations is to have meaningful sequences in the items, such as prose rather than random letters.

Reaction time and error rate are interrelated. Figure 7.18 shows that when someone reacts very quickly, the person chooses a response at random. As the person takes a longer time, he/she can take in more information before initiating a response, and there is a trade-off between time and error rate. At longer reaction times, there is a basic error rate that depends on the equipment used.

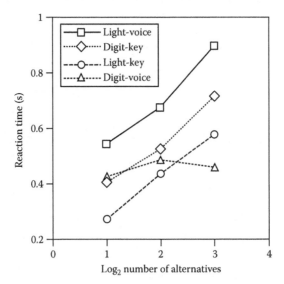

FIGURE 7.17 Effect of preview and predictability of material on response time. (Based on data in Shaffer, L.H., Latency mechanisms in transcription. In Kornblum, S. (Ed.), *Attention and Performance IV*, Academic Press, London, 1973, pp. 435–446.)

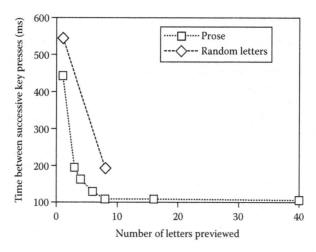

FIGURE 7.17 (continued)

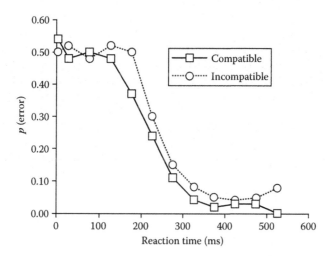

FIGURE 7.18 Speed-accuracy trade-off in two-choice reactions, and the effect of stimulus–response compatibility.

7.1.4 Action Execution

This chapter does not focus on the physical activity, but this section makes some points about the cognitive aspects of action execution. The section is divided into two parts, on acquisition movements and on continuous control or tracking movements.

The speed, accuracy, and power that a person can exert in a movement depend on its direction relative to the body position. Human biomechanics and its effects on physical performance and the implications for workplace design are vast topics, which are not reviewed here (Pheasant, 1991). Only one point is made. Workplace design affects the amount of physical effort needed to make an action and the amount of postural stress that a person is undergoing. Both these affect whether a person is willing to make a particular action or do a particular job. Thus, workplace design can affect the performance in cognitive tasks. Factors that affect what a person is or is not willing to do are discussed in detail in the section on workload.

7.1.4.1 Acquisition Movements

When someone reaches to something, or puts something in place, this is an acquisition movement. Reaching a particular endpoint or target is more important than the process of getting there. The relation between the speed and accuracy of these movements can be described by Fitts's law (Fitts, 1954), in which movement time depends on the ratio of the movement length to the target width. However, detailed studies show that all movements with the same ratio are not carried out in the same way. Figure 7.19 shows that an 80/10 movement is made with a single pulse of velocity. A 20/2.5 movement has a second velocity pulse, suggesting that the person has sent a second instruction to his or her hand about how to move. Someone making a movement gives an initial instruction to his or her muscles about the direction, force, and duration needed, and then monitors how the movement is being carried out, by vision and/or feel. If necessary, the person sends a corrected instruction to the muscles to improve the performance, and so on. This monitoring and revision represents the use of feedback. A finer movement involves the feedback to the brain and a new instruction from the brain. A less accurate movement can be made with one instruction to the hand, without needing to revise it. An unrevised movement (open-loop or ballistic) probably involves feedback within the muscles and spinal cord, but not visual feedback to the brain and a new instruction from the brain.

   Movements that are consistently made in the same way can be done without visual feedback, once learned, as mentioned in the section on location coding. Figure 7.20 shows the double use of feedback in this learning. A person chooses an action instruction that he or she expects will have the effect wanted. If the result is not as intended, then the person needs to adjust the knowledge about the expected effect of an action. This revision continues each time when the person makes an action, until the expected result is the same as the actual result. Subsequently, the person can make an action with minimal need to

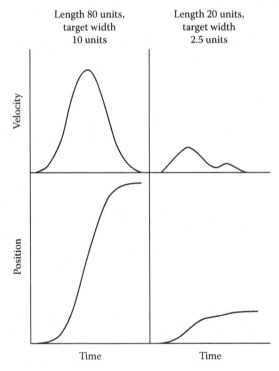

FIGURE 7.19 Execution of movements of different sizes. (From Crossman, E.R.F.W. and Goodeve, P.J., Feedback control of hand-movement and Fitts' law, Communication to the Experimental Psychology Society, University of Oxford, Oxford, U.K., 1963.)

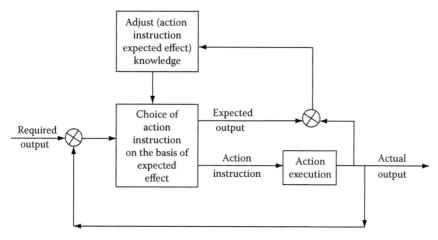

FIGURE 7.20 Double use of feedback in learning to make movements.

check that it is being carried out effectively. This reduces the amount of processing effort needed to make the movement. Knowledge about the expected results is a type of meta-knowledge. Meta-knowledge is important in activity choice, and is discussed again in the later section.

7.1.4.2 Control or Tracking Movements

Control movements are those in which someone makes frequent adjustments, with the aim of keeping some part of the external world within the required limits. They might be controlling the output of an industrial process, or keeping an aircraft straight and leveled. In industrial processes, the time lag between making an action and its full effect in the process may be anything from minutes to hours; hence, there is usually time to think about what to do. In contrast, in flying, events can happen very quickly, and human-reaction time along with neuromuscular lag adding up to half a second or more, can have a considerable effect on the performance. Hence, various factors may be important in the two types of control task.

There are two ways of reducing the human response lag (cf. Figure 7.17). Preview allows someone to prepare actions in advance and therefore, to overcome the effect of the lag. People can also learn something about the behavior of the track that they are following, and can subsequently use this knowledge to anticipate what the track will do and prepare their actions accordingly.

There are two ways of displaying a tracking task. In a pursuit display, the moving target and the person's movements are displayed separately. A compensatory display system computes the difference between the target and the person's movements, and displays this difference relative to a fixed point. Many studies show that human performance is better with a pursuit display, as shown in Figure 7.21. As mentioned earlier, people can learn about the effects of their actions and target movements, and both types of learning can lead to improved performance. On the pursuit display, the target and human movements are displayed separately, and hence, a person using this display can do both types of learning. In contrast, the compensatory display only shows the difference between the two movements. Thus, it may not be possible for the viewer to tell which part of a displayed change is owing to the target movements and which is owing to the viewer's own movements, and hence, these are difficult to learn.

A great deal is known about human fast-tracking performance (Rouse, 1980; Sheridan & Ferell, 1974). A person doing a tracking task acts as a controller. Control theory provides tools for describing some aspects of the track to be followed and how a device responds to the inputs. This has resulted in the development of a "human transfer function," a description of a human controller as if the person was an engineered control device. The transfer function contains some components that describe the human

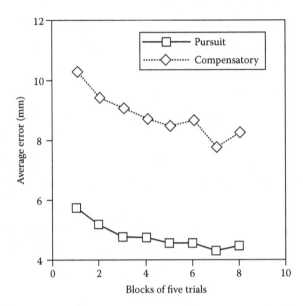

FIGURE 7.21 Errors in tracking performance using pursuit and compensatory displays. (From Briggs, G.E. and Rockway, M.R., *J. Exp. Psychol.*, 71, 165, 1966.)

performance limits, and some that partially describe the human ability to adapt to the properties of the device that the person is controlling. This function can be used to predict the combined pilot–aircraft performance. This is a powerful technique with considerable economic benefits. However, it is not relevant to this chapter as it describes the performance, and not the underlying processes, and only describes the human performance in compensatory tracking tasks. It also focuses attention on an aspect of human performance that can be poorer than that of fairly simple control devices. This encourages the idea of removing the person from the system, rather than appreciating what people can actively contribute, and designing support systems to overcome their limitations.

7.1.5 Summary and Implications

7.1.5.1 Theory

The cognitive processes underlying the classic HF/E can be relatively simple, but not so simple that they can be ignored. Cognitive processing is carried out to meet cognitive functions. Five functions are discussed in this section: distinguishing between stimuli; building up a percept of an external world containing independent entities with stable properties; naming; choosing an action; and comparison.

This section suggests that these functions could be met with simple tasks using three main cognitive processes (what happens when these processes are not sufficient has been mentioned briefly and is discussed in the next main section). The three processes are: deciding between the alternative interpretations of the evidence; integrating the data from all the sensory sources along with the knowledge about the possibilities, to an inferred percept that makes the best sense of all the information; and recoding, that is, translating from one type of code to another.

Furthermore, five other key aspects of cognitive processing have been introduced:

1. Sensory processing is relative rather than absolute.
2. The cognitive functions are not necessarily met by processes in a clearly distinct sequence. Processes that are "automated" may be carried out in parallel. The processes communicate with each other via a common "blackboard," which provides the context within which each process works, as summarized in Figure 7.22.

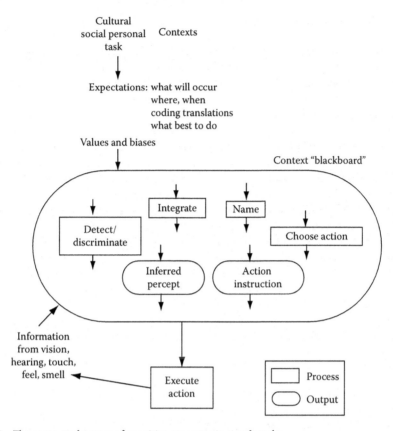

FIGURE 7.22 The contextual nature of cognitive processes in simple tasks.

As processing is affected by the context in which it is done, behavior is adaptive. However, for HF/E practitioners, this has the disadvantage that the answer to any HF/E question is always, "it depends."

3. The processing is not simply input driven: All types of processing involve the use of knowledge relevant to the context (it can therefore be misleading to use the term *knowledge-based* to refer to one particular mode of processing).
4. Preview and anticipation can improve performance.
5. Actions have associated meta-knowledge about their effects, which improves with learning.

7.1.5.2 Practical Aspects

The primary aim of classic HF/E has been to minimize unnecessary physical effort. The points made here emphasize the need to minimize unnecessary cognitive effort.

Task analysis should not only note which displays and controls are needed, but should also ask questions such as: What cognitive functions need to be carried out? By what processes? Is the information used in these processes salient?

In discrimination and integration, the following questions need to be addressed: What is the ensemble of alternatives to be distinguished? Are the items designed to maximize the differences between them? What are the probabilities and costs of the alternatives? How does the user learn these?

In recoding, questions that should addressed include: What coding vocabularies are used (shape, color, location, size, direction, alphanumeric) in each subtask, and in the task as a whole? Are the translations unambiguous, unique, consistent, and if possible, obvious? Do reaction times limit performance, and if so, can preview or anticipation be provided?

7.2 Complex Tasks

Using an interface for a simple task entails the functions of distinguishing between stimuli, integrating stimuli, naming, comparing, and choosing and making simple actions. When the interface is well-designed, these functions can be carried out by decision making, integration, and recoding processes. These processes use knowledge about the alternatives that may occur, their distinguishing features, probabilities, and costs, and the translations to be made.

More complex task needs more complex knowledge in more complex functions and processes. For example, consider that an air-traffic controller is given the two flight strips illustrated in Figure 7.23. Commercial aircraft fly from one fix point to another. These two aircrafts are flying at the same level (31,000 ft) from fix OTK to fixed LEESE7 DAL1152, and are estimated to arrive at LEESE7, 2 min after AALA19 (18–16), and are traveling faster (783 > 746). Thus, DAL1152 is closing relatively fast and the controller needs to take immediate action, to tell one of the aircrafts to change the flight level. The person telling the aircraft to change the level is doing more than simply recoding the given information. The person uses strategies for searching the displays and comparing the data about the two aircraft, along with a simple dynamic model of how an aircraft changes position in time, to build up a mental picture of the relative positions of the aircrafts, with one overtaking the other which may result in a possible collision. The person then uses a strategy for optimizing the choice of which aircraft should be instructed to change its level.

The overall cognitive functions or goals are to understand what is happening and to plan what to do about it. In complex dynamic tasks, these two main cognitive needs are met by subsidiary cognitive functions, such as

- Infer/review present state
- Predict/review future changes/events
- Review/predict task-performance criteria
- Evaluate acceptability of present or future state
- Define subtasks (task goals) to improve acceptability
- Review available resources/actions, and their effects
- Define possible (sequences of) actions (and enabling actions) and predict their effects
- Choose action/plan
- Formulate execution of action plan (including monitoring of the effects of actions, which may involve repeating all the preceding)

AAL419	OTK	16	310		+LEESE7 + KMCO	4325
MD88/R	1002	10				
T746 G722						
490 1		KMCO				

DAL1152	OTK	18	310		+LEESE7 + KMCO	3350
H/L101/R	1004	10				
T783 G759						
140 1		KMCO				

FIGURE 7.23 Two flight strips, each describing one aircraft. Column 1: (top) aircraft identification; (bottom) true airspeed/knots. Column 2: (top) previous fix. Column 3: (top) estimated time over next fix. Column 4: flight level (i.e., altitude in hundreds of feet). Column 6: next fix.

These cognitive functions are interdependent. They are not carried out in a fixed order, but are used whenever necessary. Lower level cognitive functions implement higher level ones. At the lowest levels, the functions are fulfilled by cognitive processes, such as searching for the information needed, discrimination, integration, and recoding. The processing is organized within the structure of the cognitive goals/functions.

An overview is built up in working storage by carrying out these functions. This overview represents the person's understanding of the current state of the task and the person's views about it. The overview provides the data that the person uses in later thinking, as well as the criteria for what best to do next and how best to do it. Thus, there is a cycle: Processing builds up the overview, which determines the next processing, which updates the overview, and so on (see Figure 7.24). Figure 7.22 shows an alternative representation of the context, as nested rather than cyclic (for more information about this mechanism, see Bainbridge 1993a).

The main cognitive processes discussed in the previous section were decision making, integrating stimuli, and recoding. However, additional modes of processing are needed in complex tasks, such as

- Carrying out a sequence of recoding transformations, and temporarily storing intermediate results in working memory
- Building up a structure of inference, an overview of the current state of understanding and plans, in working storage, using a familiar working method
- Using working storage to menially simulate the process of a cognitive or physical strategy
- Deciding between alternative working methods on the basis of meta-knowledge
- Planning and multitasking
- Developing new working methods

These complex cognitive processes are not directly observable. The classic experimental psychology method, which aims to control all except one or two measured variables, and to vary one or two variables so that their effects can be studied, is well-suited to investigate the discrimination and recoding processes. However, it is not well-suited to examine the cognitive activities in which many interrelated processes may occur without any observable behavior. Studying these tasks involves special techniques: case studies, videos, verbal protocols, or distorting the task in some way, perhaps slowing it down or making the person do

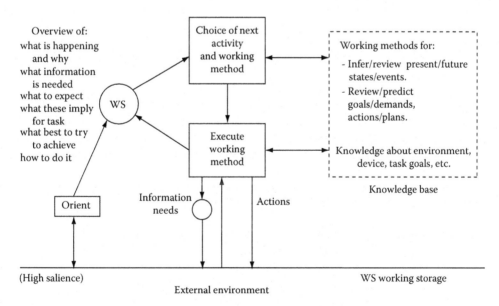

FIGURE 7.24 A sketch of the contextual cycle in relation to the knowledge base and the external environment.

extra actions to get the information (Wilson & Corlett, 1995). Both setting up and analyzing the results of such studies can take years of effort. The results tend to be as complex as the processes studied, and hence, they are difficult to publish in the usual formats. Such studies do not fit well into the conventions about how a research is to be carried out, and therefore, there are unfortunately not many studies of this type. However, the rest of this section gives some evidence about the nature of the complex cognitive processes, to support the general claims made so far. The subsections are on sequences; language understanding; inference and diagnosis; working storage; planning, multitasking, and problem solving; and knowledge.

7.2.1 Sequences of Transforms

After decision making, integrating, and recoding, the next level of complexity in cognitive processing is to carry out the sequence of recoding translations or transforms. The result of one step in the sequence acts as the input to the next step, and hence, has to be kept temporarily in working memory. Here, the notion of recoding needs to be expanded to include transforms, such as simple calculations and comparisons, and conditions leading to alternative sequences. It can be noted that in this type of processing, the goal of the behavior, the reason for doing it, is not included in the description of how it is done. Some people call this type of processing as *rule-based*. There are two typical working situations in which behavior is not structured relative to goals.

When a person is following instructions that do not give him or her any reason for why he or she has to do each action, then the person is considered to use this type of processing. This is usually not a good way of presenting instructions, as if anything goes wrong, then the person may have no reference point to identify how to correct the problem.

The second case can arise in a stable environment, in which the behavior is carried out in the same way each time. If a person has practiced often, then the behavior may be carried out without the need to check it, or to think out what to do or how to do it (see later discussion). Such overlearned sequences give a very efficient way of behaving, in the sense of using minimal cognitive effort. However, if the environment does change, then overlearning becomes maladaptive and can lead to errors (see later discussion on learning and errors).

7.2.2 Language Processing

This section covers two issues: using language to convey information and instructions, and the processes involved in language understanding. Although language understanding is not the primary task of either the pilot or air-traffic controller, it does provide simple examples of some key concepts in complex cognitive processing.

7.2.2.1 Written Instructions

Providing written instructions is often thought of as a way of making a task easy, but this is not guaranteed. Reading instructions involves interpreting the words to build up a plan of action. The way the instructions are written may make this processing more or less difficult, and videorecorder-operating manuals are notorious for this.

Various techniques have been used for measuring the difficulty of processing different sentence types. Some typical results are as follows (Savin & Perchonock, 1965):

Sentence Type	Example	% Drop in Performance
Kernel	The pilot flew the plane.	0
Negative	The pilot did not fly the plane.	−16
Passive	The plane was flown by the pilot.	−14
Negative passive	The plane was not flown by the pilot.	−34

Such data suggest that understanding negatives and passives involves two extra and separate processes. This indicates that it is usually best to use active positive forms of the sentence. However, when a negative or restriction is the important message, it should be the most salient and should come first. For example, "No smoking" is more effective than "Smoking is not permitted." Furthermore, using a simple form of sentence does not guarantee that a message makes a good sense. I recently enjoyed staying in a hotel room with a notice on which the large letters said:

Do not use the elevator during a fire.
Read this notice carefully.

Connected prose is not necessarily the best format for showing alternatives in written instructions. Spatial layout can be used to show the groupings and relations between the phrases by putting each phrase on a separate line, indenting to show the items at the same level, and using flow diagrams to show the effect of choice between the alternatives (e.g., Oborne, 1995, Chapter 4). When spatial layout is used to convey the meaning in written instructions, it is a code and should be used consistently, as discussed earlier.

Instructions also need to be written from the point of view of the reader: "If you want to achieve this, then do this." However, instruction books are often written the other way round: "If you do this, then this happens." The second approach requires the reader to have much more understanding, searching, and planning to work out what to do. It can be noted that the effective way of writing instructions is goal-oriented. In complex tasks, methods of working are, in general, best organized in terms of what is to be achieved, and this is discussed in the later section.

7.2.2.2 Language Understanding

In complex tasks, many of the cognitive processes and knowledge used are only possible, because the person has considerable experience of the task. Language understanding is the chief complex task studied by experimental psychologists (e.g., Ellis, 1993), as it is easy to find experts to test. When someone is listening to or reading a language, each word evokes learned expectations. For example:

The
can only be followed by
—a descriptor, or
—a noun
The pilot
depending on the context, either;
(a) *will be followed by the word "study" or:*
(b) *—evokes general knowledge (scenarios) about aircraft or ship pilots.*
    —can be followed by:
    —a descriptive clause, containing items relevant to living things/animals/human beings/pilots, or
    —a verb, describing possible actions by pilots

Each word leads to expectations about what will come next; each constrains the syntax (grammar) and semantics (meaning) of the possible next words. To understand the language, a person needs to know the possible grammatical sequences, the semantic constraints on what words can be applied to what types of item, and the scenarios. During understanding, a person's working storage contains the general continuing scenario, the structure of understanding built up from the words received so far, and the momentary expectations about what will come next (many jokes depend on not meeting these expectations).

The overall context built up by a sequence of phrases can be used to disambiguate alternative meanings, such as

> **The Inquiry investigated why**
> **the pilot turned into a mountain.**

or

> **In this fantasy story**
> **the pilot turned into a mountain.**

The knowledge base/scenario is also used to infer missing information. For example:

> **The flight went to Moscow.**
> **The stewardess brought her fur hat.**

Answering the question "Why did she bring her fur hat?" involves knowing that the stewardesses go on flights, and about the need for and materials used in protective clothing, which are not explicitly mentioned in the information given.

Understanding of a language does not necessarily depend on the information being presented in a particular sequence. Although it requires more effort, we can understand someone whose first language uses a different word order from English, such as

> **The stewardess her fur hat brought.**

We do this by having a general concept that a sentence consists of several types of units (noun phrases, verb phrases, etc.), and we make sense of the input by matching it with the possible types of units. This type of processing can be represented as being organized by a "frame with slots," where the frame coordinates the slots for the types of item expected, which are then instantiated in a particular case, as in

Noun phrase	Verb	Noun phrase
The stewardess	brought	her fur hat

(as language has many alternative sequences, this is by no means a simple operation; Winograd, 1972).

The understanding processes used in complex control and operation tasks show the same features that are found in language processing. The information obtained evokes both general scenarios and specific moment-to-moment expectations. The general context, as well as additional information, can be used to decide between the alternative interpretations of the given information. A structure of understanding is built up in working storage, and frames or working methods suggest the types of information that the person needs to look for to complete their understanding. These items can be obtained in a flexible sequence, and the knowledge is used to infer whatever is needed to complete the understanding, but is not supplied by the input information. Furthermore, the structure of understanding is built up to influence the state of the external world, to try to get it to behave in a particular way, which is an important addition in the control/operation tasks.

7.2.3 Inference and Diagnosis

To illustrate these cognitive processes in an aviation example, this section uses an imaginary example to make the presentation short. The later sections describe the real evidence on pilot and air-traffic controller behavior, which justifies the claims made here.

Suppose that an aircraft is flying and the "engine oil low" light goes on. What might be the pilot's thoughts? The pilot needs to infer the present state of the aircraft (cognitive functions are indicated by italics). This involves considering alternative hypotheses that could explain the light, such as whether there is an instrument fault, or there is genuinely an engine fault, and then choosing between the hypotheses according to their probability (based on previous experience of this or another aircraft) or by looking for other evidence that would confirm or disprove the possibilities. The pilot could predict the future changes that will occur as a result of the chosen explanation of the events. Experienced people's behavior in many dynamic tasks is future-oriented. A person takes anticipatory action, not to correct the present situation, but to ensure that the predicted unacceptable states or events do not occur. Before evaluating the predictions for their acceptability, the pilot needs to review the task performance criteria, such as the relative importance of arriving at the original destination quickly, safely, or cheaply. The result of comparing the predictions with the criteria will be to define the performance needs to be met. It is necessary to review the available resources, such as the state of the other engines or the availability of alternative landing strips. The pilot can then define possible alternative action sequences and predict their outcomes. A review of action choice criteria, which includes the task-performance criteria as well as others, such as the difficulty of the proposed procedures, is needed as a basis for choosing an action sequence/plan, before beginning to implement the plan. Many of these cognitive functions must be based on incomplete evidence, for example, about future events or the effects of actions, and hence, risky decision making is involved.

A pilot who has frequently practiced these cognitive functions may be able to carry them out "automatically," without being aware of the need for intermediate thought. Furthermore, an experienced pilot may not be aware of thinking about the functions in separate stages; for example, (predict + review criteria + evaluation) may be done together.

Two modes of processing have been used in this example: "automatic" processing (i.e., recoding), and using a known working method that specifies the thinking that needs to be carried out. Other modes of processing are suggested later. The mode of processing needed to carry out a function depends on the task situation and the person's experience (see later discussion on learning). An experienced person's knowledge of the situation may enable the person to reduce the amount of thinking, even when the person does need to think things out explicitly. For example, it may be clear early in the process of predicting the effects of possible actions that some will be not acceptable and hence, need not be explored further (see later discussion on planning).

Nearly all the functions and processing mentioned earlier have been acquired from the pilot's knowledge base. The warning light evokes working methods for explaining the event and choosing an action plan, as well as the knowledge about the alternative explanations of events and suggestions of relevant information to look for. Thus, the scenario is the combination of (working method + knowledge referred to in using this method + mental models for predicting events). Specific scenarios may be evoked by particular events or particular phases of the task (phases of the flight).

This account of the cognitive processes is goal-oriented. The cognitive functions or goals are the means by which the task goals are met, but are not the same. Task and personal goals act as constraints on what it is appropriate and useful to think about when fulfilling the cognitive goals.

The cognitive functions and processing build up a structure of data (in working storage) that describes the present state and the reasons for it, predicted future changes, task performance and action choice criteria, resources available, the possible actions, the evaluations of the alternatives, and the chosen action plan. This data structure is an overview that represents the results of the thinking and decisions done so far, and provides the data and context for subsequent thinking. For example, the result of reviewing task-performance criteria is not only an input to evaluation; it could also affect what is focused on in inferring the present state, in reviewing resources, or in action choice. The overview ensures that behavior is adapted to its context.

This abovementioned simple example describes the reaction to a single unexpected event. Normally, flying and air-traffic control are ongoing task. For example, at the beginning of the shift an air-traffic controller has to build up an understanding of what is happening and what actions are necessary, from the scratch. After this, each new aircraft that arrives is fitted into the controller's ongoing mental picture of what is happening in the airspace; thus, the thinking processes do not start again from the beginning. Aircrafts usually arrive according to schedule and are expected accordingly, but the overview needs to be updated and adapted to changing circumstances (see later discussion on planning and multitasking).

There are two groups of practical implications of these points. One is that cognitive task analysis should focus on the cognitive functions involved in a task, rather than simply prespecifying the cognitive processes by which they are met. The second is that designing specific displays for individual cognitive functions may be unhelpful. A person doing a complex task meets each function within an overall context, where the functions are interdependent, and the person may not think about them in a prespecified sequence. Giving independent interface support to each cognitive function or subtask within a function could make it more difficult for the person to build up an overview that interrelates the different aspects of the person's thinking.

7.2.3.1 Diagnosis

The most difficult cases of inferring that underlies the given evidence may occur during fault diagnosis. A fault may be indicated by a warning light or, for an experienced person, by a device not behaving according to the expectations. Like any other inference, fault diagnosis can be done by several modes of cognitive processing, depending on the circumstances. If a fault occurs frequently and has unique symptoms, it may be possible to diagnose the fault by visual pattern recognition, that is, pattern on interface → fault identity (e.g., Marshall, Scanlon, Shepherd, & Duncan, 1981). This is a type of recoding. However, diagnosis can also pose the most difficult issues of inference, for example, by reasoning based on the physical or functional structure of the device (e.g., Hukki & Norros, 1993).

In-flight diagnosis may need to be done quickly. Experienced people can work rapidly using recognition-primed decisions, in which situations are assigned to a known category with a known response, on the basis of similarity. The processes involved in this are discussed by Klein (1989). The need for rapid processing emphasizes the importance of training for fault diagnosis.

Amalberti (1992, Expt. 4) studied the fault diagnosis by pilots. Two groups of pilots were tested: Pilots in one group were experts on the Airbus, and those in the other group were experienced pilots beginning their training on the Airbus. They were asked to diagnose two faults specific to the Airbus, and two general problems. In 80% of the responses, the pilots gave only one or two possible explanations. This is compatible with the need for rapid diagnosis. Diagnostic performance was better on the Airbus faults, which the pilots had been specifically trained to watch out for, than on the more general faults. One of the general problems was a windshear on take-off. More American than European pilots diagnosed this successfully. American pilots are more used to windshear as a problem, and hence, are more likely to think of this as a probable explanation of an event. Thus, people's previous experience is the basis for the explanatory hypotheses that they suggest.

In the second general fault, there had been an engine fire on take-off, during which the crew forgot to retract the landing gear, which made the aircraft unstable when climbing. Most of the hypotheses suggested by the pilots to explain this instability were general problems with the aircraft, or were related to the climb phase. Amalberti suggested that when the aircraft changed the phase of flight, from take-off to climb, the pilots changed their scenario that provides the appropriate events, procedures, mental models, and performance criteria to be used in thinking. Their knowledge about the previous phase of flight became less accessible, and hence, was not used in explaining the fault.

7.2.4 Working Storage

The inference processes build up the contextual overview or situation awareness in working storage. This is not the same as the short-term memory, but short-term memory is an important limit to performance and is discussed first.

7.2.4.1 Short-Term Memory

Figure 7.25 shows some typical data on how much is retained in short-term memory after various time intervals. Memory decays over about 30 s, and is worse if the person has to do another cognitive task before being tested on what the person can remember.

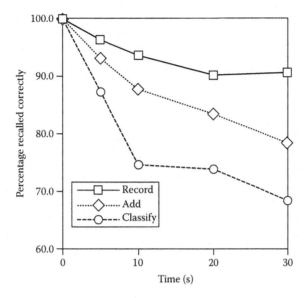

FIGURE 7.25   Decrease in recall after a time interval with different tasks during the retention interval. (From Posner, M.I. and Rossman, E., *J. Occup. Accidents*, 4, 311, 1965.)

This memory decay is important in the design of computer-based display systems in which different display formats are called up in sequence on a screen. Consider that the user has to remember an item from one display, which should be used with an item on a second display. Suppose, the second display format is not familiar, then the person has to search for the second item: This search may take about 25 s. The first item must then be recalled after doing the cognitive processes involved in calling up the second display and searching it. The memory data suggest that the person might have forgotten the first item on 30% of occasions.

The practical implication is that, to avoid this source of errors, it is necessary to have sufficient display area so that all the items used in any given cognitive processing can be displayed simultaneously. Minimizing non-task-related cognitive processes is a general HF/E aim, to increase processing efficiency. In this case, it is also necessary to reduce errors. This requirement emphasizes the need to identify what display items are used together, in a cognitive task analysis.

7.2.4.2 The Overview in Working Storage

Although there are good reasons to argue that the cognitive processes in complex dynamic tasks build up a contextual overview of the person's present understanding and plans (Bainbridge 1993a), not much is known about this overview. This section makes some points about its capacity, content, and the way items are stored.

Capacity. Bisseret (1970) asked the air-traffic area controllers, after an hour of work, about what they remembered about the aircraft that they had been controlling. Three groups of people were tested: trainee controllers, people who had just completed their training, and people who had worked as controllers for several years. Figure 7.26 shows the number of items recalled. The experienced controllers could remember on average 33 items. This is a much larger figure than the 7 ± 2 chunk capacity for static short-term memory (Miller, 1956) or the two items capacity of running memory for arbitrary material (Yntema & Mueser, 1962). Evidently, a person's memory capacity is improved by doing a meaningful task and by experience. A possible reason for this is given later.

Content. Bisseret also investigated on the items that were remembered. The most frequently remembered items were flight level (33% of items remembered), position (31%), and time at fix (14%). Leplat and Bisseret (1965) had previously identified the strategy that the controllers used in conflict identification

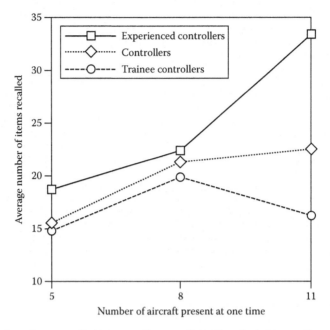

FIGURE 7.26 Number of items recalled by air-traffic controllers. (Data from Bisseret, Personal communication; based on Bisseret, A., *Ergonomics*, 14, 565, 1971.)

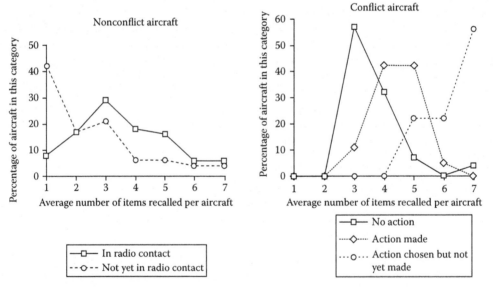

FIGURE 7.27 Recall of items about aircraft in different categories. (Based on Sperandio, J.C., Charge de travail et mémorization en contrôle d'approche (Report No. IRIA CENA, CO 7009, R24), Institut de Recherche en Informatique et Aeronautique, Paris, France, 1970.)

(checking whether aircrafts are at a safe distance apart). The frequency with which the items were remembered matched the sequence in which they were thought about: the strategy first compared the aircraft flight levels, followed by position, time at fix, and so on.

Sperandio (1970) studied another aspect (Figure 7.27). He found that more items were remembered about aircrafts involved in conflict than those that were not. With regard to nonconflict aircrafts, more was remembered about the aircrafts that had been in radio contact. With respect to conflict aircrafts, more was remembered about the aircrafts on which action had been taken, and most was remembered about the aircrafts for which an action had been chosen but not yet implemented.

These results might be explained by two classic memory effects. One is the rehearsal or repetition mechanism by which items are maintained in short-term memory. The more frequently the item or aircraft has been considered by the controllers when identifying the potential collisions and acting on them, the more likely it is to be remembered. The findings about the aircrafts in conflict could be explained by the recency effect, that items that have been rehearsed most recently are more likely to be remembered. These rehearsal and recency mechanisms make good sense as mechanisms for retaining material in real as well as laboratory tasks.

7.2.4.3 The Form in Which Material Is Retained

The controllers studied by Bisseret (1970) remembered the aircrafts in pairs or threes: "There are two flying towards DIJ, one at level 180, the other below at 160," "there are two at level 150, one passed DIJ towards BRY several minutes ago, the other should arrive at X at 22," or "I've got one at level 150 which is about to pass RLP and another at level 170 which is about 10 min behind." The aircraft were not remembered by their absolute positions, but in relation to each other. Information was also remembered relative to the future; many of the errors put the aircraft too far ahead. These sorts of data suggest that although rehearsal and recency are important factors, the items are not remembered simply by repeating the raw data, as in short-term memory laboratory experiments. What is remembered is the outcome of working through the strategy for comparing the aircrafts for potential collisions. The aircrafts are remembered in terms of the key features that bring them close together—whether they are at the same level, or flying toward the same fix point, and so on.

A second anecdotal piece of evidence is that air-traffic controllers talk about "losing the picture" as a whole, and not piecemeal. This implies that their mental representation of the situation is an integrated structure. It is possible to suggest that experienced controllers remember more, because they have better cognitive skills for recognizing the relations between aircraft, and the integrated structure makes the items easier to remember.

The only problem with this integrated structure is that the understanding, predictions, and plans can form a "whole" that is so integrated and self-consistent, that it becomes too strong to be changed. Subsequently, people may only notice information that is consistent with their expectations, and it may be difficult to change the structure of inference if it turns out to be unsuccessful or inappropriate (this rigidity in thinking is called *perceptual set*).

7.2.4.4 Some Practical Implications

Some points have already been made about the importance of short-term memory in display systems. The interface also needs to be designed to support the person in developing and maintaining an overview. It is not yet known whether an overview can be obtained directly from an appropriate display, or whether the overview can only be developed by actively understanding and planning the task, with a good display enhancing this processing but not replacing it. It is important in display systems, in which all the data needed for the whole task are not displayed at the same time, to ensure that there is a permanent overview display and that it is clear how the other possible displays are related to it.

Both control automation (replacing the human controller) and cognitive automation (replacing the human planner, diagnoser, and decision maker) can cause problems with the person's overview. A person who is expected to take over manual operation or decision making will only be able to make informed decisions about what to do after the person has built up an overview of what is happening. This may take 15–30 min to develop. The system design should allow for this sort of delay before a person can take over effectively (Bainbridge, 1983). Also, the data mentioned earlier show that a person's ability to develop a wide overview depends on experience. This indicates that, to be able to take over effectively from an automated system, the person needs to practice building up this overview. Therefore, practice opportunities should be allowed in the allocation of functions between computer and person, or in other aspects of the system design such as refresher training.

7.2.5 Planning, Multitasking, and Problem Solving

Actions in complex dynamic tasks are not simple single units. A sequence of actions may be needed, and it may be necessary to deal with several responsibilities at the same time. Organization of behavior is an important cognitive function, which depends on and is a part of the overview. This section is divided into three interrelated parts: planning future sequences of action; multitasking, dealing with several concurrent responsibilities, including sampling; and problem solving, devising a method of working when a suitable one is not known.

7.2.5.1 Planning

It may be more efficient to think about what to do in advance if there is a sequence of actions to carry out or multiple constraints to satisfy, or it would be more effective to anticipate the events. Alternative actions can be considered and the optimum ones can be chosen, and the thinking should not be done under time-pressure. The planning processes may use working storage for testing the alternatives by mental simulation and holding the plan as a part of the overview.

In aviation, an obvious example is preflight planning. Civilian pilots plan their route in relation to predicted weather. Military pilots plan their route relative to possible dangers and the availability of evasive tactics. In high-speed, low-level flight, there may be no time to think out what to do during the flight, and hence, the possibilities need to be worked out beforehand. Subsequently, the plan needs to be implemented and adjusted if changes in the circumstances make this necessary. This section is divided into two parts, on preplanning and online revision of plans.

7.2.5.1.1 Preplanning

Figure 7.28 shows the results from a study of preflight planning by Amalberti (1992, Expt. 2). Pilots anticipate the actions to take place at particular times or geographical points. Planning involves thinking about several alternative actions and choosing the best compromise with the given several constraints. Some of the constraints that the pilots consider are the level of risk of external events, the limits to maneuverability of the aircraft, and their level of expertise to deal with particular situations, as well as the extent to which the plan can be adapted, and what to do if circumstances demand major changes in the plan.

Amalberti studied four novice pilots, who were already qualified but at the beginning of their careers, and four experts. The cognitive aims considered during planning are listed on the left side of the figure. Each line on the right represents one pilot, and shows the sequence in which he thought about the cognitive functions. The results show that novice pilots took longer time to carry out their planning, and that each of the novice pilots returned to reconsider at least one point he had thought about earlier. Verbal protocols collected during the planning showed that novices spent more time mentally simulating the results of the proposed actions to explore their consequences. On the other hand, the experts did not think about the cognitive functions in the same sequence, but only one of them reconsidered an earlier point. Their verbal protocols showed that they prepared fewer responses to possible incidents than the novices.

One of the difficulties in planning is that, later in planning, the person may think of problems that may demand parts of the plan already devised to be revised. Planning is an iterative process. For example, the topics are interdependent. The possibility of incidents may affect the best choice of route to or from the objective. What is chosen as the best way of meeting any one of the aims may be affected by, or affect, the best way of meeting the other aims. As the topics are interdependent, there is no single optimum sequence for thinking about them. The results suggest that experts have the ability, when thinking about any one aspect of the flight, to take into account its implications on the other aspects, and hence, it does not need to be revised later.

The experts have better knowledge about the scenario, possible incidents, and levels of risk. They know more about what is likely to happen, and hence, they need to prepare fewer alternative responses to possible incidents. The experts also know from their experience about the results of alternative actions, including the effects of actions on other parts of the task, and hence, they do not need to mentally

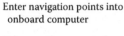

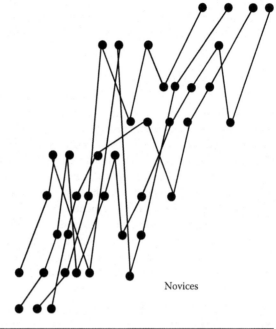

Enter navigation points into
onboard computer

Take specific account of
possible incidents

Write itinerary on map, calculate
parameters of each leg of flight
(speed, heading, altitude)

Determine route from objective
to return airfield

Determine navigation in zone of
dense enemy radar

Determine route from departure
airfield to objective

Determine navigation in zone of
poor visibility

Determine approach to
objective

General feasibility (fuel/distance
relation, weather conditions)

Novices

| Minutes | 0 | 15 | 30 | 45 | 60 |

Enter navigation points into
onboard computer

Take specific account of
possible incidents

Write itinerary on map, calculate
parameters of each leg of flight
(speed, heading, altitude)

Determine route from objective
to return airfield

Determine navigation in zone of
dense enemy radar

Determine route from departure
airfield to objective

Determine navigation in zone of
poor visibility

Determine approach to
objective

General feasibility (fuel/distance
relation, weather conditions)

Experts

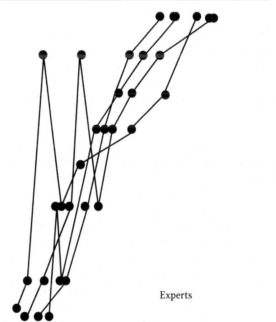

FIGURE 7.28 Preflight planning by pilots with different levels of expertise. (Translated from Amalberti, R., Modèles d'activite en conduite de processus rapides: Implications pour l'assistance á la conduite. Unpublished doctoral thesis, University of Paris, France, 1992.)

simulate the actions to check their outcomes. They also have more confidence in their own expertise to deal with given situations. All these are aspects of their knowledge about the general properties of the things that they can do, their risks, their expertise on them, and so on. This meta-knowledge was introduced in the earlier section on actions, and is also essential for multitasking as well as in workload and learning (see later discussion).

7.2.5.1.2 *Online Adaptation of Plans*

In the second part of Amalberti's study, the pilots carried out their mission plan in a high-fidelity simulator. The main flight difficulty was that they were detected by radar, and the pilots responded immediately to this. The response had been preplanned, but had to be adapted to details of the situation when it happened. The novice pilots showed much greater deviations from their original plan than the experts. Some of the young pilots slowed down before the point at which they expected to be detected, as accelerating was the only response they knew for dealing with detection. This acceleration led to a deviation from their planned course, and thus, they found themselves in an unanticipated situation. Subsequently, they made a sequence of independent, reactive, short-term decisions, because there was no time to consider the wider implications of each move. The experts made much smaller deviations from their original plan, and were able to return to the plan quickly. The reason for this was that they had not only preplanned their response to the radar, but had also thought out in advance how to recover from deviations from their original plan. Again, experience and thus, training, plays a large part in effective performance.

In situations in which events happen less quickly, people may be more effective in adapting their plans to changing events at that time. The best model for the way in which people adapt their plans to present circumstances is probably the opportunistic planning model of Hayes-Roth and Hayes-Roth (1979; see also Hoc, 1988).

7.2.5.2 Multitasking

If a person has several concurrent responsibilities, each of which involves a sequence of activities, then interleaving these sequences is called multitasking. This involves an extension of the processes mentioned under planning. Multitasking involves working out in advance what to do, along with the opportunistic response to events and circumstances at that time.

7.2.5.2.1 *Examples of Multitasking*

Amalberti (1992, Expt. 1) studied military pilots during simulated flight. Figure 7.29 shows part of his analysis, about activities during descent to low-level flight. The bottom line in this figure is a time line. The top part of the figure describes the task as a hierarchy of task goals and subgoals. The parallel double-headed arrows beneath represent the time that the pilot spent on each of the activities. These arrows are arranged in five parallel lines that represent the five main tasks in this phase of flight: maintain engine efficiency at minimum speed; control angle of descent; control heading; deal with air-traffic control; and prepare for the next phase of flight. The other principal tasks that occurred in other phases of flight were: maintain planned timing of maneuvers; control turns; and check safety. Figure 7.29 shows how the pilot allocated his time between the different tasks. Sometimes, it is possible to meet two goals with one activity. The pilot does not necessarily need to complete one subtask before changing to another. Indeed, this is often not possible in a control task, in which states and events develop over time. Usually, the pilot does one thing at a time. However, it is possible for him to do two tasks together when they use different cognitive processing resources. For example, controlling descent, which uses eyes + motor coordination, can be done at the same time as communicating with the air-traffic control, which uses hearing + speech (see later discussion on workload).

Some multitasking examples are difficult to describe in a single figure. For example, Reinartz (1989), studying a team of three nuclear power plant operators, found that they might work on 9–10 different goals at the same time. Other features of multitasking have been observed by Benson (1990):

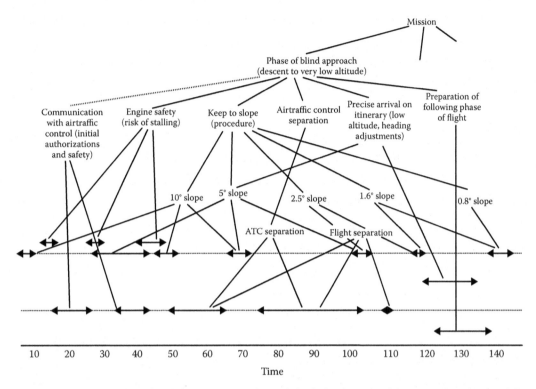

FIGURE 7.29 Multitasking by a pilot during one phase of the flight. (Translated from Amalberti, R., Modèles d'activite en conduite de processus rapides: Implications pour l'assistance á la conduite, Unpublished doctoral thesis, University of Paris, France, 1992).

- Multitasking may be planned ahead (a process operator studied by Beishon, 1974, made plans for up to 1.5 h ahead). These plans are likely to be partial and incomplete in terms of timing and detail. Planned changes in activity may be triggered by times or events. When tasks are done frequently, much of the behavior organization may be guided by habit.
- Executing the plan. Interruptions may disrupt the planned activity. As preplan is incomplete, the actual execution depends on the details of the situation at that time. Some tasks may be done when they are noticed in the process of working (Beishon, 1974, first noticed this, and called it serendipity). This is opportunistic behavior. The timing of activities of low importance may not be preplanned, but may be fitted in spare moments. The remaining spare moments are recognized as spare time.
- Effects of probabilities and costs. In a situation that is very unpredictable, or when the cost of failure is high, people may make the least risky commitment possible. If there is a high or variable workload, people may plan to avoid increasing their workload, and use different strategies in different workload conditions (see later discussion on workload).

7.2.5.2.2 A Possible Mechanism

Sampling is a simple example of multitasking, in which people have to monitor several displays to keep track of changes on them. Mathematical sampling theory has been used as a model for human attention in these tasks. In the sampling model, the frequency of attending to an information source is related to the frequency of changes on that source. This can be a useful model showing how people allocate their attention when changes to be monitored are random, as in straight and level flight; however, this model is not sufficient to account for switches in the behavior in more complex phases of flight.

Amalberti (1992) made some observations about switching from one task to another. He found that

- Before changing to a different principal task, the pilots review the normality of the situation by checking that various types of redundant information are compatible with each other.
- Before starting a task that will take some time, the pilots ensure that they are in a safe mode of flight. For example, before analyzing the radar display, pilots check that they are in the appropriate mode of automatic pilot.
- While waiting for feedback about one part of the task, pilots do part of another task that they know is short enough to fit into the waiting time.
- When doing high-risk, high-workload tasks, pilots are less likely to change to another task.

These findings suggest that, at the end of a subsection of a principal task, the pilots check that everything is all right. Subsequently, they decide (not necessarily consciously) on the next task that needs their effort, by combining their preplan with meta-knowledge about the alternative tasks, such as how urgent they are, how safe or predictable they are, how difficult they are, how much workload they involve, and how long they take (see later discussion on workload).

7.2.5.2.3 Practical Implications

Multitasking can be preplanned, and involves meta-knowledge about alternative behaviors. Both planning and knowledge develop with experience, which underlines the importance of practice and training.

The nature of multitasking also emphasizes the difficulties that could be caused by task-specific displays. If a separate display is used for each of the tasks combined in multitasking, then the user would have to call up a different display, and perhaps change the coding vocabularies, each time when the person changes to a different main task. This would require extra cognitive processing and extra memory load, and could make it difficult to build up an overview of the tasks considered together. This suggests an extension to the point made in the section on working storage. All the information used in all the principle tasks that may be interleaved in multitasking need to be available at the same time, and easily cross-referenced. If this information is not available, then coordination and opportunistic behavior may not be possible.

7.2.5.3 Problem Solving

A task is familiar to a person who knows the appropriate working methods, as well as the associated reference knowledge about the states that can occur, the constraints on allowed behavior, and the scenarios, mental models, and so on, which describe the environmental possibilities within which the working methods must be used.

Problem solving is the general term for the cognitive processes that a person uses in an unfamiliar situation, for which the person does not already have an adequate working method or reference knowledge to deal with. Planning and multitasking are also types of processing that are able to deal with situations that are not the same each time. However, both take existing working methods as their starting point, and either think about them as applied to the future, or work out how to interleave the working methods used for more than one task. In problem solving, a new working method is needed.

There are several ways of devising a new working method. Some are less formal techniques that do not use much cognitive processing, such as trial and error or asking for help. There are also techniques that do not need much creativity, such as reading an instruction book. People may otherwise use one of the three techniques for suggesting a new working method. Each of these uses working methods recursively; it uses a general working method to build up a specific working method.

1. Categorization. This involves grouping the problem situation with similar situations for which a working method is available. Thus, the working method that applies to this category of situation can then be used. This method is also called *recognition-primed decision making*. The nature of "similarity" and the decisions involved are discussed by Klein (1989).

2. Case-based reasoning. This involves thinking of a known event (a *case*) that is similar or analogous to the present one, and adapting the method used, in the present situation. This is the reason why stories about unusual events circulate within an industry. They provide people in the industry with exemplars for what they could do themselves if a similar situation arose, or with opportunities to think out for themselves what would be a better solution.

3. Reasoning from basic principles. In the psychological literature, the term problem solving may be restricted to a particular type of reasoning in which a person devises a new method of working by building it up from individual components (e.g., Eysenck & Keane, 1990, Chapters 11 and 12). This type of processing may be called *knowledge-based* by some people.

A general problem-solving strategy consists of a set of general cognitive functions that have much in common with the basic cognitive functions in complex dynamic tasks (see introduction to this section). Problem solving, for example, could involve understanding the problem situation, defining what would be an acceptable solution, and identifying what facilities are available. Meeting each of these cognitive needs can be difficult, because the components need to be chosen for their appropriateness to the situation and then fitted together. This choice could involve: identifying what properties are needed from the behavior; searching for components of behavior that have the right properties (according to the meta-knowledge that the person has about them); and then combining them into a sequence.

The final step in developing a new working method is to test it, either by mental simulation or by trial and error. This mental simulation could be similar to the techniques used in planning and multitasking. Thus, working storage may be used in problem solving in two ways: to hold both the working method for building up a working method and the proposed new method, and to simulate the implementation of the proposed working method to test whether it's processing requirements and outputs are acceptable.

7.2.6 Knowledge

Knowledge is closely involved in all modes of cognitive processing. It provides the probabilities, utilities, and alternatives considered in decision making, and the translations used in recoding. In complex tasks, it provides the working methods and reference knowledge used in thinking about cognitive functions and the meta-knowledge. Different strategies may use different types of reference knowledge. For example, a strategy for diagnosing faults by searching the physical structure of the device uses one type of knowledge, whereas a strategy that relates symptoms to the functional structure of the device uses another. The reference knowledge may include scenarios, categories, cases, mental models, performance criteria, and other knowledge about the device that the person is working with. Some knowledge may be used mainly for answering questions, for explaining why events occur, or why actions are needed. This basic knowledge may also be used in problem solving.

There are many interesting fundamental questions about how these different aspects of knowledge are structured, interrelated, and accessed (Bainbridge, 1993c), but these issues are not central to this chapter. The main questions here are the relation between the type of knowledge and how it can best be displayed, and what might be an optimum general display format.

7.2.6.1 Knowledge and Representation

Any display for a complex task can show only a subset of what could be represented. Ideally, the display should explicit the points that are important for a particular purpose, and provide a framework for thinking. The question of which display format is best for representing what aspect of knowledge has not yet been thoroughly studied, and most of the recommendations about this are assumptions based on experience (Bainbridge, 1988). For example, the following formats are often found:

Aspect of Knowledge	Form of Display Representation
Geographical position	Map
Topology, physical structure	Mimic/schematic, wiring diagram
Cause–effect, functional structure	Cause–effect network, mass-flow diagram
Task goals–means structure	Hierarchy
Sequence of events or activities	Flow diagram
Analogue variable values and limits	Scale + pointer display
Evolution of changes over time	Chart recording

Each of these aspects of knowledge might occur at several levels of detail, for example, in components, subsystems, systems, and the complete device. Furthermore, knowledge can be at several levels of distance from direct relevance; for example, it could be about a specific aircraft, about all aircrafts of this model, about aircrafts in general, about aerodynamics, or about physics.

Knowledge-display recommendations raise three sorts of question. One arises because each aspect of knowledge is one possible "slice" from the whole body of knowledge. All the types of knowledge are interrelated, but there is no simple one-to-one relation between them. Figure 7.30 illustrates some links between the different aspects of knowledge. Any strategy is unlikely to use only one type of knowledge or have no implications on the aspects of thinking that uses other types of knowledge. It might mislead the user to show different aspects of knowledge with different and separate displays that are difficult to cross-refer, as this might restrict the thinking about the task. Knowledge about cross-links is difficult to display, and is gained by experience. This emphasizes training.

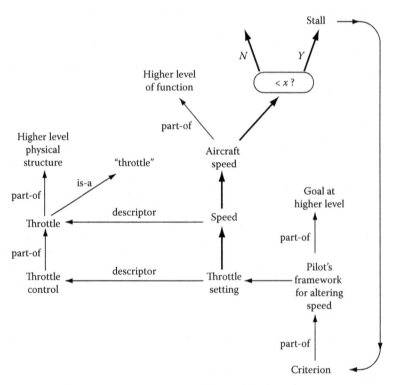

FIGURE 7.30 Some of the links in a small part of a pilot's knowledge base (thick arrow indicates cause–effect relation).

A second question is concerned with salience. Visual displays emphasize (make more salient) the aspects that can easily be represented visually (e.g., see the discussion at the end of this chapter on the limitations of Figures 7.22 and 7.24 as models of behavior). It might be unwise to make some aspects of knowledge easy to take in simply because they are easier to display, rather than because they are important in the task. There are vital types of knowledge that are not easy to display visually, such as the associations used in recoding or the categories, cases, scenarios, and meta-knowledge used in complex thinking. These are all learned by experience. The main approach to supporting nonvisual knowledge is to provide the user with reminder lists about the alternatives (see later discussion on cued recall). Display design and training are interdependent, as they are each effective at providing different types of knowledge. It could be useful to develop task-analysis techniques that identify different aspects of knowledge, as well as carry out more research on how types of knowledge and the links between them can best be presented.

The third issue about all these multiple possible display formats repeats the questions raised previously about the efficient use of codes. If a user was given all the possible display types listed earlier, each of them employing different codes possibly with different display formats using the same code with different meanings (e.g., a network with nodes could be used to represent physical, functional, or hierarchical relations between the items), then the different codes might add to the user's difficulties in making cross-connections between different aspects of knowledge.

7.2.6.2 An Optimum Format?

These issues suggest the question: Is there one or a small number of formats that subsume or suggest the others? This is a question that has not yet been much studied. A pilot study (Brennan, 1987) asked people to explain an event, given either a mimic or a cause–effect diagram of the physical device involved. The people tested, either did or did not already know how the device worked. The results suggested that people who did not know how the device worked were most helped by a cause–effect representation (which does show how it worked), whereas experts were best with the mimic representation. Contextual cues can greatly aid the memory performance (e.g., Eysenck & Keane, 1990, Chapter 6). A cue is an aid to accessing the items to be recalled. The reason for expert performance with mimic displays might be that the icons and flow links on this type of display not only give direct evidence about the physical structure of the device, but they also act as cues or reminders about other knowledge that the person has about the device—they evoke other parts of the scenario. This is an example from only one type of cognitive task, but it does point to the potential use of contextual cued recall in simplifying display systems. However, cued recall can only be effective with experienced people, who can recognize the cues and know what they evoke.

7.3 Mental Workload, Learning, and Errors

Workload, learning, and errors are all aspects of the efficiency of cognitive processing. There are limits to human-processing capacities, but these are difficult to define, because of the adaptability of the human behavior. As a result of learning, processing becomes more efficient and adapted to what is required. As efficiency increases, mental workload may decrease. Furthermore, error rates can be affected by both expertise and workload, and errors are closely involved in the processes of learning. There is a huge wealth of material that could be discussed; however, the aim here is only to give a brief survey.

7.3.1 Mental Workload

There are numerous issues involved in accounting for mental workload and how it is affected by the different aspects of a task. This section mentions three main topics: whether people can only do one task at a time; factors affecting processing capacity; and the ways in which people typically respond to overload.

7.3.1.1 Single- or Multichannel Processing

Many evidences, including the example of multitasking given in Figure 7.29, show that people usually do one task at a time. This section looks at how people attend to one source of the stimuli among many, and under what circumstances people can do more than one task at a time. Typically, the findings show how adaptable human beings are, and that there is not yet a full account of the processes involved.

7.3.1.1.1 Focused Attention

People have the ability to pick out one message against a background of the others, either visual or auditory. However, studies show that a person does not only process one of the stimulus sources, but takes in enough about the other possible signals to be able to separate them. This chapter has already used the notion of depth of processing, as in discrimination, recoding, sequences of recoding, and building up an overview. This notion is also involved here. Separation of two signal sources requires the least processing if they can be discriminated by the physical cues, such as listening to a high voice while a low voice also speaks, or reading red letters against a background of green letters. The various factors discussed earlier on discrimination affect the ability to carry out this separation. If stimuli cannot be distinguished by physical cues, then "deeper" processing may be involved. For example, Gray and Wedderburn (1960) found that the messages presented to the ears as

Left ear:	mice	5	cheese
Right ear:	3	eat	4
Were heard as:	(3 5 4)		(mice eat cheese)

In this case, the words may be grouped by recognizing their semantic category. In some tasks, deeper processing for meaning may be needed, that is, building up an overview, as in

> It is important that the subject man be car pushed house slightly boy beyond hat his shoe normal candy limits horse of tree competence pen for be only in phone this cow way book can hot one tape be pin certain stand that snaps he with is his paying teeth attention in to the empty relevant air task and hat minimal shoe attention candy to horse the tree second or peripheral task. (Lindsay & Norman, 1972)

It can be noted that if the cue used becomes ineffective, then this is disconcerting. Subsequently, it takes time and a search for clues about what would be effective is carried out, before the person can orient to a new cue and continue with the task. There is also an interplay of depths of processing: When the physical cue becomes inadequate for following the message, then the reader uses continuity of meaning as a basis for finding a new physical cue. This account fits in with several points made earlier. The person uses active attention for what the person wants to take in, and not passive reception of signals. The task setting provides the cue that can be used to minimize the effort needed to distinguish between signal sources. Thus, this cue acts as a perceptual frame for searching for relevant inputs.

The concept of depth of processing was first introduced by Crark and Lockhart (1972) to explain the results of some memory experiments. The word *depth* is here distinguished from the depth in the organization of behavior, as in goal/subgoal and so on.

7.3.1.1.2 Parallel Processing

The criteria defining whether or not people are able to do two tasks at the same time have so far proved elusive to identify. Figure 7.16 shows that, after high levels of practice, the choice of time is not affected by the number of alternatives. Such tasks are said to be automated or require no conscious attention. They can be done at the same time as something else, unless both the tasks use the same peripheral resources, such as vision or hand movement. Wickens (1984) did a series of studies showing that people can use different peripheral resources at the same time. People can also learn to do some motor tasks, so that movements are monitored by feel, rather than visually; thus, movements can be made at the same time as looking at or thinking about something else. In practice, the possibility of multiple processing means that care is needed in designing tasks. One might, for example, think that it would reduce unnecessary effort of an air-traffic controller to have the flight strips printed out, rather than expecting the controller to write the strips by hand. However, if the controller, while writing, is simultaneously thinking out how the information fits into their overview, then printing the flight strips might deprive him or her of useful attention and thinking time.

Whether or not two tasks that both involve "central" processing can be done at the same time is less clear. This is partly because what is meant by central processing has not been clearly defined. People can do two tasks at the same time if the tasks are processed by different areas of the brain—for example, a music task and a language task (Allport, Antonis, & Reynolds, 1972)—though both tasks need to be simple and perhaps done by recoding. By going to "deeper" levels of processing, a limit to the extent to which people can build up distinct overviews for two different tasks at the same time could be observed, and whether or not an overview is needed to do a task may be a part of the question. For example, people playing multiple chess games may have very good pattern-recognition skills and hence, react to each game by recognition-primed decisions as they return to it, rather than having to keep in mind a separate and continuing overview for each of the games that they are playing. Most experienced drivers can drive and hold a conversation on a different topic at the same time when the driving task is simple, but they stop talking when the driving task becomes more difficult.

This is an area in which it is challenging to identify the limits to performance, and it is probably beyond the competence of HF/E at the moment, either to define the concepts or to investigate and measure the processing involved. Fortunately, in practice, the issue can often be simplified. When predicting performance, the conservative strategy is to assume that people cannot do two tasks at the same time. This will always be the worst-case performance.

7.3.1.2 Factors Influencing Processing Capacity

The amount of mental work a person can do in a given time is not a simple quantity to specify. If it is assumed that a person can only do one thing at a time, then every factor that increases the time taken to do a unit task will decrease the number of those tasks that can be done in a given time interval, and thus, decrease the performance capacity. Hence, every factor in interface design might affect performance capacity.

Focusing on performance time emphasizes performance measures of workload effects. Other important measures of workload are physiological, such as the rate of secretion of stress chemicals, and subjective, such as moods and attitudes. Any factor could be considered a "stressor" if it deteriorates the performance levels, stress hormone secretion rates, or subjective feelings. The approach in this section is to indicate some key general topics, rather than to attempt a full review. One can obtain reviews on workload topics in the chapters on fatigue and biological rhythms, pilot performance, and controller performance (Chapters 10, 13, and 19).

Thus, the points made here are concerned with the capacities of different mental processes; extrinsic and intrinsic stressors; individual differences; and practical implications.

7.3.1.2.1 Capacities of Different Cognitive Resources

Different aspects of cognitive processing have different capacities. For a review on processing limits, see Sage (1981). The capacity of different processes may be affected differently by different factors. Figure 7.31 shows the time-of-day effects on performance in four tasks: serial search, verbal reasoning (working memory) speed, immediate retention, and alertness. The different performance trends in these tasks suggest that each task uses a different cognitive resource that responds differently to this stress. It is difficult to make reliable analyses of these differences, but some other tasks in which performance may differ in this way are coding and syllogisms (Folkhard, 1990).

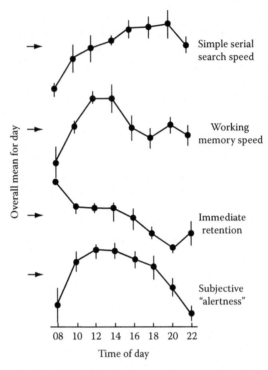

FIGURE 7.31 Cognitive processing capacities change during the day. The different patterns of change suggest that these capacities have different mechanisms. (Reproduced from Folkhard, S., Circadian performance rhythms. In Broadbent, D.E. et al. (Eds.), *Human Factors in Hazardous Situations*, Clarendon Press, Oxford, 1990, pp. 543–553.)

7.3.1.2.2 Extrinsic and Intrinsic Stressors

Extrinsic stressors are the stressors that apply to any person working in a particular environment, irrespective of the task that they are doing. Time-of-day, as in Figure 7.31, is extrinsic in this sense. Some other extrinsic stressors that can affect the performance capacity are noise, temperature, vibration, fumes, fatigue, and organizational culture.

Intrinsic stressors are factors that are local to a particular task. All the HF/E factors that affect the performance speed or accuracy come in this category. The effect of task difficulty interacts with motivation. Easy tasks may be done better with high motivation, whereas difficult tasks are done better at lower levels of motivation. This can be explained by assuming that stressors affect a person's "arousal" level, and that there is an inverted-U relation between arousal level and performance (see Figure 7.32).

Measures of stress hormones and workforce attitudes show that several factors with respect to the pacing of work and the amount of control over their work that a person feels he or she has, can be stressors (e.g., Johansson, Aronsson, & Lindström, 1978). Such aspects are of more concern in repetitive manufacturing jobs than in work, such as flying or air-traffic control.

7.3.1.2.3 Individual Differences

Individual differences affect a person's capacity to carry out a task, and the person's willingness to do it. Aspects of individual differences fall into at least five groups.

1. Personality. Many personality dimensions, such as extroversion/introversion, sensitivity to stimuli, need for achievement or fear of success, and preference for facts/ideas or regularity /flexibility, can affect a person's response to a particular task.

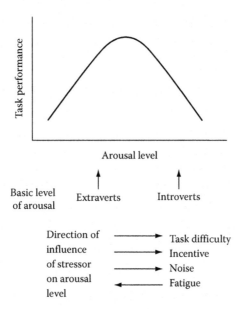

FIGURE 7.32 "Inverted U" relation between internal arousal level and performance (it is not possible to account for the effect of all stressors in this way).

2. Interests and values. A person's interests and values affect the response to various factors in the task and the organizational climate, which influence the willingness and commitment to do or learn a given task. People differ in their response to incentives or disincentives, such as money, status, or transfer to a job that does not use their skills.

3. Talent. Different people have different primary senses, different cognitive styles, and different basic performance abilities (e.g., Fleishman, 1975). For example, very few of us have the ability to fly high-speed aircraft.

4. Experience. The rest of us may be able to develop higher levels of performance though practice. Even the few who can fly high-speed aircraft have millions spent on their training. The effects of training on cognitive capacities are discussed more in the section on learning.

5. Nonwork stressors. There may be nonwork stressors on an individual which affect the person's ability to cope with work, such as illness, drugs, or home problems.

7.3.1.2.4 Practical Implications

There are so many factors affecting the amount of effort any particular individual is able or willing to devote to a particular task at a particular time, such that performance prediction might seem impossible. Actually, the practical ways of dealing with this variety are familiar. There are two groups of issues, in HF/E design and performance prediction.

Nearly all HF/E design recommendations are based on measures of performance capacity. Any factor that has a significant effect on performance should be improved, as far as it is economically justifiable. Design recommendations could be made with regard to all the intrinsic and extrinsic factors mentioned earlier, and individual differences might be considered in selection.

However, it is easier to predict that a design change will improve performance than to predict the size of the improvement. Numerical performance predictions may be made to assess whether a task can be done in the time available or with the available people, or to identify the limits to speed or accuracy on which design investment should best be concentrated. Obviously, it is not practical to include all the possible effective factors when making such predictions. Three simplifying factors can reduce the problem. One is that, although smaller performance changes may give important clues about how to optimize design, from the point of view of performance prediction, these factors may only be important if they make an order of amplitude difference to performance. Unfortunately, our data relevant to this issue are far from complete. The second point is that only conservative performance predictions are needed. For these purposes, it may be valid to extrapolate from performance in simple laboratory tasks, in which people with no relevant expertise react to random signals, which is the worst case. To predict minimum levels of performance, it may not be necessary to include the ways in which performance can improve when experienced people carry out tasks in which they know the redundancies, can anticipate, and so forth. The third point is that, in practice, many of the techniques for performance prediction that have been devised, have the modest aim of matching expert judgments about human performance in a technique that can be used by someone with less expertise, rather than attempting high levels of accuracy or validity.

7.3.1.3 Response to Overload

If people doing a simple task have too much to do, they only have the options of omitting parts of the task or accepting a lower level of accuracy in return for higher speed (Figure 7.18). People doing more complex tasks may have more scope for responding to increased workload, while maintaining acceptable task performance. This section discusses increasing efficiency, changing strategy, and practical implications.

7.3.1.3.1 Increasing Efficiency

Complex tasks often offer the possibility of increasing the efficiency with which a task is done. For example, Sperandio (1972) studied the radio messages of air-traffic approach controllers. He found that when they were controlling one aircraft they spent 18% of their time in radio communication, and when there were nine aircrafts, they spent 87% of their time on the radio. In simple model of mental workload:

$$\text{Total workload} = \text{workload in one task} \times \text{number of tasks}$$

Evidently, this does not apply here, as according to this model, the controllers would spend 162% of their time on the radio. Sperandio found that the controllers increased the efficiency of their radio messages in several ways. There were fewer pauses between the messages; redundant and unimportant information were omitted; and conversations were more efficient: The average number of conversations per aircraft decreased, but the average number of messages per conversation increased, and hence, fewer starting and ending procedures were necessary.

7.3.1.3.2 Changing Strategy

The controllers studied by Sperandio (1972) not only altered the efficiency of their messages, but the message content was also altered. The controllers used two strategies for bringing the aircrafts into the airport (this is a simplification, and hence, the description can be brief). One strategy was to treat each aircraft individually. The other was to standardize the treatment of the aircrafts by sending them all to a stack at a navigation fix point, from which they could all enter the airport in the same way. When using the individual strategy, the controllers asked an aircraft about its height, speed, and heading. In the standard strategy, they more often told an aircraft about the height and heading to be used. The standard strategy required less cognitive processing for each aircraft. Sperandio found that the controllers changed from using only the individual strategy when there were three or fewer aircrafts, to using only the standard strategy when there were eight or more aircrafts. Expert controllers changed to the standard strategy at lower levels of workload. Sperandio argued that the controllers change to a strategy that requires less cognitive processing, to keep the total amount of cognitive processing within achievable limits (Figure 7.33a).

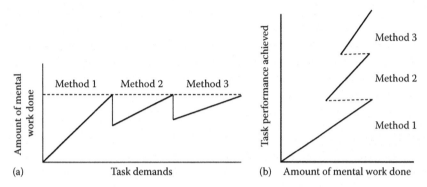

FIGURE 7.33 Effect of changing working methods on relation between mental work and task work. This figure is a simplification: in practice, the use of the methods overlaps, so there are not discontinuities. (From Sperandio, J.C., *Ergonomics*, 14, 571, 1971.)

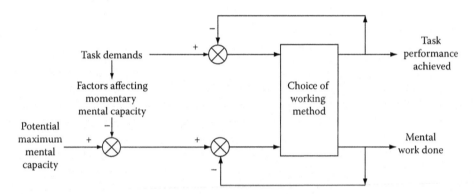

FIGURE 7.34 Choice of optimum working method depends on task and personal factors.

The relation between task performance and workload is therefore, not the same in mental work, as it is in physical work. In physical work, conservation of energy ensures there is a monotonic relation between physical work and task performance. In mental workload, if there are alternative working methods for meeting the given task demands, then there does not necessarily exists a linear relation between the task performance achieved and the amount of mental work needed to achieve it. By using different methods, the same amount of mental effort can achieve different amounts of task performance (Figure 7.33b).

In choosing an optimum working method, two adaptations are involved. A person must choose a method that meets the task demands. Furthermore, the person must also choose a method that maintains the mental workload at an acceptable level. The method chosen will affect both the task performance achieved and the mental workload experienced (Figure 7.34). There needs to be a mechanism for this adaptive choice of working method. This is another contextual effect that could be based on meta-knowledge. Suppose that the person knows, for each method, both how well it meets various task demands and what mental workload demands it poses. The person could then compare this meta-knowledge with the demands of the task and mental context, to choose the best method for the circumstances (Bainbridge, 1978).

7.3.1.3.3 Practical Implications

This flexibility of working method has several practical implications. It is not surprising that many studies have found no correlation between the task performance and subjective experience of mental workload. There are also problems in predicting the mental workload, similar to those predicting the performance capacity, mentioned earlier.

A person can only use several alternative working methods, if the performance criteria do not strictly constrain what method must be used. For example, in air-traffic control, safety has much higher priority than the costs of operating the aircraft. Task analysis could examine whether alternative methods are possible, and perhaps find out what these methods are (it may not be possible to predefine all the methods; see earlier discussion on problem solving and later discussion on learning).

Adaptive use of working methods suggests that strategy-specific displays should not be provided, as they could remove the possibility of this flexibility in dealing with varying levels of workload. It could also be useful to train people to be aware of alternative methods and the use of meta-knowledge in choosing between them.

When decision-support systems are introduced with the aim of reducing workload, it is necessary to consider a wider situation. Decision-support systems can increase rather than decrease the mental workload, if the user does not trust the system and thus, frequently checks what it is doing (Moray, Hiskes, Lee, & Muir, 1995).

7.3.2 Learning

Learning is another potentially huge topic, and all the expertise of psychology on learning, HF/E on training, and educational psychology on teaching cognitive skills and knowledge could be included in it. As this chapter focuses on cognitive processes, this section primarily discusses the cognitive skills and knowledge. The description only attempts a brief mention of some key topics, which indicate how learning interrelates with other aspects of cognitive processing, rather than being a separate phase of performance.

This section uses the word *skill* in the sense, in which it is used in psychology and in British industry. There are two key features of skilled behavior in this context. Processing can be done with increased efficiency, either because special task-related abilities have been developed which would not be expected from the average person, or because no unnecessary movements or cognitive processing are used, and behavior is adapted to the circumstances. Furthermore, choices, about what best to do next and how to do it, are adapted to the task and personal context. In this general sense, any type of behavior and any mode of cognitive processing can be skilled, and hence, it can be confusing to use the word *skill* as the name for one mode of processing.

This section has three main parts: changes in the behavior with experience; learning processes; and relations between mode of processing and appropriate training method.

7.3.2.1 Changes within a Mode of Processing

This subsection briefly surveys the modes of processing that have formed one framework of this chapter, and indicates the ways in which each can change by introducing new aspects of processing or losing inefficient ones. This is a summary of points made before and is by no means complete. Learning can also lead to changes from one mode of processing to another, as discussed later.

Physical Movement Skills. By carrying out movements in a consistent environment, people can learn

- Which movement has which effect (i.e., they develop their meta-knowledge about movements; Figure 7.20). This indicates that they do not need to make exploratory actions, and their movements do not oscillate around the target. People can then act with increased speed, accuracy, and coordination, and can reach to the correct control or make the correct size of action without checking.
- To use kinesthetic rather than visual feedback.
- The behavior of a moving target, so that its movements can be anticipated.

Changes in performance may extend over very long periods. For example, Crossman (1959) studied people doing manually dexterous task of rolling cigars, and found that performance continued to improve until people had made about 5 million items.

Perceptual Skills. These are discriminations and integrations. People learn

- Discriminations, groupings, and size, shape, and distance inferences to make
- Probabilities and biases that can be used in decision making
- Appropriate items to attend to
- Eye movements needed to locate the given displays

Recoding. The person connects from one item to another by association, without intermediate reasoning. These associations must be learned as independent facts, or there may be some general rule underlying a group of recodings, such as "choose the control with its location opposite to the location of the display." Many people need a large number of repetitions before they can learn the arbitrary associations.

Sequence of Recodings. Two aspects of learning may be involved:

- When a sequence is the same each time, so that the output of one recoding and the input of the next recoding are consistent, then a person may learn to "chunk" these recodings together, to carry them out as a single unit without using intermediate working memory.
- When a goal/function can be met in the same way each time, then choosing a working method that is adapted to circumstances is not necessary. A previously flexible working method may then reduce to a sequence of transforms that does not include goals or choice of working method.

Familiar Working Methods. People need to learn

- Appropriate working method(s).
- The reference knowledge needed while using each method. When this reference knowledge has been learned while using the method, then it may be accessed automatically, without having to think out explicitly what knowledge is needed in a particular situation.
- How to build up an integrated overview.
- Meta-knowledge about each working method, when it is used in choosing the best method for a given context.

Planning and Multitasking. People can become more skilled in these activities. They can learn a general method for dealing with a situation, and the subsidiary skills for dealing with parts of it (Samurçay & Rogalski, 1988).

Developing New Working Methods. The process of developing new working methods can itself be more or less effective. Skill here lies in taking an optimum first approach to finding a new working method. There are several possible modes of processing for doing this.

1. Recognition-primed decisions. People can only make recognition-primed decisions once they have learned the categories used. This involves several aspects of learning:
   - The features defining a category, and how to recognize an instance that has these features, as a member of the category.
   - The members of a category, and their properties (e.g., for each category of situation, what to do in it).
   - How to adapt a category method to specific circumstances.
2. Case-based reasoning. Cases (or, more distant from a particular task, analogies) provide examples as a basis for developing the knowledge or working method needed. To be able to do this, people need to know
   - Cases
   - How to recognize which case is appropriate to which circumstances
   - How to adapt the method used in one case to different circumstances

3. Reasoning from basic principles. For this sort of reasoning, people need to have acquired an adequate base of knowledge about the task and the device(s) that they are using, with associated meta-knowledge. The same type of knowledge may also be used for explaining events and actions.

7.3.2.2 Learning Processes

Little is known about how changes in processing take place. Similar processes may be involved in developing and maintaining physical and cognitive skills. This section indicates some mechanisms: repetition; meta-knowledge and feedback; independent goals–means; and changing modes of processing.

7.3.2.2.1 Repetition

Repetition is crucial for acquiring and maintaining skills. The key aspects are that, each time a person repeats a task, some aspects of the environment are the same as before, and the knowledge of the results is given. This knowledge of results has two functions: it gives information about how and how well the task was done, and it acts as a reward.

7.3.2.2.2 Meta-Knowledge and Feedback

As described in the section on movement execution, learning of motor skills involves learning both how to do an action and meta-knowledge about the action. Actions have associated expectations about their effect (meta-knowledge). Feedback about the actual effect provides information that can be used to refine the choice made next time (Figure 7.20). Thus, during learning, feedback is used both to revise the present action and the subsequent action.

Choosing an action instruction on the basis of meta-knowledge is similar in process to choosing the working method used to maintain mental workload at an acceptable level. The choice of working method involves checking the meta-knowledge about each method, to find which method has the properties best suited to the present situation. A similar process is also involved when developing a new cognitive working method: A person develops a working method, hoping (on the basis of a combination of meta-knowledge and mental simulation) that it will give the required result, and then revises the method on the basis of feedback about the actual effectiveness of what they do.

7.3.2.2.3 Independent Goals–Means

In coping with mental workload, and in developing cognitive processes while learning, several working methods may be used for meeting the same function/goal. Also, the same behavior may be used to meet several goals. Thus, the link between goal and means must be flexible. The goal and means are independent in principle, although, after learning, particular working methods may become closely linked to particular goals. In the section on workload, the goal–means link was described as a point at which a decision between working method is made on the basis of meta-knowledge.

It is generally the case (Sherrington 1906/1947) that behavior at one level of organization transfers the information about the goal to be met, and constraints on how it should be met, to the lower levels of behavior organization by which the goal is met, but not detailed instructions about how to meet it. How to carry out the function is decided locally in the context at that particular time. As behavior is not dictated from above, but has local flexibility, human beings are by nature not well suited to follow standardized procedures.

7.3.2.2.4 Changes in the Mode of Processing

Learning does not lead only to changes within a given mode of processing. A person may also change to a different mode of processing. If the task is consistent, then a person can learn to do the task in a more automatic way, that is, by using a simpler mode of processing. Inversely, when there is no fully developed working method or knowledge for meeting a given goal/function, then it is necessary to devise one. Thus, the possibility or need for developing a simpler or more complex mode of processing depends on both a person's experience with the task and the amount and types of regularity in the task. It may be possible through learning to change from any mode of processing to any other mode of processing, but two types of change are most typical: from more complex to simpler processing, or vice versa.

Someone may start a new task by developing a working method. However, once the person has had an opportunity to learn the regularities in the task, the processing may become simpler. If the task and environment are sufficiently stable, then the person may learn that making a choice between the methods to meet a goal or search for appropriate knowledge is not necessary. In familiar stable situations, the working method may become so standardized that the person using it may not be aware of the goals or choices.

Alternatively, someone may start by learning parts of a task, and gradually become capable of organizing them together into a wider overview, or become proficient in choosing the behavior that is compatible with several cognitive functions. These changes depend on the changes in processing efficiency. When someone first does a complex task, the person may start at the lowest levels of behavior organization, learning components of the task that will eventually be simple, but which initially require all the person's problem-solving, attention and other processing resources. As the processing for doing these subtasks becomes simpler with learning, this releases processing capacity. This capacity can then be used for taking in larger segments of the task at the same time, so that the person can learn about larger regularities in the task.

In general, any cognitive function and any subgoal involved in meeting it, may be met by any mode of processing, depending on the person's experience with the task, and the details of the circumstances at the moment. A task can become "automated" or flexible at any level of behavior organization, depending on the repetitions or variety of situations experienced. Thus, in some tasks, a person may learn to do the perceptual-motor components automatically, but have to rethink the task each time at a higher level, like a professional person using an office computer. In other tasks, "higher" levels of behavior organization such as planning may become automated, whereas lower levels remain flexible, as in driving to work by the same route every day. It is not necessarily the case that "higher" levels of behavior organization are only done by more complex modes of processing, such as problem solving or vice versa.

As any of the main cognitive functions in a task could become very much standardized such that they are done automatically or unconsciously, this results in the so-called shortcuts in processing. Inversely, at any moment, a change in the task situation, such as a fault, may mean that what could previously be done automatically now has no associated standard working method, and hence, problem solving is needed to find one. At any time, or at any point in the task, there is the potential for a change in the mode of processing. Hence, care is needed, if an interface design strategy is chosen to provide displays that support only one mode of processing.

7.3.2.3 Some Training Implications

Gagné (1977) first suggested the concept that different modes of processing are best developed by different training methods. It is not appropriate to survey these methods here, but some general points link to the general themes of this chapter.

7.3.2.3.1 Simple Processes

Training for simple processes needs to

- Maximize the similarity to the real task (the transfer validity) of discriminations, integrations, and recodings that are learned until they become automatic, by using high-fidelity simulation.
- Minimize the need for changes in mode of processing during learning, by presenting the task in a way that needs little problem solving to understand.
- Ensure that trainees retain a feeling of mastery, as a part of their meta-knowledge about the task activities, by avoiding training methods in which errors are difficult to recover from, and by only increasing the difficulty of the task at a rate such that trainees continue to feel in control.

7.3.2.3.2 *Complex Processes*

Tasks that involve building up an overview and using alternative strategies need more than simple repetition, if they are to be learned with least effort. The status of errors is different in learning complex tasks. In training for simple discriminations, recodings, and motor tasks, the emphasis is on minimizing the number of errors made, so that wrong responses do not get associated with the inputs. By contrast, when learning a complex task, an "error" can have positive value as a source of information about the nature and limits of the task. Thus, in learning complex tasks, the emphasis should be more on exploring the possibilities without negative consequences, to develop a variety of working methods and wide knowledge of the task alternatives. Flexibility could be encouraged by giving trainees

- Guided discovery exercises, in which the aim is to explore the task rather than to achieve the given aims
- Recovery exercises in which people practice recovering from nonoptimal actions
- Problem-solving and planning exercises, with or without real-time pressures
- Opportunities to share with other trainees the discoveries made
- Practice with considering alternative working methods and assessing the criteria for choosing between them
- Practice with thinking about alternative "hypotheses" for the best explanation of events or the best action
- Practice with multitasking
- Practice with using different methods for developing working methods and the case examples and recognition categories used

A feature of cognitive skill is to have a knowledge base that is closely linked to the cognitive processing that uses it, so that the knowledge is appropriately organized and easy to access. This suggests that knowledge is best learned as a part of doing the task, and not separately.

7.3.2.3.3 *Training as a Part of System Design*

This chapter has mentioned several ways in which training needs interact with the solutions chosen for other aspects of the system design:

- The need for training and the quality of interface or procedure design may be inversely related.
- Skills are lost if they are not maintained by practice, and hence, the amount of continuing training needed may be related to the extent of automation.

7.3.3 Difficulties and Errors

Errors occur when people are operating at the limits of modes of processing. Errors result from misuse of normally effective processes. The concept of relating error types to modes of processing was first suggested by Rasmussen (1982), although the scheme suggested here is somewhat different.

The approach to the complex tasks taken in this chapter suggests several points that should be added to most error schemes. First, the notion of error needs to be expanded. In some simple tasks such as recoding, it is possible to be wrong. However, in control tasks and complex tasks, it is useful to think in terms of difficulty or lowered effectiveness, rather than focusing on being wrong. For example, Amalberti's novice pilots (Figure 7.28) were already qualified. They completed the task; but did it less effectively than the more experienced pilots. Thus, as a basis for supporting people doing complex tasks, it is useful to look at factors that make the task more difficult, as well as those that slow the behavior down or increase the errors.

Second, many error schemes assume that task behavior can be broken down into small independent units, each of which may be right or wrong. In probabilistic risk assessment (PRA) or human reliability assessment (HRA) techniques, behavior is segmented into separate units. A probability of error is assigned to each unit, and the total probability of human error for the combined units is calculated by addition or multiplication. However, this chapter has stressed that human behavior in complex tasks does not consist of independent units. The components of complex behavior are organized into an integrated interdependent structure. This means that, although PRA/HRA techniques are useful for practical purposes, any attempt to increase their fundamental validity while retaining an "independent units" model of behavior is doomed to failure (Hollnagel, 1993).

Third, as the processes of building up and using an overview are often not included in models of human processing, the related errors are also often not discussed, and hence, they are the focus here. This section briefly suggests some of the ways in which performance can be weaker (e.g., see Bainbridge, 1993b).

7.3.3.1 Discriminations

Decisions made under uncertainty cannot always be right, and are more likely to be wrong if the evidence on which they are based is ambiguous or incomplete. Incorrect expectations about probabilities and incorrect biases about payoffs can also increase the error rates. People make errors, such as misattributing risk, importance, or urgency; ignoring a warning that is frequently a false alarm; or seeing what they expect to see. Some people when under stress may refuse to make decisions invoking uncertainty.

7.3.3.2 Recodings

There are many sorts of error that can be attributed to mistranslations. Sometimes, the person may not know the coding involved. People are more likely to make coding errors when they have to remember a specific code translation to be used in specific circumstances. Difficult codes are often ambiguous or inconsistent. Furthermore, the salience of some stimuli may give improper emphasis to them or to their most obvious meaning.

7.3.3.3 Sequences

The items that need to be retained in working memory during a sequence of behavior may be forgotten within half a minute, if other task processing distracts or interrupts the rehearsal needed to remember the items.

In an overlearned sequence, monitoring/supervision of parts of the activity may be omitted. This can lead to "slips" in performance or rigid behavior that causes difficulties when the environment changes and adaptive behavior is needed.

7.3.3.4 Overview and Behavior Organization

There may be errors in organizing the search for information. People may only attend to part of the task information, fail to keep up-to-date with changes in the environment, or look at the details without taking an overall view. They may not get information for which there is a cost for getting it. They may only look for information that confirms their present interpretation of the situation (confirmation bias). In a team work, people may assume without checking that another member of the team, particularly someone with higher status, has done something that needed to be carried out.

There may also be errors in the allocation of time between the tasks, which may lead to omissions or repetitions. People may react to events rather than anticipating events and how to deal with them. They may not apply available strategies in a systematic way. They may shift between subtasks, without relating them to the task as a whole (thematic vagabonding, Doerner, 1987). They may break the task down into subproblems in an inadequate way or fail to devise intermediate subgoals, or they may continue to do parts of the task which they know how to do (encystment, Doerner, 1987). Under high workloads, people may delay decisions in the hope that it will be possible to catch up later, or they may cycle through thinking about the task demands without taking any action.

The overview influences a person's biases about what will happen and what to do about it. If the overview is incorrect, this can lead to inappropriate behavior or expectations. People who have completed a subtask, and thus, completed a part of their own overview, may fail to tell other members of the team about this. Once people have built up a complete and consistent overview, it may be difficult to change it when it turns out to be inadequate (perceptual set). The overview may also be lost completely if a person is interrupted.

7.3.3.5 Use of Knowledge

People's knowledge of all types may be incomplete or wrong, and hence, they make incorrect inferences or anticipations. There may be problems with assumed shared knowledge in a team if the team members change.

A person may have an incorrect or incomplete representation of the device that they are using. For example, the person may not know the correct causalities or interactions, or may not be able to correctly represent the development of events over time. Or, someone may use an inappropriate category in recognition-primed decisions or in case-by-case-based reasoning.

Knowledge about probabilities may be incorrect or used wrongly, and people may be under- or over-confident. They may have a "halo effect," attributing the same probabilities to the unrelated aspects, and may give inappropriate credence to information or instructions from people of higher status. Different social groups—for example, unions, management, and the general public—may have different views on the risks and payoffs of particular scenarios.

This list of human weaknesses should not distract from two important points. One is that people can be good at detecting their errors and recovering from them, if they are given an interface and training that enable them to do this. Therefore, design to support recovery should be included in cognitive task analysis.

The second point is that care is needed with the attribution of responsibility for faults. Although it may be a given individual who makes an error, the responsibility for that error may be attributed elsewhere, to poor equipment or system design (training, workload, allocation of function, teamwork, organizational culture).

7.4 Neurotechnology-Driven Joint Cognitive Systems

This chapter has so far focused on the underlying human processing mechanisms and the human adaptation strategies used to meet the challenges of human cognitive limitations in performing complex tasks in stressful situations. Traditionally, human–machine system designers have identified bottlenecks as well-known human limitations such as short-term memory and dual-task performance. This is the standard HF/E approach whose stated aim is to design interfaces to machines that mitigate the negative impact of human limitations (Wickens, 1992). This can lead to default function allocation schemes that consider the human and machine component independently when determining which agent is best suited to accomplish some individual task. However, if one considers both the human and the machine as required elements to solve the complex problem at hand, a *joint cognitive system* (Woods, Roth, & Benett, 1990) design approach emphasizes a collaborative, complementary design, rather than the more traditional comparative approach. Such an approach depends on the ability of both the human and the machine to understand each other when performing complex tasks (Brezillin & Pomeral, 1997). Until recently, complex cognitive processes were not directly measurable. However, technologies have advanced to the state where real-time, sensor-based inferences of specific cognitive processes are feasible. Neurotechnologies, as they are collectively called, use data from physiological and neurophysiological sensors as input to algorithms to provide meaningful measurements of cognitive state—such as working memory, mental workload, attention, executive function, or other complex cognitive processes.

The goal of a joint cognitive system is to amplify the human's cognitive capability when performing complex tasks under stress, when normally human performance degrades. Neurotechnology-based measurement of the cognitive state of the individual allows the automated computational system to adapt to best mitigate human cognitive decrements, with the overall goal of modifying and mediating human cognition in order to optimize joint performance, rather than optimizing the human or the computer alone. In a truly adaptive joint cognitive system, the computational system can adapt to the current state of the user, rather than forcing the user to adapt to the system (Schmorrow & Kruse, 2002). A related field is *neuroergonomics*, an interdisciplinary approach that combines neuroscience (the study of the brain structure and function) and HF/E (the study of behavior in task environments). The focus of neuroergonomics is the study of how the brain functions during the execution of complex tasks in real domain settings (Parasuraman & Rizzo, 2006).

Neurotechnology-driven adaptive joint cognitive systems offer the opportunity to blur the line between the human and the machine, to tightly couple human and machine capabilities, and to have the machine adapt to the current state of the user. This section, while not an exhaustive overview of the collective field, will discuss the implications of the ability to measure cognitive processes on human–machine system design. A full review of joint cognitive systems, or even adaptive joint cognitive system driven by measures of cognitive state, is beyond the scope of this section. However this section will briefly introduce different measures of cognitive state and provide examples of how the measurement of underlying cognitive processes can drive adaptive joint cognitive systems in complex task domains.

7.4.1 Measuring Cognitive State

Neurophysiological- and physiological-based assessment of the cognitive state has been captured in several different ways. Methods fall into three general areas: direct measures of the brain based on cerebral hemodynamics (blood flow), those based on electromagnetic brain activity (Parasuraman & Rizzo, 2006), and indirect measures that are based on non-brain sensors. Direct sensing of blood flow measures include functional magnetic resonance imaging (fMRI), positron emission tomography (PET), transcranial Doppler sonography (TCD), and functional near-infrared (fNIR) imaging. Direct measures of brain activity include electroencephalogram (EEG), evoked-response potentials (ERPs), and magnetoencephalography (MEG). For a more detailed review of brain imaging techniques, see Cabeza and Kingstone, 2001. Finally, indirect measures include utilizing electrocardiogram (ECG), galvanic skin response (Verwey & Veltman, 1996), eyelid movement (Stern, Boyer, & Schroeder, 1994; Veltman & Gaillard, 1998; Neumann, 2002), pupil response (Beatty, 1982; Partala & Surakka, 2003), and respiratory patterns (Porges & Byrne, 1992; Backs & Seljos; Veltman & Gaillard, 1998). Some example techniques are briefly reviewed here.

EEG (through the use of cortical electrical activity from scalp electrodes) has been used extensively in the context of adaptive joint cognitive systems. It is the gold standard for providing high-resolution spatial and temporal indices of cognitive processes. Research has shown that EEG activity can be used to assess a variety of cognitive states that affect complex task performance. These include working memory (Gevins & Smith, 2000), alertness (Makeig & Jung, 1995), executive control (Garavan, Ross, Li, & Stein, 2000), and visual information processing (Thorpe, Fize, & Marlot, 1996). These findings point to the potential for using EEG measurements as the basis for driving adaptive joint cognitive systems that demonstrate a high degree of sensitivity and adaptability to human operators in complex task environments. For instance, researchers have used the engagement index, developed by NASA, in the context of mixed-initiative control of an automated system (Pope, Bogart, & Bartolome, 1995). This method uses a ratio of power in common EEG frequency bands (beta/(alpha + theta)), where cognitively alert and focused is represented in beta, wakeful and relaxed in alpha, and a daydream state in theta. Thereby higher engagement index values estimate increased levels of task engagement. The efficacy of the engagement index as the basis for adaptive task allocation has been experimentally established. For instance, under manipulations of vigilance levels (Mikulka, Hadley, Freeman, & Scerbo, 1999) and workload (Prinzel, Freeman, Scerbo, Mikulka, & Pope, 2000), an adaptive system effectively detected states where human performance was likely to fall, and took steps to allocate tasks in a manner that raised overall task performance. The results associated with the engagement index highlighted the potential benefits of a neurophysiologically triggered adaptive automation.

Evoked response potentials (ERPs) are the electrical potential recorded from the brain's neural response to a specific event or stimuli. EEG sensors are used to record the ERPs, which are detected approximately 150 milliseconds after stimulus onset (Thorpe, Fize, & Marlot, 1996). ERPs have been utilized as a measure of the underlying cognitive processes necessary for processing and coordinating a response to task-relevant stimuli (Makeig, Westerfield, Jung, Enghoff, & Townsend, 2002). The challenge in utilizing ERPs for real-time cognitive state detection is the fact that typically ERP response curves are constructed by averaging responses to hundreds of stimuli over time. However, recent advances in single-trial ERP detection, where sensor information is integrated spatially over multiple EEG sensors, rather than integrating sensor data over time, have been shown to be accurate (Parra et al., 2003; Gerson, Parra, & Sajda, 2005; Mathan et al., 2006a). In one example of joint cognitive system design, ERPs have been utilized to identify critical targets within large image sets efficiently in a triage process (Mathan et al., 2006b).

As an example of direct blood flow measures, fNIR spectroscopy conducts functional brain studies using wavelengths of light, introduced at the scalp, to measure hemodynamic changes. This type of non-invasive optical imaging has been shown to be sensitive to neuronal activity (Gratton & Fabiani, 2006), and has been used to assess cognitive workload (Izzetoglu et al., 2004).

There is an extensive research history of using cardiac, or electrocardiogram (ECG), measures to evaluate cognitive activity under a variety of task conditions. For instance, ECG has been used to measure heart rate variability in the time domain to assess mental load (Kalsbeek & Ettema, 1963), tonic heart rate to evaluate impact of continuous information processing (Wildervanck, Mulder, & Michon, 1978), variability in the spectral domain as an index of cognitive workload (Wilson & Eggemeier, 1991), and T-wave amplitude during math interruption task performance (Heslegrave & Furedy, 1979).

7.4.2 Adaptive Joint Cognitive Systems in Complex Task Domains

Function allocation involves the distribution of work between humans and automated systems. In 1951, Paul Fitts published a list of the functions best suited to humans or machines (Duncan, 1986). Function allocation decisions have been based on this paradigm ever since: compare humans to computers and assign tasks accordingly. In order to do this comparison, however, all tasks have to be reduced to a common framework, usually mathematical or technology-based framework (Jordan, 1963). Consequently, function allocation decisions have been driven more by available technology than by user needs, optimal role assignments, or an understanding of the differences between human cognition and computer algorithmic logic. Often the human roles are relegated by default, namely tasks that are too technologically difficult or expensive to automate. What is needed is a flexible, complementary, rather than comparative approach toward function allocation in the context of both the design and execution stages of human–computer systems. Figure 7.35 illustrates an adaptive joint cognitive system. The joint cognitive system is faced with task demands. The human interacts with the system to determine his or her working methods to address the

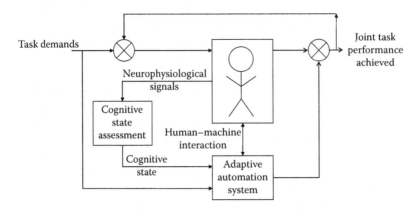

FIGURE 7.35 Closed loop system utilizes cognitive state feedback to adapt system function.

task demands. Faced with same task demands, the adaptive automation adapts its choice of working methods based on a real-time assessment of the current capabilities of the individual. Rather than traditional human–machine systems, where functions are allocated between the human and the machine at the time the system was designed (and hence by the system designers), adaptive systems dynamically determine function allocation during task execution. If the cognitive state assessment shows the human cognitive capabilities degrading, the machine can adapt to mitigate the effect on overall joint performance.

There are several categories of adaptation the joint cognitive system can utilize to respond to diminished human capabilities. The system, which adapts during execution in the current task environment, can either provide adaptive aiding, which makes a certain component of a task simpler, or can provide adaptive task allocation, which shifts an entire task from a larger multitask context to automation (Parasuraman, Mouloua, & Hilburn, 1999). Adaptive joint cognitive systems must make timely decisions on how best to use varying levels of automation to support human performance. In order for an adaptive joint cognitive system to decide when to intervene, it ideally should have some model of the context of operations, be it a functional model of system performance, a model of relevant world states, and/or a model of the operator's cognitive state. The unique qualities of adaptive joint cognitive systems attempt to capitalize on the strengths and mitigate the weaknesses of a coordinated human–computer system. By definition, their adaptive nature requires both the human and the automation to take an active role in the allocation of functions. These systems require not only the human to adapt to the situation but also the automation, thereby necessitating a high degree of coordination between the human and the computer in order to constantly inform the other about their current state. Often this coordination imposes additional tasks on the user to constantly update the system on his or her status.

Traditionally, many adaptive systems derive their inferences about the cognitive state of the operator from models, performance on the task, or from external factors related directly to the task environment (Wickens & Hollands, 2000). Neurophysiologically and physiologically triggered adaptive joint cognitive systems offer many advantages over the more traditional approaches to the human–machine system design by basing estimates of operator state on sensed data directly. These systems offer the promise of leveraging the strengths of humans and machines, augmenting human performance with automation specifically when assessed human cognitive capacity falls short of the demands imposed by task environments (Dorneich, Mathan, Ververs, & Whitlow, 2008). In addition, real-time cognitive state assessment allows the system to unobtrusively monitor state of the user, reducing or eliminating extra coordination demands placed on the user. With more refined estimates of the operator's cognitive state, measured in real-time, adaptive automation also offers the opportunity to provide aid even before the operator knows he or she is getting into trouble.

There have been a wide range of underlying cognitive processes that have been measured utilizing neurotechnology techniques, including attention, sensory memory, information-processing stages, working memory, executive function, and mental workload.

7.4.2.1 Information Processing Stages

Key cognitive bottlenecks constrain information flow and the performance of decision making, especially under stress. From an information-processing perspective, there is a limited amount of resources that can be applied to processing incoming information due to cognitive bottlenecks (Broadbent, 1958; Treisman, 1964; Kahneman, 1973; Pashler, 1994). The fusion of cognitive psychology and information theory resulted in a framework—human information processing—that considers human information-processing bottlenecks in system-oriented terms such as "input, processing, output" where each stage is limited by the nature of the subsystem that executes it (Lindsay & Norman, 1977). This approach primarily considers the limitations of the human operator independently of any emergent constraint of a joint human–machine system.

With the rapid proliferation of automation within human–machine systems, researchers now conceptualize information-processing stages as potential insertion points for automated aiding. For example,

Parasuraman, Sheridan, and Wickens (2000) proposed that automation can be applied to four broad classes of functions that correspond to the stages of human information processing: information acquisition, information analysis, decision and action selection, and action implementation, i.e., acquisition, analysis, decision, and action. This approach requires a priori, static assumptions about the relative utility of automated aiding at each stage without necessarily considering the whole system performance. In contrast, adaptive joint cognitive systems go beyond this traditional approach in several important ways. Adaptive joint cognitive systems change the task environment depending on the current state of the user, the current tasks, and the current context. Adaptive systems aim to enhance joint human-machine performance by having the system invoke varying levels of automation support in real time during task execution. Finally, adaptive automation refers to systems in which both the user and the system can initiate changes in the level of automation (Scerbo, Freeman, & Mikulka, 2003).

7.4.2.2 Attention

Attention can be broadly defined as a mechanism for allocating cognitive and perceptual resources across controlled processes (Anderson, 1995). There are many varieties of attention that need to be considered to optimize their distribution (Parasuraman & Davies, 1984): executive attention, divided attention, focused attention (both selective visual attention and selective auditory attention), and sustained attention. Breakdowns in attention lead to multiple problems: failure to notice an event in the environment, failure to distribute attention across a space, failure to switch attention to the highest priority information, or failure to monitor events over a sustained period of time.

An example of an adaptive system designed to address breakdowns in attention is the Communication Scheduler (Dorneich, et al., 2005). The system utilized EEG to determine the current mental workload of the user. The Communications Scheduler scheduled and presented messages to the user based on the user's current level of mental workload, the priority of the message, and the current task context. If an incoming aural message was of higher priority than the current task, and the user's mental workload was high, the system would aggressively interrupt the user to divert their attention to the higher-priority message. Conversely, if the incoming message was of lower priority, and the mental workload was high, the message was diverted to a text display for later reading. Care must be taken when designing such a system, since inappropriate direction of attention can greatly diminish overall performance. In this case, the Communications Scheduler showed a two-fold increase in attention allocation as measured by message comprehension, with little or no decrement in secondary tasks.

7.4.2.3 Working Memory

Interactions with complex systems require humans to review available information and integrate task-relevant information in working memory in order to have an internal representation of the problem space—one that can be manipulated and analyzed to finally reach some decision (Gentner & Franks, 1983). If the user is overloaded with information, they expend precious cognitive resources filtering out irrelevant information that takes additional time and contributes to the temporal decay of their representation in working memory (Baddeley, 1986); consequently, users are often required to make time-critical decisions based on impoverished mental models of the problem space.

An adaptive system designed to maximize working memory processes via an autonomous intelligent interruption and negotiation strategy utilized fNIR sensors to derive a diagnostic measure for verbal and spatial working memory. The system was able to dramatically increase working memory capacity for an unmanned air vehicle control task (Barker & Richards, 2005).

7.4.2.4 Mental Workload

The previous section of this chapter described how mental workload is a function of arousal, factors affecting current processing capacity, stresses, and individual responses to overload. When a joint cognitive system's assessment of user workload is high, and additional tasks or information-processing demands cannot be met, this would be a candidate time to invoke automation. However, that automation may not be appropriate when the cognitive capacity is well matched to the current task demands. Thus the adaptive system should only be invoked when the person's ability to handle the task demands breaks down. In a joint cognitive system, where automation support is provided on an as-needed basis, careful consideration must be given to the costs and benefits to determine the optimal time to invoke and disengage automation assistance.

For instance, an evaluation of an adaptive system that provided tactile navigation cues when user mental workload was overloaded demonstrated both the benefits and costs of such a system (Dorneich, Ververs, Mathan, & Whitlow, 2006). A Tactile Navigational Cueing belt used eight vibrating tactors in conjunction with position information from a GPS system and bearing information from a Dead Reckoning Module (DRM) to guide the participant to known waypoints. The system was only invoked when the cognitive state assessment indicated that the workload was higher than the available current processing capacity. Under heavy information-processing demands imposed by operational tasks (such as responding to radio communications and maintaining positional awareness), and when cognitive workload exceeded capacity, automation was triggered to assist in quickly and safely navigating a complex route by temporarily alleviating the cognitive overhead of navigating to an objective, allowing available resources to be used to complete the other tasks at hand. Thus the navigation task went from being cognitively intense (involving reading a map, mental transformation from 2D to 3D space, etc.) to one that is essentially a reactionary task to external stimuli that requires less cognitive resources. The overall benefit was to lower cognitive demands, allowing users to improve performance on the navigation task while not adversely affecting other tasks being done simultaneously. The potential cost of the adaptation is the loss of situation awareness of the surroundings, since the adapted task did not require attention to the environment. It is possible that the cost would never be realized (say, for instance, the user never returned to that area), but the potential cost must be weighed against the realized benefit when decided if the adaptation is triggered.

7.4.3 Summary and Implications

The adaptive joint cognitive system assistance, triggered by real-time classification of cognitive state, offers many advantages over traditional approaches to automation. These systems offer the promise of leveraging the strengths of humans and automation—augmenting human performance with automation specifically when human abilities fall short of the demands imposed by task environments. However, by delegating critical aspects of complex tasks to autonomous automation components, these systems run the risk of introducing many of the problems observed in many traditional human–automation interaction contexts. The pros and cons of automating complex systems have been widely discussed in the literature (e.g., Sarter, Woods, & Billings, 1997; Parasuraman & Miller, 2004) However, as widely noted, poorly designed automation can have serious negative effects. Automation can relegate the operator to the status of a passive observer, serving to limit situational awareness, and induce cognitive overload when a user may be forced to inherit control from an automated system. In addition, adaption strategies that are inappropriately applied can degrade overall performance significantly. Thus, the design of a joint cognitive adaptive system must explicitly consider the costs as well as the benefits of mitigation when deciding when and how to intervene in the decision-making process.

7.5 Conclusion

There are several integrative concepts in this chapter.

Cognitive Goals. In complex tasks people use cognitive goals when implementing task goals. A person's cognitive goals are important in organizing the person's behavior, directing attention to parts of the task, choosing the best method for meeting a given goal, and developing new working methods. The cognitive goals might be met in different ways under different circumstances, and thus, the goals and the processes for meeting them can be independent. For example, flying an aircraft involves predicting the weather, and this may be done in different ways before and during the flight.

Contextual Overview. People, and to a lesser extent automated systems, build up an overview of understanding and planning, which subsequently acts as the context for later activity. The overview provides data, expectations and values, and the criteria for deciding what would be the next best thing to do and how to do it.

Goal–Means Independence and Meta-Knowledge. Meta-knowledge is the knowledge about knowledge, such as the likelihood of alternative explanations of what is happening, or the difficulty in carrying out a particular action. Alternative working methods have meta-knowledge about their properties associated with them. Decisions about how best to meet a particular aim are based on meta-knowledge, and are involved in adapting behavior to particular circumstances, in the control of multitasking and mental workload, and in learning. Joint cognitive systems blur the line between human and machine precisely because both the human and machine employ meta-knowledge to adapt their behavior to best accomplish goals in the current context.

Modes of Processing. In addition to using different working methods, people may use different modes of processing, such as knowing the answer by association or thinking out a new working method. The mode of processing used varies from moment to moment, depending on the task and the person's experience.

7.5.1 Modeling Human Behavior

Establishing HF/E on an analysis of behavior into small independent units fits well with a "sequential stages" concept of the underlying structure of human behavior. However, a sequential-stages model does not include many of the key features of the complex tasks, such as flying and air-traffic control. Complex behavior is better described by a contextual model, in which processing builds up an overview that determines what processing is done next and how, which in turn updates the overview, and so on. In this mechanism for behavior organization, choices about what to do and how to do it depend on the details of the immediate situation interacting with the individual's nature and previous experience. It is a tenant of joint cognitive system design that knowledge of the human cognitive state is insufficient without knowledge of the current task context in which to interpret it.

The aspects missing from many sequential-stages models are

- The goal-oriented nature of behavior and the independence of goals from the means by which they are met.
- The continuing overview.
- The flexible sequencing of cognitive activity and the organization of multitasking.
- The knowledge base and the resulting predictions, anticipations, and active search for information that are part of the top-down processing.

Some of these aspects require a fundamental change in the nature of the model used. The most important aspect to add is the overview, as all cognitive processes are carried out within the context provided by this overview, and the sequence in which they are done is determined by what is in the overview.

A simple version of a contextual model has been suggested in Figures 7.22 and 7.24. These figures can act as an aide-mémoire about contextual processing, but any one-page representation can only indicate some features of what could be expressed. These simple figures do not convey explicit important aspects, such as

- Risky decision making and the effects of biases
- Goal orientation of behavior
- Typical sequences of activity
- Different modes of processing, including devising new working methods
- Use of meta-knowledge

Perhaps, the most important disadvantage of the one-page contextual model will be felt by people who are concerned with the tasks that are entirely sequential, rather than cyclic as in flying or air-traffic control. However, it can be argued that although dependencies may define the order in which some parts of a task are done, it could still be useful, when designing to support the sequential tasks, to consider the task sequence as a frame for structuring the overt behavior, while the underlying order of thinking about the task aspects may be more varied (cf. Figure 7.28).

7.5.2 The Difficulty in HF/E

Contextual processing underlies two types of difficulties in HF/E. One group of issues is concerned with HF/E techniques. As indicated earlier, the overview suggests the need for several additions to HF/E techniques:

- Considering the codings used in the task as a whole, rather than for isolated subtasks.
- Orienting cognitive task analysis toward the cognitive goals or functions to be met as an intermediary between the task goals and the cognitive processing (analysis of either goals or working methods alone is necessary but not sufficient).
- Designing the interface, training, and allocation of function between people and machines in a more dynamic way, to support the person's development and use of the contextual overview, alternative strategies, and the processes involved in the development of new working methods for both humans and machines in a joint cognitive systems design approach.
- Extending human error schemes to include difficulties with the overview and organization of the sequences of behavior.

The second group of issues is concerned with a fundamental complexity problem in human behavior and therefore, in HF/E. Human behavior is adapted to the particular circumstances in which it is carried out. This does not make it impossible to develop a general model of human behavior, but it does make it impossible to predict the human behavior in detail. Predicting human behavior is like weather prediction: It is not possible to be right, but it is possible to be useful. Any HF/E answer is always context-sensitive. The continuing complaint of HF/E practitioners that researchers do not provide them with what they need is a consequence of the fundamental nature of human behavior. Specific tests of what happens under specific circumstances will always be necessary. Furthermore, the models of human behavior need to provide, not the details, but the key issues to focus on when doing such tests or when developing and applying HF/E techniques.

References

Allport, D. A., Antonis, B., & Reynolds, P. (1972). On the division of attention: A disproof of the single channel hypothesis. *Quarterly Journal of Experimental Psychology, 24*, 225–235.

Amalberti, R. (1992). *Modèles d'activite en conduite de processus rapides: implications pour l'assistance á la conduite*. Unpublished doctoral thesis, University of Paris, France.

Anderson, J. R. (1995). *Cognitive Psychology and its Implications* (2nd Ed.). New York: Freeman.

Atkinson, W. H., Crumley, L. M., & Willis, M. P. (1952). *A study of the requirements for letters, numbers and markings to be used on trans-illuminated aircraft control; panels. Part 5: the comparative legibility of three fonts for numerals* (Report No. TED NAM EL-609, part 5). Naval Air Material Center, Aeronautical Medical Equipment Laboratory.

Backs, R. W. & Seljos, K. A. (1994). Metabolic and cardiorespiratory measures of mental effort: The effects of level of difficulty in a working-memory task. *International Journal of Psychophysiology, 16*(1), 57–68.

Baddeley, A. (1986). *Working Memory*. Oxford: Clarendon Press.

Bailey, R. W. (1989). *Human performance engineering. A guide for system designers* (2nd ed.). London: Prentice Hall.

Bainbridge, L. (l974). Problems in the assessment of mental load. *Le Travail Humain, 37*, 279–302.

Bainbridge, L. (1978). Forgotten alternatives in skill and workload. *Ergonomics, 21*, 169–185.

Bainbridge, L. (1983). Ironies of automation. *Automatica, 19*, 775–779.

Bainbridge, L. (1988). Types of representation, In L. P. Goodstein, H. B. Anderson, & S. E. Olsen (Eds.), *Tasks, errors and mental models* (pp. 70–91). London: Taylor & Francis.

Bainbridge, L. (1991). Multiplexed VDT display systems. In G. R. S. Weir & J. L. Alty (Eds.), *Human-computer interaction and complex systems* (pp. 189–210). London: Academic Press.

Bainbridge, L. (1993a). *Building up behavioural complexity from a cognitive processing element* (p. 95). London: Department of Psychology, University College London.

Bainbridge, L. (1993b). *Difficulties and errors in complex dynamic tasks.* Unpublished manuscript, University College London.

Bainbridge, L. (1993c). Typos of hierarchy imply types of model. *Ergonomics, 36*, 1399–1412.

Barker & Richards, (2005). The Boeing Team Fundamentals of Augmented Cognition. In D. Schmorrow (Ed.), *Foundations of Augmented Cognition.* Routledge.

Barnard, P. (1987). Cognitive resources and the learning of human–computer dialogues. In J. M. Carroll (Ed.), *Interfacing thought: Cognitive aspects of human–computer interaction.* Cambridge, MA: MIT Press.

Beatty, J. (1982). Task-Evoked Pupillary Responses, Processing Load, and the Structure of Processing Resources. *Psychological Bulletin, 91*(2), 276–292.

Beishon, R. J. (1974). An analysis and simulation of an operator's behaviour in controlling continuums baking ovens. In E. Edwards, & F. P. Lees (Eds.). *The human operator in process control* (pp. 79–90). London: Taylor & Francis.

Benson, J. M. (1990). *The development of a model of multi-tasking behaviour.* Unpublished doctoral thesis, Ergonomics Unit, University College London.

Bisseret, A. (1970). Mémoire opérationelle et structure du travail. *Bullets de Psychologic, XXIV*, 280–294 [English summary in *Ergonomics, 14*, 565–570 (1971)].

Brennan, A. C. C. (1987). *Cognitive support for process control: Designing system representations.* Unpublished master's thesis, Ergonomics Unit, University College London.

Brezillin, P., & Pomeral, J.-C. (1997). Joint Cognitive Systems, Cooperative Systems and Decision Support Systems: A Cooperation in Context. *Proc. of the European Conference on Cognitive Science*, Manchester, UK, April 1997, pp. 129–139.

Broadbent, D. E. (1958). *Perception and Communication.* New York: Pergamon.

Briggs, G. E., &. Rockway, M. R. (1966). Learning and performance as a function of the percentage of pursuit tracking component in a tracking display. *Journal of Experimental Psychology, 71*, 165–169.

Cabeza, R., & Kingstone, A. (Eds.) (2001). *Handbook of Functional Neuroimaging of Cognition.* Cambridge, MA: The MIT Press.

Crark, F. I. M., & Lockhart, R. S. (1972). Levels of processing: A framework for memory research. *Journal of Verbal Learning and Verbal Behaviour, 11*, 671–684.

Crossman, E. R. F. W. (1959). A theory of the acquisition of speed-skill. *Ergonomics, 2*, 153–166.

Crossman, E. R. F. W., & Goodeve, P. J. (1963, April). Feedback control of hand-movement and Fitts' law. Communication to the Experimental Psychology Society, University of Oxford, Oxford, U.K.

Davis, D. R. (1966). Railway signal!' passed at danger: The drivers, circumstances and psychological processes. *Ergonomics, 9*, 211–222.

Doerner, D. (1987). On the difficulties people have in dealing with complexity. In J. Rasmussen, K. D. Duncan, & J. Leplat (Eds.), *New technology and human error* (pp. 97–109). Chichester: Wiley.

Dorneich, M. C., Whitlow, S. D., Mathan, S., Carciofini, J., & Ververs, P. M. (2005). The Communications Scheduler: A Task Scheduling Mitigation For A Closed Loop Adaptive System. *Proceedings of the 11th International Conference on Human-Computer Interaction, (HCI International 2005)*, Las Vegas, NV, USA: Lawrence Erlbaum.

Dorneich, M. C., Ververs, P. M., Mathan, S., and Whitlow, S. D., (2006). "Evaluation of a Tactile Navigation Cueing System Triggered by a Real-Time Assessment of Cognitive State," *Proceedings of the Human Factors and Ergonomics Society Conference 2006*, San Francisco, CA.

Dorneich, M. C., Mathan, S., Ververs, P. M., and M. C., Whitlow, S. D. (2008), "Cognitive State Estimation in Mobile Environments", *Augmented Cognition: A Practitioner's Guide*, Schmorrow D., & Stanney K. (Editors), Santa Monica: HFES Press.

Duncan, J. (1986). Disorganisation of behaviour after frontal lobe damage. *Cognitive Neuropsychology, 3*, 271–90.

Easterby, R. S. (1970). The perception of symbols for machine displays. *Ergonomics, 13*, 149–158.

Ellis, A. W. (1993). *Reading, writing and dyslexia: A cognitive analysis*. Hillsdale, NJ: Lawrence Erlbaum Associates.

Eysenck, M. W., & Keane, M. T. (1990). *Cognitive psychology: A student's handbook*. Hillsdale, NJ: Lawrence Erlbaum Associates.

Fitts, P. M. (1954). The information capacity of the human motor system in controlling the amplitude of movement. *Journal of Experimental Psychology, 47*, 381–391.

Fitts, P. M., & Deininger, R. L. (1954). S-R compatibility: Correspondence among paired elements within stimulus and response codes. *Journal of Experimental Psychology, 48*, 483–492.

Fitts, P. M., & Jones, R. E. (1961a). *Analysis of factors contributing to 460 "pilot error" experiences in operating aircraft controls* (Memorandum Report No. TSEAA-694-12, Aero Medical Laboratory, Air Materiel Command, Wright Patterson Air Force Base, Dayton, OH. July 1, 1947). In H. W. Sinaiko (Ed.), *Selected papers on human factors in the design and use of control system* (pp. 332–358). New York: Dover (Original work written 1947).

Fitts, P. M., & Jones, R. E. (1961b). *Psychological aspects of instrument display. 1: Analysis of 270 "pilot error" experiences in reading and interpreting aircraft instruments* (Memorandum Report No. TSEAA-694-12A, Aero Medical Laboratory, Air Materiel Command. Wright Patterson Air Force Base, Dayton, OH. October 1, 1947). In H. W. Sinaiko (Ed.), *Selected papers on human factors m the design and use of control systems* (pp. 359–396). New York: Dover (Original work written 1947).

Fleishman, E. A. (1975). Taxonomic problems in human performance research. In W. T. Singleton & P. Spurgeon (Eds.), *Measurement of human resources* (pp. 49–72). London: Taylor & Francis.

Folkhard, S. (1990). Circadian performance rhythms: Some practical and theoretical implications. In D. E. Broadbent, L Reason, & A. Baddeley (Eds.), *Human factors in hazardous situations* (pp. 543–553). Oxford: Clarendon Press.

Gagné, R. M. (1977). *The conditions of learning* (3rd ed.). New York: Holt, Rinehart & Winston.

Gallaher, P. D., Hunt, R. A., & Williges, R. C. (1977). A regression approach to generate aircraft predictor information. *Human Factors, 19*(6), 549–555.

Garavan, H., Ross, T. J., Li, S.-J., & Stein, E. A. (2000). A parametric manipulation of central executive functioning using fMRI. *Cerebral Cortex, 10*, 585–592.

Gerson, A. D., Parra, L. C., & Sajda, P. (2005) Cortical Origins of Response Time Variability during Rapid Discrimination of Visual Objects, NeuroImage, 28(2) 326–341.

Gevins, A., & Smith, M. E. (2000). Neurophysiological measures of working memory and individual differences in cognitive ability and cognitive style. *Cerebral Cortex, 10*, 829–839.

Gibson, J. J. (1950). *The perception of the visual world*. Boston, MA: Houghton Mifflin.

Gratton, G., & Fabiani, M. (2006). Optical Imaging of Brain Function. In R. Parasuraman & M. Rizzo (eds.), *Neroergonomics* (pp. 65–81). Oxforn: Oxford University Press.

Gray, J. A., & Wedderburn, A. A. I. (1960). Grouping strategies with simultaneous stimuli. *Quarterly Journal of Experimental Psychology, 12*, 180–184.

Green, R. (1990). Human error on the flight deck. In D. E. Broadbent, J. Reason, & A. Baddeley (Eds.), *Human factors in hazardous situations* (pp. 55–63). Oxford: Clarendon Press.

Grether, W. F. (1949). The design of long-scale indicators for speed and accuracy of quantitative reading. *Journal of Applied Psychology, 33*, 363–372.

Hayes-Roth, B., & Hayes-Roth, F. (1979). A cognitive model of planning. *Cognitive Science, 3*, 275–310.

Helson, H. (1964). *Adaptation-level theory.* New York: Harper and Row.

Heslegrave, R. J., & Furedy, J. J. (1979). Sensitivities of HR and T-wave amplitude for detecting cognitive and anticipatory stress. *Physiology & Behavior, 22*(1), 17–23. The Netherlands: Elsevier Science.

Hoc, J.-M. (1988). *Cognitive psychology of planning.* London: Academic Press.

Hollnagel, E. (1993). *Reliability of cognition: Foundations of human reliability analysis.* London: Academic Press.

Howland, D., & Noble, M. (1955). The effect of physical constraints on a control on tracking performance. *Journal of Experimental Psychology, 46*, 353–360.

Hukki, K., & Norros, L. (1993). Diagnostic orientation in control of disturbance situations. *Ergonomics, 35*, 1317–1328.

Hunt, D. P., & Warrick, M. J. (1957). *Accuracy of blind positioning of a rotary control* (USAF WADC Tech. Notes 52-106). Dayton, OH: U.S. Air Force.

Izzetoglu, K., Bunce, S., Onaral, B., Pourrezaei, K., & Chance, B. (2004). Functional optical brain imaging using near-infrared during cognitive tasks. *International Journal of Human-Computer Interaction, 17*(2), 211–227.

Johansson, G., Aronsson, G., & Lindström, B. G. (1978). Social psychological and neuroendocrine stress reactions in highly mechanised work. *Ergonomics, 21*, 583–599.

Jordan, N. (1963). Allocation of functions between man and machines in automated systems. *Journal of Applied Psychology, 47*, 161–165.

Kahneman, D. (1973). *Attention and Effort.* Englewood Cliffs, NJ: Prentice-Hall.

Kahneman, D., Slovic, P., & Tversky, A. (Eds.). (1982). *Judgement under uncertainty; Heuristics and biases.* New York: Cambridge University Press.

Kalsbeek, J. W. H., & Ettema, J. H. (1963). Scored irregularity of the heart pattern and measurement of perceptual or mental load. *Ergonomics, 6*, 306–307.

Klein, G. A., (1989). Recognition-primed decisions. In W. B. Rouse (Ed.), *Advances in man-machine systems research* (Vol. 5, pp. 47–92). Greenwich, CT: JAI Press.

Leplat, J., & Bisseret, A. (1965). Analyse des processus de traitement de l'information chez le controleur de la navigation aerienne. *Bulletin dunCERP, XIV*, 51–67. [English translation in *Controller: IFACTCA Journal, 5*, 13–22 (l965)]

Lindsay, P. H., &. Norman, D. A. (1972). *Human information processing.* New York: Academic Press.

Lindsay, P. H., & Norman, D. A. (1977). *Human Information Processing: An Introduction to Psychology,* Second Edition. Academic Press, New York.

Makeig, S., Westerfield, M., Jung, T.-P., Enghoff, S., & Townsend, J. (2002). Dynamic brain sources of visual evoked responses. *Science* 295: 690–693.

Makeig, S., & Jung, T.-P. (1995). Changes in alertness are a principal component of variance in the EEG spectrum. *NeuroReport, 7*(1), 213–216.

Marshall, E. C., Scanlon, K. E., Shepherd, A., & Duncan, K. D. (1981). Panel diagnosis training for major hazard continuous process installations. *Chemical Engineer, 365*, 66–69.

Mathan, S., Whitlow, S., Erdogmus, D., Pavel, M., Ververs, P., & Dorneich, M. (2006a). Neurophysiologically Driven Image Triage. CHI '06 extended abstracts on Human factors in computing systems. *Proceedings of the 2006 Conference on Human Factors in Computing Systems.* Montreal, Canada. 1085–1090.

Mathan, S., Ververs, P., Dorneich, M., Whitlow, S., Carciofini, J., Erdogmus, Pavel, M., Huang, C., Lan, T., & Adami, A. (2006b). "Neurotechnology for Image Analysis: Searching For Needles in Haystacks Efficiently", *Proceedings of the 2nd Augmented Cognition International*, San Francisco, CA.

Mikulka, P., Hadley, G., Freeman, F., & Scerbo, M. (1999). The effects of a biocybernetic system on vigilance decrement. *Proceedings of the 43rd Annual Meeting of the Human Factors and Ergonomics Society*, Santa Monica, CA: HFES.

Miller, G. A. (1956). The magical number seven, plus or minus two: Some limits on our capacity for processing information. *Psychological Review, 63*, 81–97.

Miller, G. A., Heise, G. A., & Lichten, W. (1951). The intelligibility of speech as a function of the context of the test materials. *Journal of Experimental Psychology, 41*, 329–335.

Moray, N. P., Hiskes, D., Lee, J., & Muir, B. M. (1995). Trust and human intervention in automated systems. In J. M. Hoc, P.C. Cacciabue, & E. Hollnagel (Eds.), *Expertise and technology: Cognition and human-computer cooperation* (pp. 183–194). Hillsdale, NJ: Lawrence Erlbaum Associates.

Neumann, D. L. (2002). Effect of varying levels of mental workload on startle eyeblink modulation. *Ergonomics, 45*(8), 583–602.

Oborne, D. (1995). *Ergonomics at work* (3rd ed.). Chichester: Wiley.

Parasuraman, R., & Davies, D. R. (1984). *Varieties of Attention.* New York: Academic Press.

Parasuraman, R., & Miller, C. (2004). Trust and etiquette in high-criticality automated systems. *Communications of the Association for Computing Machinery, 47*(4), 51–55.

Parasuraman, R., Mouloua, M., & Hilburn, B. (1999). Adaptive aiding and adaptive task allocation enhance human-machine interaction. In M. W. Scerbo & M. Mouloua (Eds.), *Automation technology and human performance: Current research and trends* (pp. 129–133). Mahwah, NJ: Erlbaum.

Parasuraman, R., Sheridan, T. B., and Wickens, C. D. (2000). A model for types and levels of human interaction with automation. *IEEE Transaction on Systems, Man, and Cybernetics–Part A: Systems and Humans, 30*(3), 286–297.

Parasuraman, R., & Rizzo, M. (2006). Introduction to Neroergonomics. In R. Parasuraman & M. Rizzo (eds.), *Neroergonomics* (pp. 3–14). Oxforn: Oxford University Press.

Parra, L., Alvino, C., Tang, A., Pearlmutter, B., Yeung, N., Osman, A., and Sajda, P. (2003). Single Trial Detection in EEG and MEG: Keeping it Linear. Neurocomputing. (52–54), 177–183.

Partala, T., & Surakka, V. (2003). Pupil size variation as an indication of affective processing. *International Journal of Human-Computer Studies, 59*(1–2), 185–198.

Pashler, H. (1994). Dual-task interference in simple tasks: data and theory. *Psychological Bulletin, 116*, 220–244.

Pheasant, S. (1991). *Ergonomics: Work and health.* London: Macmillan.

Pope, A. T., Bogart, E. H., & Bartolome, D. (1995). Biocybernetic system evaluates indices of operator engagement. *Biological Psychology, 40*, 187–196.

Porges, S. W., & Byrne, E. A. (1992). Research Methods for Measurement of Heart-Rate and Respiration." *Biological Psychology, 34*(2–3), 93–130.

Posner, M. I., & Rossman, E. (1965). Effect of size and location of informational transforms upon short-term retention. *Journal of Occupational Accidents, 4*, 311–335.

Prinzel, L. J., Freeman, F. G., Scerbo, M. W., Mikulka, P. J., & Pope, A. T. (2000). A closed-loop system for examining psychophysiological measures for adaptive automation. *International Journal of Aviation Psychology, 10*, 393–410.

Rasmussen, J. (1982). Human errors: A taxonomy for describing human malfunction in industrial installations. *Journal of Occupational Accidents, 4*, 311–335.

Reinartz, S. J. (l989). Analysis of team behaviour during simulated nuclear power plant incidents. In E. D. Megaw (Ed.), *Contemporary ergonomics 1989* (pp. 188–193). London: Taylor & Francis.

Rouse, W. B. (1980). *Systems engineering models of human-machine interaction.* Amsterdam, the Netherlands: North Holland.

Rumelhart, D. E. (1977). Towards an interactive model of reading. In S. Dornic (Ed.), *Attention and performance VI.* Hillsdale, NJ: Lawrence Erldaum Associates.

Sage, A. P. (1981). Behavioural and organizational considerations in the design of information systems and processes for planning and decision support. *IEEE Transactions on Systems, Man and Cybernetics, SMC-11*, 610–678.

Samurçay, R., & Rogalski, J. (l988). Analysis of operator's cognitive activities in learning and using a method for decision making in public safety. In J. Patrick, & K. D. Duncan (Eds.), *Training, human decision making and control.* Amsterdam, the Netherlands: North-Holland.

Sarter, N. B., Woods, D. D., & Billings, C. E. (1997). "Automation surprises," *Handbook of Human Factors and Ergonomics*, 2nd Edition, Salvendy, Gavriel (Ed)., Wiley and Sons, 1997.

Savin, H. B., & Perchonock, E. (1965). Grammatical structure and the immediate recall of English sentences. *Journal of Verbal Learning and Verbal Behaviour*, 4, 348–353.

Scerbo, M. W., Freeman, F. G., & Mikulka, P. J. (2003). A brain-based system for adaptive automation Theoretical Issues in Ergonomic Science, 4, 200–219.

Schmorrow, D. D., & Kruse, A. A. (2002). Improving human performance through advanced cognitive system technology. *Proceedings of the Interservice/Industry Training, Simulation and Education Conference (I/ITSEC'02)*, Orlando, FL.

Shaffer, L. H. (1973). Latency mechanisms in transcription. In S. Kornblum (Ed.), *Attention and performance IV* (pp. 435–446). London: Academic Press.

Sheridan, T. B., & Ferell, W. R. (1974). *Man machine systems: Information, control and decision models of human performance.* Cambridge, MA: MIT Press.

Sherrington, C. (1974). *The integrative action of the nervous system.* Cambridge: Cambridge University Press (Original work published 1906).

Smith, S. L., & Thomas, D. W. (1964). Color versus shape coding in information displays. *Journal of Applied Psychology*, 48, 137–146.

Sperandio, J. C. (1970). *Charge de travail et mémorization en contrôle d'approche* (Report No. IRIA CENA, CO 7009, R24). Paris: Institut de Recherche en Informatique et Aeronautique.

Sperandio, J. C. (1972). Charge de travail et régulation des processus opératoires. *Le Travail Humain*, 35, 86–98 [English summary in *Ergonomics*, *14*, 571–577(1971)].

Stern, J. A., Boyer, D., & Schroeder, D. (1994). Blink rate: A possible measure of fatigue. *Human Factors*, *36*(2), 285–297.

Teichner, W. H., & Krebs, M. J. (1974). Laws of visual choice reaction time. *Psychological Review*, *81*, 75–98.

Thorpe, S., Fize, D., & Marlot, C. (1996). Speed of processing in the human visual system. *Nature*, *381*, 520–522.

Treisman, A.M. (1964). Verbal cues, language, and meaning in selective attention. *American Journal of Psychology*, *77*, 206–219.

Veltman, J. A., & Gaillard, A.W. K. (1998). Physiological workload reactions to increasing levels of task difficulty. *Ergonomics*, *41*(5), 656–669.

Verwey, W. B., & Veltman, H. A. (1996). Detecting short periods of elevated workload: A comparison of nine workload assessment techniques. *Journal of Experimental Psychology-Applied*, *2*(3), 270–285.

Wickens, C. D., & Hollands, J. (2000). *Engineering Psychology and Human Performance.* Prentice Hall, 3rd edition.

Wickens, C. D. (1992). *Engineering psychology and human performance* (2nd ed). New York: Harper Collins.

Wickens, C. D. (1984). Processing resources in attention. In R. Parasuraman, &. D. R. Davies (Eds.), *Varieties of attention.* London: Academic Press.

Wildervanck, C., Mulder, G., & Michon, J. A. (1978). Mapping mental load in car driving. *Ergonomics*, *21*, 225–229.

Wilson, G. F., & Eggemeier, F. T. (1991). Physiological measures of workload in multi-task environments. In D. Damos (Ed.), *Multiple-task performance* (pp. 329–360). London: Taylor & Francis.

Wilson, J. R., & Corlett, E. N. (1995). *Evaluation of human work: A practical ergonomics methodology* (2nd ed.). London: Taylor & Francis.

Winograd, T. (1972). *Understanding natural language.* Edinburgh: Edinburgh University Press (Reprinted from *Cognitive Psychology* (1972), 3, 1–195).

Woods, D. D., Roth, E. M., & Benett, K. (1990). Explorations in joint human-machine cognitive systems. In Robertson, S. Zachary, W. & Black, J. B. (Eds.) Cognition, Computing and Cooperation. Ablex (pp. 123–158).

Yntema, D. B., & Mueser, G. E. (1962). Keeping track of variables that have few or many states. *Journal of Experimental Psychology*, *63*, 391–395.

8

Automation in Aviation Systems: Issues and Considerations

Mustapha Mouloua
University of Central Florida

Peter Hancock
University of Central Florida

Lauriann Jones
University of Central Florida

Dennis Vincenzi
Navair

8.1 Introduction

Aviation continues to grow at an unprecedented rate. The Boeing Corporation is currently finalizing the Boeing Jumbo 787, which is due to make its first flight in 2008. Airbus Industries are similarly planning for their first Jumbo A380 Transatlantic flight in August of 2007. The increasing number of modern aircraft with advanced technologies can be seen in both general and commercial aviation realms. Owing to these advanced technologies, both civilian and military pilots can now fly their missions with increasing accuracy leading to greater fuel efficiency and all-weather operations. However, there have been a number of factors that continue to jeopardize flying and safety. The Boeing series 737, 747, 757, 767, and 777 represent an evolution of technological capabilities. These technological advances have changed the size and role of the aircrew throughout the years. It is now possible to fly with only two crewmembers for long-duration flights. At cruising altitudes, pilots now rely on autopilots to perform many of their flying tasks and duties. Thus, the evolution of advanced aviation covaries with the ascendancy of automated functioning.

8.2 Automation Problems

Increased automation in all transportation systems (ground, air, space, and maritime) has served to highlight the effects of incomplete specification on the performance of human operators of these systems. Although automation has brought considerable benefits to the operation and control of these critical human–machine systems, there is growing empirical evidence pointing to some negative effects, especially in relation to human-monitoring performance and system awareness (Mouloua & Koonce, 1997; Mouloua & Parasuraman, 1994; Parasuraman & Mouloua, 1996; Scerbo & Mouloua, 1998; Vincenzi, Mouloua, & Hancock, 2004; Wiener, 1988). Particularly, in aviation, a number of incidents and accidents have been attributed to system designs that led to operator overreliance on automation (Billings, 1997; Endsley & Strauch, 1997; Parasuraman & Mouloua, 1996; Wickens, 1992; Wiener, 1981). Parasuraman and Riley (1997) recently reviewed a spectrum of those human-performance costs associated with poor automation. These include problems, such as unbalanced operator mental workload, overtrust and over-reliance leading to complacency, mistrust, loss of skill, and reduced situation awareness.

In the real world, the human monitoring of automated systems for malfunctions can often be poor as a result of low frequency of automation failures when dealing with reliable automated systems (Parasuraman, 1987). In formal vigilance theory, this is considered as a low target rate. Research on vigilance has shown that the detection of low-probability events is degraded after prolonged periods on watch (Davies & Parasuraman, 1982). Therefore, one can predict that operator detections of a failure, which is very infrequent in the automated control of a task, may well be very poor after a prolonged period spent under automation control. However, most vigilance research upon which such an assumption is based has been carried out with simple, easy tasks in which the operator has little to do (Davies & Parasuraman, 1982). Hence, these findings may not apply directly to those more complex multitask environments of current automated human–machine systems.

Parasuraman, Molloy, and Singh (1993) conducted a study in which participants were required to perform a tracking task and a fuel-management task manually over four 30 min sessions. At the same time, an automated engine-status task had to be monitored for occasional automation "failures" (engine malfunctions not detected by the automation system). In another condition, the engine-status task was also performed manually. Participants detected over 75% of malfunctions on the engine-status task when they did the task manually, while simultaneously carrying out tracking or fuel management. However, when the engine-status task was under automation control, there was a marked reduction (mean = 32%) in the operator detection rate of the system malfunctions (i.e., automation failures). This substantial reduction in failure-detection sensitivity was apparent after about 20 min spent under automation control. In a separate experiment conducted in the same series, it was shown that monitoring of automation was poor under multitask, but not single-task conditions (i.e., when only the engine-status task had to be performed). In a follow-up study, experienced pilots were also found to show similar performance trends (Parasuraman, Mouloua, & Molloy, 1994). Although poor monitoring of automation appears to be a general finding under multitask-performance conditions, individuals may differ in the extent to which they exhibit this phenomenon. Singh, Molloy, and Parasuraman (1993) found evidence to suggest that individuals with high self-reported levels of "energetic arousal" showed better monitoring of automated systems, at least for short periods of time. Thus, individual differences in operator capacities are an important factor in those circumstances.

8.3 What Is Automation?

Automation can be defined as the execution of a task, function, service, or subtask by a machine agent. This mode of automation can vary from full automation to low-level control. According to the *Merriam-Webster's Dictionary*, Automation is defined as (a) "the technique of making an apparatus, a process, or a system operate automatically," (b) "the state of being operated automatically," or (c) "automatically controlled operation of an apparatus, process, or system by mechanical or electronic devices that take the place of human labor." In aviation, automation most often refers to the autopilot, flight management system (FMS), as well as other burgeoning advanced cockpit-related systems and functions. When these respective functions are shared interchangeably by the pilot(s) and automated systems, they can pose some serious behavioral problems. These problems include issues, such as human and machine error, loss of situation awareness, each acting as precursors to untoward events. In the following sections, we consider some of these issues.

8.4 Situation Awareness

Situation awareness has been defined as "the perception of the elements in the environment within a volume of time and space, the comprehension of their meaning and the projection of their status in the near future" (Endsley, 1996, pp. 165–166). Automation has been found to improve situation awareness by reducing the workload, stress, and the complexity of a system for the human operator. Recently, a research has found that adaptive automation applied to aid the human operator with the acquisition of information and implementation of actions, significantly improved situational awareness over the application of adaptive automation to cognitive functions, such as the analysis of information (see Kaber, Perry, Segall, McClernon, & Prinzel, 2006). However, when automation was applied to tasks that required the analysis of information or decision-making automation, it increased the workload.

Numerous investigators are also exploring to employ the less-utilized modalities, such as haptics, olfaction, and audition to improve the performance and direct attention when automation state changes in the adaptive automation systems (see Warren-Noell, Kaber, & Sheik-Nainar, 2006; Washburn & Jones, 2004). For example, vocal cues have been found to improve human-operator performance (workload, time-to-task completion, and situational awareness) over the standard visual cueing, and olfactory cues are being proposed to increase attention, redirect attention, and improve situation awareness and human performance in complex environments (Jones, Bowers, Washburn, Cortes, & Vijaya Satya, 2004; Washburn & Jones, 2004).

Despite what we have learned, loss of situation awareness in automated aviation systems is still a fairly common and very serious problem. If the human operators of an automated aviation system are unable to maintain their situation awareness, they are likely to be unresponsive if the automation fails (Parasuraman & Hancock, 2001; Parson, 2007; Prince, Ellis, Brannick, & Salas, 2007). They may not even detect the failed automation before the situation becomes unrecoverable.

The reason for the loss of situation awareness in a complex automated system is still an area of great concern and research. This may not be probably owing to any one reason, but to a variety of situations, such as when the pilot's role has changed from an active role to a passive monitoring role (Kaber, et al., 2006; Parasuraman & Mouloua, 1996). When a human operator's role is reduced to mere monitoring of the automated system, monitoring complacency may occur in which the human operator's monitoring performance begins to decline. Monitoring complacency is a phenomenon that may be related to vigilance decrement, low workload, or boredom, as well as theories of information processing, cognitive capacity, workload, attention, stress, and performance (Hancock & Warm, 1989; Kahneman & Treisman, 1984; Parasuraman & Mouloua, 1987; Wickens, Vidulich, & Sandry-Garza, 1984).

Another cause for the loss of situation awareness and monitoring complacency may be owing to the amount of trust that the human operator has in the system, and the sheer nature of the presence of reliable automation (Dixon & Wickens, 2006; Madhavan, Wiegmann, & Lacson, 2006; Riley, 1994). It has been posited that the operator can become over reliant on the automation, possibly trusting the automation beyond its actual reliability level. Once the operator begins to trust and rely on the automation, attention to the continuous activity of the system may begin to turn to something more active in the environment, and the operator's situation awareness may decline, possibly to dangerously low levels. A recent study by Rovira, McGarry, and Parasuraman (2007) found that decision times were significantly increased by highly reliable automation when compared with manual performance.

Situation awareness may be diminished in the design of the system itself, in which the human operator is kept *out-of-the-loop*, from the processes and activities of the automated system (Endsley, 1996). When the operator is out-of-the-loop and something in the automated system fails, the operator may not be able to detect what when wrong, where, or how to recover from it. Accordingly, when the operator is unaware of the coupling of automated functions within the system, *mode errors* and *errors of omission and commission* may occur.

8.5 Mode of Error

When *mode errors* and *errors of omission* and *errors of commission* occur, it greatly reduces the overall resilience of the system. Mode errors occur when the human operator looses the mode awareness, which may be defined as the "awareness of the status and behavior of the automation" (Sarter, 1996; Sarter & Woods, 1992). Often loss of mode awareness is described by pilots as "automation surprise" (Sarter, 1996, p. 273). Mode errors occur when the observability of the automation actions lacks in the design, often owing to inadequate feedback to the human operator (Woods, 1996). Feedback is thought to be a useful method to maintain the human operator in-the-loop and prevent some of these errors (Endsley, 1996). An automated system designed to "have high autonomy and authority but low observability" increases the risk of mode errors (Woods, 1996, pp. 8–9). Fifty-five percent of pilots said that they were still surprised by their aviation system's automation, after 1 year of experience with the aircraft. Additionally, human operators often rely on their own means, rather than on the automation system to maintain mode awareness, making their individual performance and the overall system performance, all the more less predictable (Bjorklund, Alfredson, & Dekker, 2006).

Errors of omission and commission are mode errors in which the human operator or pilot fails to act or reacts inappropriately. Errors of omission occur when the human operator fails to realize the system's status, such as if the system has changed its behavior and requires intervention, and thus, fails to act appropriately to correct the system; whereas, errors of commission occur when the human operator "take[s] an action that is appropriate for one mode of the device when it is, in fact, in a different mode" (Sarter, 1996). Such errors of information and coordination between the human operator and the automated system represent a continuing area of concern for the aviation industry's advanced cockpit systems. *Adaptive automation* is one approach to combat some of the pitfalls of a complex automated system (Mouloua & Parasuraman, 1994).

Automation may be designed into a system in many different ways or *levels*, from a no-automation manual control form to completely autonomous automation. Parasuraman, Sheridan, and Wickens (2000) examined several of these levels of automation and further categorized automation functions into four distinct types: (1) information acquisition (organizing information); (2) information analysis (information integration and summary; (3) decision and action selection (providing suggestions); and (4) action implementation (Parasuraman, et al., 2000, p. 286).

Parasuraman, et al. (2000) also categorized the degree or level of automation into 10 categories: Level (1) manual operation or no automation; Level (2) a complete set of decision/action alternatives are provided to the human operator to choose from; Level (3) only a narrowed set of decision/action alternatives are provided to the human operator; Level (4) only one decision/action alternatives is provided to the human operator; Level (5) automation makes decision and takes action, after human-operator approval; Level (6) a decision/action is provided to the human operator with a limited time to veto before automatic execution; Level (7) automatic automation action and notification to the operator; Level (8) automatic automation action and notification to the operator, unless information is requested; Level (9) automatic automation action and notification to the operator up to the discretion of the automation; and Level (10) autonomous automation. Great consideration must be given to the implications and applications of different types of automation functions and levels when designing an automation system. Kaber, et al. (2006) found adaptive automation to be particularly beneficial at levels one and four; and to increase workload, at levels two and three. It must be noted that the proper function allocation may benefit or increase the risk to the system (Parasuraman, Mouloua, & Molloy, 1996; Parasuraman, et al., 2000).

8.6 Automation Usage

Parasuraman and Riley (1997) distinguished four types of automation problems. These problems were automation use, misuse, disuse, and abuse. They defined *Use* as "the voluntary activation or disengagement of automation by human operators." Having automation designed to benefit the system is meaningless, if the human operator chooses not to use it. The use of automation is heavily influenced by the human operator's perceived trust in the system and the system's actual reliability (Riley, 1994). As people's perceptions are individual and often biased and inaccurate, especially in estimating the reliability of an automated system, the proper use of a system is difficult to predict. Misuse and disuse are Use problems (Dixon & Wickens, 2006; Madhavan, et al., 2006; Rovira, et al., 2007).

Misuse was described by Parasuraman and Riley (1997) as the "overreliance on automation, (e.g., using it when it should not be used, failing to monitor it effectively)" "which can result in failures of monitoring [i.e., monitoring complacency] or decision biases." In this case, the human operator may use the automation to the point that they practically remove themselves as active participants in the function of the system, and their performance may decline. If the automation fails, as rare as that may be, the human operator may not be prepared to recover the system and the safety of the system may be jeopardized.

Disuse is the opposite Use problem. Rather than overreliance, the human operator may not utilize the automation or ignore the automation. Parasuraman and Riley (1997) described this problem as the "neglect or underutilization of automation." Such a problem may arise when the human operator lacks trust in the automation (Riley, 1994). Repeated false alarms as well as moments of high workload may lead to a condition of disuse (Dixon & Wickens, 2006). Both disuse and misuse may also be influenced by a human operator's self-confidence in their own skills. Again, the occurrence of such states is extremely individual and therefore, often difficult to predict.

Abuse, refers to a condition that is not of the human operator's control, but rather a problem created by the humans involved in the design and implementation of automation, who do not give proper consideration to those who will be using the automation. Parasuraman and Riley (1997) defined this as "automation abuse, or the automation of functions by designers and implementation by managers without due regard for the consequences for human performance, tends to define the operator's roles as by-products of the automation."

8.7 Automation Complacency

Crew "Complacency" or automation-induced monitoring inefficiency is a major problem regarding flight performance, systems reliability, and safety. In their original paper on performance consequences of automation-induced complacency, Parasuraman, et al. (1993) examined how reliability variations in an automated monitoring system may affect the human operators' failure-detection performance. Two experiments were conducted on a revised version of the multi-attribute task (MAT) battery (Comstock & Arnegard, 1992), which allowed some systems to be automated and others to remain under manual control. The reliability of the automation was manipulated to examine its effect on the human operators' monitoring performance. It was found that consistently reliable automation was associated with poorer human-monitoring performance (a.k.a., automation-induced complacency or monitoring complacency). After approximately twenty minutes of reliable automation, the human operators' monitoring performance was found to decline, and observed this to be related to a number of other tasks allocated to the human operator. In other words, when the human operator was required to perform other tasks manually while monitoring the automation, their monitoring performance was found to decline.

8.8 Adaptive Automation

Sheridan and Parasuraman (2000) examined the problem of deciding which functions should be automated and which should be manually controlled. They compared humans and automation for superior failure detection. They used signal detection theory to analyze the probabilities of hits, misses, false alarms, and correct rejections, and then weighed the costs and benefits of various errors on the overall outcome. It was found that superiority of failure detection between human or machine depended on the weighing of the costs and benefits to each individual system. For example, in some systems, a quick response may be critical and in others, it may be the accuracy of the response. Such individual dichoto-mies will vary the outcome of the automation allocation to each individual system. The purpose of such analysis is to aid in the allocation of tasks to either machine or human, but it is suggested that such analysis could also aid in "deciding between two kinds of automation, two kinds of human interaction, or two mixes of human and automation."

A more recent study by Cummings and Guerlain (2007) examined the operators' capacity to real-locate highly autonomous tasks (in-flight missiles to time-sensitive target), while maintaining perfor-mance on other "secondary tasks of varying complexity." They found that human performance was significantly degraded when the human operator was attempting to reallocate multiple autonomous tasks. Specifically, if the human operator's busy time was 70%, then there was a significant decay in the performance. They suggested that a 70% utilization score (percentage busy time) would be a metric, generalizable, for predicting the human performance in complex environments.

Similar to Sheridan and Parasuraman (2000), Scerbo (1996) also suggested that the design of adaptive automation systems should weigh the costs and benefits of automation allocation for each individual job. He stated that "automation is neither inherently good nor bad" but it does "change the nature of work, and, in doing so, solves some problems while creating others." He suggested new theoretical approaches to the study of adaptive automation, such as a "social-centered approach" in which under-standing of team dynamics may weigh heavily on the interactions between the human operator and the automation system.

8.9 Training Issue in Aviation System

With the influx of increasingly complex aviation systems, there is an increase in complex training (Scerbo, 1996). Additional training requirements will be necessary to assure that the human opera-tor understands the complexity of the automation system, to protect against the probability of safety risks to the optimal performance of the system (e.g., mode errors of omission and commission). Scerbo (1996) compared learning to work with an adaptive automation system with "learning to work with a new team member," and suggested that team-training approaches may useful. Practice sessions are pertinent to team training with automation to increase familiarity with the strengths and weaknesses of the players in the human-automation system. Training should include (1) knowledge of pertinent systems, such as alerts, and techniques for verifying such systems; (2) practice scenarios with "what if" training; and (3) training to stop and consider alternative actions and risk analyses before respond-ing (Mouloua, Gilson, & Koonce, 1997). Today's complex aviation systems require more than mere mechanical manipulation; they demand automation management, risk management, and information management (Parson, 2007).

Automation, information, and risk management may prevent automation surprises—described as a "dangerous distraction" in a highly automated glass cockpit (Parson, 2007). For example, during an approach to land, the automated navigation system (e.g., Garmin G1000 GPS) automatically locates, identifies, and installs the course while switching the active navigation from GPS to another format to relieve the human-operator workload during the busy approach and landing phase of the flight. Without adequate training, the operator may find such changes in format surprising and distracting, or they may not be able to detect inappropriate operations by the automation and therefore, not react and recovery optimally.

The importance of understanding and monitoring automated systems cannot be overemphasized. In December 1995, an American Airlines flight to Cali, Columbia, crashed into a mountainside when the automation (autopilot) was incorrectly programmed by a sleep-deprived crewmember. The autopilot was supposed to lock onto the nearest beacon, called Rozo. By typing in the letter "R," a list of navigation beacons that start with the letter "R" was automatically displayed. The human operator selected the first name on the list, because it is usually the closest beacon, but this time it was not. The human operator failed to notice this error and the autopilot automatically slowly turned the airplane toward the new beacon. By the time the crew realized the error, it was too late to recover, and the plane crashed into a mountain, killing 149 people (Dement & Vaughan, 2000; Endsley & Strauch, 1997; Mouloua, et al., 1997). Knowing that the automation may deviate from its usual behavior and being trained to prepare for certain automation surprises is paramount to safety of the entire system.

8.10 Automation and Aging

Older adults often have trouble interacting with automated teller machines and other computer technologies (Rogers, 1996). Hence, how age factor relates to complex human-automation systems, such as aviation?

A study by Hardy, Satz, E'Elia, and Uchiyama (2007) compared the age differences in pilots' attention and executive ability, information processing and psychomotor speed, verbal and visual learning, and verbal and visual memory. They found that pilots over the age of 40 years exhibited a significant decline in performance. The study suggested that the assessment of individual differences in pilot's age may be necessary in predicting human-automation system performance.

Vincenzi and Mouloua (1998) examined age-related changes in automation-induced complacency. The study indicated that age differences in system-monitoring performance are observed only if the task demands increase to a point where resources available to perform a task are exceeded. While older adults showed a significant decrease in monitoring performance in a multitask condition, single- and dual-task performance were unaffected by adult aging. In the multitask condition, detection rate increased for the younger group as a function of time on task, whereas the older group showed a decline over time. The performance cost of automation-induced complacency was more pronounced in the older group than in the younger age group, only under high workload conditions.

One possible explanation for the differences between age groups on detection of automation failures is that younger participants have better attention allocation control, and are more capable of initiating and sustaining effortful processing. Attention allocation control refers to the extent to which one can direct one's attention to different tasks according to other task demands (Tsang & Voss, 1996). Older participants may have less attention allocation control than younger participants, and may not be able to maintain adequate attention allocation control in the presence of high workload situations involving multiple tasks being performed simultaneously. Furthermore, older participants may not be capable of maintaining constant levels of attention allocation unless a task requires active participation, as in the case of the tracking and resource-management tasks. As a result, performance in the system-monitoring task suffers. Participants who do not allocate sufficient attentional resources to the system-monitoring task may be looking, but not seeing (Molloy, 1996).

Surprisingly, subjective workload data did not appear to vary as a function of age in Vincenzi and Mouloua's (1998) study. Both the older and younger groups experienced comparably high levels of subjective workload. The fact that subjective workload expressed by both the age groups in the dual task condition was not significantly different from that expressed by both the age groups in the multitask condition indicates that subjective workload does not necessarily increase as the workload increases, even though performance on the specified tasks changes significantly. This apparent "saturation" effect suggests a dissociation between the effects of workload on subjective measures and objective performance – neither younger nor older participants experienced a significant increase in the workload from dual to multitask conditions, yet the older participants performed worse.

As mentioned earlier in this chapter, automation-induced complacency represents a human-performance cost of high-level automation. Wickens (1994) suggested that in addition to complacency, automation that replaces human decision-making functions can also lead to loss of situation awareness and skill. Collectively, he referred to these three costs as reflecting "out-of-the-loop unfamiliarity." How can such costs be mitigated? And is the solution the same for younger pilots and older pilots?

One proposed method of maintaining high levels of monitoring performance lies in the implementation of adaptive task allocation or adaptive automation, capable of dynamic, workload-triggered reallocations of task responsibility between human and machine (Hilburn, Jorna, Byrne, & Parasuraman, 1997; Hilburn, Parasuraman, & Mouloua, 1996; Parasuraman, Mouloua, & Molloy, 1996). According to proponents of adaptive systems, the benefits of automation can be maximized and the costs minimized, if tasks are allocated to automated subsystems or to the human operator in an adaptive, flexible manner, rather than in an all-or-none fashion (Rouse, 1988). An ideal situation would be the one where the operator could switch the control of a task from manual to automated, when workload conditions are high (Hilburn, et al., 1996; Parasuraman, et al., 1996). Once the operator's workload is reduced, the operator could then continue to perform the task in manual mode, thereby, maintaining familiarity with the system and preserving the operator's cognitive ability and baseline skill level. Such flexibility of an adaptive automated system could potentially optimize individual differences, such as in younger and older operators. However, one difficulty to this approach is the dissociation found between subjective workload and performance (Vincenzi & Mouloua, 1998). In a situation where the workload is presumably the greatest, older subjects did not report greater workload relative to younger subjects, although their performance was relatively worse. As the fastest growing segment of the population is the elderly population and as the proliferation of automation and technology shows no indication of slowing, further research in this area is needed across a wide spectrum of applications.

8.11 Pilots' Experience and Automation

In a human–machine system performance, operator's experience or expertise plays a major role in performance and safety. In aviation systems, this is important because pilots' experience is very critical to dealing with potential accidents/incidents.

Pilot experience is a potentially powerful moderator of the ability to monitor automation failures. For instance, experience can modulate levels of mental workload. Perhaps, age group differences in automation complacency may be reduced as a function of pilot expertise. Experience is also related to a pilot's mental model. Differences in monitoring strategy owing to mental models could alter automation complacency in both younger and older pilots. More efficient monitoring strategies should lead to better automation-failure detection. On the other hand, experienced pilots may also be more comfortable with automated systems, potentially making them more susceptible to automation complacency.

The examination of age-group differences (or similarities) in higher-order pilot skills appears to be a fruitful path of investigation (Hardy & Parasuraman, 1997). Therefore, to the degree that automation complacency is affected by experience or domain-dependent knowledge, the elucidation of age-related differences in such monitoring situations will be instructive for models of pilot performance in younger and older pilots.

8.12 Conclusions

The use of automation in advanced technical systems is no longer a choice—it is a requirement. As the number of potential operational states proliferates and the time of transition between such states diminishes, the question is not human vs. automatic control, but rather how the automated control is to be managed. In particular, it is a matter of rendering the processes and operations of the automation into "human-scale" (Hancock, 1997). Similar to the way in which any transportation system navigates the physical environment, the challenge will be for the human operator to "navigate" the operational possibilities that the system can explore in relation to the goal imposed and the environment encountered.

References

Automation. (n.d.). *The Merriam-Webster Online® Dictionary*. Retrieved April 25, 2007, from Web site: http://www.m-w.com/dictionary/automation.
Billings, C. E. (1997). *Aviation automation: The search for a human-centered approach*. Mahwah, NJ: Erlbaum.

Bjorklund, C. M., Alfredson, J., & Dekker, S. W. A. (2006). Mode monitoring and call-outs: An eye-tracking study of two-crew automated flight deck operations. *The International Journal of Aviation Psychology, 16*(3), 263–275.

Comstock, J. R., & Arnegard, R. J. (1992). *The multi-attribute task battery for human-operator workload and strategic behavior research* (NASA Technical Memorandum No. 104174). Hampton, VA: NASA Langley Research Center.

Cummings, M. L., & Guerlain, S. (2007). Developing operator capacity estimates for supervisory control of autonomous vehicles. *Human Factors, 49*(1), 1–15.

Davies, D. R., & Parasuraman, R. (1982). *The psychology of vigilance*. London: Academic Press.

Dement, W. C., & Vaughan, C. (2000). Our chronically fatigued syndrome: The not-so-friendly skies. *The promise of sleep* (pp. 217–237). New York: Dell.

Dixon. S. R., & Wickens, C. D. (2006). Automation reliability in unmanned aerial vehicle control: A reliance-compliance model of automation dependence in high workload. *Human Factors, 48*(3), 474–486.

Endsley, M. (1996). Automation and situation awareness. *Automation and human performance: Theory and applications* (pp. 163–181). Mahwah, NJ: Erlbaum.

Endsley, M., & Strauch, B. (1997). Automation and situation awareness: The accident at Cali, Columbia. In R. S. Jensen, & L. Rakovan (Eds.), *Proceedings of the 9th International Symposium on Aviation Psychology* (pp. 877–881). Columbus: Ohio State University.

Hancock, P. A. (1997). *Essays on the future of human-machine systems*. Eden Prairie, MN: Banta (second printing, 1998, CD version, 1999).

Hancock, P., & Warm, J. (1989). A dynamic model of stress and sustained attention. *Human Factors, 31*(5), 519–537.

Hardy, D. J., & Parasuraman, R. (1997). Cognition and flight performance in older pilots. *Journal of Experimental Psychology: Applied, 3*(4), 313–348.

Hardy, D. J., Satz, P., E'Elia, L. F., & Uchiyama, C. L. (2007). Age-related group and individual differences in aircraft pilot cognition. *The International Journal of Aviation Psychology, 17*(1), 77–90.

Hilburn, B., Jorna, P. G., Byrne, E. A., & Parasuraman, R. (1997). The effect of adaptive air traffic control (ATC) decision aiding on controller mental workload. In M. Mouloua, & J. Koonce (Eds.), *Human-automation interaction: Research and practice*. Mahwah, NJ: Erlbaum.

Hilburn, B., Parasuraman, R., & Mouloua, M. (1996). Effects of short- and long-cycle adaptive function allocation on performance of flight-related tasks. In N. Johnston, R. Fuller, & N. McDonald (Eds.), *Aviation psychology: Training and Selection, Proceedings of the 21st Conference of the European Association for Aviation Psychology (EAAP)* (Vol. 2, pp. 347–353). Hants, England: Avebury Aviation.

Jones, L. M., Bowers, C. A., Washburn, D., Cortes, A., & Vijaya Satya, R. (2004). The effect of olfaction on immersion into virtual environments. *Human performance, situational awareness, and automation: Issues and considerations for the 21st century* (Vol. 2, pp. 282–285). Mahwah, NJ: Erlbaum.

Kaber, D. B., Perry, C. M., Segall, N., McClernon, C. K., & Prinzel, L. J. III, (2006). Situation awareness implications of adaptive automation for information processing in an air traffic control-related task. *International Journal of Industrial Ergonomics, 36*, 447–462.

Kahneman, D., & Treisman, A. M. (1984). Changing views of attention and automaticity. In R. Parasuraman, & D. R. Davies (Eds.), *Varieties of attention*. Orlando, FL: Academic Press.

Madhavan, P., Wiegmann, D. A., & Lacson, F. C. (2006). Automation failures on tasks easily performed by operators undermine trust in automated aids. *Human Factors, 48*(2), 241–256.

Molloy, R. (1996). *Monitoring automated systems: The role of display integration and redundant color coding*. Unpublished doctoral dissertation, The Catholic University of America, Washington, DC.

Mouloua, M., Gilson, R., & Koonce, J. (1997). Automation, flight management and pilot training: issues and considerations. In R. Telfer, & P. Moore (Eds.), *Aviation training: Learners, instruction and organization* (pp. 78–86). Brookfield: Avebury Aviation.

Mouloua, M., & Koonce, J. (Eds.). (1997). *Human-automation interaction: Research and practice*. Mahwah, NJ: Erlbaum.

Mouloua, M., & Parasuraman, R. (1994). *Human performance in automated systems: Current research and trends*. Mahwah, NJ: Erlbaum.

Parasuraman, R. (1987). Human-computer monitoring. *Human Factors, 29*, 695–706.

Parasuraman, R., & Hancock, P. A. (2001). Adaptive control of mental workload. In P. A. Hancock, & P. A. Desmond (Eds.), *Stress, workload, and fatigue* (pp. 305–320). Mahwah, NJ: Erlbaum.

Parasuraman, R., Molloy, R., & Singh, I. L. (1993). Performance consequences of automation-induced complacency. *International Journal of Aviation Psychology, 3*, 1–23.

Parasuraman, R., & Mouloua, M. (1987). Interaction of signal discriminability and task type in vigilance decrement. *Perception & Psychophysics, 41*(1), 17–22.

Parasuraman, R., & Mouloua, M. (1996). *Automation and human performance: Theory and applications*. Mahwah, NJ: Erlbaum.

Parasuraman, R., Mouloua, M., & Molloy, R. (1994). Monitoring automation failures in human-machine systems. In R. Parasuraman, & M. Mouloua (Eds.), *Human performance in automated systems: Current research and trends* (pp. 45–49). Mahwah, NJ: Erlbaum.

Parasuraman, R., Mouloua, M., & Molloy, R. (1996). Effects of adaptive task allocation on monitoring of automated systems. *Human Factors, 38*(4), 665–679.

Parasuraman, R., & Riley, V. (1997). Humans and automation: Use, misuse, disuse, abuse. *Human Factors, 39*(2), 230–253.

Parasuraman, R., Sheridan, T., & Wickens, C. (2000). A model for types and levels of human interaction with automation. *IEEE Transactions on Systems, Man, and Cybernetics–Part A: Systems and Humans, 30*(3), 286–297.

Parson, S. (2007). Beyond the buttons: Mastering our marvelous flying machines. *FAA Aviation News, March/April*, 10–13. Retrieved Saturday, April 21, 2007 from the PsycINFO. http://www.faa.gov/news/aviation_news/2007/media/MarchApril2007.pdf.

Prince, C., Ellis, E., Brannick, M., & Salas, E. (2007). Measurement of team situation awareness in low experience level aviators. *International Journal of Aviation Psychology, 17*(1), 41–57.

Riley, V. (1994). A theory of operator reliance on automation. In M. Mouloua, & R. Parasuraman (Eds.), *Human performance in automated systems: Current research and trends* (pp. 8–14). Mahwah, NJ: Erlbaum.

Rogers, W. (1996). Assessing age-related differences in the long-term retention of skills. *Aging and skilled performance: Advances in theory and applications* (pp. 185–200). Mahwah, NJ: Erlbaum.

Rouse, W. B. (1988). Adaptive aiding for human/computer control. *Human Factors, 30*, 431–438.

Rovira, E., McGarry, K., & Parasuraman, R. (2007). Effects of imperfect automation on decision making in a simulated command and control task. *Human Factors, 49*(1), 76–87.

Sarter, N. B. (1996). Cockpit automation: From quantity to quality, from individual pilot to multiple agents. In R. Parasuraman, & M. Mouloua (Eds.), *Automation and human performance: Theory and applications* (pp. 267–280). Mahwah, NJ: Erlbaum.

Sarter, N., & Woods, D. (1992). Pilot interaction with cockpit automation: Operational experiences with the flight management system. *International Journal of Aviation Psychology, 2*(4), 303–321.

Scerbo, M. W. (1996). Theoretical perspectives on adaptive automation. In R. Parasuraman, & M. Mouloua (Eds.), *Automation and human performance: Theory and applications* (pp. 37–63). Mahwah, NJ: Erlbaum.

Scerbo, M., & Mouloua, M. (Eds.). (1998). *Automation technology and human performance: Current research and future trends*. Mahwah, NJ: Erlbaum.

Sheridan, T., & Parasuraman, R. (2000). Human versus automation in responding to failures: An expected-value analysis. *Human Factors, 42*(3), 403–407.

Singh, I. L., Molloy, R., & Parasuraman, R. (1993). Individual differences in monitoring failures of automation. *Journal of General Psychology, 120*, 357–373.

Tsang, P. S., & Voss, D. T. (1996). Boundaries of cognitive performance as a function of age and flight experience. *International Journal of Aviation Psychology, 6,* 359–377.

Vincenzi, D., & Mouloua, M. (1998). Monitoring automation failures: Effects of age on performance and subjective workload. In M. Scerbo, & M. Mouloua (Eds.), *Automation technology and human performance: Current research and future trends.* Mahwah, NJ: Erlbaum.

Vincenzi, D. A., Mouloua, M., & Hancock, P. A. (2004). (Eds.). *Human performance, situation awareness and automation: Current research and trends* (Vols. I and II). Mahwah, NJ: Erlbaum.

Warren-Noell, H. L., Kaber, D. B., & Sheik-Nainar, M. A. (2006). Human performance with vocal cueing of automation state changes in an adaptive system. *Proceedings of the Human Factors and Ergonomics Society 50th Annual Meeting.* Santa Monica, CA: Human Factors and Ergonomics Society.

Washburn, D. A., & Jones, L. M. (2004). Could olfaction displays improve data visualization? *IEEE Computing in Science & Engineering, 6*(6), 80–83.

Wickens, C. D. (1992). *Engineering psychology and human performance.* New York: Harper-Collins.

Wickens, C. D. (1994). Designing for situation awareness and trust in automation. In *Proceedings of the IFAC Conference on Integrated Systems Engineering.* Baden-Baden, Germany.

Wickens, C., Vidulich, M., & Sandry-Garza, D. (1984). Principles of S-C-R compatibility with spatial and verbal tasks: The role of display-control location and voice-interactive display-control interfacing. *Human Factors, 26*(5), 533–543.

Wiener, E. L. (1981). Complacency: Is the term useful for air safety? *Proceedings of the 26th Corporate Aviation Safety Seminar* (pp. 116–125). Denver, CO: Flight Safety Foundation.

Wiener, E. (1988). Cockpit automation. In E. Wiener (Ed.), *Human factors in aviation.* San Diego, CA: Academic Press.

Woods, D. D. (1996). Decomposing automation: Apparent simplicity real complexity. In R. Parasuraman, & M. Mouloua (Eds.), *Automation and human performance: Theory and applications* (pp. 3–17). Mahwah, NJ: Erlbaum.

9

Team Process

Katherine A. Wilson
University of Central Florida

Joseph W. Guthrie
University of Central Florida

Eduardo Salas
University of Central Florida

William R. Howse
U.S. Army Research Institute

9.1 Introduction

Teams have been a widely used organizational strategy within the aviation community as a means to improve safety. Team members serve as redundant systems to monitor and backup others, for example, during periods of high workload. To maintain the effectiveness of these teams, team-training strategies are used to develop key team processes. Prince and Salas (1999) offered a review on what has been done in terms of team processes and their training in aviation. Their review focused on the theoretical foundations of team training in aviation, specifically examining the input–throughput–output (IPO) models. They suggested that once a theoretical model has been selected, team training can be developed, implemented, and the performance measured. Prince and Salas discussed a number of measurement instruments, including the Cockpit Management Attitudes Questionnaire (CMAQ) developed by Helmreich (1984), and Helmreich et al. (1986) to measure the pilots' reactions to training, and the line/line-oriented flight training (LOFT) worksheet and targeted acceptable responses to generated events (TARGET) checklist to guide the observations of behaviors. Their review concluded with a discussion on the necessary components of team training, which include training tools (e.g., team task analysis, feedback, simulations), methods (e.g., lectures, videotapes, role plays), and content (e.g., knowledge, skills, abilities), as defined by Salas and Cannon-Bowers (1997).

This review takes a similar approach; however, it recognizes that crew resource management (CRM) training is the most widely used team-training strategy in aviation and therefore, will focus on this strategy throughout many of the examples presented. This chapter will be developed based on Prince and Salas (1999) review, by discussing the advancements made in terms of team processes and training in aviation, as well as expand in some areas where a deeper discussion of these issues is needed. The chapter proceeds as follows: First, we have discussed the theoretical foundations for team training in aviation, including leadership, shared cognition, team situation awareness (TSA), and multicultural teams. Second, we have reviewed the literature to determine the types of team process and the performance-measurement instruments being used by the aviation community today. Third, we have discussed a number of tools that can be used to improve the outcomes of team training. Fourth, we have discussed about CRM training specifically, and have argued for the use of other instructional strategies (e.g., scenario-based training [SBT], metacognitive training) in conjunction with CRM training to improve its effectiveness. Finally, we have presented the conclusion with several future needs for the aviation community.

9.2 Theoretical Developments

While much of the early research on team processes and training in aviation were based on general group research (e.g., IPO models), several new theories have emerged in the literature. In this chapter, we have discussed four theories, as we feel that these are the most relevant to team processes and training carried out today. Specifically, we have discussed the theories of leadership, shared cognition, TSA, and multicultural teams.

9.2.1 Leadership

Leadership in the cockpit is critical. While an authoritarian leadership style may have been more common years ago, today, the cockpit (at least in most Western countries) is more egalitarian. This leadership style can be characterized by leader–member exchange (LMX) theory. Although LMX theory has just recently been applied to the team level, its relevance to the cockpit is evident. The LMX theory posits that effective leadership occurs when the leader is able to develop and foster a mature working relationship with the individual team members (Graen & Uhl-Bien, 1995). Relationships are built on team-member competence, dependability, and interpersonal compatibility (Duchon, Green, & Tabor, 1986; Graen & Cashman, 1975; Graen & Scandura, 1987). Once the relationship is established, it is characterized by a high degree of mutual trust, respect, and open communication (e.g., Dansereau, Graen, & Haga, 1975; Dienesch & Liden, 1986; Graen & Uhl-Bien, 1995) which are critical in the cockpit. This theory is different from other leadership theories, in that it focuses on the dyadic exchange between the leader and member (e.g., captain and first officer) as opposed to the traits, behaviors, or situational styles of the leader. It could also be argued that the development of a good working relationship between the leader and the team member will foster the development of shared mental models (SMMs) and shared cognition (see Salas, Burke, Fowlkes, & Wilson, 2004). We have discussed the theories supporting these factors in the subsequent sections.

9.2.2 Shared Cognition

Pilots perform cognitive tasks every day. They detect and recognize cues in the environment, acquire knowledge, remember relevant information, plan, make decisions, and solve problems (Cooke, Salas,

Kiekel, & Bell, 2004). However, to accomplish these cognitive tasks as a team takes more than just the sum of individual cognitions. To a certain extent, team or shared cognition is necessary, which is the result of the interplay between the individual-level cognitions and team-process behaviors (Cooke et al., 2004). Shared cognition is developed through the exchange of individual cognitions by communicating and coordinating as a team (Stout, Cannon-Bowers, & Salas, 1996). The outcome of shared cognition is improved team performance.

Shared cognition among crew members is critical owing to the dynamic, ambiguous nature in the cockpit. Cannon-Bowers and Salas (2001) offered several suggestions for the use of shared cognition to explain team performance. First, shared cognition can help to explain how team members operating within effective teams interact with each other. Second, shared cognition allows the team members to interpret cues similarly, make decisions that are compatible, and take appropriate action. Finally, shared cognition can help in the diagnosis of deficiencies within a team and also provide insight on how to correct these deficiencies.

For teams to be effective, four types of information must be shared: (1) task-specific knowledge (i.e., knowledge pertaining to a specific task), (2) task-related knowledge (i.e., knowledge pertaining to a variety of tasks' processes), (3) knowledge of teammates (i.e., understanding team members' preferences, strengths, weaknesses, and tendencies), and (4) attitudes/beliefs. Task-specific knowledge allows the team members to share the expectations and take action in a coordinated manner without explicit communications (Cannon-Bowers, Salas, & Converse, 1993). Task-related knowledge, such as a shared understanding of teamwork, contributes to a team's ability to successfully complete a task (e.g., Rentsch & Hall, 1994). Knowledge of teammates allows the members to compensate for the others, predict the actions they will take, provide information without being asked, and allocate resources according to the team member's expertise. Finally, shared attitudes/beliefs allow the team members to have compatible perceptions of the task and/or environment that will lead to more effective decisions, and potentially consensus cohesion and motivation. Bressole and Leroux (1997) investigated the link between shared cognition and environmental resources used by air-traffic controllers to update a mutual cognitive environment, as well as its relation to the design and implementation of new technological tools. While research related to shared cognition is limited, its relevance to effective team performance in the cockpit (and other areas within aviation) is evident.

9.2.3 Shared Mental Models

A critical component of shared cognition in the cockpit is SMMs, which allows the crew members to anticipate and predict the needs of others (Orasanu, 1994). SMMs are the cognitive knowledge structures shared by all team members, which are related to the team (e.g., team functioning, expected behaviors) or the task (e.g., equipment needed to accomplish task; Mathieu, Heffner, Goodwin, Salas, & Cannon-Bowers, 2000). For example, SMMs can contain information pertaining to the team's goals and expectations, team members' tasks and task environment, and the method(s) by which the team will coordinate to achieve their goals (Cannon-Bowers, Tannenbaum, Salas, & Volpe, 1995). In other words, SMMs serve as a heuristic function, such that once developed, they enable the team to implicitly coordinate and more efficiently perceive, interpret, and respond while operating within a dynamic environment (Blickensderfer, Cannon-Bowers, & Salas, 1998; Cannon-Bowers et al., 1993; Entin & Serfaty, 1999; Shlecter, Zaccaro, & Burke, 1998).

Within the aviation community, it has been shown that the sharing of mental models results in improved team performance (e.g., fewer operational errors, better coordination), even under conditions of fatigue (Foushee, Lauber, Baetge, & Acomb, 1986). Additional research suggests that when team members have SMMs, this may lead to more effective communication, improved performance (e.g., Griepentrog & Fleming, 2003; Mohammed, Klimoski, & Rentsch, 2000), and a willingness to work together in the future (Rentsch & Klimoski, 2001). It has also been argued that SMMs lead to improved and more efficient communication strategies, and are thus used during periods of high workload (Stout, Cannon-Bowers, Salas, & Milanovich, 1999). This is critical in the cockpit for both routine and nonroutine situations, where errors can have severe consequences.

9.2.4 Team Situation Awareness

Much of the studies available have examined situation awareness at the individual level (see Vidulich, Dominguez, Vogel, & McMillan, 1994). Individual situation awareness is often a skill taught as a part of team-training programs in aviation, such as CRM training. However, the notion of TSA, a relatively new concept to the aviation community, is also critical and should not be overlooked. TSA is composed of two parts: (1) the combination of each team member's individual SA (based on preexisting knowledge bases and assessments of patterns and cues in the environment) and (2) the degree of shared understanding between the team members (as developed though compatible mental models and team interaction behaviors) (Salas, Muniz, & Prince, 2001). TSA is constantly in a dynamic state, and at any given time, is proposed to be affected by several factors—the situational context (e.g., aircraft type, task/mission to be performed), environmental elements (e.g., weather conditions, terrain), and temporal elements (e.g., high workload, time pressure). While research on TSA among aircrews is limited, there are few researches available. For example, Proctor, Panko, and Donovan (2004) used TARGET methodology to successfully train TSA in the cockpit. Furthermore, CRM research indicated four skills necessary for supporting TSA, namely, preparation, communication, leadership, and adaptability (Prince & Salas, 2000). By having TSA, crews will be better able to communicate and coordinate. Thus, this will foster the development of accurate SMMs and subsequent shared cognition.

9.2.5 Multicultural Teams

Multicultural teams in aviation are increasingly more common. Some have argued that the cockpit is a "culture-free zone" indicating that regardless of the nationality, pilots share the universal task of safe flight from point A to B, and conduct this task in a similar professional manner (e.g., Helmreich, Wilhelm, Klinect, & Merritt, 2001). However, some researches may suggest otherwise. For example, one study conducted by Helmreich and Merritt (1998) suggested that while many cultures agree on issues such as the importance of crew communication, coordination, and preflight briefings, there is greater disagreement on issues regarding junior crewmembers' assertiveness to speak up or question authority, the influence of personal problems on performance, and the likelihood of making errors in judgment during emergency situation. While some cultures may not see any consequences related to such attitudes, these differences severely affect the safety of flight operations. This was evident on a Korean Airlines flight when a Canadian captain and Korean copilot disagreed on the decision to land on a rain-slicked runway and physically fought for control of the aircraft (Westrum & Adamski, 1999). Though the aircraft overran the runway and caught fire, fortunately, the accident resulted in no fatalities.

There are four dimensions of culture frequently studied and discussed in the literature, which may influence teamwork in the cockpit—power distance, masculinity/femininity, individualism/collectivism, and uncertainty avoidance (Hofstede, 1991). Power distance can be defined as the degree to which individuals with less power in an organization (i.e., junior crewmembers) expect and accept their inequality. Second, masculinity is the degree to which social gender roles are distinct (e.g., male pilots are assertive and female pilots are submissive), whereas femininity is the degree to which social gender roles overlap (e.g., male and female pilots are equal). Third, individualism is the degree to which interests of individuals come first (e.g., individual pilot goals are considered more important), whereas collectivism is the degree to which the interests of the group come first (e.g., goals of crewmembers considered more important). Finally, uncertainty avoidance is defined as the degree to which individuals handle the unknown (e.g., how pilots react when faced with the potential for novel situations). Helmreich (1994) demonstrated how three of the abovementioned dimensions (i.e., power distance, collectivism, and uncertainty avoidance) contributed to one aircraft accident. In 1990, Avianca Flight 52 crashed in New York, because the Columbian pilot reported a low fuel state to air-traffic control, but failed to declare an emergency (www.airdisaster.com). As such, the urgency of the situation was not communicated and a catastrophic ending occurred. An investigation by Kuang and Davis (2001) supported the influence of culture in cockpit errors, arguing that collectivist, high power distance, and high uncertainty avoidance cultures commit more errors.

As the aviation community becomes more multicultural, especially within the organizations, the challenge thus turns to determine how culture influences teams in the cockpit and how to train teams to minimize the negative consequences. CRM training is one way to minimize the consequences of culture in the cockpit. While researches have been conducted on CRM training within the United States, relatively limited researches have looked at the influences of national culture on CRM training. The limited researches that were conducted suggest that CRM training does not apply well to other countries. Although CRM training is adapted by each nation and organization, some of the concepts (e.g., leadership, assertiveness) promoted by CRM are not advocated by some cultures (Helmreich & Merritt, 1998). For example, assertiveness is influenced by several of Hofstede's dimensions. Cultures high in power distance would probably expect the junior crewmembers to do what they are told and speak up only when asked to do so (Hofstede, 1991). However, CRM training includes assertiveness training for junior crewmembers as a result of several accidents (e.g., Air Florida). In addition, high-masculinity cultures might look negatively upon females in the assertive role of captain. These differences can have grave consequences, especially when crewmembers within the organization are multinational.

Despite the resistance to CRM by some cultures, the impact of cultural differences on the safety of flight cannot be ignored. However, some have argued that the professional culture weighs heavily in the aviation community and overrides one's national culture in the cockpit (Hutchins, Holder, & Pérez, 2002). If this is true, then the evidence supporting this is limited. Additional researches are needed which must focus on how culture (i.e., national, organizational, and professional) influences the crews' behaviors and how CRM training can help to minimize the consequences.

9.2.6 Summary of Theoretical Developments

Theoretical developments in the aviation community over the last decade have shifted its focus from team processes and behaviors to team cognitions. New theories regarding team leadership have emphasized the importance of an egalitarian atmosphere, where team members are not afraid to speak up and assist the leader when team performance is at risk. Furthermore, the advancements of research in terms of SMMs and shared cognition have helped us to better understand how these cognitive processes influence the team performance in the cockpit. By improving TSA in the cockpit, SMMs and shared cognition will be fostered. Finally, the multicultural aspect of organizations in aviation has led researchers to further investigate on how national culture impacts teamwork. However, additional researches are needed to fully understand the impact of organizational and professional cultures as well.

9.3 Team Process/Performance Measurement

For years, researchers have been trying to understand and train team processes. However, there have been several road blocks along this path. Because of the dynamic nature of teams and their tasks, determining what needs to be measured is difficult (Prince & Salas, 1999). In addition, measurement is made more difficult, because team performance is made up of both individual team-member actions and the coordinated actions of the team members. Another difficulty that researchers have encountered has been the development of valid and reliable measurement tools. Much of the researches measuring team processes utilized self-report instead of objective measures. However, Prince and Salas (1999) discussed measurement tools such as the updated Line/LOFT worksheet, Line/LOS, and TARGET as improvements made to traditional self-report measures.

In the last decade, a greater emphasis has been placed on the development and utilization of more objective measures of team process. The result of this emphasis has been the development of new measurement tools and the application of existing tools to the teams. In the subsequent sections, we have discussed a number of measurement tools currently being used in team research. While some focus on understanding the different team processes that have been trained (e.g., communication analysis), the others focus on whether the structure of the training system could be improved (e.g., line operations safety audit [LOSA]).

9.3.1 Nontechnical Skills

In the mid 1990s, the European Joint Aviation Authorities (JAA) put forth legislation to developing a generic method in which nontechnical skills trained through CRM could be evaluated (Flin et al., 2003). Nontechnical skills include the cognitive and social skills developed in flight crews which are not directly related to control of the aircraft, management of flight systems, or standard operating procedures. The nontechnical skills evaluation method (NOTECHS) was developed based on a review of the existing systems and literature, and discussions with subject experts. NOTECHS consists of four categories—two social skills (i.e., cooperation, and leadership and management) and two cognitive skills (i.e., situation awareness and decision making). Each of the four categories is made up of a subset of elements which corresponded to the behaviors expected to be demonstrated by the well-trained, high performing aircrews (see Flin et al., 2003). For example, the subelements of cooperation include team-building and maintaining, consideration of others, support of others, and conflict resolution. The reliability, validity, and usability of NOTECHS have been supported by instructors reporting high levels of satisfaction in terms of the rating system and consistency of the methodology (Flin et al., 2003). In addition, ratings provided by instructors were at an acceptable level of accuracy with those of the trained experts, and no cultural differences were found. These initial results indicate the usefulness of NOTECHS for evaluating the team processes and performance.

9.3.2 Communication Analysis

In recent years, analysis of team communication has been utilized to determine the types of communications that high performing teams, such as cockpit crews, use. Priest et al. (in preparation) suggest that the importance of understanding team communications is owing to the insight that can be gained into many different team processes, such as decision making, coordination, teamwork, leadership, back-up behavior, and situation awareness. Although much of the literature on communication focuses primarily on the amount and frequency of communication, more recent studies have focused on the actual content and sequence of communication through elements of exploratory sequential data analysis (ESDA, Bowers, Jentsch, Salas, & Braun, 1998; Priest, Burke, Guthrie, Salas, & Bowers, in preparation). ESDA is used as a means to analyze the recorded data in which temporal information has been preserved (Sanderson & Fisher, 1992). For example, Bellorini and Vanderhaegen (1995) used the ESDA techniques to analyze the communication between air-traffic control teams sequencing the landing patterns of aircraft. In addition, analysis of communication of cockpit crews by Bowers et al. (1998) investigated the differences between high- and low-performing teams in terms of the types of communications used. As communication is a focal point of CRM training, analysis of team communications can provide vital information regarding the types of communications that teams should be trained to engage in.

9.3.3 Concept Mapping

Swan (1995) described content mapping as a knowledge-elicitation method utilizing graphical representations of the knowledge structure and content of an individual within a domain. While concept mapping was introduced as a tool for use on individuals, in recent years, it has been adapted for use in teams to aid in the measurement of shared team knowledge. Concept mapping allows researchers to create representative mental models for domain-specific concepts (Cooke, Salas, Cannon-Bowers, & Stout, 2000). The different concepts contained in a map show the relationship of one concept to another,

provide infor   vidual team members and teams, and provide an
understandir   lead to successful performance of individuals and
teams. For e   ave been used to evaluate pilot knowledge acquisition over th   ɔ, Hoeft, Kochan, & Jentsch, 2005). Their research
suggests th   ent written exams can be beneficial to pilots.

9.3.4 P

Another   oncept mapping is Pathfinder. Pathfinder is used to
provide   related concepts through participant ratings of paired
compar   ith, Johnson, and Acton (1991) argued that a minimum
of 15–3   rger number of concepts used, resulting in a better structure pr   made on each of the comparisons, Pathfinder computes
an alg   the structure of the concepts and the links between them.
Previc   show changes in knowledge structure and performance
after   wers, 1995). In aviation, for example, Pathfinder analysis
has l   n novice- and expert-pilot priority ratings of information
pert   Beringer, & Lamonica, 2001).

9.   dit

L(stry to assess the effects of CRM training in the cockpit (Croft,
2   ɔ ride along in the cockpit and observe how the pilots and
c   their efforts to complete the task. The judges rate the crew's
                                       (observed performance had safety implications), (2) marginal
                                       equate), (3) good (observed performance was effective), and
                                       e was truly noteworthy). Following performance monitoring,
                                       nd their understanding of the processes at work while decisions
                                       lition to recording errors made during flight, LOSA also looks at
how the   f their errors. All these informations are combined to determine
any weaknesses or critical   hat may exist within the training program. The results of LOSA
provide the airlines with practical information to help in (1) identification of threats in the airline's
operating environment, (2) identification of threats from within the airline's operation, (3) assessing the
degree of transference of training to the airline, (4) checking the quality and usability of procedures,
(5) identification of design problems in the human–machine interface, (6) understanding the pilots'
shortcuts and workarounds, (7) assessment of safety margins, (8) providing a baseline for organizational
change, and (9) providing a rationale for allocation of resources (University of Texas, 2005). To make the
LOSA data reliable and valid, the researchers developed a list of 10 operating characteristics for LOSA
(see University of Texas, 2005).

Helmreich, Merritt, and Wilhelm (1999) discussed the use of LOSA to document threat and error,
and developed models based on LOSA data. In total, 3500 flights were observed for threats and errors to
safety. This data provided a representation of normal operations during flight and allowed the flight
community to approximate the amount of risk associated with a certain movement or a particular environment. In addition, the data not only provided strengths, where current training programs are found
to be successful in teaching cockpit behaviors, but also showed areas where the flight crews made errors
indicating where the training should be enhanced.

The different types of errors committed during flight can be traced back to decrements in CRM behaviors (Helmreich, Klinect, & Wilhelm, 1999). Procedural errors indicate the failure to use CRM behaviors,
such as monitoring and cross checking. Communication errors point to other failures in CRM behaviors, such as failing to share mental models and verifying exchanged information. Operational decision

errors can be a sign of failure "to exchange and evaluate perceptions of threat in the operating environment" (Helmreich et al., 1999, p. 680). Not only does this data highlight where current training is lacking, but it also reinforces the need for CRM training.

9.3.6 Summary of Team Process Measurement

The measurement tools discussed here follow the recommendation of Prince and Salas (1999) that tools should be (1) well designed, (2) easy to use, (3) provide reliable information on crew interactions, and (4) act as a useful guide for specific, useful feedback to the teams. NOTECHS and LOSA require rigorous training of raters and judges to ensure that the data is not biased. In addition, tools such as communication analysis, concept mapping, and Pathfinder allow for more objective measures of team processes, possibly increasing the reliability and validity of the data. The use and development of these types of measurement tools allows the researchers to gain a clearer insight into the team process and provide a better foundation to develop training.

9.4 Tools for Aviation Training

There are a variety of tools available to enhance the effectiveness of team training in aviation. Here, we have discussed three tools, simulations (full motion simulators, PC-based simulators, and part-task trainers), the rapidly reconfigurable event-set-based line-oriented evaluations generator, and distributed training tools.

9.4.1 Simulations

Simulations have been used in the aviation community for almost a century. From the development of the Link trainer in 1928 to the state-of-the-art life-size, full-motion simulators, simulations have been used to train pilots to fly in a safe environment. As such, we would argue that the aviation community is one of the biggest proponents of simulation for training. Flight simulations have been the front runner in training and evaluation technology in a number of aviation-related areas, including adaptive decision making (e.g., Gillan, 2003), performance (Aiba, Kadoo, & Nomiyama, 2002), response time (Harris & Khan, 2003), performance under workload (Wickens, Helleberg, & Xu, 2002), and team processes (Prince & Jentsch, 2001). Here, we have discussed three types of simulators used within the aviation community.

9.4.1.1 Full Motion Simulators

Full motion flight simulators are used by commercial airlines and military branches to train the pilots to develop skills necessary in the cockpit. These simulators are so realistic that pilots can become fully qualified to fly an aircraft before even setting foot in the cockpit. Simulators can recreate various locations, weather (e.g., wind, turbulence, visibility), and equipment failures. Furthermore, full motion simulators can simulate 3 or 6 degrees of freedom data (roll, pitch, yaw, x, y, and z motion) to create a realistic moving environment. While these simulators are beneficial, they are quite expensive to set up and maintain. Furthermore, the high physical fidelity of these simulators is not always necessary or more beneficial for training nontechnical skills (such as CRM skills). We have discussed the use of PC-based simulators in the subsequent section.

9.4.1.2 PC-Based Trainers

It has often been the assumption that for a simulation to be effective in helping trainees to learn, it must (as much as possible) recreate the physical task environment (i.e., have all the "bells and whistles"). However, research also suggests that this high physical fidelity (which typically gets favorable reactions from trainees) may not always be necessary to achieve learning goals (e.g., Gopher, Weil, & Bareket,

1994; Jentsch & Bowers, 1998). In one study, it was found that when a high physical fidelity simulation was used to train pilots, the training did not transfer nor had a minimal effect on on-the-job performance (Taylor, Lintern, & Koonce, 1993). Rather, learning requires the psychological (i.e., cognitive) fidelity of the simulation to be high. High psychological fidelity is essentially the notion that the simulation will require trainees to progress through the same cognitive processes that would be required to complete the task in the real world. Low physical fidelity simulators (e.g., PC-based off-the-shelf games; personal computer aviation training devices [PCATD]) may be just as effective, as long as the level of psychological fidelity meets the training needs (Bowers & Jentsch, 2001). In general, PC-based flight simulators have been successful in the transfer of flight skills to the cockpit (e.g., technical skills, Dennis & Harris, 1998; instrument flight skills, Koonce & Bramble, 1998; Taylor et al., 2002). As such, making the leap from technical-skills training to nontechnical-skills (i.e., CRM) training using these PC-based systems is natural. Several studies have proven this for training and eliciting complex skills, such as decision making (e.g., Jentsch & Bowers, 1998; Prince & Jentsch, 2001; Gopher et al., 1994). For example, studies by Stout, Cannon-Bowers, Salas, and Milanovich (1990) and Lassiter, Vaughn, Smaltz, Morgan, and Salas (1990) showed initial support for the use of low-fidelity simulations for eliciting CRM competencies when completing a task. Additional studies have been found, in which CRM skills were specifically trained (vs. used to elicit CRM) using an aviation computer game (Baker, Prince, Oser, & Salas, 1993). Ninety percent of the participants in this study agreed that the PC-based system demonstrated the importance of CRM and could be used for training CRM skills. Finally, Brannick, Prince, Salas, and Stout (2005) found that CRM skills learned using Microsoft Flight Simulator do transfer to a high-fidelity motion simulator.

Overall, the research presented here suggests that the use of PC-based simulators is a viable way to train, practice, and provide feedback regarding CRM skills. These PC-based systems (such as Microsoft's Flight Simulator; see Bowers & Jentsch, 2001 for a review on others) offer a low-cost alternative to training some of the necessary skills for flight. This evidence is not to state that physical fidelity is not important, though it is, in some cases. However, the level of simulation fidelity needed to achieve learning goals should be determined by the requirements of the task (i.e., both cognitive and behavioral) and the level needed to support learning (Salas, Bowers, & Rhodenizer, 1998).

9.4.1.3 Part-Task Trainers

Part-task trainers are similar to PC-based simulators. Part-task trainers are a method of training that teaches part of a task on a device that represents the actual equipment used (Eurocontrol, 2005). Specifically, the learning objective is broken into separate sections and practiced individually. After all the parts of the task have been mastered individually, the sections are joined in different combinations until the learning objective as a whole is completed. Another way to use part-task trainers is to allow trainees "free play" to learn the different functions of a system. Noble (2002) reported that part-task trainers that provide high fidelity but limit the equipment and controls to what is found in the cockpit and do not include other real-world factors, can be effective in training pilots ab initio. Specifically, these part-task trainers can be effective in increasing the procedural knowledge, safety, and learning from mistakes.

Part-task trainers have been especially useful as a part of integrative training programs. Martin (1993) described the Integrated Learning System that used a combination of computer-based training, part-task trainers, and flight simulators in ab initio pilot training. Ford (1997) discussed a similar training system that included part-task trainers, simulators, computer-based and classroom training. The part-task trainer provided simulated practice on the aircraft's common control unit and tactical situation display.

There are several advantages and disadvantages in using part-task trainers (Eurocontrol, 2005). The advantages include the ability to learn the functionality of a system without the use of the actual system. Part-task trainers can also be used to provide practice before the real equipment is used. In addition, they can be designed as a single component of a larger training system. Although part-task trainers can save money, as all aspects of training do not have to be completed on a full-scale simulator or the actual aircraft, they still must be designed and produced. The scenarios created in the part-task trainer must be realistic to be effective, which takes time and effort.

9.4.2 Rapidly Reconfigurable Event-Set-Based Line-Oriented Evaluations Generator

Line-oriented evaluations (LOEs) are used in the Advanced Qualification Program (AQP) to determine performance levels and proficiency. LOEs have traditionally been developed and approved on an individual basis, that is, newly constructed LOEs or changes to existing LOEs had to be approved by the Federal Aviation Administration (FAA; http://pegasus.cc.ucf.edu/~rrloe/About_RRLOE.html). Rapidly reconfigurable event-set-based line-oriented evaluations generator (RRLOE) proposes that instead of using a small number of LOEs which creates the possibility of a reduction in validity of the training or the necessity to have a new LOE approved, a basic LOE structure be developed and approved by the FAA, upon which unique LOEs could be created without the need for further approval.

For such a system to be practical, researchers have worked to overcome obstacles in four different areas: skill-based training, event-based assessment, applied research issues, and human factors aspects of the software (http://pegasus.cc.ucf.edu/~rrloe/About_RRLOE.html).

Although much of the knowledge and skills necessary for high performance have been identified for use in skill-based training, proper practice and feedback mechanisms must also be present. Therefore, one of the primary initiatives of the RRLOE has been to provide appropriate practice and feedback within the scenarios. In addition to the skill-based training, researchers have developed RRLOE with mechanisms for event-based assessment to increase the systems practicality. Therefore, performance measures are linked to events within the scripted scenarios. In addition, the RRLOE is meant to increase the speed with which LOEs are developed.

The RRLOE must also be valid and realistic. The RRLOE has adopted the "Domino-Method" to track the realism of the assessment system (see Bowers, Jentsch, Baker, Prince, & Salas, 1997), allowing better tracking of flight-relevant parameters than what could be done by a human developer. In addition to the realism of the assessment system, the RRLOE must also realistically show the impact of the weather and provide a range of difficulty in the scenarios. To overcome this obstacle, the RRLOE is equipped with a database of weather scenarios that vary in different critical elements. Furthermore, to ensure that the scenarios created by RRLOE are within the desired range of difficulty, a mathematical model was created which was executed for each LOE. Lastly, the software developed for RRLOE is based on human factors principles of computer–human interface design, and is directly related to the operational environment of the airline industry.

9.4.3 Distributed Training Tools

It could be argued that for each commercial or military flight, there are various colocated and distributed team members working together to achieve the goals of flight. For example, within commercial aviation, the team members could consist of the cockpit crew, cabin crew, ATC, and flight dispatchers. In the military, team members could comprise cockpit crew, navigation crew, ATC, and ground controllers. Some are located within the aircraft, whereas others are located at various locations, sometimes separated by a cockpit door, in separate aircraft (e.g., formation flight), or even hundreds of miles away in a ground location where face-to-face communication and coordination is not possible. To accurately simulate this environment, trainees should also be distributed. However, this poses a challenge for the trainers. There are a number of tools available to assist in the training of distributed teams. Bell (1999) defined the distributed mission training (DMT) as a set of advanced generation imaging technologies, high-resolution displays, and secure distributed networks which allow geographically dispersed virtual training platforms for mission-critical team training. Furthermore, training strategies such as guided practice and cross-training can be effectively investigated and taught in DMT systems.

The DMT is useful in aviation team training for both military and commercial pilots. For military pilots, DMT provides a tool that allows the pilots to train on necessary skills regardless of whether other teammates in their squadron are in the same location. For example, Zachary, Santarelli, Lyons, Bergondy, and Johnston (2001) discussed the SCOTT program and environment that trains the team skills by substituting the team members with human behavior representations (HBRs). The HBRs in the SCOTT program verbally interact with the trainee in real-time scenario-based environments. The HBR is programmed to give the trainee a feedback to increase learning during training. In addition to verbal interactions, the HBR is designed such that errors in decision-making occur, thereby, better simulating the performance under different levels of stress.

In addition to the SCOTT program, groups of researchers, software developers, and other scientists have been collaborating to develop Synthetic Teammates for Realtime Anywhere Training and Assessment (STRATA; Bell, 2003). The STRATA environment is specifically developed for pilot training and provides a tool that can be structured to initiate training exercises, specifically tailored to the needs of the trainee. The purpose of STRATA is to provide a simulation that utilizes both human team members and HBRs that will interact on scenario-based flight missions. By using HBRs, the STRATA program can be better equipped to train the pilots in a variety of real-time scenarios without the need for a full complement of team members. The simulation is developed so that humans joining or leaving a training scenario will either seamlessly replace an HBR or be replaced by an HBR without any failures in the structure or objective of the training.

Although current DMT systems are primarily being developed for military use, they could be easily adapted and applied for use in commercial aviation. For commercial pilots, DMT provides a way for the pilots to go through training scenarios with pilots in the surrounding aircraft and air-traffic control through the use of HBRs or a flight crew training on the DMT in a different location.

9.4.4 Summary of Tools for Training

We have discussed several tools that have been used to increase the utility and efficiency of team training in aviation. While simulations are typically used, the need for software to generate training scenarios as well as techniques to improve DMT is also useful. The continued developments of programs, such as RRLOE, SCOTT, and STRATA, as well as new and improved simulations, are providing appropriate and vital training tools to carry aviation team training into the future.

9.5 Instructional Strategies for Improving Team Performance

As noted earlier, CRM training is the most commonly used team-training strategy in the aviation community. Here, we have discussed CRM training and its origins. While CRM training has been largely successful in changing crew members' attitudes, knowledge, and behaviors (see Salas, Burke, Bowers, & Wilson, 2001; Salas, Wilson, Burke, & Wightman, in press), its impact on improving safety is less salient. Therefore, we present four instructional strategies that can be used in conjunction with CRM training to improve teamwork in the cockpit and ultimately safety.

9.5.1 Crew Resource Management Training

CRM training was introduced almost three decades ago to train aircrews to use all the available resources—people, information, and equipment. Recently, Salas et al. (1999) defined CRM training as an instructional strategy "designed to improve teamwork in the cockpit by applying well-tested training tools (e.g., performance measures, exercises, feedback mechanisms) and appropriate training methods (e.g., simulators, lectures, videos) targeted at specific content (i.e., teamwork knowledge, skills, and attitudes)" (p. 163).

Since the inception of CRM training, it has evolved and matured through five generations (Helmreich, Merritt, & Wilhelm, 1999). United and KLM airlines were the first among the commercial airlines to implement CRM training to its aircrews in the early 1980s. First-generation CRM training was based largely on existing management training approaches and was psychological in nature (i.e., focusing on psychological testing and general interpersonal behaviors, such as leadership). However, in this generation, there was a lack in the emphasis on team behaviors critical in the cockpit and also the use of games and exercises specific to the aviation community. Second-generation CRM training evolved 5 years later in 1986, and a shift was made from psychological testing to cockpit group dynamics (i.e., focusing on flight operations and team orientation). To emphasize the team nature of the second-generation CRM training, *cockpit* resource management became CRM. Nearly 10 years after CRM was introduced, in 1989, the U.S. FAA provided airlines with guidance on how to develop, implement, reinforce, and assess CRM training (i.e., Advisory Circular 120-51D). Soon after, the third-generation of CRM training evolved, and the focus shifted to the integration of technical-skills training and CRM-skills training. CRM also expanded beyond the flight deck crews to check airmen, those responsible for the training, reinforcement, and evaluation of human and technical factors, and cabin crews. In 1990, the AQP was developed (i.e., fourth generation) as a voluntary program in which airlines were able to tailor CRM training to the specific needs of its organization (Birnbach & Longridge, 1993). As of 1997, only three of the eight major U.S. airlines have chosen to implement and train pilots under the AQP guidelines (U.S. GAO, 1997). This generation also required airlines to provide CRM and Line Oriented Flight Training (LOFT) to crew members, combine nontechnical- (i.e., CRM) and technical-skills training, use Line Operational Evaluations (LOE) for all crews training in full mission simulators, and submit a detailed analysis of their training requirements for each aircraft. In 1997, CRM training shifted once again with a focus on error management. This generation served to accept human error as inevitable and argued that there are three countermeasures to error which could be taught as a part of CRM—avoiding errors, trapping incipient errors, and mitigating the consequences of errors that have occurred. Furthermore, this generation stressed on the importance of feedback and reinforcement of CRM competencies through LOFT, LOE, and in-line checking. Thus, today, a focus on human error is the main emphasis of CRM training.

9.5.2 Scenario-Based Training

SBT, also known as event-based training, is unique and may be especially helpful in training both individuals and teams to exhibit safe behaviors in the cockpit by embedding learning events into simulation scenarios (Fowlkes, Dwyer, Oser, & Salas, 1998; see the earlier discussion on types of simulations). These events are determined from critical incidents data and give the trainees a meaningful framework to learn (Salas & Cannon-Bowers, 2000; Fowlkes et al., 1998).

SBT offers several benefits over other instructional strategies. First, when applied appropriately, SBT is practice-based. With this training, trainees are provided with opportunities to practice applying critical knowledge and skills in pre-planned scenarios while their performance is being evaluated. Cues are embedded within the scenarios that trigger desired behaviors and competencies that are observed (e.g., CRM knowledge and skills), evaluated, and incorporated into feedback. By providing feedback regarding performance, trainees can correct or adjust their strategies before they become internalized. Second, SBT is flexible such that scenarios can be varied allowing a number of responses from the trainees. This allows the trainees to create a set of templates of what to expect and how to react in a variety of situations (Richman, Staszewski, & Simon, 1995). The acquisition of a template repertoire enables the trainees to rapidly recall them from memory when a similar situation is encountered and to make decisions more quickly (Klein, 1997). This can be important in situations where a delayed response could lead to catastrophic consequences. Finally, SBT is a general instructional strategy within which many competencies can be framed. The SBT, as mentioned, has been used extensively to train aircrews to operate in dynamic, decision-making environments (Smith-Jentsch, Zeisig, Acton, & McPherson, 1998).

The SBT can be combined with a number of training strategies (such as those discussed later) to encourage teamwork in the cockpit. For example, SBT could be used to supplement metacognitive training by embedding events into training that require trainees to recognize, react, and continuously reevaluate a situation and their decisions to successfully complete the scenario. Additionally, SBT can be used in conjunction with assertiveness training to recreate situations that ended disastrously when first officers did not assert themselves to captains (e.g., 1978, United Airlines Flt 173 in Portland, OR; 1982, Air Florida Flt 90 in Washington, DC).

9.5.3 Metacognitive Training

Metacognitive training will enable the teams to appropriately respond to and consciously learn from the dynamic situations faced in the cockpit. Decision making is a critical process of CRM in the cockpit. The purpose of metacognitive training is to train individuals to use general rather than specific strategies when making decisions to optimize the process (Jentsch, 1997). In addition, metacognitive training involves providing an awareness of the meta-level (or executive level) and its influence on task processes, presentation of strategies that may improve processes at the meta-level, and an opportunity to practice using the strategies presented. Learning metacognitive strategies will enable the teams to (1) recognize situations as novel or changing and to select appropriate responses, (2) monitor, evaluate, and control one's own processes, and (3) if necessary, to create new or revised strategies for responding to the situation. Jentsch (1997) found that providing junior crew members with metacognitive training improved their performance when taking on the role of pilot-not-flying and also improved specific skills (i.e., planning and prioritizing) in the cockpit.

The use of trained metacognitive strategies improves the performance, because metacognition allows the teams to recognize when they do not know an appropriate strategy for a given situation and to utilize their knowledge to develop an appropriate one (i.e., adaptive expertise). This is especially useful in nonroutine situations. Smith, Ford, and Kozlowski (1997) discuss strategies for how to build an adaptive expert. These researchers concluded that there are two critical factors necessary for adaptability (a critical skill for effective CRM). The first is that knowledge structures are required (e.g., SMMs) that provide individuals and teams with knowledge about their mission and objectives. Next, metacognition (or metacognitive strategies) are necessary to allow individuals to be aware of, understand, and control their cognitive processes (see also Osman & Hannafin, 1992). Therefore, the application of metacognitive strategies allows the team members to "refer to factual, long-term knowledge about cognitive tasks, strategies, current memory states, and conscious feelings related to cognitive activities" (Osman & Hannafin, 1992, p. 83). In other words, metacognitive strategies allow team members to consciously become aware of one's own thought processes, monitor these processes, and regulate them using one's existing knowledge. By being aware of this information, it can be shared with team members improving SMMs and shared cognition.

9.5.4 Assertiveness Training

Assertiveness training is the final instructional strategy that can be used to enhance CRM training. Assertiveness training involves teaching the team members to clearly and directly communicate their concerns, ideas, feelings, and needs to other team members (Jentsch & Smith-Jentsch, 2001). Assertiveness training is important, because it trains a less senior team member (e.g., first officers) to feel comfortable to provide input (e.g., first officer questioning captain's decision to land the aircraft during severe weather) to a more senior team member. In addition, assertiveness training teaches a junior team member to communicate this information in a way that does not demean the senior member or infringe upon their rights. However, assertiveness training is not only for junior team members. It also trains the senior team members to accept input from a team member of lower status without feeling threatened. It should be made clear that assertiveness training does not attempt to remove the authority of the team leader (e.g., captain). Rather, its purpose is to ensure that critical information (i.e., through concerns, ideas, etc.) that may impact a flight does not go unspoken owing to fear of reprimand.

Assertiveness training can be used in conjunction with SBT (discussed previously) to train junior team members to assert themselves to more senior members. One study by Smith-Jentsch, Salas, & Baker (1996) suggests that by using a behavior modeling approach to emphasize practice through role-playing and performance, feedback was more effective in training assertiveness skills than a lecture only or a lecture with demonstration format. In another study, Baldwin (1992) used behavioral reproduction (i.e., demonstrating assertiveness in a situation that was similar to the training environment) to expose trainees to positive models of assertiveness. While this proved effective, more effective was exposing trainees to both positive and negative model displays, as this achieved behavioral generalization (i.e., application of skills outside of the training simulation) 4 weeks after training. This type of role-playing not only helps the junior members to practice asserting themselves, but also creates awareness among senior members regarding the consequences of not considering the concerns and opinions of team members regardless of rank. Law (1993) found that first officers who could inquire and assert themselves were more effective than those who did not inquire or assert themselves. However, research by Orasanu et al. (1998, 1999) indicated that junior crewmembers only assert themselves in specific situations. For example, their research showed that regardless of the position, all the crewmembers would speak up if they felt another crewmember was jeopardizing the safety of the flight (e.g., procedural violation). On the other hand, in concerns related to CRM issues, the captains were more likely than the first officers or flight engineers to assert themselves. The reason cited for this is that the degree of "face threat" plays a great role in a junior crewmember's willingness to speak up.

9.6 Future Needs

Our review on the aviation team-training literature indicates that much progress has been made over the last 20 years. However, areas still exist where more and better research is needed. Subsequently, we have presented a number of team-training needs that we feel would benefit the aviation community by improving teamwork in the cockpit and flight safety.

9.6.1 Standardization of CRM Training

Our examination of the CRM literature and practice has led us to believe that CRM training needs standardization (e.g., what to train, how to train it). First, CRM training has various names associated with it. While the commercial aviation industry typically refers to training as CRM training, the U.S. military has multiple labels. Specifically, the U.S. Navy and Air Force use the term CRM training, whereas the U.S. Army calls the training program as Aircrew Coordination Training. This indicates the lack of consensus not only between the communities, but within the communities. The reviews conducted by Salas et al. (2001, in press) led us to conclude that there is a lack in the standardization of the CRM competencies (i.e., knowledge, skills, and attitudes). Their review illustrates the variability of skills being labeled as CRM skills. Specifically, 36 different skills were being trained as a part of CRM training in aviation. The most commonly trained competencies are communication, situation awareness/assessment, and decision making. Furthermore, this only constitutes the studies that actually published the CRM skills that they were teaching, while many studies failed to report them. Because of this, it makes it difficult to understand how the content of the training may be impacting CRM's success or failure. While the FAA provides guidance to airlines on how to implement CRM training (FAA, 2001), there is still a lack of an agreement on the competencies to be trained and how to implement CRM training. While most agree that team skills and performance must be focused on (Driskell & Adams, 1992), the exact skills are up for a debate. The AQP was set up to give the airlines flexibility; however, because of this, the community could not readily learn from each other. The U.S. military branches are trying work on the standardization of CRM training. A CRM working group has been organized which brings together the U.S. military personnel as well as the researchers and consultants, to make CRM training more consistent from branch to branch.

9.6.2 More and Better CRM Training Evaluations

Although CRM training has been around for nearly three decades and the results of the discussion mentioned earlier are overall encouraging (i.e., CRM training improves teamwork in the cockpit), there still exist some methodological concerns (Salas et al., 2001, in press; Wilson, Burke, Salas, & Bowers, 2001). If the aviation community wants to truly understand the impact of CRM training on safety, then these concerns need to be addressed.

First, many of the evaluation studies found in the literature evaluated the effectiveness of training using only one method of data collection. Within the studies reviewed, data was most typically collected using self-report surveys that questioned the trainees about their reactions to training as well as their attitudes towards CRM training. Furthermore, behavioral data was collected using behavioral observation forms, behavioral checklists, or analysis of crew communication. In addition, several studies utilized peer or self-evaluations. While peer/self-evaluations may appear to be effective in that they might be perceived as less threatening and provide a better viewpoint from which to evaluate crewmembers, this method can be highly prone to rating biases without proper training. Second, many of the studies used poor scientific designs (e.g., one-shot case studies and one-group pretest–posttest designs). While the one-group pretest–posttest design is argued to be "better" than the one-shot case study, both the designs have several threats to validity. Internal validity is threatened in both the designs by history and maturation, while external validity is threatened by an interaction between selection and treatment. These designs, although commonly used in organizations, pose a threat to the validity of their results.

In addition to the abovementioned methodological issues, other weaknesses were found throughout the studies examined. First, the observed studies failed to specify the guidelines used and the skills that were taught during the training programs. Next, many of the studies using self-evaluations failed to report the reliability and content validity of the evaluations used, in the literature. These weaknesses resulted in difficulty in assessing how and where the training was effective. In contrast, if the training failed, then diagnosing the cause may be just as difficult. However, it should be mentioned that this failure to discuss the specifics of the training and its evaluation might be a sign of weakness in the literature, rather than a weakness in the actual study. Finally, although more behavioral-level evaluations were found than initially expected and the behavioral evidence suggested that CRM training does have an impact on crewmember behavior, a majority of these evaluations were conducted in simulated environments rather than on the job. A reason for this might be owing to the resources administering restrictions that reduce the ease with which behavioral data can be collected on the actual job. As opposed to the simulated conditions, behavior on the actual job could provide even stronger support to the suggested effectiveness of CRM training with regard to behavioral change. Nevertheless, behavioral evaluation in simulated environments is definitely a welcomed start in the right direction.

To improve these issues, we suggest the integration of both observation and self-evaluations in order to obtain a more complete picture of the training effectiveness. In addition, the use of randomization and control groups in the experimental design may allow greater control over the studies' validity. We realized that randomization is not always feasible in naturalistic settings, and hence, other forms of control (e.g., statistical- or design-based) should be utilized. Subsequently, a thorough discussion of the skills trained and evaluation techniques may allow greater ease in determining where the training succeeded or failed. Finally, further research is needed not only on the degree to which these behaviors transfer to the actual job, but also on the stability of these behavioral changes. Evidence has shown that changes in the attitudes as a result of training might waive over time. If this is valid, then the same might be true for behaviors as well.

9.6.3 Exploration of More Robust Team-Performance Measurement Techniques

One area where research can be expanded is in the exploration of more robust techniques to measure aviation performance. Research should focus, for example, on the dynamic assessment and triangulation of measures in simulation. By dynamic assessment, we refer to a testing procedure that involves feedback and prompts to elicit learning and determination of the potential for learning (Schack & Guthke, 2003). Previous comparisons between static testing and assessment, as well as dynamic testing and assessment in aviation show the benefits of dynamic assessment (Schack & Guthke, 2003). The performance measures used in dynamic assessment should provide real-time data so that feedback can be immediate. One method of bringing about dynamic assessment in the cockpit is through triangulation of measures. Triangulation of measures involves examining research questions from multiple perspectives and collecting data using multiple independent methods to provide a result with as little uncertainty as possible (Brynam, n.d.). For example, in addition to traditional performance measures in aviation, such as the ability to stay on a given heading or within an altitude range, researchers can use additional methods like eye tracking to gain an understanding of the processes not gleaned from traditional performance measures. For example, previous research using eye tracking exactly shows the researchers the area of the instrument panel that a pilot is looking at during each part of the flight (Hook, 2005). The data gathered from the different performance measures may also provide the researchers the ability to better model performance in the cockpit, because they will have an understanding of the performance from many different perspectives. In turn, with better models of pilot performance, better scenarios can be developed for simulations increasing training effectiveness.

9.6.4 Development of a Deeper Understanding of Flight Debriefings

Without an adequate debriefing after a flight (albeit simulated or real), learning cannot occur. In simple terms, "If I am not told what I did wrong, then I must be doing it right." The dynamic nature of the aviation environment can be busy and intense. Therefore, a thorough discussion on what happened throughout the flight and why it happened is critical (Dismukes, Jobe, & McDonnell, 2000). We know that feedback is important and should be presented in a nonthreatening manner and within a timely fashion. However, some critical questions remain regarding debriefings that need to be answered if we are going to complete the learning process by providing feedback to the learner. The specific questions that we feel still need to be answered are: How long after a flight should debriefings occur? Should debriefings occur at the individual level or the team level? How detailed should debriefings be? and Should debriefings be facilitated, and if so by whom? We will not attempt to answer these questions here; however, we encourage the readers to look deeper at the process of debriefings to ensure that the community is getting the most out of them.

9.6.5 Building Better Learner-Centered Simulation Environments for Aviation

Researchers and developers should focus on building better learner-centered simulation environments in aviation. Norman and Spohrer (2005) discussed that the designs should be learner-centered such that they focus on the needs, skills, and interests of the learner. Learning under these conditions is more enjoyable and motivating, because the learning takes place while practicing, and has the potential to increase transfer of learning. The technology available today allows for the creation of better learner-centered simulation environments. Projects such as SCOTT and STRATA discussed earlier follow a learner-centered approach to simulation development, because each training scenario is specifically designed around the level of skill of the user and the particular training requirements that need to be fulfilled.

9.7 Conclusion

The purpose of this review was to provide an update on where the aviation community is in terms of team process and training research. We have discussed four new team-based theoretical drivers for team training, as well as tools and techniques for improving training and measuring team process and performance. In addition, we have focused on CRM training, as it is the most widely used team-training strategy in aviation, and have suggested several training strategies that could be used in conjunction with CRM, to further improve the teamwork in the cockpit. Finally, we concluded with various needs that the aviation community can use to guide future research programs. Much of what we have presented here serve to not only inform the readers regarding where the aviation community stands in terms of team process and training, but also to challenge the readers to critically look at where the community is and where it needs to go.

Acknowledgments

This research was partially funded by the U.S. Army Research Institute, Ft. Rucker, Alabama, and partially supported by the U.S. Army Research Institute for the Behavioral and Social Sciences, through contract DASW01-03-C-0010. The research and development reported here was also partially supported by the FAA Office of the Chief Scientific and Technical Advisor for Human Factors, AAR-100. Dr. Eleana Edens was the contracting officer's technical representative at the FAA. All the opinions stated in this chapter are those of the authors and do not necessarily represent the opinion or position of the University of Central Florida, the U.S. Army Research Institute, or the FAA.

References

Aiba, K., Kadoo, A., & Nomiyama, T. (2002). Pilot performance of the night approach and landing under the limited runway light conditions. *Reports of Aeromedical Laboratory, 42*(2–3), 19–29.

Baker, D., Prince, C., Shrestha, L., Oser, R., & Salas, E. (1993). Aviation computer games for crew resource management training. *The International Journal of Aviation Psychology, 3*(2), 143–156.

Baldwin, T. T. (1992). Effects of alternative modeling strategies on outcomes of interpersonal skills training. *Journal of Applied Psychology, 77*, 147–154.

Bell, B. (2003). On-demand team training with simulated teammates: Some preliminary thoughts on cognitive model reuse. *TTG News*, pp. 6–8.

Bell, H. H. (1999). The effectiveness of distributed mission training. *Communication of the ACM, 42*(9), 72–78.

Bellorini, A., & Vanderhaegen, F. (1995, April 24–27). *Communication and cooperation analysis in air traffic control*. Paper presented at the 8th International Symposium on Aviation Psychology, Columbus, OH.

Birnbach, R. A., & Longridge, T. M. (1993). The regulatory perspective. In E. L. Wiener, B. G. Kanki, & R. L. Helmreich (Eds.), *Cockpit resource management* (pp. 263–281). New York: Academic Press.

Blickensderfer, E., Cannon-Bowers, J. A., & Salas, E. (1998). Assessing team shared knowledge: A field test and implications for team training. In *Proceedings of the 42nd Annual Meeting of the Human Factors and Ergonomic Society* (p. 1630). Santa Monica, CA: HFES.

Bowers, C. A., & Jentsch, F. (2001). Use of commercial, off-the-shelf, simulations for team research. In E. Salas (Ed.), *Advances in human performance* (Vol. 1, pp. 293–317). Amsterdam, the Netherlands: Elsevier Science.

Bowers, C. A., Jentsch, F., Salas, E., & Braun, C. C. (1998). Analyzing communication sequences for team training needs assessment. *Human Factors, 40*(4), 672–679.

Bowers, C. A., Jentsch, F., Baker, D., Prince, C., & Salas, E. (1997). Rapidly reconfigurable event-based line operational evaluation scenarios. In *Proceedings of the 41st annual meeting to the Human Factors and Ergonomics Society* (pp. 912–915). Santa Monica, CA.

Brannick, M. T., Prince, C., Salas, E., & Stout, R. (2005). Can PC-based systems enhance teamwork in the cockpit? *The International Journal of Aviation Psychology, 15*(2), 173–187.

Bressole, M. C., & Leroux, M. (1997). Multimodality and cooperation in complex work environments—some issues for the design and validation of a decision support system in air traffic control. In *Proceedings of the Ninth International Symposium on Aviation Psychology* (pp. 212–218). Columbus: Ohio State University.

Brynam, A. (n.d). *Triangulation*. Retrieved June 23, 2005 from http://www.referenceworld.com/sage/socialscience/triangulation.pdf#search='triangulation%20of%20measures'.

Cannon-Bowers, J. A., & Salas, E. (2001). Reflections on shared cognition. *Journal of Organizational Behavior Special Issue: Shared Cognition, 22*(2), 195–202

Cannon-Bowers, J., Salas, E., & Converse, S. (1993). Shared mental models in expert team decision making. In N. J. Castellan Jr. (Ed.), *Individual and group decision making: Current issues* (pp. 221–246). Hillsdale, NJ: Lawrence Erlbaum Associates.

Cannon-Bowers, J. A., Tannenbaum, S. I., Salas, E., & Volpe, C. E. (1995). Defining team competencies and establishing team training requirements. In R. Guzzo, E. Salas, & Associates (Eds.), *Team effectiveness and decision making in organizations* (pp. 333–380). San Francisco, CA: Jossey-Bass.

Cooke, N. J., Salas, E., Kiekel, P. A., & Bell, B. (2004). Advances in measuring team cognition. In E. Salas, & S. M. Fiore (Eds.), *Team cognition* (pp. 83–106). Washington, DC: American Psychological Association.

Cooke, N. J., Salas, E., Cannon-Bowers, J. A., & Stout, R. J. (2000). Measuring team knowledge. *Human Factors, 42*(1), 151–173.

Croft, J. (2001). Research perfects new ways to monitor pilot performance. *Aviation Week & Space Technology, 155*(3), 76–81.

Dansereau, F., Graen, G., & Haga, W. (1975). A vertical dyad approach to leadership within formal organizations. *Organization Behavior and Human Performance, 13*, 46–78.

Dennis, K., & Harris, D. (1998). Computer-based simulation as an adjunct to *ab initio* flight training. *International Journal of Aviation Psychology, 8*, 261–276.

Dienesch, R. M., & Liden, R. C. (1986). Leader-member exchange model of leadership: A critique and further development. *Academy of Management Review, 11*, 618–634.

Dismukes, R. K., Jobe, K. K., & McDonnell, L. K. (2000). Facilitating LOFT debriefings: A critical analysis. In R. K. Dismukes, & G. M. Smith (Eds.), *Facilitation and debriefing in aviation training and operations* (pp. 13–25). Aldershot, U.K.: Ashgate.

Driskell, J. E., & Adams, J. A. (1992). *Crew resource management: An introductory handbook* (Report number DOT/FAA/RD-92/26). Washington, DC: Research and Development Service.

Duchon, D., Green, S. G., & Tabor, T. D. (1986). Vertical dyad linkage: A longitudinal assessment of antecedents, measures, and consequences. *Journal of Applied Psychology, 71*, 56–60.

Entin, E. E., & Serfaty, D. (1999). Adaptive team coordination. *Human Factors, 41*(2), 312–325.

Eurocontrol. (2005). *Part-task trainer or emulation*. Retrieved June 22, 2005, from http://eurocontrol.int/projects/ians/cbtguidelines/html/body_part_task.html.

Evans, A. W., Hoeft, R. M., Kochan, J. A., & Jentsch, F. G. (2005, July). *Tailoring training through the use of concept mapping: An investigation into improving pilot training.* Paper to be presented at the Human Computer Interaction International Conference, Las Vegas, NV.

Federal Aviation Administration (FAA). (2001). *Crew resource management* [Advisory Circular 120-51D]. Retrieved online February 23, 2003 from: http://www.faa.gov.

Flin, R., Martin, L., Goeters, K. M., Hormann, H. J., Amalberti, R., & Valot, C. (2003). Development of the NOTECHS (non-technical skills) system for assessing pilots' CRM skills. *Human Factors and Aerospace Safety, 3*(2), 97–119.

Ford, T. (1997). Helicopter simulation. *Aircraft Engineering and Aerospace Technology, 69*(5), 423–427.

Foushee, H. C., Lauber, J. K., Baetge, M. M., & Acomb, D. B. (1986). *Crew factors in flight operations: III. The operational significance of exposure to short-haul air transport operations* (Technical Memorandum No. 88322). Moffett Field, CA: NASA-Ames Research Center.

Fowlkes, J., Dwyer, D. J., Oser, R. L., & Salas, E. (1998). Event-based approach to training (EBAT). *International Journal of Aviation Psychology, 8*(3), 209–221.

Gillan, C. A. (2003). Aircrew adaptive decision making: A cross-case analysis. *Dissertation Abstracts International: Section B: The Sciences & Engineering, 64*(1-B), 438.

Goldsmith, T. E., Johnson, P. J., & Acton, W. H. (1991). Assessing structural knowledge. *Journal of Educational Psychology, 83*, 88–97.

Gopher, D., Weil, M., & Bareket, T. (1994). Transfer of skill from a computer game trainer to flight. *Human Factors, 36*, 387–405.

Graen, G., & Cashman, J. F. (1975). A role making model in formal organizations: A developmental approach. In J. G. Hunt, & L. L. Larson (Eds.), *Leadership frontiers* (pp. 143–165). Kent, OH: Kent State Press.

Graen, G. B., & Scandura, T. A. (1987). Toward a psychology of dyadic organizing. *Research in Organizational Behavior, 9*, 175–208.

Graen, G. B., & Uhl-Bien, M. (1995). Relationship-based approach to leadership: Development of leader-member exchange (LMX) theory of leadership over 25 years: Applying a multi-level multi-domain perspective. *Leadership Quarterly, 6*(2), 219–247.

Griepentrog, B. K., & Fleming, P. J. (2003). *Shared mental models and team performance: Are you thinking what we're thinking?* Symposium presented at the 18th Annual Conference of the Society for Industrial Organizational Psychology, Orlando, FL.

Harris, D., & Khan, H. (2003). Response time to reject a takeoff. *Human Factors & Aerospace Safety, 3*(2), 165–175.

Helmreich, R. L. (1984). Cockpit management attitudes. *Human Factors, 26*, 63–72.

Helmreich, R. L. (1994). Anatomy of a system accident: The crash of Avianca Flight 052. *International Journal of Aviation Psychology, 4*(3), 265–284.

Helmreich, R. L., & Merritt, A. C. (1998). *Culture at work in aviation and medicine: National, organizational, and professional influences.* Aldershot, U.K.: Ashgate.

Helmreich, R. L., Foushee, H. C., Benson, R., & Russini, W. (1986). Cockpit management attitudes: Exploring the attitude-performance linkage. *Aviation, Space, and Environmental Medicine, 57*, 1198–1200.

Helmreich, R. L., Klinect, J. R., & Wilhelm, J. A. (1999). Models of threat, error, and CRM in flight operations. *Proceedings of the Tenth International Symposium on Aviation Psychology* (pp. 677–682). Columbus: The Ohio State University.

Helmreich, R. L., Merritt, A. C., & Wilhelm, J. A. (1999). The evolution of crew resource management training in commercial aviation. *The International Journal of Aviation Psychology, 9*(1), 19–32.

Helmreich, R. L., Wilhelm, J. A., Klinect, J. R., & Merritt, A. C. (2001). Culture, error and crew resource management. In E. Salas, C. A. Bowers, & E. Edens (Eds.), *Improving teamwork in organizations: Applications of resource management training* (pp. 305–331). Hillsdale, NJ: Erlbaum.

Hofstede, G. (1991). *Cultures and organizations: Software of the mind.* London: McGraw-Hill.

Hook, G. T. (2005). RE: [crm-devel] Digest Number 1111. Message posted on June 23, 2005, to crm-devel@yahoogroups.com.

Hutchins, E., Holder, B. E., & Pérez, R. A. (2002). *Culture and flight deck operations.* Unpublished internal report prepared for the Boeing Company [Sponsored Research Agreement 22-5003]. San Diego: University of California.

Jentsch, F. (1997). *Metacognitive training for junior team members: Solving the "copilot's catch-22."* Unpublished doctoral dissertation, University of Central Florida, Orlando.

Jentsch, F., & Bowers, C. A. (1998). Evidence for the validity of pc-based simulations in studying aircrew coordination. *International Journal of Aviation Psychology, 8*, 243–260.

Jentsch, F., & Smith-Jentsch, K. A. (2001). Assertiveness and team performance: More than "just say no". In E. Salas, C. A. Bowers, & E. Edens (Eds.), *Improving teamwork in organizations: Applications of resource management training* (pp. 73–94). Mahwah, NJ: Lawrence Erlbaum Associates.

Klein, G. A. (1997). *Developing expertise in decision making.* Unpublished draft.

Koonce, J., & Bramble, W. (1998). Personal computer-based flight traveling devices. *International Journal of Aviation Psychology, 8*(3), 277–292.

Kraiger, K., Salas, E., & Cannon-Bowers, J. A. (1995). Measuring knowledge organization as a method for assessing learning during training. *Human Performance, 37*, 804–816.

Kuang, J. C. Y., & Davis, D. D. (2001). *Culture and team performance in foreign flightcrews.* Paper presented at the 109th Annual Convention of the American Psychological Association, San Francisco, CA.

Lassiter, D. L., Vaughn, J. S., Smaltz, V. E., Morgan, B. B., Jr., & Salas, E. (1990). A comparison of two types of training interventions on team communication performance. *Human Proceedings of the Factors Society 34th Annual Meeting* (Vol. 2, pp. 1372–1376). Orlando, FL.

Law, J. R. (1993). Position specific behaviors and their impact on crew performance: Implications for training. In *Proceedings of the Seventh International Symposium on Aviation Psychology* (pp. 8–13). Columbus: Ohio State University.

Martin, D. A. J. (1993). The use of computer based training, part task trainers and flight simulators in ab initio airline pilot training. In *Proceedings of the Annual Conference of the Flight Simulation Group* (pp. 7.1–7.5). London, England.

Mathieu, J. E., Heffner, T. S., Goodwin, G. F., Salas, E., & Cannon-Bowers, J. A. (2000). The influence of shared mental models on team process and performance. *Journal of Applied Psychology, 85*(2), 273–283.

Mohammed, S., Klimoski, R., & Rentsch, J. R. (2000). The measurement of team mental models: We have no shared schema. *Organizational Research Methods, 3*(2), 123–165.

Noble, C. (2002). The relationship between fidelity and learning in aviation training and assessment. *Journal of Air Transportation, 7*(3), 33–54.

Norman, D. A., & Spohrer, J. C. (2005). *Learner-centered education.* Retrieved June 22, 2005, from http://itech1.coe.uga.edu/itforum/paper12/paper12.html.

Orasanu, J. M. (1994). Shared problem models and flight crew performance. In N. Johnston, N. McDonald, & R. Fuller (Eds.), *Aviation psychology in practice* (pp. 255–285). Aldershot, U.K.: Ashgate.

Orasanu, J., Fischer, U., McDonnell, L. K., Davison, J., Haars, K. E., Villeda, E., et al. (1998). How do flight crews detect and prevent errors? Findings from a flight simulation study. In *Proceedings of the Human Factors and Ergonomics Society 42nd Annual Meeting* (pp. 191–195). Santa Monica, CA: HFES.

Orasanu, J., Murray, L., Rodvold, M. A., & Tyzzer, L. K. (1999). Has CRM succeeded too well? Assertiveness on the flight deck. In *Proceedings of the 10th International Symposium on Aviation Psychology* (pp. 357–361). Columbus: Ohio State University.

Osman, M. E., & Hannafin, M. J. (1992). Metacognition research and theory: Analysis and implications for instructional design. *Educational Technology Research & Development, 40*(2), 83–99.

Priest, H. A., Burke, C. S., Guthrie, J. W., & Salas, E. (in preparation). *Are all teams created equal? An examination of teamwork communication across teams.*

Prince, C., & Salas, E. (2000). Team situational awareness, errors, and crew resource management: Research integration for training guidance. In M. R. Endsley, & D. J. Garland (Eds.), *Situation awareness analysis and measurement* (pp. 325–347). Mahwah, NJ: Lawrence Erlbaum Associates.

Prince, C., & Jentsch, F. (2001). Aviation crew resource management training with low-fidelity devices. In E. Salas, C. A. Bowers, & E. Edens (Eds.), *Improving teamwork in organizations: Applications of resource management training* (pp. 147–164) Mahwah, NJ: Lawrence Erlbaum Associates.

Prince, C., & Salas, E. (1999). Team processes and their training in aviation. In D. Garland, J. Wise, & D. Hopkin (Eds.), *Handbook of aviation human factors* (pp. 193–213). Mahwah, NJ: LEA.

Proctor, M. D., Panko, M., & Donovan, S. J. (2004). Considerations for training team situation awareness and task performance through PC-gamer simulated multiship helicopter operations. *The International Journal of Aviation Psychology, 14*(2), 191–205.

Rentsch, J. R., & Hall, R. J. (1994). Members of great teams think alike: A model of team effectiveness and schema similarity among team members. In M. M. Beyerlein, & D. A. Johnson (Eds.), *Advances in interdisciplinary studies of work teams. Vol. 1. Series on self-managed work teams* (pp. 223–262). Greenwich, CT: JAI Press.

Rentsch, J. R., & Klimoski, R. J. (2001). Why do "great minds" think alike? Antecedents of team member schema agreement. *Journal of Organizational Behavior, 22*, 107–120.

Richman, H. B., Staszewski, J. J., & Simon, H. A. (1995). Simulation of expert memory using EPAM IV. *Psychological Review, 102*(2), 305–330.

Salas, E., & Cannon-Bowers, J. A. (1997). Methods, tools, and strategies for team training. In M. A. Quinones, & A. Ehrenstein (Eds.), *Training for a rapidly changing workplace: Applications of psychological research* (pp. 249–280). Washington, DC: APA.

Salas, E., & Cannon-Bowers, J. A. (2000). The anatomy of team training. In S. Tobias, & J. D. Fletcher (Eds.), *Training & retraining: A handbook for business, industry, government, and the military* (pp. 312–335). New York: Macmillan Reference.

Salas, E., Bowers, C. A., & Rhodenizer, L. (1998). It is not how much you have but how you use it: Toward a rational use of simulation to support aviation training. *International Journal of Aviation Psychology, 8*, 197–208.

Salas, E., Burke, C. S., Bowers, C. A., & Wilson, K. A. (2001). Team training in the skies: Does crew resource management (CRM) training work? *Human Factors, 43*(4), 641–674.

Salas, E., Burke, C. S., Fowlkes, J. E., & Wilson, K. A. (2004). Challenges and approaches to understanding leadership efficacy in multi-cultural teams. In M. Kaplan (Ed.), *Advances in human performance and cognitive engineering research* (Vol. 4, pp. 341–384). Oxford, U.K.: Elsevier.

Salas, E., Muniz, E. J., & Prince, C. (2001). Situation awareness in teams. In W. Karwowski (Ed.), *International encyclopedia of ergonomics and human factors* (Vol. 1, pp. 555–557). Taylor & Francis.

Salas, E., Prince, C., Bowers, C., Stout, R., Oser, R. L., & Cannon-Bowers, J. A. (1999). A methodology for enhancing crew resource management training. *Human Factors, 41*(1), 161–172.

Salas, E., Wilson, K. A., Burke, C. S., & Wightman, D. C. (2006). Does CRM training work? An update, extension, and some critical needs. *Human Factors, 48*(2), 392–412.

Sanderson, P. M., & Fisher, C. (1992). *Exploratory sequential data analysis: Traditions, techniques and tools* (CHI'92 Pre-Workshop Report). Monterey, CA: Engineering Psychology Research Laboratory, Department of Mechanical and Industrial Engineering, University of Illinois at Urbana-Champaign, Urbana.

Schack, T., & Guthke, J. (2003). Dynamic testing. *International Journal of Sport & Exercise Psychology, 1*(1), 40–60.

Shlechter, T. M., Zaccaro, S. J., & Burke, C. S. (1998). *Toward an understanding of shared mental models associated with proficient team performance.* Paper presented at the American Psychological Society Annual Meeting, Washington, DC.

Schvaneveldt, R. W., Beringer, D. B., & Lamonica, J. A. (2001). Priority and organization of information accessed by pilots in various phases of flight. *The International Journal of Aviation Psychology, 11*(3), 253–280.

Smith, E. M., Ford, J. K., & Kozlowski, S. W. J. (1997). Building adaptive expertise: Implications for training design strategies. In M. A. Quinones, & A. Ehrenstein (Eds.), *Training for a rapidly changing workplace: Applications of psychological research* (pp. 89–118). Washington, DC: APA.

Smith-Jentsch, K. A., Zeisig, R. L., Acton, B., & McPherson, J. A. (1998). Team dimensional training: A strategy for guided team self-correction. In J. A. Cannon-Bowers, & E. Salas (Eds.), *Making decisions under stress: Implications for individual and team training* (pp. 271–297). Washington, DC: American Psychological Association.

Smith-Jentsch, K., Salas, E., & Baker, D. P. (1996). Training team performance-related assertiveness. *Personnel Psychology, 49*, 909–936.

Stout, R. J., Cannon-Bowers, J. A., & Salas, E. (1996). The role of shared mental models in developing team situational awareness: Implications for training. *Training Research Journal, 2*, 85–116.

Stout, R. J., Cannon-Bowers, J. A., Salas, E., & Milanovich, D. M. (1999). Planning, shared mental models, and coordinated performance: An empirical link is established. *Human Factors, 41*(1), 61–71.

Stout, R. J., Cannon-Bowers, J. A., Morgan, B. B., Jr., & Salas, E. (1990). Does crew coordination behavior impact performance? In *Proceedings of the Human Factors Society 34th Annual Meeting* (pp. 1382–1386). Santa Monica, CA.

Swan, J. A. (1995). Exploring knowledge and cognitions in decisions about technological innovations—Mapping managerial cognitions. *Human Relations, 48,* 1241–1270.

Taylor, H. L., Lintern, G., & Koonce, J. M. (1993). Quasi-transfer as a predictor of transfer from simulator to airplane. *Journal of General Psychology, 120*(3), 257–276.

Taylor, H. L., Talleur, D. A., Bradshaw, G. L., Emanuel, T. W., Rantanen, E., Hulin, C. L., et al. (2002). *Effectiveness of personal computers to meet recency of experience requirements* (Technical Report ARL-01-6/FAA-01-1). Savoy: Aviation Research Lab, Institute of Aviation, University of Illinois at Urbana-Champaign.

University of Texas. (2005). *LOSA advisory circular.* Retrieved June 15, 2005, from http://homepage.psy.utexas.edu/homepage/group/HelmreichLAB/.

United States General Accounting Office (U.S. GAO). (1997). *Human factors: FAA's guidance and oversight of pilot crew resource management training can be improved* (GAO/RCED-98-7). Washington, DC: Author.

Vidulich, M., Dominguez, C., Vogel, E., & McMillan, G. (1994). *Situation awareness papers and annotated bibliography* (Report AL/CF-TR-1994-0085). Wright Patterson AFB, OH: Armstrong Laboratory.

Westrum, R., & Adamski, A. J. (1999). Organizational factors associated with safety and mission success in aviation environments. In D. J. Garland, J. A. Wise, & V. D. Hopkin (Eds.), *Handbook of aviation human factors.* Mahwah, NJ: Lawrence Erlbaum Associates.

Wilson, K. A., Burke, C. S., Salas, E., & Bowers, C. A. (2001, August). *Does CRM training transfer to the cockpit?: Methodological observations.* Poster presented at the 109th Annual Meeting of the American Psychological Association, San Diego, CA.

Wickens, C. D., Helleberg, J., & Xu, X. (2002). Pilot maneuver choice and workload in free flight. *Human Factors, 44*(2), 171–188.

Zachary, W., Santarelli, T., Lyons, D., Bergondy, M., & Johnston, J. (2001). Using a community of intelligent synthetic entities to support operational team training. In *Proceedings of the Tenth Conference on Computer Generated Forces and Behavioral Representation* (pp. 215–224). Norfolk, VA.

10

Crew Resource Management

Daniel E. Maurino
*International Civil
Aviation Organization*

Patrick S. Murray
Griffith University

Although there is no real possibility of a quantitative evaluation of the benefits, no airline having set up a CRM program would now consider killing it.

Pariès and Amalberti

10.1 Introduction

This rather ominous remark was made by Pariès and Amalberti in 1995. However, it would have been quite wrong to have considered it as the prophesy of doom. At that time, the comment reflected the concerns of some within the international aviation human-factors community who, while believing that crew resource management (CRM) was an essential prevention tool in the contemporary aviation system, had taken to critically review CRM and its history to ensure that there was a meaningful future for this training in aviation. In the 10 years since the comment was made, large strides have in fact been made in the evaluation of CRM—both qualitative and quantitative—although these too have not been without controversy. However, it is an undisputed fact that in reflecting on its quarter-century lifespan to date, the history of CRM is one of greatest success. Like a living being, from its inception, through its short childhood and teenage years, CRM has matured steadily, although by continuing the comparison with an adolescent, some would say—at times—that CRM causes similar angst to its parents and friends! Despite the regular healthy challenge, both from within the aviation industry and parts of the academic world, it has continued to gain impetus and has now reached a stage where it has been assigned a significant role as contributor to the safety and efficiency of the aviation system. It is perceived as a sound way to proceed by the user population and regulatory community alike. In fact, nobody would dare say that CRM does not work.

However, there is more in the successful history of the development, implementation, and operational practice of CRM, than what meets the eye. Without casting doubts about its value, there have been certain quarters that have cautioned about what the future might hold, suggesting that the relationship between CRM and improved safety is still tenuous or at least not completely proven. In these quarters, the prevailing attitude is one of critical vigil. Without endorsing optimists or skeptics, and without denying eventual merits in each relative position, it is contended that there are present-day issues that, in the best interests of CRM itself, must not be ignored.

The literature on the development and early days of CRM is abundant (Cooper, White, & Lauber, 1979; Hayward & Lowe, 1993; Orlady & Foushee, 1986; Wiener, Kanki, & Helmreich, 1993). Readers interested in early CRM development, implementation, and practice may refer to these and many other publications. However, it is noteworthy that literature on more recent developments since the late 1990s is more scattered. This chapter does not discuss CRM in itself, but it rather looks at some recent developments and assesses some of the issues that might affect its future. Such assessment is conducted within the framework provided by a historical review of the evolution of CRM and its associated safety paradigms. The position upon which this chapter builds is quite simple and should not be mistaken. CRM is too valuable a tool to be squandered by misunderstandings, misperceptions, incompetence, or plain ignorance.

10.2 Why CRM Training?

To clarify misunderstandings and misperceptions, the essential question that demands an unambiguous answer is *why* it is desirable—or necessary—to provide CRM training to operational personnel. CRM training is not an end in itself, but is a means of living for consultants, a power structure for management, or an opportunity to generate research data and statistics. In addition, CRM is not the last frontier of aviation safety, organizational efficiency, or a frontier of any kind. We provide CRM training to operational personnel so that they are better equipped to contribute to the aviation system production's goals: the safe and efficient transportation of people and goods. CRM training is a safety and production tool. In socio-technical systems, operational personnel are expected to act as the last line of defense against systemic flaws and failures. Hence, their training must be build upon both an understanding of the systemic safety and a safety paradigm that are relevant to contemporary civil aviation.

CRM training for operational personnel must evolve from a realistic understanding of how today's underspecified and unpredictable aviation system can fail. At the individual level, we must seek failures in the cognitive dimensions of human performance. At the collective level, we must seek failures in the social and organizational dimensions of the system. Once potential failures have been identified, the blueprint of a healthy system built upon proactive rather than reactive measures has been identified, and data regarding the user organization and its population have been collected, then and only then should the process of writing the CRM-training curriculum start. These fundamental steps have not always been observed, and the design of CRM training has been undertaken lightly and arguably, even unscrupulously on occasion.

CRM training for operational personnel must be relevant to their professional needs. Evidence from accident reports suggests that this has not always been the case. It is a matter of record that CRM has saved the day in many instances and averted more serious outcomes in others. It is equally a matter of record that the lack of CRM training has often been a missing defense that could have contained the consequences of flaws in human or system performance. Furthermore, flaws in equipment design and the aviation system itself are contributing realities to the breakdown of safety. Both are here to stay, at least for a long while. Changes in equipment design are very expensive, and we are only observing a slow introduction within the current generation of flight decks. Changes in the aviation system itself are also painfully slow. We can only work at the interface between pilots and equipment as well as pilots and the system through CRM training, buying time while change takes place. Therefore, the relevance of the operational context to CRM training is essential.

10.3 The Evolution of CRM Training—
Two Perspectives in Harmony

CRM training was introduced during the 1970s, not only because of the "70% factor," but also because there was a growing concern that ritualistic training—ticking boxes in a proficiency form—did not address the operational issues that eventually led to the "70% factor" (Cooper et al., 1979). Today, we know that it was not an idle concern. Over its 25 years of existence, CRM has experienced considerable evolution and change. We will initially consider the early stages of this evolution based on two perspectives: European and North American.

Paries and Amalberti (1995) asserted that "…No training is suspended in a vacuum: any training policy is built on a specific understanding of the system's potential failures, and follows a particular safety paradigm, including an accident causation model and some remedial action principles." Thus, they tried to establish a connection between changes in the understanding of aviation safety, their allied prevention strategies, and the evolution of CRM training. They suggested that what started as a "cockpit crisis" prevention and management program has been gradually shifting toward a macrosystem education: "…CRM corresponds to a revolution in accident causation models. It was a shift from 'training for the abnormal' to 'training for the normal.' It was a dramatic acceptance that the prevailing bipolar equation—pilot proficiency plus aircraft reliability equals safe flights—had been proven wrong by hard real life reasons." Pariès and Amalberti distinguished four generations of CRM training.

- First generation aimed at individual attitudes, leadership, and communication. The objective was to prevent accidents owing to flawed flight-crew team performance. The safety paradigm was that safety was a function of flight-crew performance exclusively, and that there were individuals with either the "right" or the "wrong stuff."

The resistance to initial CRM training by segments of the pilot community (Helmreich, 1992) led to a revision of the original approach, thus, giving birth to the second generation of CRM training. In an attempt to overcome resistance, this second generation essentially distanced itself from the notion of the "right/wrong stuff."

- Second generation aimed at individual attitudes, leadership, and communication, but expanded to include situation awareness, the error chain, stress management, and decision making. Like the first, the second generation of CRM training aimed at preventing accidents through improved crew performance, and its underlying safety paradigm was that safety was a consequence of improved crew synergy.

Both first and second generation of CRM training relied intensively on role-playing and nonaviation-related games, and they resorted to repetitive accident case studies. A distinct characteristic of first and second generation of CRM training programs was that they appeared to consciously and purposefully introduce and maintain a clear separation between technical and CRM training.

The introduction of "glass cockpits" led to the development of a third generation of CRM training, with a broadened human-factors knowledge base, and with particular attention to the cognitive dimensions of small teams acting in dynamic environments as well to the importance of shared mental models (Orasanu, 1993). The third generation of CRM programs also revisited human–machine interface issues, in the relationship between pilots and computers. It was during this third generation that Cockpit Resource Management became CRM.

- Third generation aimed at improving the overall system performance through improved performance of the system's basic flight operational units (aircraft/crew system; flight/cabin crew system). It added the concepts of mental models, stress and fatigue management, automation management, vigilance, and human reliability to the basic issues included in the two first generations of CRM programs. It further included discussions intended to develop not only *skills*, but also *understanding* and *knowledge*.

The major step forward in the third generation of CRM was a change in its underlying safety paradigm: safety was now considered to be a proactive rather than reactive endeavor, and the consequence of a healthy system and its effective performance. Two consequences of adopting a proactive stance also represented a major change: the integration of human-factors knowledge into the training, thus, resolving the prevailing dichotomy of "technical" versus "nontechnical" training; and the gradual shift from nonaviation games and role-playing toward the realities of aviation, such as the justification of operational doctrines.

The recognition that safety as an outcome is the consequence of the global health of the system, and that training is a tool to help the process and therefore, to influence the outcome, led to the development of the fourth generation of CRM training.

- Fourth generation aims at improving the overall system performance through improved performance of as many system's components as possible. It includes topics, such as interaction among teams, shared mental models, role and status, and organizational synergy. The safety paradigm corresponds to the shift in the safety thinking observed since the beginning of the 1990s: safety is one positive outcome of the system's health.

Fourth generation of CRM training includes maintenance (Robertson, Taylor, Stelly, & Wagner, 1995; Taggart, 1995), air-traffic control (ATC) (Baker et al., 1995), and flight dispatch (Chidester, 1993; FAA, 1995). Furthermore, it recognizes that management actions or inactions influence the outcome of aviation operations (Maurino, 1992). In this aspect, Pariès and Amalberti advocated the term cross-corporate or Company Resource Management to reflect that the benefits of CRM extend beyond safety to include cost efficiency, service quality, and job satisfaction. The term Organizational Resource Management (ORM) reflects the same line of thinking (Heinzer, 1993), fully developed by Smith (1992), who viewed CRM as an organizational development.

On the North American side of the Atlantic, Helmreich (1994) agreed with Pariès and Amalberti and acknowledged the existence of four generations of CRM training. With slight differences with regard to the milestones and emphasis—probably reflecting American empiricism *vis-à-vis* French encyclopaedism—Helmerich drew a "road map" of the evolution of CRM which is consistent with that of his colleagues across the Atlantic. According to Helmreich, the distinctive features of each of the four generations of CRM were:

- First generation: derived from classical management development; focused on management styles and interpersonal skills; aimed at fixing the "wrong stuff."
- Second generation: focused on concepts (situational awareness, stress management) and modular in conception (error chain, decision-making models).
- Third generation: observes a systems approach, with a focus on specific skills and behaviors. It places emphasis on team building and in the integration of CRM with technical performance. This generation includes the first attempts to assess CRM training; therefore, to allow such assessment, special training is designed for check airmen and instructors. Lastly, the training transcends beyond the cockpit door to include flight attendants, maintenance personnel, dispatchers, and air-traffic controllers.
- Fourth generation: addresses specialized curriculum topics, including automation and fatigue, joint training between flight and cabin crew, crew-performance training directly derived from the incident data and, very importantly, includes an added focus on cultural issues, including national and organizational culture, and the particular issues of multinational crews.

The consensus of opinion among Paries, Amalberti, and Helmreich regarding the evolution and *raison d'être* of CRM was significant. Each "road map" would fit into the blueprint proposed by the other without major difficulties. There was a coincidence with regard to the two fundamental changes experienced by CRM: the blending of CRM within technical training, and the expansion of CRM beyond the cockpit to become a systems concept. In terms of conceptual preferences, while Paries and

Amalberti made it a clear point to present CRM espoused to prevailing safety paradigms, Helmreich seemed less concerned about this relationship (while not ignoring it). Instead, and by virtue of the globalization of CRM, Helmreich placed greater emphasis on cross-cultural issues as they affect CRM training.

10.4 CRM Fifth and Sixth Generations

Since this time of broad Trans-Atlantic consensus within a context of increasing awareness of cultural issues, there has been a further and most significant milestone. This was the narrowing of the goals of CRM from the broad strategic concept of "safer flight," to specifically what CRM was attempting to achieve on a tactical basis, on every flight. Threat and Error Management, as a concept, was born in the late 1990s, based on the studies at the University of Texas and their close liaison with Continental Airlines in the development of the line operations safety audit (LOSA) (Helmreich et al. 1993, 1994, 1996, 1997, 1999; Jones & Tesmer 1999; Klinect et al. 1999, 2001; 2003). LOSA, quite simply, was a methodology whereby trained observers were placed on the flightdeck during normal line flights with a view to collecting data on what flight crews did to fly safely from A to B. All the data were de-identified and collected on a strictly anonymous and nonjeopardy basis. In its inception in the early 1990s, LOSA—in its earliest form—was conceived to analyze CRM behaviors in avoiding error. LOSA was related to earlier research that was conducted on the development of the Line/LOSS checklist (Helmreich, Klinect, Wilhelm & Jones, 1999), and as a result, two major findings emerged. It not only became apparent that human error was inevitable, but it was also apparent that some pilots, at both individual and crew level, were employing specific countermeasures and strategies to good effect. These defenses were observed at three levels: The first was to avoid or minimize error. The second was the trapping of incipient errors in their early stages before becoming consequential. The third was mitigating the effects of those errors that had not been trapped. These strategies were closely linked with certain classic CRM "skills," and taken together, the countermeasures were called "Error Management."

While human error was the original impetus for the first generation of CRM, the realization and articulation of this was less than perfect. Even when the training advocated certain behaviors, the reasons for them were not always explicit. Error management provided a more sharply defined justification that was accompanied by proactive organizational support that was gaining ground in the area of systemic improvement (Reason, 1990, 1997).

Without wishing to perpetuate the comparison with the "generations" of a popular space-science fiction series, and continuing the "Generations of CRM" analogy, the so-called fifth generation of CRM involved the development and adoption of these concepts of error management. Whilst accurate in its philosophy, the emphasis on "pilot error" and its avoidance and management—was less welcomed by some pilots than others. There was a nagging feeling that this was still not the whole story. During the continuing LOSA research on airlines conducted by Helmreich's team at the University of Texas, the last pieces of the jigsaw quickly came together, with the importance and variability of the operational context being highlighted. LOSA showed quite clearly that operational context played an enormous role in terms of error management. The concept of "threats," as being situations or events outside the influence of the flight crew but needing their immediate attention to manage, was a breakthrough. Although intuitively pilots knew that this was, in effect, their job description, it had not previously been articulated as such—particularly, in the context of managing error. By incorporating the complexities of managing operational context, "Error Management" became "Threat and Error Management" or TEM. Thus, TEM became the framework for the "6th Generation of CRM." This framework provided context to earlier CRM training and reframed conventional CRM skills as countermeasures for effective TEM. Currently, TEM is growing rapidly in acceptance and popularity, interestingly demonstrating acceptance within disparate cultures. The following sections of this chapter discuss TEM within a context of increasing cultural awareness, as being the current focus for contemporary CRM implementation.

10.5 Where CRM and Culture Meet

The International Civil Aviation Organization (ICAO) is the specialized agency of the United Nations tasked with establishing standards to ensure uniformity of procedures and practices in international civil aviation. ICAO started its Flight Safety and Human Factors Programme in 1989 (Maurino, 1991) and given its international nature, obtained sensible knowledge about the influence of cross-cultural issues in aviation safety and efficiency. This information suggests the need to properly take cross-cultural issues into account when developing CRM training.

The most important lesson that ICAO learned is that any kind of human endeavor has strong cultural components (Maurino, 1995a, 1995b). An important and practical consequence of the cross-cultural issues is that there are no solutions valid "across the board." Problems in different contexts might *seem* similar on the surface, but they encode important cultural biases. Therefore, they require different, culturally calibrated solutions. It is naive to assume that by simply exporting a solution that worked in one culture will bring the same benefits to another. Indeed, such attempt might only set a collision course between the respective biases of the different cultures involved, making the problem worst. Another important lesson that ICAO learned is that there is a tendency to deal with cultural issues from an "us/right" versus "them/wrong" position. The perceived "qualities" or "defects" involved in rating different cultures are not only without foundation, but they create barriers in understanding the implications of the cross-cultural issue (Phillips, 1994). Difficult as they may be, the implications of cultural differences in cross-border endeavors are worth considering in depth (Phelan, 1994).

Culture can be defined as "the values, beliefs and behaviors we share with other people and which help define us as a group, especially in relation to other groups. Culture gives us cues and clues on how to behave in normal and novel situations" (Merritt & Helmreich, 1998). CRM originated in North America as a solution to the intricacies of human interrelationship on flight decks. It is based on social psychology, a practice scarcely known in large geographical areas, including Africa, Latin America, the Middle and Far East, Asia, and largely across the Pacific, except for the English-speaking enclaves. In its beginnings and for some time, it was considered that "…any culture, whether it is Japanese, American, or any other—fits in with the cockpit environment. In this sense, CRM is culture free" (Yamamori, 1986). This opinion was shared by many—if not most—proponents of CRM during the 1980s. In fact, the shared perception was that with few minor cosmetic changes—or even without them—CRM could fit within the operational practices of any airline around the world. This view is still held by some even today.

We now know that, well-intentioned and honest as it might have been, the myth of culture-free CRM is a fallacy. Not only does culture influence CRM, but available evidence suggests that CRM is not a good traveler unless culturally calibrated. In fact, implementing off-the-shelf CRM training in an organization without due consideration to national, corporate, and even pilot subgroup cultures may generate aberrant situations (Helmreich & Merritt 1998; Maurino, 1994; Merritt, 1993, 1995).

For example, Johnston (1993a) expressed concern about the universal understanding of the concepts of CRM. He questions the effectiveness of simple translation, arguing that meaning is also provided by the cultural and environmental context. Examples of cognitive incompatibilities and limitations of translation are everyday occurrences in organizations that work in several languages—such as ICAO—where interpretation and translation are frequent victims of cross-cultural perceptions. It is a logical corollary to this state of affairs that led Johnston to question "…In this general context, consider how the specialist vocabulary of CRM might be received in different cultures; indeed, might it not be the case that existing CRM training may be unsuitable, or misdirected in some cultures?"

Helmreich and Merritt continued to lead the research to provide solutions to the challenge of making CRM a universal, albeit culturally calibrated concept. They built their research on culture in the cockpit on two basic tools: the research in culture by the Dutch anthropologist Gert Hofstede, and the Flight Management Attitudes Questionnaire (FMAQ).

Hofstede (1980) developed a survey measure to profile the dimensions of national cultures. He defined four dimensions of culture:

- Power distance (PD)—the extent to which people accept that power is distributed unequally. In high PD cultures, subordinates do not question superiors. Those in authority are seldom accountable, as it is assumed that authority "belongs" to them.
- Individualism–collectivism (IC)—the extent to which individual costs and rewards or the group is the dominant factor that motivates the behavior.
- Uncertainty avoidance (UA)—the extent to which people feel threatened by uncertain or unknown situations or conditions; the need for defined structures and highly proceduralized protocols.
- Masculinity—the extent to which people are concerned with achievement rather than quality of life and interpersonal relationships.

Of these four dimensions, PD and IC have been found to be highly relevant to cross-cultural research in aviation, while UA and masculinity have been found to be of lesser relevance.

The FMAQ (Helmreich, Merrit, Sherman, Gregorich, & Wiener, 1993) is an extension of a previous tool, the Cockpit Attitudes Management Questionnaire (CMAQ) (Helmreich, Wilhelm, & Gregorich, 1988), developed to assess crewmembers' attitudes about group performance, leadership, and susceptibility to stress among flight crews. The FMAQ is an iteration which combines the CMAQ with Hoftede's measures to capture cross-cultural differences in flight-deck management attitudes and to provide a measure of organizational climate.

This research produced evidence that CRM practices that were considered to be of universal application are indeed culturally influenced. Thus, it would not seem reasonable to apply training designed for a group in one country to other groups of different nationalities, without first understanding the receiver group's attitudes toward leadership and followership, communication styles, work values, team concept, and the use and flexibility of regulations (Merritt, 1993). CRM is a *process*, the output of which is improved safety and efficiency of aviation operations. Culture is among the many input factors, and cultural preferences influence CRM practices.

Helmreich and Merritt (1998) explained that while Anglo-speaking countries rank high on individualism, Asian and Latin countries are at the opposite side of the dimension. Individualism is inversely correlated to the PD: individualist cultures tend to be egalitarian, while collectivist cultures tend to be hierarchical. Merritt summed up the implications of these cultural preferences on communication styles, leadership, coordination, conflict resolution, and role expectations as follows:

- People in individualist cultures consider the implications of their behaviors within narrowly defined areas of personal costs and benefits; while those in collectivist cultures consider the implications of their behavior in a context that extends beyond their immediate family.
- Communication in individualist cultures is direct, succinct, personal, and instrumental; communication in collectivist cultures is indirect, elaborate, contextual, and affective. Differences in communication styles affect the feedback and monitoring: people in individualist cultures provide precise, immediate, and verbal feedback; feedback in collectivist cultures tends to be vague, delayed, nonpersonal, and nonverbal.
- People in individualist cultures place more emphasis on resolving conflicts than being agreeable; in collectivist cultures, it is more important to be agreeable than to be right. Further implications in conflict resolution arise from the fact that while collectivist cultures address the broad context and avoid focus on specific detail, individualist cultures concentrate on the key issues and play down the broad context.
- Conflict resolution in collectivist cultures follows strategies based on compromise, avoidance, collaboration and, lastly, competition. The order is almost reversed in individualist cultures: collaboration, competition, compromise, and lastly, avoidance.
- In collectivist cultures, the superiors have considerable authority and discretion. In individualist cultures, the leaders are bound by rules that have been previously negotiated in detail by all involved. While leaders in individualist cultures "hold office" and are therefore accountable, superiors in collectivist cultures "hold the power" and are not questioned.

Failure to consider cultural issues may result in what Merritt (1993) called "cultural imperialism," which in turn produces "cultural mutiny": "...first the consultants promote a model of optimal CRM behaviors. Masquerading as universally applicable, the behaviors are in reality only 'optimal' in their host culture. Tolerance and polite deference for the trainers' odd ideas give way finally to open disagreement in the face of being asked to provide direct specific feedback (in essence, criticism) to members of their own in-group. In what will be a face-saving strategy for both senior management and the consultant I suspect the remainder of the training will be provided as specified in the contract."

The CRM training that is the product of "cultural imperialism" may be at odds not only with the national culture, but also with the organizational culture. Merritt and Helmreich (1996) also explored the relationship between organizational culture and organizational climate that is, the pilots' appraisal of the organizational culture. A positive organizational climate reflects an organizational culture that is in harmony with the pilots' values; a negative organizational climate indicates conflict between the organizational culture and pilots' values. Merritt and Helmreich drew an interesting correlation between pilots' attitudes and negative organizational climate: pilots who endorse organizational climates considered negative by the majority of the peer group demonstrate less positive CRM attitudes ("macho" attitudes, hierarchical, less interactive cockpits). This is the case when the airline's management is hierarchical, uncommunicative, and unrealistic in its appraisal of stressors. In such cases, pilots with less positive CRM attitudes do not perceive a conflict between their own personal style and the way the airline functions. However, pilots with positive CRM attitudes will experience a conflict between their professional style and the airline's management style.

Having drawn this connection, Merritt and Helmreich explained that, nonetheless, a positive organizational climate is positively correlated to the CRM attitudes, and that airlines with poor morale have weaker attitudes toward CRM, while those with high morale support positive CRM attitudes. They concluded that: "...managers can play a strong and influential role in shaping CRM within their company, simply by the way they choose to manage." This conclusion supports with data, a contention, first expressed by Bruggink (1990): pilots will model their own behavior after the behavior that they observe in the organization, because they believe that it is the behavior that the management expects from them.

Cross-cultural issues in aviation are here to stay. As aviation becomes a global village and as CRM expands beyond the flight deck and across national boundaries, intra- and cross-cultural issues will position themselves at the leading edge of research, design, prevention, and training endeavors. There is too much at stake to engage in simplistic reasoning and cosmetic solutions or, worse still, in denial. The proclaimed objectives of the global aviation system are the same across the community—the safe and efficient transportation of people and goods. However, there are as many different approaches to implement these objectives similar to the cultures and subcultures that exist.

10.6 The Link between CRM and Accidents

The relationship between improper CRM behaviors, *vis-à-vis* accident causation seems to have persuasively been established in the United States as far back as 1979 (Cooper et al., 1979). On the other hand, regional safety statistics provided by Boeing (Russell, 1993) suggest differences that might encourage generalizations between CRM and safety. Johnston (1993a) suggested caution when qualitatively linking aviation accidents and CRM. Care must be exercised when drawing fast conclusions and extending generalizations across different aviation contexts. Statistical analyses show a sequence of cause/effect relationships that reflect agreed categorizations largely determined by prevailing beliefs. For all the scientific rigor and impartiality that statistics usually reflect, the benefit of hindsight plays a role in defining the numbers. Those involved in statistical analysis know that the "slings and arrows" of temptation are real. An analyst's major and deliberate effort must be to stick to absolute impartiality and to resist "seeing what one wants to see" in the data evaluated.

Moreover, while revealing relationships perceived to be prevalent in accident causation, statistics do not reveal the processes involved in such relationships. These processes—as well as their underlying and supporting beliefs—are influenced by cultural factors and bounded by environmental and contextual constraints. It is contended that the answers to the safety questions lie not in the numbers, but in the understanding of the processes. It has been asserted that 70% of aviation accidents in a particular context—the United States, for example—owing to human error might involve processes that relate to the complexities of human interaction in small teams in dynamic environments. However, the North American aviation system is arguably fit and healthy, and supports flight crews in discharging their duties. In this case, where the system is beyond serious challenge, CRM training is an answer to operational human error. However, in a developing aviation system, the "70% argument" might simply attest to the impossibility of humans to achieve the system's production goals with existing tools, and to the fact that the system generates risk-taking behaviors as normal practice. In this case, CRM, although a palliative, would be far from a solution.

Commenting with insider's knowledge about aviation safety performance of Third World countries, Faizi (1994) reflected that: "Safety culture in a society is the end product of many ingredients: literacy rate, socioeconomic conditions, level of available technical education, natural resources, industrial base and last, but not least, political stability." He prefaced a candid, yet thorough analysis explaining the poor safety record of developing regions of the world by warning "...what follows cannot be considered the story of any one single country. The majority of Third World countries are tormented by similar problems of varying intensity or gravity." After advocating for the need to collect exclusive data from the Third World countries to address safety deficiencies, and comparing it with prescribing the medicine appropriate to the patient's and not somebody else's symptoms, Faizi summed up the deficiencies of the developing regions of the world:

- Slackness of regulatory functions
- Inadequate professional training
- Nonprofessional safety management
- Funds—mismanagement and scarcity
- Aging fleet

What Faizi enumerated are flawed organizational processes (Maurino, Reason, Johnston, & Lee, 1995; Reason, 1990) that contribute to the differences in the safety statistics among different regions. Interestingly, CRM training only deserves a passing remark in Faizi's analysis: "...Words like Crew Resource Management (CRM) and Line-Oriented Flight Training (LOFT), which have become household words in the 'NORTH', are alien to the crews of the 'SOUTH'. In these countries, it is still 'stick and rudder' skills which determine the efficiency and merit of an airman..."

The foregoing would seem to validate Johnston's concerns and suggests caution when proceeding forward. While the symptoms of certain safety deficiencies may appear similar on the surface, the underlying cultural and contextual factors may dictate radically different solutions. In fact, safety deficiencies that could be addressed by CRM training in North America might not be effectively addressed at all by training in other regions of the world (Maurino, 1994).

10.7 Latest Developments

There is a continued concern about the reactive application of human-factors knowledge largely favored in the past; most often than not after the investigation of an accident uncovered flaws in human performance owing to inherent human limitations and/or fostered by deficient human–technology interfaces. Reactive approaches tend to focus on immediate rather than on the root causes of problems, because of the emotional contexts within which they take place. Improving safety through the application of human-factors knowledge requires the proactive application of the knowledge to anticipate and minimize the consequences of human error and deficient human–technology interfaces.

A proactive stance must start from the foundations. Human factors must progress beyond the "knobs and dials" of long ergonomic standing. It is essential to incorporate existing human-factors knowledge at the stage of system design, before the system is operational and not after—that is, during the certification process of equipment, procedures, and personnel. Likewise, it is essential to the provide end-users (flight crews, air-traffic controllers, mechanics, and dispatchers) with relevant knowledge and skills related to human capabilities and limitations.

This guiding philosophy led ICAO to develop human-factors-related standards and recommended practices (SARPs) for inclusion in the Annexes to the Chicago Convention and associated documentation. Annex 1 (Personnel Licensing) and Annex 6 (Operation of Aircraft) have been amended to include human-factors training standards. The human-factors training requirements for flight crews in Annexes 1 and 6 carry important consequences for trainers and training developers, regulators and human-factors researchers. However, a considerable number of organizations are yet to implement such training. The onus is on trainers and training developers to see that initial CRM training is optimized, and that recurrent CRM training is operationally relevant and observes the philosophy underlying the requirements in Annexes 1 and 6, which are discussed in this chapter.

The regulatory community will have the responsibility of developing an appropriate regulatory framework for a field in which, notwithstanding a major educational campaign by ICAO, there are still misconceptions about the aim and objective behind human-factors regulations. However, the real onus falls upon the research community. The evaluation of CRM has been the focus of research in different countries. Nevertheless, universally accepted tools for CRM student evaluation still does not exist. While progress has been made, the ICAO requirement dictates the need for ongoing research. Until an objective evaluation tool is designed and accepted by consensus of opinion, we can evaluate human-factors *knowledge*, but we must be very cautious when it comes to the evaluation of human-factors *skills*.

10.8 A Tale of Two Continents

The response of the aviation community to ICAO's requirements and to the challenge of an integrated approach to aviation safety is biased—it could not be otherwise—by cultural preferences.

In Europe, the European Joint Aviation Authority (JAA) responded promptly to ICAO's requirements to include human-factors training within operational personnel training curricula. The JAA is an associated body representing the civil aviation regulatory authorities of European countries who have agreed to cooperate in developing and implementing common safety and regulatory standards—the Joint Aviation Regulations (JARs) (Pearson, 1995). All JARs are based on ICAO standards, and the primary aim is to reduce national variance. Flight-crew licensing is governed by the JAR-Flight Crew Licensing (FCL) provisions.

In terms of human-factors training aimed at developing knowledge, JAR-FCL provisions are consistent with developments in aviation human-factors training. Furthermore, applicants for pilot's licenses are required to pass a written examination in basic human performance and limitations (HPL) knowledge. A number of books and educational materials have been published and are easily accessible (Campbell & Bagshaw, 1991; Green, Muir, James, Gradwell, & Green, 1991; Hawkins 1993; Trollip & Jensen, 1991). The United Kingdom has been the first authority among the JAA member states which made HPL training and examination mandatory. In 1993, some 10,000 license applicants had taken the examination, with an initial pass rate that rose from 30% to 70%–80% between 1991 and 1993 (Barnes, 1993).

JAR-FCL also requires that any applicant to a commercial pilot license and instrument rating (CPL/IR) who intends to operate in multicrew aircraft types must satisfactorily complete a training course aimed at developing skills, referred to as MCC (multicrew cooperation) course. The objective of the MCC course is "...to enable pilots to become proficient in instrument flying and multicrew cooperation..." Technical exercises are mandated in detail, including the type-specific training exercises, such as engine failure and fire and emergency descents. JAR-FCL also includes the requirement for skills evaluation after completion of MCC training.

According to Johnston (1993b), the MCC course represents an innovative departure of existing licensing and training practices. Initial JAR-FCL guidance material on human-factors issues within the MCC course implied a basic CRM course. The MCC included 30 hours of simulator training; however, it neither included mandatory groundschool training nor any guidance as to the possible content of theoretical training. The human-factors skills evaluation embedded in the MCC course was based upon information in JAR-FCL which related to the general conduct of checks and assessment of acceptable pilot performance. Tests were conducted in a multicrew environment, and there was the requirement to assess the "management of crew cooperation" and "maintaining a general survey of the airplane operation by appropriate supervision" on a pass/fail basis.

In North America, the response to the challenge of the integration of human-factors knowledge into operational personnel training took the form the Advanced Qualification Programme (AQP) (FAA, 1991). Rather than establishing a programmed number of hours for a training course, AQP is based on the concept of training to proficiency. The AQP applicant rather than the regulatory authority develops a set of proficiency objectives that substitute the traditional training and checking requirements, thus, assuring that valuable training resources are not expended in training and checking activities of marginal relevance.

AQP is an attempt to download the complex regulatory maze of Federal Aviation Regulation (FARs). Regulatory standards have traditionally addressed technical performance in individual positions. This approach was effective for the generation of aircraft to which it was aimed. However, as technology was introduced, the relevance of existing regulatory standards became under scrutiny. Moreover, existing standards were silent regarding proficiency in crew functions. After the National Transportation Safety Board (NTSB) recommended in 1988 that CRM training should be included in simulator or aircraft training exercises, airlines gradually embraced the recommendation. However, given the absence of a regulatory requirement, widely different CRM training programs could be observed. Another deciding factor in developing AQP was that existing regulatory requirements did not reflect advances in aircraft technology or the changing function of the pilot from a controller to a systems manager. All these developments carry significant training implications, and AQP provides a systematic methodology to accommodate them.

AQP is type-specific. The first step in AQP is to conduct an aircraft type-specific task analysis that includes the identification of CRM behaviors pertinent to the execution of piloting tasks, within the context in which the task will be developed. Thus, proficiency objectives determined are supported by enabling objectives, which prepare the crews for further training in operational flight-deck environments. For recurrent training, certain proficiency objectives may be categorized as currency items, that is, activities on which proficiency is maintained by virtue of frequent exposure during routine line operations. Although verification that proficiency in such items is maintained remains as a requirement, they need not be addressed for training or evaluation during periodic training.

Proficiency objectives can be further categorized as critical or noncritical, based on the significance of the consequences of operational errors, which in turn determines the interval within which training and evaluation of these items must take place. Noncritical items can be distributed over a longer period of time than critical objectives. In this way, training resources may be devoted with greater emphasis on abnormal conditions, emergencies, CRM, and other skills that line experience may identify as important and relevant to pilots' operational needs (Birnbach & Longridge, 1993).

AQP mandates the evaluation of proficiency objectives that reflect both technical and CRM performance in a crew-oriented environment. In its first AC on AQP, the Federal Aviation Administration (FAA) warned that CRM issues and measures are not completely developed at this writing. AQP is expected to support further development of CRM. Collection and analysis of anonymous data (not identified with a named individual) will validate the CRM factors as well as the overall crew performance. This quest, at least in part, was one of the drivers for the development of LOSA.

The implementation of AQP has not been free of problems and has been considered by some, as almost an obsession in the United States. Concern has been expressed that AQP may in some cases result in "no formal CRM training at all in the classroom; it is all done via the sim ride"

(Komich, personal communication, April 5, 1995). Whether simulator training can effectively address all CRM issues that have contributed to aircraft accidents in the past remains an issue open to debate. Real-line operations are situations of unmonitored human performance while, for all its realism, flight crews undergoing simulator training are under monitored conditions and crews might arguably act differently after knowing that they are being scrutinized. While simulation can be an ideal tool to combat flight-crew errors, it might not be equally efficient in dealing with violations that take place within the social, regulated environment, and which include a motivational component (Maurino et al., 1995; Reason, Parker, & Lawton, 1996). Personal and organizational factors influence motivation and thus foster or discourage violations, and such factors are hardly replicable through simulation. While the simulator remains a vital tool, it is not sufficient to address certain CRM issues that might more effectively be addressed in a classroom.

Practitioners involved in AQP development have also expressed concern that the continuing development and validation of CRM issues and measures might become an academic exercise and therefore, loose track of the real issues at hand. Accounts abound of lengthy discussions regarding issues, such as whether a briefing should be evaluated as standard or substandard if the individual involved looks away and breaks the eye contact. The immediate question should be "How does this help in reducing human error in the cockpit"?

10.9 Apples and Oranges: An Interpretation

Readers might perceive the previous section as a comparison between the MCC and the AQP; if so, it would be the proverbial comparison of apples and oranges, because of the differences in cultural perceptions and understandings between Europe and North America. The previous section should simply be considered as a presentation of facts and differences at the developmental phase. However, why are there such differences when pursuing identical goals? Is there such a difference in philosophy? This section is an attempt to briefly answer these questions.

The differences in European and North American approaches to implement CRM training as a system safety tool are a case of cultural issues involved in the transfer of technology. CRM was conceived in a small PD society with low degree of UA. These cultural dimensions are reflected both in CRM and in the caution with which the United States pursued its implementation. Large PD societies, or societies with a high degree of UA, will instead try to "regulate" themselves out of problems. It is a way of building the authority system, evading accountability, and passing the "buck" to the front-line operators in case of safety breakdowns or system failures, so that authority remains intact. Cultural dimensions are clearly reflected in the issue of CRM evaluation: AQP, observing collective-based thinking, warns about the dangers involved in evaluating individual CRM behaviors, advocates for nonjeopardy training, and imposes the requirement for feedback only to improve the system. On the other hand, MCC, following individual-based thinking, regulates the student evaluation on a pass/fail basis. Neither approach is "better" than the other; both the approaches are as good as their respective effectiveness within the systems in which they are implemented. However, we must not miss the crucial point in terms of understanding the differences: if we import bits and pieces—usually those which we like—of a solution that worked in one context to another without due consideration to the social givens and without an understanding of the original context, we may generate aberrant results within the receiving context (Maurino, 1995a).

The JAA favors a highly regulated approach with the attendant lack of flexibility. Such an approach builds almost exclusively upon assuring high levels of individual competence and optimum operational behaviors, with scant consideration for the system within which humans discharge their responsibilities. Consistent with this approach, physicians play a central role in keeping humans fit, supported by psychologists. Owing to the reliance upon individual rather than system performance, individuals deemed unfit by prevailing standards must immediately be removed; hence, there exists emphasis on evaluation and the arguments surrounding psychological testing. The relevance of the knowledge to operational realities is of no great consequence, as knowledge is only a tool for the authority to dictate what operational personnel must and must not do. Across the Atlantic, the FAA also favors a regulated, yet flexible approach. In this case, system performance is the objective, although substandard individual performance is expected to be addressed. As system operation is paramount, operational personnel must be brought into the decision-making loop to support regulators and academicians in defining the blueprint of a healthy system. Knowledge belongs to the community, as the community plays a role in defining the desired standards. As system performance is the objective, individual evaluation is of lesser concern. Likewise, because knowledge is an essential tool for all involved, its relevance to operational realities is of prime importance.

As long as both the approaches remain independent, there is no conflict. However, if we try to embed unmodified CRM that builds upon a social/collective, nonjeopardy, flexible, agreement-by-consensus approach into individualistic, accountable, and inflexible top-down contexts, then we will be laying grounds for a clash of objectives. Thus, proper consideration of differences and cultural calibration of ideas, no matter how good they may be, is essential.

10.10 Threat and Error Management

Having outlined earlier in this chapter that the genesis of TEM was in an analysis of what flight crews did to fly safely from A to B, some may be amazed by the thought that this simple question had not been asked earlier. The previous section highlighted that the traditional pilot training, on both sides of the Atlantic, focuses on error avoidance. It was the belief that high standards of knowledge and technical proficiency would lead to fewer errors. Whilst there were fundamental differences in the approach to the issue, the underpinning philosophy was that with appropriate amounts of knowledge and skill, coupled with maximum zeal on the part of the pilot, flight that was free of errors could be achieved. Sadly, such an utopian ideal is not only fundamentally flawed, but by perpetuation of the notion, flight was arguably being made less safe as pilots were not being taught to deal with error when it occurred.

The research indicated that whilst error minimization by sound training, etc. was important, the need for a major paradigm shift was required in the fact that, irrespective of the training or the zeal of the pilot, if human beings were involved, errors *would* occur. Once this is accepted (and it still meets with resistance from some traditional aviation trainers), the next step is logical. That is, if error is inevitable, then we can develop strategies only by studying the nature of the error in the operational contexts, to best identify *and* manage these errors to prevent them from becoming consequential. Accidents are not caused by error alone, but by mismanaged error.

The operational context in which pilots fly is complex and continuously variable. Planning plays a very important role, but owing to the real-time nature of changing conditions during flight, flexibility and the ability to almost continuously review and modify plans are also essential skills. Issues that individually make up this operational context are labeled as threats. They can be overt or latent. Overt threats are those that are observable or tangible to the crew. Examples include poor weather, aircraft malfunctions, terrain, etc. Overt threats are certain in aviation and little can be done at the crew level to completely avoid them. Nevertheless, under certain operational circumstances they can pose significant risks to the safety, and need to be actively managed. Latent threats are often not readily observable at a crew level, but are concealed within the fabric of the particular operation or the aviation system. Examples of latent threats include ATC practices, industrial issues, poor procedures, and/or manuals and operational pressure owing to conflicting goals between commercial and safety objectives. The growth of LOSA has provided large quantities of data in this area and, for example, has disproved the line of thought that the majority of errors originate in mismanagement of a threat. Whilst superficially attractive, the notion that pilots only err when dealing with difficult circumstances has been quickly dispelled. LOSA data from normal operations indicate that though some threats, if mismanaged, can lead to bad outcomes, approximately 50% of errors are independent of the presence of threats.

Clearly, there will be an element of task saturation present if the threat environment is very complex making error more likely, but the relationship between the presence of threats and the increased likelihood of error is complex and certainly not linear. Significant errors are randomly committed, and are often undetected in extremely benign, low-threat environments. Monitoring and cross-checking of actions and particularly in terms of monitoring and cross-checking of actions and information is strongly reinforced as being amongst the most vital of CRM countermeasures.

Whilst purists from certain disciplines of psychology may argue with the categorization, for the sake of simplicity, ease of use, and "observability," the TEM model lists error under four areas. These are

1. Intentional noncompliance. These are intentional deviations from regulations and/or operators' procedures.
2. Procedural. This is where the intention is correct, but the execution is flawed. They also include errors where the crew simply forgot to do something that was intended—the so-called slips and lapses.
3. Communication error. This includes missing, misinterpreting, or failing to communicate pertinent information. It can be between crewmembers or between the crew and external agencies (e.g., ATC, maintenance personnel, etc.).
4. Operational decision error. These are decision-making errors in areas which are not standardized by regulations or operator procedures and compromise safety. To be categorized as a decision error in the TEM framework, at least one of three conditions must exist. First, the crew must have had other more conservative options available and decided not to take them. The second condition is that the decision was not discussed between the crew members. Third is that the crew had time available but did not use it to evaluate the decision.

Procedural, communication, and operational decision errors may be proficiency-based, that is, they may indicate a lack of knowledge or psychomotor skills, for example, "Stick and Rudder" skills or automation-mode misunderstanding.

Intentional noncompliance is worthy of further consideration. By definition, it indicates no lack of proficiency, but instead signifies a deliberate deviation from the procedures. Such willful violations automatically conjure up images of "rogue pilots" roaming our skies, which may be particularly concerning as LOSA data indicate that it occurs on approximately 35% of all flights (Klinect et al 2001). The truth is more comforting, although it raises other concerns in itself. It is in fact likely that much intentional noncompliance is committed in honest attempts by crews to optimize the operation in a less than perfect system. Whilst on occasion there would be examples of individual pilots believing that they are above the rules, "optimization" is the most frequent cause of intentional noncompliance and is perceived by the crews as being necessary, because the rules and the tasks are often incompatible and sometimes mutually exclusive.

To complete the TEM model, we must examine what happens if the crew is unsuccessful in its threat and/or error management. The result is an undesired aircraft state where the aircraft is in a position in space, time, or configuration that is not intended by the crew. Typical examples are an unstable approach, an altitude "bust," or speed excursion. The presence of an undesired aircraft state, whilst indicating that prior error management has broken down, also indicates the need for immediate action to manage that state. Data show that there is a temptation among the crew to become "locked in" to analyzing why the error occurred, rather than quickly switching to the recovery mode. This is particularly prevalent in errors caused by incorrect interaction with automation.

It is well beyond the scope of this chapter to discuss in detail the specific skill sets needed for effective TEM, but again for simplicity, TEM categorizes the skills and behaviors under three broad areas:

1. Planning skills
2. Execution skills
3. Review/modify skills

In a typical sixth generation of CRM training, the various TEM skill sets needed in each of these areas are discussed—ideally, in the context of actual incidents or LOSA data from that specific operator. TEM training, whilst classically introduced in the classroom, needs to be adopted as a holistic philosophy within an airline. Clearly, it would provide the basis not only for LOFT, but also the philosophy for training and checking in line operations.

10.11 Conclusion

Despite the popular perceptions amongst some aviation practitioners as being at the cutting edge of innovation, in fact, the opposite is probably closer to the mark. The majority of people in aviation are conservative in nature and the inertia involved in bringing about regulatory change perpetuates the relatively slow pace of such change. Therefore, in view of the scope and nature of recent changes in the CRM world, it might be argued by some that these are the best of the times and yet the worst of the times for CRM. CRM has established itself as a central protagonist of the aviation system, and it has become an international training requirement mandated by ICAO. However, as the warning flags raised in this chapter attest, there are very good reasons to plan the future of CRM with critical intelligence. Effective CRM training requires dropping the piecemeal strategies largely preferred in the past in favor of a systems approach. Implementing CRM does not only mean training pilots, controllers, or mechanics, but developing the organization. The notion that CRM is another tool to improve organizational effectiveness has transcended "traditional" operational boundaries, and is gradually becoming acknowledged by those in "high places" (Harper, 1995). This must continue to gain momentum.

Whilst TEM in the context of improving cultural awareness has undoubtedly improved the global picture, the imbalance of priorities and possibilities among different contexts continues to remain a serious obstacle to the globalization of CRM as it exists today. Nothing could be more distant from reality than assuming that because CRM—as we know it—has worked in the Western world, it will work everywhere. This does not imply that the basic principles underlying CRM are not of universal value—CRM is indeed a global concept. It is merely a reminder that the contents of the package with the basic principles might need to be substantially different. TEM, in its intuitive acceptance by the pilots, appears to be helping to breech this chasm.

Airlines within the "industrial belt" of the aviation community are dealing with "rocket science CRM," while other not fortunate enough to have their headquarters located within this belt are still struggling with "CRM 101." Again, the sixth-generation CRM (TEM training) provides a good starting point as well as a focus for improvement of existing programs. Will "American"-designed CRM/TEM find breeding grounds within this organizational context? Faizi's plea for consideration of contexts, Johnston's doubts about the suitability of early CRM to some cultures, and Merritt's concerns about cultural imperialism, are clear reminders that frontiers exist not only to provide a means of living for customs and immigration officers. Let us hope that we have learned from lessons of the past.

Any discussion on cultural issues associated with the transfer of technology has sensitive overtones, some of which have been discussed in this chapter. Airlines challenge aircraft manufacturers' statistics on regional accident rates, arguing that in spite of having an individual "clean record," owing to an airline being based in a poor-record region, such statistics may damage its corporate image. While opinions regarding the fairness of this complaint (or the fairness of the manufacturers' statistics for that matter) will surely differ, this position reflects, beyond doubt, the perceived legitimacy of regional comparisons originated "at the top." It is only logical that less-developed regions might perceive cultural research on the transfer of technology as an indictment against their cultures. However, as the discussion on the implementation of CRM training in Europe reminds us, cultural issues are also involved in the transfer of technology between industrialized societies! If lessons are to be drawn from these skirmishes, they are that cultural research can become extremely vulnerable, there are potential dangers in loose interpretations of cross-cultural research data, and there are definite dangers in going beyond the data, assuming evidence in attempting to support such data.

Airline pilots are often heard lamenting that in every situation in which the professional group was asked to make concessions in the name of safety (concessions, which they say that other groups would have thought unacceptable), the result was a slap in the face. The pilots contended that such was the case with periodic reading of the Flight Data Recorder (FDR) for quality assurance purposes. The file cases of International Federation of Airline Pilots' Associations (IFALPA) report that FDR exceedances were used as the pretext to get rid of industrially "undesirable" pilots. The frightening perspective that CRM might be used for similar purposes is very real, according to pilot groups. They are deeply concerned that, despite the fact that human-factors experts have demanded for a *bona fide* approach to develop CRM, individuals have been disqualified (i.e., fired) through CRM and research, personal data have been used unethically, whilst some human-factors experts have kept a conspicuous silence. ICAO (Maurino, 1995a, 1995c) has repeatedly warned against the considerable damage potentially involved in misunderstandings and misperceptions about CRM. Helmreich (1995) was the first researcher to take a position in denouncing what he has dubbed "The European Crisis." In the best interests of the future of CRM, it remains a matter of hope that more human-factors experts will join Helmreich and Klinect in proving that the pilots are wrong in their fears about further beatings. LOSA, with its underpinning characteristics of ensuring nonjeopardy through completely de-identifying data, continues to expand rapidly as a popular and highly effective safety tool, not only within the United States, but globally. In particular, the growth within South America and the Asia Pacific regions has been nothing short of spectacular. It is encouraging that, at the time of writing, it is also gaining greater acceptance in Europe. LOSA uses TEM as its data collection framework and whilst it is still in its early days, TEM philosophy appears to be easily accepted by the pilots from the most varied of national cultures. Whilst the breakthrough is now happening, it is intriguing to conjecture why the acceptance of TEM and LOSA has been slower off the mark in Europe than the rest of the world. If nothing else, this will provide the basis for much animated discussion amongst human-factors practitioners on perceptions and social/cultural stereotyping!

The real issue is that we must not forget the past. Since the early days of CRM at the NASA Workshop at San Francisco in 1979 all the way to the present day, the "founding fathers" in North America have taken considerable efforts to present CRM as fact and not an ideal. The involvement of operational personnel, the use of operational events, maintaining a healthy distance from the notion of cheap group therapy, and the clear differentiation between changing attitudes (whilst deliberately never addressing personality) were all attempts to convey the notion of a training tool without mystical connotations. CRM is a process and not an outcome. Whilst research efforts in North America to assess CRM have been aimed toward the process, a better definition of desired outcomes in terms of TEM has not only assisted the analysis of CRM skill sets, but has also proven to be intuitively attractive to pilots. As a model, or more correctly—a taxonomy, TEM accurately describes what pilots "do" under that intangible heading of "Airmanship." Over time, all of these givens remain essentially the same. It would nonetheless seem that, simple and sensible as they are, they continue to find difficulty in obtaining an outbound clearance from U.S. Immigration owing not only to actual, but also perceived cultural origins. The most recent problem has not been the outward clearance from the United States, but the inward clearance to Europe. However, there is light on the horizon and many of these barriers are beginning to dissolve. The eminently practical outcomes of the research by Helmreich and Merritt, developed on the seminal work by Hofstede have been timely. It has allowed the development of the concept of TEM with an awareness of cultural issues. The outcome has been that TEM is making inroads across many cultures at a pace that is extremely encouraging. The recent adoption and endorsement of TEM by both ICAO and the International Air Transport Association will undoubtedly speed up this process.

However, complacency has no place anywhere in the aviation system and we must learn from mistakes in the past. We must focus on CRM as a practical tool. It is about developing skills to help better manage threats and errors that pilots face during every flight. It is nothing more and nothing less. We must be cautious of those who wish to measure CRM performance in isolation—outside the technical context. If we forget that assessment of social activities involves a lot more than assessing one hundred feet/five degree/ten knot tolerances, CRM will become extremely vulnerable to those who have been taught to measure costs in a similar isolationist manner.

References

Baker, S., Barbarino, M., Baumgartner, M., Brown, L., Deuchert, I., Dow, N., et al. (1995). *Team resource management in ATC—interim report*. Brussels, Belgium: EUROCONTROL.

Barnes, R. M. (1993). Human performance and limitations requirements: The United Kingdom experience. In International Civil Aviation Organization (Ed.), *Proceedings of the Second ICAO Global Flight Safety and Human Factors Symposium*, Washington DC, April 12–15, 1993 (pp. 23–29). Montreal, Canada: Author.

Birnbach, R. A., & Longridge, T. M. (1993). The regulatory perspective. In E. L. Wiener, B. G. Kanki, & R. L. Helmreich (Eds.), *Cockpit resource management* (pp. 263–281). San Diego, CA: Academic Press Inc.

Blixth, J. (1995). Description of the SAS flight academy crew resource management program. In Ministere de l'Equipment, des Transports et du Tourisme (Ed.), *Proceedings of an International Seminar on Human Factors Training for Pilots*. Paris, France: Author.

Bruggink, G. M. (1990). Reflections on air carrier safety. *The Log, British Airline Pilot Association (BALPA)* (pp. 11–15), July 1990 . London, England: Author.

Campbell, R. D., & Bagshaw, M. (1991). *Human performance and limitations in aviation*. London, England: BSP Professional Books.

Cooper, G. E., White, M. D., & Lauber, J. K. (Eds.). (1979). *Resource management on the flight deck* (NASA Conference Publication 2120). Moffett Field, CA: NASA Ames Research Center.

Chidester, T., (1993). Role of dispatchers in CRM training. In R. S. Jensen, & D. Neumeister (Eds.), *Proceedings of the Seventh International Symposium on Aviation Psychology* (pp. 182–185). Columbus: The Ohio State University.

Dahlburg, J. T. (1995, June 30). Battle of the sexes taken to the skies over India. *The Gazette, Montreal.*

Damos, D. (1995). Pilot selection batteries: A critical examination. In R. Fuller, N. Johnston, & N. McDonald (Eds.), *Applications of Psychology to the Aviation System* (pp. 165–169). Hants, England: Averbury Technical.

Faizi, G. H. (1994). Aviation safety performance of Third World Countries. In:

Federal Aviation Administration (1991). *Advanced qualification program* (Advisory Circular AC 120-54). Washington, DC: U.S. Department of Transportation.

Federal Aviation Administration (1995). *Dispatch resource management* (Advisory Circular AC 121-32). Washington, DC: U.S. Department of Transportation.

Goeters, K. M. (1995). Psychological evaluation of pilots: The present regulations and arguments for their application. In R. Fuller, N. Johnston, & N. McDonald (Eds.), *Applications of psychology to the aviation system* (pp. 149–156). Hants, England: Averbury Technical.

Green, R. G., Muir, H., James, M., Gradwell, D., & Green, R. L. (1991). *Human factors for pilots*. Hants, England: Averbury Technical.

Harper, K. (1995, June 30). Pilots given too much blame for crashes: IATA chief Jeanniot. *The Gazette, Montreal.*

Hawkins, F. H. (1993). *Human factors in flight* (2nd ed.). Hants, England: Ashgate Publishing Limited.

Hayward, B. J., & Lowe, A. R. (Eds.). (1993). *Towards 2000—future directions and new solutions*. Albert Park, Victoria, Australia: The Australian Aviation Psychology Association.

Heinzer, T. H. (1993). Enhancing the impact of human factors training. In International Civil Aviation Organization (Ed.), *Proceedings of the Second ICAO Global Flight Safety and Human Factors Symposium*, Washington DC, April 12–15, 1993 (pp. 190–196). Montreal, Canada: Author.

Helmreich, R. L., Wilhelm, J. A., & Gregorich, S. E. (1988). *Revised versions of the cockpit management attitudes questionnaire (CMAQ) and CRM seminar evaluation form* (Technical Report 88-3). Austin: NASA/University of Texas.

Helmreich, R. L. (1992). Fifteen years of CRM wars: A report from the trenches. In B. J. Hayward, & A. R. Lowe (Eds.), *Towards 2000—future directions and new solutions* (pp. 73–88). Albert Park, Victoria, Australia: The Australian Aviation Psychology Association.

Helmreich, R. L., Merrit, A., Sherman, P., Gregorich, S., & Wiener, E. (1993). *The flight management attitude questionnaire* (NASA/UT/FAA Technical Report 93-4). Austin, TX.

Helmreich, R. L. (1994). New developments in CRM and LOFT training. In International Civil Aviation Organization (Ed.), *Report of the Seventh ICAO Flight Safety and Human Factors Seminar*, Addis Ababa, Ethiopia, October, 18–21, 1994 (pp. 209–225). Montreal, Canada: Author.

Helmerich, R. L. (1995). Critical issues in developing and evaluating CRM. In International Civil Aviation Organization (Ed.), *Seminar Manual—First Regional Human Factors Training Seminar*, Hong Kong, September 4–6, 1995 (pp. 112–117). Montreal, Canada: Author.

Helmreich, R. L., & Fouhee, H. C. (1993). Why crew resource management? Empirical and theoretical bases of human factors training in aviation. In E. Weiner, B. Kanki, & R. Helmreich (Eds.), *Cockpit resource management* (pp. 3–45). San Diego, CA: Academic Press.

Helmreich, R. L., & Merritt, A. C. (1998). *Culture at work in aviation and medicine: National organizational and professional influences*. Aldershot, U.K.: Ashgate.

Helmreich, R. L., Wilhelm, J. A., Klinect, J. R., & Merritt, A. C. (1999). Culture error and crew resource management. In E. Salas, C. A. Bowers, & E. Edens (Eds.), *Improving teamwork in organisations: Applications of resource management training* (pp. 305–331). Hillsdale, NJ: Erlbaum.

Helmreich, R. L., Klinect, J. R., & Wilhelm, J. A. (1999). Models of threat, error and CRM in flight operations. *Proceedings of the Tenth International Symposium on Aviation Psychology* (pp. 677–682). Columbus: The Ohio State University.

Hofstede, G. (1980). Motivation, leadership and organizations: Do American theories apply abroad? *Organizational Dynamics*, Summer 1980, 9, 42–63.

Johnston, A. N. (1993a). CRM: Cross-cultural perspectives. In E. L. Wiener, B. G. Kanki, & R. L. Helmreich (Eds.), *Cockpit resource management* (pp. 367–398). San Diego, CA: Academic Press Inc.

Johnston, A. N. (1993b). Human factors training for the new joint European (JAA) pilot licences. In R. S. Jensen (Ed.), *Proceedings of the Seventh International Symposium on Aviation Psychology* (pp. 736–742). Columbus: The Ohio State University.

Johnston, A. N. (1994). Human factors perspectives; the training community. In International Civil Aviation Organization (Ed.), *Report of the Sixth ICAO Flight Safety and Human Factors Seminar*, Amsterdam, Kingdom of The Netherlands, May 16–19, 1994 (pp. 199–208). Montreal, Canada: Author.

Jones, S. G., & Tesmer, B. (1999). A new tool for investigating and tracking human factors issues in incidents. In *Proceedings of the Tenth International Symposium on Aviation Psychology*. Columbus: The Ohio State University.

Klinect, J. R., Wilhelm, J. A., & Helmreich, R. L. (2001). Threat and error management: Data from line operations safety audits (LOSA). In *Proceedings of the Tenth International Symposium on Aviation Psychology*. Columbus: The Ohio State University.

Klinect, J. R., Murray, P. S., Merritt, A. C., & Helmreich, R. L. (2003). Line operations safety audits (LOSA): Definitions and operating characteristics. In *Proceedings of the Eleventh International Symposium on Aviation Psychology*. Columbus: The Ohio State University.

Maurino, D. E. (1991). The ICAO flight safety and human factors programme. In R. S. Jensen (Ed.), *Proceedings of the Sixth International Symposium on Aviation Psychology* (pp. 1014–1019). Columbus: The Ohio State University.

Maurino, D. E. (1992). Shall we add one more defense? In R. Heron (Ed.), *Third Seminar in Transportation Ergonomics*, October 7, 1992. Montreal, Canada: Transport Canada Development Centre.

Maurino, D. E. (1994). Crosscultural perspectives in human factors training: Lessons from the ICAO human factors programme. *The International Journal of Aviation Psychology*, 4(2), 173–181.

Maurino, D. E. (1995a). The future of human factors and psychology in aviation from the ICAO's perspective. In R. Fuller, N. Johnston, & N. McDonald (Eds.), *Applications of psychology to the aviation system* (pp. 9–15). Hants, England: Averbury Technical.

Maurino, D. E. (1995b). Cultural issues and safety: Observations of the ICAO human factors programme. *Earth Space Review*, 4(1), 10–11.

Maurino, D. E. (1995c). Human factors training for pilots: A status report. In Ministere de l'Equipment, des Transports et du Tourisme (Ed.), *Proceedings of an International Seminar on Human Factors Training for Pilots*. Paris, France: Author.

Maurino, D. E., Reason, J., Johnston, A. N., & Lee, R. (1995). *Beyond aviation human factors*. Hants, England: Averbury Technical.

Merritt, A. C. (1993). The influence of national and organizational culture on human performance. *Invited paper at an Australian Aviation Psychology Association Industry Seminar*, October 25, 1993. Sydney, Australia: The Australian Aviation Psychology Association.

Merritt, A. C. (1994). Cultural issues in CRM training. In International Civil Aviation Organization (Ed.), *Report of the Sixth ICAO Flight Safety and Human Factors Seminar*, Amsterdam, Kingdom of The Netherlands, May 16–19, 1994 (pp. 326–243). Montreal, Canada: Author.

Merritt, A. C. (1995). Designing culturally sensitive CRM training: CRM in China. *Invited paper at the Training and Safety Symposium, Guangzhou, People's Republic of China*, May 1995. Guangzhou, People's Republic of China: Civil Aviation Authority of China.

Merritt, A. C., & Helmreich, R. L. (1996). Culture in the cockpit: A multi-airline study of pilot attitudes and values. In R. S. Jensen (Ed.), *Proceedings of the Eight International Symposium on Aviation Psychology*. Columbus: The Ohio State University.

Murphy, E. (1995). JAA psychological testing for pilots: Objection and alarms. In R. Fuller, N. Johnston, & N. McDonald (Eds.), *Applications of psychology to the aviation system* (pp. 157–164). Hants, England: Averbury Technical.

Orasanu, J. M. (1993). Shared problem models and flight crew performance. In A. N. Johnston, N. McDonald, & R. Fuller (Eds.), *Aviation psychology in practice* (pp. 255–285). Hants, England: Averbury Technical.

Orlady, H. W., & Foushee, H. C. (Eds.). (1986). *Cockpit resource management training* (NASA Conference Publication 2455). Moffett Field, CA: NASA Ames Research Center.

Paries, J., & Amalberti, R. (1995). Recent trends in aviation safety: From individuals to organizational resources management training. In Risoe National Laboratory-Systems Analysis Department (Eds.), *Technical report* (Risoe I series). Roskilde, Denmark: Author.

Pearson, R. A. (1995). JAA psychometric testing: The reasons. In R. Fuller, J. Johnston, & N. McDonald (Eds.), *Applications of psychology to the aviation system* (pp. 139–140). Hants, England: Averbury Technical.

Phelan, P. (1994, August 24–30). Cultivating safety. *Flight International, 146*, 22–24.

Phillips, D. (1994, August 23). Aviation safety vs. national culture? Boeing takes a flyer. *Paris Herald Tribune*.

Reason, J. (1990). *Human error*. England: Cambridge University Press.

Reason, J. (1997). *Managing the risks of organizational accidents*. Aldershot, U.K.: Ashgate.

Reason, J., Parker, D., & Newton, R. (1995). Organisational controls and the varieties of rule-related behaviour. In Economic & Social Research Council (Eds.), *Proceedings of a Conference on Risk in Organisational Settings*, London, May 16–17, 1995. University of York, Heslington, York, England: Author.

Robertson, M. M., Taylor, J. C., Stelly, J. W., & Wagner, R. H. (1995). Maintenance CRM training: Assertiveness attitudes effect on maintenance performance in a matched sample. In R. Fuller, N. Johnston, & N. McDonald (Eds.), *Human factors in aviation operations* (pp. 215–222). Hants, England: Averbury Technical.

Russell, P. D. (1993). Crew factors accidents; regional perspective. In International Air Transport Association (Ed.), *Proceedings of the 22nd Technical Conference: Human Factors in Aviation* (pp. 45–61). Montreal, Canada: Author.

Smith, P. M. (1992). Some implications of CRM/FDM for flight crew management. *Flight Deck*, Autumn 1992,(5) 38–42. Heathrow, England: British Airways Safety Services.

Taggart, W. R. (1995). Implementing human factors training in technical operations and maintenance. *Earth Space Review, 4*(1), 15–18.

Trollip, S. R., & Jensen, R. S. (1991). *Human factors for general aviation.* Englewood, CO: Jeppesen Sanderson, Inc.

Vandenmark, M. J. (1991). Should flight attendants be included in CRM training? A discussion of a major air carrier's approach to total crew training. *The International Journal of Aviation Psychology, 1*(1), 87–94.

Wiener, E. L., Kanki, B. G., & Helmreich, R. L. (Eds.). (1993). *Cockpit resource management.* San Diego, CA: Academic Press Inc.

Yamamori, H. (1986). Optimum culture in the cockpit. In H. W. Orlady, & H. C. Foushee (Eds.), *Cockpit resource management training* (pp. 75–87) (NASA Conference Publication 2455). Moffett Field, CA: NASA Ames Research Center.

11

Fatigue and Biological Rhythms

Giovanni Costa
Università di Milano

This chapter is concerned with the temporal factors affecting human performance and work efficiency. The aim is to emphasize that a careful consideration of the temporal structure of the body functions and, consequently, a proper timing of work activities can be of paramount importance to ensure high levels of performance efficiency, decreasing fatigue, and enhancing health and safety.

11.1 Biological Rhythms of Body Functions

The temporal organization of the human biological system is one of the most remarkable characteristics of living organisms. In the last few decades, chronobiology has highlighted the importance of this aspect for human life, revealing the complex mechanisms underlying the temporal interactions among the various components of the body (systems, organs, tissues, cells, subcellular structures).

These are characterized by a large spectrum of rhythms having different frequencies and amplitudes. According to their periodicity (τ), three main groups of biological rhythms have been defined: (a) ultradian rhythms ($\tau \leq 20\,h$), such as heart rate, respiration, and electric brain waves; (b) circadian rhythms ($20\,h \leq \tau\ 28\,h$), such as sleep/wake cycle or temperature; and (c) infradian rhythms ($\tau \geq 28\,h$), among which circaseptan (weekly), circatrigintan (monthly), and circannual rhythms can be found, like immunological response, the menstrual cycle, and seasonal mood/hormonal changes, respectively. Circadian (Latin: *circa diem* = about a day) rhythms are the most extensively studied owing to their great influence on everyday life (Czeisler & Jewett, 1990; Minors & Waterhouse, 1986).

11.1.1 Circadian Rhythms and Their Mechanisms

Humans are day-light creatures; in the course of evolutionary adaptation, the human species has associated its own state of wakefulness and activity (ergotropic phase) with the day-light period and its sleep and rest state (trophotropic phase) with the night-dark period.

Although in modern society artificial lighting makes it possible to have light for the whole 24 h span, body functions (hormonal, metabolic, digestive, cardiovascular, mental, etc.) are still influenced mainly by the natural light/dark cycle, showing periodic oscillations that have, in general, peaks (acrophases) during the daytime and troughs at night.

For example, body temperature, the main integrated index of body functions, decreases during the night sleep, reaching a minimum of 35.5°C–36°C at about 4 a.m., and increases during the day up to a maximum of 37°C–37.3°C at around 5 p.m. This reflects the increased basal level of body arousal during the day to be fit for activity, whereas it decreases at night to restore the body and recover from fatigue.

After experiments with subjects living in isolation chambers or caves without any reference to external time cues ("free-running"), circadian rhythms have been proved to be sustained by an endogenous timing system, located in the suprachiasmatic nuclei of the hypothalamus in the brain, acting as an autonomous oscillator or "body clock." Its inherent oscillation shows an effective period of 25 h that can be shortened up to 22.5 h or lengthened up to 27 h ("entrainment") by varying the light/dark cycle.

In normal living conditions, this endogenous clock is entrained to the 24 h period ("masking effect") by the rhythmic changes of the external socioenvironmental synchronizers or *zeitgebers*, such as the light period, habitual sleep and meal times, and the timing of work and leisure activities.

As a result, a multitude of circadian rhythms of psychological and physiological variables, having different phases and amplitudes, interact with each other and harmonize on the 24 h period to sustain normal body functioning. For example, during the day, increasing temperature levels are associated with high sympathetic nervous-system activity, higher metabolic rate, increased alertness, better vigilance performance, and physical fitness; during the night, lower temperature levels are associated with increased parasympathetic activity, low metabolic rate, increased sleepiness, and poorer work efficiency. With regard to hormones, cortisol shows its acrophase in the early morning, adrenaline around noon, and melatonin around midnight.

In subjects synchronized to the normal living routine, the loss of this harmonization, or the disruption of some circadian rhythms, can be a premonitory symptom of health impairment as well as one of the clinical manifestations of a disease (Reinberg & Smolenski, 1994).

Two work-related conditions are the main causes of disruption of this circadian rhythmicity: (a) shift and night work that requires a periodic rotation of the duty period around the 24 h span; and (b) long transmeridian flights that impose rapid time-zone transitions and irregular changes in the light/dark regimen.

Therefore, respect for the rhythmic organization of body functions is of paramount importance for people engaged in aviation jobs. In fact, most of them have to face continuous interference with biological and social rhythms in relation to their specific work activity: Flying personnel have to cope with time-zone transitions ("jet lag") and, like ground personnel (e.g., air-traffic control, maintenance, services), do shift and night work.

11.1.2 Circadian Rhythms of Vigilance and Performance

The circadian oscillation in the physiological state plays an important role in influencing human performance and fatigue. It is common knowledge that work efficiency during the night is not the same as during the day. There is in fact clear evidence that it can be significantly influenced by the time of day in which the task is performed, owing to the interaction between the homeostatic (time since awake) and circadian processes that regulate sleep and wakefulness (Turek & Zee, 1999).

Under normal conditions, alertness and the efficiency of many psychomotor functions generally show a progressive increase after waking with peaks 8–10 h later, in the afternoon; after that, they progressively worsen, with troughs at night. They appear roughly influenced by the wake/sleep cycle and parallel the body-temperature rhythm.

Moreover, after well-controlled laboratory experiments, these fluctuations have been shown to vary according to the task demands, suggesting a different weight of endogenous (multioscillators?) and exogenous components on mental activities. For example, performance in simple perceptual motor tasks shows an improvement over the day with higher levels in the late afternoon, whereas in tasks with high short-term memory load, performance decreases from morning to evening. On the other hand, performance having a high cognitive or "working memory" component (e.g., verbal reasoning and information processing) shows an alternate trend with better levels around midday. These fluctuations can vary up to 30% and also reflect the interactions of many other factors, in which the effect of the time of the day can be seen as well, both in terms of phase and amplitude.

The relationship between mental efficiency and basal arousal (the inverse of sleepiness) is described by an inverted U-shaped curve; furthermore, the optimal arousal level for a task depends on its structure, with more complex tasks having lower optimal arousal levels. Increased arousal levels, owing to higher motivation or work demand, decrease the circadian oscillation of performance efficiency, which is counteracted by an extra effort that results in a reduction of alertness and increased fatigue. However, lack of interest and boredom have the opposite effect, as fluctuations increase when motivation is low (Folkard, 1990; Johnson et al., 1992; Monk, 1990).

A "post-lunch dip" in performance has been documented in both laboratory and field conditions (Monk & Folkard, 1985). Although it can be related to the exogenous masking effect of food ingestion (with consequent "blood steal" from the brain and increased sleepiness), there is some evidence that it can be also owing to an endogenous temporal component. In fact, after experiments with no meals and in temporal isolation conditions, it appears that at least part of it is owing to a decrease in the arousal level that does not parallel the underlying increase in body temperature. According to some authors, the post-lunch dip reflects a bimodality of circadian rhythms owing to a 12 h harmonic of the circadian system, probably related to two oscillator systems (Monk, Buysse, Reynolds, & Kupfer, 1996). On the other hand, others support the assumption that the sleepiness/alertness cycle has a 4 h rhythm (Zulley & Bailer, 1988), so that a high level of performance cannot be maintained for more than 2 consecutive hours. Besides, Lavie (1985) indicated the existence of 90 min cycles (ultradian rhythms) in alertness and, consequently, in perceptual-motor performance ("biological working units").

Moreover, performance impairment and negative oscillations increase with prolonged working hours and/or sleep deprivation, particularly in more complex tasks (Babkoff, Mikulincer, Caspy, & Sing, 1992; Doran, Van Dongen, & Dinges, 2001). The maximum decrement during the same extended duty period can be twice as severe when work starts at midnight rather than at midday (Klein & Wegmann, 1979a). Naps inserted during work periods, especially at night, reduce the negative fluctuations, although a temporary grogginess ("sleep inertia") might appear if the person is suddenly awakened during the deep-sleep phase (Gillberg, 1985; Naitoh, Kelly, & Babkoff, 1993; Tassi & Muzet, 2000).

With regard to individual factors, the circadian-phase position of biological rhythms seems to have a relevant influence on performance efficiency and fatigue.

The morning active types (or "larks") appear to have more difficulties in coping with night work when compared with the evening active types (or "owls"), probably because of their earlier psychophysical activation. In fact, they show an advanced phase position of alertness and body temperature, with acrophase 2 h earlier during the day, and a consequent quicker decrease during the night hours. They have fewer difficulties in coping with early morning activities, as their temporal structure facilitates an early awakening and higher vigilance in the first part of the day (Foret, Benoit, & Royant-Parola, 1982; Kerkhof, 1985; Ostberg, 1973).

Introverted people appear to act mainly as morning types, whereas extroverts tend to behave as evening types; the differences in operating behavior become more evident when these characteristics are associated with neurotic instability, with the neurotic introverts showing a worse phase adjustment of circadian rhythms (Colquhoun & Folkard, 1978).

Age also appears to correlate with morningness (Åkerstedt & Torsvall, 1981; Carrier, Parquet, Morettini, & Touchette, 2002); moreover, aging shows a greater susceptibility to the occurrence of rhythm disturbances, sleep disorders, and psychic depression, which can in turn contribute to causing performance impairment.

It is worth mentioning that more than 50% of the variance in the neurobehavioral response to sleep deprivation can be owing to the systematic interindividual differences in vulnerability to performance impairment (Van Dongen, Maislin, & Dinges, 2004).

11.2 Problems Connected with Shift Work and Transmeridian Flights

11.2.1 Shift and Night Work

Shift work interferes with biological and social rhythms, forcing people to adapt themselves to unusual work schedules. On night work, in particular, they must change their normal sleep/wake pattern and adjust their body functions to inverted rest/activity periods, having to work when they should sleep and sleep when they should stay awake.

Such adjustment entails a progressive phase shift of body rhythms, which increases with the number of successive night shifts, oriented forward or backward, respectively, according to the advanced (afternoon–morning–night) or delayed (morning–afternoon–night) rotation of the duty periods. The circadian system is exposed to a continuous stress in the attempt to adjust as quickly as possible to the new working hours, while at the same time, being invariably frustrated by the continuous "changeover" imposed by the alternating shift schedules. Therefore, people seldom or never adjust completely or reach a total inversion, even in cases of permanent night work or slowly rotating shift systems (7–15 consecutive nights shifts), because family and social cues are diurnal and workers immediately go back to their normal habits during rest days.

In many cases, a flattening of the amplitude or a delinking ("desynchronization") among the different rhythmic functions, having different speeds and directions of readjustment, can be observed (Åkerstedt, 1985; Nesthus et al., 2001; Reinberg & Smolenski, 1994).

Such a perturbation of the rhythmic structure plays an important role in influencing health and work capacity. People can suffer to a greater or lesser extent from a series of symptoms ("shift lag" syndrome), characterized by feelings of fatigue, sleepiness, insomnia, digestive problems, poorer mental agility, and impaired performance (Comperatore & Krueger, 1990).

Sleep is the main function altered, being decreased both in quantity and quality. A significant reduction of hours of sleep occurs during the morning-shift period in relation to the early wakeup, which also causes a reduction of REM (rapid eye movement) sleep. On night shifts, diurnal sleep is perturbed both with regard to "circadian reasons," owing to difficulty in falling asleep during the rising phase of the body temperature, and the interferences connected with unfavorable environmental conditions (light and noises in particular). Consequently, it is more fragmented and perturbed in its ultradian components (less REM sleep and phase 2), thus losing some of its restorative properties (Åkerstedt 1996, 2003; Kogi, 1982; Tepas & Carvalhais 1990).

Such conditions in the long run not only can give rise to permanent and severe disturbances of sleep, but also can cause chronic fatigue and changes in behavior patterns, characterized by persistent anxiety or depression, which often requires medical treatment with the administration of psychotropic drugs (Cole, Loving, & Kripke, 1990; Gordon, Cleary, Parker, & Czeisler, 1986).

Shiftworkers' health impairment can include other psychosomatic disorders of the gastrointestinal tract (colitis, gastroduodenitis, and peptic ulcer) and cardiovascular system (hypertension, ischemic heart diseases), as well as metabolic disturbances, that are influenced by other time- and work-related factors and behaviors (Costa, 1996; Knutsson, 2003; Waterhouse, Folkard, & Minors, 1992). In fact, eating habits are often disrupted in shiftworkers, who are forced to change their timetables (according to the shift schedules) and food quality. Moreover, they tend to increase the consumption of stimulating substances (coffee, tea, alcohol, tobacco smoke) that have negative effects on digestive and cardiovascular functions. Furthermore, shiftworkers may well experience more difficulties in keeping normal relationships at the family and social levels, with negative impacts on marital relations, care of children, and social contacts, which can further contribute to the development of the already mentioned psychosomatic disorders.

Besides, we have to consider the multifactorial characteristic of such disorders, and their chronic-degenerative trend. In fact, they are quite common among the general population and show the influence of several factors concerning genetic and family heritage, personality, lifestyles, and social and working conditions. Therefore, shift work has to be seen as one of the many risk factors that favor the development of such disorders, which are more likely to become apparent after long-term exposure.

11.2.2 Jet Lag

Air crews operating on long transmeridian flights have to cope with a shift of external time besides the shift of the working period. Therefore, the individual's biological rhythms have to adjust to abnormal working hours in a changed environmental context, in which a shift of time has occurred as well. The short-term problems arising from these conflicts are similar to those of normal shift work, but are often aggravated by the fatigue owing to the extended duty periods and by loss of the usual external time cues.

After a long transmeridian flight, the circadian system does not adjust immediately to the new local time but requires several days in relation to the number of time zones crossed; the greater the number is, the longer is the time required, considering that the human circadian system can adjust to no more than 60–90 min per day (Wegmann & Klein, 1985).

The speed of resynchronization can differ among individuals (e.g., aged people adjust more slowly than youths) and between variables, leading to an "internal dissociation" (i.e., heart rate and catecholamines adjust more quickly than body temperature and cortisol; the same is true for simple mental tasks when compared with complex psychomotor activities). The adjustment is generally more rapid in westbound (about 1 day per hour of shift) than eastbound flights (about 1.5 day per hour of shift) (Ariznavarreta et al., 2002; Gander, Myhre, Graeber, Andersen, & Lauber, 1989; Gundel & Wegmann, 1987; Minors, Akerstedt, & Waterhouse, 1994; Suvanto, Partinen, Härmä, & Ilmarinen, 1990); in the former case, there is a progressive phase delay of the circadian rhythms in relation to the extended personal day, whereas in the latter, there is a phase advance owing to the compressed day ("directional asymmetry"). A complete readjustment after transitioning six time zones was found to take 13 days in eastward and 10 days in westward flights (Wegmann & Klein, 1985). The reason for the quicker adjustment of body rhythms to a phase delay seems to be related to the natural lengthening of the "biological" day, which arises when people are kept in complete isolation ("free running"), showing a period of 25 h.

However, the recovery pattern is not necessarily linear, but can present a "zigzag" trend, with some of the postshift days showing worse levels of adjustment than those on the day immediately preceding (Monk, Moline, & Graeber, 1988). Furthermore, in several cases, particularly after more extended time-zone transitions, a splitting of rhythms may occur, that is, some adjust by phase advance whereas others adjust by phase delay ("resynchronization by partition") (Gander et al., 1989; Klein & Wegmann, 1979b).

During this period, the person suffers from the so-called jet lag syndrome, characterized by a general feeling of malaise with sleepiness, fatigue, insomnia, hunger, and digestive disturbances with constipation or diarrhea. Sleep times and patterns appear much more variable and disrupted after eastward than westward flights across an equivalent number of time zones, in particular with regard to a reduction in REM sleep and an increase in SWS (slow wave sleep) phases (Graeber, 1994; Lowden & Akerstedt, 1998).

As a result of the desynchronization (and in association with sleep deficit and fatigue), performance on many psychomotor activities (e.g., reaction time, hand–eye coordination, logical reasoning, vigilance) also shows an acute 8%–10% decrement that can last for 3–5 days (in the case of six or more time zones crossed). This appears more pronounced in the afternoon and early night hours after a westbound flight, and during the morning and early afternoon after an eastbound flight (Klein & Wegmann, 1979b; Wegmann & Klein, 1985) and corresponds to the effect of moderate alcohol consumption (Klein, Wegmann, & Hunt, 1972).

This decrement also depends on the nature of the task and can lead to a lower work efficiency that must be compensated by higher motivation and extra effort. Moreover, in air crews operating on long distance flights, jet lag is one of the components of impaired well-being and fitness that is also considerably affected by the great irregularity of the rest/activity patterns and prolonged duty periods. Therefore, the resulting fatigue is owing to two components: (a) desynchronization of the circadian rhythms and (b) prolonged physical and mental effort; the latter is usually compensated by an adequate night's rest, which occurs after long north/south flights without time-zone transition.

With regard to long-term effects on health, it is still questionable whether people engaged for many years on transmeridian routes (and therefore subjected to frequent and irregular perturbations of the body temporal structure) have more negative consequences. The complaints mostly reported are concerned with sleep and nervous disturbances, chronic fatigue, and digestive problems, like normal shift-workers; however, clinical findings do not report an incidence significantly different from the general population. Nevertheless, this comparison can be affected by some confounding factors: (1) air crews are a highly selected and supervised population; (2) possible masking behaviors, such as a negative assessment of medical fitness for work may have important economic consequences; and (3) the presence of high interindividual variability in terms of years spent on long-haul flights during working life (Haugli, Skogtad, & Hellesøy, 1994; Wegmann & Klein, 1985).

11.2.3 Errors and Accidents Owing to Performance Decrement and Fatigue in Workers on Irregular Work Schedules

The circadian fall in psychophysical activation at certain hours of the day, often aggravated by disruption of the biological rhythms, sleep deficit, and fatigue, certainly decreases the work efficiency and increases the possibility of errors and accidents (Ashberg, Kecklund, Akerstedt, & Gamberale, 2000; Dinges, 1995).

Significant decrements in work performance at night (up to 100%) with consequent errors or accidents have been reported in many groups of workers engaged on irregular work schedules, such as switchboard operators, gas-work log-keepers, car drivers, spinners, ship workers, nurses, and train and truck drivers (Monk & Folkard, 1985). In all reports, a less pronounced "post-lunch dip" has also been documented, the possible causes of which have already been mentioned.

Electroencephalographic recordings (Torsvall & Åkerstedt, 1987) have clearly demonstrated the occurrence of dramatic episodes of sleepiness on the job in train drivers during night work, particularly in the second half, leading to potentially hazardous errors (failure to respond to signals). Moreover, the too early starting hours of morning shifts have been documented to influence higher frequencies of errors and accidents in train and bus drivers (Hildebrandt, Rohmert, & Rutenfranz, 1975; Pokorny, Blom, & Van Leeuwen, 1981).

Lapses in performance have also been described among air-traffic controllers (Folkard & Condon, 1987) owing to the so-called night-shift paralysis, such as a sudden immobility of the voluntary muscles during consciousness, which can last from a few seconds to a few minutes, occurring in about 6% of the subjects with peaks around 5 a.m.; this seems related to the number of successive night shifts worked and sleep deprivation.

These are important aspects to consider, particularly, when high and sustained levels of performance efficiency are required as public health is at stake, and its failure may be very costly both from the social and economic points of view; and such is the case with aviation activities. Besides, new technologies that increase cognitive tasks and require more alertness and vigilance are often more vulnerable to errors than manual work activities. On the other hand, automated systems may increase monotony and boredom, thus decreasing vigilance and safety, particularly in case of an emergency. It is worth mentioning that many relevant accidents and disasters that occurred in recent years, for example, both the two main nuclear-reactor accidents at Three Mile Island (1979) and Chernobyl (1986), the Bhopal chemical plant disaster (1984), and the Exxon-Valdez shipwreck (1989), started during the night hours (at 0400, 0125, 0057, and 0020 h, respectively): in all the situations, "human error," in which sleep deprivation, fatigue, and abnormal shift schedules were major determinants, has been claimed as an important factor.

On long-haul flights, cockpit crews' periods of decrement in vigilance, feelings of extreme sleepiness, and involuntary sleep are more likely to occur during the monotonous cruise, at night hours, and occur more frequently during eastbound than westbound flights (Cabon, Coblentz, Mollard, & Fouillot, 1993; Cullen, Drysdale, & Mayes, 1997; Graeber, 1994; Wright & McGown, 2004).

On reviewing some of the main airline accidents that occurred during the period of 1967–1988, Price and Holley (1990) underlined that chronic fatigue, sleep loss, and desynchronosis were three "human factors" that contributed significantly to the unfavorable events. In most cases, they were the consequence of improper work scheduling that imposed prolonged duty periods and irregular wake times in the previous hours, not allowing sufficient time to rest and sleep. In other cases, the influence of circadian desynchronization owing to time-zone changes or night work (in one case, concerning maintenance personnel) appeared evident. The negative effect of sleep loss and shift work on mental performance has also been claimed for the Space Shuttle Challenger accident (1986).

On the other hand, the epidemiological studies concerning work accidents among normal shiftworkers are quite controversial; some investigations reported more accidents on night shifts, others found more on day shifts, and still others reported less frequent but more serious events on night shifts. This can be explained by considering the different tasks and work sectors examined (at major or minor risk of accident) and the daily fluctuations in work demands, usually with lower levels at night (i.e., night interruption of higher risk jobs, slowing down of work pace, added automation), which can compensate for the reduction in psychophysical performance.

However, the macro-analysis of some well-controlled studies concerning accidents in road transport, maritime operations, and industrial situations showed a common trend of accident risk, which appears to parallel the mean trend in sleep propensity (Lavie, 1991) over the 24 h day: it is highest in the early hours of the day (02.00–04.00), showing a second minor peak in the early afternoon (14.00–16.00) corresponding to the post-lunch dip, and lowest in the late morning (10.00–12.00) and late afternoon (18.00–20.00) (Folkard, 1997).

Besides time of day, two other temporal factors can have a significant effect on fatigue and accident risk: (a) hours on duty: in the twelfth hour, the risk is more than double than that during the first 8 h; and (b) number of successive night shifts: for example, in the fourth night, the risk is 36% higher when compared with the first one (Folkard & Akerstedt, 2004; Folkard & Tucker, 2003; Hänecke, Tiedemann, Nachreiner, & Grzech-Sukalo, 1998; Nachreiner, 2000; Sammel, Veijvoda, Maaβ, & Wenzel, 1999).

According to these findings, in recent years, several biomathematical models have been developed, aimed at predicting times of reduced alertness and performance impairment, owing to the cumulative effects of time of day and time on duty period, as well as establishing times that are most suitable for restful recovery sleep and napping, and developing ergonomic shift schedules that are both safe and productive (Mallis, Mejdal, Nguyen, & Dinges, 2004; Van Dongen, 2004).

11.2.4 Specific Problems for Women

It is legitimate to presume that irregular working hours and altered sleep/wake cycles may have increasing specific adverse effects on women's health, above all in relation to their peculiar periodical hormonal activity (the menstrual cycle) and reproductive function.

Moreover, female shiftworkers may have to face more stressful living conditions in relation to higher work loads and time pressure connected with their additional domestic duties. For example, women with children show greater sleep problems with consequent higher cumulative fatigue (Dekker & Tepas, 1990; Estryn-Behar, Gadbois, Peigne, Masson, & Le Gall, 1990). Therefore, these biological and socio-cultural factors may affect women and men differently, not only in terms of social life, but also with respect to adaptation to and tolerance of irregular working hours.

With regard to sexual differences in circadian rhythms, studies carried out in well-controlled exper-imental conditions did not report significant differences between males and females, although women seemed more likely to desynchronize by a shortening of their sleep/wake cycle (Wever, 1985). Also, in real working conditions, many studies carried out widely on nurses did not report differences in adjustments of biochemical parameters and psychophysical functions, including vigilance and per-formance (Folkard, Monk, & Lobban, 1978, 1979; Minors & Waterhouse 1986; Smith, 1979; Suvanto et al., 1990).

In relation to the menstrual cycle, it has been frequently reported that arousal and mood tend to worsen in the premenstrual and menstrual phases, with increases in mental tension, anxiety, and depression; this can determine the changes in alertness and negatively affect work performance. It is still an open question on whether this is related to hormonal mechanisms or more to psychological and social factors (Patkai, 1985).

However, women engaged on irregular shift schedules and night work show a higher incidence of menstrual disorders (irregular cycles and menstrual pains) and problems of fertility, such as lower rates of pregnancies, more abortions/miscarriages or preterm deliveries, and low-birth-weight infants (Nurminen, 1998).

Also, several studies reported high frequencies of irregular menstrual cycles and dysmenorrhea (Cameron, 1969; Preston, Bateman, Short, & Wilkinson, 1973) and a possible increased risk of sponta-neous fetal loss, but no other adverse outcomes on pregnancy among airline flight attendants (Daniell, Vaughan, & Millies, 1990; Lyons, 1992).

Moreover, some recent reports suggest the possibility of an increased risk of breast cancer for women shiftworkers: a potential mechanism would be via impairment of the secretion of melatonin by the pineal gland, owing to exposure to light at night (Swerdlow, 2003). That is also valid for cabin attendants, who can have other concomitant risk factors, such as cosmic radiation and pesticides sprayed (Lancet Oncology, 2002).

11.2.5 Interindividual Differences

It is generally recognized that about 20% of workers cannot stand irregular working schedules, includ-ing night work, because of manifested intolerance, particularly dealing with difficult rhythm adjust-ment and sleep deprivation. On the other hand, only 10% of them do not complain at all, whereas the remaining withstand shift and night work with different levels of discomfort and health disorders. In fact, the effects of such stress conditions can vary widely among people in relation to many inter-vening variables concerning individual factors as well as working situations and social conditions (see Costa, 2003b, for a review).

It is well known that ageing per se is associated with tendencies toward flattening and instability of circadian rhythms; deterioration of sleep restorative efficiency; and a decrease in overall psychophysi-cal fitness and capacity to cope with stressors. In shiftworkers, aging has been found to decrease the ability for circadian adjustment to night work and increase sleep disturbances (Åkerstedt & Torsvall, 1981; Härmä & Kandolin, 2001; Matsumoto & Morita, 1987), so that the effort involved in adjusting to jet lag and from day to night work may become increasingly difficult with advancing age (Petrie, Powell, & Broadbent, 2004). Furthermore, aged (over 50) workers have been observed to have more difficulties in passing from 8 to 12 h shifts, which was reported by both Conrad-Betschart (1990) and Aguirre, Heitmann, Imrie, Sirois, and Moore-Ede (2000).

Gander, De Nguyen, Rosekind, and Connell (1993) documented a significant increase in daily sleep loss and a decline in the amplitude of baseline temperature rhythm with increasing age of air-crew members, particularly those involved in long-haul flights: subjects aged 50–60 years showed on average 3.5 times more sleep loss per day than those aged 20–30 years. Other authors recorded more difficulties in adjustment of psychophysical parameters in both flight attendants (Suvanto et al. 1990) and pilots (Ariznavarreta et al., 2002).

Subjects having greater amplitude of oral-temperature circadian rhythm, who show a more stable circadian structure and a slow adjustment after a phase change, seem to have a better long-term tolerance to shift and night work. Contrastingly, those who exhibit low amplitudes are more prone to a persisting internal desynchronization (Reinberg & Smolenski, 1994).

Moreover, the already-mentioned "evening types," who show delayed circadian peaks of temperature and alertness, have fewer sleeping problems and a better adaptation to night work than "morning types" (Folkard & Monk, 1985). The same appears with respect to the adjustment to jet lag after transmeridian flights (Colquhoun, 1979).

Other authors have stressed the influence of some personality and behavioral aspects, such as the characteristics of "rigidity of sleeping habits" and "languidity to overcome drowsiness" (Costa, Lievore, Casaletti, Gaffuri, & Folkard, 1989; Folkard et al., 1979; Iskra-Golec & Pokorski, 1990; Kaliterna,Vidacek, Prizmic, & Radosevic-Vidacek, 1995), as well as extroversion and neuroticism (Costa, Schallenberg, Ferracin, & Gaffuri, 1995; Gander et al., 1989; Härmä, Ilmarinen, & Knauth, 1988; Iskra-Golec, Marek, & Noworol, 1995), in negatively conditioning both short-term adjustment and long-term tolerance to irregular work schedules and jet lag. However, it is also possible that increased neuroticism could be more a consequence than a cause of long-term shiftwork intolerance (Nachreiner, 1998).

On the other hand, good physical fitness, as a result of physical-training interventions, lessens fatigue and increases performance on night shifts (Härmä et al., 1988; Shapiro, Helsegrave, Beyers, & Picard, 1997).

A good sleep hygiene is also one of the most important factors capable of counteracting sleepiness and fatigue. The more a worker is able to adopt proper sleeping regimens and avoid external sleep disturbances, the more he or she can compensate for the sleep disruption owing to irregular working hours (Åkerstedt, 1996; Kogi, 2000; Tepas & Carvalhais, 1990; Wedderburn, 1991a).

Moreover, the level of commitment to (shift) work, that is, the degree to which the person is able to structure his or her life around it (e.g., avoiding moonlighting or other stressful extra-job activities, adopting more stable sleep/wake regimens, and regular living routine), can favor better adjustments and tolerance (Folkard et al., 1978; Minors & Waterhouse, 1983; Rosa, 1990). This can be enhanced by a vigorous personality or high motivation.

11.3 Preventive Measures

11.3.1 Personal Coping Strategies

As a countermeasure against jet lag, air crews are sometimes advised to try to maintain their home-base time as much as possible. This can be possible in case of a short stay and quick return from home (in 2–3 days), if provided with an adequate sleep length in proper bedrooms shielded from light and noise. Otherwise, increased drowsiness is likely to occur during the subsequent long-haul flight.

In case of a prolonged work period in a different time zone, it is advisable to force the speed of adjustment by immediate adaptation to the local time and engagement in social activities. This can be facilitated by a proper use of exposure to bright light, both through outdoor physical activities and increasing indoor artificial lighting, or by using light visors. Bright light (>1000 lux), in fact, besides having a direct stimulating effect on mental activity, influences the pineal gland and suppresses the secretion of melatonin, a hormone that plays an important role in the circadian system. Therefore, proper timing of light exposure can help in resetting the phase, and affect the direction and magnitude of the entrainment of circadian rhythms: for example, light exposure in the morning causes a phase advance, whereas light exposure in the evening causes a phase delay (Arendt & Deacon, 1996; Bjorvatn, Kecklund, & Akerstedt, 1999; Boulos et al., 2002; Czeisler et al., 1989; Eastman, 1990; Eastman & Martin 1999; Khalsa et al., 1997; Lewy & Sack, 1989; Samel & Wegmann, 1997; Wever, 1989).

These effects also have useful implications on shiftwork, provided that bright light could be used during the night shift (and wearing dark sunglasses while traveling home to avoid natural sunlight), which results not only in short-term adjustment but also long-term tolerance (Costa, Ghirlanda, Minors, & Waterhouse, 1993; Crowley & Eastman, 2001; Czeisler & Dijk, 1995; Eastman, 1990). In fact, bright light can reduce the symptoms of seasonal affective disorders, and some of the negative effects of night work can be linked to a mild form of endogenous depression.

In recent years, oral administration of melatonin has also been tested to counteract both shift lag (Folkard, Arendt, & Clark, 1993) and jet lag (Arendt, 1999; Comperatore, Lieberman, Kirby, Adams, & Crowley, 1996; Croughs & De Bruin, 1996; Herxheimer & Petrie, 2001). It has been proven to be useful in inducing sleep and hastening the resetting of circadian rhythms, reducing feelings of fatigue and sleepiness, and increasing sleep quality and duration, without impairing performance and causing negative effects on health (although long-term effects have not been fully assessed). Similar effects have been recorded after the administration of some short-acting hypnotic agents (Paul et al., 2004a, 2004b; Suhner et al., 2001).

Moreover, proper timing and composition of meals can help in the adaptation. In principle, people should try to maintain stable meal times, which can act as cosynchronizers of body functions and social activities. In cases when full resynchronization of circadian rhythms is required, some authors propose special diet regimens, assuming that meals with high carbohydrate contents facilitate sleep by stimulating serotonin synthesis, whereas meals with high protein contents, which stimulate catecholamines secretion, favor wakefulness and work activity (Ehret, 1981; Romon-Rousseaux, Lancry, Poulet, Frimat, & Furon, 1987). During night work, in particular, it would be preferable that shiftworkers have the meal before 0100 h (also to avoid the coincidence of the post-meal dip with the alertness trough), then take only light snacks with carbohydrates and soft drinks, and not later than 2 h before going to sleep (Waterhouse et al., 1992; Wedderburn, 1991a).

These strategies can help in reducing or avoiding the use of many drugs currently taken to alleviate jet lag symptoms. In fact, the assumption that hypnotics induce sleep (usually benzodiazepines) actually has no effect on the process of resynchronization and may even retard it by interacting with neurotransmitters and receptors; moreover, they can cause a transient (up to 12 h) impairment in psychomotor performance (e.g., visuomotor coordination). Furthermore, in the case of prolonged stays in different time zones, forcing the sleep recovery can also disturb the slow physiological realignment of the other circadian functions, taking into consideration the "zigzag" pattern of the readjustment process (Monk et al., 1988; Walsh, 1990).

On the other hand, the use of stimulating substances, such as xanthines (contained in coffee, tea, and cola drinks) or amphetamines to fight drowsiness and to delay the onset of sleep, in addition to having a potential influence on the adjustment of the circadian system at high doses only, may also disrupt sleep patterns and have negative effects on the digestive system (Walsh, Muehlbach, & Schweitzer, 1995; Wedderburn, 1991a), as well as on performance efficiency if the proper dosage is not taken (Babkoff, French, Whitmore, & Sutherlin, 2002; Wesensten, Belenky, Thorne, Kautz, & Balkin, 2004).

Good sleep strategies and relaxation techniques should also be adopted to help to alleviate desynchronosis and fatigue. People should try to keep a tight sleeping schedule while on shiftwork and avoid disturbances (e.g., by arranging silent and dark bedrooms, using ear plugs, making arrangements with family members and neighbors). The timing of diurnal sleep after a night duty should also be scheduled taking into consideration that sleep onset latency and length can be influenced more by the phase of the temperature rhythm than by prior wakefulness, so that sleep starting in the early morning, during the rising phase of the temperature rhythm, tends to have longer latency and shorter duration than that commencing in the early afternoon (Åkerstedt, 1996; Peen & Bootzin, 1990; Shapiro et al. 1997; Wedderburn, 1991a).

Furthermore, the proper use of naps can be very effective in compensating for sleep loss, improving alertness, and alleviating fatigue, and the length of the nap seems irrelevant (20 min and 2 h may have the same value), but rather its temporal position in relation to duty period and kind of task is significant. Useful naps can be taken before night shift or extended operations ("prophylactic naps"), during night as "anchor sleep" (Minors & Waterhouse, 1981) to alleviate fatigue ("maintenance naps"), or after early morning and night shifts, to integrate normal sleep ("replacement naps") (Åkerstedt, 1998; Åkerstedt & Torsvall, 1985; Bonnet, 1990; Bonnet & Arand, 1994; Naitoh, Englund, & Ryman, 1982; Rosa, 1993; Rosa et al., 1990; Sallinen, Härmä, Åkerstedt, Rosa, & Lillquist, 1998).

11.3.2 Compensatory Measures

Many kinds of interventions, aimed at compensating for shift- and night-work inconveniences, have been introduced in recent years, usually in a very empirical way according to different work conditions and specific problems arising in different companies, work sectors, and countries. Such interventions can act as *counterweights*, aimed only at compensating for the inconveniences, or as *countervalues*, aimed at reducing or eliminating the inconveniences (Thierry, 1980; Wedderburn, 1991b).

The main counterweight is monetary compensation, adopted as a worldwide basic reward for irregular work schedules and prolonged duty periods. It is a simple monetary translation of the multidimensional aspects of the problem, and can have a dangerous masking function. Other counterweights may be represented by interventions aimed at improving work organization and environmental conditions.

With regard to countervalues, most are aimed at limiting the consequences of the inconveniences, for example, medical and psychological health checks; the possibility of early retirement or transfer from night work to day work; availability of extra time off and/or more rest periods at work; canteen facilities; and social support (transports, housing, children care). One important preventive measure can be the exemption from shiftwork for transient periods during particular life phases, owing to health impairments or significant difficulties in family or social life (Rutenfranz, Haider, & Koller, 1985). Andlauer et al. (1982) pointed out that "6 weeks of unbroken rest per year is a minimum requirement to compensate the adverse effects of shift work," thus, allowing an effective recovery of biological functions.

The possibility, or the priority, for transfer to day work after a certain number of years on night shifts (generally 20 years) or over 55 years of age, has been granted by collective agreements in some countries. Passing from shift work that includes night work to schedules without night work brought an improvement in physical, mental, and social well-being (Åkerstedt & Torsvall, 1981). Moreover, some national legislation and collective agreements enable the night workers having a certain amount of night work to their credit (at least 20 years), to retire some years earlier (from 1 to 5 years) than the normal age of retirement (International Labour Office, 1988).

Some countervalues are aimed at reducing the causes of inconveniences, that is, reduction of working hours, night work in particular; adoption of shift schedules based on physiological criteria (see later discussion); rest breaks; reduced work load at night; and sleep strategies and facilities. For example, the introduction of supplement crews is a positive measure that constitutes reduction in the amount of night work of the individual worker by sharing it with a larger number of workers. This also makes it possible to reduce the number of hours on night shift to 7 or 6 or even less, particularly when there are other stress factors, such as heavy work, heat, noise, or high demands on attention.

11.3.3 Some Guidelines for the Arrangement of Shift Work Schedules According to Ergonomic Criteria

Designing shift systems based on the psychophysiological and social criteria also has a positive effect on shiftworkers' performance efficiency and well-being. In recent years, many authors gave some recommendations aimed at making shift schedules more respectful of human characteristics, in particular, the biological circadian system (Knauth, 1998; Knauth & Hornberger, 2003; Monk, 1988; Rosa et al., 1990; Wedderburn, 1991b).

They deal with the following points in particular: the number of consecutive night duties, speed and direction of shift rotation, timing and length of each shift, regularity and flexibility of shift systems, and distribution of rest and leisure times. The most relevant can be summarized as follows.

The number of consecutive night shifts should be reduced as much as possible (preferably one or two at most); this prevents accumulation of sleep deficit and fatigue, and minimizes the disruption of the circadian rhythms. Consequently, rapidly rotating shift systems are preferable to slowly rotating shifts (weekly or fortnightly) or permanent night work. This also helps to avoid prolonged interferences with social relations, which can be further improved by keeping the shift rotation as regular as possible and inserting some free weekends. Moreover, at least one rest day should be scheduled after the night-shift duty.

The forward or "clockwise" rotation of the duty periods (morning–afternoon–night) must be preferred to the backward one (afternoon–morning–night), because it allows a longer rest interval between the shifts, and parallels the "natural" tendency of phase delay of circadian rhythms over 24 h, as in "free-running" conditions. Therefore, shift systems including fast changeovers or doublebacks (e.g., morning and night shifts in the same day), which are very attractive for the long blocks of time off, should be avoided as they do not leave sufficient time for sleeping between the duty shifts.

Morning shift should not start too early, to allow a normal sleep length (as people go to bed at the usual time) and to save the REM sleep, which is more concentrated in the second part of the night sleep. This can decrease fatigue and risk of accidents on the morning shift, which often has the highest workload. A delayed start of all the shifts (e.g., 07.00–15.00–23.00 or 08.00–16.00–24.00 h) could favor a better exploitation of leisure time in the evening also for those working on night shift.

The length of the shifts should be arranged according to the physical and mental load of the task. Therefore, a reduction in the duty hours can become a necessity in job activities requiring high levels of vigilance and performance for their complexity or safety reasons (e.g., fire fighters, train and aircraft drivers, pilots and air-traffic controllers, workers in nuclear and petrochemical plants). For example, Andlauer et al. (1982), after the Three Mile Island accident, proposed doubling up the night shift with two teams and providing satisfactory rest facilities for the off-duty team, so that no operator should work more than 4.5 h in the night shift.

On the other hand, extended work shifts of 9–12 h, which are generally associated with compressed working weeks, should only be contemplated if the nature of work and the workload is suitable for prolonged duty hours, the shift system is designed to minimize accumulation of fatigue and desynchronization, and when there are favorable environmental conditions (e.g., climate, housing, commuting time) (Rosa, 1995).

Besides, in case of prolonged or highly demanding tasks, it may be useful to insert short nap periods, particularly during the night shift. As mentioned earlier, this has been found to have favorable effects on performance (Costa et al., 1995; Gillberg, 1985; Rogers, Spencer, Stone, & Nicholson, 1989; Rosekind et al., 1994), physiological adjustment (Matsumoto, Matsui, Kawamori, & Kogi, 1982; Minors & Waterhouse, 1981), and tolerance of night work (Costa, 1993; Kogi, 1982).

After an extensive review, Kogi (2000) concluded by stating that "napping can only be effective when it is combined with improved work schedules and detailed consultations about improving work assignments, work environment, and other shift working conditions." Therefore, the use of naps during the night shift should be promoted and negotiated officially, taking into consideration that night workers in many cases take naps or "unofficial" rest periods during the night shifts, through informal arrangements among colleagues and under the tacit agreement of the management.

Furthermore, it is important to give the opportunity to maintain the usual meal times as fixed as possible, by scheduling sufficiently long breaks and providing hot meals.

Anyway, it is quite clear that there is no "optimal shift system" in principle, as each shift system has advantages and drawbacks, or in practice, as different work sectors and places have different demands.

Therefore, there may be several "best solutions" for the same work situation, and flexible working time arrangements appear to be very useful strategies in favoring adaptation to shift work (Costa et al., 2003; Knauth, 1998).

11.3.4 Some Suggestions for Air-Crew Scheduling and Crew Behavior

A proper strategy in flight schedules arrangement as well as in timing rest and sleep periods can be of paramount importance in counteracting performance impairment and fatigue owing to desynchronosis and prolonged duty period. This can be achieved by restricting flight-duty periods of excessive length and/or reducing maximum flight time at night and/or extending the rest periods prior to or after long-haul flights.

It is obviously impossible to fix rules to deal with all the possible flight schedules and routes all over the world, but it seems right and proper to consider these aspects and try to incorporate some indications from chronobiological studies on transmeridian flights in flight scheduling (Graeber, 1994; Klein & Wegmann, 1979b; Wegmann, Hasenclever, Christoph, & Trumbach, 1985).

In general, night time between 22.00 and 06.00 h is the least efficacious time to start a flight, as it coincides with the lowest levels of psychophysical activation.

The resynchronization on a new time zone should not be forced, but the crew should return as soon as possible to their home base and be provided with a sufficient rest time to prevent sleep deficits (e.g., 14 h of rest is considered the minimum after crossing four or more time zones).

After returning home from transmeridian flights, the length of the postflight rest period should be directly related to the number of time zones crossed. According to Wegmann, Klein, Conrad, and Esser (1983), the minimum rest period should be as long as the number of time zones crossed multiplied by 8, to avoid a residual desynchronization of no more than 3 h (which seems to have no operational significance) before beginning a new duty period.

The final section of long transmeridian flights should be scheduled to avoid its coincidence with the nocturnal trough of alertness and performance efficiency (Klein & Wegmann, 1979a; Wright & McGown, 2004). For example, the most advantageous time for departure of eastward flights would be in the early evening, as this allows a nap beforehand, which can counteract sleepiness during the first part of the flight; moreover, the circadian rising phase of psychophysiological functions, occurring in correspondence to the second part of the flight, may support a better performance for approach and landing.

Preadjustment of the duty periods in 2–3 days preceding long and complex transmeridian flights, to start work either progressively earlier or later according to the direction of the flight, can avoid abrupt phase shifts and increase the performance efficiency.

Rest and sleep schedules should be carefully disciplined to help compensating for fatigue and desynchronosis. For example, in case of prolonged layover after eastward flights, it would be advisable to limit sleep immediately after arrival and prolong the subsequent wake according to the local time. This would increase the likelihood of an adequate duration of sleep immediately preceding the next duty period.

In the case of flights involving multiple segments and layovers in different time zones, sleep periods should be scheduled based on the two troughs of the biphasic (12 h) alertness cycle, such as a nap of 1 or 2 h plus a sleep of 4–6 h. This would allow better maintenance of performance levels during the subsequent periods of higher alertness, in which work schedules might be optimally adjusted (Dement, Seidel, Cohen, Bliwise, & Carskadon, 1986).

To post the entire crews overseas for prolonged periods of time would be the best for chronobiological adjustment, but not for family and social relationships.

Naps may be very helpful; they pay an essential role in improving alertness. They can be added at certain hours of the rest days to integrate sleep periods, and can be inserted during flight duty (Nicholson et al., 1985; Robertson & Stone, 2002). After several studies on long-haul and complex flights showing that circadian rhythms remain close to home time for about the first 2 days, Sasaki, Kurosaki, Spinweber,

Graeber, and Takahashi (1993) suggested that crew members should schedule their sleep or naps to correspond to early morning and afternoon of home time, to reduce sleepiness and minimize the accumulation of sleep deficit. On the other hand, it could be preferable to permit and schedule flight-deck napping for single crew members, if operationally feasible, instead of letting it happen unpredictably (Petrie et al., 2004).

Planning rest breaks during the flight is also a good measure to reduce physiological sleepiness and avoid unintended sleep. They are more effective in proximity of the nocturnal circadian nadir of alertness and in the middle and latter portion of the flight (Neri et al., 2002; Rosekind et al., 1994).

For air crews not involved in long transmeridian flights, the general guidelines suggested for rapid rotating shiftworkers may be followed, but they should be further adapted in relation to the more irregular patterns of duty sections during the working day.

Finally, it may be advisable to try to take advantage from some individual chronobiological characteristics. It could be useful to consider the different activation curve between morning and evening types, as already mentioned, when scheduling flight timetables, to allow people to work in periods when they are at their best levels. For example, morning-type crew members would certainly be fitter on flights scheduled on the first part of the day, whereas evening types would show a lower sleepiness on evening and night flights. Some suggestions on this are presented in the study by Sasaki, Kurosaki, Mori, and Endo (1986).

11.3.5 Medical Surveillance

Good medical surveillance is essential to ensure that operators are in good health and able to carry out their job without excessive stress and performance impairment. Besides the careful application of precise norms and recommendations given by international authorities (European JAA, 2002; FAA, 1996; ICAO, 1988) for the medical certification of license holders, medical checks should be oriented toward preserving physical and mental health with regard to the temporal organization of body functions (Dinges, Graeber, Rosekind, Samel, & Wegmann, 1996).

In the light of the possible negative consequences connected with desynchronization of the biological rhythms, both selection and periodical checks of workers engaged on irregular work schedules should take into consideration some criteria and suggestions proposed by several authors and institutions (Costa, 2003a; International Labour Office, 1988; Rutenfranz et al., 1985; Scott & LaDou, 1990).

Work at night and on irregular shift schedules should be restricted for people suffering from severe disorders that are associated with or can be aggravated by shift lag and jet lag, in particular: chronic sleep disturbances; important gastrointestinal diseases (e.g., peptic ulcer, chronic hepatitis, and pancreatitis); insulin-dependent diabetes, as regular and proper food intake and correct therapeutic timing are required; hormonal pathologies (e.g., thyroid and suprarenal gland), because they demand regular drug assumption strictly associated with the activity/rest periods; epilepsy, as the seizures can be favored by sleep deprivation and the efficacy of treatment can be hampered by irregular wake/rest schedules; chronic psychiatric disorders, depression in particular, as they are often associated with a disruption of the sleep/wakefulness cycle and can be influenced by the light/dark periods; and coronary heart diseases, severe hypertension, and asthma, as exacerbations are more likely to occur at night and treatment is less effective at certain hours of the day.

Moreover, occupational health doctors should very carefully consider those who may be expected to encounter more difficulty in coping with night work and jet lag on the basis of their psychophysiological characteristics, health, and living conditions, such as age over 50 years; low amplitude and stability of circadian rhythms; excessive sleepiness; extreme morningness; high neuroticism; long commuting and unsatisfactory housing conditions; and women with small children but lacking social support.

Therefore, medical checks have to be focused mainly on sleeping habits and troubles, eating and digestive problems, mood disorders, psychosomatic complaints, drug consumption, housing conditions, transport facilities, work loads, and off-job activities, preferably using standardized questionnaires, for example, the Standard Shiftwork Index (Barton et al., 1995), as well as checklists and rating scales, to monitor the worker's behavior throughout the years.

Besides this, permanent education and counseling should be provided for improving self-care strategies for coping, in particular, with regard to sleep, smoking, diet (e.g., caffeine), stress management, physical fitness, and medications. On the latter, a careful medical supervision has to be addressed to people who are taking medications that can affect the central nervous system, such as antihistaminics, antihypertensives, and psychotropic drugs, either as stimulants (e.g., amphetamines, modafinil) or antidepressants (e.g., monoamino-oxidase and serotonin reuptake inhibitors, triyciclic compounds), as well as hypnotics (including melatonin) and anxiolitics, to avoid any abuse or misuse (Arendt & Deacon 1997; Caldwell, 2000; Ireland, 2002; Jones & Ireland, 2004; Nicholson, Stone, Turner, & Mills, 2000; Nicholson, Roberts, Stone, & Turner, 2001; Wesensten et al., 2004).

The adoption of these criteria could also improve the efficacy of preemployment screenings, to avoid allocating some people who are more vulnerable in circadian rhythmic structure and psychophysical homeostasis, to jobs that require shift and night work, and/or frequent time-zone transitions.

References

Aguirre, A., Heitmann, A., Imrie, A., Sirois, W., & Moore-Ede, M. (2000). Conversion from an 8-H to a 12-H shift schedule. In S. Hornberger, P. Knauth, G. Costa, & S. Folkard (Eds.), *Shiftwork in the 21st century* (pp. 113–118). Frankfurt, Germany: Peter Lang.

Åkerstedt, T. (1985). Adjustment of physiological circadian rhythms and the sleep-wake cycle to shiftwork. In S. Folkard, & T. H. Monk (Eds.), *Hours of work. Temporal factors in work scheduling* (pp. 185–197). Chichester: John Wiley & Sons.

Åkerstedt, T. (1998). Is there an optimal sleep-wake pattern in shift work? *Scandinavian Journal of Work Environment Health, 24*(Suppl. 3), 18–27.

Åkerstedt, T. (1996). *Wide awake at odd hours. Shift work, time zones and burning the midnight oil* (pp. 1–116). Stockholm, Sweden: Swedish Council for Work Life Research.

Åkerstedt, T. (2003). Shift work and disturbed sleep/wakefulness. *Occupational Medicine, 53,* 89–94.

Åkerstedt, T., & Torsvall, L. (1981). Age, sleep and adjustment to shift work. In I. W. P. Koella (Ed.), *Sleep 80* (pp. 190–194). Basel: Karger.

Åkerstedt, T., & Torsvall, L. (1985). Napping in shift work. *Sleep, 8,* 105–109.

Andlauer, P., Rutenfranz, J., Kogi, K., Thierry, H., Vieux, N., & Duverneuil, G. (1982). Organization of night shifts in industries where public safety is at stake. *International Archives of Occupational and Environmental Health, 49,* 353–355.

Arendt, J. (1999). Jet lag and shift work: (2) Therapeutic use of melatonin. *Journal of the Royal Society of Medicine, 92,* 402–405.

Arendt, J., & Deacon, S. (1996). Adapting to phase shifts. I. An experimental model for jet lag and shift work. *Physiology & Behaviour, 59,* 665–673.

Arendt, J., & Deacon, S. (1997). Treatment of circadian rhythm disorders—melatonin. *Chronobiology International, 14*(2), 185–204.

Ariznavarreta, C., Cardinali, D. P., Villanua, M. A., Granados, B., Martìn, M., Cjiesa, J. J., et al. (2002). Circadian rhythms in airline pilots submitted to long-haul transmeridian flights. *Aviation, Space and Environmental Medicine, 73*(5), 445–455.

Ashberg, E., Kecklund, G., Akerstedt, T., & Gamberale, F. (2000). Shiftwork and different dimensions of fatigue. *International Journal Industrial Ergonomics, 26,* 457–465.

Babkoff, H., French, J., Whitmore, J., & Sutherlin, R. (2002). Single-dose bright light and/or caffeine effect on nocturnal performance. *Aviation, Space and Environmental Medicine, 73,* 341–350.

Babkoff, H., Mikulincer, M., Caspy, T., & Sing, H. C. (1992). Selected problems of analysis and interpretation of the effects on sleep deprivation on temperature and performance rhythms. *Annals of the New York Academy of Sciences, 658*, 93–110.

Barton, J., Spelten, E., Totterdell, P., Smith, L., Folkard, S., & Costa, G. (1995). The standard shiftwork index: A battery of questionnaires for assessing shiftwork related problems. *Work & Stress, 9*, 4–30.

Bjorvatn, B., Kecklund, G., & Akerstedt, T. (1999). Bright light treatment used for adaptation to night work and re-adaptation back to day life. A field study at an oil platform in the North Sea. *Journal of Sleep Research, 8*, 105–112.

Bonnet, M. H. (1990). Dealing with shift work: Physical fitness, temperature, and napping. *Work & Stress, 4*, 261–274.

Bonnet, M. H., & Arand, D. L. (1994). The use of prophylactic naps and caffeine to maintain performance during a continuous operation. *Ergonomics, 37*, 1009–1020.

Boulos, Z., Macchi, M., Stürchler, M. P., Stewart, K. T., Brainard, G. C., Suhner, A., et al. (2002). Light visor treatment for jet lag after westward travel across six time zones. *Aviation, Space and Environmental Medicine, 73*, 953–963.

Cabon, P. H., Coblentz, A., Mollard, R. P., & Fouillot, J. P. (1993). Human vigilance in railway and long-haul flight operation. *Ergonomics, 36*, 1019–1033.

Caldwell, J. L. (2000). The use of melatonin: An information paper. *Aviation, Space and Environmental Medicine, 71*, 238–244.

Cameron, R. G. (1969). Effect of flying on the menstrual function of air hostesses. *Aerospace Medicine, 40*, 1020–1023.

Carrier, J., Parquet, J., Morettini, J., & Touchette, E. (2002). Phase advance of sleep and temperature circadian rhythms. *Neuroscience Letters, 320*, 1–4.

Cole, R. J., Loving, R. T., & Kripke, D. F. (1990). Psychiatric aspects of shiftwork. *Occupational Medicine: State of Art Reviews, 5*, 301–314.

Colquhoun, W. P. (1979). Phase shift in temperature rhythm after trasmeridian flights, as related to pre-shift phase angle. *International Archives of Occupational and Environmental Health, 42*, 149–157.

Colquhoun, W. P., & Folkard, S. (1978). Personality differences in body temperature rhythm, and their relation to its adjustment to night work. *Ergonomics, 21*, 811–817.

Comperatore, C. A., & Krueger G. P. (1990). Circadian rhythm desynchronosis, jet-lag, shift lag, and coping strategies. *Occupational Medicine: State of Art Reviews, 5*, 323–341.

Comperatore, C. A., Lieberman, H. R., Kirby, A. W., Adams, B., & Crowley, J. S. (1996). Melatonin efficacy in aviation missions requiring rapid deployment and night operations. *Aviation Space and Environmental Medicine, 67*, 520–524.

Conrad-Betschart, H. (1990). Designing new shift schedules: Participation as a critical factor for an improvement. In G. Costa, G. C. Cesana, K. Kogi, & A. Wedderburn (Eds.), *Shiftwork: Health, sleep and performance* (pp. 772–782). Frankfurt, Germany: Peter Lang.

Costa, G. (1993). Evaluation of work load in air traffic controllers. *Ergonomics, 36*, 1111–1120.

Costa, G. (1996). The impact of shift and night work on health. *Applied Ergonomics, 27*, 9–16.

Costa, G. (2003a). Shift work and occupational medicine: An overview. *Occupational Medicine, 53*, 83–88.

Costa, G. (2003b). Factors influencing health and tolerance to shift work. *Theoretical Issues in Ergonomics Sciences, 4*, 263–288.

Costa, G., Åkerstedt, T., Nachreiner, F., Carvalhais, J., Folkard, S., Frings Dresen, M., et al. (2003). *As time goes by—flexible work hours, health and wellbeing* (Working Life Research in Europe Report No. 8). Stockholm, Sweden: The National Institute for Working Life.

Costa, G., Ghirlanda, G., Minors, D. S., & Waterhouse, J. (1993). Effect of bright light on tolerance to night work. *Scandinavian Journal of Work Environment and Health, 19*, 414–420.

Costa, G., Lievore, F., Casaletti, G., Gaffuri, E., & Folkard, S. (1989). Circadian characteristics influencing interindividual differences in tolerance and adjustment to shiftwork. *Ergonomics, 32*, 373–385.

Costa, G., Schallenberg, G., Ferracin, A., & Gaffuri, E. (1995). Psychophysical conditions of air traffic controllers evaluated by the standard shiftwork index. *Work & Stress, 9,* 281–288.

Croughs, R. J. M., & De Bruin, T. W. A. (1996). Melatonin and jet lag. *Netherlands Journal of Medicine, 49,* 164–166.

Crowley, S. J., & Eastman, C. I. (2001). Black plastic and sunglasses can help night workers. *Shiftwork International Newsletter, 18,* 65.

Cullen, S. A., Drysdale, H. C., & Mayes, R. W. (1997). Role of medical factors in 1000 fatal aviation accidents: Case note study. *British Medical Journal, 314,* 1592.

Czeisler, C. A., & Jewett, M. E. (1990). Human circadian physiology: Interaction of the behavioural rest-activity cycle with the output of the endogenous circadian pacemaker. In M. J. Thorpy (Ed), *Handbook of sleep disorders* (pp. 117–137). New York: Marcel Dekker Inc.

Czeisler, C. A., Kronauer, R. E., Allan, J. S., Duffy, J. F., Jewett, M. E., Brown, E. N., et al. (1989). Bright light induction of strong (type O) resetting of the human circadian pacemaker. *Science, 244,* 1328–1333.

Czeisler, C. H. A., & Dijk, D. J. (1995). Use of bright light to treat maladaptation to night shift work and circadian rhythm sleep disorders. *Journal of Sleep Research, 4*(Suppl. 2), 70–73.

Daniell, W. E., Vaughan, T. L., & Millies, B. A. (1990). Pregnancy outcomes among female flight attendants. *Aviation Space and Environment Medicine, 61,* 840–844.

Dekker, D. K., & Tepas, D. I. (1990). Gender differences in permanent shiftworker sleep behaviour. In G. Costa, G. C. Cesana, K. Kogi, & A. Wedderburn (Eds.), *Shiftwork: Health, sleep and performance* (pp. 77–82). Frankfurt, Germany: Verlag Peter Lang.

Dement, W. C., Seidel, W. F., Cohen, S. A., Bliwise, N. G., & Carskadon, M. A. (1986). Sleep and wakefulness in aircrew before and after transoceanic flights. In R. C. Graeber (Ed.), *Crew factors in flight operations: IV. Sleep and wakefulness in international aircrews* (pp. 23–47) [Technical Memorandum 88231]. Moffett Field, CA: NASA Ames Research Center.

Dinges, D. F. (1995). An overview of sleepiness and accidents. *Journal of Sleep Research, 4*(Suppl. 2), 4–14.

Dinges, D. F., Graeber, R. C., Rosekind, M. R., Samel, A., & Wegmann, H. M. (1996). *Principles and guidelines for duty and rest scheduling in commercial aviation* [Technical memorandum No. 11040]. Moffett Field, CA: NASA Ames Research Center.

Doran, S. M., Van Dongen, H. P. A., & Dinges, D. F. (2001). Sustained attention performance during sleep deprivation: Evidence of state instability. *Archives Italian Biology, 139,* 253–267.

Eastman, C. I. (1990). Circadian rhythms and bright light: Recommendations for shiftwork. *Work & Stress, 4,* 245–260.

Eastman, C. I., & Martin, S. K. (1999). How to use light and dark to produce circadian adaptation to night shift work. *Annals of Medicine, 31,* 87–98.

Ehret, C. F. (1981). New approaches to chronohygiene for the shift worker in the nuclear power industry. In A. Reinberg, A. Vieux, & P. Andlauer (Eds.), *Night and shift work: Biological and social aspects* (pp. 263–270). Oxford: Pergamon Press.

Estryn-Behar, M., Gadbois, C., Peigne, E., Masson, A., & Le Gall, V. (1990). Impact of night shifts on male and female hospital staff. In G. Costa, G. C. Cesana, K. Kogi, & A. Wedderburn (Eds.), *Shiftwork: Health, sleep and performance* (pp. 89–94). Frankfurt, Germany: Verlag Peter Lang.

European Joint Aviation Authorities (JAA). (2002). *Joint aviation requirements* [JAR-FCL 3.205 and 3.325, Appendix 10]. Hoofddorp: The Netherlands.

Federal Aviation Administration. (1996). *Guide for aviation medical examiners.* Washington, DC: Department of Transportation.

Folkard, S. (1990). Circadian performance rhythms: Some practical and theoretical implications. *Philosophical Transactions of the Royal Society of London, B327,* 543–553.

Folkard, S. (1997). Black times: Temporal determinants of transport safety. *Accident Analysis & Prevention, 29/4,* 417–430.

Folkard, S., & Akerstedt, T. (2004). Trends in the risk of accidents and injuries and their implications for models of fatigue and performance. *Aviation, Space and Environmental Medicine, 75*, A161–A167.

Folkard, S., Arendt, J., & Clark, M. (1993). Can Melatonin improve shift workers' tolerance of the night shift? Some preliminary findings. *Chronobiology International, 10*, 315–320.

Folkard, S., & Condon, R. (1987). Night shift paralysis in air traffic control officers. *Ergonomics, 30*, 1353–1363.

Folkard, S., & Monk, T. H. (1985). Circadian performance rhythms. In S. Folkard, & T. Monk (Eds.), *Hours of work. Temporal factors in work scheduling* (pp. 37–52). Chichester: John Wiley & Sons.

Folkard, S., Monk, T. H., & Lobban, M. C. (1978). Short and long-term adjustment of circadian rhythms in "permanent" night nurses. *Ergonomics, 21*, 785–799.

Folkard, S., Monk, T. H., & Lobban, M. C. (1979). Towards a predictive test of adjustment to shift work. *Ergonomics, 22*, 79–91.

Folkard, S., & Tucker, P. (2003). Shift work, safety and productivity. *Occupational Medicine, 53*, 95–101.

Foret, J., Benoit, O., & Royant-Parola, S. (1982). Sleep schedules and peak times of oral temperature and alertness in morning and evening "types." *Ergonomics, 25*, 821–827.

Gander, P. H., De Nguyen, B. E., Rosekind, M. R., & Connell, L. J. (1993). Age, circadian rhythms, and sleep loss in flight crews. *Aviation Space and Environmental Medicine, 64*, 189–195.

Gander, P. H., Myhre, G., Graeber, R. C., Andersen, H. T., & Lauber, J. K. (1989). Adjustment of sleep and circadian temperature rhythm after flights across nine time zones. *Aviation Space and Environmental Medicine, 60*, 733–743.

Gillberg, M. (1985). Effects of naps on performance. In S. Folkard, & T. Monk (Eds.), *Hours of work. Temporal factors in work scheduling* (pp. 77–86). Chichester: John Wiley & Sons.

Gordon, N. P., Cleary, P. D., Parker, C. E., & Czeisler, C. A. (1986). The prevalence and health impact of shiftwork. *American Journal of Public Health, 76*, 1225–1228.

Graeber, R. C. (1994). Jet lag and sleep disruption. In M. H. Kryger, T. Roth, & W. C. Dement (Eds.), *Principles and practice of sleep medicine* (pp. 463–470). London: W. B. Saunders Co.

Gundel, A., & Wegmann, H. (1987). Resynchronisation of the circadian system following a 9-hr advance or a delay zeitgeber shift: Real flights and simulations by a Van-der-Pol oscillator. *Progress in Clinical and Biological Research, 227B*, 391–401.

Hänecke, K., Tiedemann, S., Nachreiner, F., & Grzech-Sukalo, H. (1998). Accident risk as a function of hour at work and time of day as determined from accident data and exposure models for the German working population. *Scandinavian Journal of Work Environment Health, 24*(Suppl. 3), 43–48.

Härmä, M., Ilmarinen, J., & Knauth, P. (1988). Physical fitness and other individual factors relating to the shiftwork tolerance of women. *Chronobiology International, 5*, 417–424.

Härmä, M., & Kandolin, I. (2001). Shiftwork, age and well-being: Recent developments and future perspectives. *Journal of Human Ergology, 30*, 287–293.

Haugli, L., Skogtad, A., & Hellesøy, O. H. (1994). Health, sleep, and mood perceptions reported by airline crews flying short and long hauls. *Aviation Space and Environmental Medicine, 65*, 27–34.

Herxheimer, A., & Petrie, K. J. (2001). *Melatonin for preventing and alleviating jet lag*. Oxford: The Cochrane Library, Issue 4.

Hildebrandt, G., Rohmert, W., & Rutenfranz, J. (1975). The influence of fatigue and rest period on the circadian variation of error frequency of shift workers (engine drivers). In W. P. Colquhoun, S. Folkard, P. Knauth, & J. Rutenfranz (Eds.), *Experimental studies of shiftwork* (pp. 174–187). Opladen: Westdeutscher Verlag.

ICAO Standards and Recommended Practices. (1988). *Personnel licensing* [Annex 1, Chapter 6, Medical requirements]. Montreal, Canada: ICAO.

International Labour Office. (1988). *Night work*. Geneva.

Ireland, R. R. (2002). Pharmacologic considerations for serotonin reuptake inhibitor use by aviators. *Aviation, Space and Environmental Medicine, 73*, 421–429.

Iskra-Golec, I., Marek, T., & Noworol C. (1995). Interactive effect of individual factors on nurses' health and sleep. *Work & Stress, 9*, 256–261.

Iskra-Golec, I., & Pokorski, J. (1990). Sleep and health complaints in shiftworking women with different temperament and circadian characteristics. In G. Costa, G. C. Cesana, K. Kogi, & A. Wedderburn (Eds.), *Shiftwork: Health, sleep and performance* (pp. 95–100). Frankfurt, Germany: Peter Lang.

Johnson, M. P., Duffy, J. F., Dijk, D. J., Ronda, J. M., Dyal, C. M., & Czeisler, C. A. (1992). Short-term memory, alertness and circadian performance: A reappraisal of their relationship to body temperature. *Journal of Sleep Research, 1*, 24–29.

Jones, D. R., & Ireland, R. R. (2004). Aeromedical regulation of aviators using selective serotonin reuptake inhibitors for depressive disorders. *Aviation, Space and Environmental Medicine, 75*, 461–470.

Kaliterna, L., Vidacek, S., Prizmic, S., & Radosevic-Vidacek, B. (1995). Is tolerance to shiftwork predictable from individual differences measures? *Work & Stress, 9*, 140–147.

Kerkhof, G. (1985). Individual differences in circadian rhythms. In S. Folkard, & T. Monk (Eds.), *Hours of work. Temporal factors in work scheduling* (pp. 29–35). Chichester: John Wiley & Sons.

Khalsa, S. B., Jewett, M. E., Klerman, E. B., Duffy, J. F., Rimmer, D. K., Kronauer, R., et al. (1997). Type 0 resetting of the human circadian pacemaker to consecutive bright light pulses against a background of very dim light. *Sleep Research, 26*, 722.

Klein, E. K., & Wegmann, H. M. (1979a). Circadian rhythms of human performance and resistance: Operational aspects. In *Sleep, wakefulness and circadian rhythm* (pp. 2.1–2.17). London: AGARD Lectures Series No. 105.

Klein, E. K., & Wegmann, H. M. (1979b). Circadian rhythms in air operations. In *Sleep, wakefulness and circadian rhythm* (pp. 10.1–10.25). London: AGARD Lectures Series No. 105.

Klein, E. K., Wegmann, H. M., & Hunt, B. I. (1972). Desynchronization of body temperature and performance circadian rhythm as a result of outgoing and homegoing transmeridian flights. *Aerospace Medicine, 43*, 119–132.

Knauth, P. (1998). Innovative worktime arrangements. *Scandinavian Journal of Work Environment Health, 24*(Suppl. 3), 13–17.

Knauth, P., & Hornberger, S. (2003). Preventive and compensatory measures for shift workers. *Occupational Medicine, 53*, 109–116.

Knutsson, A. (2003). Health disorders of shift workers. *Occupational Medicine, 53*, 103–108.

Kogi, K. (1982). Sleep problems in night and shift work. *Journal of Human Ergology, 11*(Suppl.), 217–231.

Kogi, K. (2000). Should shiftworkers nap? Spread, roles and effects of on-duty napping. In S. Hornberger, P. Knauth, G. Costa, & S. Folkard (Eds.), *Shiftwork in the 21st century* (pp. 31–36). Frankfurt, Germany: Peter Lang.

Lancet Oncology (2002). Editorial. Hormonal resynchronization—an occupational hazard. *Lancet Oncology, 3*, 323.

Lavie, P. (1985). Ultradian cycles in wakefulness. Possible implications for work-rest schedules. In S. Folkard, & T. Monk (Eds.), *Hours of work. Temporal factors in work scheduling* (pp. 97–106). Chichester: John Wiley & Sons.

Lavie, P. (1991). The 24-hour sleep propensity function (SFP): Practical and theoretical implications. In T. H. Monk (Ed.), *Sleep, sleepiness and performance* (pp. 65–93). Chichester: Wiley.

Lewy, A. J., & Sack, R. L. (1989). The dim light melatonin onset as a marker for circadian phase position. *Chronobiology International, 6*, 93–102.

Lowden, A., & Akerstedt, T. (1998). Sleep and wake patterns in aircrew on a 2-day layover on westward long distance flights. *Aviation, Space and Environmental Medicine, 69*, 596–602.

Lyons, T. J. (1992). Women in the fast jet cockpit—Aeromedical considerations. *Aviation, Space and Environmental Medicine, 63*, 809–818.

Mallis, M. M., Mejdal, S., Nguyen, T. T., & Dinges, D. F. (2004). Summary of the key features of seven biomathematical models of human fatigue and performance. *Aviation, Space and Environmental Medicine, 75*, A4–A14.

Matsumoto, K., Matsui, T., Kawamori, M., & Kogi, K. (1982). Effects of nighttime naps on sleep patterns of shiftworkers. *Journal of Human Ergology, 11*(Suppl.), 279–289.

Matsumoto, K., & Morita, Y. (1987). Effects of night-time nap and age on sleep patterns of shiftworkers. *Sleep, 10,* 580–589.

Minors, D., Akerstedt, T., & Waterhouse, J. (1994). The adjustment of the circadian rhythm of body temperature to simulated time-zone transitions: A comparison of the effect of using raw versus unmasked data. *Chronobiology International, 11,* 356–366.

Minors, D. S., & Waterhouse, J. M. (1981). Anchor sleep as a synchronizer of rhythms on abnormal routines. In L. C. Johnson, D. I. Tepas, W. P. Colquhoun, & M. J. Colligan (Eds.), *Advances in sleep research. Vol. 7. Biological rhythms, sleep and shift work* (pp. 399–414). New York: Spectrum.

Minors D. S., & Waterhouse, J. M. (1983). Circadian rhythms amplitude—is it related to rhythm adjustment and/or worker motivation? *Ergonomics, 26,* 229–241.

Minors, D. S., & Waterhouse, J. M. (1986). Circadian rhythms and their mechanisms. *Experientia, 42,* 1–13.

Monk, T. (1988). *How to make shift work safe and productive.* Pittsburgh, PA: University of Pittsburgh School of Medicine.

Monk, T. (1990). Shiftworker performance. *Occupational Medicine: State of Art Reviews, 5,* 183–198.

Monk, T. H., Buysse D. J., Reynolds, C. F., & Kupfer, D. J. (1996). Circadian determinants of the postlunch dip in performance. *Chronobiology International, 13,* 123–133.

Monk, T. H., & Folkard, S. (1985). Shiftwork and performance. In S. Folkard, & T. Monk (Eds.), *Hours of work. Temporal factors in work scheduling* (pp. 239–252). Chichester: John Wiley & Sons.

Monk, T. H., Moline, M. L., & Graeber R. C. (1988). Inducing jet-lag in the laboratory: Patterns of adjustment to an acute shift in routine. *Aviation Space and Environmental Medicine, 59,* 703–710.

Nachreiner, F. (1998). Individual and social determinants of shiftwork tolerance. *Scandinavian Journal of Work Environment Health, 24*(Suppl. 3), 35–42.

Nachreiner, F. (2000). Extended working hours and accident risk. In T. Marek, H. Oginska, J. Pokorski, G. Costa, & S. Folkard (Eds.), *Shiftwork 2000. Implications for science, practice and business* (pp. 29–44). Krakow, Poland: Institute of Management, Jagiellonian University.

Naitoh, P., Englund, C. E., & Ryman, D. (1982). Restorative power of naps in designing continuous work schedules. *Journal of Human Ergology, 11*(Suppl.), 259–278.

Naitoh, P., Kelly, T., & Babkoff, H. (1993). Sleep inertia: Best time not to wake up? *Chronobiology International, 10,* 109–118.

Neri, D. F., Oyung, R. L., Colletti, L. M., Mallis, M. M., Tam, P. Y., & Dinges, D. F. (2002). Controlled breaks as a fatigue countermeasure on the flight deck. *Aviation, Space and Environmental Medicine, 73,* 654–664.

Nesthus, T., Cruz, C., Boquet, A., Detwiler, C., Holcomb, K., & Della Rocco, P. (2001). Circadian temperature rhythms in clockwise and counter-clockwise rapidly rotating shift schedules. *Journal of Human Ergology, 30,* 245–249.

Nicholson, A. N., Pascoe, P. A., Roehrs, T., Roth, T., Spencer, M. B., Stone, B. M., et al. (1985). Sustained performance with short evening and morning sleep. *Aviation Space and Environmental Medicine, 56,* 105–114.

Nicholson, A. N., Roberts, D. P., Stone, B. M., & Turner, C. (2001). Antihypertensive therapy in critical occupations: Studies with an angiotensin II agonist. *Aviation, Space and Environmental Medicine, 72,* 1096–1101.

Nicholson, A. N., Stone, B. M., Turner, C., & Mills, S. L. (2000). Antihistamines and aircrew: Usefulness of fexofenadine. *Aviation, Space and Environmental Medicine, 71,* 2–6.

Nurminen, T. (1998). Shift work and reproductive health. *Scandinavian Journal of Work Environment and Health, 15,* 28–34.

Ostberg, O. (1973). Circadian rhythms of food intake and oral temperature in "morning" and "evening" groups of individuals. *Ergonomics, 16,* 203–209.

Patkai, P. (1985). The menstrual cycle. In S. Folkard, & T. Monk (Eds.), *Hours of work. Temporal factors in work scheduling* (pp. 87–96). Chichester: John Wiley & Sons.

Paul, M. A., Gray, G., Sardana, T. M., & Pigeau, R. A. (2004a). Melatonin and Zopiclone as facilitators of early circadian sleep in operational air transport crews. *Aviation, Space and Environmental Medicine, 75*, 439–443.

Paul, M. A., Gray, G., MacLellan, M., & Pigeau, R. A. (2004b). Sleep-inducing pharmaceuticals: A comparison of Melatonin, Zaleplon, Zopiclone and Temazepam. *Aviation, Space and Environmental Medicine, 75*, 512–519.

Peen, P. E., & Bootzin, R. R. (1990). Behavioural techniques for enhancing alertness and performance in shift work. *Work & Stress, 4*, 213–226.

Petrie, K. J., Powell, D., & Broadbent, E. (2004). Fatigue self-management strategies and reported fatigue in international pilots. *Ergonomics, 47*, 461–468.

Pokorny, M., Blom, D., & Van Leeuwen, P. (1981). Analysis of traffic accident data (from bus drivers). An alternative approach (I). In A. Reinberg, A. Vieux, & P. Andlauer (Eds.), *Night and shift work: Biological and social aspects* (pp. 271–278). Oxford: Pergamon Press.

Preston, F. S., Bateman, S. C., Short, R. V., & Wilkinson, R. T. (1973). Effects of time changes on the menstrual cycle length and on performance in airline stewardesses. *Aerospace Medicine, 44*, 438–443.

Price, W. J., & Holley, D. C. (1990). Shiftwork and safety in aviation. *Occupational Medicine: State of Art Reviews, 5*, 343–377.

Reinberg, A., & Smolenski, M. H. (1994). Night and shift work and transmeridian and space flights. In Y. Touitou, & E. Haus (Eds.), *Biologic rhythms in clinical laboratory medicine* (pp. 243–255). Berlin: Springer-Verlag.

Robertson, K. A., & Stone, B. M. (2002). *The effectiveness of short naps in maintaining alertness on the flightdeck: A laboratory study* (Report No. QINETIQ/CHS/P&D/CR020023/1.0). Farnborough, U.K.: QinetiQ.

Romon-Rousseaux, M., Lancry, A., Poulet, I., Frimat, P., & Furon, D. (1987). Effects of protein and carbohydrate snacks on alertness during the night. In A. Oginski, J. Pokorski, & J. Rutenfranz (Eds.), *Contemporary advances in shiftwork research* (pp. 133–141). Krakow, Poland: Medical Academy.

Rogers, A. S., Spencer, M. B., Stone, B. M., & Nicholson, A. N. (1989). The influence of a 1 H nap on performance overnight. *Ergonomics, 32*, 1193–1205.

Rosa, R. (1990). Editorial: Factors for promoting adjustment to night and shift work. *Work & Stress, 4*, 201–202.

Rosa, R. (1993). Napping at home and alertness on the job in rotating shift workers. *Sleep, 16*, 727–735.

Rosa, R. (1995). Extended workshifts and excessive fatigue. *Journal of Sleep Research, 4*(Suppl. 2), 51–56.

Rosa, R. R., Bonnet, M. H., Bootzin, R. R., Eastman, C. I., Monk, T., Penn, P. E., et al. (1990). Intervention factors for promoting adjustment to nightwork and shiftwork. *Occupational Medicine: State of Art Reviews, 5*, 391–414.

Rosekind, M. R., Graeber, R. C., Dinges, D. F., Connel, L. J., Rountree, M. S., Spinweber, C. L., et al. (1994). *Crew factors in flight operations IX: Effects of planned cockpit rest on crew performance and alertness in long-haul operations* (Technical memorandum No. 108839). Moffet Field, CA: NASA Ames Research Center.

Rutenfranz, J., Haider, M., & Koller, M. (1985). Occupational health measures for nightworkers and shiftworkers. In S. Folkard, & T. Monk (Eds.), *Hours of work. Temporal factors in work scheduling* (pp. 199–210). Chichester: John Wiley & Sons.

Sallinen, M., Härmä, M., Åkerstedt, T., Rosa, R., & Lillquist, O. (1998). Promoting alertness with a short nap during a night shift. *Journal of Sleep Research, 7*, 240–247.

Samel, A., & Wegmann, H. M. (1997). Bright light: A countermeasure for jet lag? *Chronobiology International, 14*, 173–183.

Sammel, A., Veijvoda, M., Maaß, H., & Wenzel, J. (1999). Stress and fatigue in 2-pilot crew long-haul operations. *Proceedings of CEAS/AAF Forum Research for safety in civil aviation*, Oct. 21–22, Paris (Chapter 8.1, p. 9).

Sasaki, M., Kurosaki, Y. S., Mori, A., & Endo, S. (1986). Patterns of sleep-wakefulness before and after transmeridian flight in commercial airline pilots. In R. C. Graeber (Ed.), *Crew factors in flight operations: IV. Sleep and wakefulness in international aircrews* (Technical Memorandum 88231). Moffett Field, CA: NASA Ames Research Center.

Sasaki, M., Kurosaki, Y. S., Spinweber, C. L., Graeber, R. C., & Takahashi, T. (1993). Flight crew sleep during multiple layover polar flights. *Aviation Space and Environmental Medicine, 64,* 641–647.

Scott, A. J., & LaDou, J. (1990). Shiftwork: Effects on sleep and health with recommendations for medical surveillance and screening. *Occupational Medicine: State of Art Reviews, 5,* 273–299.

Shapiro, C. M., Helsegrave, R. J., Beyers, J., & Picard, L. (1997). *Working the shift. A self-health guide.* Thornhill, Ontario: JoliJoco Publications.

Smith, P. (1979). A study of weekly and rapidly rotating shift workers. *International Archives of Occupational and Environmental Health, 46,* 111–125.

Suhner, A., Schlagenauf, P., Höfer, I., Johnson, R., Tschopp, A., & Steffen, R. (2001). Effectiveness and tolerability of melatonin and zolpidem for the alleviation of jet-lag. *Aviation, Space and Environmental Medicine, 72,* 638–646.

Suvanto, S., Partinen, M., Härmä, M., & Ilmarinen, J. (1990). Flight attendant's desynchronosis after rapid time zone changes. *Aviation Space and Environmental Medicine, 61,* 543–547.

Swerdlow, A. (2003). *Shift work and breast cancer: A critical review of the epidemiological evidence* (p. 26) [Research report 132]. Sudbury, U.K.: HSE Books.

Tassi, P., & Muzet, A. (2000). Sleep inertia. *Sleep Medicine Reviews, 4,* 341–353.

Tepas, D. I., & Carvalhais, A. B. (1990). Sleep patterns of shiftworkers. *Occupational Medicine: State of Art Reviews, 5,* 199–208.

Thierry, H. K. (1980). Compensation for shiftwork: A model and some results. In W. P. Colquhoun, & J. Rutenfranz (Eds.), *Studies of shiftwork* (pp. 449–462). London: Taylor & Francis.

Torsvall, L., & Åkerstedt, T. (1987). Sleepiness on the job: Continuously measured EEG changes in train drivers. *Electroencephalography and Clinical Neurophysiology, 66,* 502–511.

Turek, F. W., & Zee, P. C. (Eds.) (1999). *Regulation of sleep and circadian rhythms.* Basel: Marcel Dekker Inc.

Van Dongen, H. P. A. (2004). Comparison of mathematical model predictions to experimental data of fatigue and performance. *Aviation, Space and Environmental Medicine, 75,* A15–A36.

Van Dongen, H. P. A., Maislin, G., & Dinges, D. F. (2004). Dealing with inter-individual differences in the temporal dynamics of fatigue and performance: Importance and techniques. *Aviation, Space and Environmental Medicine, 75,* A147–A154.

Walsh, J. K. (1990). Using pharmacological aids to improve waking function and sleep while working at night. *Work & Stress, 4,* 237–243.

Walsh, J. K., Muehlbach, M. J., & Schweitzer, P. K. (1995). Hypnotics and caffeine as countermeasures for shift work-related sleepiness and sleep disturbance. *Journal of Sleep Research, 4*(Suppl. 2), 80–83.

Waterhouse, J. M., Folkard, S., & Minors D. S. (1992). *Shiftwork, health and safety. An overview of the scientific literature 1978–1990.* London: Her Majesty's Stationery Office.

Wedderburn, A. (1991a). Guidelines for shiftworkers. *Bulletin of European Studies on Time* (No. 3). Dublin: European Foundation for the Improvement of Living and Working Conditions.

Wedderburn, A. (1991b). Compensation for shiftwork. *Bulletin of European Shiftwork Topics* (No. 4). Dublin: European Foundation for the Improvement of Living and Working Conditions.

Wegmann, H. M., Hasenclever, S., Christoph, M., & Trumbach, S. (1985). Models to predict operational loads of flight schedules. *Aviation Space and Environmental Medicine, 56,* 27–32.

Wegmann, H. M., & Klein, K. E. (1985). Jet-lag and aircrew scheduling. In S. Folkard, & T. Monk (Eds.), *Hours of work. Temporal factors in work scheduling* (pp. 263–276). Chichester: John Wiley & Sons.

Wegmann, H. M., Klein, K. E., Conrad, B., & Esser, P. (1983). A model of prediction of resynchronization after time-zone flights. *Aviation Space and Environmental Medicine, 54,* 524–527.

Wesensten, N. J., Belenky, G., Thorne, D. R., Kautz, M. A., & Balkin, T. J. (2004). Modafinil vs. caffeine: Effects on fatigue during sleep deprivation. *Aviation, Space and Environmental Medicine, 75,* 520–525.

Wever, R. A. (1985). Man in temporal isolation: Basic principles of the circadian system. In S. Folkard, & T. Monk (Eds.), *Hours of work. Temporal factors in work scheduling* (pp. 15–28). Chichester: John Wiley & Sons.

Wever, R. A. (1989). Light effects on human circadian rhythms: A review of recent Andechs experiments. *Journal of Biological Rhythms, 4,* 161–185.

Wright, N., & McGown, A. (2004). Involuntary sleep during civil air operations: Wrist activity and the prevention of sleep. *Aviation, Space and Environmental Medicine, 75,* 37–45.

Zulley, J., & Bailer, J. (1988). Polyphasic sleep/wake patterns and their significance to vigilance. In J. P. Leonard (Ed.), *Vigilance: Methods, models, and regulations* (pp. 167–180). Frankfurt, Germany: Verlag Peter Lang.

12

Situation Awareness in Aviation Systems

Mica R. Endsley
SA Technologies

In the aviation domain, maintaining a high level of situation awareness (SA) is one of the most critical and challenging features of an aircrew's job. SA can be considered as an internalized mental model of the current state of the flight environment. This integrated picture forms the central organizing feature from which all decision making and action takes place. A vast portion of the aircrew's job is involved in developing SA and keeping it up-to-date in a rapidly changing environment. Consider the following excerpt demonstrating the criticality of SA for the pilot and its frequent elusiveness.

> Ground control cleared us to taxi to Runway 14 with instructions to give way to two single-engine Cessnas that were enroute to Runway 5. With our checklists completed and the Before Takeoff PA [public announcement] accomplished, we called the tower for a takeoff clearance. As we called, we noticed one of the Cessnas depart on Runway 5. Tower responded to our call with a "position

and hold" clearance, and then cleared the second Cessna for a takeoff on Runway 5. As the second Cessna climbed out, the tower cleared us for takeoff on Runway 5.

Takeoff roll was uneventful, but as we raised the gear we remembered the Cessnas again and looked to our left to see if they were still in the area. One of them was not just in the area, he was on a downwind to Runway 5 and about to cross directly in front of us. Our response was to immediately increase our rate of climb and to turn away from the traffic.... If any condition had prevented us from making an expeditious climb immediately after liftoff, we would have been directly in each other's flight path. (Kraby, 1995)

The problem can be even more difficult for the military pilot who must also maintain a keen awareness of many factors pertaining to enemy and friendly aircraft in relation to a prescribed mission, in addition to the normal issues of flight and navigation, as illustrated by this account.

We were running silent now with all emitters either off or in standby... We picked up a small boat visually off the nose, and made an easy ten degree turn to avoid him without making any wing flashes...

Our RWR [radar warning receiver] and ECM [electronic counter measures] equipment were cross checked as we prepared to cross the worst of the mobile defenses. I could see a pair of A-10's strafing what appeared to be a column of tanks. I was really working my head back and forth trying to pick up any missiles or AAA [anti-aircraft artillery] activity and not hit the ground as it raced underneath the nose. I could see Steve's head scanning outside with only quick glances inside at the RWR scope. Just when I thought we might make it through unscathed, I picked up a SAM [surface to air missile] launch at my left nine o'clock heading for my wingman!... It passed harmlessly high and behind my wingman and I made a missile no-guide call on the radio....

Before my heart had a chance to slow down from the last engagement, I picked up another SAM launch at one o'clock headed right at me! It was fired at short range and I barely had time to squeeze off some chaff and light the burners when I had to pull on the pole and perform a last ditch maneuver... I tried to keep my composure as we headed down towards the ground. I squeezed off a couple more bundles of chaff when I realized I should be dropping flares as well! As I leveled off at about 100 feet, Jerry told me there was a second launch at my five o'clock.... (Isaacson, 1985)

To perform in the dynamic flight environment, aircrew must not only know how to operate the aircraft and the proper tactics, procedures and rules for flight, but they must also have an accurate, up-to-date picture of the state of the environment. This is a task that is not simple in light of the complexity and sheer number of factors that must be taken into account to make effective decisions. SA does not end with the simple perception of data, but also depends on a deeper comprehension of the significance of that data based on an understanding of how the components of the environment interact and function, and a subsequent ability to predict future states of the system.

Having a high level of SA can be seen as perhaps the most critical aspect for achieving successful performance in aviation. Problems with SA were found to be the leading causal factor in a review of military aviation mishaps (Hartel, Smith, & Prince, 1991). In a study of accidents among major air carriers, 88% of those involving human error could be attributed to problems with SA (Endsley, 1995a). Owing to its importance and the significant challenge that it poses, finding new ways of improving SA has become one of the major design drivers for the development of new aircraft systems. Interest has also increased within the operational community in finding ways to improve SA through training programs. The successful improvement of SA through aircraft design or training programs requires the guidance of a clear understanding of SA requirements in the flight domain, the individual, the system and environmental factors that affect SA, and a design process that specifically addresses SA in a systematic fashion.

12.1 Situation Awareness Definition

SA is formally defined as "the perception of the elements in the environment within a volume of time and space, the comprehension of their meaning and the projection of their status in the near future" (Endsley, 1988). Thus, SA involves perceiving critical factors in the environment (Level 1 SA), understanding what those factors mean, particularly when integrated together in relation to the aircrew's goals (Level 2), and at the highest level, an understanding of what will happen with the system in the near future (Level 3). These higher levels of SA allow the pilots to function in a timely and effective manner.

12.1.1 Level 1 SA: Perception of the Elements in the Environment

The first step in achieving SA is to perceive the status, attributes, and dynamics of the relevant elements in the environment. A pilot needs to perceive important elements, such as other aircraft, terrain, system status, and warning lights along with their relevant characteristics. In the cockpit, just keeping up with all of the relevant system and flight data as well as other aircraft and navigational data can be quite taxing.

12.1.2 Level 2 SA: Comprehension of the Current Situation

Comprehension of the situation is based on the synthesis of disjointed Level 1 elements. Level 2 SA goes beyond simply being aware of the elements that are present, to include an understanding of the significance of those elements in light of one's goals. The aircrew puts together Level 1 data to form a holistic picture of the environment, including a comprehension of the significance of the objects and events. For example, upon seeing warning lights indicating a problem during take-off, the pilot must quickly determine the seriousness of the problem in terms of the immediate air worthiness of the aircraft, and combine this with the knowledge on the amount of runway remaining to know whether it is an abort situation or not. A novice pilot may be capable of achieving the same Level 1 SA as more experienced pilots, but may fall far short of being able to integrate various data elements along with pertinent goals to comprehend the situation.

12.1.3 Level 3 SA: Projection of Future Status

It is the ability to project the future actions of the elements in the environment, at least in the very near term, which forms the third and highest level of SA. This is achieved through knowledge of the status and dynamics of the elements and a comprehension of the situation (both Level 1 and Level 2 SA). Amalberti and Deblon (1992) found that a significant portion of experienced pilots' time was spent in anticipating the possible future occurrences. This gives them the knowledge (and time) necessary to decide on the most favorable course of action to meet their objectives.

12.2 Situation Awareness Requirements

Clearly understanding SA in the aviation environment rests on a clear elucidation of its elements (at each of the three levels of SA), identifying what the aircrew needs to perceive, understand, and project. These are specific to individual systems and contexts, and, as such, must be determined for a particular class of aircraft and missions (e.g., commercial flight deck, civil aviation, strategic or tactical military aircraft, etc.). However, in general, across many types of aircraft systems, certain classes of elements are needed for SA.

Geographical SA—location of own aircraft, other aircraft, terrain features, airports, cities, waypoints, and navigation fixes; position relative to designated features; runway and taxiway assignments; path to desired locations; climb/descent points.

Spatial/Temporal SA—attitude, altitude, heading, velocity, vertical velocity, G's, flight path; deviation from flight plan and clearances; aircraft capabilities; projected flight path; projected landing time.

System SA—system status, functioning and settings; settings of radio, altimeter, and transponder equipment; air-traffic control (ATC) communications present; deviations from correct settings; flight modes and automation entries and settings; impact of malfunctions/system degrades and settings on system performance and flight safety; fuel; time and distance available on fuel.

Environmental SA—weather formations (area and altitudes affected and movement); temperature, icing, ceilings, clouds, fog, sun, visibility, turbulence, winds, microbursts; instrument flight rules (IFR) vs. visual flight rules (VFR) conditions; areas and altitudes to avoid; flight safety; projected weather conditions.

In addition, for military aircraft, elements relative to the military mission will also be important.

Tactical SA—identification, tactical status, type, capabilities, location and flight dynamics of other aircraft; own capabilities in relation to other aircraft; aircraft detections, launch capabilities, and targeting; threat prioritization, imminence, and assignments; current and projected threat intentions, tactics, firing, and maneuvering; mission timing and status.

Determining specific SA requirements for a particular class of aircraft is dependent on the goals of the aircrew in that particular role. A methodology for determining SA requirements has been developed and applied to fighter aircraft (Endsley, 1993), bomber aircraft (Endsley, 1989), commercial pilots (Endsley, Farley, Jones, Midkiff, & Hansman, 1998), and air-traffic controllers (Endsley & Rodgers, 1994).

12.3 Individual Factors Influencing Situation Awareness

To provide an understanding of the processes and factors that influence the development of SA in complex settings such as aviation, a theoretical model describing the factors underlying SA was developed (Endsley, 1988, 1994, 1995c). The key features of the model will be summarized here and are shown in Figure 12.1 (the reader is referred to Endsley (1995c) for a full explanation of the model and supporting research). In general, SA in the aviation setting is challenged by the limitations of human attention and working memory. The development of relevant long-term memory stores, goal-directed processing, and automaticity of actions through experience and training are seen as the primary mechanisms used for overcoming these limitations to achieve high levels of SA and successful performance.

12.3.1 Processing Limitations

12.3.1.1 Attention

In aviation settings, the development of SA and the decision process are restricted by limited attention and working-memory capacity for novice aircrew and those in novel situations. Direct attention is needed for perceiving and processing the environment to form SA, and for selecting actions and executing responses. In the complex and dynamic aviation environment, information overload, task complexity, and multiple tasks can quickly exceed the aircrew's limited attention capacity. As the supply of attention is limited, more attention to some information may mean a loss of SA on other elements. The resulting lack of SA can result in poor decisions leading to human error. In a review of National Transportation Safety Board (NTSB) aircraft-accident reports, poor SA resulting from attention problems in acquiring data accounted for 31% of accidents involving human error (Endsley, 1995a).

Pilots typically employ a process of information sampling to circumvent attention limits, attending to information in rapid sequence following a pattern dictated by long-term memory concerning the relative priorities and the frequency with which information changes. Working memory also plays an important role in this process, allowing the pilot to modify attention deployment on the basis of other information perceived or active goals. For example, in a study of pilot SA, Fracker (1990) showed that a limited supply of attention was allocated to environmental elements on the basis of their ability to contribute to task success.

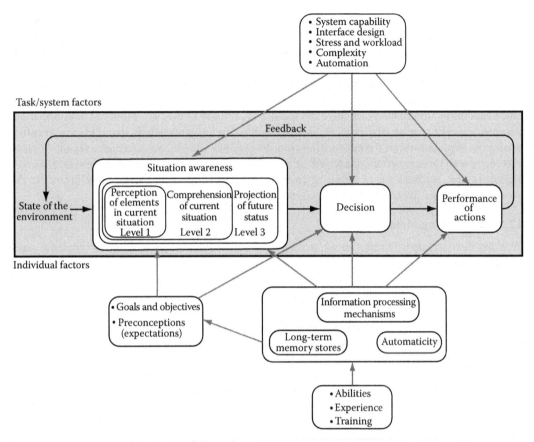

FIGURE 12.1 Model of SA. (From Endsley, M.R., *Hum. Factors*, 37(1), 32, 1995c.)

Unfortunately, people do not always sample information optimally. Typical failings include: (1) forming nonoptimal strategies based on a misperception of the statistical properties of elements in the environment, (2) visual dominance—attending more to visual elements than information coming through competing aural channels, and (3) limitations of human memory, leading to inaccuracy in remembering statistical properties to guide sampling (Wickens, 1984). In addition, owing to information overload, which is a frequent occurrence, pilots may feel that the process of information sampling is either insufficient or inefficient, in which case the pilot may choose to attend to certain information, and neglect other information. If the pilot is correct in this selection, all is well. However, in many instances, this is not the case.

As a highly visible example, reports on controlled descent into the terrain by high-performance fighter aircraft are numerous (McCarthy, 1988). While various factors can be implicated in these incidents, channelized attention (31%), distraction by irrelevant stimuli (22%), task saturation (18%), and preoccupation with one task (17%) have all been indicated as significant causal factors (Kuipers, Kappers, van Holten, van Bergen, & Oosterveld, 1990). Some 56% of the respondents in the same study indicated a lack of attention for primary flight instruments (the single highest factor) and having too much attention directed toward the target plane during combat (28%), as major causes. Clearly, this demonstrates the negative consequences of both intentional and unintentional disruptions of scan patterns. In the case of intentional attention shifts, it is assumed that attention was probably directed to other factors that the pilots erroneously felt to be more important, because their SA was either outdated or incorrectly perceived in the first place. This leads to a very important point. To know which information to focus on and which information to be temporarily ignored, the pilot must have, at some level, an understanding about all of it—that is, "the big picture."

The way in which information is perceived (Level 1 SA) is affected by the contents of both working memory and long-term memory. Advanced knowledge of the characteristics, form, and location of information, for instance, can significantly facilitate the perception of information (Barber & Folkard, 1972; Biederman, Mezzanotte, Rabinowitz, Francolin, & Plude, 1981; Davis, Kramer, & Graham, 1983; Humphreys, 1981; Palmer, 1975; Posner, Nissen, & Ogden, 1978). This type of knowledge is typically gained through experience, training, or preflight planning and analysis. One's preconceptions or expectations about the information can affect the speed and accuracy of the perception of the information. Repeated experience in an environment allows people to develop expectations about future events that predispose them to perceive the information accordingly. They will process information faster, if it is in agreement with those expectations and will be more likely to make an error if it is not (Jones, 1977). As a classic example, readback errors, repeating an expected clearance instead of the actual clearance to the air-traffic controller, are common (Monan, 1986).

12.3.1.2 Working Memory

Working-memory capacity can also act as a limit on SA. In the absence of other mechanisms, most of a person's active processing of information must occur in working memory. The second level of SA involves comprehending the meaning of the data that is perceived. New information must be combined with the existing knowledge and a composite picture of the situation must be developed. Achieving the desired integration and comprehension in this fashion is a very taxing proposition that can seriously overload the pilot's limited working memory, and will draw even further on limited attention, leaving even less capacity to direct toward the process of acquiring new information.

Similarly, projections of future status (Level 3 SA) and subsequent decisions as to the appropriate courses of action will draw upon working memory as well. Wickens (1984) stated that the prediction of future states imposes a strong load on working memory by requiring the maintenance of present conditions, future conditions, rules used to generate the latter from the former, and actions that are appropriate to the future conditions. A heavy load will be imposed on working memory if it is taxed with achieving the higher levels of SA, in addition to formulating and selecting responses and carrying out subsequent actions.

12.3.2 Coping Mechanisms

12.3.2.1 Mental Models

In practice, however, experienced aircrew may use long-term memory stores, most likely in the form of schemata and mental models, to circumvent these limits for learned classes of situations and environments. These mechanisms help in the integration and comprehension of information and the projection of future events. They also allow for decision making on the basis of incomplete information and under uncertainty.

Experienced aircrews often have internal representations of the system that they are dealing with—a mental model. A well-developed mental model provides (a) knowledge of the relevant "elements" of the system that can be used in directing attention and classifying information in the perception process, (b) a means of integrating elements to form an understanding of their meaning (Level 2 SA), and (c) a mechanism for projecting future states of the system based on its current state and an understanding of its dynamics (Level 3 SA). During active decision making, a pilot's perceptions of the current state of the system may be matched to the related schemata in memory that depict prototypical situations or states of the system model. These prototypical situations provide situation classification and understanding, and a projection of what is likely to happen in the future (Level 3 SA).

A major advantage of these mechanisms is that the current situation does not need to be exactly like the one encountered before owing to the use of categorization mapping (a best fit between the characteristics of the situation and the characteristics of known categories or prototypes). The matching process can be almost instantaneous owing to the superior abilities of human pattern-matching mechanisms. When an individual has a well-developed mental model for the behavior of particular systems or domains, it will provide (a) the dynamic direction of attention to critical environmental cues, (b) expectations regarding future states of the environment (including what to expect as well as what not to expect), based on the projection mechanisms of the model, and (c) a direct, single-step link between recognized situation classifications and typical actions, providing very rapid decision making.

The use of mental models also provides useful default information. These default values (expected characteristics of elements based on their classification) may be used by aircrew to predict system performance with incomplete or uncertain information, providing more effective decisions than novices who will be more hampered by missing data. For example, experienced pilots are able to predict within a reasonable range about how fast a particular aircraft is traveling just by knowing what type of aircraft it is. Default information may furnish an important coping mechanism for experienced aircrew in forming SA in many situations, where information is missing or overload prevents them from acquiring all the information that they need.

Well-developed mental models and schema can provide the comprehension and future projection required for the higher levels of SA almost automatically, thus, greatly off-loading working memory and attention requirements. A major advantage of these long-term stores is that a great deal of information can be called upon very rapidly, using only a limited amount of attention (Logan, 1988). When scripts have been developed and tied to these schemas, the entire decision-making process can be greatly simplified, and working memory will be off-loaded even further.

12.3.2.2 Goal-Driven Processing

In the processing of dynamic and complex information, people may switch between data-driven and goal-driven processing. In a data-driven process, various environmental features are detected whose inherent properties determine which information will receive further focalized attention and processing. In this mode, cue salience will have a large impact on which portions of the environment are attended to and thus, SA. People can also operate in a goal-driven fashion. In this mode, SA is affected by the aircrew's goals and expectations, which influence how attention is directed, how information is perceived, and how it is interpreted. The person's goals and plans direct which aspects of the environment are attended to; that information is then integrated and interpreted in light of these goals to form level 2 SA. On an on-going basis, one can observe trade-offs between top-down and bottom-up processing, allowing the aircrew to process information effectively in a dynamic environment.

With experience, aircrew may develop a better understanding of their goals, which goals should be active in which circumstances, and how to acquire information to support these goals. The increased reliance on goal-directed processing allows the environment to be processed more efficiently than with purely data-driven processing. An important issue for achieving successful performance in the aviation domain lies in the ability of the aircrew to dynamically juggle multiple competing goals effectively. They need to rapidly switch between pursuing information in support of a particular goal to responding to perceived data activating a new goal, and back again. The ability to hold multiple goals has been associated with distributed attention, which is important for performance in the aviation domain (Martin & Jones, 1984).

12.3.2.3 Automaticity

SA can also be affected by the use of automaticity in processing information. Automaticity may be useful in overcoming attention limits, but may also leave the pilot susceptible to missing novel stimuli. Over time, it is easy for actions to become habitual and routine, requiring a very low level of attention. However, when something is slightly different, for example, a different clearance than usual, the pilots may miss it and carry out the habitual action. Developed through experience and a high level of learning, automatic processing tends to be fast, autonomous, effortless, and unavailable to conscious awareness in that it can occur without attention (Logan, 1988). Automatic processing is advantageous in that it provides good performance with minimal attention allocation. While automaticity may provide an important mechanism for overcoming processing limitations, thus allowing people to achieve SA and make decisions in complex, dynamic environments like aviation, it also creates an increased risk of being less responsive to new stimuli, because automatic processes operate with limited use of feedback. When using automatic processing, a lower level of SA can result in nontypical situations, decreasing decision timeliness and effectiveness.

12.3.2.4 Summary

In summary, SA can be achieved by drawing upon a number of internal mechanisms. Owing to limitations of attention and working memory, long-term memory may be heavily relied upon to achieve SA in the highly demanding aviation environment. The degree to which these structures can be developed and effectively used in the flight environment, the degree to which aircrew can effectively deploy goal-driven processing in conjunction with data-driven processing, and the degree to which aircrew can avoid the hazards of automaticity will ultimately determine the quality of their SA.

12.4 Challenges to Situation Awareness

In addition to SA being affected by the characteristics and processing mechanisms of the individual, many environmental and system factors may have a large impact on SA. Each of these factors can act to seriously challenge the ability of the aircrew to maintain a high level of SA in many situations.

12.4.1 Stress

Several types of stress factors exist in the aviation environment which may affect SA, including (a) Physical stressors—noise, vibration, heat/cold, lighting, atmospheric conditions, boredom, fatigue, cyclical changes, G's and (b) Social/Psychological stressors—fear or anxiety, uncertainty, importance or consequences of events, self-esteem, career advancement, mental load, and time pressure (Hockey, 1986; Sharit & Salvendy, 1982). A certain amount of stress may actually improve performance by increasing the attention to important aspects of the situation. However, a higher amount of stress can have extremely negative consequences, as accompanying increases in autonomic functioning and aspects of the stressors can act to demand a portion of a person's limited attentional capacity (Hockey, 1986).

Stressors can affect SA in a number of different ways, including attentional narrowing, reductions in information intake, and reductions in working-memory capacity. Under stress, a decrease in the attention has been observed for peripheral information, those aspects which attract less attentional focus (Bacon, 1974; Weltman, Smith, & Egstrom, 1971), with an increased tendency to sample dominant or probable sources of information (Broadbent, 1971). This is a critical problem for SA, leading to the neglect of certain elements in favor of others. In many cases, such as in emergency conditions, it is those factors outside the person's perceived central task that prove to be lethal. An L-1011 crashed in the Florida Everglades killing 99 people, when the crew became focused on a problem with a nose-gear indicator and failed to monitor the altitude and attitude of the aircraft (National Transportation Safety Board, 1973). In military aviation, many lives are lost owing to controlled flight into terrain accidents, with attentional narrowing being a primary culprit (Kuipers, et al., 1990).

Premature closure, that is, arriving at a decision without exploring all the available information, has also been found to be more likely under stress (Janis, 1982; Keinan, 1987; Keinan & Friedland, 1987). This includes considering less information and attending more to negative information (Janis, 1982; Wright, 1974). Several authors have also found that scanning of information under stress is scattered and poorly organized (Keinan, 1987; Keinan & Friedland, 1987; Wachtel, 1967). A lowering of attention capacity, attentional narrowing, disruptions of scan patterns, and premature closure may all negatively affect Level 1 SA under various forms of stress.

A second way in which stress may negatively affect SA is by decreasing working-memory capacity and hindering information retrieval (Hockey, 1986; Mandler, 1979). The degree to which working-memory decrements will impact SA depends on the resources available to the individual. In tasks where achieving SA involves a high-working memory load, a significant impact on SA Levels 2 and 3 (given the same Level 1 SA) would be expected. However, if long-term memory stores are available to support SA, as in more well-learned situations, less effect can be expected.

12.4.2 Overload/Underload

High mental workload is a stressor of particular importance in aviation that can negatively affect SA. If the volume of information and number of tasks are too great, SA may suffer as only a subset of information can be attended to, or the pilot may be actively working to achieve SA, yet suffer from erroneous or incomplete perception and integration of information. In some cases, SA problems may occur from an overall high level of workload, or, in many cases, owing to a momentary overload in the tasks to be performed or in information being presented.

Poor SA can also occur under low workload. In this case, the pilot may be unaware of what is going on and not be actively working to find out owing to inattentiveness, vigilance problems, or low motivation. Relatively little attention has been paid to the effects of low workload (particularly on long haul flights, for instance) on SA; however, this condition can pose a significant challenge for SA in many areas of aviation and deserves further study.

12.4.3 System Design

The capabilities of the aircraft for acquiring needed information and the way in which it presents that information will have a large impact on aircrew SA. While a lack of information can certainly be seen as a problem for SA, too much information poses an equal problem. Improvements in the avionics capabilities of aircraft in the past few decades have brought a dramatic increase in the sheer quantity of information available. Sorting through this data to derive the desired information and achieve a good picture of the overall situation is no small challenge. Overcoming this problem through better system designs that present integrated data is currently a major design goal aimed at alleviating this problem.

12.4.4 Complexity

A major factor creating a challenge for SA is the complexity of the many systems that must be operated. There has been a boom in the avionics systems, flight management systems, and other technologies on the flight deck that have greatly increased the complexity of the systems that aircrew must operate. System complexity can negatively affect both the pilot workload and SA through an increase in the number of system components to be managed, a high degree of interaction between these components, and an increase in the dynamics or rate of change of the components. In addition, the complexity of the pilot's tasks may increase through an increase in the number of goals, tasks, and decisions to be made with regard to the aircraft systems. The more complex the systems are to be operated, the greater is the increase and the mental workload that is required to achieve a given level of SA. When that demand exceeds human capabilities, SA will suffer.

System complexity may be somewhat moderated by the degree to which the person has a well-developed internal representation of the system to aid in directing attention, integrating data, and developing higher levels of SA. These mechanisms may be effective for coping with complexity; however, developing those internal models may require a considerable amount of training. Pilots have reported significant difficulties in understanding what their automated flight management systems are doing and why (Sarter & Woods, 1992; Wiener, 1989). McClumpha and James (1994) conducted an extensive study on nearly 1000 pilots from across varying nationalities and aircraft types. They found that the primary factor explaining the variance in pilots' attitudes toward advanced technology aircraft was their self-reported understanding of the system. Although pilots eventually develop a better understanding of the automated aircraft with experience, many of these systems do not appear to be well designed to meet their SA needs.

12.4.5 Automation

SA may also be negatively impacted by the automation of the tasks, as it is frequently designed to put the aircrew "out-of-the-loop." System operators working with automation have been found to have a diminished ability to detect system errors and subsequently perform tasks manually in the face of automation failures when compared with the manual performance on the same tasks (Billings, 1991; Moray, 1986; Wickens, 1992; Wiener & Curry, 1980). In 1987, a Northwest Airlines MD-80 crashed on take-off at Detroit Airport owing to an improper configuration of the flaps and slats, killing all but one passenger (National Transportation Safety Board, 1988). A major factor in the crash was the failure of an automated take-off configuration warning system on which the crew had become reliant. They did not realize that the aircraft was improperly configured for take-off and had neglected to check manually (owing to other contributing factors). When the automation failed, they were not aware of the state of the automated system or the critical flight parameters, and depended on the automation to monitor these. While some of the out-of-the-loop performance problem may be owing to the loss of manual skills under automation, loss of SA is also a critical component for this accident and many similar ones.

Pilots who have lost SA through being out-of-the-loop may be slow in detecting problems and additionally, may require extra time to reorient themselves to relevant system parameters to proceed with the problem diagnosis and assumption of manual performance when automation fails. This has been found to occur for a number of reasons, including (a) a loss of vigilance and increase in complacency associated with becoming a monitor for the implementation of automation, (b) being a passive recipient of information rather than an active processor of information, and (c) a loss of or change in the type of feedback provided to the aircrew concerning the state of the system being automated (Endsley & Kiris, 1995). In their study, Endsley and Kiris found evidence for SA decrement accompanying automation of a cognitive task which was greater under full automation than under partial automation. Lower SA in the automated conditions corresponded to a demonstrated out-of-the-loop performance decrement, supporting the hypothesized relationship between SA and automation.

However, SA may not suffer under all forms of automation. Wiener (1993) and Billings (1991) stated that SA may be improved by systems that provide integrated information through automation. In commercial cockpits, Hansman, et al. (1992) found that automated flight-management system input was superior to manual data entry, producing better error detection of clearance updates. Automation that reduces unnecessary manual work and data integration required to achieve SA may provide benefits to both workload and SA. However, the exact conditions under which SA will be positively or negatively affected by automation needs to be determined.

12.5 Errors in Situation Awareness

Based on this model of SA, a taxonomy for classifying and describing errors in SA was created (Endsley, 1994; Endsley, 1995c). The taxonomy, presented in Table 12.1, incorporates factors affecting SA at each of its three levels. Endsley (1995a) applied this taxonomy to an investigation of causal factors underlying aircraft accidents involving major air carriers in the United States, based on NTSB accident investigation reports over a 4-year period. Of the 71% of the accidents that could be classified as having a substantial human-error component, 88% involved problems with SA. Of the 32 SA errors identified in these accident descriptions, 23 (72%) were attributed to problems with Level 1 SA, a failure to correctly perceive some pieces of information in the situation. Seven (22%) involved a Level 2 error in which the data was perceived but not integrated or comprehended correctly, and two (6%) involved a Level 3 error in which there was a failure to properly project the near future, based on the aircrew's understanding of the situation.

More recently, Jones and Endsley (1995) applied this taxonomy to a more extensive study of SA errors, based on voluntary reports in NASA's Aviation Safety Reporting System (ASRS) database. This provided some indication on the types of problems and the relative contribution of the causal factors leading to SA errors in the cockpit, as shown in Figure 12.2.

TABLE 12.1 SA Error Taxonomy

Level 1: Failure to correctly perceive information

- Data not available
- Data hard to discriminate or detect
- Failure to monitor or observe data
- Misperception of data
- Memory loss

Level 2: Failure to correctly integrate or comprehend information

- Lack of or poor mental model
- Use of incorrect mental model
- Over-reliance on default values
- Other

Level 3: Failure to project future actions or state of the system

- Lack of or poor mental model
- Overprojection of current trends
- Other

General

- Failure to maintain multiple goals
- Habitual schema

Source: Adapted from Endsley, M.R., A taxonomy of situation awareness errors, in Fuller, R. et al. (Eds.), *Human Factors in Aviation Operations*, Avebury Aviation, Ashgate Publishing Ltd., Aldershot, England, 1995a, 287–292.

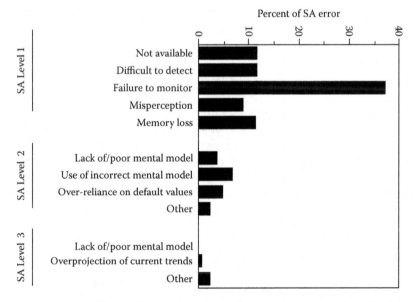

FIGURE 12.2 SA error causal factors. (From Jones, D.G. and Endsley, M.R., *Proceedings of the 8th International Symposium on Aviation Psychology*, The Ohio State University, Columbus, OH, 1995.)

12.5.1 Level 1: Failure to Correctly Perceive the Situation

At the most basic level, important information may not be correctly perceived. In some cases, the data may not be available to the person, owing to a failure of the system design to present it or a failure in the communications process. This factor accounted for 11.6% of SA errors, most frequently occurring owing to a failure of the crew to perform some necessary task (such as resetting the altimeter) to obtain the correct information. In other cases, the data are available, but are difficult to detect or perceive, accounting for another 11.6% of SA errors in this study. This included problems owing to poor runway markings and lighting, and those owing to noise in the cockpit.

Often the information is directly available, but for various reasons, is not observed or included in the scan pattern, forming the largest single causal factor for SA errors (37.2%). This is owing to several factors, including simple omission—not looking at a piece of information, attentional narrowing, and external distractions that prevent them from attending to important information. High taskload, even momentary, is another major factor that prevents information from being attended to.

In other cases, information is attended to, but is misperceived (8.7% of SA errors), frequently owing to the influence of prior expectations. Finally, in some cases, it appears that a person initially perceives some piece of information but then forgets about it (11.1% of SA errors), which negatively affects SA, as it relies on keeping information about a large number of factors in the memory. Forgetting has been found to be frequently associated with disruptions in normal routine, high workload, and distractions.

12.5.2 Level 2 SA: Failure to Comprehend the Situation

In other cases, information is correctly perceived, but its significance or meaning is not comprehended. This may be owing to the lack of a good mental model for combining information in association with pertinent goals. The lack of a good mental model is attributed to 3.5% of the SA errors that are most frequently associated with an automated system.

In other cases, the wrong mental model may be used to interpret information, leading to 6.4% of the SA errors in this study. In this case, the mental model of a similar system may be used to interpret information, leading to an incorrect diagnosis or understanding of the situation in areas where that system is different. A frequent problem is where aircrews have a model of what is expected and then interpret all the perceived cues into that model, leading to a completely incorrect interpretation of the situation.

In addition, there may also be problems with over-reliance on defaults in the mental model used, as was found for 4.7% of the SA errors. These defaults can be thought of as general expectations about how parts of the system function which may be used in the absence of real-time data. In other cases, the significance of perceived information relative to operational goals is simply not comprehended, or several pieces of information are not properly integrated. This may be owing to the working-memory limitations or other unknown cognitive lapses. Miscellaneous factors, such as these are attributed to 2.3% of the SA errors.

12.5.3 Level 3 SA: Failure to Project Situation into the Future

Finally, in some cases, individuals may be fully aware of what is going on, but may be unable to correctly project what that means for the future, accounting for 2.9% of the SA errors. In some cases, this may be owing to a poor mental model or over projection of the current trends. In other cases, the reason for not correctly projecting the situation is less apparent. Mental projection is a very demanding task at which people are generally poor.

12.5.4 General

In addition to these main categories, two general categories of causal factors are included in the taxonomy. First, some people are poor at maintaining multiple goals in memory, which could impact SA across all the three levels. Second, there is evidence that people can fall into a trap of executing habitual schema, doing tasks automatically, which render them less receptive to important environmental cues. Evidence for these causal factors was not apparent in the retrospective reports analyzed in the ASRS or NTSB databases.

12.6 SA in General Aviation

While much SA research has been focused on military or commercial aviation pilots, many of the significant problems with SA occur in the general aviation (GA) population. GA accidents account for 94% of all U.S. civil aviation accidents and 92% of all fatalities in civil aviation (National Transportation Safety Board, 1998). The pilot was found to be a "broad cause/factor" in 84% of all GA accidents and 90.6% of all fatal accidents (Trollip & Jensen, 1991). They attributed 85% of GA accidents to pilot error, with faulty decision making cited as the primary cause. However, SA problems appear to underlie the majority of these errors.

Endsley et al. (2002) conducted an in-depth analysis of SA problems in low-time GA pilots. They examined 222 incident reports at a popular flight school that contained reported problems with SA. Overall, a number of problems were noted as particularly difficult, leading to the SA problems found across this group of relatively inexperienced GA pilots.

1. Distractions and high workload. Many of the SA errors could be linked to problems with managing task distractions and task saturation. This may reflect the high workload associated with tasks that are not learned with regard to high levels of automaticity, problems with multitasking, or insufficiently developed task-management strategies. These less-experienced pilot groups had significant problems in dealing with distractions and high workload.
2. Vigilance and monitoring deficiencies. While associated with task overload in about half of the cases, in many incidents, vigilance and monitoring deficiencies were noted without these overload problems. This may reflect insufficiently learned scan patterns, attentional narrowing, or an inability to prioritize information.
3. Insufficiently developed mental models. Many errors in both understanding perceived information, and projecting future dynamics could be linked to insufficiently developed mental models. In particular, the GA pilots had significant difficulties with operations in new geographical areas, including recognizing landmarks and matching them to maps, and understanding new procedures for flight, landings, and departures in unfamiliar airspace. They also had significant difficulties in understanding the implications of many environmental factors on aircraft dynamics/behaviors. Pilots at these relatively low levels of experience also exhibited problems with judging relative motion and rates of change in other traffic.
4. Over-reliance on mental models. Reverting to habitual patterns (learned mental models) when new behaviors were needed was also a problem for the low-experience GA pilots. They failed to understand the limits of the learned models and how to properly extend these models to new situations.

In a second study, Endsley et al. (2002) conducted challenging simulated flight scenario studies with both inexperienced and experienced GA pilots. Those pilots who were scored as having better SA (in both the novice and experienced categories) all received much higher ratings for aircraft handling/psychomotor skills, cockpit task management, cockpit task prioritization, and ATC communication/coordination than those who were rated as having lower SA.

A step-wise regression model, accounting for 91.7% of the variance in SA scores across all the pilots, included aircraft handling/psychomotor skill and ATC communication and coordination. Aircraft handling might normally be considered as a manual or psychomotor task, and not one significantly involved in a cognitive construct like SA. However, other studies have also found a relationship between psychomotor skills and SA, presumably because of issues associated with limited attention (Endsley & Bolstad, 1994; O'Hare, 1997). The development of higher automaticity for physically flying the aircraft ("stick skills") helps to free-up attention resources needed for SA. Keeping up with ATC communications was also challenging for many of the novice GA pilots. They requested numerous repeats of transmissions, which used up their attentional resources.

However, not all experienced GA pilots were found to have high SA. Among the experienced pilots with high SA, good aircraft-handling skills and good task prioritization were frequently noted. Their performance was not perfect, but this group appeared to detect and recover from their own errors better than the others. Many were noted as flying first and only responding to ATC clearances or equipment malfunctions when they had the plane under control.

The experienced pilots who were rated as having only moderate SA were more likely to have difficulty in controlling the simulated aircraft and poorer prioritization and planning skills. Thus, in addition to physical performance (aircraft handling), skills associated with task prioritization appear to be important for high levels of SA in aviation.

12.7 SA in Multicrew Aircraft

While SA has primarily been discussed at the level of the individual, it is also relevant for the aircrew as a team (Endsley & Jones, 2001). This team may comprise a two- or three-member crew in a commercial aircraft to as many as five- to seven-member crew in some military aircraft. In some military settings, several aircraft may also be deployed as a flight, forming a more loosely coupled team in which several aircraft must work together to accomplish a joint goal.

Team SA has been defined as "the degree to which every team member possesses the SA required for his or her responsibilities" (Endsley, 1989). If one crew member has a certain piece of information, but another who needs it does not, then the SA of the team may suffer and their performance may suffer as well, unless the discrepancy is corrected. In this light, a major portion of inter-crew coordination can be seen as the transfer of information from one crew member to another, as required for developing SA across the team. This coordination involves more than just sharing of data. It also includes sharing of higher levels of SA (comprehension and projection), which may vary widely between individuals depending on their experiences and goals.

The process of providing shared SA can be greatly enhanced by shared mental models that provide a common frame of reference for crew-member actions, and allow team members to predict each other's behaviors (Cannon-Bowers, Salas, & Converse, 1993; Orasanu, 1990). A shared mental model may provide more efficient communications by providing a common means of interpreting and predicting actions based on limited information, and therefore, may be important for SA. For instance, Mosier and Chidester (1991) found that better-performing teams actually communicated less than poorer-performing teams.

12.8 Impact of CRM on SA

Crew resource management (CRM) programs have in the last few years received a great deal of attention and focus in aviation, as a means of promoting better teamwork and use of crew resources. Robertson and Endsley (1995) investigated the link between SA and CRM programs, and found that CRM can have an effect on crew SA by directly improving individual SA, or indirectly, through the development of shared mental models and by providing efficient distribution of attention across the crew. They hypothesized that CRM could be used to improve team SA through various behaviors measured by the Line/LOS Checklist (LLC), as shown in Figure 12.3, which are positively impacted by CRM (Butler, 1991; Clothier, 1991).

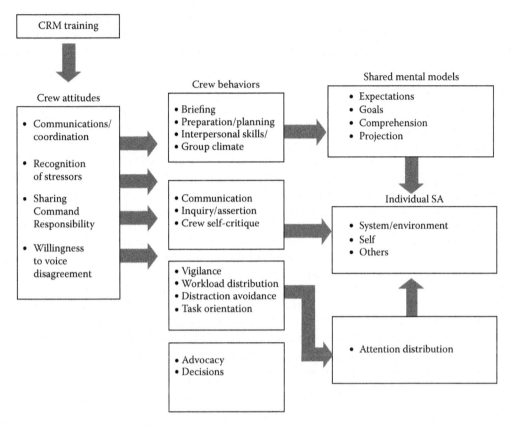

FIGURE 12.3 CRM factors affecting SA. (From Robertson, M.M. and Endsley, M.R., The role of crew resource management (CRM) in achieving situation awareness in aviation settings, in Fuller, R. et al. (Eds.), *Human Factors in Aviation Operations*, Avebury Aviation, Ashgate Publishing Ltd., Aldershot, England, 1995, 281–286.)

12.8.1 Individual SA

Improved communication between crew members can obviously facilitate effective sharing of needed information. In particular, improved inquiry and assertion behaviors by crew members helps to insure the needed communication. In addition, an understanding of the state of the human elements in the system (inter-crew SA) also forms a part of SA. The development of good self-critique skills can be used to provide an up-to-date assessment of one's own and other team member's abilities and performance, which may be impacted by factors such as fatigue or stress. This knowledge allows the team members to recognize the need for providing more information and taking over functions in critical situations, an important part of effective team performance.

12.8.2 Shared Mental Models

Several factors can help to develop shared mental models between the crew members. The crew briefing establishes the initial basis for a shared mental model between the crew members, providing shared goals and expectations. This can increase the likelihood that two crew members will form the same higher levels of SA from low level information, improving the effectiveness of communications. Similarly, prior preparation and planning can help to establish a shared mental model. Effective crews tend to "think ahead" of the aircraft, allowing them to be ready for a wide variety of events. This is closely linked to

Level 3 SA—projection of the future. The development of interpersonal relationships and group climate can also be used to facilitate the development of a good model of other crew members. This allows individuals to predict how others will act, forming the basis for Level 3 SA and efficient functioning teams.

12.8.3 Attention Distribution

The effective management of the crew's resources is extremely critical, particularly in high task load situations. A major factor in effectively managing these resources is ensuring that all aspects of the situation are being attended to—avoiding attentional narrowing and neglect of important information and tasks. CRM programs that improve task orientation and the distribution of tasks under workload can directly impact how the crew members are directing their attention, and thus their SA. In addition, improvements in vigilance and the avoidance of distractions can directly impact SA.

Thus, there are a number of ways in which existing CRM programs can affect SA at the crew level, as well as within individuals. Programs have been developed to specifically train for factors that are lacking in team SA. Endsley and Robertson (2000) developed a two-day course for AMTs, which was built on the previous CRM training for this group. The course focused on: (1) shared mental models, (2) verbalizations of decisions, (3) shift meetings and teamwork, (4) feedback, and (5) dealing with SA challenges. Robinson (2000) developed a 2 days program for training SA at British Airways as its CRM II program. This program combined training on the three levels of SA in an inspired combination with error management research (in terms of avoidance, trapping, and mitigation) from the work of Helmreich, Merritt, and Sherman (1996) and Reason (1997). In addition to very positive subjective feedback on the training (78% strongly agreed that the program had practical value), the pilots who received the training were rated as having significantly better team skills, and showed a significant increase in operating at Level 3 SA (as compared with Level 1 or 2 SA).

12.9 Building SA

12.9.1 Design

Cockpit design efforts can be directed toward several avenues for improving SA, including searching for (a) ways to determine and effectively deliver critical cues, (b) ways to ensure accurate expectations, (c) methods for assisting pilots in deploying attention effectively, (d) methods for preventing the disruption of attention, particularly under stress and high workload, and (e) ways to develop systems that are compatible with pilot goals. Many ongoing design efforts are aimed at enhancing SA in the cockpit by taking advantage of new technologies, such as advanced avionics and sensors, datalink, global positioning systems (GPS), three-dimensional visual and auditory displays, voice control, expert systems, helmet-mounted displays, virtual reality, sensor fusion, automation, and expert systems. The glass cockpit, advanced automation techniques, and new technologies, such as traffic alert/collision avoidance system (TCAS) have become a reality in today's aviation systems.

Each of these technologies provides a potential advantage: new information, more accurate information, new ways of providing information, or a reduction in crew workload. However, each can also affect SA in unpredicted ways. For instance, recent evidence showed that automation that is often cited as being potentially beneficial for SA through the reduction of workload, can actually reduce SA, thus, contributing to the out-of-the-loop performance problem (Carmody & Gluckman, 1993; Endsley & Kiris, 1995). Three-dimensional displays, also touted as beneficial for SA, have been found to have quite negative effects on pilots' ability to accurately localize other aircrafts and objects (Endsley, 1995b; Prevett & Wickens, 1994).

The SA-Oriented Design Process was developed to address the need for a systematic design process that builds on the substantial body of SA theory and research that has been developed. The SA-Oriented Design Process (Endsley, Bolte, & Jones, 2003), given in Figure 12.4, provides a key methodology for

FIGURE 12.4 SA-Oriented Design Process. (From Endsley, M.R. et al., *Designing for Situation Awareness: An Approach to Human-Centered Design*, Taylor & Francis, London, 2003.)

developing user-centered displays by focusing on optimizing SA. By creating designs that enhance the pilot's awareness of what is happening in a given situation, decision making and performance can improve dramatically.

SA requirements are first determined through a cognitive task analysis technique called Goal-Directed Task Analysis (GDTA). A GDTA identifies the major goals and subgoals for each job. The critical decisions that the individual must make to achieve each goal and subgoal are then determined, and the SA needed for making these decisions and carrying out each subgoal is identified. These SA requirements focus not only on the data that the individual needs, but also on how that information is integrated or combined to address each decision. This process forms the basis for determining the exact information (at all three levels of SA) that needs to be included in display visualizations.

Second, 50 SA-Oriented Design principles have been developed based on the latest research on SA. By applying the SA-Oriented Design principles to SA requirements, user-centered visualization displays can be created which organize information around the user's SA needs and support key cognitive mechanisms for transforming captured data into high levels of SA. These principles provide a systematic basis, consistent with human cognitive processing and capabilities, for establishing the content of user displays.

The final step of the SA-Oriented Design Process emphasizes on the objective measurement of SA during man-in-the-loop simulation testing. The Situation Awareness Global Assessment Technique (SAGAT) provides a sensitive and diagnostic measure of SA that can be used to evaluate new interface technologies, display concepts, sensor suites, and training programs (Endsley, 1995b, Endsley, 2000). It has been carefully validated and successfully used in a wide variety of domains, including army infantry and battle command operations.

The Designer's Situation Awareness Toolbox (DeSAT) was created to assist designers in carrying out the SA-Oriented Design Process (Jones, Estes, Bolstad, & Endsley, 2004). It includes (1) a software tool for easily creating, editing, and storing effective GDTAs, (2) a GDTA Checklist Tool, to aid designers in evaluating the degree to which a display design meets the SA requirements of the user, (3) an SA-Oriented Design Guidelines Tool, which guides the designers in determining how well a given design will support the user's SA, and (4) an SAGAT tool, which allows the designers to rapidly customize SAGAT queries to the relevant user domain and SA requirements, and which administers SAGAT during user testing, to empirically evaluate display designs.

As many factors surrounding the use of new technologies and design concepts may act to both enhance and degrade SA, significant care should be taken to evaluate the impact of the proposed concepts on SA. Only by testing new design concepts in carefully controlled studies, can the actual impact of these factors can be identified. This testing needs to include not only an examination of how the technologies affect the basic human processes, such as accuracy of perception, but also how they affect the pilot's global state of knowledge when used in a dynamic and complex aviation scenario, where multiple sources of information compete for attention and must be selected, processed, and integrated in light of dynamic goal changes. Real-time simulations employing the technologies can be used to assess the impact of the system by carefully measuring the aircrew performance, workload, and SA. Direct measurement of SA during design testing is recommended for providing sufficient insight into the potential costs and benefits of design concepts for aircrew SA, allowing the determination of the degree to which the design successfully addresses these issues. Techniques for measuring SA within the aviation system design process are covered in more detail in the study by Endsley and Garland (2000).

12.9.2 Training

In addition to improving SA through better cockpit designs, it may also be possible to find new ways of training aircrew to achieve better SA with a given aircraft design. The potential role of CRM programs in this process has already been discussed. It may also be possible to create "SA-oriented training programs" that seek to improve SA directly in individuals. This may include programs that provide aircrew with better information needed to develop mental models, including information on their components, the dynamics and functioning of the components, and projection of future actions based on these dynamics. The focus should be on training aircrew to identify prototypical situations of concern associated with these models by recognizing critical cues and what they mean in terms of relevant goals.

The skills required for achieving and maintaining good SA also need to be formally taught in training programs. Factors such as how to employ a system to best achieve SA (when to look, for what, and where), the appropriate scan patterns, or techniques for making the most of the limited information, need to be determined and explicitly taught in the training process. A focus on aircrew SA would greatly supplement the traditional technology-oriented training that concentrates mainly on the mechanics of how a system operates.

For example, a set of computer-based training modules was designed to build the basic skills underlying SA for new general-aviation pilots (Bolstad, Endsley, Howell, & Costello, 2002). These modules include training in time-sharing or distributed attention, checklist completion, ATC communications, intensive preflight planning and contingency planning, and SA feedback training, which were all found to be problems for new pilots. In tests with low-time general-aviation pilots, the training modules were generally successful in imparting the desired skills. Some improvements in SA were also found in the follow-on simulated flight trials, but the simulator was insensitive to detect flight-performance differences. More research is warranted to track whether this type of skills training can improve SA in the flight environment.

In addition, the role of feedback as an important component of the learning process should be more fully exploited. It may be possible to provide feedback on the accuracy and completeness of pilot SA as a part of training programs. This would allow the aircrew to understand their mistakes and better assess and interpret the environment, leading to the development of more effective sampling strategies and better schema for integrating information. Riley et al. (2005), for example, developed a system for assessing SA in virtual reality simulators that provided feedback to participants as a means of training SA. Techniques like this deserve more exploration and testing, as a means of developing higher levels of SA in aircrew.

12.10 Conclusion

Maintaining SA is a critical and challenging part of an aircrew's job. Without good SA, even the best trained crews can make poor decisions. Numerous factors that are a constant part of the aviation environment make the goal of achieving a high level of SA at all times quite challenging. In the past decade, enhancement of SA through better cockpit design and training programs has received considerable attention, and will continue to do so in the future.

References

Amalberti, R., & Deblon, F. (1992). Cognitive modeling of fighter aircraft process control: A step towards an intelligent on-board assistance system. *International Journal of Man-Machine Systems, 36,* 639–671.

Bacon, S. J. (1974). Arousal and the range of cue utilization. *Journal of Experimental Psychology, 102,* 81–87.

Barber, P. J., & Folkard, S. (1972). Reaction time under stimulus uncertainty with response certainty. *Journal of Experimental Psychology, 93,* 138–142.

Biederman, I., Mezzanotte, R. J., Rabinowitz, J. C., Francolin, C. M., & Plude, D. (1981). Detecting the unexpected in photo interpretation. *Human Factors, 23,* 153–163.

Billings, C. E. (1991). *Human-centered aircraft automation: A concept and guidelines* (NASA Technical Memorandum 103885). Moffett Field, CA: NASA Ames Research Center.

Bolstad, C. A., Endsley, M. R., Howell, C., & Costello, A. (2002). General aviation pilot training for situation awareness: An evaluation. *Proceedings of the 46th Annual Meeting of the Human Factors and Ergonomics Society* (pp. 21–25). Santa Monica, CA: Human Factors and Ergonomics Society.

Broadbent, D. E. (1971). *Decision and stress.* London: Academic Press.

Butler, R. E. (1991). Lessons from cross-fleet/cross airline observations: Evaluating the impact of CRM/LOS training. In R. S. Jensen (Ed.), *Proceedings of the Sixth International Symposium on Aviation Psychology* (pp. 326–331). Columbus: Department of Aviation, the Ohio State University.

Cannon-Bowers, J. A., Salas, E., & Converse, S. (1993). Shared mental models in expert team decision making. In N. J. Castellan (Ed.), *Current issues in individual and group decision making* (pp. 221–247). Hillsdale, NJ: Lawrence Erlbaum.

Carmody, M. A., & Gluckman, J. P. (1993). Task specific effects of automation and automation failure on performance, workload and situational awareness. In R. S. Jensen, & D. Neumeister (Eds.), *Proceedings of the Seventh International Symposium on Aviation Psychology* (pp. 167–171). Columbus: Department of Aviation, the Ohio State University.

Clothier, C. (1991). Behavioral interactions in various aircraft types: Results of systematic observation of line operations and simulations. In R. S. Jensen (Ed.), *Proceedings of the Sixth International Conference on Aviation Psychology* (pp. 332–337). Columbus: Department of Aviation, the Ohio State University.

Davis, E. T., Kramer, P., & Graham, N. (1983). Uncertainty about spatial frequency, spatial position, or contrast of visual patterns. *Perception and Psychophysics, 5,* 341–346.

Endsley, M. R. (1988). Design and evaluation for situation awareness enhancement. In *Proceedings of the Human Factors Society 32nd Annual Meeting* (pp. 97–101). Santa Monica, CA: Human Factors Society.

Endsley, M. R. (1989). *Final report: Situation awareness in an advanced strategic mission* (NOR DOC 89-32). Hawthorne, CA: Northrop Corporation.

Endsley, M. R. (1993). A survey of situation awareness requirements in air-to-air combat fighters. *International Journal of Aviation Psychology, 3*(2), 157–168.

Endsley, M. R. (1994). Situation awareness in dynamic human decision making: Theory. In R. D. Gilson, D. J. Garland, & J. M. Koonce (Eds.), *Situational awareness in complex systems* (pp. 27–58). Daytona Beach, FL: Embry-Riddle Aeronautical University Press.

Endsley, M. R. (1995a). A taxonomy of situation awareness errors. In R. Fuller, N. Johnston, & N. McDonald (Eds.), *Human factors in aviation operations* (pp. 287–292). Aldershot, England: Avebury Aviation, Ashgate Publishing Ltd.

Endsley, M. R. (1995b). Measurement of situation awareness in dynamic systems. *Human Factors, 37*(1), 65–84.

Endsley, M. R. (1995c). Toward a theory of situation awareness. *Human Factors, 37*(1), 32–64.

Endsley, M. R. (2000). Direct measurement of situation awareness: Validity and use of SAGAT. In M. R. Endsley, & D. J. Garland (Eds.), *Situation awareness analysis and measurement* (pp. 147–174). Mahwah, NJ: LEA.

Endsley et al. (2002). *Situation awareness training for general aviation pilots* (Final report (No. SATECH 02-04). Marietta, GA: SA Technologies.

Endsley, M. R., & Bolstad, C. A. (1994). Individual differences in pilot situation awareness. *International Journal of Aviation Psychology, 4*(3), 241–264.

Endsley, M. R., & Garland, D. J. (Eds.). (2000). *Situation awareness analysis and measurement.* Mahwah, NJ: Lawrence Erlbaum.

Endsley, M. R., & Jones, W. M. (2001). A model of inter- and intrateam situation awareness: Implications for design, training and measurement. In M. McNeese, E. Salas, & M. Endsley (Eds.), *New trends in cooperative activities: Understanding system dynamics in complex environments* (pp. 46–67). Santa Monica, CA: Human Factors and Ergonomics Society.

Endsley, M. R., & Kiris, E. O. (1995). The out-of-the-loop performance problem and level of control in automation. *Human Factors, 37*(2), 381–394.

Endsley, M. R., & Robertson, M. M. (2000). Situation awareness in aircraft maintenance teams. *International Journal of Industrial Ergonomics, 26*, 301–325.

Endsley, M. R., & Rodgers, M. D. (1994). *Situation awareness information requirements for en route air traffic control* (DOT/FAA/AM-94/27). Washington, DC: Federal Aviation Administration Office of Aviation Medicine.

Endsley, M. R., Bolte, B., & Jones, D. G. (2003). *Designing for situation awareness: An approach to human-centered design.* London: Taylor & Francis.

Endsley, M. R., Farley, T. C., Jones, W. M., Midkiff, A. H., & Hansman, R. J. (1998). *Situation awareness information requirements for commercial airline pilots* (No. ICAT-98-1). Cambridge, MA: Massachusetts Institute of Technology International Center for Air Transportation.

Fracker, M. L. (1990). Attention gradients in situation awareness. In *Situational Awareness in Aerospace Operations (AGARD-CP-478)* (Conference Proceedings #478) (pp. 6/1–6/10). Neuilly Sur Seine, France: NATO-AGARD.

Hansman, R. J., Wanke, C., Kuchar, J., Mykityshyn, M., Hahn, E., & Midkiff, A. (1992, September). *Hazard alerting and situational awareness in advanced air transport cockpits.* Paper presented at the 18th ICAS Congress, Beijing, China.

Hartel, C. E., Smith, K., & Prince, C. (1991, April). *Defining aircrew coordination: Searching mishaps for meaning.* Paper presented at the Sixth International Symposium on Aviation Psychology, Columbus, OH.

Helmreich, R. L., Merritt, A. C., & Sherman, P. J. (1996). Human factors and national culture. *ICAO Journal, 51*(8), 14–16.

Hockey, G. R. J. (1986). Changes in operator efficiency as a function of environmental stress, fatigue and circadian rhythms. In K. Boff, L. Kaufman, & J. Thomas (Eds.), *Handbook of perception and performance* (Vol. 2, pp. 44/1–44/49). New York: John Wiley.

Humphreys, G. W. (1981). Flexibility of attention between stimulus dimensions. *Perception and Psychophysics, 30*, 291–302.

Isaacson, B. (1985). A lost friend. *USAF Fighter Weapons Review, 4*(33), 23–27.

Janis, I. L. (1982). Decision making under stress. In L. Goldberger, & S. Breznitz (Eds.), *Handbook of stress: Theoretical and clinical aspects* (pp. 69–87). New York: The Free Press.

Jones, R. A. (1977). *Self-fulfilling prophecies: Social, psychological and physiological effects of expectancies.* Hillsdale, NJ: Lawrence Erlbaum.

Jones, D. G., & Endsley, M. R. (1995). Investigation of situation awareness errors. In *Proceedings of the 8th International Symposium on Aviation Psychology.* Columbus: The Ohio State University.

Jones, D., Estes, G., Bolstad, M., & Endsley, M. (2004). *Designer's situation awareness toolkit (DESAT)* (No. SATech-04-01). Marietta, GA: SA Technologies.

Keinan, G. (1987). Decision making under stress: Scanning of alternatives under controllable and uncontrollable threats. *Journal of Personality and Social Psychology, 52*(3), 639–644.

Keinan, G., & Friedland, N. (1987). Decision making under stress: Scanning of alternatives under physical threat. *Acta Psychologica, 64*, 219–228.

Kraby, A. W. (1995). A close encounter on the Gulf Coast. Up front: The flight safety and operations publication of Delta Airlines, 2nd Quarter, 4.

Kuipers, A., Kappers, A., van Holten, C. R., van Bergen, J. H. W., & Oosterveld, W. J. (1990). Spatial disorientation incidents in the R.N.L.A.F. F16 and F5 aircraft and suggestions for prevention. In *Situational awareness in aerospace operations (AGARD-CP-478)* (pp. OV/E/1–OV/E/16). Neuilly Sur Seine, France: NATO-AGARD.

Logan, G. D. (1988). Automaticity, resources and memory: Theoretical controversies and practical implications. *Human Factors, 30*(5), 583–598.

Mandler, G. (1979). Thought processes, consciousness and stress. In V. Hamilton, & D. M. Warburton (Eds.), *Human stress and cognition: An information-processing approach.* Chichester: Wiley and Sons.

Martin, M., & Jones, G. V. (1984). Cognitive failures in everyday life. In J. E. Harris, & P. E. Morris (Eds.), *Everyday memory, actions and absent-mindedness* (pp. 173–190). London: Academic Press.

McCarthy, G. W. (1988, May). Human factors in F16 mishaps. *Flying Safety*, pp. 17–21.

McClumpha, A., & James, M. (1994). Understanding automated aircraft. In M. Mouloua, & R. Parasuraman (Eds.), *Human performance in automated systems: Current research and trends* (pp. 183–190). Hillsdale, NJ: LEA.

Monan, W. P. (1986). *Human factors in aviation operations: The hearback problem* (NASA Contractor Report 177398). Moffett Field, CA: NASA Ames Research Center.

Moray, N. (1986). Monitoring behavior and supervisory control. In K. Boff (Ed.), *Handbook of perception and human performance* (Vol. II, pp. 40/1–40/51). New York: Wiley.

Mosier, K. L., & Chidester, T. R. (1991). Situation assessment and situation awareness in a team setting. In Y. Queinnec, & F. Daniellou (Eds.), *Designing for everyone* (pp. 798–800). London: Taylor & Francis.

National Transportation Safety Board. (1973). *Aircraft Accidents Report: Eastern Airlines 401/L-1011*, Miami, FL, December 29, 1972. Washington, DC: Author.

National Transportation Safety Board. (1988). *Aircraft Accidents Report: Northwest Airlines, Inc., McDonnell-Douglas DC-9-82, N312RC, Detroit Metropolitan Wayne County Airport, August, 16, 1987 (NTSB/AAR-99-05)*. Washington, DC: Author.

National Transportation Safety Board. (1998). *1997 U.S. Airline fatalities down substantially from previous year; general aviation deaths rise*. NTSB press release 2/24/98 (No. SB 98-12). Washington, DC: Author.

O'Hare, D. (1997). Cognitive ability determinants of elite pilot performance. *Human Factors, 39*(4), 540–552.

Orasanu, J. (1990, July). *Shared mental models and crew decision making*. Paper presented at the 12th Annual Conference of the Cognitive Science Society, Cambridge, MA.

Palmer, S. E. (1975). The effects of contextual scenes on the identification of objects. *Memory and Cognition, 3*, 519–526.

Posner, M. I., Nissen, J. M., & Ogden, W. C. (1978). Attended and unattended processing modes: The role of set for spatial location. In H. L. Pick, & E. J. Saltzman (Eds.), *Modes of perceiving and processing* (pp. 137–157). Hillsdale, NJ: Erlbaum Associates.

Prevett, T. T., & Wickens, C. D. (1994). *Perspective displays and frame of reference: Their interdependence to realize performance advantages over planar displays in a terminal area navigation task (ARL-94-8/NASA-94-3)*. Savoy, IL: University of Illinois at Urbana-Champaign.

Reason, J. (1997). *Managing the risks of organizational accidents*. London: Ashgate Press.

Riley, J. M., Kaber, D. B., Hyatt, J., Sheik-Nainar, M., Reynolds, J., & Endsley, M. (2005). *Measures for assessing situation awareness in virtual environment training of infantry squads* (Final Report No. SATech-05-03). Marietta, GA: SA Technologies.

Robertson, M. M., & Endsley, M. R. (1995). The role of crew resource management (CRM) in achieving situation awareness in aviation settings. In R. Fuller, N. Johnston, & N. McDonald (Eds.), *Human factors in aviation operations* (pp. 281–286). Aldershot, England: Avebury Aviation, Ashgate Publishing Ltd.

Robinson, D. (2000). The development of flight crew situation awareness in commercial transport aircraft. *Proceedings of the Human Performance, Situation Awareness and Automation: User-Centered Design for a New Millennium Conference* (pp. 88–93). Marietta, GA: SA Technologies, Inc.

Sarter, N. B., & Woods, D. D. (1992). Pilot interaction with cockpit automation: Operational experiences with the flight management system. *The International Journal of Aviation Psychology, 2*(4), 303–321.

Sharit, J., & Salvendy, G. (1982). Occupational stress: Review and reappraisal. *Human Factors, 24*(2), 129–162.

Trollip, S. R., & Jensen, R. S. (1991). *Human factors for general aviation*. Englewood, CO: Jeppesen Sanderson.

Wachtel, P. L. (1967). Conceptions of broad and narrow attention. *Psychological Bulletin, 68*, 417–429.

Weltman, G., Smith, J. E., & Egstrom, G. H. (1971). Perceptual narrowing during simulated pressure-chamber exposure. *Human Factors, 13*, 99–107.

Wickens, C. D. (1984). *Engineering psychology and human performance* (1st ed.). Columbus, OH: Charles E. Merrill Publishing Co.

Wickens, C. D. (1992). *Engineering psychology and human performance* (2nd ed.). New York: Harper Collins.

Wiener, E. L. (1989). *Human factors of advanced technology ("glass cockpit") transport aircraft* (NASA Contractor Report No. 177528). Moffett Field, CA: NASA-Ames Research Center.

Wiener, E. L. (1993). Life in the second decade of the glass cockpit. In R. S. Jensen, & D. Neumeister (Eds.), *Proceedings of the Seventh International Symposium on Aviation Psychology* (pp. 1–11). Columbus: Department of Aviation, the Ohio State University.

Wiener, E. L., & Curry, R. E. (1980). Flight deck automation: Promises and problems. *Ergonomics, 23*(10), 995–1011.

Wright, P. (1974). The harassed decision maker: Time pressures, distractions, and the use of evidence. *Journal of Applied Psychology, 59*(5), 555–561.

III

Aircraft

13

Personnel Selection and Training

D. L. Pohlman
Institute for Defense Analyses

J. D. Fletcher
Institute for Defense Analyses

13.1 Introduction

This chapter focuses on the selection and training of people who work in aviation specialties. Aviation work encompasses a full spectrum of activity from operators of aircraft (i.e., pilots), to flight attendants, dispatchers, flight controllers, mechanics, engineers, baggage handlers, ticket agents, airport managers, and air marshals. The topic covers a lot of territory. For manageability, we concentrated on three categories of aviation personnel: pilots and aircrew, maintenance technicians, and flight controllers.

One problem shared by nearly all aviation specialties is their workload. Workload within most categories of aviation work has been increasing since the beginning of aviation. In the earliest days, available technology limited what the aircraft could do, similarly limiting the extent and complexity of aircraft operations. Pilots flew the airplane from one place to another, but lacked instrumentation to deal with poor weather conditions—conditions that were simply avoided. Maintainers serviced the airframe and engine, but both of these were adapted from relatively familiar, non-aviation technologies and materials. Flight controllers, if they were present at all, were found standing on the airfield waving red and green flags.

Since those days, aircraft capabilities, aircraft materials, and aviation operations have progressed remarkably. The aircraft is no longer a limiting factor. Pilots, maintainers, and controllers are no longer pushing aviation technology to its limits, but are themselves being pushed to the edge of the human performance envelope by the aircraft that they operate, maintain, and control.

To give an idea about the work for which we are selecting and training people, it may help to discuss the workloads that different specialties impose on aviation personnel. The following is a short discussion about each of the three selected aviation specialties and the workloads that they may impose.

13.1.1 Pilots

Control of aircraft in flight has been viewed as a challenge from the beginning of aviation—if not before. McRuer and Graham (1981) reported that in 1901, Wilbur Wright addressed the Western Society of Engineers as follows:

> Men already know how to construct wings or aeroplanes, which when driven through the air at sufficient speed, will not only sustain the weight of the wings themselves, but also that of the engine, and of the engineer as well. Men also know how to build screws of sufficient lightness and power to drive these planes at sustaining speed.... Inability to balance and steer still confronts students of the flying problem.... When this one feature has been worked out, the age of flying machines will have arrived, for all other difficulties are of minor importance (p. 353).

The "age of flying machines" has now passed the century mark. Many problems of aircraft balance and steering—of operating aircraft—have been solved, but, as McRuer and Graham concluded, many remain. A pilot flying an approach in bad weather with most instruments nonfunctional or a combat pilot popping up from a high-speed ingress to roll over and deliver ordnance on a target while dodging surface to air missiles and ground fire, is working at the limits of human ability. Control of aircraft in flight still "confronts students of the flying problem."

To examine the selection and training of pilots, it is best, as with all such issues, to begin with the requirements. What are pilots required to know and do? The U.S. Federal Aviation Administration (FAA) tests for commercial pilots reflect the growth and current maturity of our age of flying machines. They cover the following areas of knowledge (U.S. Department of Transportation, 1995b):

1. FAA regulations that apply to commercial pilot privileges, limitations, and flight operations
2. Accident reporting requirements of the National Transportation Safety Board (NTSB)
3. Basic aerodynamics and the principles of flight
4. Meteorology to include recognition of critical weather situations, wind sheer recognition and avoidance, and the use of aeronautical weather reports and forecasts
5. Safe and efficient operation of aircraft
6. Weight and balance computation
7. Use of performance charts
8. Significance and effects of exceeding aircraft performance limitations
9. Use of aeronautical charts and magnetic compass for pilotage and dead reckoning
10. Use of air navigation facilities
11. Aeronautical decision-making and judgment
12. Principles and functions of aircraft systems
13. Maneuvers, procedures, and emergency operations appropriate to the aircraft
14. Night and high altitude operations
15. Descriptions of and procedures for operating within the National Airspace System

Despite the concern for rules and regulations reflected by these knowledge areas, there remains a requirement to fly the airplane. All pilots must master basic airmanship, operation of aircraft systems, and navigation. Military pilots must add to these basic skills the operation of weapons systems while meeting the considerable workload requirements imposed by combat environments.

13.1.1.1 Basic Airmanship

There are four basic dimensions to flight: altitude (height above a point), attitude (position in the air), position (relative to a point in space), and time (normally a function of airspeed). A pilot must control these four dimensions simultaneously. Doing so allows the aircraft to take off, remain in flight, travel from point A to point B, approach, and land.

Basic aircraft control is largely a psychomotor task. Most student pilots need 10–30 h of flying time to attain minimum standards in a slow moving single engine aircraft. Experience in flying schools suggests that of the four basic dimensions listed above, time is the most difficult to master. A good example might be the touchdown portion of a landing pattern. Assuming that the landing target is 500 ft beyond the runway threshold, and that the aircraft is in the appropriate dimensional position as it crosses the threshold at about 55 miles per hour, a student in a single-engine propeller aircraft has about 6.2 s to formulate and implement the necessary decisions to touch the aircraft down. A student in a military trainer making a no flap, heavy weight landing at about 230 miles per hour has approximately 1.5 s to formulate and implement the same necessary decisions. The requirement to make decisions at 4 times the pace of slower aircraft prevents many student pilots from graduating to more advanced aircraft, and is the cause of a large number of failures in military flight schools. The problem of flying more powerful aircraft than those used to screen pilot candidates is compounded by the steadily increasing complexity of aircraft systems.

13.1.1.2 Aircraft Systems

Pilots must operate the various systems found in aircraft. These systems include engine controls, navigation, fuel controls, communications, airframe controls, and environmental controls, among others. Some aircraft have on-board systems that can be run by other crew members, but the pilot remains responsible for them and must be aware of the status of each system at all times. For instance, the communications system can be operated by other crew members, but the pilot must quickly recognize from incessant radio chatter, the unique call sign in use that day and respond appropriately. Increases in the number and complexity of aircraft systems, faster and more capable aircraft, and increased airway system density and airport traffic all combine to increase the difficulty of operating aircraft. Increasing difficulty translates to an increased demand on the pilot's already heavy workload. These systems make it possible for aircrews to perform many tasks that would be impossible in their absence, but the systems also increase appetite, demand, and expectations for higher levels of performance that reach beyond the capabilities afforded by emerging aircraft systems. The result is a requirement for remarkable levels of performance, as well as serious increases in aircrew workload.

13.1.1.3 Navigation

Once pilots master basic airmanship and the use of basic aircraft systems, they must learn to navigate. Navigating in four dimensions is markedly different from navigating in two dimensions. Flying in the Federal Airway system requires pilots to know and remember all the five different types of airspace while maintaining the aircraft on an assigned course, at an assigned airspeed, on an assigned altitude, and on an assigned heading. Pilots must also be prepared to modify the assigned parameters at an assigned rate and airspeed (i.e., pilots may be required to slow to 200 knots and descend to 10,000 ft at 500 ft per min). They must accomplish all these tasks, while acknowledging and implementing new instructions over the radio. They may further be required to perform all these tasks under adverse weather conditions (clouds, fog, rain, or snow) and turbulence.

13.1.1.4 Combat Weapons Systems

Combat aircraft confront pilots with all the usual problems of "balance and steering" and systems operation/navigation, but add to them the need to contend with some of the most complex and advanced weapons systems and sensors in the world. Each weapon that the aircraft carries, affects flight parameters in different ways. Combat pilots must understand how each weapon affects the aircraft when it is aboard and when it is deployed. They must understand the launch parameters of the weapons, their in-flight characteristics, and any additional system controls that the weapons require. These controls include buttons, switches, rockers, and sliders located on the throttles, side panels, instrument panel, and stick grip. Some controls switch between different weapons, others change the mode of the selected weapons, while others may manipulate systems such as radar and radios. The pilot must understand, monitor, and properly operate (while wearing flight gloves) all the controls belonging to each weapon system. It is not surprising to find that the capabilities of state-of-the-art fighter aircraft often exceed the pilots' capabilities to use them. But, we have yet to get our overloaded pilot into combat.

13.1.1.5 Combat Workload

The task of flying fighter aircraft in combat is one of the most complex cognitive and psychomotor tasks imaginable. "Fifty feet and the speed of heat" is an expression that military fighter pilots use to describe an effective way to ingress a hostile target area. A fighter pilot in combat must be so versed in the flying and operation of the aircraft that nearly all of the tasks just described are assigned to background, or "automatic," psychomotor and cognitive processing. The ability to operate an aircraft in this manner is described as *strapping the aircraft on*. A combat pilot must:

- Plan the route through space in relation to the intended target, suspected threats, actual threats, other known aircraft, wingmen, weather, rules of engagement, and weapons
- Monitor the aircraft displays for electronic notification of threats
- Differentiate among threat displays (some systems can portray 15 or more different threats)
- Plan ingress to and egress from the target
- Set switches for specific missions during specific periods of the flight
- Monitor radio chatter on multiple frequencies for new orders and threat notification
- Monitor progress along the planned route
- Calculate course, altitude, and airspeed corrections
- Plan evasive maneuvers for each type of threat and position during the mission
- Plan and execute weapons delivery
- Execute battle damage assessment
- Plan and execute safe egress from hostile territory
- Plan and execute a successful recovery of the aircraft

This workload approaches the realm of the impossible. However, other aviation specialties also present impressive workloads. One of the most highly publicized of these workloads is that of flight controllers.

13.1.2 Flight Controllers

In semiformal terms, flight controllers are responsible for the safe, orderly, and expeditious flow of air traffic on the ground at airports and in the air where service is provided using instrument flight rules (IFR) and visual flight rules (VFR), depending on the airspace classification. In less formal terms, they are responsible for reducing the potential for chaos around our airports, where as many as 2000 flights a day may require their attention.

In good conditions, all airborne and ground-based equipments are operational and VFR rules prevail. However, as weather deteriorates and night approaches, pilots increasingly depend on radar flight controllers to guide them and keep them at a safe distance from obstacles and other aircraft. Radar images used by controllers are enhanced by computers that add to each aircraft's image such information as the call sign, aircraft type, airspeed, altitude, clearance limit, and course. If the ground radar becomes unreliable or otherwise fails, controllers must rely on pilot reports and "raw" displays, which consist of small dots (*blips*), with none of the additional information provided by computer-enhanced displays. During a radar failure, controllers typically calculate time and distance mechanically, drawing pictures on the radarscope with a grease pencil. The most intense condition for flight controllers occurs when all ground equipment is lost except radio contact with the aircraft. To exacerbate this situation there may be an aircraft that declares an emergency during IFR conditions with a complete radar failure. This condition is rare, but not unknown in modern aircraft control.

Using whatever information is available to them, flight controllers must attend to the patterns of all aircraft (often as many as 15) in the three-dimensional airspace under their control. They must build a mental, rapidly evolving image of the current situation and project it into the near future. Normally, controllers will sequence aircraft in first-in, first-out order so that the closest aircraft begins the approach first. The controller changes courses, altitudes, aircraft speeds, and routing to achieve "safe, orderly, and expeditious flow of aircraft." During all these activities, the controller must prevent aircraft at the same altitude from flying closer to each other than three miles horizontally.

The orderly flow of aircraft may be disrupted by emergencies. An emergency aircraft is given priority over all aircraft operating normally. The controller must place a bubble of safety around the emergency aircraft by directing other aircraft to clear the airspace around the emergency aircraft and the path of its final approach. The controller must also determine the nature of the emergency so that appropriate information can be relayed to emergency agencies on the ground. If the ground equipment fails, the only separation available for control may be altitude with no enhanced radar image feedback to verify that the reported altitude is correct. The controller must expedite the approach of the emergency aircraft while mentally reordering the arriving stack of other aircrafts.

Knowledge and skill requirements for aircraft controller certification include (US Department of Transportation, 1995c):

1. Flight rules
2. Airport traffic control procedures
3. En-route traffic-control procedures
4. Communications procedures
5. Flight assistance services
6. Air navigation and aids to air navigation
7. Aviation weather and weather reporting procedures
8. Operation of control tower equipment
9. Use of operational forms
10. Knowledge of the specific airport, including rules, runways, taxiways, and obstructions
11. Knowledge of control zones, including terrain features, visual checkpoints, and obstructions
12. Traffic patterns, including use of preferential runways, alternate routes and airports, holding patterns, reporting points, and noise abatement procedures
13. Search and rescue procedures
14. Radar alignment and technical operation

The stress levels during high traffic volume periods in Air Traffic Control (ATC) are legendary. At least, however, ATC controllers are housed in environmentally controlled towers and buildings. This is not necessarily the case for aircraft maintenance technicians (AMTs).

13.1.3 Aircraft Maintenance Technicians

A typical shift for an AMT may consist of several calls to troubleshoot and repair problems ranging from burnt-out landing lights to finding a short in a cannon plug that provides sensor information to an inertial navigation system. To complicate matters, some problems may only be present when the aircraft is airborne—there may be no way to duplicate an airborne problem on the ground. The inability to duplicate a reported problem greatly complicates the process of isolating the malfunction. For example, the problem may be that one of many switches indicates that the aircraft is not airborne when it actually is, or the malfunction may arise from changes in the aircraft frame and skin due to temperature variations and condensation or intermittent electrical shorts due to vibration, all of which may occur only in flight. Also, of course, the variety and the rapidly introduced, constantly changing materials and the underlying technologies applied in aviation increase both the workload for AMTs and their continuing need for updated training and education.

Despite these complications, the AMT is usually under pressure to solve problems quickly because many aircraft are scheduled to fly within minutes after landing. Additionally, an AMT may have to contend with inadequate descriptions of the problem(s), unintelligible handwriting by the person reporting the problem, and weather conditions ranging from 140°F in bright sun to −60°F with 30 knots of wind blended with snow. All these factors combine to increase the challenge of maintaining modern aircraft.

Although some research on maintenance issues had been performed earlier for the U.S. military, until about 1985 most human factors research in aviation, including research on selection and training, was concerned with cockpit and ATC issues. Concern with maintenance as a human factors issue was almost nonexistent. However, this emphasis has evolved somewhat in recent years (Jordan, 1996). Although the selection, training, and certification of maintenance technicians have lagged behind increases in the complexity and technological sophistication of modern aircraft, they also have been evolving. Appreciation of aviation maintenance as a highly skilled, often specialized profession requiring training in institutions of higher learning has been developing, albeit slowly (Goldsby, 1996).

Current FAA certification of AMTs still centers on mechanical procedures involving the airframes and power plants. The AMTs are required to possess knowledge and skills concerning (U.S. Department of Transportation, 1995a):

1. Basic electricity
2. Aircraft drawings
3. Weight and balance in aircraft
4. Aviation materials and processes
5. Ground operations, servicing, cleaning, and corrosion control
6. Maintenance publications, forms, and records
7. Airframe wood structures, coverings, and finishes
8. Sheet metal and nonmetallic structures
9. Welding
10. Assembly and rigging
11. Airframe inspection
12. Hydraulic and pneumatic power systems
13. Cabin atmosphere control systems
14. Aircraft instrument systems
15. Communication and navigation systems
16. Aircraft fuel systems
17. Aircraft electrical systems
18. Position and warning systems
19. Ice and rain systems
20. Fire protection systems
21. Reciprocating engines
22. Turbine engines
23. Engine inspection
24. Engine instrument systems
25. Lubrication systems
26. Ignition and starting systems
27. Fuel and fuel metering systems
28. Induction and engine airflow systems
29. Engine cooling systems
30. Engine exhaust and reverser systems
31. Propellers

This is a long list, but still more areas of knowledge need to be covered if maintenance training and certification are to keep pace with developments in the design and production of modern aircraft. The list needs to include specialization in such areas as: (a) aircraft electronics to cover the extensive infusion of digital electronics, computers, and fly-by-wire technology in modern aircraft, (b) composite structures, which require special equipment, special working environments, and special precautions to protect the structures themselves and the technicians' own health and safety, and (c) nondestructive inspection technology, which involves sophisticated techniques using technologies such as magnetic particle and dye penetrants, x-rays, ultrasound, and eddy currents.

Even within more traditional areas of airframe and power-plant maintenance, current business practices and trends are creating pressures for more extensive and specialized training and certification. Goldsby (1996) suggested that these pressures arise from increasing use of: (a) third parties to provide increasing amounts of modification and repair work; (b) aging aircraft; (c) leased aircraft requiring greater maintenance standardization and inspection techniques; (d) noncertified airframe specialists; and (e) second- and third-party providers of noncertified technicians.

The most important problem-solving skills for AMTs may be those of logical interpretation and diagnostic proficiency. These higher-order cognitive skills can only be developed by solving many problems provided by extensive and broad experience in working on actual aircraft or by long hours spent with appropriately designed and employed maintenance simulators. Talent and logical thinking help, which is say to that, personnel selection and classification remain relevant, but they increasingly need to emphasize problem solving and judgment in addition to the usual capacities for learning and systematically employing complex procedural skills. There appears to be no real substitute for experience in developing troubleshooting proficiency, but the time to acquire such experience has been considerably shortened by the availability of simulations used in maintenance training and the need for training can be lightened through the use of portable, hand-held maintenance-aiding devices (Fletcher & Johnston, 2002).

13.2 Personnel Recruitment, Selection, and Classification for Aviation

How people are recruited from the general population pool, selected for employment, and classified for occupational specialties affects the performance and capabilities of every organization. Effective recruitment, selection, and classification procedures save time, materiel, and funding in training, and improve the quality and productivity of job performance. They help ensure worker satisfaction, organizational competence, productivity, and, in military circles, operational readiness.

Among personnel recruitment, selection, and classification, recruitment is the first step—people are first recruited from a general or selected population pool, then selected for employment and subsequently classified into specific jobs or career paths. In civilian practice, personnel selection and classification are often indistinguishable; individuals with the necessary pretraining are identified and recruited to perform specific jobs. Selection is tantamount to classification. In large organizations such as the military services, which provide appreciable amounts of training to their employees, the processes of recruitment, selection, and classification are more separate. For instance, people are recruited from the general population by the various recruiting services within the military. They are then selected for military service based on general, but well-observed standards. Those people selected are then classified and assigned for training to one of many career fields with which they may have had little or no experience. These efforts pay off. Zeidner and Johnson (1991) determined that the U.S. Army's selection and classification procedures save the Army about $263 million per year.

There are more pilots currently available than there are flying jobs, in both the military and civilian sectors. Radar controllers, aviation mechanics, air marshals, and many other specialties do not enjoy the same situation. People entering the aviation-mechanics field fell to 60%, from 1991 to 1997 (Phillips, 1999). In May 2003, the United States Air Force (USAF) needed 700 ATC controllers, and the National Air Traffic Controllers Association, the union that represents 15,000 controllers, reported that the Federal Aviation Administration (FAA) needs to immediately begin hiring and training the next generation of ATCs who would fill the gaps created by upcoming retirements, increased traffic growth, and system capacity enhancements (McClearn, 2003). The FAA Controller training facility is preparing to increase training from 300 controllers a year in 2001 to 1600 a year in 2009 (Nordwall, 2003), but aviation must compete with many other industries requiring similar skill levels, such as the electronics industry and the automotive industry, most of which pay better and impose less personal liability.

It should be noted that classification may matter as much as selection, as pointed out by Zeidner and Johnson (1991). Researchers have found that how well people are classified for specific jobs or career paths has a major impact on job performance, job satisfaction, and attrition, regardless of how carefully they are selected for employment. One study found that personnel retention rates over a 5 year period differed by 50% for well-classified versus poorly-classified individuals (Stamp, 1988). Zeidner and Johnson suggested that the Army might double the $263 million it saves through proper selection by paying equal attention to classifying people into the occupation specialties for which they are best suited by ability, interest, and values. This is to say nothing of the increases in productivity and effectiveness that could result from early identification and nurturing of potential "aces" across all aviation specialties—mechanics and controllers as well as pilots.

Because of the expense, complexity, and limited tolerance for error in aviation work, more precise selection and classification have been sought almost from the beginning of the age of flying machines (at the very beginning, the Wright bothers just flipped a coin). Hunter (1989) wrote that "almost every test in the psychological arsenal has been evaluated at one time or another to determine its applicability for aircrew selection" (p. 129). Hilton and Dolgin (1991) wrote that there may be no other "occupation in the world that benefits more from personnel selection technology than that of military pilot" (p. 81).*

13.2.1 A Brief Historical Perspective

Aviation and many personnel management procedures began their systematic development at about the same time. This fact is not entirely coincidental. The development of each increased the requirement for the other.

Recruitment is necessary when the voluntary manpower pool is insufficient to provide the necessary personnel flow to fill the current and future job requirements. In the history of most aviation careers, the issue of recruitment is a relatively new phenomenon. When aviation began in the early 1900s it was a glamorous endeavor. At the beginning of World War I, many Americans left the safety of the United States and volunteered to fight for France if they could fly aeroplanes. Flying was high adventure, not only for the military, but also for the commercial carrier personnel. During this period, it was the U.S. Air Mail Service that laid the foundation for commercial aviation worldwide. With the cooperation of the U.S. Air Service, the U.S. Post Office flew the mail from 1918 to 1927 (http://www.airmailpioneers.org/).

Aviation matured rapidly during World War I and World War II. By 1945, the fledgling air industry in America was beginning to gain momentum. Excess post-war transport aircraft initially filled the need for equipment. Pilots and mechanics, and other service personnel who entered the job market after the war's end provided the labor. Even though there remained an air arm in the military, the U.S. Mail routes precipitated the aviation revolution in America. For the most part, volunteers provided sufficient manpower to populate the military and its aviation requirements. With the end of the Vietnam-era draft and the initiation of the All Volunteer Force in 1973, the Armed Services began a systematic recruiting drive that has continued to fulfill the nation's military and most of its civilian requirements for aviation personnel, but the pressure on the Services to do so has increased steadily. The U.S. Army began recruiting only high-school graduates with Armed Forces Vocational Aptitude Battery (ASVAB) scores in the upper 50th percentile in 1978, resulting in an entry-level training reduction of 27% (Oi, 2003).

---

* The history of recruiting in aviation has not always been honorable. The term fly-by-night comes from early aviators who would descend on a town, "recruit" (through not entirely scientific means) individuals who were proclaimed to have a talent for flying, collect a fee for training these individuals, and fly out at night before the lessons were to begin (Roscoe, Jensen, & Gawron, 1980).

Currently, selection and classification procedures are applied across the full range of aviation personnel, but the development of systematic personnel management procedures in aviation initially focused on selection of pilots, rather than aviation support personnel. These procedures grew to include physical, psychomotor, mental ability, and psychological (personality) requirements, but they began with self-selection.

13.2.1.1 Self-Selection

Probably, from the time of Daedelus and certainly from the time of the Wright Brothers, people have been drawn to aviation. In the early days of World War I, many pilots were volunteers who came from countries other than the one providing the training (Biddle, 1968). Some of these early pilots could not even speak the language of the country for which they flew, but they wanted to fly. Among them, the Americans established the base of America's early capabilities in aviation during and after that war.

Self-selection continues to be a prominent factor in pilot and aircrew selection in both military and civilian aviation. Only people with a strong desire to fly civil aircraft are likely to try and obtain a license to fly. Advancement past the private pilot stage and acquiring the additional ratings required of commercial pilots is demanding, time-consuming, and expensive. The persistence of a prospective pilot in finishing training and pursuing an aviation career beyond a private pilot license constitutes a form of natural selection. That aviation continues to attract and hold so many able people who select themselves for careers in aviation attests to its strong and continuing appeal.

Early it was observed that training pilots was an expensive undertaking, and selection for aircrew personnel soon evolved from self-selection alone to more systematic and formal procedures. The arguments for this evolution frequently cite the costs of attrition from flight training. These costs have always been high, and they have risen steadily with the cost and complexity of aircraft. Today, it costs more than $1M to train a jet pilot, and the current cost to the Air Force for each failed aviation student is estimated to be $50,000 (Miller, 1999). This latter expense excludes the very high cost of aircraft whose loss might be prevented by improved selection and classification procedures. As a consequence, research, development, implementation, and evaluation of procedures to select and classify individuals for aviation training have been a significant investment and a major contribution of the world's military services. These procedures began with those used for the general selection and classification of military personnel—physical qualifications.

13.2.1.2 Physical Qualification Selection

With World War I, the demand for flyers grew, and the number of applicants for flying training increased. Military organizations reasonably assumed that physical attributes play a significant role in a person's ability to successfully undertake flight training and later assume the role of pilot. Flight physicals became a primary selection tool.

At first, these physicals differed little from the standard examinations of physical well-being used to select all individuals for military service (Brown 1989; Hilton & Dolgin, 1991).[*] Soon, however, research aimed specifically to improve selection of good pilot candidates began in Italy and France (Dockeray & Isaacs, 1921). Needs for balance in air, psychomotor reaction, appropriate concentration and distribution of attention, emotional stability, and rapid decision-making were assumed to be greater than those for non-aviation personnel, and more stringent procedures were established for selecting aviation personnel.

Italian researchers, who may have initiated this line of research, developed measures of reaction time, emotional reaction, equilibrium, attention, and perception of muscular effort and added them to the

---

[*] Vestiges of early physical standards for military service held on long after the need for them was gone. As late as the Korean War, fighter pilots were required to have opposing molars. This requirement was eventually traced to the Civil War era need to bite cartridges before they could be fired. Only when fighter pilots became scarce in the early 1950s did anyone question its enforcement.

standard military physical examinations specifically used to select pilots. Other countries, including the United States, undertook similar research and development efforts.

Rigorous flight physicals continue to be used today to qualify and retain individuals in flight status, for both military and civilian pilots. The FAA defines standards for first-, second-, and third-class medical certificates covering eyesight, hearing, mental health, neurological conditions (epilepsy and diabetes are cause for disqualification), cardiovascular history (annual electrocardiograph examinations are required for people over 40 years with first-class certificates), and general health as judged by a certified federal air surgeon (U.S. Department of Transportation, 1996).

13.2.1.3 Mental Ability Selection

During World War I, the military services also determined that rigorous flight physicals for selecting pilots were not sufficient. Other methods were needed to reduce the costs and time expended on candidates who were washing out of training despite being physically qualified. A consensus developed that pilots need to make quick mental adjustments using good judgment in response to intense, rapidly changing situations. It was then assumed that pilot selection would benefit from methods that would measure mental ability. These methods centered on use of newly developed paper-and-pencil tests of mental ability. What was new about these tests was that they could be inexpensively administered to many applicants all at the same time.

Assessment procedures administered singly to individuals by specially trained examiners had been used in the United States at least as early as 1814 when both the Army and the Navy used examinations to select individuals for special appointments (Zeidner & Drucker, 1988). In 1883, the Civil Service Commission initiated the wide use of open, competitive examinations for appointment in government positions. Corporations, such as General Electric and Westinghouse, developed and implemented employment testing programs in the early 1900s. However, it took the efforts of the Vineland Committee working under the supervision of Robert Yerkes in 1917, to develop reliable, parallel paper-and-pencil tests that could be administered by a few individuals to large groups of people using simple, standardized procedures (Yerkes, 1921).

The Vineland Committee developed a plan for the psychological examination of the entire U.S. Army. It produced the Group Examination Alpha (the Army Alpha), which was "an intelligence scale for group examining… [making] possible the examination of hundreds of men in a single day by a single psychologist" (Yerkes, 1921, p. 310). The Army Alpha provided the basis for many paper-and-pencil psychological assessments that were developed for group administration in the succeeding years. It was used by the United States Committee on Psychological Problems of Aviation to devise a standard set of tests and procedures that were adopted in 1918 and used to select World War I pilots (Hilton & Dolgin, 1991).

The Army Alpha test laid the foundation for psychological assessment of pilots performed by the U.S. Army in World War I, by the Civil Aeronautics Authority in 1939, and after that by the U.S. Army and Navy for the selection of aircrew personnel in World War II. Reducing the number of aircrew student washouts throughout this period saved millions of dollars that were thereby freed to support other areas of the war effort (U.S. Department of the Air Force, 1996). It is also likely that the aircrews selected and produced by these procedures were of higher quality than they might have been without them, thereby significantly enhancing military effectiveness. However, the impact of personnel selection and classification on the ultimate goal of military effectiveness—or on productivity in nonmilitary organizations—then and now has received infrequent and limited attention from researchers (Kirkpatrick, 1976; Zeidner & Johnson, 1991).

After World War I, there was a flurry of activity concerning psychological testing and pilot selection. It differed from country to country (Dockeray & Isaacs, 1921; Hilton & Dolgin, 1991). Italy emphasized psychomotor coordination, quick reaction time, and constant attention. France used vasomotor reactions during apparatus testing to assess emotional stability. Germany concentrated on the use of apparatus tests to measure individual's resistance to disorientation. Great Britain emphasized physiological signs as indicators of resistance to altitude effects. Germany led in the development of personality measures for pilot selection. The United States, Japan, and Germany all used general intelligence as an indicator of aptitude for aviation.

In the United States the rapid increase of psychological testing activity was short-lived. The civil aircraft industry was embryonic, and there was a surplus of aviators available to fly the few existing civil aircraft. Only in the mid-1920s, when monoplanes started to replace postwar military aircraft, did civil air development gain momentum and establish a growing need for aviation personnel. Hilton and Dolgin reported that the pattern of reduced testing was found in many countries, consisting of a rigorous physical examination, a brief background questionnaire, perhaps a written essay, and an interview.

In the 1920s and 1930s, as aircraft became more sophisticated and expensive, the selection of civilian pilots became more critical. The development of a United States civilian aviation infrastructure was first codified through the Contract Mail Act (the Kelly Act) of 1925 (Hansen & Oster, 1997). This infrastructure brought with it requirements for certification and standardized management of aviation and aviation personnel. It culminated in the Civil Aeronautics Act in 1938, which established the Civil Aeronautics Authority, later reorganized as the Civil Aeronautics Board in 1940.

Another world war and an increased demand for aviation personnel both appeared likely in 1939. For these reasons the Civil Aeronautics Authority created a Committee on Selection and Training of Aircraft Pilots, which immediately began to develop qualification tests for screening civilian aircrew personnel for combat duty (Hilton & Dolgin, 1991). This work formed the basis for selection and classification procedures developed by the Army Air Force Aviation Psychology Program Authority. Viteles (1945) published a comprehensive summary description of this program and its accomplishments at the end of World War II.

The procedures initially developed by the Aviation Psychology Program were a composite of paper-and-pencil intelligence and flight aptitude tests. They were implemented in 1942 as the Aviation Cadet Qualifying Examination and used thereafter by the U.S. Army Air Force to select aircrew personnel for service in World War II (Flanagan, 1942; Hilton & Dolgin, 1991; Hunter, 1989; Viteles, 1945). These procedures used paper-and-pencil tests, motion picture tests, and apparatus tests. The Army's procedures were designed to assess five factors that had been found to account for washouts in training: intelligence and judgment, alertness and observation including speed of decision and reaction, psychomotor coordination and technique, emotional control and motivation, and ability to divide attention. The motion picture and apparatus tests were used to assess hand and foot coordination, judgment of target speed and direction, pattern memory, spatial transposition, and skills requiring timed exposures to visual stimuli.

Flanagan (1942) discussed the issues in classifying personnel after they had been selected for military aviation service. Basically, he noted that pilots need to exhibit superior reaction speed and the ability to make decisions quickly and accurately, bombardiers need superior fine motor steadiness under stress (for manipulating bomb sights), concentration and ability to make mental calculations rapidly under distracting conditions, and navigators need superior ability to grasp abstractions, such as those associated with celestial geometry and those required to maintain spatial orientation, but not the high level of psychomotor coordination needed by pilots and bombardiers.

In contrast, the U.S. Navy relied primarily on physical screening, paper-and-pencil tests of intelligence and aptitude (primarily mechanical comprehension), the Purdue Biographical Inventory, and line officer interviews to select pilots throughout World War II (Fiske, 1947; Jenkins, 1946). The big differences between the two Services were that the Army used apparatus (we might call them simulators today), whereas the Navy did not and that the Navy used formal biographical interviews, whereas the Army did not. The Army studied the use of interviews and concluded that even those that were reliable contributed little to reductions in time, effort, and costs (Viteles, 1945).

Today, the military services depend on a progressive series of selection instruments. These include academic performance records, medical fitness, a variety of paper-and-pencil tests of general intelligence and aptitude, possibly a psychomotor test such as the Air Force's Basic Abilities Test (BAT), and flight screening (flying lessons) programs. Newer selection methods include the use of electroencephalography to test for epileptiform indicators of epilepsy (Hendriksen & Elderson, 2001). Commercial airlines rarely hire a pilot who has no experience. They use flight hours to determine if candidates will be able to acclimate to the life of airline pilots. They capitalize on aviation personnel procedures developed by the military to hire large numbers of pilots, maintainers, controllers, and others who have been selected, classified, and trained by the military Services.

Three conclusions may be drawn from the history of selection for aircrew personnel. First, most research in this area has focused on the selection of individuals for success in training, and not on performance in the field, in operational units, or on the job. Nearly all validation studies of aircrew-selection measurements concern their ability to predict performance in training.* This practice makes good monetary sense—the attrition of physically capable flight candidates is very costly. Trainers certainly want to maximize the probability that individuals selected for aircrew training will successfully complete it. Also, it is not unreasonable to expect some correlation between success of individuals in training and their later performance as aircrew members. However, over 100 years into the age of flying machines, information relating selection measures to performance on the job remains scarce.† It would still be prudent to identify those individuals who, despite their successes in training, are unlikely to become good aviators on the job. And we would like to identify, earlier than we can now, those exceptional individuals who are likely to become highly competent performers, if not aces, in our military forces and master pilots in our civilian aircraft industry.

The second and third conclusions were both suggested by Hunter (1989). His review of aviator selection concludes that there seems to be little relationship between general intelligence and pilot performance. It is certainly true that tests of intelligence do not predict very well either performance in aircrew training or on the job. These tests largely measure the verbal intelligence that is intended to predict success in academic institutions—as these institutions are currently organized and operated. Newer multifaceted measures of mental ability (e.g., Gardner, Kornhaber, & Wake, 1996) may more successfully identify aspects of general intelligence that predict aviator ability and performance. Also, by limiting variability in the population of pilots, our selection and classification procedures may have made associations between measures of intelligence and the performance of pilots difficult to detect. In any case, our current measures of intelligence find limited success in accounting for pilot performance.

Hunter also suggested a third conclusion. After a review of 36 studies performed between 1947 and 1978 to assess various measures used to select candidates for pilot training, Hunter found that only those concerned with instrument and mechanical comprehension were consistent predictors of success—validity coefficients for these measures ranged from 0.20 to 0.40. Other selectors, assessing factors such as physical fitness, stress reactivity, evoked cortical potentials, age, and education were less successful. A follow-up study by Hunter and Burke (1995) found similar results. The best correlates of success in pilot training were job samples, gross dexterity, mechanical understanding, and reaction time. General ability, quantitative ability, and education were again found to be poor correlates of success.

In brief, selection for aircrew members currently centers on predicting success in training, and includes measures of physical well-being, general mental ability, instrument and mechanical comprehension, and psychomotor coordination, followed by a brief exposure to flying an inexpensive, light airplane and/or a simulator. Attrition rates for training by the military services range around 22% (Duke & Ree, 1996).

The best hope for reducing attrition rates further and for generally increasing the precision of our selection and classification procedures may be the use of computer-based testing. Early techniques of

---

* Notably, they are concerned with the prediction of success in training, given our current training procedures. Different training procedures could yield different "validities."
† There are exceptions. See for example the efforts discussed by Carretta and Ree (1996) to include supervisory performance ratings in the assessment of selection and classification validities.

computer-based testing were innovative in using the correct and incorrect responses made by individuals to branch rapidly among pools of items with known psychometric characteristics and difficulty until they settled on a level of ability within a sufficiently narrow band of confidence. Newer techniques may still use branching, but they go beyond the use of items originally developed for paper-and-pencil testing (Kyllonen, 1995). These tests capitalize on the multimedia, timing, and response-capturing capabilities that are only available through the use of computers. These computerized tests and test items have required and engendered new theoretical bases for ability assessment. For a more complete discussion of assessment for pilot training see O'Neil & Andrews (2000). Most of the theoretical bases that are emerging are founded on information processing models of human cognition. These models are discussed, briefly and generically, in the next section.

13.2.2 A Brief Theoretical Perspective

Over the years, work in aviation has changed. The leather-helmeted, white-scarfed daredevil fighting a lone battle against the demons of the sky, overcoming the limited mechanical capabilities of his aircraft, and evading the hostile intent of an enemy at war is gone. The problems remain: The sky must, as always, be treated with respect, maintenance will never reach perfection, and war is still with us, but the nature of aviation work and the requisite qualities of people who perform it have evolved with the evolution of aviation technology.

Today, in place of mechanical devices yoked together for the purposes of flight and requiring mostly psychomotor reflexes and responses, we have computer-controlled, highly-specialized, integrated aviation systems requiring judgment, abstract thinking, abstract problem-solving, teamwork, and a comprehensive grasp of crowded and complex airspaces along with the rules and regulations that govern them (Driskell & Olmstead, 1989; Hansen & Oster, 1997). Aviation work has evolved from the realms of the psychomotor to include those of information processing and from individual dash and élan to leadership, teamwork, and managerial judgment. With an evolution toward information processing, and the resulting increase in the demands on both the qualitative and quantitative aspects of human performance in aviation, it is not surprising to find information-processing models increasingly sought and applied in the selection, classification, assignment, training, and assessment of aviation personnel.

The complexity of human performance in aviation has always inspired similarly complex models of human cognition. Primary among the models to grow out of aviation psychology in World War II was Guilford's (1967) well-known and wonderfully heuristic "Structure of the Intellect" which posited 120 different ability factors based on all combinations of 5 mental operations (memory, cognition, convergent thinking, divergent thinking, and evaluation), 6 types of products (information, classes of units, relations between units, systems of information, transformations, and implications), and 4 classes of content (figural, symbolic, semantic, and behavioral). An appropriate combination and weighting using "factor pure" measures of these 120 abilities would significantly improve the selection and classification of individuals for work in aviation.

Despite the significant research and substantial progress that these abilities engendered in understanding human abilities, Guilford's ability factors—or perhaps our ability to assess them—failed to prove as independent and factor, pure as hoped, and the psychological research community moved on to other, more dynamic models. These models center on notions of human information processing and cognition and are characterized by Kyllonen's (1995) Cognitive Abilities Measurement approach.

Information processing encompasses a set of notions, or a method, intended to describe how people think, learn, and respond. Most human information-processing models use stimulus-thought-response as a theoretical basis (Bailey, 1989; Wickens & Flach, 1988). The information-processing model depicted in Figure 12.1 differs from that originally developed by Wickens and Flach, but it is derived from and based on their model. Figure 12.1 covers four major activities in information processing: short-term sensory store, pattern recognition, decision-making, and response execution.

13.2.2.1 Short-Term Sensory Store

The model presented here is an extension, shown in Figure 13.1, of the Wickens and Flack model. It assigns stimuli input received by the short-term sensory store into separate buffers, or registers, for the five senses. Input from internal sensors for factors such as body temperature, heart and respiration rates, blood chemistry, limb position and rates of movement, and other internal functions could be added (Bailey, 1989), but are not needed in this summary discussion.

Visual and auditory sensory registers have been fairly well supported as helpful constructs that account for research findings (e.g., Paivio, 1991; Crowder & Surprenant, 2000). Evidence to support the other sensory registers is more limited, but as Crowder and Surprenant suggested, it is not unreasonable to posit these as constructs in a human information processing model. They have been added and included here.

13.2.2.2 Pattern Recognition

Over the past 30 years general theories of perception and learning have changed. They have evolved from the fairly strict logical positivism of behavioral psychology, which emphasized the study of directly observable and directly measurable actions, to consideration of the internal, mediating processes that have become the foundation of what is generally called cognitive psychology. Cognitive psychology gives more consideration to these internal, less observable processes. They are posited as bases for human learning and the directly observable behavior that is the subject of behaviorist investigations.

The keynote of these notions, which currently underlies our understanding of human perception, memory, and learning, may have been struck by Neisser (1967) who stated, "The central assertion is that seeing, hearing, and remembering are all acts of *construction,* which may make more or less use of stimulus information depending on circumstances." (p. 10). These ideas were, of course, prevalent long before Neisser published his book. For instance, while discussing what he called the general law of perception, William James stated in 1890 that "Whilst part of what we perceive comes through our senses from the object before us, another part (and it may be the larger part) always comes out of our mind" (p. 747, 1890/1950). After many years of wrestling with strictly behaviorist models, which only reluctantly considered internal processes such as cognition, Neisser's book seems to have freed the psychological research community to pursue new, more "constructivist" approaches to perception, memory, learning, and cognition.

Neisser was led to this point of view by a large body of empirical evidence showing that many aspects of human behavior, such as seeing and hearing, simply could not be accounted for by external physical cues reaching human perceptors, such as eyes and ears. Additional processes had to be posited to account for well-established and observable human abilities to detect, identify, and process physical stimuli.

Human cognition, then, came to be viewed as an overwhelmingly constructive process (Dalgarno, 2001). Perceivers and learners are not viewed as blank slates, passively recording bits of information transmitted to them over sensory channels, but as active participants who use the fragmentary cues permitted them by their sensory receptors to construct, verify, and modify their own cognitive simulations of the outside world. Human perception, cognition, and learning are understood to be enabled through the use of simulations of the world that the perceiver constructs and modifies based on sensory cues received from the outside world. In attempting to perform a task, a student will continue to act on an internal, cognitive simulation until that simulation no longer agrees with the sensory cues he/she is receiving from the physical world. At this point the student may modify the internal simulation so that it is more nearly in accord with the cues being delivered by his/her perceptual sensors. Even memory has come to be viewed as constructive with recollections assumed to be reconstructed in response to stimuli rather than retrieved whole cloth from long-term storage.

13.2.2.3 Attention Processes

For a stimulus to be processed, it must be detected by the information-processing system. Stimulus detection and processing distribute human ability to attend to the stimuli. When there is little or no workload, attention resources are distributed in an unfocused random pattern (Huey & Wickens, 1993).

As more sensory input becomes available, the individual must begin to prioritize what stimuli are going to be selected for interpretation. The attention process, based on pattern recognition from both long-term and working memory resources, decides the stimuli to be processed further. The selection of signals that should receive attention may be guided by the following (Wickens & Flach, 1988):

- Knowledge: Knowing how often a stimulus is likely to be presented, and if that stimulus is likely to change enough to affect a desired outcome, will influence the attention it receives.
- Forgetting: Human memory will focus attention on stimuli that have already been adequately sampled, but lost to memory.
- Planning: A plan of action that is reviewed before an activity is to take place will focus attention on some stimuli at the expense of others.
- Stress: Stress reduces the number of stimuli that can receive attention. Stress can also focus attention on stimuli that are of little consequence. For instance, fixating on a minor problem (a burnt out light) while ignoring a major problem (aircraft on a collision course).

Stimuli attended to may not be the brightest, loudest, or most painful, but they will be those deemed most relevant to the situation (Gopher, Weil, & Siegel, 1989). The likelihood that a stimulus will be detected depends at least partly on the perceived penalty for missing it. Klein (2000) offered a constructive view of attention. He stated that the decision-maker judges the situations as either typical or atypical, and, if judged as typical, or "recognition primed," the decision-maker then knows what the relevant cues are through experience extracted from long-term memory.

13.2.2.4 Working Memory

In an unpublished study, Pohlman and Tafoya (1979) investigated the fix-to-fix navigation problem in a T-38 instrument simulator. They found two primary differences between student pilots and instructor pilots. First, the accuracy of student in solving a fix-to-fix problem was inconsistent, whereas the instructor pilots were consistently accurate. Second, student pilots used a classic geometric approach to solve the problem in contrast to the instructors who used a rate-of-change comparison approach. Notably, almost every instructor denied using rate-of-change comparison until it was demonstrated they were in fact doing that, showing once again that experts may be unaware of the techniques that they use (Gilbert, 1992).

Although students were working geometry problems in the cockpit, instructors were merely comparing the rates at which the distance and bearing were changing, and flew the aircraft so that the desired range and desired bearing were arrived at simultaneously. A real bonus was that the rate of change comparison method automatically accounted for wind. Since current rate-of-change information is kept in working memory rather than in long-term memory (Wickens & Flach, 1988), the use of current rate-of-change information by these experts indicates that working memory is integral and essential to the distribution of attention. Observations such as this support the inclusion of a working-memory interface between the attention process and the long-term memory used primarily for pattern matching.

13.2.2.5 Long-Term Memory

Long-term memory becomes relevant in pattern matching and perception when the signal attended to requires interpretation. Long-term memory is the primary repository of patterns and episodic information. Patterns of dark and light can be converted into words on a page or pictures remembered and linked to names, addresses, and events. Memory that is linked to the meaning of the patterns is usually called *semantic memory*. Memory relating to events and the people, places, things, and emotions involved in them is usually called *episodic memory*. It is primarily semantic memory that is used in psychomotor tasks such as piloting an aircraft, fixing a landing gear, or sequencing an aircraft in the traffic pattern.

13.2.2.6 Automaticity

Humans are capable of different types of learning. One of these learning types involves choosing responses at successively higher levels of abstraction. For instance, in learning to read one may first attend to individual letters, then, with increased practice and proficiency, one may attend to individual words, then to phases, and finally, perhaps, to whole ideas. There are different levels of automaticity imposed by individual talents and abilities. As a boy, Oscar Wilde often demonstrated (for wagers) his ability to read both facing pages of a book at the same time and complete entire three-volume novels in 30 min or less (Ellmann, 1988). Clearly, there are levels of automaticity to which most of us can only aspire. In general, automaticity is more likely to be attained in situations where there are strict rules governing the relationship between stimuli and responses as in typing (Huey & Wickens, 1993).

The key for aviation tasks, with all their time pressures and demands for attention, is that automatic processing frees up attention resources for allocation to other matters such as perceiving additional stimuli (Bailey, 1989; Shiffrin & Schneider, 1977). As Figure 12.1 suggests, automatic responses are evoked by patterns abstracted from many specific situations and then stored in long-term memory.

13.2.2.7 Situation Awareness

Situation awareness is a product of the information processing components shown in Figure 13.1. It has become a topic of particular interest in discussions of aircrew skill. Situation awareness is not a matter limited to aviation—it transcends issues directly related to aviation skills and knowledge—but it arises out of discussions concerning those flying skills that distinguish average from exceptional pilots. It concerns the ability of individuals to anticipate events and assess their own progress through whatever environmental conditions they may encounter.

Researchers have emphasized measuring and modeling situation awareness and then using their findings to develop individual situation-awareness skill and instrumentation intended to enhance it. As a foundation for this work, Endsley devised a widely-accepted three-level definition of situation awareness as (1) perception of the elements in the environment, (2) comprehension of the current situation, and (3) projection of future status (Endsley, 2000). This framework has proven heuristic and helpful, but researchers still find situation awareness difficult to measure with sufficient precision to provide prescriptive reliability and validity. They have developed techniques for quantifying situation awareness such as structured interviews, testable responses, online probes, and error tracking (Endsley & Garland, 2000; Pritchett, Hansman, & Johnson, 1996; Wickens & McCarley, 2001). These techniques have proven helpful in assessing Endsley's first two levels—perceiving elements that are present in the environment and comprehending their impact on the current situation.

However, the third level—projecting future environmental status on the basis of what is currently noted and understood—has proven more difficult, possibly because it involves so many of the components shown in Figure 13.1, and their interactions. Once working memory, with some help from

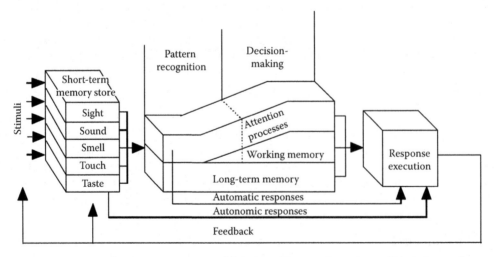

FIGURE 13.1   Generic information processing model. (Adapted from Wickens, C.D. and Flach, J.M., Information processing, in Wiener, E.L. and Nagel, D.C. (Eds.), *Human Factors in Aviation*, Academic Press, New York, 1988.)

long-term memory and its pattern recognition capabilities, has constructed a model—an environmental pattern—from the items presented to it by stimuli and attention processes, it must "run" the model as a cognitive simulation of what the future may bring. This simulation must take into account many possibilities and their interactions that must be identified and then prioritized with respect to their impact on future status. This requirement presents working memory with a problem. It must decide on which environmental possibilities or parameters to enter first into its simulation without information from that simulation indicating their impact on the future. Experience and pattern recognition seem essential in solving this problem, but in a complex fashion not yet well-informed by empirical research findings. Their contributions may have much to do with successful situation awareness and may provide its foundation.

Overall, situation awareness remains an important target for research. The difficulties encountered may be worth the effort. Being able to develop situation awareness training for novice operators may produce expert behavior in much less time than it would take by simply relying on happenstance experience to stock long-term memory with the necessary patterns and behaviors.

Of course, the story of human performance does not end with situation awareness. Perceiving and understanding the current environment and being able to project various possibilities into the future may be necessary, even essential, but it does not fully describe competent human performance. Knowing what is and what might be is a good start, but deciding what to do remains to be done. Situation awareness must be complemented by situation competence, which primarily involves decision-making. It brings us more directly back to the model depicted in Figure 13.1.

13.2.2.8 Decision-Making

Once stimuli have been detected, selected, and pattern matched, a decision must be made. As the process proceeds, cues are sought to assist the decision-maker in gathering information that will help with the decision. These cues are used to construct and verify the simulation, or runnable model, of the world that an individual constructs, verifies, and modifies to perceive and learn. As each situation is assessed, the individual chooses among possible responses by first "running" them in the simulation. This constructivist approach is markedly different from the highly formal, mathematical approaches that have been taught for decades. These rational approaches are designed using an engineering rationale. While they work well in relatively static environments, they are less useful and less effective in more dynamic environments such as flying or radar controlling where time constraints may reign (Klein, 2000).

Lack of time is a significant problem in aviation decision-making. Unlike other vehicles, an aircraft cannot stop in mid-air and shut down its systems to diagnose a problem. Decision-making is often stressed by this lack of time combined with the inevitable uncertainty and incompleteness of relevant sensory input. Another problem is that stress may be increased when sensory input is increased because of the greater workload placed on pattern recognition to filter out what is relevant and what is not. A pilot, controller, or maintenance technician may have too little time and too much sensory input to adapt to new situations or recognize cues needed for problem solution.

An individual may also miss relevant cues because they do not support his/her simulation of the situation. If the cues do not fit, an individual can either modify the underlying model or ignore them, with the latter leading to faulty decision-making. These factors influence what cues are available to long-term and working memory for situation assessment. Tversky and Kahneman (1974) discussed a variety of these interference factors as biases and interfering heuristics in the decision-making processes. Zsambok and Klein (e.g., 1997) described what they called naturalistic decision-making, which focuses on how people use experience and pattern recognition to make decisions in real-world practice.

Determining the ways that prospective aviation personnel process information and their capacities for doing so should considerably strengthen our procedures for selecting, classifying, and training them. For instance, the ability to filter sensory cues quickly and accurately may be critical for aircrew personnel, especially combat pilots, and flight controllers who must frequently perform under conditions of sensory overload. Creative, accurate, and comprehensive decision-making that takes account of all the salient cues and filters out the irrelevant ones may be critical for AMTs. Rapid decision-making that quickly adjusts situation assessment used to select among different decision choices may be at a premium for pilots and controllers. A large working-memory capacity with rapid access to long-term memory may be especially important for combat pilots whose lives often depend on the number of cues they process rapidly and accurately.

Emerging models of human information processing are, in any case, likely to find increasing application in the selection, classification, and training of aviation personnel. The dynamic nature of these models requires similarly dynamic measurement capabilities. These measurement capabilities are now inexpensive and readily available. Computer-based assessment can measure the aspects of human cognitive processes that were heretofore inaccessible, given the military's need for inexpensive, standard, procedures to assess hundreds of people in a single day by a single examiner. Development of computerized measurement capabilities may be as important a milestone in selection and classification as the work of the Vineland Committee in producing the Army Alpha Test. These possibilities were until recently, being pursued by Air Force laboratory personnel performing leading research in this area (Carretta, 1996; Carretta & Ree, 2000; Kyllonen, 1995; Ree & Carretta, 1998).

Finally, it should be noted that improvements in selection and classification procedures are needed for many aviation personnel functions, not just for potential aircrew members. Among U.S. scheduled airlines, domestic passenger traffic (revenue passenger enplanements) increased by 83% over the years 1980–1995, and international passenger traffic doubled in the same period (Aviation & Aerospace Almanac, 1997). Despite the 9/11 attack, aircraft passenger enplanements increased an additional 18% from 1996 through 2002 (U.S. Department of Transportation, 2004). Thousands of new aviation mechanics and flight controllers are needed to meet this demand. They are needed to operate and maintain the new digital equipment and technologies being introduced into modern aircraft and aviation work, and to satisfy the expansion of safety inspection requirements brought about by policies of deregulation.

The FAA has stated that there is an unacceptably high attrition rate in ATC controller training, costing the FAA about $9000 per washout. Therefore, both modernized training and more precise selection and classification are necessary (U.S. Department of Transportation, 1989). The plan is to introduce more simulation into the processes of selection and classification. It raises significant questions about the psychometric properties—the reliability, validity, and precision—of simulation used to measure human capabilities and performance (Allessi, 2000). These questions are by no means new, but they remain inadequately addressed by the psychometric research community.

Although these procedures fall short of perfection, they provide significant savings in funding, resources, and personnel safety over less systematic approaches. Still, our current selection and classification procedures rarely account for more than 25% of the variance in human performance observed in training and on the job (e.g., U.S. Department of the Air Force, 1996). There remains plenty of leverage to be gained by improving the effectiveness and efficiency of other means for securing the human competencies needed for aviation. Prominent among these means is training. As the age of flying machines has developed and grown, so too has our reliance on improving safety and performance through training.

13.3 Training for Aviation

13.3.1 A Little Background

Training and education may be viewed as opposite ends of a common dimension that we might call *instruction*. Training may be viewed as a means to an end—as preparation to perform a specific job. Education, on the other hand, may be viewed as an end in its own right and as preparation for all life experiences—including training. The contrast matters because it affects the way we develop, implement, and assess instruction—especially with regard to trade-offs between costs and effectiveness. In education, the emphasis is on maximizing the achievement—the improvements in human knowledge, skills, and performance—returned from whatever resources can be brought to bear on it. In training, the emphasis is on the other side of the cost-effectiveness coin—on preparing people to perform specific, identifiable jobs. Rather than maximize learning of a general sort, in training, we seek to minimize the resources that must be allocated to produce a specified level of learning—a specifiable set of knowledge, skills, and attitudes determined by the job to be done.

These distinctions between education and training are, of course, not hard and fast. In military training, as we pass from combat systems support (e.g., depot maintenance, hospital care, finance and accounting), to combat support (e.g., field maintenance, field logistics, medical evacuation), and to combat (i.e., warfighting), the emphasis in training shifts from a concern with minimizing costs toward one of maximizing capability and effectiveness. In education, as we pass from general cultural transmission to programs of professional preparation and certification, the emphasis shifts from maximizing achievement within given cost constraints toward minimizing the costs to produce specifiable thresholds of instructional accomplishment.

These considerations suggest that no assessment of an instructional technique for application in either education or training is complete without some consideration of both effectiveness and costs. During early stages of research, studies may honestly be performed to assess separately the cost or effectiveness of an instructional technique. However, once the underlying research is sufficiently complete to allow implementation, evaluations to effect change and inform decision-makers will be incomplete unless both costs and effectiveness considerations are included in the data collection and analysis.

It may also be worth noting that recruitment, selection, classification, assignment, training, human factoring, and job and career design, are all components of systems designed to produce needed levels of human performance. As in any system, all these components interact. More precise selection and classification reduce requirements for training. Embedded training in operational equipment will reduce the need for ab initio (from the beginning) training and either ease or change standards for selection and classification. Addition of job performance aids will do the same, and so on. Any change in the amount and quality of resources invested in any single component of the system is likely to affect the resources invested in other components—as well as the return to be expected from these investments.

The problem of completely understanding the interaction of all recruiting, selection, classification, and training variables has yet to be successfully articulated, let alone solved. What is the return to training from investments in recruiting or selection? What is the return to training or selection from investment in ergonomic design? What is the impact on training and selection from investment in electronic performance support systems? What, even, is the impact on training, selection, and job design from investments in spare parts? More questions could be added to this list. These comments are just to note the context within which training, in general, and aviation training, in particular, operate to produce human competence. Properly considered, training in aviation and elsewhere does not occur in a vacuum, separate from other means used to produce requisite levels of human competence.

13.3.2 Learning and Training

At the most general level, training is intended to bring about human learning. Learning is said to take place when an individual alters his/her knowledge and skills through interaction with the environment. Instruction is characterized by the purposeful design and construction of that environment to produce learning. Theories of learning, which are mostly descriptive, and theories of instruction, which are mostly prescriptive, help to inform the many decisions that must be made to design, develop, and implement training environments and the training programs that use them.

Every instructional program represents a view of how people perceive, think, and learn. As discussed earlier, these views have evolved over the past 30 years to include more consideration of the internal processes that are assumed to mediate and enable human learning. These cognitive, constructive notions of human learning are reflected in our current systems of instruction. They call into question the view of instruction as straightforward information transmission.

Instead, these constructive views suggest that the role of instruction is to supply appropriate cues for learners to use in constructing, verifying, and modifying their cognitive simulations—or runable models—of the subject matter being presented. The task of instruction design is not so much to transmit information from teacher to student to create environments in which students are enabled and encouraged to construct, verify, and correct these simulations. A learning environment will be successful to the extent that it also is individualized, constructive, and active. Systems intended to bring about learning, systems of instruction, differ in the extent to which they assist learning by assuming some of the burdens of this individualized, constructive, and active process for the student.

13.3.3 Training-Program Design and Development

These considerations do not, however, lead to the conclusion that all instruction, especially training, is hopelessly idiosyncratic and thereby beyond all structure and control. There is still much that can and should be done to design, develop, and implement instructional programs beyond simply providing opportunities for trial and error with feedback. Systematic development of instruction is especially important for programs intended to produce a steady stream of competent individuals, an intention that is most characteristic of training programs. All aspects of the systematic development of training are concerns of what is often called as Instructional System Design (ISD) (Logan, 1979) or the Systems Approach to Training (SAT) (Guptill, Ross, & Sorenson, 1995). ISD/SAT approaches apply standard systems engineering to the development of instructional programs. They begin with the basic elements of systems engineering, which are shown in Figure 13.2. These are the generic steps of analysis, design, production, implementation, and evaluation. ISD/SAT combines these steps with theories of learning and instruction to produce systematically designed and effective training programs.

Training *analysis* is based on systematic study of the job and the task(s) to be performed. It identifies training inputs and establishes training objectives to be accomplished in the form of student flow and the knowledge, skill, and attitude outcomes to be produced by the training. Training *design* devises the instructional interactions needed to accomplish the training objectives identified by training analysis. It is also used to select the instructional approaches and media used to present these interactions. Training *production* involves the development and preparation of instructional materials, which may include hardware such as simulators, software such as computer programs and audiovisual productions,

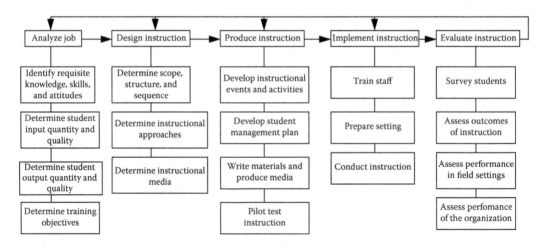

FIGURE 13.2 Example procedures for instructional system development.

and databases for holding information such as subject content and the performance capabilities of weapon systems. Training *implementation* concerns the appropriate installation of training systems and materials in their settings and attempts to ensure that they will perform as designed. Training *evaluation* determines if the training does things correctly (verification) and if it does the right things (validation). As discussed by Kirkpatrick (1976), it provides verification that the training system meets its objectives (Kirkpatrick's Level II) and the validation that meeting these objectives prepares individuals to better perform the targeted tasks or jobs (Kirkpatrick's Level III), and improves the operation of the organization overall (Kirkpatrick's Level IV). Notably, evaluation provides formative feedback to the training system for improving and developing it further.

Many ISD/SAT systems for instructional design have been devised—Montemerlo and Tennyson (1976) found that manuals for over 100 such systems had been written as of 1976, more doubtless exist now—but all these systems have some version of the basic steps for systems engineering in common. An ISD/SAT approach seeks to spend enough time on the front end of the system life cycle to reduce its costs later on. It is a basic principle of systems development that down-line modifications are substantially more expensive than designing and building something properly the first time. The same is true for training systems. It is more efficient to develop and field a properly designed training system than simply to build the system and spend the rest of its life fixing it. But the latter approach is pursued far more frequently than the former. For that matter, many training systems currently in use have never been evaluated, let alone subjected to Kirkpatrick's four levels of assessment. To some extent, training for aviation is an exception to these very common, seemingly haphazard approaches.

13.3.4 Training in Aviation

An aircraft pilot performs a continuous process of what Williams (1980) described as discrimination and manipulation. A pilot must process a flood of stimuli arriving from separate sources, identify which among them to attend to, generate from a repertoire of discrete procedures an integrated plan for responding to the relevant stimuli, and perform a series of discrete acts, such as positioning levers, switches, and controls, and continuous manual control movements requiring small forces and adjustments based on counter pressures exerted in response to the control movements. Williams suggested that the heart of these actions is decision-making and that it concerns: (a) when to move the controls; (b) which controls to move; (c) which direction to move the controls; (d) how much to move the controls; and (e) how long to continue the movement. It is both straightforward and complicated.

The task of flight controllers might be described in the same way. Both pilots and controllers must contend with significant time pressures and with the possibilities of severe consequences for error. Both require psychomotor responses, and both properly involve some degree of artistry and personal expression. No two people will perform psychomotor activities in precisely the same way, and these activities may be most effectively accomplished in ways that are consonant with other aspects of personal style (Williams, 1980). So, while the movements, decisions, and responses of aviation personnel can be catalogued, those actions cannot be prescribed since each individual has a different framework underlying the rule set. This framework does not filter what stimuli are available, but how the person attends to and interprets those stimuli.

Responses to the flood of incoming stimuli involve performance of pretrained procedures, but the procedures must be assembled into an integrated, often unique, response. As described by Roscoe, Jensen, and Gawron (1980), the performance of aviation personnel concerns procedural, decisional, and perceptual-motor responses. Responses chosen are generative and created to meet the demands of the moment. They involve the sensing, transforming, recollecting, recognizing, and manipulating of concepts, procedures, and devices. These responses are controlled by decision-making that is basically cognitive, but with emotional overtones. Responses made by pilots and controllers key on this decision-making, but the decision-making is more tactical than strategic. The decisions may be guided by general principles, but they are made under significant time pressures and resemble those of a job shop or a military-command post, more than those of an executive suite. These issues are discussed in more detail by Klein (2000).

Aviation training is just now beginning to evolve from the World War I days of the Lafayette Escadrille, as described by Charles Biddle, an American who was enlisted in the French Foreign Legion Aviation Section in 1917. Biddle was later commissioned in the U.S. Army Air Force where he performed with distinction as a fighter pilot* and a squadron commander. He was also a prolific letter writer. His letters, which were collected and published, provide a grass-roots description of training for pilots in World War I (Biddle, 1968).

This early training consisted mostly of an accomplished (hence, instructor) pilot teaching each student one-on-one in the aircraft. Ground training consisted of academic classes and some small group sessions with an instructor pilot. Each individual was briefed on what to do and then allowed to practice the action under the guidance of a monitor. Flying began, as it does today, with students taxiing the aircraft around on the ground, learning to balance, and steer.† As subsequent steps were mastered and certified by the instructor, the student proceeded to actual flight, and new, more difficult, and often more specialized stages of learning with more capable aircraft to fly and more complex maneuvers to complete.‡ Today's flight instruction follows the same basic pattern—probably because it works. It leads trainees reliably to progressively higher levels of learning and performance.

This "building block" approach has led to a robust set of assumptions concerning how aircrew training must be done. It emphasizes one-on-one student instruction for both teaching and certification, a focus on the individual, the use of actual equipment (aircraft, radar, airframe/powerplant) to provide the training, and hours of experience to certify proficiency. Each of these assumptions deserves some discussion.

13.3.4.1 One-on-One Instruction

One-on-one instruction receives somewhat more emphasis in aviation training than elsewhere. For an activity as complex and varied as piloting an airplane, it is difficult to imagine an alternative to this approach. One-on-one instructor to student ratios have long been recognized as effective, perhaps the most effective, format for instruction. Bloom (1984) found that the difference between students taught in classroom groups of 30 and those taught one-on-one by an individual instructor providing individualized instruction was as large as two standard deviations in achievement. Recent research into constructivist teaching methods (Alesandrini & Larson, 2002) supports the typical method used for one-on-one instruction. It involves teaching the semantic knowledge necessary for the mission, mental rehearsal (constructing mental models), and finally, practicing the mission with help from the instructor to correct inaccuracies in performance. The next step is to allow the student to practice the mission alone to further refine the performance.

It may be worth noting that many benefits of one-on-one instruction can be lost through improper implementation—with no reductions in their relatively high cost. Instructors who have not themselves received instruction in how to teach and then assess student progress may do both poorly despite their own high levels of proficiency and best intentions (Semb, Ellis, Fitch, & Matheson, 1995). Roscoe et al. (1980) stated that "there is probably more literal truth than hyperbole in the frequent assertion that the flight instructor is the greatest single source of variability in the pilot training equation" (p. 173). Instructors must both create an environment in which students learn and be able to assess and certify students' learning progress.

---

* He attributed much of his success in air combat to his earlier experience with duck hunting—learning how to track and lead moving targets in three-dimensional space.
† This is the so-called "penguin system" in which a landborne airplane, in Biddle's case, a Bleriot monoplane with reduced wingspan, is used to give students a feel for its controls.
‡ As early as 1915 in World War I, these maneuvers included aerobatics, which Biddle credits with saving the lives of many French-trained aviators—some of whom were, of course, Americans.

Much can be done to simplify and standardize the subjective assessment of student achievement accomplished during flight checks. Early on, Koonce (1974) found that it is possible to achieve inter-rater reliabilities exceeding 0.80 in flight checks, but these are not typical. In practice, instructors still, as reported earlier by Roscoe and Childs (1980), vary widely in their own performance of flight maneuvers and the indicators of competence that they consider in assessing the performance of their students. Despite variance in instructional quality, one-on-one instruction is still the bulwark of initial pilot training, in both the civilian and military schools.

Unfortunately, one-on-one instruction is also very expensive. One-on-one teaching has been described as both an instructional imperative and an economic impossibility (Scriven, 1975). Data-based arguments have been made (e.g., Fletcher, 2004) that technology, such as computer-based instruction that tailors the pace, content, sequence, difficulty, and style of presentations to the needs of individual students, can help to fill this gap between what is needed and what is affordable. Technology can be used more extensively in aviation training,* and FAA efforts have been made to encourage and increase not just the use of technology, but also the use of relatively inexpensive personal computers in aviation training. The discussion surrounding the correct mix of different training delivery devices has yet to be fully defined, much less solved.

For instance, a successful line of research was undertaken at Embry Riddle University to develop PC-based training that emphasizes less the number of flight hours in aircraft and more the knowledge and competencies of the trainees, and improved validity for FAA certification (e.g., Williams, 1994). Hampton, Moroney, Kirton, and Biers (1993) found that students trained using PC-based training devices needed fewer trials and less time to reach pilot test standards for eight maneuvers performed in an aircraft. They also found that the per-hour operating costs of the PC-based devices were about 35% less than those of an FAA-approved generic training device costing about $60,000 to buy.

The Air Force Human Resources Laboratory (now a part of the Human Effectiveness Directorate of the Air Force Research Laboratory) pursued some of this work and found that PC-based approaches produced superior achievement compared to paper-based approaches (programmed instruction) used in F-16 weapons control training (Pohlman & Edwards, 1983). The same laboratory developed a Basic Flight Instruction Tutoring System (BFITS) using a PC equipped with a joystick and rudder petals, intended for ab initio flight training (Benton, Corriveau, Koonce, & Tirre, 1992). Koonce, Moore, and Benton (1995) reported positive transfer of BFITS training to subsequent flight instruction. More recent work has shown effectiveness for modified commercial games and simulators in aircrew training (Pratt & Henninger, 2002).

Despite the expense and difficulty of one-on-one instruction and despite the technology-based opportunities for providing means that are both more effective and less costly for achieving many aviation training objectives, the use of individual instructors is likely to remain a key component of aviation training for some time to come.

13.3.4.2 Focus on Aircrew Teams

The days of barnstorming, ruggedly individualistic pilots are mostly gone. Even combat pilots fly under the tightening control of attack coordinators and radar operators, and they must coordinate their actions with wingmen. Commercial airline pilots must deal with an entire crew of people who are specialists in their fields and whose knowledge of specific aspects of aviation may well exceed that of the aircraft captain. However, the culture of the individual master of the craft still remains. This cultural bias may be less than ideal in an age of aircrews and teams. It represents a challenge for training.

Foushee and Helmreich (1988), among others, have pointed out that group performance has received little attention from the aviation training community and the attention it has received has been stimulated by unnecessary and tragic accidents. Generally these accidents seem to occur because of

---

* One of the first applications of speech recognition technology in technology-based instruction was for training naval flight controllers (Breaux, 1980).

a failure to delegate tasks (attention being focused on a relatively minor problem, leaving no one to mind the store) or an unwillingness to override the perceived authority of the aircraft captain. Still, it is interesting to note that the 1995 areas of knowledge listed earlier and required by the FAA for pilot certification are silent with regard to crew, team, and group abilities.

Communication skills are particularly important in successful crew interaction. Roby and Lanzetta (1958) and Olmstead (1992) reported empirical studies in which about 50% of team performance was accounted for by the presence and timing of particular kinds of communications. These were problem-solving teams placed under the sort of time pressures that are likely to occur in aviation. An interesting study reported by Foushee and Helmreich compared the performance of preduty (rested) with postduty (fatigued) crews. The study is notable because the postduty crews performed better than the preduty crews on operationally significant measures—and others—despite their fatigue. This relative superiority may be attributed to learning by the postduty crews to perform as a team, something that the preduty crews were yet to accomplish. Communication patterns were the key to these differences.

In brief, communications and other crew skills can and probably should be both taught and certified in aviation-training programs. These issues are currently addressed under the heading of cockpit resource management (Wiener, Kanki, & Helmreich, 1993). They deserve the attention of the military and civilian aviation communities and are discussed in detail in this *Handbook*. This is not to suggest that a focus on individuals is undesirable in aviation training. Rather it suggests that crew and team communication, management, and behavior should be added to current aviation training and certification requirements. However, more is required to bring this about. As recently as 2002, Nullmeyer and Spiker (2002) argued that there is little empirical data to guide the development of crew resource management instruction.

13.3.4.3 Aircraft versus Simulators

To a significant extent, the study of aviation training is the study of training simulators. This is true in training of aircrew members, flight controllers, and AMTs. Simulation is a sufficiently important topic on its own to deserve a separate chapter in this book. Comments here are of a general nature and focused on the use of simulation in training.

Rolfe and Staples (1986), Caro (1988), and others have provided useful and brief histories of flight simulators. The first flight simulators were developed early in the age of flying machines and were often aircrafts tethered to the ground, but capable of responding to aerodynamic forces. The Sanders Teacher, one of the first of these, was introduced in 1910. Some of these devices depended on natural forces to provide the wind needed to give students an experience in learning to balance and steer, and some, like the Walters trainer, also introduced in 1910, used wires and pulleys manipulated by flight instructors to give students this experience. Motion for flight simulators was made possible through the use of compressed air actuators developed for aviation simulators by Lender and Heidelberg in 1917 and 1918. However, the use and value (cost-effectiveness and training effectiveness) of motion in flight simulation was as much a matter of discussion then as it is now (e.g., Alessi, 2000; Hays, Jacobs, Prince, & Salas, 1992; Koonce, 1979; Pfeiffer & Horey, 1987; Waag, 1981).

As instrumentation for aircraft improved, the need to include instruments coupled with simulated flight characteristics increased. The Link Trainers succeeded in doing this. By the late 1930s, they were able to present both the instrument layout and performance of specific aircraft to students. Simulators using electrical components to model characteristics of flight were increasingly used as World War II progressed. In 1943, Bell Telephone Laboratories produced an operational flight trainer/simulator for the U.S. Navy's PBM-3 aircraft using electrical circuitry to solve flight equations in real time and display their results realistically, using the complete system of controls and instruments available in the aircraft. Modern simulators evolved further with the incorporation of computers that could not only respond to controls in simulators and display the results of flight equations on aircraft instruments, but also could provide motion simulation and generate out the window visual displays as well. Today, following the lead of Thorpe (1987), groups of aircraft simulators are linked together, either locally or over wide area computer networks to provide training in air combat tactics and distributed mission operations (Andrews & Bell, 2000).

Rolfe and Staples (1986) pointed out that a faithful simulation requires: (a) a complete model of the response of the aircraft to all inputs, (b) a means of animating the model (rendering it runnable in real time), and (c) a means of presenting this animation to the student using mechanical, visual, and aural responses. They noted that the degree to which all this is necessary is another question. The realism, or "fidelity" needed by simulation to perform successful training of all sorts is a perennial topic of discussion. Much of this discussion is based either in actuality or on the intuitive appeal of Thorndike's (1903) early argument for the presence and necessity of identical elements in training to ensure successful transfer of what is learned in training to what is needed on the job. Thorndike suggested that such transfer is always specific, never general, and keyed to either substance or procedure.

Not knowing precisely what will happen on the job leads naturally to the desire to provide as many identical elements in training as possible. In dynamic pursuits such as aviation, where unique situations are frequent and the unexpected is expected, this desire may lead to an insistence on maximizing simulator fidelity in training. Unfortunately, fidelity does not come free. As fidelity increases, so do costs, reducing the number, availability, and/or accessibility of training environments that can be provided to students. If the issue ended here, we might solve the problem by throwing more money at it—or not as policy dictated.

However, there is another issue involving fidelity, simulation, and training. Simulated environments permit the attainment of training objectives that cannot or should not be attempted without simulation. As discussed by Orlansky et al. (1994) among many others, aircraft can be crashed, expensive equipment ruined, and lives hazarded in simulated environments in ways that range from impractical to unthinkable without simulators. Simulated environments provide other benefits for training. They can make the invisible visible, compress or expand time, and repeatedly reproduce events, situations, and decision points. Training using simulation is not just a degraded, less-expensive reflection of the realism that we would like to provide, but enables the attainment of training objectives that are otherwise inaccessible.

Training using simulation both adds value and reduces cost. Evidence of this utility comes from many sources. In aircrew training the issue keys on transfer are the skills and knowledge acquired in simulation of value in flying actual aircraft? Do they transfer from one situation to the other? Many attempts to answer this question rely on transfer effectiveness ratios (TER) (Roscoe & Williges, 1980). These ratios may be defined for pilot training in the following way:

$$\text{TER} = \frac{A_C - A_S}{S}$$

where
   TER is the transfer effectiveness ratio
   A_C is the aircraft time required to reach criterion performance, without access to simulation
   A_S is the aircraft time required to reach criterion performance, with access to simulation
   S is the simulator time

Roughly, this TER is the ratio of aircraft time savings to the expenditure of simulator time—it tells us how much aircraft time is saved for every unit of simulator time invested. If the TER is small, a cost-effectiveness argument may still be made for simulation since simulator time is likely to cost much less than aircraft time. Orlansky and String (1977) investigated precisely this issue in a now-classic and often-cited study. They found (or calculated, as needed) 34 TERs from assessments of transfer performed from 1967 to 1977 by military, commercial, and academic organizations. The TERs ranged from −0.4 to 1.9, with a median value of 0.45. Orlansky, Knapp, and String (1984) also compared the cost to fly actual aircraft with the cost to "fly" simulators. Very generally, they found that (1) the cost to operate a flight simulator is about one-tenth the cost to operate a military aircraft; (2) an hour in a simulator saves about one-half hour in an aircraft; so that (3) use of flight simulators is cost-effective if the TER is 0.20 or greater.

At a high level of abstraction, this finding is extremely useful and significant. Because nothing is simple, a few caveats may be in order. First, as Provenmire and Roscoe (1973) pointed out, not all simulator hours are equal—early hours in the simulator appear to save more aircraft time than later ones. This consideration leads to learning curve differences between cumulative TERs and incremental TERs with diminishing returns best captured by the latter. Second, transfer is not a characteristic of the simulator alone. Estimates of transfer from a simulator or simulated environment must also consider what the training is trying to accomplish—the training objectives. This issue is well illustrated in a study by Holman (1979) who found 24 TERs for a CH-47 helicopter simulator ranging from 2.8 to 0.0, depending on which training objective was under consideration. Third, there is an interaction between knowledge of the subject matter and the value of the simulation alone. Gay (1986) and Fletcher (1991) found that the less the student knows about the subject matter, the greater is the need for tutorial guidance in simulation. The strategy of throwing a naive student into a simulator with the expectation that learning will occur does not appear to be viable. Kalyuga, Ayres, Chandler, and Sweller (2003) summarized a number of studies demonstrating an "expertise reversal effect" indicating that high levels of instructional support are needed for novice learners, but have little effect on experts and may actually interfere with their learning. Fourth, the operating costs of aircraft differ markedly and will create quite different trade-offs between the cost-effectiveness of training with simulators and without them. In contrast to the military aircraft considered by Orlansky, Knapp, and String where the cost ratio was about 0.10, Provenmire and Roscoe were concerned with flight simulation for the Piper Cherokee, where the cost ratio was 0.73.

Nonetheless, many empirical studies have demonstrated the ability of simulation to both increase effectiveness and lower costs for many aspects of flight training. Hays et al. (1992) reviewed 26 studies of transfer from training with flight simulators to operational equipment. They found that there was significant positive transfer from the simulators to the aircraft, that training using a simulator and an aircraft was almost always superior to training with a simulator alone, and that self-paced simulator training was more effective than lock-step training. Also the usual ambiguities about the value of including motion systems in flight simulators emerged. Beyond this, the findings of Orlansky and String (1977), Orlansky, Knapp, and String (1984), and Hammon and Horowitz (1996) provided good evidence of lowered costs in flight training obtained through the use of simulators.

The value of simulation is, of course, not limited to flight. From a broad review of interactive multimedia capabilities used for simulation, Fletcher (1997) extracted 11 studies in which simulated equipment was used to train maintenance technicians. These studies compared instruction with the simulators to use of actual equipment, held overall training time roughly equal, and assessed the final performance using actual (not simulated) equipment. Over the 11 studies, the use of simulation yielded an effect size (which is the measure of merit in such meta-analyses) of 0.40 standard deviations, suggesting an improvement from 50th percentile to about 66th percentile achievement among students using simulation. Operating costs using simulation were about 0.40 of those without it, because the equipment being simulated did not break and could be presented and manipulated on devices costing 1–2 orders of magnitude less than the actual equipment that was the target of the training. Although simulators are an expected component of any aircrew program of instruction, they may deserve more attention and application in the nonflight components of aviation training (Hemenway, 2003).

13.3.4.4 Distributed Training/Distance Learning

According to the United States Distance Learning Association, distance learning is an education program that allows students to complete their work in a geographical location separate from the institution hosting the program (http://www.usdla.org/html/resources/dictionary.htm). The students may work alone or in groups at home, workplace, or training facility. They may communicate with faculty and other students via e-mail, electronic forums, videoconferencing, chat rooms, bulletin boards, instant messaging, and other forms of computer-based communication. Most distance learning programs are synchronous, requiring students and teachers to be engaged in instructional activities at the same time, albeit at different locations. Video teletraining and teleconferencing are typically used in distance learning.

Distributed learning programs are primarily asynchronous. They typically include computer-based training (CBT) and communications tools to produce a virtual classroom environment. Because the Internet and World Wide Web are accessible from so many computer platforms, they serve as the foundation for many distributed learning systems although local area networks and intranets are also commonly found in distributed training settings. There have been major increases in both the technologies and the use of distributed training in the last 5 years. These applications are beginning to incorporate more exotic technologies such as virtual reality (Weiderhold and Weiderhold, 2005). As distributed training becomes an operational reality, more attention needs to be focused on instructional design and defining performance outcomes.

13.3.4.5 Embedded Training

Most embedded training is based on training software installed in operational systems. Ideally, the only prior information an individual would need to operate a system with embedded training would be how to turn it on—the rest would be handled by the system itself. Such training systems can be installed in command and control facilities, radar, aircraft, ship, ground vehicles, and many other operational systems. In effect, embedding training in the actual equipment allows it to be used as an operational simulator while leaving the equipment in its intended theater of operations. Embedded training, intended to enhance the behavioral repertoire of its user(s), can readily be used as a Performance Support System (PSS) intended to help user(s) apply the system to aid decision-making and solve problems. This capability is enabled by the underlying knowledge structures, which are nearly identical for both training and performance aiding.

13.3.4.6 Performance Support Systems

A PSS is an integrated group of tools used to assist an individual or group in the performance of a specific task (Gery, 1991). A PSS can include a wide variety of media including computer-based training and electronic manuals (Seeley & Kryder, 2003). Its primary function is to help the users to solve a problem or make a decision, not to effect a persistent change in capability or behavior, an objective that is more characteristic of training than of a PSS, but the same knowledge structures underlie and can be used for both. For this reason, the "Janis Principle" states that learning and PSSs should coexist and not be separated even though a PSS does not require training elements to accomplish its role (Eitelman, Neville, & Sorensen, 2003). Most PSSs today include both learning and performance support.

PSS can be used for a wide variety of performance tasks, from Space Operations to aircraft maintenance. PSS design and development considerations begin with target performance task analyses, but they must also consider the overall integration of the users (humans), the PSS, and the target system (Seeley & Kryder, 2003). In order to enhance performance effectively, the PSS must be obtrusive and avoid interfering with the performance of the target system. A PSS that is difficult to use can obviate any potential gains it might provide. A PSS that integrates task performance criteria, the target system, and the human can expand and broaden the learning that takes place in classrooms. It can also allow the operator to experiment without jeopardizing the target system. These capabilities may create an atmosphere where the operator can learn to create innovative solutions to new and unexpected difficulties (Kozlowski, 1998).

Fletcher and Johnston (2002) summarized empirical findings from use of three hand-held, computer-controlled PSS: Computer-Based Maintenance Aids System (CMAS), Portable Electronic Aid for Maintenance (PEAM), and Integrated Maintenance Information System (IMIS). CMAS and IMIS were pioneering efforts by the Air Force to support the performance of AMTs. Evaluation of CMAS found that technicians using CMAS compared those using paper-based technical manuals took less than half the time to find system faults, checked more test points, made fewer (i.e., no) false replacements, and solved more problems correctly. Evaluation of IMIS concerned fault-isolation problems for three F-16 avionics subsystems—fire control radar, heads-up display, and inertial navigation system. Technicians in the evaluation study used paper-based technical manuals for half of the problems and IMIS for the other half. Technicians using IMIS when compared with those using Task Orders found more correct solutions in less time, used fewer parts to do so, and took less time to order them. Findings also showed that technicians with limited avionics training performed as well as avionics specialists when they used IMIS. Analysis of costs found net savings of about $23 million per year in maintaining these three avionics subsystems for the full Air Force fleet of F-16s.

PSS research findings suggest that a strong cost-effectiveness case can be made for using them, optimal trade-offs between training and performance aiding should be sought, PSS can benefit from the individualization capabilities developed for intelligent tutoring systems, and more effort is needed to ensure that the state of practice in maintenance operations advances along with the state of the art.

13.3.4.7 Process Measurement versus Performance Measurement

Experience is a thorough teacher and especially valuable when the nuances and enablers of human proficiency are ill-defined and incompletely captured by instructional objectives. However, one hour of experience will produce different results in different people. Personnel and instructional strategies based solely on the assumption that time in the aircraft or working with actual equipment (in the case of flight controllers and maintenance technicians) equates to learning, are limited. Training and the certification that it bestows may be better served by increased emphasis on performance assessment in place of process measurements such as hours of experience. New technologies incorporated into aviation, such as the individually configurable cockpit displays of the F-35 Joint Strike Fighter, may require new teaching methodologies. Current training paradigms often neglect processes for training a user regarding how to configure a piece of operational equipment, so that it will optimize the performance produced by the user and the equipment working together.

These comments are not to suggest that traditional instructional strategies such as one-on-one instruction, use of actual aircraft, and hours of experience should be eliminated from training programs. They do suggest that by simply doing things the way they have always been done sooner or later leads to inefficiency, ineffectiveness, and stagnation. All assumptions that are emphasized in aviation training should be routinely subjected to analytical review and the possibility of change.

13.3.5 Pathways to Aviation Training

According to the Department of Transportation, in 2001 (the last year statistics are available), there were 129,000 pilots and navigators and 23,000 ATC controllers working in the transportation industry (U.S. Department of Transportation, 2004). There are five types of pilot certificates: student, private, commercial, airline transport, and instructor. Except for student pilot, ratings for aircraft category (airplane, rotorcraft, glider, and lighter-than-air), aircraft class (single-engine land, multi-engine land, single-engine sea, and multi-engine sea), aircraft type (large aircraft, small turbojet, small helicopters, and other aircraft), and aircraft instruments (airplanes and helicopters) are placed on each certificate to indicate the qualification and limitations of the holder. AMTs are certified for two possible ratings (airframe and power plant combined) and repairman. As discussed earlier, the number of maintenance certifications may be increased to meet the requirements posed by modern aircraft design. Separate certification requirements also exist for ATC controllers, aircraft dispatchers, and parachute riggers. The aviation workforce is large and both technically and administratively complex.

In response to Congressional concerns, the National Research Council (NRC) undertook a study (Hansen & Oster, 1997) to assess our ability to train the quantity and quality of people needed to sustain this workforce. The NRC identified five "pathways" to aviation careers:

1. Military training has been a major source of aviation personnel in the past and its diminution provided a major impetus for the NRC study. The military is likely to become much less prominent and civilian sources are likely to become substantially more important as the military Services continue to downsize and the air-transport industry continues to expand and replace its aging workers.
2. Foreign hiring has been used little by U.S. airlines and is not expected to increase in the future. In fact, many U.S.-trained pilots are expected to seek employment in other countries when U.S. openings are scarce.
3. On-the-job training allows individuals to earn FAA licenses and certificates by passing specific tests and without attending formal training programs. U.S. airlines prefer to hire people who have completed FAA certificated programs, and on-the-job training is not likely to grow as a source of training in the future.
4. Collegiate training is offered by about 280 postsecondary institutions tracked by the University Aviation Association currently located at Auburn University. Collegiate training is already the major source for AMTs, and the NRC report suggested that it will become successively more important as a source of aircrew personnel. However, the report also points out that pilots, even after they complete an undergraduate degree in aviation, must still work their way up through nonairline flying jobs before accumulating the hours and ratings certifications currently expected and required by the airlines for placement.
5. Ab initio ("from the beginning") training is offered by some foreign airlines to selected individuals with no prior flying experience. As yet, U.S. airlines have not considered it necessary to provide this form of training.

The NRC study concluded that civilian sources will be able to meet market demand, despite the downsizing of the military. However, they stressed the need to sustain and develop the professionalization and standardization of collegiate aviation programs—most probably by establishing an accreditation system similar to that in engineering and business and supported by the commercial aviation industry and the FAA. As described earlier in this paper, the U.S. aviation industry continues to grow, as it does worldwide. The next 5 to 10 years will be both interesting and challenging to those concerned with the support and growth of the aviation workforce. The NRC study suggests some means for accomplishing these ends successfully. The community concerned with human competence in aviation has been given a significant opportunity to rise to the challenge.

References

Alesandrini, K., & Larson, L. (2002). Teachers' bridge to constructivism. *The Clearing House: Educational Research, Controversy, and Practices, 75*(3), 118–121.

Alessi, S. (2000). Simulation design for training and assessment. In H. F. O'Neil Jr. & D. H. Andrews (Eds.), *Aircrew training and assessment* (pp. 197–222). Mahwah, NJ: Lawrence Erlbaum Associates, Inc.

Andrews, D. H., & Bell, H. H. (2000). Simulation-based training. In S. Tobias & J. D. Fletcher (Eds.), *Training & retraining: A handbook for business, industry, government, and the military* (pp. 357–384). New York: Macmillan Reference.

Aviation & Aerospace Almanac 1997 (1997). New York: Aviation Week Group, McGraw-Hill.

Bailey, R. W. (1989). *Human performance engineering.* Englewood Cliffs, NJ: Prentice-Hall.

Benton, C., Corriveau, P., Koonce, J. M., & Tirre, W. C. (1992). *Development of the basic flight instruction tutoring system (BFITS)* (AL-TP-1991-0060). Brooks Air Force Base, TX: Armstrong Laboratory Human Resources Directorate (ADA 246 458).

Biddle, C. J. (1968). *Fighting airman: The way of the eagle.* Garden City, NY: Doubleday & Company.

Bloom, B. S. (1984). The 2-sigma problem: The search for methods of group instruction as effective as one-to-one tutoring. *Educational Researcher, 13,* 4–16.

Breaux, R. (1980). Voice technology in military training. *Defense Management Journal, 16,* 44–47.

Brown, D. C. (1989). Officer aptitude selection measures. In M. F. Wiskoff & G. M. Rampton (Eds.), *Military personnel measurement: Testing, assignment, evaluation.* New York: Praeger.

Caro, P. W. (1988). Flight training and simulation. In E. L. Wiener & D. C. Nagel (Eds.), *Human factors in aviation.* New York: Academic Press.

Carretta, T. R. (1996). *Preliminary validation of several US Air Force computer-based cognitive pilot selection tests* (AL/HR-TP-1996-0008). Brooks Air Force Base, TX: Armstrong Laboratory Human Resources Directorate.

Carretta, T. R., & Ree, M. J. (1996). Factor structure of the air force officer qualifying test: Analysis and comparison. *Military Psychology, 8,* 29–43.

Carretta, T. R., & Ree, M. J., (2000). *Pilot selection methods* (AFRL-HE-WP-TR-2000-0116). Wright-Patterson Air Force Base, OH: Human Effectiveness Directorate, Crew Systems Interface Division.

Crowder, R. G., & Surprenant, A. M. (2000). Sensory stores. In A. E. Kazdin (Ed.), *Encyclopedia of psychology* (pp. 227–229). Oxford, U.K.: Oxford University Press.

Dalgarno, B. (2001). Interpretations of constructivism and consequences for computer assisted learning. *British Journal of Educational Technology, 32,* 183–194.

Dockeray, F. C., & Isaacs, S. (1921). Psychological research in aviation in Italy, France, England, and the American Expeditionary Forces. *Comparative Psychology, 1,* 115–148.

Driskell, J. E., & Olmstead, B. (1989). Psychology and the military: Research applications and trends. *American Psychologist, 44,* 43–54.

Duke, A. P., & Ree, M. J. (1996). Better candidates fly fewer training hours: Another time testing pays off. *International Journal of Selection and Assessment, 4,* 115–121.

Eitelman, S., Neville, K., & Sorensen, H. B. (2003). Performance support system that facilitates the acquisition of expertise. In *Proceedings of the 2003 Interservice/Industry Training System and Education Conference (I/ITSEC)* (pp. 976–984). Arlington, VA: National Security Industrial Association.

Ellmann, R. (1988). *Oscar Wilde.* New York: Vintage Books.

Endsley, M. (2000). Situation awareness in aviation systems. In D. Garland, J. Wise, & V. Hopkin (Eds.), *Handbook of aviation human factors* (pp. 257–276). Mahwah, NJ: Lawrence Erlbaum Associates.

Endsley, M. R., & Garland, D. J. (2000). Situation Awareness Analysis and Measurement. Mahwah, NJ: Lawrence Erlbaum Associates.

Fiske, D. W. (1947). Validation of naval aviation cadet selection tests against training criteria. *Journal of Applied Psychology, 5,* 601–614.

Flanagan, J. C. (1942). The selection and classification program for aviation cadets (aircrew—bombardiers, pilots, and navigators). *Journal of Consulting Psychology, 6,* 229–239.

Fletcher, J. D. (1991). Effectiveness and cost of interactive videodisc instruction. *Machine Mediated Learning, 3,* 361–385.

Fletcher, J. D. (1997). What have we learned about computer-based instruction in military training? In R. J. Seidel & P. R. Chatelier (Eds.), *Virtual reality, training's future?* New York: Plenum.

Fletcher, J. D. (2004). Technology, the Columbus effect, and the third revolution in learning. In M. Rabinowitz, F. C. Blumberg, & H. Everson (Eds.), *The design of instruction and evaluation: Affordances of using media and technology* (pp. 139–157). Mahwah, NJ: Lawrence Erlbaum Associates.

Fletcher, J. D., & Johnston, R. (2002). Effectiveness and cost benefits of computer-based aids for maintenance operations. *Computers in Human Behavior, 18,* 717–728.

Foushee, H. C., & Helmreich, R. L. (1988). Group interaction and flight crew performance. In E. L. Wiener & D. C. Nagel (Eds.), *Human factors in aviation* (pp. 189–227). New York: Academic Press.

Gardner, H., Kornhaber, M., & Wake, W. (1996). *Intelligence: Multiple perspectives*. Fort Worth, TX: Harcourt Brace.

Gay, G. (1986). Interaction of learner control and prior understanding in computer-assisted video instruction. *Journal of Educational Psychology, 78*, 225–227.

Gery, G. (1991). *Electronic performance support systems*. Boston, MA: Weingarten.

Gilbert, T. F. (1992). Foreword. In H. D. Stolovitch & E. J. Keeps (Eds.), *Handbook of human performance technology*. San Francisco, CA: Jossey-Bass.

Goldsby, R. (1996). Training and certification in the aircraft maintenance industry: Technician resources for the twenty-first century. In William T. Shepherd, *Human factors in aviation maintenance—phase five progress report* (DOT/FAA/AM-96/2) (pp. 229–244). Washington, DC: Department of Transportation, Federal Aviation Administration (ADA 304 262).

Gopher, D., Weil, M., & Siegel, D. (1989). Practice under changing priorities: An approach to the training of complex skills. *Acta Psychologica, 71*, 147–177.

Guilford, J. P. (1967). *The nature of human intelligence*. New York: McGraw-Hill.

Guptill, R. V., Ross, J. M., & Sorenson, H. B. (1995). A comparative analysis of ISD/SAT process models. In *Proceedings of the 17th Interservice/Industry Training System and Education Conference (I/ITSEC)* (pp. 20–30). Arlington, VA: National Security Industrial Association.

Hammon, C. P., & Horowitz, S. A. (1996). *The relationship between training and unit performance for naval patrol aircraft—revised* (IDA Paper P-3139). Alexandria, VA: Institute for Defense Analyses.

Hampton, S., Moroney, W., Kirton, T., & Biers, W. (1993). *An experiment to determine the transfer effectiveness of PC-based training devices for teaching instrument flying* (CAAR-15471-93-1). Daytona Beach, FL: Center for Aviation/Aerospace Research, Embry-Riddle Aeronautical University.

Hansen, J. S., & Oster, C. V. (Eds.) (1997). *Taking flight: Education and training for aviation careers*. Washington, DC: National Research Council, National Academy Press.

Hays, R. T., Jacobs, J. W., Prince, C., & Salas, E. (1992). Flight simulator training effectiveness: A meta-analysis. *Military Psychology, 4*, 63–74.

Hemenway, M. (2003). Applying learning outcomes to media selection for avionics maintenance training. In *Proceedings of the 2003 Interservice/Industry Training System and Education Conference (I/ITSEC)* (pp. 940–951). Arlington, VA: National Security Industrial Association.

Hendriksen, I. J. M., & Elderson, A. (2001). The use of EEG in aircrew selection. *Aviation Space Environmental Medicine, 72*, 1025–1033.

Hilton, T. F., & Dolgin, D. L. (1991). Pilot selection in the military of the free world. In R. Gal & A. D. Mangelsdorff (Eds.), *Handbook of military psychology* (pp. 81–101). New York: Wiley.

Holman, G. J. (1979). *Training effectiveness of the CH-47 flight simulator* (ARI-RR-1209). Alexandria, VA: Army Research Institute for the Behavioral and Social Sciences (ADA 072 317).

Hunter, D. R. (1989). Aviator selection. In M. F. Wiskoff & G. M. Rampton (Eds.), *Military personnel measurement: Testing, assignment, evaluation* (pp. 129–167). New York: Praeger.

Hunter, D. R., & Burke, E. F. (1995). Predicting aircraft pilot training success: A meta-analysis of published research. *International Journal of Aviation Psychology, 4*, 297–313.

Huey, B. M., & Wickens, C. D. (1993). *Workload transition*. Washington, DC: National Academy Press.

James, W. (1890/1950). *Principles of Psychology: Volume I*. New York: Dover Press.

Jenkins, J. G. (1946). Naval aviation psychology (II): The procurement and selection organization. *American Psychologist, 1*, 45–49.

Jordan, J. L. (1996). Human factors in aviation maintenance. In W. T. Shepherd (Ed.), *Human factors in aviation maintenance—phase five progress report* (DOT/FAA/AM-96/2) (pp. 251–253). Washington, DC: Department of Transportation, Federal Aviation Administration (ADA 304 262).

Kalyuga, S., Ayres, P., Chandler, P., & Sweller, J. (2003). The expertise reversal effect. *Educational Psychologist, 38*, 23–31.

Kirkpatrick, D. L. (1976). Evaluation of training. In R. L. Craig (Ed.), *Training and development handbook*. New York: McGraw-Hill.

Klein, G. (2000). How can we train pilots to make better decisions? In H. F. O'Neil Jr. & D. H. Andrews (Eds.), *Aircrew training and assessment* (pp. 165–195). Mahwah, NJ: Lawrence Erlbaum Associates.

Koonce, J. M. (1974). *Effects of ground-based aircraft simulator motion conditions upon prediction of pilot proficiency* (AFOSR-74-1292). Savoy: Aviation Research Laboratory, University of Illinois (AD A783 256/257).

Koonce, J. M. (1979). Predictive validity of flight simulators as a function of simulation motion. *Human Factors, 21*, 215–223.

Koonce, J. M., Moore, S. L., & Benton, C. J. (1995). Initial validation of a Basic Flight Instruction tutoring system (BFITS). Columbus, OH: *8th International Symposium on Aviation Psychology*.

Kozlowski, S. W. J. (1998). Training and developing adaptive teams: Theory, principles, and research. In J. A. Cannon-Bowers & E. Salas, (Eds.), *Making decisions under stress: Implications for individual and team training* (pp. 115–153). Washington, DC: American Psychological Association.

Kyllonen, P. C. (1995). CAM: A theoretical framework for cognitive abilities measurement. In D. Detterman (Ed.), *Current topics in human intelligence, Volume IV, Theories of intelligence*. Norwood, NJ: Ablex.

Logan, R. S. (1979). A state-of-the-art assessment of instructional systems development. In H. F. O'Neil Jr. (Ed.), *Issues in instructional systems development* (pp. 1–20). New York: Academic Press.

McClearn, M. (2003). Clear skies ahead. *Canadian Business, 76*, 141–150.

McRuer, D., & Graham, D. (1981). Eighty years of flight control: Triumphs and pitfalls of the systems approach. *Journal of Guidance and Control, 4*(4), 353–362.

Miller, W. D. (1999). The pre-pilots fly again. *Air Force Magazine, 82*(6), available at http://www.airforce-magazine.com

Montemerlo, M. D., & Tennyson, M. E. (1976). *Instructional systems development: Conceptual analysis and comprehensive bibliography* (NAVTRAEQUIPCENIH 257). Orlando, FL: Naval Training Equipment Center.

Neisser, U. (1967). *Cognitive psychology*. New York: Appleton-Century-Crofts.

Nordwall, B. D. (2003). Controller attrition could intensify air traffic woes. *Aviation Week & Space Technology, 158*, 49.

Nullmeyer, R. T., & Spiker, V. A. (2002). Exploiting archival data to identify CRM training needs for C-130 aircrews. In *Proceedings of the 2002 Interservice/Industry Training System and Education Conference (I/ITSEC)* (pp. 1122–1132). Arlington, VA: National Security Industrial Association.

Olmstead, J. A. (1992). *Battle Staff Integration* (IDA Paper P-2560). Alexandria, VA: Institute for Defense Analyses (ADA 248-941).

O'Neil, H. F., Jr., & Andrews, D. H. (Eds.). (2000). *Aircrew training and assessment*. Mahwah, NJ: Lawrence Earlbaum Associates.

Oi, W. (2003). The virtue of an all-volunteer force. *Regulation Magazine, 26*(2), 10–14.

Orlansky, J., Dahlman, C. J., Hammon, C. P., Metzko, J., Taylor, H. L., & Youngblut, C. (1994). *The value of simulation for training* (IDA Paper P-2982). Alexandria, VA: Institute for Defense Analyses (ADA 289 174).

Orlanksy, J., Knapp, M. I., & String, J. (1984). *Operating costs of military aircraft and flight simulators* (IDA Paper P-1733). Alexandria, VA: Institute for Defense Analyses (ADA 144 241).

Orlansky, J., & String, J. (1977). *Cost-effectiveness of flight simulators for military training* (IDA Paper P-1275). Alexandria, VA: Institute for Defense Analyses (ADA 051801).

Paivio, A. (1991). *Images in mind: The evolution of a theory*. Hempstead, Herfordshire, U.K.: Harvester Wheatshaft.

Pfeiffer, M. G., & Horey, J. D. (1987). *Training effectiveness of aviation motion simulation: A review and analyses of the literature* (Special Report No. 87-007). Orlando, FL: Naval Training Systems Center (ADB 120 134).

Phillips, E. H. (1999). *Aviation Week & Space Technology, 151*, 41.

Pohlman, D. L., & Edwards, D. J. (1983). Desk-top trainer: Transfer of training of an aircrew procedural task. *Journal of Computer-Based Instruction, 10,* 62–65.

Pohlman, D. L., & Tafoya, A. F. (1979). *Perceived rates of motion in a cockpit instruments as a method for solving the fix to fix navigation problem.* Unpublished Technical Paper, Williams Air Force Base, AZ: Air Force Human Resources Laboratory.

Pratt, D. R., & Henninger, A. E. (2002). A case for micro-trainers. In *Proceedings of the 2002 Interservice/ Industry Training System and Education Conference (I/ITSEC)* (pp. 1122–1132). Arlington, VA: National Security Industrial Association.

Pritchett, A., Hansman, R., & Johnson, E. (1996). Use of testable responses for performance-based measurement of situation awareness. In *International Conference on Experimental Analysis and Measurement of Situation Awareness,* Daytona Beach, FL. Available from http://web.mit.edu/aeroastro/www/labs/ASL/SA/sa.html#contents

Provenmire, H. K., & Roscoe, S. N. (1973). Incremental transfer effectiveness of a ground-based aviation trainer. *Human Factors, 15,* 534–542.

Ree, M. J., & Carretta, T. R. (1998). Computerized testing in the U. S. Air Force. *International Journal of Selection and Assessment 6,* 82–89.

Roby, T. L., & Lanzetta, J. T (1958). Considerations in the analysis of group tasks. *Psychological Bulletin, 55,* 88–101.

Rolfe, J. M., & Staples, K. J. (1986). *Flight simulation.* Cambridge, England: Cambridge University Press.

Roscoe, S. N., & Childs, J. M. (1980). Reliable, objective flight checks. In S. N. Roscoe (Ed.), *Aviation psychology* (pp. 145–158). Ames, IA: Iowa State University Press.

Roscoe, S. N., Jensen, R. S., & Gawron, V. J. (1980). Introduction to training systems. In S. N. Roscoe (Ed.), *Aviation psychology* (pp. 173–181). Ames, IA: Iowa State University Press.

Roscoe, S. N., & Williges, B. H. (1980). Measurement of transfer of training. In S. N. Roscoe (Ed.), *Aviation psychology* (pp. 182–193). Ames, IA: Iowa State University Press.

Scriven, M. (1975). Problems and prospects for individualization. In H. Talmage (Ed.), *Systems of individualized education* (pp. 199–210). Berkeley, CA: McCutchan.

Seeley, E., & Kryder, T. (2003). Evaluation of human performance design for a task-based training support system. In *Proceedings of the 2003 Interservice/Industry Training System and Education Conference (I/ITSEC)* (pp. 940–951). Arlington, VA: National Security Industrial Association.

Semb, G. B., Ellis, J. A., Fitch, M. A., & Matheson, C. (1995). On-the job training: Prescriptions and practice. *Performance Improvement Quarterly, 8,* 19.

Shiffrin, R. M., & Schneider, W. (1977). Controlled and automatic human information processing II: Perceptual learning, automatic attending, and a general theory. *Psychological Review, 84,* 127–190.

Stamp, G. P. (1988). *Longitudinal research into methods of assessing managerial potential* (Tech. Rep. No. 819). Alexandria, VA: U.S. Army Research Institute for the Behavioral and Social Sciences (ADA 204 878).

Thorndike, E. L. (1903). *Educational psychology.* New York: Lemcke and Buechner.

Thorpe, J. A. (1987). The new technology of large scale simulator networking: Implications for mastering the art of warfighting. In, *Proceedings of the Ninth InterService/Industry Training Systems Conference* (pp. 492–501). Arlington, VA: American Defense Preparedness Association.

Tversky, A., & Kahneman, D. (1974). Judgment under uncertainty: Heuristics and biases. *Science, 185,* 1124–1131.

U.S. Department of the Air Force (1996). *New World Vistas: Air and Space Power for the 21st Century: Human Systems and Biotechnology.* Washington, DC: Department of the Air Force, Scientific Advisory Board.

U.S. Department of Transportation (1989). *Flight Plan for Training: FAA Training Initiatives Management Plan.* Washington, DC: U.S. Department of Transportation, Federal Aviation Administration.

U.S. Department of Transportation (1995a). *Aviation Mechanic General, Airframe, and Powerplant Knowledge and Test Guide* (AC 61-28). Washington, DC: U.S. Department of Transportation, Federal Aviation Administration.

U.S. Department of Transportation (1995b). *Commercial Pilot Knowledge and Test Guide* (AC 61–114). Washington, DC: U.S. Department of Transportation, Federal Aviation Administration.

U.S. Department of Transportation (1995c). *Control Tower Operator (CTO) Study Guide* (TS-14-1). Washington, DC: U.S. Department of Transportation, Federal Aviation Administration.

U.S. Department of Transportation (1996). *Commercial Flight Regulations Chapter 1, Part 67* (14 CFR). Washington, DC: U.S. Department of Transportation, Federal Aviation Administration.

U.S. Department of Transportation (2004). *BTS—Airline Information—Historical Air Traffic Data 2003.* Washington, DC: U.S. Department of Transportation, Federal Aviation Administration.

Viteles, M. S. (1945). The aircraft pilot: 5 years of research, a summary of outcomes. *Psychological Bulletin, 42,* 489–521.

Waag, W. L. (1981). *Training effectiveness of visual and motion simulation* (AFHRL-TR-79-72). Brooks Air Force Base, TX: Air Force Human Resources Laboratory (ADA 094 530).

Wickens, C. D., & Flach, J. M. (1988). Information processing. In E. L. Wiener & D. C. Nagel (Eds.), *Human factors in aviation.* New York: Academic Press.

Wickens, C., & McCarley, J. (2001). *Attention-situation awareness (A-SA) model of pilot error* (ARL-01-13/NASA-01-6). Moffett Field, CA: NASA Ames Research Center.

Wiener, E. L., Kanki, B. J., & Helmreich, R. L. (Eds.). (1993). *Cockpit resource management.* San Diego, CA: Academic Press.

Williams, A. C. (1980). Discrimination and manipulation in flight. In S. N. Roscoe (Ed.), *Aviation psychology* (pp. 11–30). Ames, IA: Iowa State University Press.

Williams, K. W. (Ed.). (1994). *Summary Proceedings of the Joint Industry-FAA Conference on the Development and Use of PC-based aviation training devices* (DOT/FAA/AM-94/25). Washington, DC: Office of Aviation Medicine, Federal Aviation Administration, U.S. Department of Transportation (ADA 286-584).

Yerkes, R. M. (Ed.). (1921). *Memoirs of the national academy of sciences* (Vol. 15). Washington, DC: National Academy of Sciences.

Zeidner, J., & Drucker, A. J. (1988). *Behavioral science in the army: A corporate history of the army research institute.* Alexandria, VA: U.S. Army Research Institute for the Behavioral and Social Sciences.

Zeidner, J., & Johnson, C. (1991). *The economic benefits of predicting job performance: Volume 3, estimating the gains of alternative policies.* New York: Praeger.

Zsambok, C. E., & Klein, G. (Eds.) (1997). *Naturalistic decision making.* Mahwah, NJ: Lawrence Erlbaum Associates.

14

Pilot Performance

Lloyd Hitchcock*
Hitchcock & Associates

Samira Bourgeois-
Bougrine
Clockwork Research

Phillippe Cabon
Université Paris Descartes

The determination of pilot performance and the efforts to maximize it are central to aviation safety. It is generally conceded that two out of three aviation accidents are attributable to inappropriate responses of the pilot or crew. Although the catch phrase "pilot error" is all too often laid on the pilot who is guilty only of making a predictable response to "mistakes waiting to happen" that are intrinsic to the design of his cockpit controls or displays or to the work environment surrounding him (or her), there is no question that the greatest improvement in flight safety can be achieved by eliminating the adverse elements of the human component in the aircraft system. Being the most important contributor to aviation safety, the pilot is also the most complicated, variable, and least understood of the aviation "subsystems." Pilot performance refers to both technical flying skills and nontechnical skills related to interpersonal communications, decision-making, and leadership. Pilot performance has been shown to be affected by everything from eating habits to emotional stress, both past and present. Scheduling decision can disrupt the pilots' sleep-and-rest cycle and impose the requirement for pilots to execute the most demanding phase of flight at the point of their maximum fatigue. Illness and medication can degrade the performance markedly, as can the use of alcohol and tobacco. Although a complete exposition of all the factors that serve to determine or delimit pilot performance is impossible within the constraints of a

---

* It should be noted that our friend and colleague Lloyd Hitchcock died since the publication of the first edition and his input was sincerely missed. The chapter was updated by the second and third authors.

single chapter, it is hoped that the following will at least make the reader aware of many of the variables that have an impact on the skill and the ability of the commercial and general aviation pilot.

14.1 Performance Measurement

Before the role played by any factor in determining pilot behavior can be objectively assessed, we must first be able to quantitatively measure the performance within the cockpit environment. The purpose of performance measurement is to provide an assessment of the pilot's knowledge, skills, or decision-making. It depends on overt actions that are produced by internal complex processes such as decision-making, which are not directly observable. In aviation's infancy, the determination of pilot performance was simple and direct. Those who flew and survived were considered adequate aviators. Since that time, the increased complexity and demands of the airborne environment have continued to confound the process of evaluating the performance of those who fly. Incident and accident investigation remain the most used tool to obtain information on operational human performance and define remedial countermeasures.

14.1.1 Subjective Evaluation of Technical Skills

The earliest measures of pilot technical skills were the subjective ratings of the pilot's instructors. The "up-check" was the primary method of evaluation used by the military flight training programs through World War II and, to a great extent, remains the dominant method of pilot assessment today. The general aviation pilot receives his or her license based on the subjective decision of a Federal Aviation Administration (FAA) certified flight examiner. Despite the relative ease of subjective measure implementation, this approach depends on the expertise and the skills of the evaluator and therefore remains prone to the problems of inter- and intra-raters' reliability. Additionally, the limitation of human observation capabilities restricts the capture of the "whole" flying tasks such as the use of aids and equipment, the interpersonal and interface communications, and the performance on secondary tasks. It is highly recommended to use a standardized checklist where all the items to be evaluated are explicitly defined and to provide sufficient training to the evaluator, who must have an intimate knowledge of the appropriate procedures and the pitfalls and most common mistakes, to achieve reasonable inter and intra-raters' reliability (Rantanen, 2001). A proactive approach based on the observation of crew performance called Line Operations Safety Audit (LOSA) has been developed by the University of Texas and endorsed by ICAO (ICAO, 2002;[*] Klinect, 2002[†]). LOSA uses highly trained expert observers who record all threats and errors, how they were managed and their outcomes. The criteria used for observation are defined and inter-observer reliability are conducted at the end of the training session. According to ICAO document, data from LOSA provide a real-time picture of system operations and that can guide organizational strategies in regard to safety, training and operations.

14.1.2 Objective Evaluation of Technical Skills

The appearance of flight simulations not only has enhanced the training of aviators but has made possible a level of quantitative assessment of pilot performance that was not possible before the age of the simulator. In their exhaustive literature review, Johnson and Rantanen (2005) found 19 flight parameters and 17 statistical or mathematical metrics based on these (Table 14.1). Among flight parameters, altitude, airspeed, roll, control inputs, heading, and pitch accounting for 65% of all parameters measured in the literature. The basic statistical measures most frequently applied to flight data are: root mean square error (RMSE),

---

[*] ICAO. Line operations safety audit (LOSA). Montreal, Canada: International Civil Aviation Organisation; 2002.
[†] Klinect JR. LOSA searches for operational weaknesses while highlighting systemic strengths. International Civil Aviation Organisation (ICAO) Journal 2002; 57:8–9, 25.

TABLE 14.1 Flight Parameters and Derivative Measures
Used in the Literature

Parameters		Derivative Metric	
Altitude	Glide slope	RMSE	Autocorrelation
Airspeed	Tracking	Std. Dev	Time outside
Roll	Flaps	Max/min	tolerance
Control inputs	Trim	Mean	Median
Heading	Speed brakes	Frequency analyses	ND
Pitch	Sideslip	Range	Boolean
Vertical speed	Landing gear	Deviation from	Correlation
VOR tracking	Acceleration	Criterion	Moments
Yaw	Position	Time on target	MTE
Turn rate	NDB tracking	Mean absolute error	

Source: Adapted from Johnson, N.R. and Rantanen, E.M., Objective pilot performance measurement: A literature review and taxonomy of metric, in *The 13th International Symposium on Aviation Psychology.* Dayton, OH, 2005.
Notes: VOR, very high frequency omnidirectional range; NDB, nondirectional beacon.

standard deviation (SD), maximum and minimum values, and mean. A small SD is usually indicative of good performance in case of piloting an aircraft, but does not provide any information about the possible error relative to a given flight parameter. RMSE, used for tracking performance, summarizes the overall position error, but does not contain the information about the direction and the frequency of the deviation. To overcome these limitations, additional measures were developed such as the number of deviations (ND) outside the tolerance, the total time spent outside the tolerance for a given flight segment (TD), and the mean time to exceed tolerance (MTE: time the aircraft will remain in the tolerance region, Rantanen et al., 2001). Low ND and TD or Large MTE is indicative of good performance.

In addition, several attempts have been made to reduce the number of measures into something manageable and interpretable by combining individual flight parameter measure into an index of pilot performance. Hitchcock and Morway (1968) developed a statistical methodology allowing them to place probability values on the occurrence of given magnitudes of variation in airspeed, angle-of-attack, roll angle, altitude, and G-load as a function of aircraft weight, penetration altitude, storm severity, and the use of a penetration programmed flight director. This technique permitted the combination of several variables (e.g., G-loading, angle-of-attack variation, and airspeed deviation) into a multidimensional probability surface that described the statistical boundaries of the sampled population of simulated turbulence penetrations. Bortolussi and Vidulich (1991) developed a figure of merit (FOM) of pilot performance from the mean and standard deviation of different flight parameters such as control inputs, altitude, airspeed, and heading. Total FOM and specific flight parameter FOMs (an altitude FOM, for example) were studied to evaluate their sensitivity to flight scenario difficulty. Another approach to help in data reduction and interpretation is based on the use of natural linking of flight parameters through the hierarchical structure of pilot goals and control order (Johnson & Rantanen, 2005). Such hierarchy offers a promising framework for the choice, analysis, and interpretation of objective metrics available from different maneuvers. As pointed out by De Maio, Bell, and Brunderman (1983), automated performance measurement systems (APMS) are generally keyed to quantitative descriptions of aircraft state (e.g., altitude, airspeed, bank angle, etc.), which are usually plotted as a function of elapsed flight time. This time-referenced methodology can ignore the variable of pilot intention and can result in the averaging of performance inputs that may well have been made to accomplish totally different objectives but were grouped together solely because they occurred at the same temporal point in the task sequence. Some widely divergent measures of pilot performance in the course of simulations are found in the literature.

Objective measurement based on flight data represents an alternative or complementary approach to pilot performance measures. However, flying is a complex task, which can yield a vast number of measures, and simply considering a single flight parameter may not provide a complete picture of the performance. Johnson and Rantanen (2005) concluded that the major problem is the lack of unifying theoretical foundation of pilot performance that defines what must be measured, the relative importance of each measure, and the interactions with other measures under the given circumstances.

14.1.3 Evaluation of Nontechnical Skills

Human error in air crashes has been identified as the failure of interpersonal communications, decision-making, and leadership. Therefore, new crew training program, crew resource management (CRM), was applied to reduce pilot error by making good use of the human resource in the flightdeck (Helmreich & Wilhelm, 1991). In the early 1990s, the FAA introduced the Advanced Qualification Program (AQP), which requires commercial aircrews to be trained and evaluated on both their technical flying skills and teamwork skills prior to being certified to fly. Helmreich et al., (1994) developed a checklist of performance markers of specific behaviors associated with more or less effective CRM (NASA/UA/FAA Line LOS checklist). It includes a list of 16 performance markers concerning different behavioral categories: Team Management & crew communication, Automation Management, Situational Awareness & decision-making, attitudes toward special situations, and technical proficiency. Overall performance of crews is classified as "poor," "minimum expectation," "standard," or "outstanding" by a trained observer. The nature of the CRM training has changed over the last two decades and the latest fifth generation of CRM deals with the management of error (Helmreich et al., 1999). This approach defines behavioral strategies as error countermeasures employed to avoid errors, to trap incipient errors, and to mitigate the consequences of errors.

14.2 Workload

During the past 30 years, owing to the evolution of cockpit design, mental workload of aircrews and air traffic control operators have received more and more attention. If task demands are over the capabilities of the operators, errors may occur. These errors might become critical and detrimental for safety. Moreover, workload assessment may also have economic benefits, in saving resources with a better work organization.

The psychophysiological approach (called "psychophysiological engineering") of the evaluation for human–machine interaction has been developed during the past years with a large amount of work on the area of workload (Cabon & Mollard, 2002).

14.2.1 Definition of Workload

The workload could be simply defined as a required demand for the human. However, this definition limits exclusively workload to an external source (the task difficulty) although the internal source (the operator state) should be included. Therefore, Human Factors defines workload as follows:

Workload is the part of the resources for the attention used for the perception, the reasonable decision-making, and action. As resources are limited, the resources needed for a specific task can exceed the available resources. Workload can also be defined as the ratio of the available resources and the required resources during the task. This means that a given task will not produce the same workload level for different operators (depending on their experience of this task) or even for the same operator (depending on his state during the task). Therefore, workload is an individual experience and thus specific methods that take into account this dimension should be applied.

14.3 Measurement of Workload

Over the past years, three kinds of workload measurements have been the most used for human–machine interface design: performance, subjective ratings, and physiological parameters.

14.3.1 Aircrew Performance Measures

As shown by the De Waard model (De Waard, 1996) at certain levels of task difficulty, performance is not correlated with effort. Therefore, it would not be suitable to use performance as the only indicator of workload. However, it could be used as a complementary measure during the evaluations.

There are three types of measurement of the performance related to workload.

14.3.1.1 Primary-Task Measures

In laboratory tasks, motor, or tracking performance, the number of errors, the speed of performance, or reaction time measures can be used as the primary performance measures (Brookhuis, Louwerans, & O'Hanlon, 1985, Green, Lin, & Bagian, 1993). On the field, primary-task performance is, by its nature, very task-specific. However, in this project, specific simulator data and a structured observation of aircrews should be used as a complement of direct workload measurements.

14.3.1.2 Secondary-Task Measures

When another task is added to the primary task, secondary-task measures can be taken. The instruction to maintain primary-task performance is given. Consequently, secondary-task performance varies with difficulty and indicates "spare capacity," provided that the secondary is sufficiently demanding (O'Donnel, 1976; Bortollussi, Hart, & Shively, 1987). However, this method has been criticized because of the possible interference of the secondary task on the primary task.

14.3.1.3 Reference Task

Reference tasks are standardized laboratory tasks that measure performance before and after task under evaluation and they mainly serve as a checking instrument for trend effects. The changes of performance on reference tasks indicate effects of mental load of the primary task. If subjective and physiological measures are added to the reference tasks, the costs for maintaining performance on the primary task could also be inferred, particularly when the operator's state is affected. The use of standard reference task batteries is very common in organizational and occupational psychology (see, e.g., Van Ouerkerk, Meijman, & Mulder, 1994).

14.3.2 Subjective Measures

The most frequently used self-reports of mental workload in aviation are the Subjective Workload Assessment Technique (SWAT) (Papa & Stoliker, 1988) and the NASA-Task Load indeX (TLX) (Bittner, Byers, Hill, Zaklad, & Christ, 1989). The disadvantage of self-reports is that operators are sometimes unaware of internal changes or that the results could be biased by other variables than workload (e.g., psychosocial environment). Therefore, it is not recommended to use them as a unique measure of workload.

14.3.3 Physiological Measures

These categories of workload measure are those derived from the operator's physiology. Probably, the most frequently applied measure in applied research is the Electrocardiogram (ECG) (Cabon, Bourgeois-Bougrine, Mollard, Coblentz, & Speyer, 2000; Cabon & Mollard, 2002; David et al., 1999, 2000).

TABLE 14.2 Summary of Several Studies Where HR Has Been Measured in Aviation

Key Words	Authors	Context	Results
HR and stress	Koonce (1976, 1978), Smith (1967), Hasbrook et al. (1975), Nicolson et al. (1973)	Flight simulator	HR is considered as one of the best indicators of physical stress during flight
HR and workload	Hart and Hauser (1987), Roscoe (1976), Wilson (1993)	Laboratory	HR is high when mental workload is high
HR and experience	Billings et al. (1973)	Flight simulator	HR activation during flight depends on not only flight task, but also the experience of the pilots
HR and responsibility	Roman (1965), Roscoe (1976, 1978), Wilson (1993)	Flight simulator	Risk plus responsibility is more potent in evoking HR than risk alone

For the cardiac-related recording, there are several parameters used for the workload evaluation studies:

Heart rate (HR), expressed in beat per minute. Table 14.2 summarizes several works where HR has been measured in aviation.

14.3.3.1 Heart Rate Variability

Heart rate variability (HRV) in the time domain is also used as a measure of mental workload. The basic assumption is that the higher the workload, the lower the HRV. In other terms, the more the operator exerts an effort, the more regular is the HR.

In the past years, numerous studies have used the spectral analysis of HR, and therefore expressed the HRV in the frequency domain. Three frequency bands have been identified:

- A low frequency band (0.02–0.06 Hz) related to the regulation of the body temperature
- A mid frequency band (0.07–14 Hz) related to the short-term blood-pressure regulation
- A high frequency band (0.15–0.50 Hz) influenced by respiratory-related fluctuations

A decrease in power in the mid frequency band, also called the 0.10 Hz component, has been shown to be related to mental effort and task demand (Vicente, Thorton, & Moray, 1987; Jorna, 1992; Paas, Van Merriënboe, & Adam, 1994). One of the main limitations of this parameter is that it can be used only with an accurate task observation and analysis because this measure is very sensitive to slight variations of workload. Table 14.3 compares the advantages and drawbacks of three workload measures mentioned here.

This section shows that the evaluation method should comprise multidimensional evaluation techniques to capture the complexity of factors involved in workload.

TABLE 14.3 Comparison of the Advantages and Disadvantages of Three Workload Measures

Types of Measures	Advantages	Disadvantage
Subjective	Cheap Assesses the perception of the individual	Can be biased by motivation or other factors
Performance	*Primary task*: No additional measures are required *Secondary task*: Provides the residual resource available	*Primary task*: Not sensitive *Secondary task*: Low ecological validity
Physiological	Sensitive Provides a continuous measure of workload	Can be expensive and needs expertise to perform

14.4 Rest and Fatigue

Pilot's fatigue is a genuine concern in terms of safety, health, efficiency, and productivity. Fatigue is recognized as one of the major factors that can impair performance, and has been often cited as a cause of accidents and incidents in industry and transport. In 1993, it was the first time that fatigue was officially recognized as a contributing factor in DC-8 crash in Guantanamo Bay. In 1999, fatigue was also cited in the crash of Korea Air Flight 801 at Guam international airport (228 deaths), and the crash of American Airline Flight 1420 (11 deaths). Extended duty and sleep loss were the root causes of fatigue. In 1981, Lyman and Orlady showed that fatigue was implicated in 77 (3.8%) of 2006 incidents reported by pilots to the Aviation Safety Reporting System. When the analysis was expanded to include all factors that could be directly or indirectly linked to fatigue, incidents potentially related to fatigue increased to 426 (21.2%). Over 50 years ago, Drew (1940) published a seminal study showing that such measured aspects of precision pilotage as deviations in airspeed, sideslip, course heading, and altitude holding were all markedly affected by flight duration. In his book *Fatal Words*, Cushing (1994) cited the role that fatigue can play in missed or misunderstood communications.

The major problem with fatigue issues is the lack of a coherent definition of fatigue itself, and of a reliable and valid assessment tool to measure it. Therefore, fatigue was and still is generally difficult to investigate on a systematic basis and to code in accidents and incidents databases. However, the main causal factors of pilots' fatigue are well known and could be used to improve work schedules or to assess fatigue implications in accidents and incidents analysis. In addition, there are a number of major efforts that focus on the elaboration and the application of predictive biomathematical models of fatigue and performance. The causal factors and the predictive models of pilot's fatigue are described in the following section.

14.4.1 The Causes and Manifestations of Pilot Fatigue

Fatigue in aviation refers to decreases in alertness and feeling tired, sleepy, and/or exhausted in both short- and long-range flights. The work of Gander et al. (Gander et al., 1985, 1986, 1987, 1989, 1991) and Foushee et al., (1986) described the negative impact of changes in the pilot's day–night cycles on their sleep and rest patterns. A recent survey (Bourgeois-Bougrine et al., 2003a) confirmed that night flights and jet lag are the most important factors that generated fatigue in long-range flights. In SRF, multileg flights and early wake-ups are the main causes of fatigue (Bourgeois-Bougrine et al., 2003a,b). In addition, time constraints, high numbers of legs per day, and consecutive work days seemed to increase fatigue, suggesting that flight and duty time limitations have to take into account the flight category (Cabon et al., 2002). When considering themselves, pilots cited the manifestations of fatigue caused by sleep deprivation as a reduction in alertness and attention, and a lack of concentration (Bourgeois-Bougrine et al., 2003a). However, for their cockpit crewmembers, they reported mental manifestations (increased response times, small mistakes) and verbal manifestations (reduction of social communications, bad message reception). In addition, these pilots reported that when they are tired, all the flying tasks seemed to be more difficult than usual, particularly supervisory or monitoring activities. Among nontechnical skills, attitude toward conflicts is the most affected by fatigue.

The need to minimize personnel costs by pilot reduction has further constrained the operations manager's crew scheduling options. Indeed, the current trend to the use of two-person flight crews, as opposed to the three- and sometimes four-person crews of the past, has removed the option of carrying a "rested" pilot along in the cockpit in case one were needed. Using physiological recordings on 156 flights, a previous study showed that reductions in alertness were frequent during flights, including the descent and approach phases (Cabon et al., 1993). Most decreases in alertness occurred during the monotonous part of the cruise and were often observed simultaneously in both pilots in two-person crews. Based on these results, specific operational recommendations were designed. These recommendations have been validated in further studies (Cabon et al., 1995a) and they were extended to cover all long-haul flight schedules (around the clock) and all time zone transitions (±12). The recommendations were gathered into a booklet for the use of long-haul aircrews and a software is now available that enables crewmembers to simply enter their flight details to obtain a detailed set of recommendations (Cabon et al., 1995b, Mollard et al., 1995, Bourgeois-Bougrine et al., 2004).

14.4.2 Fatigue Predictions

Several research groups have developed models for estimating the work-related fatigue associated with work schedules. Seven of these models were discussed at a workshop held in 2002 in Seattle and compared in a number of scenarios: the Sleepwake Predictor (Akerstedt and Folkard); the Fatigue Audit Interdyne (FAID; D. Dawson et al.); the two-process model (P. Achermann, A.A. Borbely); Fatigue Avoidance Scheduling Tool (FAST; S. Hursh), The Circadian Alertness Simulator (CAS; M. Moore-Ede et al.); the Interactive Neurobehavioral Model (M.E. Jewett, R.E. Kronauer); the System for Aircrew Fatigue Evaluation (SAFE; M. Spencer). The detailed description of these models is available in the preceding workshop (Aviation Space and Environmental Medicine Vol. 75 No 3, Section II, March 2004). Most of these models are based on the two-process model of sleep regulation first proposed by Borbely. Sleep inertia is included in some models, as are time on task, cumulative fatigue, and effect of light and workload. The majority of the models seek to predict some aspects of subjective fatigue or sleepiness (six models), performance (five models), physiological sleepiness or alertness (four models), or the impact of countermeasure such as naps and caffeine (five models). But, there are only two models concerned with predicting accident risk, three by optimal work/rest schedules, two by specific performance task parameters, and three by circadian phase. The required inputs are mainly work hours and/or sleep–awake time. Despite their differences, these models have a fundamental similarity and can be used as tools to anticipate and predict the substantial performance degradation related to fatigue that often accompanies around the clock operations, transmeridian travel, and sustained or continuous military operations.

   Predictive models of fatigue risk are mainly based on the results of simple cognitive tasks such as the Psychomotor Vigilance Test (PVT) focusing on individuals rather than a multi-pilot crew performance. Human performance on PVT has proven to be an effective method for measuring sleepiness due to sleep restriction and the effectiveness of countermeasures against fatigue such as cockpit napping. However, flight simulator-based studies suggest that fatigue has a complex relationship with aircrew operational performance (Foushee, 1986; Thomas, Petrilli, Lamond, Dawson, & Roach, 2006). Crew familiarity was seen to improve crew communication in non-rested crew leading to less operational errors (Foushee, 1986). More recently, Thomas et al. (2006) suggested that fatigue is associated with increased monitoring of performance as an adaptive strategy to compensate for the increased likelihood of errors in fatigued crew.

14.5 Stress Effects

14.5.1 Acceleration

The dominant impact of linear acceleration on the pilot is a reduction in peripheral vision and ultimate loss of consciousness associated with sustained high levels of positive G-loadings (+Gz).* Such effects are of great importance to the military combat aviator pilot and the aerobatic pilot, but are far less of a

---

* Traditionally, the direction in which acceleration is imposed on the body is defined in terms of the change in weight felt by the subject's eyeballs. Thus, positive acceleration (+Gz) such as that felt in tight high-speed turns or in the pullout from a dive is known as "eyeballs down." The forward acceleration (+Gx) associated with a dragster or an astronaut positioned on his or her back during launch would be "eyeballs in." Accelerations associated with sharp level turns (+ or −Gy) would result in "eyeballs right" during a left turn and "eyeballs left" while in a flat right turn. The negative loading (−Gx) associated with a panic stop in an automobile would be "eyeballs out" and the loading (−Gz) associated with an outside loop would be "eyeballs up."

challenge for the commercial or general aviation pilots, who, hopefully, will never experience the acceleration levels necessary to bring about such physical consequences. These acceleration effects are the result of two factors, the pooling of blood in the lower extremities and the increase in the effective vertical distance (hemodynamic column height) that the heart must overcome to pump blood to the brain. Chambers and Hitchcock (1963) showed that highly motivated pilots would voluntarily sustain up to 550 s of +Gx (eyeballs in), and even the most determined would tolerate exposures of approximately 160 s of +Gx (eyeballs down). The seminal work on acceleration-induced loss of vision (grayout) was done by Alice Stoll in 1956. She demonstrated that grayout, blackout, and subsequent unconsciousness are determined not only by the magnitude of the acceleration level but also by the rate of onset (the time required to reach the programmed G-level). More rapid rates of onset apparently do not allow the body time to adapt to the acceleration imposed changes in blood flow. A great deal of effort has been expended in the development of special suiting to constrain blood pooling and the use of grater reclining angles as ways in which the pilot's tolerance to acceleration can be enhanced. In addition, the work of Chambers and Hitchcock (1963) demonstrated the roles that variables like control damping, cross-coupling, balancing, and number of axes being controlled have in the impact of acceleration of a pilot's tracking control precision, with well-damped, balanced, and moderately cross-coupled controls achieving the best performance. A general review of the effects of sustained acceleration is available in Fraser's chapter on sustained linear acceleration in the NASA *Bioastronautics Data Book* (NASA, 1973). More recent work has focused not just on the physical effects of acceleration but also on its impairment of a pilot's cognitive capabilities. Research by Deaton, Holmes, Warner, and Hitchcock (1990) and Deaton and Hitchcock (1991) has shown that the seatback angle of centrifuge subjects has a significant impact on their ability to interpret the meaning of four geometric shapes even though the variable of back angle did not affect the subjects' physical ability to perform a psychomotor tracking task. A much earlier unpublished study by Hitchcock, Morway, and Nelson (1966) showed a strong negative correlation between acceleration level and centrifuge subjects' performance on a televised version of the Otis Test of Mental Abilities. Such findings are consistent with the pilot adage that states the "all men are morons at 9G."

14.5.2 Vibration

The boundaries of acceptable human body vibration are established by the International Standards Organization *Guide for the Evaluation of Human Exposure to Whole Body Vibration* (1985) and the Society of automotive engineers *Measurement of Whole Body Vibration of the Seated Operator of Off-Highway Work Machines* (1980). The dynamic vibration environment experienced by the pilot is the product of many factors including maneuver loads, wing loading, gust sensitivity, atmospheric conditions, turbulence, aircraft size, structural bending moments, airframe resonant frequency, and the aircraft's true airspeed. A clear picture of the impact of vibration on pilot performance is not easily obtained. Investigations of vibration stress have used so many diverse tasks involving such a variety of control systems and system dynamics that it is difficult to integrate their findings. Ayoub (1969) found significant (40%) reduction in a single-axis side-arm controller compensatory racking task during a 1-h exposure to a ±.2 g* sinusoidal vibration at 5 Hz (hertz) or cycles per second.[2] Recovery had not been completed for at least 15 min after exposure. Hornick and Lefritz (1966) exposed subject pilots to 4-h simulation of three levels of a terrain following task using a two-axis side-stick controller. The vibration spectrum used ranged from 1 to 12 Hz with the peak energy falling between 1 and 7 Hz and with g loadings of .10, .15, and .20 g. There was no tendency for error to increase as a function of exposure time for the two easier task levels, although performance degraded after 2.5 h of exposure to the heaviest loading. Further, these researchers found that reaction time to a thrust change command was almost

---

* Although the uppercase G is used to denote steady-state acceleration, convention dictates that the lowercase g should be used to designate the level of vibration exposure.

four times long during vibration exposure than during the nonvibratory control period. In general, the effects of vibration on pilot performance, as measured by tracking performance during simulation, can be summarized as:

- Low-frequency (5 Hz) sinusoidal vibrations from .2 to .8 g can reduce tracking proficiency up to 40%.
- When vibration-induced performance decrement is experienced, the effect can persist for up to 0.5 h after exposure.
- Higher levels of random vibration exposure are required to affect performance than are required for sinusoidal exposure.
- For gz exposure, vertical tracking performance is more strongly affected than is horizontal.

Under sufficiently high levels of vibration exposure, visual capabilities and even vestibular functioning can be impaired. Although the role of vibration exposure in determining pilot performance should not be ignored, the level of exposure routinely experienced in the commercial aviation environment would not generally be expected to introduce any significant challenge to pilot proficiency.

14.5.3 Combined Stresses

The appearance of other stressors in the flight environment raises the possibility of interactive effects between the individual variables. For example, heat tends to lower acceleration tolerance, whereas cold, probably owing to its associated vascular constriction, tends to raise G tolerance. In the same vein, pre-existing hypoxia reduces the duration and magnitude of acceleration exposure required to induce peripheral light loss (Burgess, 1958). The nature of stress interactions is determined by (a) their order of occurrence, (b) the duration of their exposure, (c) the severity of exposure, and (d) the innate character of their specific interaction. Any analysis of the flight environment should include a consideration of the potential for synergy between any stressors present. An excellent tabulation of the known interactions between environmental stresses is contained in Murray and McCalley's chapter on combined environmental stresses in the NASA *Bioastraunautics Data Book* (NASA, 1973).

14.6 Physical Fitness

14.6.1 Aging

The interactive role of the potentially negative impact of the aging process and the safety enhancements that are assumed to accompany the gaining of additional operational experience has been assessed in a comprehensive overview of the subject by Guide and Gibson (1991). These authors cite the studies of Shriver (1953), who found that the physical abilities, motivation, skill enhancement, and piloting performance (cognitive) and physical capabilities of pilots deteriorated with age. More recently, it was found that the ability to respond to communication command and time-sharing efficiency in complex, multitask environments declines with age (Morrow, Ridolfo, Menard, Sanborn, & Stine-Morrow, 2003). However, the prevalence and the pattern of crew errors in air carrier accidents do not seem to change with pilot age (Guohua, Grabowski, Baker, & Rebok, 2006). In large part, the FAA imposition of the so-called Age 60 Rule, which prohibits anyone from serving as pilot or copilot of an aircraft heavier than 7500 lb after their 60th birthday is based on a concern for the potential for "sudden incapacitation" by the older pilot (General Accounting Office, 1989). However, a number of studies have shown that this concern is most probably misplaced. Buyley (1969) found that the average pilot experiencing sudden inflight incapacitation resulting in an accident was 46 years old. This finding was subsequently confirmed by Bennett (1972), who found that most incapacitation accidents were not related to age. However, age does have an observable impact on aviation safety in that the accident rate for private pilots aged 55–59 (4.97/1000) is almost twice that for the 20–24

(2.63/1000) age group) (Guide & Gibson, 1991). On the other hand, the accident rate of airline transport rated (ATR) pilots aged 55–59 (3.78/1,000) is approximately one-third of that of pilots with the same rating who are aged 20–24 (11.71/10,000). This difference between the age effects for the private and ATR pilot population is most likely the result of two factors. The first is the far more stringent physical and check ride screening given to the airline pilots. Downey and Dark (1990) found that the first-class medical certificate failure rate of ATR pilots went from 4.3/1000 for the 25–29 age group to 16.2/1000 for pilots in the 55–59 age group. Thus, many of those age-related disabilities that are seen in the private pilot population appear to have been successfully eliminated from the airline pilot group before they have had a chance to impact safety. The second factor is proposed by Kay et al. (1994), who found that the number of recent flight hours logged by a pilot is a far more important determinant of flight safety than is the age of the pilot. The Kay study authors concluded that their "analyses provided no support for the hypothesis that the pilots of scheduled carriers had increased accident rates as they neared the age of 60" (p. 42). To the contrary, pilots with more than 2000 h total time and at least 700 h of recent flight time showed a significant reduction in accident rate with increasing age. These findings replicate and confirm the conclusions of Guide and Gibson (1991), who also found that the recent experience gained by the aviator was, at least for the mature ATR-rated pilot population, a major determinant of flight safety. According to the comprehensive analyses of flight safety records performed by these researchers, pilots flying more than 400 h per year have fewer than a third of the accidents per hour flown than do those with less than 400 h annually. In addition, though the senior pilots would appear to be slightly less safe than those in their 40s, they are "safer" than the younger (25–34) pilots who would be most apt to replace them when they are forcibly retired by the Age 60 Rule. Hultsch, Hertzog, and Dixon (1990) and Hunt and Hertzog (1981) also point out that extensive recent experience enables many individuals to develop compensatory mechanisms and thus significantly reduce the negative effects of many of the more general aspects of aging. Stereotyping may play a part in the perception of the aging pilot. Hyland et al. (Hyland, Kay, & Deimler, 1994), in an experimental simulation study of the role of aging in pilot performance, found that the subjective ratings given to the subject pilots by the evaluating check pilots declined as a function of the age of the pilots are routinely subjected. Tsang (1992), in her extensive review of the literature on the impact of age on pilot performance, pointed out that much of the information on the impact of aging comes from the general psychological literature due to the "sparcity of systematic studies with pilots." She cautioned against the uncritical transfer of findings from the general literature to the tasks of the pilot because most laboratory studies on the effects of aging on cognitive and perceptual processes tend to concentrate on a single isolated function, but the act of flying involves integration of interactive mental and physical functions.

A corollary of aging that is critical to flight safety is the degradation in vision that all too often afflicts the mature aviator. Whether the problem is an impairment of the ability to focus on near object (presbyopia) or on far objects (myopia), the result is a need for the pilot to rely on some form of corrective lenses for at least some portion of his or her visual information acquisition. Using a hand to remove and replace glasses as the pilot switches back and forth between the view out of the cockpit to the instrument panel is less than desirable, to say the least. The use of bifocal or trifocal glasses imposes a potentially annoying requirement for the wearer to tilt the head forward and backward to focus through the proper lens. In addition, a representative study by Birren and Shock (1950) determined that the aviator's dark adaptation ability can be expected to degrade progressively from about the age of 50.

The older pilot (40 and above) also shows a marked degradation in auditory sensitivity. The older pilot can show a decline of 15 decibels or more when compared with that of the typical 25-year-old. In earlier days, Graebner (1947) reported that the age-related decline of auditory sensitivity, particularly at the high frequencies (200 cps [cycles per second] and above), was more pronounced for pilots than for the general population. This was attributed to the high cockpit noise levels associated with the reciprocal engines in use at the time. It is reasonable to assume that the transition to the jet engine would have significantly reduced this effect.

Those who are interested in a more comprehensive study and detailed evaluation of the role of age in determining flight safety are referred to two recent studies supported by the FAA Office of Aviation Medicine. This first is an annotated bibliography of age-related literature performed by Hilton systems, Inc. (1994), under contract to the civil Aeromedical Institute in Oklahoma. The second is an analytic review of the scientific literature, compiled by Hyland, Kay, Deimler, and Gurman (1994), relative to aging and airline pilot performance.

14.6.2 Effects of Alcohol

A number of general reviews of the impact of alcohol on both psychological and physiological performance are available (Carpenter, 1962; McFarland, 1953; Ross & Ross, 1995; Cook CC*, 1997). In general, the documented effects of alcohol are all deleterious, with alcohol consumption adversely affecting a wide range of sensory, motor, and mental functions. The drinker's visual field is constricted, which could affect both instrument scan and the detection of other aircraft (Moskowitz & Sharma, 1974). Alcohol reduces a pilot's ability to see at night or at low levels of illumination, with the eye of one who has consumed ingestion of the alcohol. In addition, the intensity of light required to resolve flicker has been found to be a direct function of the observer's blood alcohol concentration. Alcohol consumption has also been found to reduce the sense of touch. The effects of alcohol ingestion on motor behavior are considered to be the result of its impairment of nervous functions rather than as direct degradation of muscle action. Such activities as reflex actions, steadiness, and visual fixation speed and accuracy are adversely affected by the consumption of even a small amount of alcohol. The consumption of sufficient quantities of alcohol can result in dizziness, disorientation, delirium, or even loss of consciousness. However, at the levels that would most often be encountered in the cockpit, the most significant effects would most likely be in the impairment of mental behavior rather than a degradation of motor response. A detailed review of the literature by Levine, Kramer, and Levine (1975) confirmed the alcohol-induced performance deterioration in the area of cognitive domain, perceptual-motor processes, and psychomotor ability, with the psychomotor domain showing the greatest tolerance for alcohol effects. Alcohol has been also found to degrade memory, judgment, and reasoning. More recent work by Barbre and Price (1983) showed that alcohol intake not only increased search time in a target detection task but also degraded touch accuracy and hand travel speed. In addition, alcohol was found to reduce the subject's motivation to complete a difficult task. Both Aksnes (1954) and Henry, Davis, Engelken, Triebwasser, and Lancaster (1974) demonstrated the negative effect of alcohol on Link Trainer performance. Billings, Wick, Gerke, and Chase (1973) showed similar alcohol-induced performance decrements in light aircraft pilots. Studies by Davenport and Harris (1992) showed the impact of alcohol on pilot performance in a landing simulation. Taylor, Dellinger, Schillinger, and Richardson (1983) found similar degradation of both holding pattern performance and instrument landing system (ILS) approaches as a function of alcohol intake. Ross and Mundt (1988) evaluated the performance of pilots challenged with simulated very high frequency omnidirectional range (VOR) tracking, vectoring, traffic avoidance, and descent tasks. Using a multiattribute modeling analysis, pilot performance was evaluated by flight instructor judgments under 0.0% and 0.04% blood alcohol concentrations (BACs). The multiattribute approach was sufficiently sensitive to reveal "a significant deleterious effect on overall pilot performance" associated with alcohol consumption of even this rather low level, which is the maximum allowable by FAA regulation in 1985 and 1986. Ross, Yeazel, and Chau (1992) using light aircraft simulation studies of pilots under BACs ranging from 0.028% to 0.037% challenged pilots with the demands of simulated complicated departures, holding patterns, and approaches under simulated instrument meteorological conditions (IMC) or instrument landing approaches involving turbulence, cross winds, and wind shear. Significant alcohol-related effects were found at the higher levels of works. Of particular significance for those interested in the effects of alcohol on pilots is the synergistic relationship between alcohol

---

* Cook CC Alcohol and Aviation, *Addition*, 1997, 92:539–55.

and the oxygen lack associated with altitude. Early studies by McFarland and Forbes (1936), McFarland and Barach (1936), and Newman (1949) established the facts that, even at altitudes as low as 8000 ft, the ingestion of a given amount of alcohol results in a greater absorption of alcohol into the blood than at sea level and that, at altitude, it takes the body significantly longer to metabolize the alcohol out of the blood and spinal fluid. More recent studies by Collins et al. (Collins & Mertens, 1988; Collins, Mertens, & Higgins, 1987) confirmed the interaction of alcohol and altitude in the degradation in the perception of professional pilots of the seriousness of the alcohol usage problem. The average overall level of concern over pilot drinking was found to be just below 3 on a scale of 0 (no problem) to 10 (a very serious problem). Noncarrier pilots rated usage as a more serious problem for the scheduled airline pilot than did the major carrier pilots themselves. The majority of commercial pilots approved of the proposal to enact laws making drinking and flying a felony and also approved of random blood alcohol concentration testing, although they were almost evenly divided on the potential effectiveness of such testing and expressed significant concern about the possibility that such a testing program could violate the pilots' rights.

A recent study Guohua, Baker, Qiang, Rebok, and McCarthy (2007) analyzed data from the random alcohol testing and post-accident alcohol testing programs reported by major airlines to the Federal Aviation Administration for the years 1995 through 2002. During the study period, random alcohol testing yielded a total of 440 violations with a prevalence rate of 0.03% for flight crews, and without any significant increase of the risk of accident involvement. The authors concluded that alcohol violations among U.S. major airline are rare, and play a negligible role in aviation accidents.

14.6.3 Drug Effects

In 1953, McFarland published one of the first and most comprehensive descriptions of the potential negative effects of commonly used pharmaceuticals on flight safety. Some of the more common antibiotic compounds have been found to adversely affect the aviator's tolerance to altitude-induced hypoxia and therefore psychomotor performance. Of course, those antihistamines that advise against the operation of machinery after use should be avoided by the pilot, as should any use of sedatives prior to or during flight operations. The use of hyoscine (scopolamine) as a treatment of motion sickness was found to reduce visual efficiency in a significant number of users. In general, the use of common analgesics, such as aspirin, at the recommended dosage levels, does not appear to be a matter of concern. However, because any medication has the potential for adverse side effects in the sensitized user, the prudent pilot would be well advised to use no drug except under the direction of his flight surgeon.

14.6.4 Tobacco

The introduction of nicotine into the system is known to have significant physiological effects. HR is increased by as much as 20 beats per minute, systolic blood pressure goes up by 1020 mm Hg, and the amount of blood flowing to the extremities is reduced. Although these effects have clear significance for the pilot's potential risk of in-flight cardiac distress, perhaps the most significant impact of smoking on flight safety lies in the concomitant introduction of carbon monoxide into the pilot's blood stream. Human hemoglobin has an affinity for carbon monoxide that is over 200 times as strong as its attraction to oxygen (O_2). Hemoglobin cannot carry both oxygen and carbon dioxide molecules. Therefore, the presence of carbon monoxide will degrade the body's capability to transport oxygen, essentially producing a temporary state of induced anemia. McFarland, Roughton, Halperin, and Niven (1944) and Sheard (1946) demonstrated that the smoking-induced level of carboxyhemoglobin (COHb) of 5%–10%, the level generally induced by smoking a single cigarette, can have a significant negative effect on visual sensitivity although this CO content is well below the 20% or more COHb considered necessary to induce general physiological discomfort. Trouton and

Eysenck (1960) reported some degradation of limb coordination at 2%–5% COHb levels. Schulte (1963) found consistent impairment of cognitive and psychomotor performance at this same COHb level. Putz (1979) found that CO inhalation also adversely affected dual-task performance. These findings are not unanimously accepted. Hanks (1970) and Stewart et al. (1970) found no central nervous system functions at COHb levels below 15%. The carbon monoxide anemia induced by smoking synergizes with the oxygen deficits imposed by altitude. According to McFarland et al. (1944), by both decreasing the effectiveness of the oxygen transport system and increasing the metabolic rate, and thus the need for oxygen, smoking can raise the effective altitude experienced by the pilot by as much as 50%, making the physiological effect of 100,000 ft on the smoker equivalent to those felt by the nonsmoker at 15,000 ft. Although most commercial flights now restrict the occurrence of smoking in flight, the uncertainties about the rate with which the effects of smoking prior to flight are dissipated will cause the issue of smoking to continue to be of concern for those interested in optimizing pilot performance. The in-flight use of tobacco by the general aviation pilot will remain as a potential concern. To date, no studies defining the role of second-hand smoke inhalation on pilot performance were located.

14.6.5 Nutrition

Perhaps the earliest impact of nutrition on pilot performance was reported by McFarland, Graybiel, Liljencranz, and Tuttle (1939) in their description of the improvement in vision brought about by vitamin A supplementation of the diet of night-vision-deficient airmen. Hecht and Mendlebaum (1940) subsequently confirmed this effect by experimentally inducing marked degradation in the dark-adaptation capability of test subjects fed a vitamin A-restricted diet. Currently, the ready availability of daily vitamin supplements and the general level of nutrition of the population as a whole have tended to virtually eliminate any concern about a lack of vitamin C on the health of skin, gums, and capillary system or a degradation in the pilot's nervous system, appetite, or carbohydrate metabolism due to a deficiency in the B vitamin complex. However, the intrinsic nature of airline operations inevitably results in some irregularity in the eating habits of the commercial pilot. Extended periods without eating can result in low blood sugar (hypoglycemia). Although the effects of long-term diet deficiency are generally agreed on (marked reduction in endurance and a correspondingly smaller degradation of physical strength), the exact relationship between immediate blood sugar level and performance is less well established. Keys (1946) demonstrated that reaction time was degraded at blood sugar levels below 64–70 mg%.

14.7 Summary

The importance of each variable described in this section is sufficient for all are the subjects of book chapters and, in many cases, the entire texts in their own right. The best that can be hoped is that the foregoing will create sensitivity to the complexity of the topic field of pilot performance. There is much work that remains to be done in developing more objective methods for measuring the essential components of piloting skill. Even more challenging is the pressing need to define and quantify the cognitive components of the concept of pilot workload. Because of the economic and safety implications of aging on both the airline industry and the pilot ranks, the issue of aging will remain a major topic of interest and concern. Because age does not seem to be a prime determinant of sudden in-flight incapacitation, additional effort is clearly needed to determine the physical factors that can be effective in predicting such occurrences. We already know enough to be certain of the negative impacts of alcohol, smoking, and controlled substances on pilot performance. In short, it is unfortunately clear that although pilot performance is unquestionably the most critical element in flight safety, it is the aircraft system area about which we know far less than we should.

References

Aksnes, E. G. (1954). Effects of small does of alcohol upon performance in a link trainer. *Journal of Aviation Medicine, 25*, 680–688.

Barbre, W. E., & Price, D. L. (1983). Effects of alcohol and error criticality on alphanumeric target acquisition. In *Proceedings of the Human Factors Society 27th Annual Meeting* (Vol. 1, pp. 468–471). Santa Monica, CA: Human Factor Society.

Bennett, G. (1972, October). Pilot incapacitation. *Flight International*, pp. 569–571.

Billings, C. E., Wick, R. L., Gerke, R. J., & Chase, R. C. (1973). Effects of ethyl alcohol on pilot performance. *Aerospace Medicine, 44*, 379–382.

Birren, J. E., & Shock, N. W. (1950). Age changes in rate and level of visual dark adaptation. *Applied Physiology, 2*(7), 407–411.

Bittner, A. C., Byers, J. C., Hill, S. G., Zaklad, A. L., & Christ, R. E. (1989). Generic workload ratings of a mobile air defense system. In *Proceedings of the Human Factors Society 33rd Annual Meeting* (pp. 1476–1480). Santa Monica, CA: Human Factor Society.

Bortollussi, M. R., Hart, S. G., & Shively, R. J. (1987). Measuring moment-to-moment pilot workload using synchronous presentations of secondary tasks in a notion-base trainer . In *Proceedings of the 4th Symposium on aviation Psychology*. Columbus: Ohio State University.

Bortolussi, M. R., & Vidulich, M. A. (1991). An evaluation of strategic behaviours in a high fidelity simulated flight task. Comparing primary performance to a figure of merit. In *Proceedings of the 6th ISAP* (Vol. 2, pp. 1101–1106).

Bourgeois-Bougrine, S., Cabon, P., Gounelle, C., Mollard R., Coblentz, A., & Speyer, J. J. (2003a). Fatigue in aviation: Point of view of French pilots. *Aviation Space and Environmental Medicine, 74*(10), 1072–1077.

Bourgeois-Bougrine, S., Cabon, P., Mollard, R., Coblentz, A., & Speyer, J. J. (2003b). Fatigue in aircrew from short-haul flights in civil aviation: The effects of work schedules. *Human Factors and Aerospace Safety. An International Journal, 3*(2), 177–187.

Bourgeois-Bougrine, S., Cabon, P., Folkard, S., Normier, V., Mollard, R., & Speyer, J. J. (2004, July). In Blagnac (Ed.), *Getting to grips with Fatigue and Alertness Management* (Issue III, 197 p.). France: Airbus Industry.

Brookhuis, K. A., Louwerans, J. W., & O'Hanlon, J. F. (1985). The effect of several antidepressants on EEG and performance in a prolonged car driving task. In W. P. Koella, E. Rüther, & H. Schulz (Eds.), *Sleep' 84* (pp. 129–131). Stuttgart: Gustav Fisher Verlag.

Burgess, B. F. (1958). The effect of hypoxia on tolerance to positive acceleration. *Journal of Aviation Medicine, 29*, 754–757.

Buyley, L. E. (1969). Incidence, causes, and results of airline pilot incapacitation while on duty. *Aerospace Medicine, 40*(1), 64–70.

Cabon, P., Bourgeois-Bougrine, S., Mollard, R., Coblentz, A., & Speyer, J. J. (2002). Flight and duty time limitations in civil aviation and their impact on crew fatigue: A comparative analysis of 26 national regulations. *Human Factors and Aerospace Safety: An International Journal, 2*(4), 379–393.

Cabon, P., Coblentz, A., Mollard, R., & Fouillot, J. P. (1993). Human vigilance in railway and long-haul flight operation. *Ergonomics, 36*(9), 1019–1033.

Cabon, P., Mollard, R., Coblentz, A., Fouillot, J.-P., & Speyer, J.-J. (1995a). Recommandations pour le maintien du niveau d'éveil et la gestion du sommeil des pilotes d'avions long-courriers. *Médecine Aéronautique et Spatiale, XXXIV*(134), 119.

Cabon, P., Mollard, R., Bougrine, S., Coblentz, A., & Speyer, J.-J. (1995b, November). In Blagnac (Ed.), *Coping with long range flying. Recommendations for crew rest and alertness* (215 p.). France: Airbus Industry.

Cabon, P., Farbos, B., Mollard, R., & David, H. (2000). Measurement of adaptation to an unfamiliar ATC interface. Ergonomics for the new millennium. *Proceedings of the 14th Triennial Congress of the International Ergonomics Association and 44th Annual Meeting of the Human Factors and Ergonomics Society*. San Diego, CA, July 29–August 4, 2000. Human Factors and Ergonomics Society Santa Monica, CA, Volume 3, pp. 212–215.

Cabon, P., & Mollard, R. (2002). Prise en compte des aspects physiologiques dans la conception et l'évaluation des interactions homme-machine (pp. 99–138). L'Ingénierie Cognitive: IHM et Cognition/G. Boy dir, Paris: Hermes.

Carpenter, J. A. (1962). Effects of alcohol on some psychological processes. *Quarterly Journal of Studies on Alcohol, 24*, 284–314.

Chambers, R. M., & Hitchcock, L. (1963). *The effects of acceleration on pilot performance* (Tech. Rep. No. NADC-MA-6219). Warminster, PA: Naval Air Development Center.

Collins, W. E., & Mertens, H. W. (1988). Age, alcohol, and simulated altitude: Effects on performance and breathalyzer scores. *Aviation, Space, and Environmental Medicine, 59*(11), 1026–1033.

Crabtree, M. S., Bateman, R. P., & Acton, W. H. (1984). Benefits of using objective and subjective workload measures. *Proceedings of the Human Factors Society 28th Annual Meeting* (Vol. 2, pp. 950–953). Santa Monica, CA: Human Factor Society.

Cushing, S. (1994). *Fatal words* (p. 71). Chicago: University of Chicago Press. Davenport, M., & Harris, D. (1992). The effect of low blood alcohol levels on pilot performance in a series of simulated approach and landing trials. *International Journal of Aviation Psychology, 2*(4), 271–280.

David, H., Caloo, F., Mollard, R., Cabon, P., & Farbos, B. (2000). Eye point-of-gaze, EEG and ECG measures of graphical/keyboard interfaces. In P. T. McCabe, M. A. Hanson, & S. A. Robertson (Eds.), *Simulated ATC. Contempory Ergonomics 2000* (pp. 12–16). London: Taylor & Francis.

David, H., Caloo, F., Mollard, R., & Cabon, P. (1999). Trying out strain measures on a simulated simulator. *Proceedings of the Silicon Valley Ergonomics Conference and Exposition—ErgoCon'99.* San Jose, CA, June 4, 1999. Silicon Valley Ergonomics Institute, San Jose State University, San Jose, CA, pp. 54–59.

De Maio, J., Bell, H. H., & Brunderman, J. (1983). Pilot oriented performance measurement. In *Proceedings of the Human Factors Society 27th Annual Meeting* (Vol. 1, pp. 463–467). Santa Monica, CA: Human Factor Society.

De Maio, J., Bell, H. H., & Brunderman, J. (1983). Pilot oriented performance measurement. *Proceedings of the Human Factors Society 27th Annual Meeting* (Vol. 1, pp. 463–467). Santa Monica, CA: Human Factor Society.

Deaton, J. E., & Hitchcock, E. (1991). Reclined seating in advanced crew stations: Human performance considerations. *Proceedings of the Human Factors Society 35th Annual Meeting* (Vol. 1, pp. 132–136). Santa Monica, CA: Human Factor Society.

Deaton, J. E., Holmes, M., Warner, N., & Hitchcock, E. (1990). *The development of perceptual/motor and cognitive performance measures under a high G environment* (Tech. Rep. No. NADC-90065-60). Warminster, PA: Naval Air Development Center.

Downey, L. E., & Dark, S. J. (1990). *Medically disqualified airline pilots in calendar years 1987 and 1988* (Report No. DOT-FAA-AM-90-5). Oklahoma City, OK: Office of Aviation Medicine.

Drew, G. C. (1940). *Mental fatigue* (Report 227). London: Air Ministry, Flying Personnel Research Committee.

Foushee, H. C. (1986). Assessing fatigue. A new NASA study on short-haul crew performance uncovers some misconceptions. Airline Pilot.

Foushee, H. C., Lauber, J. K., Baetge, M. M., & Acombe, D. B. (1986). *Crew factors in flight operations: III, The operational significance of exposure to short-haul air transport operations* (Technical Memorandum 88322). Moffett Field, CA: National Aeronautics and Space Administration.

Gander, P. H., Nguyen, D., Rosekind, M. R., & Connell, L. J. (1993). Age, circadian rhythm and sleep loss in flight crews. *Aviation, Space, and Environmental Medicine, 64*, 189–195.

Gander, P. H., & Graeber, R. C. (1987). Sleep in pilots flying short-haul commercial schedules. *Ergonomics, 30*, 1365–1377.

Gander, P. H., Connell, L. J., & Graeber, R. C. (1986). Masking of the circadian rhythms of body temperature by the rest-activity cycle in man. *Journal of Biological Rhythms Research, 1*, 119–135.

Gander, P. H., Kronauer, R., & Graeber, R. C. (1985). Phase-shifting two coupled circadian pacemakers: Implications for jetlag. *American Journal of Physiology, 249*, 704–719.

Gander, P. H., McDonald, J. A., Montgomery, J. C., & Paulin, M. G. (1991). Adaptation of sleep and circadian rhythm to the Antarctic summer: A question of zeit-geber strength. *Aviation, Space, and Environmental Medicine, 62,* 1019–1025.

Gander, P. H., Myrhe, G., Graeber, R. C., Anderson, H. T., & Lauber, J. K. (1989). Adjustment of sleep and the circadian temperature rhythm after flights across nine time zones. *Aviation, Space, and Environmental Medicine, 60,* 733–743.

General Accounting Office. (1989). *Aviation safety: Information on FAA's Age 60 Rule for pilots* (GAO-RCED-90-45FS). Washington, DC: Author.

Graebner, H. (1947). Auditory deterioration in airline pilots. *Journal of Aviation Medicine, 18*(1), 39–47.

Green, P., Lin, B., & Bagian, T. (1993). *Driver workload as a function of road geometry: A pilot experiment.* (Report UMTRI-93-39), Ann Arbor, MI: The University of Michigan Transportation Research Institute.

Guide, P. C., & Gibson, R. S. (1991). An analytical study of the effects of age and experience on flight safety. *Proceedings of the Human Factors Society 35th Annual Meeting* (Vol. 1, pp. 180–184). Santa Monica, CA: Human Factor Society.

Guohua, L., Grabowski, J. G., Baker, S. P., & Rebok, G. W. (2006). Pilot error in air carrier accidents: Does age matter? *Aviation, Space, and Environmental Medicine, 77*(7), 737–741.

Guohua, L., Baker, S. P., Qiang, Y., Rebok, G. W., & McCarthy, M. L. (2007, May). Alcohol violations and aviation accidents: Findings from the U.S. mandatory alcohol testing program. *Aviation Space and Environmental Medicine, 78*(5): 510–513.

Hancock, P. A., & Meshkati, N. (Eds.) (1988). *Human mental workload.* Amsterdam, the Netherlands: North-Holland.

Hanks, T. H. (1970, February). *Analysis of human performance capabilities as a function of exposure to carbon monoxide.* Paper presented at Conference on the Biological Effects of Carbon Monoxide, New York Academy of Sciences. New York.

Hecht, S., & Mendlebaum, J. (1940). Dark adaptation and experimental human vitamin A deficiency. *Journal of General Physiology, 130,* 651–664.

Helmreich, R. L., Butler, R. E., Taggart, W. R., & Wilhelm, J. A. (1994). *The NASA/University of Texas/FAA Line/LOS Checklist: A behavioural marker-based checklist for CRM skills assessment* (NASA/UT/FAA Technical Report 94–02. Revised 12/8/95). Austin, TX: The University of Texas.

Helmreich, R. L., Merritt, A. C., & Wilhelm, J. A. (1999). The evolution of crew resource management training in commercial aviation. University of Texas at Austin Human Factors Research Project: 235. *International Journal of Aviation Psychology, 9*(1), 19–32.

Helmreich, R. L., & Wilhelm, J. A. (1991). Outcomes of crew resource management training. *International Journal of Aviation Psychology, 1*(4), 287–300.

Henry, P. H., Davis, T. Q., Engelken, E. J., Triebwasser, J. A., & Lancaster, M. C. (1974). Alcohol-induced performance decrements assessed by two Link Trainer tasks using experienced pilots. *Aerospace Medicine, 45,* 1180–1189.

Hilton Systems, Inc. (1994). *Ago 60 rule research, Part I: Bibliographic database* (Report No. DOT/FAA/AM-94/20). Oklahoma City, OK: Civil Aeromedical Institute, FAA.

Hitchcock, L., & Morway, D. A. (1968). *A dynamic simulation of the sweptwing transport aircraft in severe turbulence* (Tech. Rep. No. NADC-MR-6807, FAA Report No. FAA-DS-68-12). Warminster, PA: Naval Air Development Center.

Hitchcock, L., Morway, D. A., & Nelson, J. (1966). *The effect of positive acceleration on a standard measure of intelligence.* Unpublished study, Aerospace Medical Acceleration Laboratory, Naval Air Development Center, Warminster, PA.

Hornick, R. J., & Lefritz, N. M. (1966). A study and review of human response to prolonged random vibration. *Human Factors, 8*(6), 481–492.

Hultsch, D. F., Hertzog, C., & Dixon, R. A. (1990). Ability correlates of memory performance in adulthood and aging. *Psychology and Aging, 5,* 356–358.

Hunt, E., & Hertzog, C. (1981). *Age related changes in cognition during the working years (final report).* Department of Psychology, University of Washington, Seattle.

Hyland, D. T., Kay, E. J., & Deimler, J. D. (1994). *Age 60 study, Part IV: Experimental evaluation of pilot performance* (Office of Aviation Medicine Report No. DOT/FAA/AM-94/23). Oklahoma City, OK: Civil Aeromedical Institute, FAA.

Hyland, D. T., Kay, E. J., Deimler, J. D., & Gurman, E. B. (1994). *Age 60 study, Part II: Airline pilot age and performance—A review of the scientific literature* (Office of Aviation Medicine Report No. DOT/FAA/AM-94/21). Washington, DC: FAA.

International Standards Organization. (1985). Guide for the evaluation of human exposure to whole body vibration (ISO 2631). Geneva: Author.

Johnson, N. R., & Rantanen, E. M. (2005). Objective pilot performance measurement: A literature review and taxonomy of metric. In *The 13th International Symposium on Aviation Psychology.* Dayton, OH.

Jorna, P. G. A. (1992). Spectral analysis of heart rate and psychological state: A review of its validity as a workload index. *Biological Psychologie, 34,* 1043–1054.

Kay, E. J., Hillman, D. J., Hyland, D. T., Voros, R. S. Harris, R. M., & Deimler, J. D. (1994). *Age 60 study, Part III: Consolidated data experiments final report* (Office of Aviation Medicine, Report No. DOT/FAA/AM-94/22). Washington, DC: FAA.

Keys, A. (1946). Nutrition and capacity for work. *Occupational Medicine, 2*(6), 536–545.

Levine, J. M., Kramer, G. G., & Levine, E. N. (1975). Effects of alcohol on human performance: An integration of research findings based on an abilities classification. *Journal of applied Psychology, 60,* 285–293.

Lyman, E. G., & Orlady, H. W. (1981). *Fatigue and associated performance decrements in air transport operations.* National Aeronautics and Space Administration: NASA CR 166167.

McFarland, R. A., & Barach, A. L. (1936). The relationship between alcoholic intoxication and anoxemia. *American Journal of Medical Science, 192*(2), 186–198.

McFarland, R. A., & Forbes, W. H. (1936). The metabolism of alcohol in man at high altitudes. *Human Biology, 8*(3), 387–398.

McFarland, R. A., Graybiel, A., Liljencranz, E., & Tuttle, A. D. (1939). An analysis of the physiological characteristics of two hundred civil airline pilots. *Journal of Aviation Medicine, 10*(4), 160–210.

McFarland, R. A., Roughton, F. J. W., Halperin, M. H., & Niven, J. I. (1944). The effect of carbon monoxide and altitude on visual thresholds. *Journal of Aviation Medicine, 15,* 382–394.

Mollard R., Coblentz A., Cabon P., & Bougrine S. (1995, Octobre). Vols long-courriers. Sommeil et vigilance des équipages. Guide de recommandations. Volume I: *Fiches de recommandations.* Volume II: *Synthèse des connaissances de base* (202 p.). Paris: DGAC ed.

Morrow, D. G., Ridolfo, H. E., Menard, W. E., Sanborn, A., & Stine-Morrow, E. A. L. (2003). Environmental support promotes expertise-based mitigation of age differences in pilot communication tasks. *Psychology and Aging, 18,* 268–284.

Moskowitz, H., & Sharma, S. (1974). Effects of alcohol on peripheral vision as a function of attention. *Human Factors, 16,* 174–180.

National Aeronautics and Space Administration. (1973). *Bioastronautics data book.* Washington, DC: Scientific and Technical Information Office, NASA.

Newman, H. W. (1949). The effect of altitude on alcohol tolerance. *Quality Journal of Studies on Alcohol, 10*(3), 398–404.

O'Donnel, R. D. (1976). Secondary task assessment of cognitive workload in alternative cockpit evaluation. In B. O. Hartman (Ed.), Higher mental functioning in operational environments, *AGARD Conference Proceeding Number 181* (pp. C10/1–C10/5). Neuilly sur Seine, France: Advisory Group for aerospace Research and Development.

Papa, R. M. & Stoliker, J. R. (1988). *Pilot workload assessment: A flight test approach.* Washington, DC: American Institute of Aeronautics and Astronautics, 88–2105.

Pass, F. G. W. C., Van Merriënboe, J. G. J., & Adam, J. J. (1994). Measurement of cognitive load instructional research. *Perceptual and Motor Skills, 79*, 419–430.

Putz, V. R. (1979). The effects of carbon monoxide on dual-task performance. *Human Factors, 21*, 13–24.

Rantanen, E. M., & Talleur D. A. (2001). Measurement of pilot performance during instrument flight using flight data recorders. *International Journal of Aviation Research and Development, 1*(2), 89–102.

Ross, L. E., & Mundt, J. C. (1988). Multiattribute modeling analysis of the effects of low blood alcohol level on pilot performance. *Human Factors, 30*(3), 293–304.

Ross, L. E., & Ross, S. M. (1995). Alcohol and aviation safety. In R. R. Watson (Ed.), *Drug and alcohol abuse reviews* (Vol. 7: Alcohol, cocaine, and accidents). Totowa, NJ: Humana.

Ross, L. E., & Ross, S.M. (1992). Professional pilots' evaluation of the extent, causes, and reduction of alcohol use in aviation. *Aviation and Space Environment Medicine, 63*, 805–808.

Ross, L. E., Yeazel, L. M., & Chau, A. W. (1992). Pilot performance with blood alcohol concentration below 0.04%. *Aviation Space and Environmental Medicine, 63*, 951–956.

Schulte, J. H. (1963). Effects of mild carbon monoxide intoxication. *AMA Archives of Environmental Medicine, 7*, 524.

Sheard, C. (1946). The effect of smoking on the dark adaptation of rods and cones. *Federation Proceedings, 5*(1–2), 94.

Society of Automotive Engineers. (1980). Measurement of whole body vibration of the seated operator of off-highway work machines (Recommended Practice J1013). Detroit, MI: Author.

Stewart, R. D., Peterson, J. E., Baretta, E. D., Blanchard, R. T., Hasko, M. J., & Herrmann, A. A. (1970). *AMA Archives of Environmental Medicine, 21*, 154.

Stoll, A. M. (1956). Human tolerance to positive G as determined by the physiological endpoints. *Journal of Aviation Medicine, 27*, 356–359.

Taylor, H. L., Dellinger, J. A., Schillinger, R. F., & Richardson, B. C. (1983). Pilot performance measurement methodology for determining the effects of alcohol and other toxic substances. *Proceedings of the Human Factors Society 27th Annual Meeting* (Vol. 1, pp. 334–338). *Proceedings of the Human Factors Society 35th Annual Meeting.*

Thomas, M. J. W., Petrilli, R. M., Lamond, N., Dawson, D., & Roach, G. D. (2006). Australian long haul fatigue study. *Proceedings of the 59th annual IASS. Enhancing Safety Worldwide.* Paris, France.

Trouton, D., & Eysenck, H. J. (1960). The effects of drugs on behavior. In H. J. Eysenck (Ed.), *Handbook of abnormal psychology.* London: Pitman Medical.

Tsang, P. S. (1992). A reappraisal of aging and pilot performance. *International Journal of Aviation Psychology, 2*(3), 193–212.

Van Ouerkerk, R., Meijman, T. F., & Mulder, G. (1994). *Arbeidspsychologie taakanalyse. Het onderzoek van cognitieveen emotionele aspecten van arbeidstaken* (*Workpsychological task analysis. Research on cognitive and emotional aspects of tasks of labour*). Utrecht, the Netherlands: Lemma.

Vicente, K. J., Thorton, D. C., & Moray, N. (1987). Spectral analysis of sinus arrhythmia: A measure of mental effort. *Human Factors, 29*, 171–182.

Waard, D. (1996). *The Measurement of Drivers' Mental Workload.* The traffic Research Center VSC, University of Groningen, p. 125.

15

Controls, Displays, and Crew Station Design

Kristen K. Liggett
U.S. Air Force Research Laboratory

15.1 Introduction

Aircraft control and display (C/D) technologies have changed dramatically over the past 30 years. The advent of compact, high power, rugged digital devices has allowed the onboard, real-time processing of data electronically. The digital impact has allowed a major shift from electromechanical to electro-optical devices and has also had a far-reaching effect on the way in which C/D research is being conducted. Since electro-optical C/Ds are computer controlled, and, therefore, multifunctional, there has been a shift away from experiments concerned with the optimal arrangement of physical instruments within the crew stations, and an added emphasis has been placed on the packaging of the information that appears on the display surface. The reason for this shift is that multifunction displays can show many formats on the same display surface and portray the same piece of information in many different ways. Also, with the advent of such technologies as touch-sensitive overlays and eye control, the same physical devices serve both as control and display, blurring the previously held careful distinction between the two. Section 15.1.1 discusses the history of crew station technology from the mechanical era through the electro-optical era. Subsequent sections will discuss the applications and impact of the new technology on the military environment.

15.1.1 Transition of Crew Stations with Time and Technology

The history of crew station technology is divided into a number of different eras. For this chapter, we chose three mechanization eras—mechanical, electromechanical (E-M), and electro-optical (E-O)—because they have a meaningful relationship with instrument design changes. Although we can, and will, discuss these as separate periods, the time boundaries are very vague, even though design boundaries are clear (Nicklas, 1958). Mechanical instruments, of course, were used first. Nevertheless, the use of E-M instruments can be traced to the very early days of flight, around 1920. E-O instruments were investigated in the 1930s. For example, in 1937, a cathode ray tube (CRT)-based E-O display called the

Sperry Flightray was evaluated on a United Air Lines "Flight Research Boeing" (Bassett & Lyman, 1940). The fact that all operators, private, commercial, and military, have flown with instruments incorporating all three designs also makes the era's boundaries fuzzy.

For the purpose of this section, we shall consider the mechanical era as that time from the beginning of flight until the introduction of the Integrated Instrument System by the Air Force in the late 1950s (Klass, 1956). The E-M era extends from that point until the introduction of the U.S. Navy's F-18 aircraft, which makes extensive use of multipurpose CRT displays. The issues of the E-O era, and beyond, comprise the primary subject matter of this chapter.

15.1.1.1 The Mechanical Era

The importance of instrumenting the information needed to fly an airplane was recognized by the Wright brothers very early in their flying adventures. The limitations of measuring airspeed by the force of the wind on one's face were not very subtle. From the time these famous brothers first installed an anemometer, a mechanical device used to measure wind velocity, and a weather vane to measure the angle of incidence, aviators and designers have been concerned about crew station instrument issues such as weight, size, shape, accuracy, reliability, and environmental effects (Nicklas, 1958). As aviators gained more flying experience, they recognized the need for additional pieces of information in the crew station, which in turn meant that there was a need for some kind of instrument. It did not take many engine failures before the need for data that would warn of an impending failure became obvious.

The requirement for displaying most pieces of information in a crew station can be traced to the need to identify or solve a problem. So the research process during most of the mechanical era was to invent a device or improvise from something that already existed in the nonaviation world. Any testing was generally done in flight. Simulators, as we have come to know them over the past 35 years, were virtually nonexistent during the mechanical era. The first simulators were modified or upgraded flight trainers, and were not generally regarded as an adequate substitute for flight trials. During this era, it was not unusual for a potential solution to progress from conception to a flight trial in a matter of weeks as opposed to the years it currently takes.

It would certainly be wrong to leave one with the impression that the mechanical era was one of only simple-minded evolutionary changes in the crew station. On the contrary, the history of instrument flying, even as we know it today, can be traced back to the early flying days of Lt. James Doolittle of the Army Air Corps (Glines, 1989). In 1922, he performed the first crossing of the United States accomplished in less than 24 hours. Hampered by darkness and considerable weather, he claimed that the trip would have been impossible without the "blessed bank and turn indicator," an instrument invented in 1917 by Elmer Sperry. In his Gardner Lecture, Doolittle claimed that it was the "blind flying" pioneering exploits of a number of other aviators that provided the "fortitude, persistence, and brains" behind the blind flying experiments of the 1920s and early 1930s (Doolittle, 1961). In 1929, Doolittle accomplished the first flight that was performed entirely on instruments. It was obvious to these pioneers that instrument flying, as we know it today, was going to become a pacing factor in the future of all aviation.

Although many milestones in the development of instrument flying technology took place in the mechanical era, technology had advanced sufficiently by 1950 to begin to shift the emphasis from mechanical instruments to instruments powered by electricity.

15.1.1.2 The Electromechanical Era

As mentioned earlier, this era began when the United States Air Force (USAF) introduced the Integrated Instrument System, often simplistically referred to as the "T-line" concept, for high performance jet aircraft. This was the first time that the USAF had formed an internal team of engineers, pilots, and human factors specialists to produce a complete instrument panel. The result was a revolutionary change in how flight parameters were displayed to pilots. These changes were necessitated because aircraft were flying faster and weapons systems were becoming more complex. This complexity reduced the time available for the pilot to perform an instrument cross check, and the fact that each parameter was displayed on

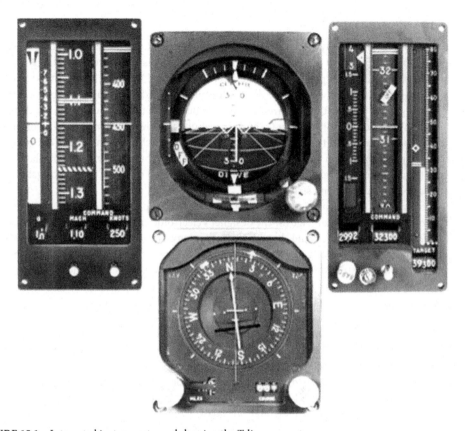

FIGURE 15.1   Integrated instrument panel showing the T-line concept.

a dedicated 3–4 in. round dial compounded the problem. The solution was to display all air data, i.e., angle of attack, Mach, airspeed, altitude, and rate of climb, on vertical moving tapes that were read on a fixed horizontal lubber line that extended continuously across all of the tape displays and the Attitude Director Indicator (ADI). In addition, lateral navigation information was read on a vertical reference line that traversed through the center of the ADI and the Horizontal Situation Indicator (HSI). The two reference lines thus formed the "T" (Figure 15.1). Manually selectable command markers were added to the tape displays to provide easily noticeable deviations from a desired position.

Again, flight trials provided the "proof of the pudding" and were critical to the design and development process. Ideas were incorporated, flown, changed, flown, changed again, and flown, until all of the design team members were satisfied. Seemingly simple questions, such as which direction the individual tapes should move, and how they should move in relation to each other, were answered through many flying hours. In the end, a system emerged that was easier to read and cross check than the old mechanical round dials. Though the displays were simpler, the electromechanization was orders of magnitude more complex. The servomechanisms, with their tremendously complex mechanical gearing, were a watchmaker's nightmare but, even so, the data was processed in a relatively simple fashion within the constraints imposed by analogue processing of electrical signals and mechanical gear mechanisms. The concept, although mechanically complex, has stood the test of time and can be seen on many of this era's aircraft. However, the pure economics of maintaining this type of instrumentation fueled the transition to solid-state displays. For example, both the new Airbus A-380 on the commercial side and the F-35 Joint Strike Fighter (JSF) on the military side have multifunction displays that cover the vast majority of the front instrument panel. A major reason for this trend is the increasing cost to maintain and support E-M instruments (Galatowitsch, 1993).

15.1.1.3 The Electro-Optical Era

The advent of the F-18 is generally regarded as a watershed in cockpit display design, and can be considered as the beginning of the E-O era. The crew station displays of this era are composed largely of CRTs presenting data that is digitally processed by the aircraft's onboard systems. An unintended but very real impact of this digital processing was the design flexibility of the displays, and the ability to vary the display according to the information required by the user. Because of this characteristic, the displays are generally known as multifunction displays (MFDs). The ability to show large amounts of information on a limited display surface shifted the emphasis of crew station research from packaging of instrumentation to packaging of information. Specifically, the concern was how best to format the displays and how to structure the control menus so that the user did not drown in an overflow of data, or get lost in the bowels of a very complicated menu structure.

The F-18 cockpit truly broke new ground, but its introduction represented only the tip of a technological iceberg in terms of the challenge for the designer's electronic crew stations. While the MFD gave a degree of freedom over what it could display, the technology of the CRT (size, power consumption, and weight) still posed some serious limitations on the positioning of the display unit itself. Since then, there has been a continual struggle to reduce the bulk of the display devices while increasing the display surface area. The goal is to provide the operator with a display that covers all of the available viewing area with one contiguous, controllable display surface. This would enable the ultimate in "designability," but are we in a position to adapt to this amount of freedom?

The problem given to the crew station designer by the MFD is—"how does one show the air crew the massive amount of data now available without their becoming swamped?" The answer is to present only that information required for the current phase of a mission and to configure the format of the display accordingly, which in turn requires the ability for displays to be changed, or controlled, during the course of a mission. Initially, this change was performed by the operator who decided what display was needed to suit the particular phase of flight. Unfortunately, extensive operator involvement was counter productive in terms of reducing operator workload. The response to this problem is to develop continually more sophisticated decision aids to predict the requirements of the user and then display recommendations (Reising, Emerson, & Munns 1993). This subject will be addressed later in this chapter.

The current generation of display devices is typically 6″ × 8″ or 8″ × 8″, although the F-35 JSF will employ two 8″ × 10″ displays. This is a halfway house to our ultimate goal, but already we are confronting some of the problems associated with freedom of design. There is a continual struggle between the mission planners who wish to use the now flexible displays for the portrayal of tactical, mission-oriented data and those designers concerned with the safe operation of the aircraft from a fundamental instrument flying point of view. The latter see the real estate previously dedicated to primary flight instrumentation now being shared with, or usurped by, "secondary" displays. There are still many questions to be answered concerning the successful integration of the various display types. It is essential that the operator maintains situational awareness both from a battle management perspective and from a basic flight control standpoint.

Further freedom is offered by the advent of Uninhabited Aerial Vehicles (UAVs) in that the operator need no longer be positioned in the aircraft. [The term "uninhabited" was chosen deliberately; the authors think it is more accurate than the term "unmanned," which implies only a male crewmember would be the operator.]

Systems onboard the UAV are capable of taking real world images, which in turn can be combined with a variety of information from various sources. The entire information package can then be displayed to the operator at the ground station. In addition, in many UAVs the operator does not fly the vehicle, but rather uses supervisory control to watch the vehicle's activities and intervene if necessary. Indeed, if we can supply the operator with an enhanced, or even virtual, view of the world, and the operator is not flying the vehicle, do we need instruments in the conventional sense?

It is clear that there are a great many paradigms to be broken. To a large extent, we have followed the design precedents set when displays were constrained by mechanical limitations. This will change as a greater body of research is developed to indicate the way in which the human will respond to the E-O technology. Indeed, in the same way that the advent of faster aircraft forced the display designer's hand at the start of the E-M era, it could well be the introduction of the new generation of high agility fighters, capable of sustained angles of attack in excess of 70°, which will force the full exploitation of electronic media. Time will also see the growth of a population of operators not steeped in the traditional designs, thus allowing a more flexible approach and less of a penalty in terms of retraining.

As always, the role of the designer is to provide the operator the information needed, in the most intuitive and efficient manner. The difference now is that the interface can be designed to meet the requirements of the human, without the human having to be redesigned to meet the engineering constraints of the system to be controlled.

15.1.2 Displays and Controls

As the E-O era unfolds, flat panel display technologies (anything from current thin-film-transistor active matrix liquid crystal displays [TFT AMLCD] to futuristic active matrix organic light emitting diode displays [AMOLED]) dominates the visual display market because of their reliability, lighter weight, smaller volume, and lower power consumption, as compared to CRTs and E-M displays (Desjardins & Hopper, 2002). Coupled with advances in visual displays is growth in alternative display and control technologies, such as three-dimensional (3-D) audio displays, tactile displays, touch control, and voice control. These display and control technologies have the potential of providing a substantial increase in the operator's efficiency. Translating that potential into actuality is, however, another matter and is a challenge for display and control designers. This section is comprised of descriptions of current and future C/D technologies, as well as examples of research studies, which address a major issue in the crew station design world, that is, how the operator might take advantage of the unique opportunities offered by these new technologies. All of the controls and displays discussed in the subsequent part of this section can be used by a number of different types of operators, such as pilots, soldiers, and UAV operators. Specific examples in this section focus on pilot applications, but the issues apply to the whole host of potential users of this technology.

15.1.2.1 Current and Future Displays

Although the majority of visual displays in the crew station are head down, there are more and more aircraft hosting head-up displays (HUDs) and helmet-mounted displays (HMDs). For instance, HUDs are found in most fighter aircraft and are making their way into transport aircraft as well. Additionally, HMDs, most popular to date in helicopters, are finding there way into legacy fighter aircraft and will provide the primary flight reference in the F-35 JSF. Also resident in the F-35 JSF is a 3-D audio display for presenting targeting information. These, as well as other controls and displays, which are not yet planned for operational employment, will be discussed.

15.1.2.1.1 Head-Up Displays

A HUD is "a virtual-image display in which the symbology typically appears to be located at some distance beyond the cockpit" (Weintraub & Ensing, 1992, p. 1). Basically, it is a piece of glass on which symbols are projected. The glass is positioned such that the operator has to look through it when looking straight ahead. The advantage to this type of display is in its name—it allows pilots to receive information on a display that keeps their head up during operations. The HUD evolved out of a need for a display that referenced the outside world and could be used for weapon-aiming purposes. At first, this consisted of a simple reticule, but it quickly developed into a more sophisticated projection device through which the user could correlate the position or vector of the airframe or weapon with the outside world.

Although the HUD started its evolution with a very limited function, it did not take long for the community to realize that a great deal of information could also be displayed to aid with the basic control of the aircraft. This brought on a new challenge.

15.1.2.1.1.1 Military Standardization As display designers become increasingly confronted by the advent of new C/D technologies, display designs have become abundant. Every airframer has its own version of tactics displays, situational awareness displays, map displays, and HUD symbology. On the one hand, the copious formats allow for creativity and invention of new ways to display important information; on the other hand, pilots are unable to transfer training from one aircraft to the next. Each new crew station poses new display formats for the pilot to learn and become proficient with in a short period of time. Because of this dilemma, there has been an emphasis on standardizing certain display formats—especially the HUD format, because it is capable of being used as a primary flight instrument. The standardization of the HUD symbology will allow pilots to maintain familiarity with the symbology regardless of the aircraft they fly.

As the HUD matured over the years, data was added to the HUD in a piecemeal fashion without any central coordination or philosophy. This haphazard growth resulted in a great deal of diversity in the design. In 1991, the USAF started a program to develop and test baseline formats for its electronic displays. The first phase of work led to a published design of HUD symbology for fighter-type aircraft (Mil-Std 1787B) (U.S. Department of Defense, 1996). Mil-Std 1787 Version C begins to address standard formats for HMD use; Version D includes rotary wing displays. The aim is to define tested designs for all electronic media in USAF aircraft to form the basis for any future development work.

15.1.2.1.1.2 Transport Aircraft HUDs Although developed originally for use in fighter aircraft, HUDs have recently been incorporated into transport aircraft, both military and civilian. In the civilian transport arena, the primary reason for including a HUD was to enable takeoffs and landings in low-visibility conditions. Alaska Airlines lead the way with the incorporation of HUDs into their 727s. "With the HUDs, Alaska can go down to Cat IIIa landing minima on a Cat II ILS beam" (Adams, 1993, p. 27). Now, Southwest has HUDs in all of their fleet, Delta has HUDs in their 767s, and a number of other airlines are following suit (i.e., Crossair, easyJet, Horizon, United Postal Systems, etc.) (Wiley, 1998).

As far as military transports are concerned, the C-17 is the only current transport that employs a HUD, but plans for the C-130J aircraft modernization program include incorporating a HUD (Rockwell Collins, 2003). The primary use of a HUD in these aircraft is to aid in visual approaches to austere fields that possess little or no landing guidance. An additional use is to aid the pilot in low-altitude parachute extraction maneuvers that require steep angles of descent.

15.1.2.1.2 Helmet-Mounted Displays

The advantage of a HUD is that it does not require users to bring their eyes into the cockpit to obtain pertinent information. It also provides information correlated with the real world. However, one of the limitations of the HUD is its limited field of view (FOV). Pilots can benefit from the HUD's information only when they are looking through the glass. Because of this limitation, there has been a push for the incorporation of HMDs, so pilots can constantly benefit from information superimposed on the real world—regardless of where they are looking. The HUD's FOV is typically 30° horizontal. It is thus not possible for information (or weapon-aiming reticules) to be presented to the operator outside this limited FOV. Clearly, the FOV limitations of a conventional HUD are raised to a new level of significance where the aircraft is capable of moving sideways and even in reverse (as in the case of the AV-8B Harrier)!

Helmet- or head-mounted displays, which project onto the visor or onto a combining glass attached to the helmet, have been developed to overcome this problem. By using miniature display technology to produce a display for each eye, combined with accurate head, and in some cases, eye-pupil tracking, it is theoretically possible to present a stereoscopic, full color image to the user in any direction (Adam, 1994). This could be anything from a simple overlay of information on the outside scene to a totally artificial virtual image.

15.1.2.1.2.1 HMD Issues Two of the challenges still facing HMD manufacturers are the image source used to produce and project the symbology, and head-tracking fidelity. Head tracking is important because different informations can be displayed based on where the pilot is looking. For instance, when a pilot is looking straight ahead, primary flight information is important. However, when a pilot is looking for targets, different symbology is needed to enhance performance on this task. Certainly, some attitude information may be present when the pilot is not looking straight ahead (referred to as off-boresight), but most of the FOV of the HMD would be displaying targeting information. Typically, the pilot is not looking forward during these times, and the use of a head tracker can change the symbology presented to the pilot based on the head position. This brings up two important issues— latency and accuracy. Certainly, if the change in symbology lags the head movement, disorientation can occur. Also, in the targeting case, the information about the target must be accurate. The accuracy must be at least equivalent to that of a HUD. Both of these issues will drive pilot acceptability of this new technology.

As mentioned earlier, flat panel display technology is dominating in the head-down display arena, and the same is true for HMD image sources. Traditional HMDs use an image source to project a picture onto a piece of glass that resides in front of the user's eye(s). Like a HUD, pilots look through the glass to obtain information while simultaneously viewing the real world. However, there is a new technology that eliminates the need for the glass or visor presentation. A retinal-scanning display (RSD) is a head- or helmet-mounted display that uses a scanning beam that actually "paints" or projects images directly on the retina of the eye. Although this may sound a bit risky, these systems meet safety rules set by the American National Standards Association and the International Electrotechnical Committee (Lake, 2001).

The advantages of this type of display are that it provides head-up information and hands-free control in full color with daylight readability in a variety of ambient settings. The RSD is based on open standards, so it can receive television signals and graphics formats, which can be displayed on an 800 pixel wide by 600 pixel high image. With the advent of wearable computers, this type of HMD is not only suited for military applications (such as for cockpits, command and control centers, soldiers, etc.), but is finding uses in a variety of commercial applications including firefighters viewing floor plans during a rescue, technicians viewing manuals during a repair, drivers viewing moving maps during a trip, or surgeons viewing patient's vital statistics during surgery.

15.1.2.1.2.2 Military HMDs The first military group to embrace HMD technology was the rotary-wing community. When the idea of using an HMD to aim the turret-mounted gun on the UH-60s caught on, helicopters that were previously tasked simply with airborne transport were suddenly employed as attack helicopters. The AH-64 Apaches were the first helicopters to integrate an HMD (developed in1976 by Honeywell), and these displays are still flown today (Williams, 2004). While the original HMD was a somewhat crude monocular display with a limited FOV, the Comanche HMD was, before the aircraft's cancellation, slated to have a binocular, large FOV (52° horizontal by 30° vertical), high resolution (1280 × 1024) full color display (Desjardins & Hopper, 2002).

On the fixed-wing side, the Joint Helmet-Mounted Cueing System (JHMCS) is a combination head tracker and HMD that is scheduled to be incorporated into the existing F-15s, F-16s, F/A-18s, and F-22s. Although the symbology set to be displayed on JHMCS for each aircraft is different, there is 95% commonality among the systems (Fortier-Lozancich, 2003). The advantage of JHMCS is that it provides a high off-boresight targeting tool that will provide the slaving of weapons and sensors to the pilot's head position. This allows for more effective air-to-air and air-to-ground missions. The hardware consists of a single monochrome CRT image source that projects symbology on the inside of the pilot's helmet visor.

Finally, the F-35 JSF will not have a HUD, but in fact, an HMD for its primary flight reference. The specifications for the F-35 are similar to the Comanche in that the image source (provided by Kopin) will provide a visor-projected wide FOV, high resolution binocular view containing primary flight information as well as critical-mission-, threat-, and safety-related information. This HMD system will also allow the steering of weapons and sensors (Adams, 2003).

15.1.2.1.3 Audio Displays

In addition to visual displays, audio displays are showing their value in increasing applications within the crew station environment. More recently, attention has shifted to localized audio (commonly referred to as 3-D audio), which are tones or cues presented at a fixed position in the external environment of the listener. This is accomplished with the use of localization systems that utilize digital signal-processing technologies to encode real-time directional information for presentation over headphones. Head tracking is used to position the tone relative to the listener's external environment regardless of his/her head position. The tone placement can vary in azimuth (left and right), elevation (up and down), and range (distance from the listener).

There are numerous applications of this technology in the crew station. The addition of localized audio to visual displays has been shown to significantly reduce the time required to search and detect targets as compared to visual-only times (with 50 distractors, target identification time averaged 15.8 s with visual only, compared to 1.5 s with visual plus localized audio) (Simpson, Bolia, McKinley, & Brungart, 2002). Also, localized audio cues have been shown to effectively redirect gaze (Perrott, Cisneros, McKinley, & D'Angelo, 1996), and have demonstrated an increase in communication intelligibility and a decrease in pilot workload when operating multiple channels for command and control tasks (Bolia, 2003).

15.1.2.1.4 Tactile Displays

Tactile displays are another up-and-coming display systems that show promise for portraying information to operators, especially those who are visually saturated. Tactile systems include anything from basic stick shakers, to vibrating wrist bands, to full vests which employ an array of tactors. The Navy's Tactile Situation Awareness System (TSAS), one of the most well-known tactile displays, is an example of the latter. TSAS incorporates a number of pneumatic and E-M tactors that vibrate in specific areas on the user's torso to convey various types of information (Institute for Human and Machine Cognition, 2000). In a fixed-wing aircraft application, TSAS can be used to present attitude information by using the various tactors to represent the horizon. For example, as the pilot maneuvers the aircraft, tactors vibrate to indicate where the horizon is with respect to the aircraft. If pilots perform a loop, the tactile sensation experienced would be vibrations that move up their back as the plane climbs, vibrations that are present on their shoulders when the plane is inverted, and then vibrations that come down the front of their vest as the loop continues. In a rotary-wing aircraft application, TSAS has been shown to improve hover capability by providing significantly increased total time on target (Raj, Kass, & Perry, 2000). TSAS has also been shown to be effective for a number of applications, including augmenting visual display information for high altitude, high-opening parachute operations in the air, and navigating on the ground for U.S. military Special Forces (Chiasson, McGrath, & Rupert, 2002). Along those same lines, researchers at TNO Human Factors Research Institute in the Netherlands have been investigating the use of a vibro-tactile vest for human–computer interactions and provide some guidelines for its incorporation into many interfaces (van Erp, 2002).

Wrist tactors are a simpler form of the tactile display. Basically, one vibro-tactor is incorporated into a wrist band to portray information in a variety of applications. These include enhanced situational awareness for altered-gravity environments (Traylor & Tan, 2002), alerting pilots of automation interventions (Sarter, 2000), and for operators detecting faults in a multitask environment (Calhoun, Draper, Ruff, & Fontejon, 2002).

15.1.2.1.5 Summary of Displays

The future holds much promise for the efficient display of information. Head-down visual displays, once the only way to convey important data, will be complemented and augmented with head-up, head- or helmet-mounted, and multisensory displays. The advantages of these head-up visual displays are obvious and the auditory and tactile displays can provide much needed attentional guidance in environments that are overtasking the visual channel. This trend is true in the aviation environment, as well as in other areas, such as medical applications, automobile applications, and virtual reality for entertainment.

15.1.2.2 Current and Future Controls

Control technology is also advancing beyond the common buttons and switches, which are standard in traditional crew stations. No longer are pilots required to "learn how to play the piccolo" to be proficient in executing the correct button sequences on the stick and throttle to control the aircraft and its displays. Some the technologies discussed in this section are ready to be incorporated today; others still need research and development before they are ready for operational employment.

15.1.2.2.1 Voice Control/Speech Recognition

Voice control has various applications in crew stations. The cognitive demands on military pilots will be extremely high because of the very dynamic environment within which they operate. The pilot has limited ability to effectively manage available onboard and offboard information sources using just hands and eyes. Because workload is high and the ability to maintain situation awareness is imperative for mission success, voice control is ideal for military crew station applications.

Speech recognition has long been advocated as a natural and intuitive method by which humans could potentially communicate with complex systems. Recent work in the area of robust speech recognition, in addition to advances in computational speed and signal processing techniques, has resulted in significant increases in recognition accuracy, spawning a renewed interest in the application of this technology. Just recently, speech recognition systems have advanced to the point where 98% accuracy in a laboratory environment is obtainable (Williamson, Barry, & Draper, 2004). This high accuracy is essential to acceptance of the technology by the user community.

15.1.2.2.2 Gesture-Based Control

There are a variety of sensing techniques (optical, magnetic, and ultrasonic) to read body movements directly (Sturman & Zeltzer, 1994). Since the operator's body and hands can be involved in other activities, gesture-based control may best involve detecting defined movements of the face or lips. In one implementation, a headset boom located in front of the speaker's lips contains an ultrasonic signal transmitter and receiver. A piezoelectric material and a 40 KHz oscillator were used to create a continuous wave ultrasonic signal (Jennings & Ruck, 1995). The transmitted signal was reflected off the speaker's mouth, creating a standing wave that changes with movements of the speaker's lips. The magnitude of the received signal was processed to produce a low-frequency output signal that can be analyzed to produce lip-motion templates.

In one candidate application of lip-motion measurement, lip movements were processed during speech inputs to provide "lip reading." An experiment using an ultrasonic lip-motion detector in a speaker-dependent, isolated word recognition task demonstrated that the combination of ultrasonic and acoustic recognizers enhanced speech recognition in noisy environments (Jennings & Ruck, 1995). An alternate application approach would be to translate symbolic lip gestures into commands that are used as control inputs.

15.1.2.2.3 Summary of Controls

Controls in future crew stations are likely to be multifunctional and designed to enable the operator to attend to primary tasks, while minimizing overall workload. In the case of aviation, this means control technologies that enable pilots to keep their hands on the stick and throttle and their heads up, out of the cockpit. Additionally, there will be more frequent use of multimodal (employing more than one sense) controls for a variety of reasons (Calhoun & McMillan, 1998; Hatfield, Jenkins,

Jennings, & Calhoun, 1996). First, mapping several control modalities to a single control action provides the operator with increased flexibility: (a) the operator may have individual preferences, (b) a temporary task or environmental condition may deem one controller more efficient than another, and (c) should one control device malfunction, the operator can use a different control. A multimodal approach is also useful when two or more controls are integrated such that they are used together to perform a task. Additionally, it will be likely that controls in the future will be adaptive depending on several potential triggers. This will be explained more in Section 15.2.2.3.1.

15.1.2.3 Controls and Display Research

This section will highlight some research that has been conducted on traditional and nontraditional controls and displays. The first study deals with the flexibility afforded to display designers with the advent of the E-O era. Not only have HUDs and HMDs become more prevalent, but head-down displays have become larger, providing an electronic blackboard upon which almost any display format can be drawn. For instance, the F-35 JSF will have two 8 × 10 in. projection displays, which can support various sized windows for displaying information. Because of their versatility, the head-down displays can be configured in nontraditional ways. Although the duplication of E-M instrumentation on E-O display formats is possible, the flexibility of E-O displays allows designers to explore new formats. The research described next gives an example of an investigation aimed at taking advantage of the digitally based displays.

15.1.2.3.1 Background Attitude Indicator

This study dealt with one of the basic aspects of flying—maintaining flight safety when there is no dedicated head-down primary attitude indicator. If one grants the premise that the more mission-related information the better, the logical conclusion is that all the glass displays in a modern cockpit should contain this type of information, with the baseline Mil-Std 1787 HUD (U.S. Department of Defense, 1996) or HMD used as the primary flight display. Because of this idea, the elimination of a dedicated head-down primary attitude indicator would free up head-down real estate for mission-related glass displays. Loss of attitude awareness (a potential flight safety problem) could result when the pilot is focusing his/her head down to do mission-related tasks.

This problem was investigated by researchers at Lockheed—Ft. Worth (Spengler, 1988) who created a background attitude indicator (BAI) using only a 3/4 in. "electronic border" around the outer edge of the display (Figure 15.2). The three displays on the front instrument panel presented mission-related information on the central rectangular portion of each, and presented, on the background border, a single attitude display format, which extended across all three displays. The attitude information, in essence, framed the mission-essential display information and acted as one large attitude indicator (Figure 15.3). The BAI consisted of a white horizon line with blue above it to represent positive pitch, and brown below

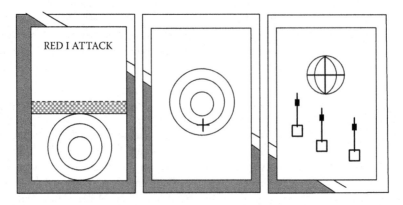

FIGURE 15.2 Spengler background attitude indicator.

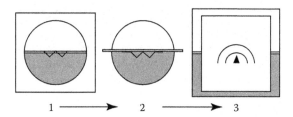

FIGURE 15.3 Evolution from attitude director indicator to background attitude indicator.

it to represent negative pitch. This display worked very well for detecting deviations in roll, but was less successful in showing deviations in pitch, because once the horizon line left the pilot's field of view, the only attitude information present in the BAI was solid blue (sky) or brown (ground). Because the concept was effective in showing roll deviations but lacked in the pitch axis, enhancing the pitch axis became the focus of work conducted at the Wright Laboratory's Cockpit Integration Division, Wright Patterson Air Force Base, Ohio, now known as the Human Effectiveness Directorate of the Air Force Research Laboratory.

The Lab's initial work began by enhancing the pitch cues for a BAI, which framed one display format only (as opposed to framing three display formats as in the original Lockheed work) (Liggett, Reising, & Hartsock, 1992). The Lab's BAI contained wing reference lines, digital readouts, and a ghost horizon (a dashed horizon line that appeared when the true horizon left the pilot's field of view, and that indicated the direction of the true horizon) (Figure 15.4). The BAIs also contained variations of color shading, color patterns, and pitch lines with numbers.

Experimental results revealed that the combination of color shading and color patterns (Figure 15.5) was the format that provided the pilot with the best performance when recovering from unusual attitudes. When using this format, the pilots moved the control stick to begin their successful recoveries more quickly than when using any other format. This measure of initial stick-input time relates to the interpretability of the format because the pilots looked at the format, determined their attitude via the cues on the BAI, and began their recovery as quickly as possible.

The design ideas from the initial Wright Lab study were transferred to framing three displays as in the original Lockheed work to provide the pilot with one large attitude indicator, which pilots highly favored. This display provided effective peripheral bank cues, as well as two types of pitch cues—the shaded patterns supplied qualitative cues while the pitch lines with numbers gave quantitative indications of both the degree of pitch and pitch rate information. Based on the results of these simulation studies, BAIs appear to be a viable means of enabling the pilot to recover from unusual attitudes.

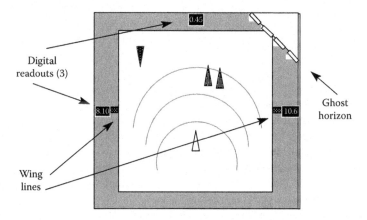

FIGURE 15.4 Wright laboratory's background attitude indicator.

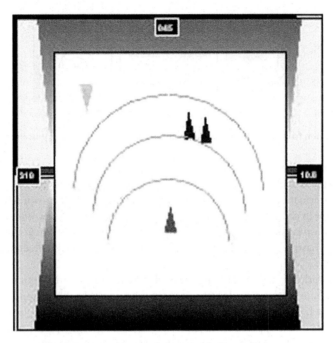

FIGURE 15.5   Background attitude indicator with color shading and patterns.

This research does indeed proclaim a paradigm shift from the old way of displaying attitude informa-tion head down on a dedicated piece of real estate for an ADI, to an innovative new way of displaying the same information. Another prime example of a paradigm shift is the use of 3-D stereo display formats. MFD displays with 3-D computer graphics have the potential of creating map formats that closely match the essential 3-D aspects of the real world. The next study deals with how pilots would control objects within a 3-D map.

15.1.2.3.2 Cursor Control within 3-D Display Formats

Mental models play an important role in the efficient operation of systems (Wickens, 1992). A mental model is the picture operators have in their heads of the way a system works. Since direct views of the inner workings of a system are often not possible (e.g., the flow of electrons inside the avionics system), displays are a major means of conveying the operation of a system. Given that the user's mental model is correct, the closer the display formats conform to the user's mental model, the more beneficial they are. In the airborne arena, the pilot is operating in a 3-D world; consequently, the more accurately a display can portray this 3-D aspect, the more accurately it can conform to the pilot's mental model.

A perspective view of terrain features for low-altitude missions should aid pilots, since this view should conform very well to their 3-D mental model of the world. Perspective map views, however, only contain monocular depth cues. Adding 3-D stereo cues can enhance agreement between a pilot's mental model and the actual display by making it more representative of the real world.

Given designers can create this 3-D perspective map, an obvious question is, "How does the operator manipulate a cursor in the 3-D map world?" Moving a cursor to mark items is one of the most important tasks involved in using map displays. The operator may be required to mark geographic features such as hill tops or river bends, as well as man-made features such as dams or bridges. The 3-D perspective view can be interpreted as *X*, *Y*, and *Z* coordinates. The problem now arises as to how to move a cursor to areas of interest in these displays.

The Lab's research in this area has focused on two types of continuous cursor controllers (a joystick and a hand tracker) and one discrete controller (a voice control system) to manipulate a cursor in 3-D space so as to designate targets on a map. The joystick and hand tracker had been used in previous 3-D research (Ware & Slipp, 1991), while voice control was chosen based on researchers' experience with it in the two-dimensional (2-D) arena.

Based on previous research in the cursor control area (Reising, Liggett, Rate, & Hartsock, 1992), it was determined that using aiding techniques with continuous controllers could enhance the pilot's performance when designating targets. This study investigated two types of aiding. Contact aiding provided participants with position feedback information via a color change in the target once the cursor came in contact with it (Figure 15.6). This aiding eliminated some of the precise positioning necessary when using a cursor to designate targets. Proximity aiding (Osga, 1991) used the Pythagorean theorem to calculate the distance between the cursor and all other targets on the screen. The target in closest proximity to the cursor was automatically selected; therefore, the requirement for precise positioning was completely eliminated.

The display formats consisted of a perspective-view map containing typical features, targets, and terrain. The targets could be presented in different depth volumes within the 3-D scene (Figure 15.7).

Participants designated targets significantly faster with proximity aiding (with the hand tracker or joystick) than when using either voice or contact aiding (with the hand tracker or joystick) (Figure 15.8). When using a continuous controller, there are two components to positioning: gross and precise movements. The addition of proximity aiding to both continuous controllers greatly reduced gross positioning and eliminated precise positioning. Contact aiding, on the other hand, did not affect gross positioning but decreased the amount of precise positioning.

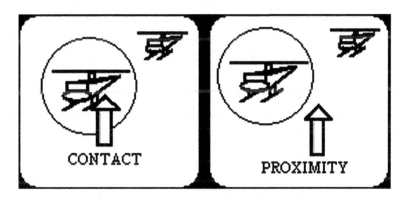

FIGURE 15.6   Types of aiding. Solid circle indicates selected target.

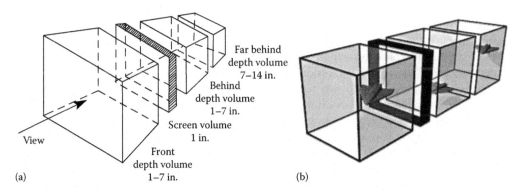

FIGURE 15.7   Depth volumes within the 3-D scene.

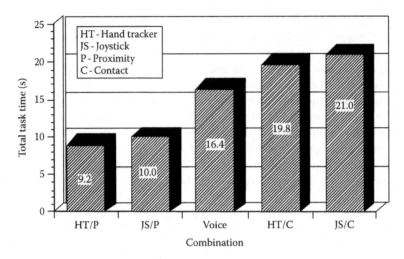

FIGURE 15.8   Effect of proximity and contact aiding on target-designation times.

Another interesting finding was that the voice control system performed significantly better than either of the continuous controllers with contact aiding. The reason for superior performance of the voice control system relates to the components of the positioning task. Both the continuous controllers with contact aiding had gross and fine positioning to deal with. The voice control system and the controllers with proximity aiding, however, eliminated the fine positioning factor to a large extent. Since the target was large enough to visually identify in all cases, the movement to the target was basically reduced to a gross-positioning task, and fine adjustment was eliminated. Because the results were positive, voice control was pursued in the Lab.

15.1.2.3.3 Voice Recognition Flight Test

The potential use of voice control as a natural, alternative method for the management of aircraft subsystems has been studied by both the Air Force and Navy for over 10 years, but because recognition accuracies had not attained acceptable levels for use in the cockpit, this technology has not yet become operational. Now that speech recognition performance is adequate and reliable, and has shown value as a cockpit control mechanism, it was an optimal time to verify that performance would not deteriorate in the operational flight environment due to high noise, acceleration, or vibration.

The objective of this experiment (Williamson, Barry, & Liggett, 1996) was to measure word recognition accuracy of the ITT Voice Recognizer Synthesizer (VRS)-1290 speech recognition system in an OV-10A test aircraft both on the ground and in 1G and 3G flight conditions. A secondary objective was the collection of a speech database that could be used to test other speech recognition systems.

Sixteen participants were involved in this study. All participants were tested in the laboratory, in the hangar sitting in the aircraft cockpit with no engines running, and in flight. During flight, participants experienced a 1G data-collection session (referred to as 1G1), followed by a 3G data-collection session, and then another 1G data-collection session (referred to as 1G2), to test for possible fatigue effects.

Participation was divided into two separate sessions. The first session consisted of generating the participants' templates in a laboratory setting and collecting some baseline performance data. Participants were briefed on the nature of the experiment and performed template enrollment. An identical system to the one in the aircraft was used as the ground-support system for template generation. The participants used the same helmet and boom-mounted microphone that was used in the aircraft. Template training involved the participants' speaking a number of sample utterances. Once the template generation was completed, a recognition test followed that consisted of reciting the utterances to collect baseline recognition data.

The first aircraft-test session was performed in the hangar to provide a baseline on the aircraft in quiet conditions. This consisted of each participant's speaking the 91 test utterances twice, for a total of 182 utterances. During both ground and airborne testing, participants needed little or no assistance from the pilot of the aircraft. The participants sat in the rear seat of the OV-10A and were prompted with a number of phrases to speak. All prompts appeared on a 5 × 7 in. monochromatic LCD in the instrument panel directly in front of the participants. Their only cockpit task was to reply to the prompts. Close coordination was required, however, between the pilot and participants while the 3G maneuvers were being performed since the pilot had to perform a specific maneuver in order to keep the aircraft in a 3G state.

Three comparisons of word recognition accuracy were of primary interest:

1. Ground (Lab + Hangar) versus air (1G1 + 3G + 1G2)
2. 1G (1G1 + 1G2) versus 3G
3. 1G1 versus 1G2

Orthogonal comparisons were done to make each of these comparisons. No significant differences were found for any of the comparisons (Figure 15.9).

Results showed that the ITT VRS-1290 system performed very well, achieving over 97% accuracy over all flight conditions. The concept of speech recognition in the fighter cockpit is very promising. Any technology that enables an operator to stay head-up and hands-on will greatly improve flight safety and situation awareness.

This flight test represented one of the most extensive in-flight evaluations of a speech recognition system ever performed. Over 5,100 utterances comprised of over 25,000 words or phrases were spoken by the 12 participants in flight (4 of the 16 participants' flight-test data was not useable). This combined with the two ground conditions resulted in a test of over 51,000 words and phrases. The audio database of Digital Audio Tape (DAT) recordings has been transferred onto CD-ROM and has been used to facilitate laboratory testing of other speech recognition systems. The DAT recordings have proven to be extremely valuable since many new voice recognition systems have been produced after this study was conducted. With this database, new systems can be tested against speech recorded in an extremely harsh environment (the participants' crew station was directly in line with the noisy engines) without requiring additional flight tests. The CD-ROM database has been made available for distribution to the speech recognition research community. Finally, the example study illustrates the importance of flight-testing controls and displays in the environment in which they will be used.

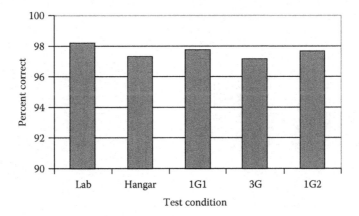

FIGURE 15.9 Mean word accuracy for each test condition.

15.2 Overall Thoughts on the Benefits of New Crew Station Technologies

New crew station technologies have the potential for enhancing the human–machine interface that is essential for effectively operating in a complex environment. The research discussed highlights the potential benefits of some of these new technologies in application-oriented studies. However, these technologies by themselves are no panacea; in fact, if not implemented in an intelligent manner, they could become a detriment to the operator. The designers still need to spend the majority of their time figuring out how the subcontrol modes, coupled with the myriad of possible formats, "play" together to present pilots with a clear picture of what the aircraft is doing and how to change its subsystems, if required. These new technologies are a two-edged sword—they offer the designers virtually unlimited freedom to present information to operators; on the other hand, these technologies also give designers the opportunity to swamp operators in data. The clever application of these C/D technologies will be the key to ensure that they help, rather than hinder operators.

The intelligent design of these controls and displays, and their integration into crew stations, can be facilitated by using a structured design process and taking advantage of the computer-aided design tools that complement the process. The next section will cover the design process and its supporting design tools.

15.2.1 Current Crew Station Design

The overall design process invoked in human–machine systems is well documented (Gagne, 1962). A paradigm specifically related to the crewstation design process for aircraft is shown in Figure 15.10. It consists of five steps: mission analysis, preliminary design, prototype-level evaluation, simulation

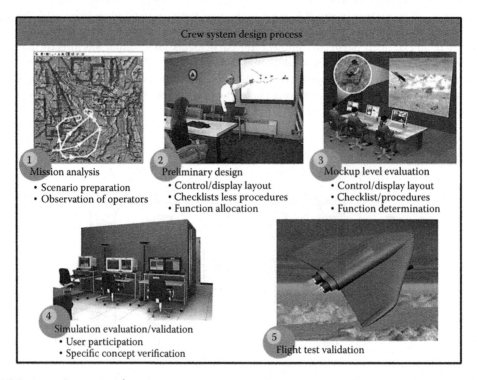

FIGURE 15.10   Crew system design process.

evaluation/validation, and flight-test validation (Kearns, 1982). The steps in the figure are numbered numerically to show the order in which they should be addressed. The order should be followed to ensure a good design. Before the process is described in detail, the design team, or players who participate in the design process, will be discussed.

15.2.1.1 The Team and the Process

15.2.1.1.1 The Design Team

To be successful, each step in the process needs strong user involvement. A multidisciplined design team is formed to follow the design from birth to implementation. Certain players take the lead during different steps of the process. The team should include, as a minimum, operators, design engineers, avionics specialists, human factors engineers, computer engineers, hardware specialists, and software specialists. Participation from each of the players throughout the process will allow for a more thorough design of the system. The ultimate goal of the design team is to get it "right the first time."

15.2.1.1.2 Mission Analysis

The first step, mission analysis, is often referred to as problem definition because it specifies a problem with the current system that needs to be solved, or it identifies deficiencies in the crewstation where a problem may occur without the incorporation of a new system. This step is initiated with a thorough examination of the intended operational use of the system to be designed. This examination is followed by a derivation and documentation of the total system and individual component requirements. The requirements document published by the future user of the system provides important baseline material for this step. Typically, the documentation produced during this step includes a mission profile describing a sequential listing of all the operations the system must perform in order to be effective in the flight environment. This profile is decomposed from a very generic state of aircraft operations to a very detailed state that includes all of the specific tasks performed by the aircraft, its systems, and each of the crew members during the mission profile (ORLOC, 1981). With modern crew stations becoming increasingly decision centered, the design team should also perform a cognitive task analysis to determine the decisions that have to be made by the crewmembers as the mission progresses. An essential output of this step is the identification of the information that the crew needs to perform its mission. The product of this phase is a specification of system requirements to include a set of alternatives for accomplishing these requirements. The alternatives must be defined in terms of their anticipated effects on human performance.

15.2.1.1.3 Preliminary Design

The second step in the crew station design process, as depicted in Figure 15.10, is preliminary design. This step is often referred to as "develop a solution." During this part of the process, most of the activity is devoted to generating a design. The requirements generated in the first step are reviewed, and decisions are made regarding how the functions necessary to complete the mission will be performed. The functions can be allocated to the operator, the computer, or a combination of both. Because modern aircraft have a great deal of automation, supervisory control has a high potential for becoming a key function of today's crew station operator. An example of current supervisory control involves the use of the flight management system that navigates the aircraft automatically through the airspace without direct pilot hands-on control. A series of trade studies are often performed to (1) determine who will do what, (2) determine applicable decision aids, and (3) establish the governing logic of these "smart systems." A further discussion of automation takes place in Section 15.2.2.3 of this chapter. The results of these trade studies will play a major role in the crew station design.

The crew station design will also be driven by the information requirements determined from step one. The intuitive presentation of information in the crew station will govern the success of the design. A key element in the evolving design is operator and user involvement. The sustained participation of operators with relevant experience results in fewer false starts, better insight in how and why the mission is performed, and a great savings in time, as well as money, in the latter steps of the process. By getting the operator involved from the beginning, the costly problem of making design changes further down the road is avoided.

The dividing line between problem definition and solution development is often vague. Specific designs may affect task sequencing during the mission profile. This change in sequencing can reveal workload problems within the crew station. Because of this overtasking, the operator may shed tasks, which in turn alter the mission profile. Once the profile has changed, the designs may affect the tasks in a different way, and thus, the cycle continues. The design process is indeed an iterative process.

15.2.1.1.4 *Prototype Evaluation, Simulation Evaluation/Validation, Flight Test*

The last three steps are interdependent and very critical to the successful completion of an effective and proven crew station design. These three steps all work synergistically to "prove the solution." Prototype evaluation marks the initial introduction of the implemented design concepts to the user. Although the users should be involved in the preliminary design step, the actual implementation into a prototype design will show the design in a whole new light. The design concepts are evaluated in a limited context, and suggestions are made by the user as to which designs should move forward to simulation. This step weeds out unfeasible design concepts. Human-in-the-loop simulation evaluation provides a more realistic and robust testing of the design concepts. In simulation evaluation, it is recommended that the new design concept be compared to an existing design in order to measure the "goodness" of the design concept. This step provides the final recommendation of a design concept for flight test.

Traditionally, this process involved human-in-the-loop simulations, or virtual simulation as they are referred to today. At present, constructive simulation, which involves the use of models in simulated environments, is becoming a required part of the evaluation process as a low-cost alternative to conducting trade studies. Modeling specific systems, such as structures, engines, sensors, etc., for use in constructive simulation has been very successful (Aviation Week and Space Technology, 2003). However, one of the current challenges is modeling human behavior. Certainly, to determine the benefits of different technologies in this step of the design process, the simulation must not only model the technology, but also how the operator interacts with it. The Combat Automation Requirements Testbed (CART) program is developing an architecture that allows human behavior/performance models to interface with various constructive simulation environments to determine the "goodness" of various cockpit designs and how the operator interfaces with them.

CART has been used to integrate such models successfully (Martin, Barbato, & Doyal, 2004). In one example, CART was used to model human tasks performed during an air-to-ground segment in a strike-fighter mission using a human performance model integrated with the Joint Integrated Mission Model aircraft model. Once the integrated model was run, results from the constructive simulation were compared with pilot performance from a virtual simulation in which real pilots performed the same tasks as the model. The human performance model was shown to predict the pilot performance with fairly high accuracy (correlation of 0.78 between the model-dependent measures and the pilot-dependent measures) (Brett et al., 2002). Once the human performance models are validated, using constructive simulation prior to virtual simulation can save time and money by providing a quick way of thoroughly testing design concepts and advancing only the most promising one(s) to virtual simulation studies.

Flight testing often involves only one design to be tested in operational use; however, in the case of the F-16, F-22, and the F-35 JSF, two prototypes were involved in a "fly-off." For the purpose of this discussion, these final steps are combined to provide "Solution Evaluation." Once again, there may not be a clear break between the solution evaluation and the solution definition step. It has been observed that most designers design, evaluate, redesign, etc., as they go. The transition from solution definition to solution evaluation occurs when formal, total-mission, total-system, human-in-the-loop evaluations begin. But even then, decisions made during the problem and solution definition steps are often revisited, changes made, and simulation sessions (or even flight tests) rescheduled—all resulting in, as previously suggested, a very iterative or cyclic process.

15.2.1.1.5 Traceability

As the process evolves, it is important that the design team maintain an accurate record of the changes that have taken place along the way, the decisions that were made that influenced the design, and the rationale behind their decisions. This information provides traceability of the design from requirements to final product. Traceability is important because the design process can take a long time, and it is helpful to know why things were done the way they were. The traceability document provides a record of past decisions, which may be reviewed periodically, so the design flows in an evolutionary manner, as opposed to a revolutionary manner, and thus, avoids regression. Also, the design of a new product can benefit from the traceability information of previous products, thus saving time and effort. This discipline of documenting the design is (or should be) a MUST feature of the design process, not a "nice to have" feature.

15.2.1.2 Facilitating the Crew Station Design Process with Computer Support

The above discussion of the crew station design process serves as a guideline for crew station designers. The process has been in existence for a long time and has been complimented over the years with a variety of computer-aided design tools. These new tools allow designers to visualize and modify their design ideas much easier than the traditional way of hand-drawing design concepts. There are various categories of tools that support this process, including physical/anthropometric tools, cognitive modeling tools, and overall system design tools. The goal of each of these will be discussed and some specific tools will be highlighted.

15.2.1.2.1 Physical/Anthropometric Tools

The purpose of these types of tools is to ensure that the crew station properly "fits" the operator. The common questions to be answered by these tools are (1) can the controls be reached by the operator's arms and legs, (2) can the visual displays be seen, and (3) do the knees fit under the instrument panel (especially in cockpits where ejection is an option). Jack is one such software package that addresses the first two issues. It includes a detailed human model capable of interacting in a 3-D environment to assess reach envelopes, strength, leg clearance, seat angles, eye and head position for visibility analyses, etc. (Engineering Animation, Inc., 2000). To address the third question, the Articulated Total Body model can be used to determine human body dynamics during hazardous events, e.g., ejection or crashes (Pellettiere, 2002). It predicts the motion and forces on the human body to determine the safety of restraint systems and ejection seats. ManneQuin is another anthropometric tool that features 3-D human figures for a number of populations, percentiles, and body types. These "humanoid" figures can interact with various systems, which are imported from graphics software packages (i.e., AutoCAD) for testing (NexGen Ergonomics, Inc., 2003).

15.2.1.2.2 Cognitive Modeling Tools

In addition to physical modeling, cognitive modeling is also important to determine the "goodness" of a crew station design. This is still a new area of research, but there are a few cognitive models available for use. One such tool, the Applied Cognitive Task Analysis tool, assists the designer in identifying the cognitive skills necessary for performing a given task (Klein Associates, Inc., 2000). For instance, it determines what the critical cues or patterns of cues are necessary for the operator to make decisions and solve problems. Another interesting tool is Active Control of Thought—Rational (ACT—R), which is a framework constructed on assumptions about human cognition (Budiu, 2003). Researchers can add to the human cognition model by introducing their own assumptions about conducting a specific task. These assumptions can be tested by comparing the results of the model (time and accuracy of performing a task) to human-in-the-loop testing results, as was mentioned earlier with the CART case study (Brett et al., 2002).

15.2.1.2.3 System Design Tools

System design tools often integrate some of the previously discussed tools to achieve a more thorough test of the system. One of the most popular design tools is the Computer-Aided Three-Dimensional Interactive Application (CATIA). CATIA can assist with all stages of product design while improving product quality and saving money. Dassault Systemes, Paris, France, designed and developed CATIA, and the system is marketed and supported worldwide by IBM. The latest, Version 5.0, includes an integrated suite of Computer-Aided Design (CAD), Computer-Aided Engineering (CAE), and Computer-Aided Manufacturing (CAM) applications. CATIA has an integrated approach to the entire product design, and because of this, is internationally recognized as an industry leader (EDGE, 1993). A key aspect of this tool is that it allows everyone on the design team access to the same data in a common format with all updates. This facilitates concurrent activity among the design team, which speeds up the entire process.

Not only has CATIA played a major part in the design process in the 1990s (i.e., the development of Boeing's 777; [Hughes, 1994]), it continues to be an essential part of modern aircraft design. For instance, both the Airbus A380 and the Boeing 7E7 utilize CATIA (Sparaco, 2003; Mecham, 2003). CATIA is used by designers to check the physical layout of parts of the aircraft. CATIA uses its 3-D human models to test and evaluate these procedures. Additionally, CATIA facilitates the use of digital mock-ups that can eliminate the need for physical mock-ups of sections of the aircraft, which results in a significant cost-savings (Rich, 1989).

For a more in-depth model-based design of the crew station, the Man–Machine Integration Design and Analysis System (MIDAS) is available. "MIDAS contains tools to describe the operating environment, equipment, and mission of manned systems, with embedded models of human performance/behavior to support static and dynamic "what-if" evaluations of the crewstation design and operator task performance" (Smith & Hartzell, 1993, p. 13).

15.2.1.2.4 Summary of Design Tools

The tools described, as well as others available, all have the same goal—to assist the designers during the crew system design process. This section was meant to introduce the reader to some available products. Obviously, the list of tools described in this section is not all inclusive. A good source for design support tools and links to specific tool information is http://www.dtic.mil/dticasd/ddsm (MATRIS, 2004).

15.2.1.3 Research Examples Using Crew Station Design Tools

This section is provided so the reader can gain a better understanding of how the process and tools have been used in previous design projects. The examples provided will describe the use of the process and/or support tools for the development of a system from scratch, as well as for upgrading existing systems.

15.2.1.3.1 Navy Example: Multimodal Watch Station

The Navy's Multimodal Watch Station (MMWS) (Osga, 2000) is a classic example of designing a brand new system using the crew station design process. In an attempt to reduce costs for future navy ship operations, the plan was laid to design a new ship with a control center that would support a reduction in the operational crew size, while maintaining mission effectiveness. However, advancements in new systems, such as sensors and weapons, provided even more tasks for the new crew. Because of these factors, it became obvious that a certain level of automation would have to be supported to achieve these goals.

Using a task-centered workstation design process to determine information requirements for the total workstation, human factors engineers were able to effectively design the MMWS. They used this process to define task characteristics that drove the design requirements. By taking into account the operator's future role of multitasking and supervisory control, effective human–computer interactions were established. The focus was not only on the mission-specific requirements, but also on the computer interface requirements and work management task requirements. For example, operators in this new

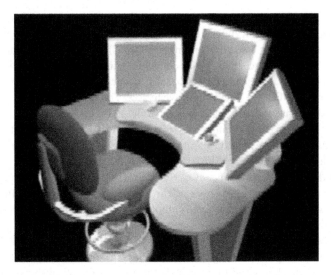

FIGURE 15.11 Multimodal watch station.

role will require a larger visual space within an anthropometrically comfortable environment that supports these new tasks (Figure 15.11). The design process used for the MMWS supported the design of a workstation that allowed the operator to easily shift between tasks without overloading his/her physical and cognitive resources. "Without regard to careful study of tasks and their information needs, display technologies will present increased opportunities for a designer to overload the user with more visual stimuli than currently possible. With proper design, however, this increased visual space can allow the user to visually shift between tasks with minimum control effort." (Osga, 2000, p. 1–706).

Testing of the MMWS has shown that the design was successful when the performance of operators using the MMWS was compared to Aegis crewmembers using traditional equipment. For instance, Aegis crews used last-second response methods when combating attacks from the air. MMWS operators were prepared for the attacks and, even with a significantly smaller crew size (50% smaller than the Aegis crew size), reported lower workload throughout the entire test (Osga, Van Orden, Kellmeyer, & Campbell, 2001).

15.2.1.3.2 Air Force Example: Work-Centered Support System

Linking computers together through machine-to-machine communication has become an essential part of achieving network-centric systems, and great progress is being made in this arena. However, just because the machines can communicate with each other electronically does not mean they can communicate with the operator efficiently—they each can have unique interfaces for the operator to understand. In addition, the operators cannot easily move among the various interface types. An analogy of this can be represented by the following example. Suppose software engineers wanted to electronically integrate three different computer systems, one of which only had a word-processing software package, the second had only a graphics software package, and the third only had a spreadsheet package. The operator would have to understand the "language" of each of these packages. And, on top of all that, the operator could not copy, cut, or paste information among the three packages.

What is needed in addition to the machine-to-machine communication is the ability for the interface to focus on the work that the operator is to achieve in this network-centric system. By first performing a cognitive work analysis, the proper information required by the operator can be determined. The next step addresses how to acquire the information from the electronically integrated machines. The software integrating the machines is called middleware. By using intelligent software agents that achieve appropriate information from the middleware, the customized operator interface can be created.

One program that employs this approach to operator console design is called the Work-Centered Support System (WCSS) (Eggleston, 2003). This approach has been successfully applied to the design of operators' consoles at USAF Air Mobility Command's (AMC) Tanker Airlift Control Center (TACC). The purpose of the TACC is to schedule flights for AMC's aircraft throughout the world. The job of the mission planners can get quite complicated because of such factors as weather changes, diplomatic clearances, and aircraft availability. They often have to access multiple databases in order to solve these problems. Also, the different databases have their own unique languages and menu structures; therefore, the mission planner has to learn the unique system's characteristics to complete the task. The bottom line is that the amount of time the mission planner spends on learning the language of each system is not really helping him/her get the job done. The real purpose of his/her job is to make sure the aircraft can efficiently travel to their final destination—everything else, such as learning unique languages, diverts them from their primary task. The purpose of the WCSS was to maximize time on the essential task—scheduling flights (Young, Eggleston, & Whitaker, 2000). An example of a work-centered display, the Port Viewer, is shown in Figure 15.12.

The purpose of the Port Viewer is to enable the mission planners to see, in one display, all the important parameters relative to a particular airfield (port). This is in contrast to the mission planners' having to go through multiple databases and then compile the parameters. With the WCSS software, agents obtain the appropriate information from the middleware and present it in the unified display. The Port Viewer display reduces the cognitive load on the operators by relieving them of the task of going through multiple databases.

15.2.1.3.3 FAA Examples: Air Traffic Control Consoles

The Federal Aviation Administration (FAA) has a complete virtual reality laboratory capable of recreating a variety of environments that users can interact with dynamically and in three dimensions to facilitate design work. By using a combination of hardware (head-mounted displays, data gloves, and trackers), with software (3-D graphics packages and Jack) hosted on sophisticated computing machines, several prototype systems have been developed.

One example of the use of this technology is the development of the next generation air traffic control display system (a replacement to the existing system). This system was initially designed and evaluated using only virtual reality tools and techniques. This allowed for a quick preliminary design of the system. The process was successful in identifying and fixing problems with a design that would have been expensive to change at a later point in the project (Virtual Reality Laboratory, Display System Replacement, n.d.).

Another successful upgrade to an existing system was achieved when the FAA used its virtual reality laboratory to redesign the Area Supervisors Workstation. This is the station that air traffic supervisors use to manage operations. The design process resulted in detailed drawings that became the specifications for the final workstation design (Virtual Reality Laboratory, Area Supervision Position, n.d.). The system was mocked-up and installed at an FAA facility where the mock-up was employed to determine user acceptability.

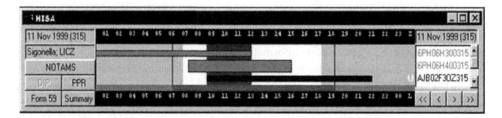

FIGURE 15.12 Port viewer.

15.2.2 What Will the Future Hold?

The U.S. Navy is depending very heavily on the versatile F/A-18 Super Hornet as the mainstay of its carrier fighter/attack force in the foreseeable future. In addition, an electronic attack version is also planned to augment the attack force, with deliveries starting in 2009. Aircraft will have either one or two crew stations depending on the version. On the Air Force side, the F/A-22 Raptor and the F-35 JSF are the latest aircraft. Both will have a single person in the crew station. The Navy and Marine Corps also plan to purchase the F-35. The first deliveries of the Air Force and Marine Corps versions of the F-35 will be in 2008, with the Navy's first deliveries starting in 2010. The bottom line is that these three aircraft will provide the two services' fighter/attack force well into the future (Schweitzer, 2003). But what type of aircraft will we have beyond these? And what type of crew station will they have?

One of the issues currently being addressed is the role of future long-range bombers within the Air Force. "The Air Force is rethinking long-range strike, a term that used to mean only one thing: big bombers. As the service adjusts to the Pentagon's new capabilities-based strategy and focuses on desired effects rather than the platforms needed to achieve them, the eventual successor to today's bomber fleet remains intentionally unsettled" (Tirpak, 2002, p. 29). The various versions being studied include not only conventional bombers as we think of them, but also various types of space planes. Another interesting aspect of these long-range strike vehicles is whether they will have a crew onboard or on the ground. Among the options being considered are systems with no airborne crew, which means it may become a UAV (Hebert, 2003).

UAVs have become well-known based on the conflict in Afghanistan. They served to give the command and control authorities continuous pictures of possible targets, and also enabled a dramatic reduction in the time from which the target was identified until it could be engaged.

A number of NATO countries are now using UAVs to augment their forces, especially in performing tasks that are dull (long-range reconnaissance), dirty (chemical or radiation problems), or dangerous (behind enemy lines). Force augmentation issues relevant to the human operator exist on several levels, including individual UAV control station design, vehicle interoperability by different organizations, and integration of UAVs with manned systems. Human interface issues associated with individual UAV control station design include guaranteeing appropriate situation awareness for the task, minimizing adverse effects of lengthy system time delays, establishing an optimum ratio of operators to vehicles, incorporating flexible levels of autonomy (manual through semiautonomous to fully automatic), and providing effective information presentation and control strategies. UAV interoperability requires development of a standard set of control station design specifications and procedures to cover the range of potential UAV operators and applications across military services and countries.

Finally, for UAVs to be successful, they must be fully integrated with manned systems so as to enhance the strength of the overall force. Human factors considerations in this area include how manned systems should best collaborate with UAVs, deconfliction concerns, operation with semiautonomous systems, and command and control issues. The essence of this paragraph can be summarized by the following statement: What is the proper role for the operator of UAVs? The operator's role can be defined in terms of three key factors: advanced UAV operator control/display interface technologies, supervisory control and decision support concepts, and trust and levels of automation. Each of these factors will be discussed in detail in the next few sections.

15.2.2.1 Factor 1: Advanced UAV Operator Control/Display Interface Technologies

The operators' stations for the U.S. Air Force's Predator and Global Hawk UAVs are mounted in vans with the operators sitting at command and control stations. The ground-based operators of these two vehicles control them quite differently. The Predator, at least in the landing in takeoff phase, uses teleoperation with the operator actually flying the vehicle from a distance. The Global Hawk, on the other hand, takes off and lands automatically and is largely autonomous during its mission. The operator, using supervisory control, "flies" the Global Hawk by using a mouse and keyboard, not stick and throttle. Different

(a)   (b)

FIGURE 15.13   Predator operator station (left) and Dragon Eye operator station (right).

UAVs require different control stations. For example, the operator station for the U.S. Marine Corps's Dragon Eye UAV is the size of a small suitcase, which makes it easily transportable; the Predator operator station is contained in a large van (Figure 15.13).

Research efforts with the Predator console have addressed a number of C/D features. Two examples are: head-coupled head-mounted display applications (Draper, Ruff, Fontejon, & Napier, 2002) and tactile system alerts (Calhoun, Draper, Ruff, Fontejon, & Guilfoos, 2003). Two additional efforts will be discussed in more detail.

As an example of a display enhancement, Draper, Geiselman, Lu, Roe, and Haas (2000) examined four different display formats that would aid the abilities of the Air Vehicle Operator (AVO) and the Sensor Operator (SO) to determine target location. If the AVO located a target in the wide field-of-view camera, it was often difficult to communicate the location to the SO who had a narrow FOV camera. Four different formats were examined to improve communication between the two crewmembers (Figure 15.14). The results showed that the two formats utilizing the locator line allowed participants to achieve statistically significantly better performance than the other formats. "Time to designate targets was reduced to an average of almost 50% using the telestrator [locator line]..." (Draper et al., 2000, p. 388). The reason for the superiority of the locator line was that, once the AVO designated the target it gave the SO a direct bearing to the target, thereby providing a very efficient means of exchanging information between the two operators.

As an example of control research, Draper, Calhoun, Ruff, Williamson, and Barry (2003) compared manual versus speech-based input involving the use of menus to complete data entry tasks. Pilots also performed flight and navigation tasks in addition to the menu tasks. Results showed that speech input was significantly better than manual for all eight different kinds of data entry tasks. The overall reduction

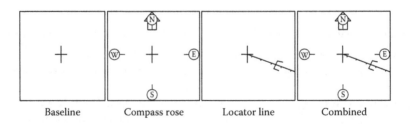

Baseline   Compass rose   Locator line   Combined

FIGURE 15.14   Locator line symbology from Predator.

was approximately 40% in task time for voice entry when compared with manual input. The operators also rated manual input as more difficult and imposing higher workload than the speech method. The reason for the superiority of the voice system was that it enabled the operator to go directly to the proper command without having to manually drill down through a number of menu sublevels in order to find the proper command.

Different types of control modes for operators' consoles were discussed in a recent conference (Association of Unmanned Vehicle Systems International, 2002). One recurring theme was a strong desire to move away from teleoperation of the UAVs and progress toward a combination of semiautonomous and fully autonomous operation of these vehicles—regardless of the type of operator console. In order to achieve this goal, a significant amount of automation will be required, especially, when coupled with the desire, in the case of UAVs, to move from a situation where a number of operators control one vehicle to one operator controlling a number of vehicles.

Research exploring the issues of one operator controlling multiple vehicles is important. Barbato, Feitshans, Williams, and Hughes (2003) examined a number of operator console features that would aid the operator in controlling four Uninhabited Combat Aerial Vehicles (UCAVs). The mission was to carry out a Suppression of Enemy Air Defenses. The operator's console contained three liquid crystal displays onto which was presented a situation awareness (SA) map, UCAV status, and multifunction information. The SA format presented the overall geographical situation along with, among other information, the flight routes of the four aircraft. The participants were required to manage the flight routes in two ways: manual versus semiautomatic using a route planner. Although the operators where favorable toward the real-time route planner, they did want information regarding what the real-time planner was actually doing (its intent) and they wanted both the original route and the planned route displayed in order to evaluate the two against each other. In essence, the study showed that one operator could manage four UCAVs when everything went as planned, and even when a single, unexpected event occurred.

15.2.2.2 Factor 2: Supervisory Control and Decision Support Concepts

In the case of UAVs, the avionics will be partly contained in the flying platform and partly incorporated into the operator's console, whether airborne or ground-based. In either case, because of present day capabilities in computers and intelligent agent software, the resulting product can be much closer to a true team. Operator–machine relationships are being created that emulate those occurring between two human crewmembers—mutual support and assistance. A diagram depicting this overall relationship is shown in Figure 15.15.

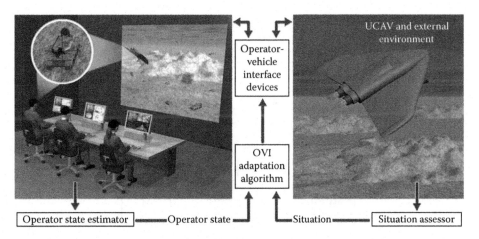

FIGURE 15.15 Operator—UAV system diagram.

A major component in achieving this mutual support and assistance is through software-entitled associate systems. Associate systems are "knowledge-based systems that flexibly and adaptively support their human users in carrying out complex, time-dependent problem-solving tasks under uncertainty" (Paterson & Fehling, 1992). Geddes (1997) lists three very important rules for associate systems and their relationship with the human operator.

- Mixed initiative—both the human operator and decision aid can take action
- Bounded discretion—the human operator is in charge
- Domain competency—decision aid has broad competency, but may have less expertise than the human operator

Because of the mixed initiative aspects of an associate system, function allocation, which assigns roles to the operator and the computer based on their abilities, has to be looked at in an entirely new light. The idea of function allocation has been around since the 1950s and had as its basic premise that the role of operator and the machine (computer), once assigned, would stay relatively constant during the operation of the system. However, this premise does not hold for modern computers since they contain associate systems that can have varying levels of automation at different times during a particular mission; therefore, static-function allocation is no longer applicable (Hancock & Scallen, 1996.). Rather, dynamic-function allocation is a key feature of associate systems with varying levels of automation.

Taylor (1993) illustrates how dynamic-function allocation changes the working relationship between the human operator and the machine (with associate-system-based automation); this changing relationship is shown in Figure 15.16. Cooperative Functionings indicates how the operator and automation would work together in an associate system. It is quite different from both manual control and supervisory control. In manual control, the human operator specifies the goals and functions to be accomplished and the machine carries out the tasks. In the next level, supervisory control, the human operator still specifies the goals, but the machine carries out both the tasks and functions. In the cooperative functionings (associate system), the human operator and machine interact at all levels, and either can specify the goals, functions, and tasks. It is through this dynamic sharing of authority that the operator and the associate can begin to operate as a team—an operator and a type of electronic crewmember (EC). However, to function as a team, the operator must trust the EC.

15.2.2.3 Factor 3: Trust and Levels of Automation

One means of establishing operator trust in the EC is to allow the operator to decide how much authority or autonomy, called levels of automation (LOA), to give the EC. "LOA defines a small set

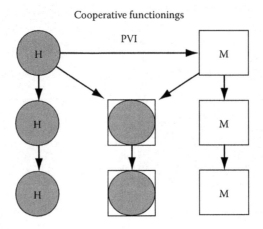

FIGURE 15.16 Systems authority concept.

("levels") of system configurations, each configuration specifying the degree of automation or autonomy (an "operational relationship") at which each particular subfunction performs. The pilot sets or resets the LOA to a particular level as a consequence of mission planning, anticipated contingencies, or in-flight needs" (Krobusek, Boys, & Palko, 1988, p. 124). While originally conceived for a piloted aircraft, LOAs apply equally well to UAV consoles and their operators. One question that must be answered is how many levels of automation should be assigned to the associate? A number of researchers have examined this issue. The result is as many as 10 (Sheridan, 1980) and as few as 5 (Endsley, 1996).

In order to create an effective team, once the levels are determined, the next task is to determine how they relate to the way humans process information. A further expansion of LOA was proposed by Parasuraman, Sheridan, and Wickens (2000); they matched levels of automation with a four-stage human information-processing model (information acquisition, information analysis, decisions selection, and action implementation). The 10 LOAs proposed by Parasuraman et al. are based on a model proposed by Sheridan (1980), which also contained an original set of 10 LOA's. They then illustrate how various systems could have different levels of automation across the four portions of the information-processing model. This work is very important because it begins to blend levels of automation with human information-processing capabilities. The authors realize that the model is not finalized, "We do not claim that our model offers comprehensive design principles but a simple guide" (Parasuraman et al., 2000, p. 294). However, it certainly is in the right direction toward achieving an optimal matching between automation and human capabilities for particular systems.

Using automation levels and having an indication of the information-processing workload of the mission, the operators could establish a "contract" with the EC in the premission phase. They could, through a dialogue at a computer workstation, define what autonomy they wish the EC to have as a function of flight phase and system function. As an example, weapon consent would always remain exclusively the operator's task, but reconfiguration of the UAVs flight control surfaces to get the best flight performance in the event of battle damage would be the exclusive task of the EC.

15.2.2.3.1 Adaptive Automation

Although the premission contract with the EC helps to establish roles for it and the human operator, the functions allocated to each crewmember remain static throughout the mission. However, missions are highly dynamic, and, as stated before, it would be desirable to change the function allocation during the mission. This dynamic-function allocation is achieved through adaptive automation (AA). "In AA, the level or mode of automation or the number of systems that are automated can be modified in real time. Furthermore, both the human and the machine share control over changes and the state of automation" (Scerbo, 1996, p. 43).

Two of the key aspects of AA are *when* to trigger the shift and for *how long*. The *when* aspect is discussed by Scerbo, Parasuraman, Di Nocero, and Prinzel, (2001, p. 11) who list a number of methods for triggering the shifting tasks between the operator and the automation: critical events, operator modeling, performance measurement, psychophysiological measurement, and hybrid methods. A diagram of how many of these allocation methods can be used in a system is shown in Figure 15.17.

As an example of how psychophysiological measurement is used to determine operator state, Wilson and Russell (2003) required USAF air traffic controllers, in a simulation, to manage air traffic around the Los Angeles airport. The task loading was manipulated by the number of aircraft they had to manage (volume) and the different kinds of aircraft they had to manage (complexity). The tasks were first given to subject-matter experts (SMEs), and the difficulty was increased until the SMEs verified that they were in an overload condition and could not effectively handle the traffic. The participants were then given the same type of task and their physiological data was processed by a computer-generated neural net. The result was the neural net could identify the nonoverload condition 99% of the time and the overload condition 96% of the time. These results indicate that psychophysiological measures may potentially be very useful in determining operator overload in real-world applications.

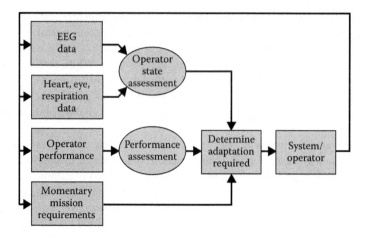

FIGURE 15.17 Adaptive automation system diagram.

Once the state of the operator can be reliably assessed, the next question is, can the workload be shifted quickly between the operator and the automation? Wilson, Lambert, and Russell (2000) addressed this question in a study using NASA's Multi-Attribute Test Battery (MATB). There are four tasks in the MATB: tracking, systems monitoring, resource management, and communications. As in the air traffic control study previously discussed, pretest conditions were defined to discover when the operators were overloaded, and the neural nets were used to identify this condition. In one experimental condition, the participants managed all four of the tasks, regardless of the difficulty. In the other condition, when the participants reached the overload condition, the systems monitoring and communications tasks were handed off to the automation. The operator continued controlling the tracking and resource management tasks. The results showed that, relative to the manual condition, the adaptive-aiding condition resulted in a 44% reduction in tracking error and a 33% error reduction in resource management tasks.

The psychophysiological triggering of adaptation appears to be very promising; however, researchers are still very early in applying this technology to real-world settings. "At present, however, there is not enough existing psychophysiological research to provide adequate information on which to base adaptive-allocation decisions" (Prinzel, Freeman, Scerbo, & Mikulka, 2000, p. 407). Although the shifting of tasks from the operator to the automation by psychophysiological methods (the *when* aspect) resulted in successful performance in the Wilson et al. study (2000), there does not appear to be any general consensus as to *how long* the automation should keep the transferred task in order to optimize overall systems performance. The *how long* aspect has been examined by a number of authors, and the answer appears to be task specific. For example, Scallen and Hancock (2001) utilized AA in a study which required pilots to perform tracking, monitoring, and targeting tasks while flying a simulator. After a target was presented, the tracking task was automated for a 20 s interval, after which it was returned to the pilot. Conversely, in another research effort (Scallen & Duley, 1995), which looked at three different cycle times between the operator and the automation (15, 30, or 60 s), the 15 s switching time resulted in the best tracking performance. However, three of the five pilots who took part in the study reported that the switching back and forth was distracting. As a result, the author states that "In the case of adaptive allocation systems we propose a moratorium strategy in which there is a minimum frequency with which the system can either assume or relinquish task control" (Scallen et al., 1995, p. 402).

15.2.2.3.2 Putting It Together

With all of the levels of automation, human information processing models, and AA, things are getting complicated. How do we make sense of all this? Kaber, Prinzel, Wright, and Claman (2002) addressed two of the three components in a study which looked at the issue of AA relative to the four stages of the information-processing model. Besides a manual control condition where there was no AA, it was applied to all the stages of the four-stage model: information acquisition, information analysis, decision making, and action implementation.

The participants used Multitask that created a simulated air traffic control environment. Their task was to provide a landing clearance to various aircraft depicted on the radar scope. The aircraft were flying from the periphery to the center of the display. An error occurred if the aircraft reached the center of the display, or collided with another aircraft, before the clearance was issued. A secondary task was also used. If the participant's performance on the secondary task fell below a predetermined level, the primary task would be automated. NASA's Task Load Index (TLX) was used to measure workload.

Although the performance utilizing AA was superior to the manual control condition, the results showed that AA was most effective when applied to the information acquisition and action, implementation, and information-processing stages. It was not effective in the information-analysis and decision-making stages. The authors conclude, "All these results suggest that humans are better able to adapt to AA when applied to lower-level sensory and psychomotor functions, such as information acquisition and action implementation, as compared to AA applied to cognitive (analysis and decision making) tasks" (Kaber et al., 2002, p. 23).

The Kaber et al. (2002) study began to give some insight into the interaction of two components: information processing and AA. But, as mentioned at the beginning of this section, there are three components, the third being levels of automation. How do they *all* fit together? Kaber and Endsley (2004) attempted to show the relationship among all three factors. They also constructed 10 levels of automation and an information-processing model similar to Parasuraman et al. (2000), with the stages being monitoring, generating, selecting, and implementing. In addition, they incorporated AA. They then conducted a study utilizing six levels of automation: manual, action support, batch processing, decision support, supervisory control, and full automation (numbers 1, 2, 3, 5, 9, and 10, in Figure 15.18). Manual and Full Automation are self-explanatory. Action Support is similar to teleoperation. Batch Processing requires the human to create and decide the options to implement, and the computer carries these out. Decision Support involves the computer suggesting options and once the operator selects one of these

	Level of Automation	Monitoring	Generating	Selecting	Implementing
1.	Manual control	Human	Human	Human	Human
2.	Action support	Human/computer	Human	Human	Human/computer
3.	Batch processing	Human/computer	Human	Human	Computer
4.	Shared control	Human/computer	Human/computer	Human	Human/computer
5.	Decision support	Human/computer	Human/computer	Human	Computer
6.	Blended decision making	Human/computer	Human/computer	Human/computer	Computer
7.	Rigid system	Human/computer	Computer	Human	Computer
8.	Automated decision making	Human/computer	Human/computer	Computer	Computer
9.	Supervisory control	Human/computer	Computer	Computer	Computer
10.	Full automation	Computer	Computer	Computer	Computer

FIGURE 15.18 LOA taxonomy for human–computer performance in dynamic Multitask scenarios.

options (or one self-generated), it is then put into operation by the computer. In Supervisory Control the computer generates and carries out the options. The operator monitors and gets involved if necessary. These six levels were then combined with three levels of AA cycle time (AACT) (20%, 40%, and 60%).

For example, in a 20 min trial the task would be allocated to the automation either 4, 8, or 12 min. The results showed that "The best combination of LOA and AACT involved human strategizing combined with computer implementation (Batch processing (LOA 3)) during high automation cycle times (12 min on cycle and 8 min off cycle)" (Kaber & Endsley, 2004, p. 147).

This result is a big step forward, but also illustrates the difficulty in implementing AA, levels of automation, and human information processing. If we put this research on a time scale relative to more than 80 years of research in the design of aircraft crew stations, we are just beginning to explore this area. So, we cannot expect instant answers to these very difficult questions. To make matters even more interesting, there are also plans to place varying levels of automation within the airborne platform.

15.2.2.3.3 Levels of Automation within the Air Vehicle

Earlier in this section it was mentioned that there would be intelligent software both in the operator's console as well as within the UAV itself. The airborne computing system enables varying levels of autonomy called autonomous control levels (ACLs) within the UAV (OSD, 2002) At first glance, it would seem logical to assume that these 10 levels (Figure 15.19) map onto Sheridan's 10 levels of autonomy mentioned in Factor 3: Trust and Levels of Automation. Sheridan's levels deal with the interaction between the operator and the UAV. However, these ACLs are referring to autonomy levels within the aircraft only and not between the aircraft and the operator. One thing to note about this chart is that the lower levels of the chart refer to the ACLs within each aircraft in, for example, a flight of four. But, from levels five and higher, they refer to how the entire flight works together as a group. They range from Level 1: Remotely Guided (teleoperation) to Level 10: Fully Autonomous Swarms where the vehicles are acting in concert with one another to achieve a common goal. Teleoperation has already been discussed in Factor 1: Advanced UAV Operator Control/Display Interface Technologies, and will not be further enumerated

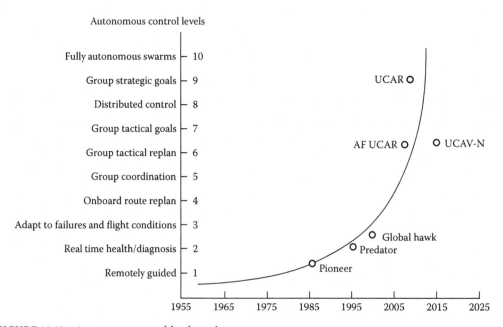

FIGURE 15.19 Autonomous control-level trend.

upon here. But Level 10: Swarms, which offer a whole new level of control both within a group of aircraft and between that group and the operator, will be examined in more detail.

The fascinating thing about swarms is that there does not appear to be any central controller telling the swarm what to do. If you observe a school (swarm) of fish, they just appear to act as one with no central leader fish giving them directions. The same is true for flocks of birds, groups of ants, and swarms of bees. "'Swarming' itself is a type of emergent behavior, a behavior that is not explicitly programmed, but results as a natural interaction of multiple entities" (Clough, 2002, p. 1). As an example of forming a swarm, consider how ants communicate that they have found a source of food. The ants lay down a pheromone trail (chemical markers) that other ants can follow. The strength of the pheromones, however, decays over time; therefore, the ant that finds the closest food supply and returns with it will have the strongest pheromone trail. Other ants will then follow this trail with no central commander ant directing them to do this (Bonabeau & Theraulaz, 2000).

So, what does this have to do with UAVs? Think of the possibilities if a flight of UAVs could act as a swarm. For example, instead of an operator giving the UAVs explicit, detailed instructions on the location of surface-to-air missile batteries, the UAVs could be directed to just loiter about a certain area of enemy territory. Then, if they come across the missiles, they could destroy them. Of course, they would be acting within the level of responsibility given to them by the human operator. Creating digital pheromones for UAVs is one way the UAVs could communicate within such a swarm. These types of pheromones are not based on chemicals, but rather on the strength of electrical fields. In a computer-based (constructive) simulation, a UAV swarm using digital pheromones significantly outperformed the nonswarm case (Parunak, Purcell, & O'Connell, 2002).

15.2.2.3.4 Conclusion

UAVs have a wide range of avionics sophistication, from the relatively basic Dragon Eye to very complex Global Hawks and UCAVs. Many of the UAVs used at the small unit level will have limited automation although; for example, they will be able to plan their own flight route. However, most future aircraft, whether inhabited or not, will contain associate systems that will incorporate varying levels of autonomy and AA as basic operating principles. These principles will enable the UAV operator and the associate to form a team consisting of two crewmembers—one human and one electronic. In order to function effectively, the operator and the EC must work together as a close-knit team, and the EC may not only supervise one aircraft but the entire swarm. One essential feature of a successful team is human trust in the associate partner. Currently, this is not a two-way street—ECs cannot trust at this level of sophistication or have the ability to trust—they are not Lieutenant Commander Data of the Starship Enterprise. In the meantime, guidelines to create such trust must include specifying the EC's level of autonomy. By using these guidelines, the operator can achieve a high-quality trusting relationship with the EC. This internal trust will, in turn, lead to an efficient and effective team, which can operate successfully in a system of systems environment.

15.2.3 Conclusions

Aircraft crewstations have progressed from those of Doolittle's days containing a myriad mechanical devices to those of today based almost entirely on E-O devices where the distinction between controls and displays continues to blur. In addition, automation has progressed from simple autopilots to flight management systems with numerous software decision aids. Computerized design tools are being used to both create and perform evaluations of conceptual crew stations before they are turned into hardware. With the increasing emphasis on UAVs, there is discussion in the military environment as to how many future airborne systems will posses human crew members. No matter how this issue is resolved, so long as there are operators involved either in the air or on the ground, crew stations will offer one of the most interesting and challenging areas of work for the human factors professional.

References

Adam, E. C. (1994). Head-up displays vs. helmet-mounted displays: The issues. In *Proceedings of SPIE Vol. 2219: Cockpit Displays* (pp. 13–21). Bellingham, WA.

Adams, C. (1993, November). HUDs in commercial aviation. *Avionics*, pp. 22–28.

Adams, C. (2003, September). JSF integrated avionics par excellence. *Avionics*, pp. 18–24.

Association of Unmanned Vehicle Systems International. (2002). Unmanned systems in a new era, *AUVSI Second Annual Ground, Sea and Air Conference*, February 12–14. Washington, DC.

Aviation Week and Space Technology. (2003, June 23). 7E7 Tunnel Tests Begin, p. 24.

Barbato, G., Feitshans, G., Williams, R., & Hughes, T. (2003). Operator vehicle interface laboratory: Unmanned combat air vehicle controls and displays for suppression of enemy air defenses. In *Proceedings of the 12th International Symposium on Aviation Psychology*. Dayton, OH.

Bassett, P., & Lyman, J. (1940, July). The flightray, a multiple indicator. *Sperryscope*, p. 10.

Belt, R., Kelley, J., & Lewandowski, R. (1998). Evolution of helmet mounted display requirements and honeywell HMD/HMS systems. In *Proceedings of SPIE Vol 3362: Helmet and Head-Mounted Displays III* (pp. 373–384). Bellingham, WA.

Bolia, R. S. (2003). Spatial intercoms for air battle managers: Does visually cueing talker location improve speech intelligibility? In *Proceedings of the 12th International Symposium on Aviation Psychology* (pp. 136–139). Dayton, OH.

Bonabeau, E., & Theraulaz, G. (2000, March). Swarm smarts. *Scientific American*, pp. 72–79.

Brett, B. E. et al. (2002). *The Combat Automation Requirements Testbed (CART) Task 5 Interim Report: Modeling a Strike Fighter Pilot Conducting a Time Critical Target Mission, U.S. Air Force Research Laboratory Technical Report* (AFRL-HE-WP-TR-2002-0018). Wright-Patterson AFB, OH: Air Force Research Laboratory.

Budiu, R. (2003). About ACT-R. Retrieved May 20, 2004, from Carnegie Mellon University, Department of Psychology, ACT-R Research Group Web site: http://act-r.psy.cmu.edu/about/

Calhoun, G. L., & McMillan, G. R. (1998). Hands-free input devices for wearable computers. *HICS 4th Annual Symposium on Human Interaction with Complex Systems* (pp. 118–123). Dayton, OH.

Calhoun, G. L., Draper, M. H., Ruff, H. A., & Fontejon J. V. (2002). Utility of a tactile display for cueing faults. In *Proceedings of the Human Factors and Ergonomics Society 46th Annual Meeting* (pp. 2144–2148). Santa Monica, CA.

Calhoun, G. L., Draper, M. H., Ruff, H. A., Fontejon, J. V., & Guilfoos, B. J. (2003). Evaluation of tactile alerts for control station operation. In *Proceedings of the Human Factors and Ergonomics Society 47th Annual Meeting* (pp. 2118–2122). Denver, CO.

Chiasson, J., McGrath, B. J., & Rupert, A. H. (2002). Enhanced situation awareness in sea, air and land environments. In *Proceedings of the RTO HFM Symposium on Spatial Disorientation in Military Vehicles: Causes, Consequences, and Cures* (pp. 32-1–32-10). La Coruna, Spain.

Clough, B. (2002). UAV's swarming? So what are those swarms, what are the implications, and how we handle them? In *AUVSI Unmanned Systems 2002 Proceedings* (pp. 1–15), July 2002. Baltimore, MD: Association for Unmanned Vehicle Systems International.

Desjardins, D. D., & Hopper, D. G. (2002). Military display market: Third comprehensive edition. In *Proceedings of SPIE Vol 4712: Cockpit Displays IX* (pp. 35–47). Orlando, FL.

Doolittle, J. (1961). Blind flying Gardner Lecture Aerospace Engineering,

Draper, M. H., Calhoun, G. L., Ruff, H. A., Williamson, D. T., & Barry, T. P. (2003). Manual versus speech input for unmanned aerial vehicle control station operations. In *Proceedings of the Human Factors and Ergonomics Society 47th Annual Meeting* (pp. 109–113). Denver, CO.

Draper, M. H., Geiselman, E. E., Lu, L. G., Roe, M. M., & Haas, M. W. (2000). Display concepts supporting crew communication of target location in unmanned air vehicles. In *Proceedings of the Human Factors and Ergonomics Society 44th Annual Meeting* (pp. 385–388). Santa Barbara, CA.

Draper, M. H., Ruff, H. A., Fontejon, J. V., & Napier, S. (2002). The effects of head-coupled control and head-mounted displays (HMDs) on large-area search tasks. In *Proceedings of the Human Factors and Ergonomics Society 46th Annual Meeting* (pp. 2139–2143). Baltimore, MA.

EDGE. (1993). CAD/CAM/CAE solutions: New IBM offerings for industry-leading CATIA software further speed product design, manufacture, and delivery). *Work-Group Computing Report, 4*(178), 28.

Eggleston, R. G. (2003). Work center design: A cognitive engineering approach to system design. In *Proceedings of the Human Factors and Ergonomics Society 47th Annual Meeting* (pp. 263–267). Denver, CO.

Endsley, M. R. (1996). Automation and situational awareness. In R. Parasuraman, & M. Mouloua (Eds.), *Automation and human performance: Theory and applications* (pp. 163–181). Mahwah, NJ: Erlbaum.

Engineering Animation, Inc. (2000). Jack' human simulation software. Retrieved May 20, 2004, from MATRIS Web site: http://dtica.dtic.mil/ddsm/srch/ddsm106.html

Fortier-Lozancich, C. (2003). The JHMCS Operational Flight Program Is Usable on Three Tactical Aircraft. *Crosstalk—The Journal of Defense Software Engineering*, July http://www.stsc.hill.af.mil/crosstalk/2003/07/top5jhmcs.html

Gagne, R. M. (1962). *Psychological principles in systems development*. New York: Holt, Rinehart, and Winston.

Galatowitsch, S. (1993, May). Liquid crystals vs. cathode ray tubes. *Defense Electronics*, p. 26.

Glines, C. V. (1989, September). Flying blind. *Air Force Magazine*, pp. 138–141.

Geddes, N. (1997). Associate systems: A framework for human-machine cooperation. In M. Smith, G. Salvendy, & R. Koubek (Eds.), *Designing of computing systems: Social and ergonomic considerations*. Amsterdam, the Netherlands: Elsevier.

Hancock, P. A., & Scallen, S. F. (1996). The future of function allocation. *Ergonomics in Design, Q4*, 24–29.

Hatfield, F., Jenkins, E. A., Jennings, M. W., & Calhoun, G. L. (1996). Principles and guidelines for the design of eye/voice interaction dialogs. *HICS 3rd Annual Symposium on Human Interaction with Complex Systems* (pp. 10–19). Dayton, OH.

Hebert, A. (2003, November). The long reach of heavy bombers. *Air Force Magazine*, pp. 24–29.

Hughes, D. (1994, January). Aerospace sector exploits CAD/CAM/CAE. *Aviation Week and Space Technology*, pp. 56–58.

Institute for Human and Machine Cognition (2000). Tactile situation awareness system. Retrieved May 19, 2004, from University of West Florida Web site: http://www.coginst.uwf.edu/projects/tsas/main.html

Jennings, D. L., & Ruck, D. W. (1995). Enhancing automatic speech recognition with an ultrasonic lip motion detector. In *Proceedings of the IEEE International Conference on Acoustics, Speech and Signal Processing* (pp. 868–871). Detroit, MI.

Kaber, D. B., & Endsley, M. R. (2004, March–April). The effects of level of automation and adaptive automation on human performance, situational awareness in workload in a dynamic control task. *Theoretical Issues in Ergonomics Science, 5*(2), 113–153.

Kaber, D. B., Prinzel, L. J., Wright, M. C., & Claman, M. P. (2002, September). Workload-matched adaptive automation support of air traffic-control or information processing stages (NASA/TP-2002-211932). Hampton, VA: NASA Langley Center.

Kearns, J. H. (1982). *A systems approach for crew station design and evaluation* (Tech. Report AFWAL-TR-81-3175). Wright-Patterson Air Force Base, OH: Flight Dynamics Laboratory.

Klass, P. (1956, July 23). USAF reveals new instrument concept. *Aviation Week*, p. 62.

Klein Associates, Inc. (2000). Applied cognitive task analysis. Retrieved May 20, 2004, from MATRIS Web site: http://dtica.dtic.mil/ddsm/srch/ddsm83.html

Krobusek, R. D., Boys, R. M., & Palko, K. D. (1988). Levels of autonomy in a tactical electronic crewmember. In *Proceedings of The Human—Electronic Crew: Can They Work Together?* (pp. 124–132) (Tech. Rep. WRDC-TR-89-7008). Wright-Patterson Air Force Base, OH: Cockpit Integration Directorate.

Lake, M. (2001). How it works: retinal displays add a second data layer. Retrieved May 29, 2003 from http://www.telesensory.com/nomad_press_releases.html

Liggett, K. K., Reising, J. M., & Hartsock, D. C. (1992). The use of a background attitude indicator to recover from unusual attitudes. In *Proceedings of the 36th Annual Meeting of the Human Factors Society* (pp. 43–46). Santa Monica, CA: Human Factors and Ergonomics Society.

Martin, E. A., Barbato, G. J., & Doyal, J. A. (2004). A tool for considering human behaviour within design trade-study constructive simulations: Lessons learned from two case study applications. In D. de Waard, K. Brookhuis, & C. Weikert (Eds.), *Human factors in design*. Maastricht NL: Shaker Publishing.

MATRIS. (2004). Directory of design support methods. Retrieved May 19, 2004, from Defense Technical Information Center A MATRIS Resources Web site: http://rdhfl.tc.faa.gov/VR/VRLAreasup.html

Mecham, M. (2003, October 27). Betting on suppliers. *Aviation Week and Space Technology*, pp. 51–54.

NexGen Ergonomics, Inc. (2003). MannequinPRO. Retrieved May 20, 2004, from MATRIS Web site: http://dtica.dtic.mil/ddsm/srch/ddsm125.html

Nicklas, D. (1958). *A history of aircraft cockpit instrumentation 1903–1946* (WRDC Technical Report No. 57-301). Wright-Patterson Air Force Base, OH: Wright Air Development Center.

ORLOC. (1981). *KC-135 crew system criteria* (Tech. Rep. No. AFWAL-TR-81-3010). Wright-Patterson Air Force Base, OH: Flight Dynamics Laboratory.

OSD. (2002). Unmanned aerial vehicles roadmap, 2002–2007, p. 84. Retrieved 16 Jun, 2004, http://www.acq.osd.mil/usd/uav_roadmap.pdf

Osga, G. A. (1991). Using enlarged target area and constant visual feedback to aid cursor pointing tasks. In *Proceedings of the Human Factors Society 35th Annual Meeting* (pp. 369–373). Santa Monica, CA: Human Factors Society.

Osga, G. A. (2000). 21st century workstations—active partners in accomplishing task goals. In *Proceedings of the IEA 2000/HFES 2000 Congress* (pp. 1-704–1-707). Santa Monica, CA.

Osga, G. A., Van Order, K. F., Kellmeyer, D., & Campbell, N. L. (2001). "Task-Managed" Watchstanding: Providing Decision Support for Multi-Task Naval Operations. SSC San Diego TD 3117, SSC San Diego Biennial Review August 2001, pp. 176–185.

Parasuraman, R., Sheridan, T. B., & Wickens, C. D. (2000, May). A model for types and levels of human interaction with automation. *IEEE Transactions on Systems, Man, and Cybernetics—Part A: Systems and Humans, 30*(3), 286–297.

Parunak, H., Purcell, M., & O'Connell, R. (2002). Digital pheromones for autonomous coordination of swarming UAVs. In *Proceedings of First AIAA Unmanned Aerospace Vehicles, Systems, Technologies, and Operations Conference* (Paper No. AIAA 2002-3446, pp. 1–9). Portsmouth, VA.

Paterson, T., & Fehling, M. (1992). Decision methods for adaptive task-sharing in associate systems. In *Proceedings of 8th Conference on Uncertainty in Artificial Intelligence*. San Mateo, CA.

Pellettiere, J. (2002). Articulated total body model. Retrieved May 20, 2004, from MATRIS Web site: http://dtica.dtic.mil/ddsm/srch/ddsm35.html

Perrott, D. R., Cisneros, J., McKinley, R. L., & D'Angelo, W. R. (1996). Aurally aided visual search under virtual and free-field listening conditions. *Human Factors, 38*, 702–715.

Prinzel, L. J., Freeman, F. G., Scerbo, M. W., & Mikulka, P. J. (2000). A closed-loop system for examining psychophysiological measures for adaptive task allocation. *The International Journal of Aviation Psychology, 10*(4), 393–410.

Raj, A. K., Kass, S. J., & Perry, J. F. (2000). Vibrotactile displays for improving spatial awareness. In *Proceedings of the IEA 2000/HFES 2000 Congress* (pp. 1-181–1-184). Santa Monica, CA.

Reising, J. M., Emerson, T. J., & Munns, R. C. (1993). Automation in military aircraft. In *Proceedings of the HCI International '93: 5th International Conference of Human-Computer Interaction* (pp. 283–288). Orlando, FL.

Reising, J. M., Liggett, K. K., Rate, C., & Hartsock, D. C. (1992). 3-D target designation using two control devices and an aiding technique. In *Proceedings of the SPIE/SPSE Symposium on Electronic Imaging Science and Technology*. San Jose, CA.

Rich, M. A. (1989). Digital mockup (airplane design and production using computer techniques). In *Proceedings for the AIAA, AHS, and ASEE, Aircraft Design, Systems and Operations Conference*. Seattle, WA: AIAA.

Rockwell Collins Press Release October 29, 2003. Rockwell Collins awarded $20 million to provide dual HUD system for U.S. Air Force and Marine Corps C-130J aircraft. http://www.rockwellcollins.com/news/page3094.html

Sarter, N. B. (2000). The need for multisensory interfaces in support of effective attention allocation in highly dynamic event-driven domains: The case of cockpit automation. *The International Journal of Aviation Psychology, 10*(3), 231–245.

Scallen, S. F., & Hancock, P. A. (2001). Implementing adaptive function allocation. *The International Journal of Aviation Psychology, 11*(2), 197–221.

Scallen, H., & Duley, (1995). Pilot performance and preference for short cycles of automation in adaptive function allocation. *Applied Economics, 26*(6), 397–403.

Scerbo, M. W. (1996). Theoretical perspectives on adaptive automation. In R. Parasuraman, & M. Mouloula (Eds.), *Automation and human performance* (pp. 37–63). Mahwah, NJ: Erlbaum.

Scerbo, M. W., Parasuraman, R., Di Nocero, F., & Prinzel, L. J. (2001). The efficacy of physiological measures for implementing adaptive technology (NASA/TP-2001-211018, p. 11). Hampton, VA: NASA Langley Technical Report Server.

Schweitzer, R. (2003, June). Big bucks for the best there is. *Armed Forces Journal*, 24–28.

Sheridan, T. B. (1980, October). Computer control and human alienation. *Technology Review*, 61–73.

Simpson, B. D., Bolia, R. S., McKinley, R. L., & Brungart, D. S. (2002). Sound localization with hearing protectors: Performance and head motion analysis in visual search task. In *Proceedings of the Human Factors and Ergonomics Society 46th Annual Meeting* (pp. 1618–1622). Baltimore, MD.

Smith, B. R., & Hartzell, E. J. (1993). A³I: Building the MIDAS touch for model-based crew station design. *CSERIAC Gateway, IV*(3), 13–14.

Sparaco, P. (2003, October 27). Weight watchers. *Aviation Week and Space Technology*, 48–50.

Spengler, R. P. (1988). *Advanced fighter cockpit* (Report No. ERR-FW-2936). Fort Worth, TX: General Dynamics.

Sturman, D. J., & Zeltzer, D. (1994). A survey of glove-based input. *IEEE Computer Graphics and Applications, 23*, 30–39.

Taylor, R. (1993). Human factors of mission planning systems: Theory and concepts. *AGARD LS 192* (New Advances in Mission Planning and Rehearsal Systems), 2-1-2-22.

Tirpak, J. (2002, October). Long arm of the air force. *Air Force Magazine*, 28–34.

Traylor, R., & Tan, H. Z. (2002). Development of a wearable haptic display for situation awareness in altered-gravity environment: Some initial findings. In *Proceedings of the 10th Symposium on Haptic Interfaces for Virtual Environments and Teleoperator Systems* (pp. 159–164). Orlando, FL: IEEE Computer Society.

U.S. Department of Defense. (1996). *Department of Defense interface standard, aircraft display symbology* (MIL-STD-1787B). Washington DC: Author.

van Erp, J. B. F. (2002). Guidelines for the use of vibro-tactile displays in human computer interaction. In S. A. Wall, B. Riedel, A. Crossan, & M. R. McGee (Eds.), *Proceedings of EuroHaptics* (pp. 18–22). Edinburgh.

Virtual Reality Laboratory, Display System Replacement. (n.d.). Retrieved May 19, 2004, from FAA Web site: http://rdhfl.tc.faa.gov/VR/VRLDsr.html

Virtual Reality Laboratory, Area Supervisor Position. (n.d.). Retrieved May 19, 2004, from FAA Web site: http://rdhfl.tc.faa.gov/VR/VRLAreasup.html

Ware, C., & Slipp, L. (1991). Using velocity control to navigate 3-D graphical environments: A comparison of three interfaces. In *Proceedings of the 35th Annual Meeting of the Human Factors Society* (pp. 300–304). Santa Monica, CA: Human Factors Society.

Weintraub, D. J., & Ensing, M. (1992). *Human factors issues in head-up display design: The book of HUD*. Wright-Patterson AFB, OH: Human Systems Information Analysis Center.

Wickens, C. (1992). *Engineering psychology and human performance* (2nd ed.). New York: HarperCollins.

Wiley, J. (1998, September 7). HUD sales remain strong. *Aviation Week*. http://www.awgnet.com/shownews/day1/hardwr10.htm

Williams, D. (2004, May). Deep attacks from the air. *Armed Forces Journal*, 8–9.

Williamson, D. T., Barry, T. P., & Draper, M. H. (2004). Commercial speech recognition technology in the military domain: Results of two recent research efforts. In *Proceedings of Applied Voice Input/Output Society's Speech TEK Spring 2004 Conference (CDROM)*. San Jose, CA: AVIOS.

Williamson, D. T., Barry, T. P., & Liggett, K. K. (1996). Flight test performance optimization of ITT VRS-1290 speech recognition system. In *Proceedings of the 1996 AGARD Aerospace Medical Panel Symposium on Audio Effectiveness in Aviation*. Neuilly Sur Seine, France: North Atlantic Treaty Organization.

Wilson, G. F., Lambert, J. D., & Russell, C. A. (2000). Performance enhancement with real-time physiologically controlled adaptive aiding. In *Proceedings of the IEA 2000/HFES 2000 Congress, Human Factors and Ergonomics Society* (pp. 3-61–3-64). Santa Monica, CA.

Wilson, G. F., & Russell, C. A. (2003). Operator functional state classification using multiple psychophysiological features an air traffic control task. *Human Factors, 45*(3), 381–389.

Young, M., Eggleston, R. G., & Whitaker, R. (2000). Direct manipulation interface techniques for users interacting with software agents. In *Proceedings of the NATO/TRO Symposium on Usability of Information and Battle Management Operations*, Oslo, Norway 10–13 April, 2000.

16

Flight Deck Aesthetics and Pilot Performance: New Uncharted Seas

Aaron J. Gannon
Honeywell Aerospace

16.1 Aesthetics: Adrift in Aerospace

Travelers are always discoverers, especially those who travel by air. There are no signposts in the sky to show a man has passed that way before. There are no channels marked. The flier breaks each second into new uncharted seas.

Anne Morrow Lindbergh
North to the Orient

Although the human factors discipline has made substantial headway in aviation since the 1940s, the development of our scientific knowledge of cockpit aesthetics largely remains adrift in shallow waters. Specifically, there is a lack of research regarding designed aesthetics and their interaction with the crew's performance. Furthermore, the disciplines that are most concerned with the interface and relationship between the human and the machine, namely, industrial design and human factors, have not integrated in a meaningful way in aerospace to give direction and progress to the research track.[*][†]

Historically, with few exceptions, these two disciplines were segregated at the cabin door: in general, human factors engineers turned to the left to design the cockpit, and industrial designers turned to the right to design the cabin. Currently industrial design is enjoying a surge of activity forward of the flight deck door, particularly in the Very Light Jet (VLJ) market segment. This has not, however, broadly changed the nature of the segregation between industrial design and human factors, but rather has revealed it.

Specifically, the separation can now be seen to extend beyond the simple station lines defining cockpit and cabin, and rather manifests as a partitioning of form and function. Even as industrial design has enjoyed an increasing role in cockpit development in recent years, there is little evidence that the effort is regularly integrated in a meaningful way with functional systems and human factors engineering—in practice, industrial designers are often given the task from the marketing department to make it attractive, while the engineering team is assigned separately to make it work.[‡] And when schedules and budgets run short, making it work takes precedence and the segregation is magnified. Louis Sullivan summarized that form ever follows function, but a typical aerospace engineering implementation of that tenet might be more accurately stated: optimal form is naturally guaranteed out of a fixed focus on function alone. But we need to ask, is this a valid assumption, and if perhaps not, are we unnecessarily segregating form and function and eliciting some unknown impact on pilot performance, perhaps even preventing optimization of function?[§]

> Form follows function –
> that has been misunderstood.
> Form and function should be one, joined in a spiritual union.

> **Frank Lloyd Wright**

The segregation was made clear to me recently as I talked with an engineer about a project with tasking addressing not just the usability but also the aesthetics of a cockpit control under development. Even as I espoused the importance of a balanced approach to form and function, the engineer replied flatly, "I don't care how it looks, I only care how it works."

---

[*] It is notable that closer ties between human factors and industrial design (and their societies) is evident in other application areas, such as medical device design.

[†] While this discussion largely refers to industrial design, a broader consideration of the design arts and sciences is in order. For instance, integration of graphic design in a graphical user interface development process is analogous to industrial design in a physical user interface development process. The industrial design terminology is used here as an efficient means to refer to a creative design discipline concerned with functional things that connect with a human, having a mobile quality, and an intention for mass-production manufacturing.

[‡] While this discussion places aesthetics in the domain of industrial design, I am in no way suggesting that industrial design's only, or even primary, concern is aesthetics; appropriate aesthetics is one of several concerns of industrial design; for instance, design for contextual meaning, innovation, manufacturability, sustainability, value, and ease of use are typical concerns to industrial designers. However, of industrial designers and human factors practitioners, it is likely that only the industrial designer will have the skill and interest to attend to the aesthetics of a design as important in their own right.

[§] For simplicity, the present discussion at times equates aesthetics and form, which is of course an oversimplification—the "presence" of a designed thing goes beyond just its aesthetics (Gilles, 1999). Furthermore, Bill Rogers (personal communication, June 7, 2007) proposed a model whereby form (visual appearance) can be divided into how well form naturally affords function (from Gibson's direct perception theory)—or "functhetics," and how pleasing and perceptually beautiful the form is—or aesthetics. In this model, functhetics feeds cognition, aesthetics feeds emotion, and emotion feeds up into cognition. The cognitions and emotions (built up from functhetics and aesthetics, respectively) then yield performance effects. Bill also pointed out that upon measurement, the functhetics and aesthetics of a design artifact might not be particularly highly correlated.

This sentiment helped me understand why many industrial designers are so hesitant to claim any affection for aesthetics, at least openly when flight deck engineers are in the room. Industrial design is so much more than aesthetics, the designers tell me (and, I should emphasize, I believe–one has only to look at the pioneering work of Dreyfuss and its influence on the industrial design discipline to recognize this truth). Yet, I also think the industrial designers protest unnecessarily. Beauty in and of itself is important and may influence performance.

While accepting that industrial design is indeed more than aesthetics, let us take this issue of aesthetics head-on. Aesthetics, I will assert, is worthy of discussion, research, and understanding in the context of flight deck usability and crew performance. The purpose of this chapter is to open a dialogue on flight deck form and function, to provide preliminary evidence of the importance of aesthetics and crew performance, and to suggest starting points for bringing together utility, usability, look, and feel in practice and in product.

16.2 The Hard Sell of Flight Deck Industrial Design

It is not a surprise that considerations of aesthetics on the flight deck are minimized. Aviation is about physics and engineering, and design considerations that appear to do little more than pretty up a flight deck are easily set aside during the cost–benefit assessments early in system definition. This is not new. Speaking on the state of industrial design in the 1950s, inventor and designer Richard Buckminster Fuller predicted that airframe manufacturers would not suffer the inclusion of industrial design at all:

> industrial design is a very tarnished affair…I assure you that no aircraft company will let an industrial designer through its engineering front door. Industrial designers are considered to be pure interior and exterior decorators. And yet, I've listened to industrial designers assert that they designed the steamship, *United States*. If you were to exhibit schematically all the items that the industrial designers created for the *United States*, you would have sailing down New York Harbour an array of window curtains, chairs, paint clouds and bric-a-brac floating in space, with nothing to really hold it together. (quoted in Woodham, 1997, pp. 75–76)

Suspending judgment on its broad characterization of industrial design for the moment, Fuller's criticism captures the suspicion of typical flight deck engineering regarding industrial design—it does not produce the essence of the engineered system, but is rather all bric-a-brac floating in space—and, by way of extension, has no place in the hard engineering of the cockpit. Zukowsky (1997) confirmed that flight deck design "…usually is the preserve of engineers at aerospace companies…" (p. 67). And an engineering manager at an aircraft company recently told me, "In my experience, things that look better always work worse!"

That statement stuck with me as I struggled to understand it. It conflicted directly with a recently articulated theory in the human–computer interaction literature; specifically, that attractive things work better (Norman, 2004a). And yet there it was, a parsimonious hypothesis, representing years of professional aerospace design practice and expertise: attractive things work worse. These two positions, attractive things work better, and attractive things work worse, summarize the current ambivalent state of our understanding of aesthetics in flight deck design.

16.3 Design and Disappointment

A colleague in product marketing relayed a story to me on why a looks better, works worse relationship makes sense in terms of expectation and contrast. On a business trip, he was choosing a car to rent. There in the lot were the traditional beige and cream sedans, four doors, and a trunk. But then there was the standout—four doors and a trunk still, but now with windows cut in high up, an imposing chrome plated grille, low profile tires with huge rims, and an aesthetic visage that said one thing: I am the boss.

But with the accelerator down, the experience reverted to the beige rental sedan—there just was not much under this hood. And worse, my colleague was stuck with windows that now seemed smallish and difficult to see through. The aesthetic styling or form had set his expectations high, while the performance or function then let him down. In general, many things are designed to look good, and yet they are soon discovered to be in violation of our expectations and even our use of the product. The effect is a basic loss of trust in the immediate product, and worse, a general loss of trust in the brand.

Beyond the concept of expectation and contrast, things that look better may actually work worse because the design that is gauche and provocative might be not only disconnected with the function, but worse, might be unusable or even dangerous to the user. Chapanis revealed just such a case, wherein a purely aesthetic feature in the center of a steering wheel—so designed to "sell"—amounted to "…a spear aimed at the driver's heart" (quoted in Lavietes, 2002). That type of designed disloyalty, wherein something that looks so good ends up treating the user so badly, concentrates the type of distrust that Fuller described decades ago, and absolutely will not fly at any aircraft company.*

16.4 Tailfins and Tailspins

Frequently in consumer products, a form of beauty serves only the function of selling. This is usually why many human factors people become deeply turned off to a focus on aesthetics, because there would seem to be little tie to intended function. What sells one year is not often perceived as provocative enough to sell the next year, and so there is a continuing cycle of planned obsolescence to introduce new and exciting forms to bolster sales. Indeed, revisiting my marketing colleague's experience in the rental car lot, the renter's choice is in part about trying out the new and the exciting (which may or may not be connected with an improvement in function).

To ground our present aviation-centric discussion, let us take a short roadtrip back in time to 1948, when the first tailfins began to take off. At Cadillac, Franklin Quick Hershey's design team introduced tailfins on almost all of the 1948 models (except for the Series 75, Headrick, 2008, pp. 6–7). Cadillac management wanted the tailfins removed for 1949, but public acceptance grew with exposure, and Harley Earl made the decision to keep the tailfins. Over the next decade, the tailfin generally continued its climb, until 1959 when it set an altitude record at about 38 in. above the tarmac (see Figure 16.1), 2.5 in. above the same model year Chrysler New Yorker, and 3.75 in. above the previous model year 1958 Cadillac (Headrick, 2008, p. 66).

What was behind the rise of the tailfin? While it is not possible to tie its climb rate to just one influence, the clear inspiration from aerospace is striking. Starting with 1948, the tailfin drew apparent influence from the P-38 lightning, and continued its climb along with the jet age and the beginning of the space age in the United States.

Anecdotally, shortly after the space age overtook the jet age, the tailfin saw its apogee (about a year after NASA was founded) and began its descent back into the side slab of the automobile. While there were management changes and design battles inside automotive companies that influenced the descent of the tailfin, perhaps it was the very height and audacity of the tailfin that hastened its fall; the trajectory of the tailfin was tied almost entirely to the environment outside of the automotive industry and this became more obvious as tailfins became more prominent and aggressive in the late-1950s. Tailfins were, and still are, a formal sign of the times, with little if any tie to function. Roughly, the tailfin saw its rise and fall over about a decade and a half, and was finally memorialized, critiqued, grounded, and buried in the Ant Farm performance art work, *Cadillac Ranch* in 1974 (Figure 16.2).

---

* While Fuller's critique names industrial design, it is unfair to suggest that modern industrial design is obsessed with bric-a-brac or ornament. In fact, entire design movements, such as the modernist movement, railed against ornamentation, so much so that one of its pioneers, Adolf Loos appropriately titled his work "Ornament und Verbrechen" or "Ornament and Crime," and Le Corbusier proclaimed "Trash is always abundantly decorated…." (quoted in Woodham, 1997, p. 33).

FIGURE 16.1 Tailfin apogee: Inspiration from the sky.

FIGURE 16.2 Cadillac Ranch 1974 Ant Farm (Lord Marquez Michaels). (Photo courtesy Wyatt McSpadden Photography, www.wyattmcspadden.com).

FIGURE 16.3   Porsche 911 in 1977 and 2006.

This is not to say that finding inspiration or innovation from an alternative environment is a misstep (such as automotive looking to aviation), or that function necessarily must come before form. While the tailfin was not lasting, it was not useless either. It served a reflective design purpose for consumers, and a business purpose for the corporation. Beautiful forms, sometimes especially the provocative, temporary forms, bolster sales. But is it necessary that beautiful forms and useful function cannot simultaneously coexist and mature over long periods of time to serve business goals? To discover an answer and adopt a balanced approach to formal and functional integration, we need to avoid the extreme assumptions. For instance, one extreme assumption is that provocation is the only way to sell. Another extreme assumption is that any attention to aesthetics is provocation and nothing more.

In an interesting example of balance, the Porsche 911, first introduced in 1964, has kept its same basic form as well as its seductive appeal for over 40 years (with, of course, many significant design refinements). There are most likely many reasons and interpretations as to why, but one argument is that the 911's original design was an expression of integrated form and function, and that basic structure was sustainable, extensible, and desirable over an almost unfathomable number of years for automotive design (Figure 16.3).

Encouraging form and function to grow together may be a way to generate designs that avoid the microbursts embedded in the storms of style. Moreover, it is very likely that this integrated approach will generate products that retain long-term value and meaning in retrospect that goes far beyond a sign of the times.

Unlike consumer products that may be tossed aside in a few years, flight decks must be lasting in their appearance and their use—they can easily last for 25 years with little upgrade. Tracing back to my marketing colleague, to the extent that any form is disconnected from the function, it runs the risk of being part of violated expectations (which, incidentally, influences what we believe to be beautiful). The Porsche 911 touchstone gives us confirmation that great function and seductive form can and should go together. Aesthetics, usability, and function, they all can be in harmony. Moreover, if managed carefully and tied to core values, the added benefit of this integration is a discernable brand that can last, produce an enviable product, and return sustainable profit.

16.5 Should Human Factors Care about Appearance?

Selling lasting, meaningful, and profitable products is a fine business goal, but human factors practitioners tend to care more about intended function and usability in delivering that function. The basics of physics and physiology in aviation—weight, power, heat, thrust, strength, time, error, performance—magnify the importance of function and thus make it quite impossible for us to loosely assert that an aesthetic detail has a functional role without some hard data behind it. Moreover, linking the form of a design to usability performance benefits is not a traditional role for human factors, since the discipline tends toward function without much attention to form, and further tends toward cognition without much attention to emotion. In other words, it may be acceptable to allow good looks, so long as these looks do not get in the way of the basic system function. The assumption is that the appearance or styling is separate from the function and usability of the design, or alternatively, that a purely functionalist approach automatically yields good form. And therefore human factors specialists do not concern themselves with the appearance of things as an end unto itself. Industrial design, of course, does. Lindgaard and Whitfield (2004) asserted,

> Aesthetics is even ignored in areas of applied scientific research where its impact would seem self-evident. Thus, research in affective computing overlooks it (Picard, 1997), as largely does Human Factors Research, despite the existence of an entire industry—the design industry— increasingly devoted to it. This oversight is inevitably due to the failure to identify aesthetics as either a cognition, an emotion, or both. Aesthetics, therefore, lacks an affinity with the main paradigms of psychological research and, therefore, has no secure theoretical attachment point: it lacks a home. However, aesthetics is so prevalent that it must be something. (p. 74)

Thus, the overt split in perspective between human factors and industrial design is not unique to aviation. Only recently has the discussion of aesthetics and emotion come into vogue as a legitimate topic in the human factors literature, under labels such as hedonomics (Hancock, Pepe, & Murphy, 2005) and emotional design (Norman, 2004a). For instance, the Association for Computing Machinery devoted a special issue in *Communications of the ACM* to address human computer etiquette (Miller, 2004), the journal *Interacting with Computers* provided a special issue on emotion and human–computer interaction (Cockton, 2004), the journal *Human-Computer Interaction* published a special section on beauty, usability, and goodness (Norman, 2004b), and the ACM's *Interactions* magazine published a special section on funology (Blythe, Hassenzahl, & Wright, 2004). In sum, more and more research in the human–computer interaction domain is suggesting that human factors professionals need to be concerned about aesthetics for more than just the usual reasons of market acceptance and sex appeal, specifically, for reasons more closely aligned with the performance concerns of human factors. For flight deck design, it is very likely that this illumination of the aesthetics–performance relationship could come from a fusion of industrial design and aviation human factors.[*]

16.6 Some Evidence of Industrial Design on the Flight Deck

Integration of industrial design and human factors has happened in consumer product and information technology systems design, and industrial design of course has substantial representation in automotive design. Examples of its inclusion in flight deck design, however, are harder to find. There is a first problem that industrial design may be misunderstood and viewed with suspicion, and therefore may not have substantial inclusion or representation on the flight deck.[†] Confounding the problem is the fact that resulting documentation from flight deck industrial design that *does* occur is either so scarce or so secret as to prohibit any meaningful dialogue that could illuminate the relationship of flight deck aesthetics and crew performance—that is, the historical record is very poor. Zukowsky (1997) lamented,

---

[*] While this discussion generally is cast in the context of attractive and desirable things, a thoughtful approach to emotional design actually considers the appropriate function of design, aesthetics, and emotion more broadly. There are certain elements of flight deck design that are very invasive, very uncomfortable, and generate a visceral reaction that is quite intense and unwelcome. Items on the non-normal, versus normal, checklist provide good examples for consideration. A no takeoff alert, for instance, accomplishes its purpose when it focuses attention like a laser and calls the crew to action. Making the alert "attractive" or "desirable" as it were, would be a misstep indeed. This raises the greater question, what are the *appropriate* aesthetics for emotional design? Norman (2004a) discussed this issue of emotional design for a given context, and Noam Tractinsky (personal communication, May–June 2007) pointed out its tie to the situation, for instance, a high-workload versus low-workload phase of flight and autopilot mode engagement or disengagement.

[†] In writing about industrial design in a chapter that is limited to aesthetics, there is obviously the risk of inadvertently deepening this misunderstanding of industrial design; the reader is asked to recognize that the limitations of this chapter's scope are not indicative of any limitations on the scope of industrial design.

Aerospace museums celebrate aircraft as designed objects as well as engineering accomplishments, and they are included in important museum collections of design. Yet, the work of industrial designers for the aviation industry is among the least known, and least researched of topics related to transportation in our era. For whatever reason, there is no real chronicle of industrial design for aviation, and what we know of the subject is filled with inaccuracies. While preparing an exhibition and catalog on architecture and design for commercial aviation, I encountered numerous difficulties in even finding basic documentation on designers and what they did. (p. 66)

Despite these general difficulties, a few instances of documented flight deck industrial design give us at least a small sample to consider. A few contemporary air transport flight decks have demonstrated particular attention to industrial design. For instance, the Boeing 777 (Figure 16.4) received in 1993 an Industrial Design Excellence Award (IDEA) for its flight deck (Boeing, 2008), and in its marketing artifacts, Boeing acknowledged the role its long-time industrial design partner Teague played in the design of the 777 flight deck (Boeing, 2003, p. 30). It is important to restate that the holistic effect of human factors and industrial design is, not surprisingly, far beyond surface aesthetics, and assists in making a substantial and meaningful connection between pilot and flight deck interaction and brand in this example.

Further, Airbus's common cockpit concept not only sets the expectation for pilots in terms of function—such that operational experience in one Airbus flight deck sets the stage for operations in another Airbus flight deck, but also in terms of form—such that the feel of brand familiarity is communicated in an instant and throughout the interaction: the design language and details of one Airbus flight deck are recognizable across the Airbus fleet. Even as it was developing its first cockpits, such as that for the A310, Airbus consulted with Porsche on the design of the flight deck (Zukowsky, 1996). Zukowsky (1997) asserted, "this kind of flexibility in design contrasts with the rigidity of most aerospace companies, who often consider the overall aircraft form and flight deck the domain of their aerospace engineers, and not the more aesthetically inclined industrial designers" (p. 77).

For a plane to fly well, it must be beautiful.

Marcel Dassault

FIGURE 16.4   The Boeing 777 Flight deck won the 1993 Gold IDSA IDEA. (Photo courtesy Boeing.)

FIGURE 16.5   HondaJet VLJ. (Photo courtesy George Hall/Check Six 2007.)

While we currently have just a handful of flight deck industrial design examples, the opportunity to see more industrial design in cockpits is on the increase, particularly with the advent of the VLJ (e.g., Eclipse 500, HondaJet, Citation Mustang, Grob sp^n, Embraer Phenom 100). Major design consultancies such as Porsche Design Group and BMW DesignWorks have contributed both cabin and cockpit designs for VLJs, and IDEO recently won an IDEA for the interaction design of the Eclipse 500 cockpit (Scanlon, 2007). It is important to pause here and emphasize that the IDEO work is a noteworthy example of an interdisciplinary approach, which considered human factors and industrial design together (as well as other disciplines). Further, the HondaJet (Figure 16.5) comes from a corporate heritage that values and routinely integrates human factors and industrial design in meaningful ways, particularly in automotive design. While the final cockpit is still under development, it is very likely that the integration of form and function will receive substantial attention in the design of this flight deck.

That said, it is still generally the case that industrial designers are not commonly found within flight deck design groups at *avionics* manufacturers, where functional operation of the flight deck is defined, distilled, and manufactured, and where integration with human factors activities would be most meaningful.

In addition to VLJs, light general aviation aircraft cockpits are being more informed by industrial design, as demonstrated by levels of fit and finish that strive toward automotive standards of refinement (although at times simply taking the rather ornament-only form of a burl walnut appliqué). Also in this market, the use of external design consultants that traditionally work in automotive or consumer product spaces is on the increase.

There is a reason for the growing interest in industrial design. The light general aviation and VLJ product lines can remove the difference between the jet aircraft buyer and the jet aircraft pilot. Now, the person with the principal financial stake in the jet aircraft may actually spend time piloting that jet—and design *is* important to this person. The market positioning targets a demographic that expects industrial design integration in its automobiles, its mobile communication devices, its wristwatches—everything that is designed for it. Having a cockpit panel design that could have been executed just as well by a plastic bucket manufacturer as an airframer disconnects with the VLJ's target buying and flying demographic.*

---

* John Zukowsky (personal communication, June 1, 2007) found a historical willingness to involve designers in light aircraft design, where the product line is treated more as a consumer product than a commodity. "This was true, I found, with Charles Butler hiring a car designer, Peter Cambridge, to completely design the Bell Jet Ranger, inside and out. That was in the 1960s and other manufacturers like Hiller (w/Raymond Loewy) followed suit. Likewise, Richard Ten Eyck, after working on the interior/flight deck of the first Beech Model 35 Bonanzas after WWII, was hired by Cessna to style, inside and out, their personal lightplanes…just as an auto designer would style a car, inside and out."

Yet even as we will continue to see an increase in flight deck industrial design activity over the next decade as VLJs grow up, it is questionable how much industrial design will actually be allowed to touch, and have a major say on, the core functional elements of the flight deck, including controls, displays, and graphical user interfaces, versus simply owning the general interior design aesthetic of the overall flight deck environment. That is, how much will industrial designers be allowed to address the tasks within the aviate, navigate, communicate, and manage systems functions?

At a basic level, even if industrial designers get the opportunity to interact with more and more functional aspects of flight deck design, a misunderstanding of industrial design and the rift with flight deck or human factors engineering groups may increase as flight deck engineers feel that something that should be purely functional is getting too much aesthetic treatment. This is in part because our knowledge of industrial design is superficial and our understanding of aesthetics and pilot performance is paltry.[*] Consequently, today we are still facing the fundamental problem: is it looks better, works better, or looks better, works worse? Or is there more complexity to the aesthetics–usability relationship? This is a compelling problem, because until we understand the interaction and appropriately plan for integrated form and function, we are unlikely to find ways to reliably create designs that are as usable as they are beautiful. And we will continue to experience great difficulty in integrating industrial design and human factors effectively when such an integration could be enlightening and beneficial to both disciplines.

16.7 Looks Better, Works Better

Why should a human factors professional care about beauty? Norman (2004a) proposed that attractive things work better, which puts beauty on a footing to drive human performance. Taking the antithesis, unattractive things should work, well, worse. The implication then is if we do not pay attention to aesthetics, our designs may deliver unintended performance effects based on their relative attractiveness.[†]

> When I am working on a problem,
> I never think about beauty.
> I think only of how to solve the problem.
> But when I have finished, if the solution is not beautiful, I know something is wrong.

Richard Buckminster Fuller

So, we have a few gaps becoming apparent already. First, industrial design is poorly represented on the flight deck, and its research is poorly documented. Second, the flight deck human factors literature is very nearly silent on the topics of aesthetics and emotion. A third, and most specific gap, is in the hypothesis itself that links aesthetics and performance—the attractive things work better theory needs more articulation and evidence.

---

[*] Beith and Jasinski (2006) noted that "the underlying difference between HF/E [human factors/ergonomics] types and most researchers is that HF/E specialists are both researchers and designers. This last orientation is often forgotten or lost in translation, perhaps due to the inability of many HF/E researchers to draw a straight line or a recognizable stick figure. It is also one of the underlying reasons for the separation between industrial design and HF/E that should never have occurred." (p. 26)

[†] For the sake of this discussion, we will limit attractive to mean visually attractive, even though it could certainly refer to attractiveness felt on any of the senses (tactile, auditory), or more importantly, to something beyond a purely visceral reaction, for instance, the cognitive-emotional response to a socio-technical experience. Noam Tractinsky (personal communication, May–June 2007) pointed out that the aesthetics of reflective design impacts both cognition and emotion.

16.8 Clarifying the Hypotheses

Even limiting attractiveness to only the visual sense, we are faced with a substantial problem of definition. Attractive things may indeed work better, but it is important to understand what is really meant by attractive, and what is meant by work better. For instance, "attractive" could mean beauty in the eye of the beholder, in which every individual maintains a personal definition for beauty for every environmental object. Alternatively, an attempt could be made to define beauty objectively according to some measurable universal characteristic or group-accepted standard. The concept of the golden section, for instance (in which the rectangular ratio of 1.618:1, about the aspect ratio of a wide format HDTV or laptop computer LCD screen), has been explored by psychologists such as Fechner for its application as a universal indicator of beauty. While the golden section notably lacks any conclusive support for a clear relationship with beauty, it is exemplary in the attempt to create a notion of objective, universal beauty, in which a dimension of nature is used as a rule for beauty in the designed environment, in a reducible, measurable way. The idea of distilling objective definitions for beauty may have value, particularly where a specific user population is able to be specified and understood sufficiently. Beauty, therefore, could be considered subjectively to the individual or objectively to some group criterion.

Similarly, "work better" could mean more usable according to a subjective measure (such as perceived usability or self-reported workload) or an objective standard (such as response time and error magnitude measures). Usability, then, can of course also be subjective or objective.

To date, most research has dealt with the looks better, works better hypothesis in one specific, and entirely subjective form:

Things perceived to look better → **are perceived to perform better**

Stated another way, if we design an object such that beauty is perceived by the user, that user will also perceive better usability in the interaction with the designed object. But there are at least three other ways to think of the looks better, works better hypothesis. We can visualize the crossing of objective and subjective aesthetics with objective and subjective usability to generate a matrix of the four hypotheses as shown in Figure 16.6.

Let us consider how one might operationalize beauty subjectively and objectively. For simplicity, we will return for the moment to the example of the golden section, and setting aside its lack of empirical support, use it as the hypothetical measure for objective beauty. We could take a range of visual data displays, some being 16:9 aspect ratio, some being 4:3 aspect ratio, and easily rank which better fits the

		Usability	
		Subjective	Objective
Aesthetics	Subjective	Things perceived to look better → are perceived to perform better	Things perceived to look better → perform better on an objective standard
	Objective	Things that look better by an objective standard → are perceived to perform better	Things that look better by an objective standard → perform better on an objective standard

FIGURE 16.6 Subjective and objective linkages between aesthetics and usability.

objective definition of beauty (as defined as 1.618:1) on physical dimensions alone—in this case, the displays trending toward the widescreen 16:9 aspect ratio are more inherently beautiful than the displays trending toward the 4:3 aspect ratio according to our chosen beauty metric.*

Alternatively, if we wished to get at subjective beauty of the interfaces, we could take the same sample of data displays and through a series of pairwise comparisons, ask users to complete an eye test for beauty (e.g., "do you prefer A or B? A or B? A or B?"). From the results of the comparisons, we could then rank the user's definition of beauty based on personal preferences. In addition to the ranking technique, we could implement some survey form that allowed users to rate elements of beauty (such as those dimensions and items discovered by Lavie & Tractinsky, 2004). Which one is more beautiful? The definition of beauty is, in this case, defined in the eye of the user.†

The concepts of objective and subjective usability are quite likely much more familiar to human factors professionals than the concepts of objective and subjective beauty. We could conceptualize a rudimentary experiment in which users have to complete a series of data manipulation tasks using the most beautiful or least beautiful display interfaces. In measuring the usability of the displays, we could use objective metrics like response time and error rate.

Alternatively, we could employ subjective metrics like NASA Task Load Index (TLX) workload wherein users self-report their workload on scales of physical demand, mental demand, temporal demand, performance, effort, and frustration. We could even use a single rating scale item called usability, with endpoints that represent low usability and high usability and simply have users rate the overall usability.

Assuming a purely scientific lens, what we would really wish to do is to move from the top left box in Figure 16.6 to the lower right box, from the subjective and perhaps more variable to the objective and more repeatable. This is because the lower right box means that we have been able to reduce the aesthetics to a repeatable quality, and further can show its usability effects on a measurable performance scale. While perhaps a bit sterile for the artist, to the engineer this is the most meaningful use of the looks better–works better hypothesis, because it is one that can be used in a production environment and delivers effects that satisfy quantifiable aesthetic and usability targets. Also of note, the ability to study, measure, and quantify is important in the aerospace environment for human factors and certification testing.

16.9 A Skin Study

In a study on aesthetics and performance performed at Honeywell (Gannon, 2005), I asked a sample of 24 pilots (mostly airline transport and business jet pilots) to rate and rank the visual appeal of different Primary Flight Displays or PFDs (see Figure 16.7), and to hand-fly a series of instrument approaches in a part-task simulator, collecting subjective workload and objective performance data from each run. While only the aesthetics of the PFD interface skin was manipulated, this singular difference was not disclosed to pilots until after the test was complete.

The quantification of aesthetics was measured pilot by pilot (i.e., beauty in the eye of the beholder) using pairwise comparisons and the rating dimensions developed by Lavie and Tractinsky (2004). Thus, aesthetics was subjective and not operationalized as an objective quality. I did, however, consider

---

* For designed objects that need to be mass-produced, eventually beauty has to be standardized to a repeatable quality. It may not be important to demonstrate an object's beauty with reference to an absolute objective criterion such as the golden section. What may be more important is that a preponderance of the perceptual evidence supports the assertion that a designed object is indeed beautiful to the target user group, in this case, pilots of a particular category and class of aircraft.

† As further research is conducted, we may find that eventual "objective" measures for beauty are more accurately described as group preferences, where the mean of internal, subjective individual preferences are translated into a group preference, which becomes the external objective standard, a sort of objectivity via summed subjectivity. More apparently objective, perhaps, are those elements of aesthetics that have nearly hard-wired emotional tie-outs, the types of things that would fit into Norman's (2004a) visceral level of design (e.g., unfamiliar, loud, fast, visually expanding, dark objects elicit the same emotional response in most any human).

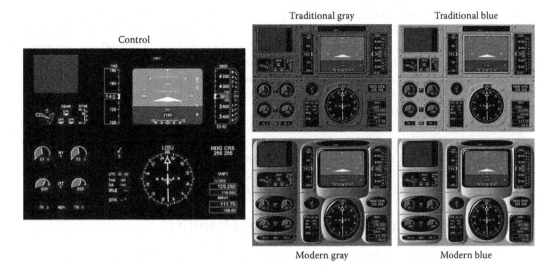

FIGURE 16.7 PFDs and skins. (From Gannon, A.J., The effects of flight deck display interface aesthetics on pilot performance and workload, doctoral dissertation, Arizona State University, Tempe, AZ (UMI No. 3178236), 2005.)

usability as both a subjective and an objective quality, using the NASA TLX as the subjective workload metric and using flight technical error and alert message response time as objective measures. Therefore, I tested the hypotheses in the top row of Figure 16.6.

Upon analyzing those PFDs that individual pilots considered to be the most and least attractive, I found a linkage between aesthetics and NASA TLX workload. Specifically, I found significantly ($p < 0.05$) lower overall workload associated with most attractive PFDs, and significantly higher overall workload associated with least attractive PFDs. That is, if pilots believed that a PFD was most attractive, they also associated it with significantly lower workload than a PFD that they perceived to be least attractive. So there it is: things perceived to look better are perceived to perform better.

What I did not find, however, was the elusive aesthetics–performance linkage. That is, in terms of objective flight technical error and alert message response time, there were no significant performance differences in using the most attractive versus the least attractive PFDs.* Thus, the importance is made evident of clarifying what we mean when we say looks better, works better.

It is critical to emphasize that the skin study of Gannon (2005) was in no way intended to suggest an appropriate method for integrating form and function. To be sure, this research was the epitome of spray-on aesthetics, applied entirely after the functional details were complete, that is, a flagrant segregation of form and function (though notably, perhaps not unlike the approach of many real-world product development efforts). The study's purpose was theory building, with the stimuli created to test the looks better, works better hypothesis with frugality, and we found that while pilots thought their most attractive interfaces were easier to use (i.e., lower workload), their objective performance with these interfaces was no better than their objective performance with their least attractive interfaces.

---

* An interesting question is whether objective performance can be driven by particularly good or particularly poor aesthetics alone, and we did not test for this. Rather, the interfaces were colored in likely flight deck colors, and the skins selected were considered to be within the boundaries of reasonable skins for a flight deck (to prevent intentionally skewing the results). Could a departure from reasonable skins drive performance, for instance a hot pink base color with a mohair texture? It seems likely that at some point, aesthetics could be made to drive performance, at which point the semantic lines between what we define as pure aesthetics and pure usability are likely to blur...and pointing out a transition zone that is extremely important and not well understood, specifically, the boundaries and qualities of that zone wherein aesthetics and usability are inseparable qualities.

The segregation of form and function at design time (i.e., spray-on aesthetics separate from the underlying function) is identified in Gannon (2005) as the reason that the looks better, works better hypothesis (as focused on objective performance) unravels in the first place. The hypothesis itself stove-pipes industrial design and human factors, for human factors practitioners will not be creating "looks better" at design time (the first half of the hypothesis), and industrial designers will not be measuring "works better" at evaluation time (the second half of the hypothesis). So, the disciplines are even kept at arms length in the very words of the hypothesis itself!

The central thesis of this chapter, and Gannon (2005), is that industrial design and human factors should be integrated at design time to enjoy the benefits at evaluation time. In other words, as an inter-disciplinary (rather than a multidisciplinary) theory statement, integrating designed aesthetics and designed usability will yield a functional thing measurably better than the sum of the benefits from beauty and usability design contributions made separately.

16.10 Aesthetics as Cover up for Poor Usability

The Gannon (2005) study largely focused on the potential benefits of appropriate aesthetics, but there is another reason to believe that flight deck aesthetics are important. In a study of automated teller machines, Tractinsky, Katz, and Ikar (2000) found that study participants believed that attractive interfaces were significantly more usable than unattractive interfaces, even in cases where the researchers had degraded the function of the attractive interfaces through errors and latencies. In other words, users believed that the attractive interfaces were more usable, even though the unattractive interfaces without the usability problems would perform better on purely objective metrics. Here, we see fascia serving as a cover up, not only to the physical ugliness of a system, but more importantly as a cover up to the inherent un-usability of the system. This latter cover up is much more worrisome, since it suggests a lack of transparency of the system to the user. Using aesthetics as a cover up to poor function or usability is at best insincere, and at worst seditious. Karvonen (2000) noted:

> What may be surprising in this world of ours that so admires the assumed rationality of the *homo sapiens,* is that such an issue as trusting a service to be reliable and secure is, at least partially, based on something as irrational as aesthetics seems to be. If a Web site strikes me beautiful, I will gladly give away my credit card number—is that how it goes? With our Swedish users, this is exactly what we experienced: users admitted to making intuitive, and rather emotional, on-the-spot decisions to trust a service provider when shopping online. A user comment included 'if it looks pleasant, I just trust it.'" (p. 87).

However, note on the other hand that Lindgaard and Dudek (2003) found that website beauty and usability were not necessarily always positively correlated, and Karvonen (2000) suggested (based on Norman's *The Psychology of Everyday Things*) that beauty as a cover-up to an unusable system turned against that system, "ugliness connotes effectiveness" (p. 88). Further, Murphy, Stanney, and Hancock (2003) found that attractiveness only enhanced the perception of usability when the underlying usability was good. Are some specific users using aesthetics as a rule of thumb to inversely correlate usability? Or is the effect one of realizing that the pretty system is cheating versus not being aware?[*]

---

[*] Noam Tractinsky (personal communication, May–June 2007) noted that sometimes beautiful things will work better and sometimes they will work worse, and what is important is teasing out the processes or factors that can drive each case. Further, Tractinsky cited the importance of personality variables (and the need to identify these factors) in predisposing one to believe that beauty either connotes good or poor functionality.

Whatever the effect is, aesthetics are serving a purpose. Regardless of *how* aesthetics are interacting with usability, for now the more important point is understanding that aesthetics *are* interacting with usability.* Engineering psychologists have for several decades been concerned about mode confusion on the flight deck, wherein the novel and complex and highly integrated aspects of the system design conspire to create a lack of transparency among the system's states and the crew's perceptions. Similarly, with the evidence that beautiful form can mask poor function, we can plot the relationship using the terms of signal detection theory or hypothesis testing as illustrated in Figure 16.8.

> "Beauty is truth, truth beauty" – that is all
> Ye know on earth, and all ye need to know.
>
> **John Keats**
> *Ode on a Grecian Urn*

Assuming that we have a system with degraded function but good form, and the user does not perceive the degraded function (as in Tractinsky et al., 2000), we are using aesthetics as cover up for a degraded system, and are in the lower right quadrant, a miss for the user. Particularly on the flight deck, it is critical that the true state of the system is being conveyed to the pilot, and that there is nothing hiding this state (be it poorly designed modes or aesthetics or some other factor entirely).

Clearly, aesthetics of form applied inappropriately and segregated from the function is not the right answer. Even if the system creates just annoyances and no real safety danger to the user by hiding the system state, the user will eventually find out about the infidelity. In highly interactive systems, this infidelity breeds mistrust. Just as in our human-to-human relationships, in human-to-machine relationships, it is not okay to have a beautiful design that cheats on the user. Transparency is the basis for trust.

		True system functional state	
H_0: No system degradation present i.e., No effect difference between desired and present system state H_1: System degradation present i.e., Effect difference between desired and present system state		Function not degraded (H_0 True)	Function degraded (H_0 False)
Perceived system functional state	Degradation detected (Reject H_0)	α/Type I Error "False Alarm" False Positive	Correct detection "Hit" True Positive
	Degradation not detected (Accept H_0)	Correct nondetection correct rejection True Negative	β/Type II error "Miss" False Negative

FIGURE 16.8 Use of a hypothesis testing or signal detection matrix to explore aesthetics masking usability.

---

* Moreover, Tractinsky (personal communication, May–June 2007) emphasized the importance of conceptualizing aesthetics as a multidimensional concept, with some dimensions that are also highly correlated with usability. Specifically, "Two of the significant contributions of Lavie and Tractinsky (2004) are: (1) that we empirically demonstrated that such [multidimensional] conceptualizations exist, and (2) that at least one such dimension is highly correlated with usability....Some principles of usability (e.g., order, grouping, clarity) can be regarded as aesthetic principles as well. The bottom line is that not only are usability and aesthetics not two opposites, but they may not be orthogonal as well. (I think that this is one more argument in favor of the integration of HF and ID.)"

So, now we have a few reasons to integrate human factors and industrial design activities. First, aesthetics can mask usability problems—if not indefinitely, at least for a while. Second, people translate aesthetics as usability. And we need a means to integrate form and function, because by identifying this means, we may be able to realize real performance benefits, which is a very appealing, final reason to link industrial design and human factors. Tractinsky et al. (2000) clarified:

> ...the advantage of aesthetic interfaces might transcend the mere (yet important) perceptions of the system's quality to the domain of actual performance. As demonstrated by Isen [46], positive effect is likely to improve decision making and creativity. Thus, the positive effect created by aesthetically appealing interfaces may be instrumental in improving users' performance as well. (p. 141, original emphasis)

It is interesting that the researchers emphasize decision-making and creativity. While much of human factors work has centered on the interface details like appropriate text size and color assignment, how to help pilots become better decision-makers and make better judgments to avoid the omnipresent human error accident remains elusive. Aesthetics are not a panacea, but dealing with the emotional underpinnings of decision-making and creativity could be a route to improved accident statistics. Even the Federal Aviation Administration's (1991) list of hazardous attitudes (e.g., invincibility, macho, resignation) manifest as emotions, do they not?

16.11 Beauty with Integrity

To realize the potential of integrated form and function, what we are after in our designs is a beauty with integrity that can grow and be sustained through the years. Even as our definition of useful and usable functions matures with age and technology progression, our definition of beauty can also progress with time and context. Let us consider how this progression can manifest within the context of seeking out design integrity.

Passengers generally favor turbofans to turboprops. Turboprops are considered noisy, slow, and ugly, while "jets" are thought of as quiet, fast, and sleek. Yet, new turboprops can provide better fuel economy, and in an era of inflating petroleum costs, a functional linkage to this underlying green characteristic could drive a trend back toward turboprops. But would this possible move be at the expense of passenger desirability? That is, must we accept that sometimes function must trump form, that fuel economy, and choosing green means we must select an option that is inherently ugly or affectively undesirable?

Consider the Piaggio Avanti as an example (Figure 16.9). While the aircraft uses twin turboprop powerplants, the aesthetic classification of the aircraft is not foremost as a "turboprop" (i.e., a statement of specification on basic function). Rather, the impression is more generally that of a sleek aircraft, a striking aircraft, a distinctive aircraft, even a beautiful aircraft. The Piaggio is fast, it is efficient, it has a quiet cabin, and it is beautiful, all at once. Not only does the aircraft's form influence what we think can be beautiful, e.g., "turboprops are beautiful again," but more fundamentally, it influences what we think of as being a modern airplane. Perhaps, the large wing does belong in back. Perhaps, the propellers should face the other way. Perhaps, that is what a turboprop engine looks like.

It is notable that in new airplane designs, sometimes an aesthetic detail (window design, for instance) will be changed in initial visualizations to give an impression of a more modern aircraft, and only thereafter are the aerodynamic analyses done to determine to what degree the aesthetic changes conflict with aerodynamic efficiency.* Yet the case of the Piaggio suggests that we need not sacrifice form at the

---

* Interestingly, window design has a history of creating issues between designers and engineers. John Zukowsky (personal communication, June 1, 2007) noted, "...I recall that Norman Bel Geddes had a run-in with engineers at Martin about the placement of the windows in the M130 (and he lost, as I recall)...Raymond Loewy fought with NASA engineers to include one in Skylab, America's first space station." Zukowsky proposed that engineers view windows as penetrations that weaken the structure, while designers view windows as a means for giving a point of reference to the human user.

FIGURE 16.9   Italian-designed Piaggio Avanti. (Photo courtesy George Hall/Check Six 2007.)

expense of function nor sacrifice function at the expense of form. Indeed, we can enjoy both. Meeting new functions, such as the green function, in fact can help us see new definitions of what beautiful can mean and what the depth and breadth of beautiful should be. And considering how to endow beauty on a system can help us imagine new ways of implementing function or generating new, useful functions entirely. That is, form informs function and function informs form. Function with ignorance to beauty is not the goal, because it will produce a design that is less informed and less valuable than it could be (often in its very functional aspects). Likewise, beauty as fascia or cover up is not the goal, and indeed does not fit a long term, or meaningful basic definition of beauty.

At the core of this approach is the idea of integrity rather than primacy. That is, it is not so important which came first—form or function, but rather that form and function should coexist and grow, magnifying the goodness of each other. It does not so much matter whether it was form or function that was there first in the beginning, so long as both are there in the end.

16.12 Interdisciplinarity Yields Skill Diversity

Setting aside the notion of the designed product's characteristics for a moment, there is a design process and skill-based reason for integrating industrial design and human factors. Using Bloom and Krathwohl's (1956) taxonomy of educational objectives for the cognitive domain,[*] we find very different approaches to the skills of analysis, synthesis, and evaluation among human factors and industrial design practitioners. While both are concerned with human interfaces and interactions, the traditional processes toward developing the product can be very different.

For instance, a human factors practitioner using the synthesis skill might assemble data from specifications and past programs and write design constraints for the new product. An industrial designer, on the other hand, might take their knowledge of the design context and purpose, and put pencil to paper or hand to clay, and generate new design options based on a creative brief.

Analysis in human factors means task analysis, in industrial design it means ethnography. Synthesis in human factors means reaching for a specification, in industrial design it means reaching for a sketching pencil and paper. Evaluation in human factors means getting statistical significance, in industrial

---

[*] Incidentally, these researchers also understood the importance of emotion and affect in education (vs. design), cf. Krathwohl, Bloom, and Masia (1964).

design it means getting formal critiques from experienced peers to learn if the concept aligns with the context. Fusing the different approaches to analysis–synthesis–evaluation is where the interdisciplinary value lies. Beith and Jasinski (2006) stated,

> The integration and interaction of industrial design and human factors can be challenging. Because the players are so different, there can be a natural angst associated with the interactions and ultimate compromises needed to find the best solutions and the best designs. However, when combined effectively, industrial design and human factors can produce dramatic and very successful results. (p. 29)

The integration of human factors and industrial design activities has become a topic of discussion and conference across both the Human Factors and Ergonomics Society (HFES), and the Industrial Designers Society of America (IDSA). There is an increasing appreciation of the overlap between the disciplines, as well as a growing respect for their respective and unique skill contributions as well as their inherent connectedness and compatibility. A handful of design schools are including formal human factors content in their curriculums, while somewhat fewer engineering psychology departments are also integrating industrial design.

In realizing this interdisciplinarity, the implications for a much deeper, much richer product are far beyond the simple aesthetics focus of this chapter. We have two disciplines with fundamentally different approaches to analysis–synthesis–evaluation, and the outcome of their integration on the flight deck is the opportunity for formal and functional integrity.

16.13 Summary and Next Steps

We have broadly assumed for many years that emotion does not belong in the flight deck—humans in the cockpit are meant to do their jobs rationally and efficiently, serving as an extension of the aircraft system. In fact, pilots themselves may be the first to claim that accommodation of emotion is bunk, the rational pilot is meant to have nerves of steel, and to be emotionless. It is not the purpose of this chapter to suggest that this basic discipline of emotion is misplaced—indeed it is a necessity in aviation. But even the concept of nerves of steel is conveyed as an emotion, and we can certainly consider the design's role in facilitating the generation of appropriate feeling with formal and functional integrity. Moreover, we are faced with the probability that emotions can easily drive and constrain cognitions.

At its core, electing to fly airplanes is not a terribly rational choice. There are certainly easier ways to make a living, and the risks and rewards of aviation are not always logically connected. We often hear of the love or romance of flying, and are well acquainted with the inspiration and wonder that human flight engenders. Yet, all of these characteristics are predominately emotional in nature, even as the job of flying is mostly, we think, cognitive and physical in nature. Thus, we have a substantial rift— the emotionally disciplined individuals who excel at flying an aircraft most likely resonate with the emotions that flying gives them.

Accounting for the emotional aspects of flight deck design, whether through visual aesthetics as discussed here, or through feelings of aircraft integrity as suggested by integrated form and function together, or by some yet unresearched aspect of design, seems an obvious area in which to extend our knowledge. What feeling does a pilot need to receive from a flight deck? A confirmation of control? A feeling of precision? An emotion of engagement? A validation of power? And what are the consequences when the design adversely affects, puts the pilot out of control, or engenders distrust? And what are the appropriate modalities and interactions of aesthetics and emotional design for flight decks— visual, auditory, tactile? And how do they interact with basic spatial and temporal perception, cognition, and decision-making?

A number of questions like these are yet unanswered. My hope is that by moving beyond the fundamental block of "I don't care how it looks, I only care how it works," we can move on to the larger issues of understanding: all of the design thinking that is uniquely industrial design, all of the functional thinking that is uniquely human factors, and the intersection of the disciplines that signals greater understanding and interchange among the disciplines.

A sampling of future recommendations for research includes:

Remove the Catch-22. A core problem is that industrial designers do not frequently get invited to work on flight deck functional problems, because they do not have experience, which they could only gain by being invited to work on them. At a very practical level, getting experience in emotional design on the flight deck means putting industrial designers to work on the interdisciplinary teams working on functional flight deck design. Well-known industrial design firms are contributing to flight deck design for an increasing number of jet aircraft. We are truly at an exciting juncture, for the industrial design for tomorrow's cockpits has the potential to go well beyond surface aesthetics, and extend to the functions and interactivity of the system.

Show the integrated process. Practically, how do we set about doing the work of integrated flight deck aesthetics? This research should explore the practical aspects of assembling a team that can address form and function together, and answer how integrating the team's disciplines helps in integrating a design of deep formal and functional integrity. Articulating an integrated design process for aerospace may be informed by those nonaviation companies that are already doing this as a given necessity.

Describe beauty in the eyes of the pilot. Define appropriate aesthetics—what is objective beauty to a pilot? Is there some definition that a majority can agree on? Can it be characterized by type of aircraft, type of mission, and type of pilot? Does it boil down to function alone? And what connotes function to the pilot? While we may find that pilots find function to be the greatest beauty, the way we have implemented function may be something very unattractive to pilots. Beautiful function to a pilot may be very different than beautiful function to the engineer.[*]

Use technology as a means to align form to the human. Technology often results in a change of form. For instance, the microprocessor miniaturization and transition from cathode ray tube to LCD technology that has taken place in the past two decades has allowed the portable computer to move from the form factor of a large briefcase to that of a fashion magazine. Technology progression, then, often acts as an enabler to align the form to the human rather than to the underlying system. This is a core discussion for the industrial designer and the human factors practitioner, because it is simultaneously a discussion of form, function, and technology.

Develop integrated models of cognition and emotion in aviation. In addition to creating general models of integrated emotion and cognition (e.g., Lindgaard & Whitfield, 2004), models of cognition and emotion in aviation need to be articulated. Stephane (2007), for instance, proposed a model of integrated cognition and emotion appropriate for human-centered design in aerospace, and further suggested human factors topic areas in which emotion might be considered: the relationship of the system to the user (e.g., trust and security), design of the graphical user interface (e.g., emotion with the real vs. the virtual), training (e.g., emotion, anxiety, risk, and expertise), and team collaboration (e.g., nonverbal communication, normal vs. nonnormal operations).

Develop the metrics. For both analytic and evaluative skills, it is important to define and develop integrated metrics that are meaningful to industrial designers and human factors professionals. For instance, what aesthetic measures are appropriate for rating flight deck beauty? How do they tie out to usability measures? How can a task analysis and an ethnographic study be used together as a force multiplier?

---

[*] In addition to task analysis and ethnography, one path might be to take a psychological approach and explore the dimensions of pilot personality (which trace back to the personnel selection research prior to WWII). Another approach could be more philosophical, as John Zukowsky (personal communication, June 1, 2007) suggests, and review the writings of Antoine de St. Exupery (e.g., *Wind, Sand and Stars*) and William Langewiesche (e.g., *Inside the Sky*) to more clearly define what it is that pilots find beautiful and meaningful.

Extend the theory. We started with the multidisciplinary hypothesis, attractive things work better, and extended it to an interdisciplinary hypothesis, integrate designed aesthetics and usability to yield a functional thing measurably better than the sum of the designed beauty or usability parts alone. The hypothesis is untested, and needs further articulation, critique, extension, and falsification.

> After midnight the moon set and I was alone with the stars.
> I have often said that the lure of flying is the lure of beauty, and I need no other flight to convince me that the reason flyers fly, whether they know it or not, is the esthetic appeal of flying.
>
> **Amelia Earhart**
> *Last Flight*

16.14 Conclusion

After 100 years into this experiment of powered flight, we may be entering a time when beauty and emotion deserve a renaissance. Inadvertently, perhaps, the inspiration and the wonder have been slowly weeping out of aviation. In a flying age that is characterized by commoditization, congestion, and complexity, returning meaningful, functional, usable beauty to the crew might go a long way to return attention to the emotional rewards that at one time naturally accompanied flying. Moreover, in a powerful linkage with human factors, addressing aesthetics may reduce workload for the pilot, and if the present theory is correct when applied to objective metrics, may improve performance.

Even as aerospace has led the way in integrating and giving headway to human factors, now recognizing the potential to broaden our approach, there is the opportunity to give headway to an integrated approach to flight deck form and function and begin moving the interface and interaction research into deeper waters. Eventually, we should be simultaneously and collaboratively designing for integrity among the useful, the usable, and the beautiful. This is important, because an integration of appropriate form and function sets the stage for an effective mission.

Acknowledgments

I sincerely thank Bill Rogers, Technical Fellow with Honeywell Aerospace, and John Hajdukiewicz, Senior Analyst with Honeywell Corporate Strategy, for reviewing and commenting on this manuscript. I further benefited from analysis and critique from the design team at Worrell, Inc.: Bob Worrell, President, Pete Madson, Director of Design, Worrell Asia, and Dan Clements, Designer. I am grateful to John Zukowsky, Chief Curator at the Intrepid Sea, Air, and Space Museum, for his assessment of the text, particularly in the context of design history. I appreciated the expert criticism spanning across the usability–aesthetics divide from Noam Tractinsky, Senior Lecturer at the Department of Information Systems Engineering, Ben-Gurion University of the Negev. Finally, I thank my partner and design Muse, Michele Gannon, for her extensive review and substantial improvement on the chapter's concept and content.

References

Beith, B. H., & Jasinski, J. E. (2006, Summer). The integration of industrial design and human factors: Bricks and mortar. *Innovation, 25*(2), 26–29.

Bloom, B. S., & Krathwohl, D. R. (1956). *Taxonomy of educational objectives: The classification of educational goals, by a committee of college and university examiners. Handbook 1: Cognitive domain.* New York: Longmans.

Blythe, M., Hassenzahl, M., & Wright, P. (2004, September–October). More funology. *Interactions, 11*(5), 37.

Boeing. (2003). *Crucial committed competitive: Boeing commercial airplanes.* Chicago, IL: Author.

Boeing. (2008). *Boeing 777 program awards, records and firsts.* Retrieved July 23, 2008, from http://www.boeing.com/commercial/777family/pf/pf_awards.html

Cockton, G. (2004, August). Doing to be: Multiple routes to affective interaction. *Interacting with Computers, 16*(4), 683–691.

Federal Aviation Administration. (1991). *Aeronautical decision making* (Advisory Circular No. 60-22). Washington, DC: Author.

Gannon, A. J. (2005). *The effects of flightdeck display interface aesthetics on pilot performance and workload.* Doctoral dissertation, Arizona State University, Tempe, AZ (UMI No. 3178236).

Gilles, W. (1999). *The context of industrial product design.* Ottawa, Ontario, Canada: Carleton University.

Hancock, P. A., Pepe, A. A., & Murphy, L. L. (2005, Winter). Hedonomics: The power of positive and pleasurable ergonomics. *Ergonomics in Design,* 8–14.

Headrick, R. J., Jr. (2008). *Cadillac: The tailfin years.* Hudson, WI: Iconografix.

Karvonen, K. (2000). The beauty of simplicity. In *Proceedings of the Conference on Universal Usability* (pp. 85–90). New York: Association for Computing Machinery.

Krathwohl, D. R., Bloom, B. S., & Masia, B. B. (1964). *Taxonomy of educational objectives: Handbook II: Affective domain.* New York: David McKay Co.

Lavie, T., & Tractinsky, N. (2004, March). Assessing dimensions of perceived visual aesthetics of web sites. *International Journal of Human Computer Studies, 60*(3), 269–298.

Lavietes, S. (2002, October 15). Alphonse Chapanis dies at 85; Was a founder of ergonomics. *New York Times.* Retrieved July 23, 2008, from http://query.nytimes.com/gst/fullpage.html?res=9C06E0D E1E3AF936A25753C1A9649C8B63

Lindgaard, G., & Dudek, C. (2003, June). What is this evasive beast we call user satisfaction? *Interacting with Computers, 15*(3), 429–452.

Lindgaard, G., & Whitfield, T. W. A. (2004). Integrating aesthetics within an evolutionary and psychological framework. *Theoretical Issues in Ergonomics Science, 5*(1), 73–90.

Miller, C. A. (2004, April). Human-computer etiquette: Managing expectations with intentional agents. *Communications of the ACM, 47*(4), 30–34.

Murphy, L., Stanney, K., & Hancock, P. A. (2003). The effect of affect: The hedonomic evaluation of human-computer interaction. In *Proceedings of the Human Factors and Ergonomics Society 47th Annual Meeting* (pp. 764–768). Denver, CO.

Norman, D. A. (2004a). *Emotional design: Why we love (or hate) everyday things.* New York: Basic Books.

Norman, D. A. (2004b, February). Introduction to this special section on beauty, goodness, and usability. *Human-Computer Interaction, 19*(4), 311–318.

Scanlon, J. (2007, July 30). The Eclipse: Safety by design. *BusinessWeek.* Retrieved July 23, 2008, from http://www.businessweek.com/magazine/content/07_31/b4044416.htm?chan=innovation_innovation+%2B+design_top+stories

Stephane, L. (2007). Cognitive and emotional human models within a multi-agent framework. *Proceedings of the 12th International Conference, HCI International 2007, 4562,* 609–618. Beijing, China.

Tractinsky, N., Katz, A. S., & Ikar, D. (2000, December). What is beautiful is usable. *Interacting with Computers, 13*(2), 127–145.

Woodham, J. M. (1997). *20th century design.* Oxford: Oxford University Press.

Zukowsky, J. (1996). *Building for air travel: Architecture and design for commercial aviation.* New York: Prestel.

Zukowsky, J. (1997, Autumn). Design for the jet age: Charles Butler and Uwe Schneider. *Design Issues, 13*(3), 66–81.

17

Helicopters

Bruce E. Hamilton
Boeing

Helicopters are just like fixed-wing aircraft except that helicopters are different. The differences are not in the men and women who fly helicopters, for they can be, and sometimes are, the same men and women who fly fixed-wing aircraft. Their abilities and limitations are the same regardless of the kind of aircraft they fly. Helicopters and fixed-wing aircraft differ in how the crew makes flight control inputs, the information required to decide the necessary control movements, and the missions assigned to the crew. There are many areas of similarities, such as in navigation, communication, subsystem management, monitoring vehicle status, coordination between crew members, and interaction between the helicopter and other aircraft. Helicopters and fixed-wing aircraft follow, for the most part, the same flight rules and procedures. Minor differences exist in the flight rules, mostly about the minimum visual ranges and decision heights. Although rotary- and fixed-wing flight are mostly the same, the differences are important and often overshadow the similarities.

One difference is in how helicopters fly. Fixed- and rotary-wing aircraft all obey the same laws of physics and use the same principle of differential pressure caused by air flowing across and under a shaped surface to generate lift. The difference is that the rotary wing, as the name implies, rotates the wing about a mast to generate airflow while the fixed wing moves forward through the air. The difference in the method of generating lift accounts for the helicopter's ability to hover and move at slow speeds in any direction. Figure 17.1 illustrates the method by which the helicopter balances opposing forces to fly. In short, the rotating blades (rotor disk) generate lift. Tilting the rotor disk provides thrust with the resultant vector—a function of how much lift (pitch of the blades) and thrust (degree of tilt) are commanded. This resultant vector counters the force of gravity acting on the mass of the helicopter and payload, and the drag of the fuselage, as it moves through the air. Increasing the pitch of the blades (more lift) without tilting the rotor disk (thrust constant) causes the helicopter to rise, whereas

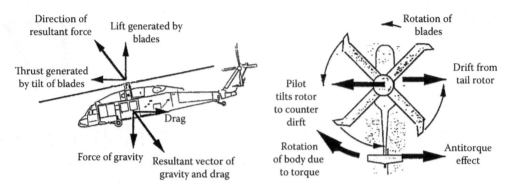

FIGURE 17.1 Example of forces that must be balanced to fly a helicopter.

increasing pitch and tilting the disk causes movement in the direction of tilt. When hovering, the resulting vector is vertical (without thrust) to balance the force of gravity. However, because the engines rotate the blades, the body of the helicopter tends to rotate in the opposite direction owing to torque effects. A small set of blades is mounted on a tail boom and oriented so that its lift counters the torque of the main rotor blades. However, because the torque effect is rotational and the antitorque tail rotor applies lateral force, the tail rotor tends to push the helicopter across the ground. This is countered by tilting the main rotor disk to counter the tail thrust. In some helicopters, two sets of main blades are used, rotating in opposite directions to counter torque rather than the more frequent main blade and tail rotor configuration. Changing the amount of pitch of main or tail rotor or changing the tilt of the main rotors determines the flight of the helicopter; however, any change in a force results in imbalances, which may or may not have to be corrected by the pilot.

The controls of a helicopter manipulate the aircraft's airfoils differently than in fixed wings but, in many respects, the result is functionally the same, especially at higher speeds. For instance, the cyclic provides pitch and roll control as does the stick or yoke in a fixed wing, the collective controls "power" as does the fixed-wing throttle, and the pedals control lateral forces about the tail just as does the rudder. However, rotary-wing flight requires more frequent adjustments, and each control interacts with the other controls as indicated earlier. As a result, special attention is paid to control placement in the helicopter cockpit.

The ability to generate and maintain lift in various directions leads to a second significant difference between fixed- and rotary-wing aircraft, namely, the missions they fly. Helicopters are able to move slowly, at very low altitudes and hover stationary over a point on the earth. This allows the helicopter to be used in a variety of unique missions. These unique missions have an impact on the design of the helicopter and the way the crew uses its ability to fly.

Besides differences in controlling flight and the flight missions, maintenance of helicopters is more demanding than that of fixed-wing aircraft. The need for more frequent maintenance, more complicated control systems, and limited access within compartments makes the time as well as the costs required for helicopter maintenance high relative to fixed wing. Recognition and consideration of the human factors of maintenance early in the design process will be significantly rewarded in the cost of ownership.

17.1 Issues Unique to Helicopters

Control of the helicopter is different from that of fixed-wing aircraft based on the way in which lift is generated. In the helicopter, lift is generated by rotating blades (airfoils) and varying the angle, or pitch, of the blade as it rotates. The act of increasing pitch causes the blade to rotate about its long axis and increases the lift generated but at the cost of more power requirement. Adjusting pitch is accomplished using a control called the collective. The collective is located on the pilot's left and arranged so that pulling up on the collective increases pitch and pushing down decreases pitch.

Where the fixed wing has rudder pedals to induce lateral forces against the tail of the fuselage, the helicopter has pedals that control lateral force by varying the pitch of blades mounted at the end of the tail boom. Together, the collective, cyclic, and pedals are used to control and stabilize lift in varying directions, thereby bestowing the helicopter with its freedom of movement. Controlling flight and following a flight path are a matter of balancing lift (via collective pitch), thrust of the rotor disk (via cyclic), and antitorque (via pedals), with the balance point changing with every control input (in helicopters without engine speed governors, engine speed must also be adjusted with each control input). To fly, the pilot must make continuous control adjustments with both hands and feet. This imposes severe restrictions on nonflight tasks such as the tuning of radios or subsystem management. This must be compensated for in the design of the crew station. Advances in flight control and handling qualities are reducing much of the demands on the crew by automating the balancing act.

As a result of the differences in the controls, visual requirements in helicopters differ from those in fixed-wing units, especially during low speed or hover flight. In these modes, constant visual contact with the outside world is used to determine minor changes in position (fore/aft, left/right, up/down, rotation), to compensate and station keep. At hover, the pilot maintains awareness by focusing at distant visual points with quick crosschecks close by to sense small movement. A rock or bush may be used to determine whether the helicopter is moving forward/backward or to a side (which is why hovering at altitude, in bad weather, or over water is so difficult). In addition, the pilot must visually check to insure that there is clearance between the main rotors and antitorque rotors and objects such as trees, wires, and so on. During take offs, the body of the helicopter can pitch down as much as 45 degrees, as the pilot shifts the rotor disk severely forward. Similarly, on approaches, the body may pitch up by 45 degrees. The need for unobstructed vision determines how much and where in the cockpit glass is required. The loop between visual cues, control movements, and compensation is continuous and demanding and is one of the primary components of pilot workload.

The helicopter gains freedom of movement by adjustments of rotating blades (airfoils) overhead. This has the undesirable side effects of causing vibration and noise. As each blade moves through space, its form causes differential airflow between the top and bottom surfaces, which then merges at the rear of the blade, resulting in turbulent air. This fact coupled with normal disturbances of the air mass, and the additional fact that the blade is always alternatively advancing toward and retreating (and stalling) from the flight path on each revolution leads to vibration. Vibrations are transmitted along each blade to the mast and then into the airframe to be added to transmission and airframe vibrations. At low airspeeds, the blades act as individual airfoils, whereas at higher airspeeds the blades act like a unified disk. The transition from individual to group behavior is another contributor to vibration.

All the movement, vibration, and blade stall contribute to noise in the cockpit. The vibrating environment created by the rotor dynamics also affects display readability and control (switch, knobs, and dials) movements. Noise also interferes with effective communication and contributes to fatigue. Light passing through the rotor system is intermittently blocked by the blades and causes some flicker. Certain combinations of blade size, blade number, rotation speed, and color of transparencies can cause visual disturbances and are to be avoided. All of these impacts are results of the helicopter's method of generating lift.

The freedom with which helicopters fly leads to unique challenges in that the missions of helicopters vary widely. The same airframe, with minimal modification, may be used as an air taxi, air ambulance, search and rescue, air-to-ground attack, air-to-air attack, antisubmarine warfare, heavy lift of goods and raw materials, aerial fire fighting, police surveillance, sightseeing, and aerial film platform, to name a few. Any of these missions might be conducted during the day or at night, sometimes using night vision devices. The same helicopter can be expected to fly under visual meteorological conditions or under instrument meteorological conditions, under either visual flight rules or instrument flight rules (IFR). These different missions involve the addition of equipment to the helicopter and some minor cockpit modifications. The cockpit usually retains its original configuration with some controls and displays moved around to make room for the control panels of the add-on equipment. The pilot–vehicle interface of the helicopter must be unusually flexible.

Another issue for helicopter pilots is the frequent operation at the power limits of the engines. For instance, desired payloads, coupled with density altitude and wind conditions, may tax the helicopter's lift capability to such an extent that a "standard" take off cannot be accomplished. The pilot has to recognize this condition and fly forward at only a few feet in altitude to build up forward airspeed. This allows the pilot to use ground effect for additional lift. Ground effect is the additional lift the helicopter gains from the blade downwash being trapped beneath the rotors and acting as a cushion. A rough analog is the lift generated by a hovercraft sitting on trapped air. After airspeed increases, the blades transition from acting as separate airfoils to acting as a coherent disk occurs, which increases their efficiency, and more lift for the same power is available. The margin provided by this flight technique is relatively small and difficult to judge. Failure to correctly judge the margin may mean the difference between a successful edge-of-the-envelope take off and possible disaster. Once aloft, environmental conditions may change (such as density altitude, owing to the arrival at a touchdown point at higher elevation), and adequate power margin may no longer be available. Again, the safety margin may be small and difficult to judge, and numerous accidents have occurred in which helicopters crashed after several failed attempts at landing or take off. The pilots were apparently convinced that the helicopter was within its operating envelope or, perhaps, outside the published envelope but within the mystical "extra design margin" that all pilots believe engineers give them.

Another major mission with unique human factors impact is the requirement to fly in close formation with the earth. Helicopters are routinely used in low-altitude missions such as aerial survey, installation and maintenance of power lines. In a military setting, helicopters are expected to fly using trees for concealment from detection. In nap of the earth flight, the pilot flies slowly (often at speeds at which the aircraft is unstable), and as close to the ground as conditions permit. This might be below treetop level or behind small hills. Confined area landings and hover above short trees but next to tall trees can be expected. All of this is expected during day or night. At night, limited night vision is augmented by vision aids such as night vision goggles (which amplify available star/moonlight) or infrared sensors (which sense minute differences in temperature in a scene and encode the differences at various grayscales on a cathode ray tube [CRT]). Night vision aids all change the visual world, usually by reducing the field of view, changing the color of the world into shades of green, reducing visual acuity, and, in the case of the infrared image displayed on a helmet mounted display, reducing vision to a single eye. If a pilot who applied for a license was colorblind, could see out of only one eye, and had only 20/60 visual acuity, he would be laughed at and denied a license. However, as a helicopter pilot, he may be reduced to that visual capability and told to ignore the fog, go hover behind a tree, and land in small, impromptu landing zones.

Other missions routinely expected of helicopters impose their own unique challenges. For example, some helicopters are expected to fly in high winds, and land on a spot on a ship's deck scarcely bigger than the rotor disk, while all the time the ship is pitching and rolling. Those helicopters have standard cockpits without specialized flight controls or displays. The message is that the designers of helicopter cockpits should design it in such a way that it is suitable for almost any mission.

Another way in which helicopters present challenges is the maintenance and support of the helicopter itself. The engine of the helicopter (the number may range from one to three engines) drives a transmission that reduces the high speed of the engine to the relatively low speed of the blades and provides drive for the tail rotor. The engines must have fuel flow constantly adjusted to keep the blades turning at a constant speed against the forces trying to slow or speed the blades. The transmission also drives hydraulic and electrical generators for use by other systems. The flight control system must translate cyclic, collective, and pedal movements into adjustment of blade pitch, while the blade rotates around the mast. This requires a fairly complex mixing of control by rotor position all superimposed upon the normal requirement to compensate for the fact that the blade generates lift during the forward portion of the rotation and stalls when retreating. In older helicopters, this control system is completely mechanical, whereas in newer helicopters hydraulic systems are used to provide the required forces to move the blades. The electrical system powers the avionics suite, which can range from minimal to extremely complex systems including radar, infrared, or satellite communications, among other systems. All of these systems, mechanical and electronic, require maintenance. Providing easy access, with simple, quick, low workload procedures, is a major challenge with direct impact on the cost of ownership and safety of operation.

Many human factors challenges are posed by the method of vertical flight and missions to which vertical flight is applied. Those mentioned here are important ones but not the only ones. The pilots of helicopters face major challenges in the control and monitoring of the aircraft health and systems status, just as the fixed-wing pilots do. Communications between the helicopter and other aircraft and the ground represent significant workload. The unique issues derive from how vertical flight is achieved and what it is used for.

17.2 The Changing Nature of Helicopter Design

Human factors engineers traditionally provide information on human capabilities and limitations to the design community and serve as a "check" function once the design was completed. The role of the human factors engineer has been to provide specific details such as how tall are people when sitting, what size should the characters on the labeling be, and what color the warning indicator should be. Often the human factors engineers found themselves helping to select which vendor's display should be selected, trying to change completed designs that failed to take something into account, or answering why a design was not acceptable to the customer. These roles were generally all that was required when the issues were primarily those of pilot fit and arrangement of selected displays in general utility helicopters. In October 2000 the U.S. Government, the primary developer and purchaser of helicopters, revised its rules governing development and procurement. The new DoD Directive 5000.1 changes the mandatory procedures and formalizes the trend to eliminate government specifications in favor of industry practice and performance-oriented procurements. This change in the nature of the procurement business came at a time when the impact of the latest generation of computers was being felt in military hardware development. The advent of high-speed miniature computers and their impact on aviation, in general, has lead to a change in the development rules and the way in which the human factors engineer interacts with the design process. Human System Integration has become a required element of the design process (Booher, 2003).

The impact of the computer on the cockpit and the way in which cockpits are developed has been significant in many areas but has created the highest change in two areas. The first is that what used to be discrete hardware functions have become integrated into a computer system. The whole infrastructure of the aircraft was assumed to be "given" with only simple crosscheck required before actions were taken. In a computerized system, not only does consideration have to be given to the status from a health monitor, but also to the health of the monitoring devices. The meaning of the information provided must also be considered. For instance, a chip detector used to be a simple design, in which the metallic chip physically completed an electrical circuit illuminating a light on the caution/warning panel. Now the size of chips, the number of chips, and the rate of chip accumulation can be monitored and passed to a central processor through a digital data bus. Besides a greater range of information and prognosis about the developing situation, the health of this detection and reporting network has to be considered. That is, is the detector performing its function? Is it talking to the digital bus? Is the software that uses the data alerts the crew functioning? The pilots of the emerging generation of helicopters are becoming managers of systems designed to free them from the workload of housekeeping the aircraft, but managers also have information and control needs.

The second major area of computer impact is that the displays being purchased are blank and can display almost anything desired in any format, whereas formerly displays were built function specific and limited. For example, attitude displays were selected by comparing the vendor's marketing material and specification sheet that fitted the available space, interfaced with the sensor system, and appeared most pleasing to the customer. Now displays are picked for size, weight, type (CRT, liquid crystal, etc.), pixel density, interface protocol, color, and so forth. What can be shown on a display is a function of the capability of the display generator, throughput of the processors, and data rates of the available sensors. Often the decision is whether to use one vendor's flight management system (including displays, controls, information, moding, capabilities, etc.) or another. The other option is to develop a purpose-built cockpit by mixing and matching or developing components. Widely varying styles of interface within the same cockpit may result from piecing together off-the-shelf systems. This second impact has surfaced in the Human Systems Integration movement as a requirement for early task analysis to assist the system engineering activity to develop more mature requirements for software development and an emphasis on workload and mission effectiveness. Human Systems Integration is tasking the human factors engineer to help articulate software requirements and verify that the resultant system will be mission effective.

The cost to buy a helicopter is usually equivalent to or less than a comparable fixed-wing aircraft; however, the costs of support and maintenance are significantly more for helicopters. The primary cost drivers are the person-hours required and the necessity to replace parts after a fixed number of hours of use. Human factors engineers can help reduce the cost of ownership by reducing the complexity of the maintenance and support tasks. Generally speaking, the time to complete safety and maintenance inspections is short compared with the time required to open and close inspection hatches and access that which needs to be inspected. Careful attention during the design phase to how the inspection process is conducted can significantly reduce the person-hours required. Once the determination is made to conduct maintenance, how the task is completed should be engineered to reduce the number of steps, number of tools, and number of person-hours required. When conducting these analyses, the human factors engineer should consider the education level, training requirements, and job descriptions of the population of maintainers and supporters.

The next generation of helicopters will be incorporating computerized maintenance support devices in an effort to reduce the time required to determine what maintenance must be done and replace parts as a function of usage or performance rather than time. Most avionics will be designed with built-in test (BIT) capability that will continuously report on the health of the device. The trend is toward a hand-held maintenance aid that can access a storage device located on the airframe that holds data taken during the flight. This portable device will be capable of diagnostics, cross-reference to parts lists, and display of maintenance procedures. It will also interact with ground-based systems for record keeping, updates, and prognostics. There exists an urgent need for human factors engineers to assist in the development of the electronic manuals, diagnostic procedures, and record-keeping interfaces of these electronic devices.

17.3 The Role of Human Factors in Future Helicopter Design

Future helicopter crew station design will require many of the human factors subspecialties. The particular skills will depend on the phase of development. The phases of a development program are shown generically in Figure 17.2. While these phases are no longer mandated by DoD directive, the emphasis upon mission effectiveness and workload does not significantly change the need for stepwise progression toward design and development. The airframe development starts with general requirement definition and the creation of basic data. In this phase, the outer lines of the helicopter are established along with size, weight, capabilities, etc., to form the general arrangements. This phase ends with a review of the data and conceptual design. Once the requirements are defined, the preliminary design phase begins. During preliminary design, interior lines are established and the general arrangements are refined. Analysis is conducted to define technical parameters and to determine how well the design meets its goals. Prototype components are built and tested to reduce the risks associated with new designs. This phase ends with a preliminary design review and an accounting of how well the original

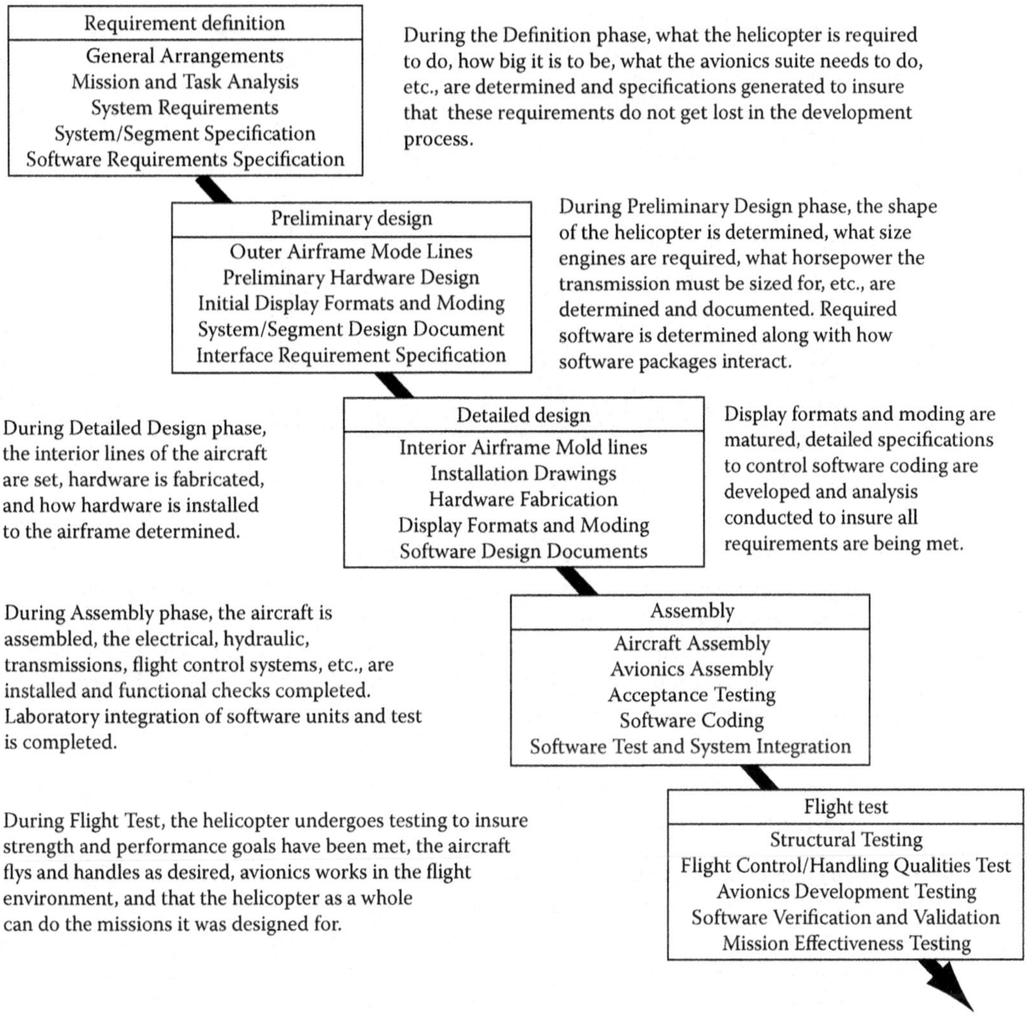

Requirement definition
General Arrangements
Mission and Task Analysis
System Requirements
System/Segment Specification
Software Requirements Specification

During the Definition phase, what the helicopter is required to do, how big it is to be, what the avionics suite needs to do, etc., are determined and specifications generated to insure that these requirements do not get lost in the development process.

Preliminary design
Outer Airframe Mode Lines
Preliminary Hardware Design
Initial Display Formats and Moding
System/Segment Design Document
Interface Requirement Specification

During Preliminary Design phase, the shape of the helicopter is determined, what size engines are required, what horsepower the transmission must be sized for, etc., are determined and documented. Required software is determined along with how software packages interact.

During Detailed Design phase, the interior lines of the aircraft are set, hardware is fabricated, and how hardware is installed to the airframe determined.

Detailed design
Interior Airframe Mold lines
Installation Drawings
Hardware Fabrication
Display Formats and Moding
Software Design Documents

Display formats and moding are matured, detailed specifications to control software coding are developed and analysis conducted to insure all requirements are being met.

During Assembly phase, the aircraft is assembled, the electrical, hydraulic, transmissions, flight control systems, etc., are installed and functional checks completed. Laboratory integration of software units and test is completed.

Assembly
Aircraft Assembly
Avionics Assembly
Acceptance Testing
Software Coding
Software Test and System Integration

During Flight Test, the helicopter undergoes testing to insure strength and performance goals have been met, the aircraft flys and handles as desired, avionics works in the flight environment, and that the helicopter as a whole can do the missions it was designed for.

Flight test
Structural Testing
Flight Control/Handling Qualities Test
Avionics Development Testing
Software Verification and Validation
Mission Effectiveness Testing

FIGURE 17.2   Generic phases of a helicopter development.

requirements are being met, as well as progress on meeting weight and cost goals. The next phase after the preliminary design phase is the detailed design phase. In this phase, the helicopter design matures into something that can be built and assembled. The design is evaluated in a critical design review-again to verify that all systems are meeting design goals and components are ready to be manufactured. This is followed by assembly of the helicopter and acceptance testing of components. Tests of the helicopter are conducted to insure strength requirements are met and operations of the subsystems (hydraulics, engines, transmissions, etc.) are evaluated. This phase ends with an airworthiness review and safety of flight release. Flight tests are conducted following the assembly phase. The flight test program is conducted to allow a safe, orderly demonstration of performance and capabilities. The purpose of the flight test is to validate the "envelope" of performance and insure safe flight under all specified conditions.

A human factors program that supports analysis, design, integration, and test of the pilot–vehicle interface should be conducted in support of the development and validation of hardware and software requirements. This program, depicted in Figure 17.3, should be iterative and interactive with the other disciplines involved in the helicopter development. Within this program, the human factors engineer

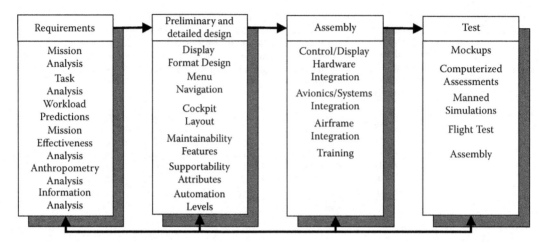

FIGURE 17.3 Features of notional human factors engineering program supporting helicopter development.

analyzes the proposed missions to the task level to determine what the pilot and aircraft must do. This is done during the requirements definition phase and is updated during preliminary and detailed design. In other words, the mission and task analysis is first conducted independent of specific implementation to define the human information requirements and required tasks. Later in the development cycle, the early mission and task analysis will be revisited using the proposed designs and finally the actual controls and display formats. The message is that although the name is the same, the products during the three phases are different and serve different purposes.

17.3.1 Requirements Definition Phase

The beginning of any program, whether a new aircraft development, a variant, or a retrofit, includes analysis to document and disseminate the requirements of the program. This is done by generating a system/segment specification (system if a large program or segment if more restricted in scope). This document informs the engineering staff of what the design is expected to do. At the system level, the specification contains requirements for capabilities that typically can be met only through the contributions of individual segments of the system design. When working with a segment specification, higher-level requirements would have been decomposed into requirements for the specific segment that are "segment pure", that is, can be met by design solely within the purview of the specific segment. For the cockpit, this often consists of creating a list of instruments required for flight and support of navigation, communication, and mission peculiar equipment at the system level with decomposed requirements in the supporting segments, such that flight data comes from one segment and communications information comes from a separate segment. The kind of requirements generated during this stage of a new helicopter program might be that the aircraft is to be single engine, dual crew, side-by-side seating, IFR certified, three radios with waveforms in the regions of very high frequency-frequency modulation (VHF-FM), very high frequency-amplitude modulation (VHF-AM), and ultra high frequency (UHF), and so on. These requirement specifications are then provided to the avionics system and airframe developers. The airframe developers, using approximate sizes from the vendors, begin the task of arranging the cockpit. At this point, tradition is most often the moving force behind the development of the cockpit.

The impact of computers on the airframe and the cockpit has changed this portion of the development process. The list of requirements provided to the airframe engineers includes specifications for the number and sizes of displays. The airframe developers, however, cannot practically talk to display vendors about weight, space, and power because the vendors wish to know the type of display, pixel density, video interfaces, and so on. The group generating the system/segment specification will still specify the requirements as before, but the requirements list must be "interpreted" into usable information for vendors prior to making the traditional information available to airframe designers. In turn, the data needed to create the attitude display, the number of displays required, the rate at which data must be displayed, the colors used, the resolution required, and so on, must be identified and provided to the avionics system designers so that they can determine the mix of displays and controls required. These decisions are documented in the software requirement specifications. Human factors engineers are well qualified to help the avionics system designers address these issues as well as to assist the airframe designers in arranging the cockpit, given that a mission and task analysis has been conducted. If this process is not followed, when detailed design of the software begins, what can be achieved will be limited by what hardware was selected and where it is placed.

17.3.2 Preliminary Design Phase

Once the helicopter's general arrangements have been created and the outer mold lines established, work begins on preliminary design of the crew station. This typically results in generating two-dimensional drawings specifying instrument panel size, location, control placement, and so on. It is during this phase that the size and location of glass is fixed, controls positioned, and more. Human factors engineers are involved to insure that the location of the supporting structure allows adequate vision down, up, and to the sides. However, the problem of vision is more than just placement of the glass. Designers must consider the allowable distortion of vision through the glass, the angle of viewing through the glass, transmissivity of the glass, and internal reflections, among other design issues. Other topics to be addressed by the human factors engineers working in conjunction with the airframe developers during this phase include how the crew gets in and out of the aircraft normally and during emergency exits, reach requirements for nominal and emergency operations, location of display surfaces and control panels, and safety requirements.

The preliminary design of helicopter cockpits must also take into consideration the fact that helicopters crash in a way that differs from that of fixed wings. Helicopters tend to impact the ground with high vertical descent rates, whereas fixed-wing aircraft tend to have a lot of forward speed at impact. As a result, helicopter cockpit designers need to be concerned about instrument panel entrapment (panel tearing away and dropping onto the legs), attenuating vertical movement of the crew, and keeping the cyclic and collective from striking the crew during the crash sequence. Information about how the human body moves during the crash sequence can help guide the airframe developer during cockpit layout. A tradeoff usually results in that if controls and displays are placed out of the way for crash environments, they usually are too far away to be used during normal operations, and if they are too close they may pose a hazard during crash.

During the preliminary design phase, the software system is also being designed. Software development starts with the general requirements defined in the requirements phase and derives requirements that must be satisfied to fulfill the original requirements. This process is called requirements decomposition. For instance, a general requirement might be the ability to store and use flight plans from a data transfer device. A requirement derived from this general requirement might be that a menu is to be presented to allow the pilot to select one flight plan from up to five stored plans. The decomposition process continues until all high-level requirements have been decomposed into specific discrete requirements that allow a programmer to write and test code for discrete functions. Part of this process is the assigning of responsibility for functions to specific software groups, along with how fast the task is to be completed, how often the task must be done, what information is needed, and in what format the product must be. In this way, the software system is laid out analogous to the physical layout of the aircraft. These decisions are documented in the system/segment design documents and the interface requirement specification.

The human factors engineer assists this process by updating the mission and task analysis created in the first phase based on the increased understanding of implementation and probable uses of the helicopter. Specification of the display formats at this point in the program provides insight into the information required and the software processes that will be required to produce that information. Menu structure can be designed to control number of key presses, to reduce confusion in menu navigation, and to provide a low-workload interface. This information allows the software system to provide what is needed by the crew and in an optimal format while simultaneously easing the task of software decomposition.

If the proposed interface is defined in this manner during this phase, there are additional benefits besides the flow down of requirements. For instance, the proposed interface and its operation can be simulated from this information on any one of the number of devices ranging from desktop to high-fidelity motion simulators. Proposed users of the helicopter can be brought in to evaluate the software design before coding. The result is a mature design and less probability of costly recoding due to mistakes in understanding, failure to identify requirements, and misconceptions by the customer as to what is technically feasible for a given cost and schedule.

The effort of the human factors engineer to create task and workload analyses and to conduct manned simulations is costly. It is always difficult to persuade program planners to expend money on these tasks, especially during the beginning phase of the program when the demand is the transition of almost ready technology from the laboratory to production. The primary selling point for the early human factors analysis and simulation is cost avoidance. Significant changes that occur after software has been validated for flight are extremely expensive in both time and money. Avoiding a single major problem will pay for all the upfront expenditures. This is the same rationale for doing wind tunnel testing on a blade design before flight test. No one would actually propose that a blade design could be drawn and would work as desired without testing (either real or virtual) and redesign. The same is true of the pilot–vehicle interface and the information content and menu structure.

17.3.3 Detailed Design Phase

Reality sets in during the detailed design phase. The cockpit layout during preliminary design is clean and tidy, but during detailed design a million and one little details have to be accommodated. For example, everything that floated freely in earlier drawings must now have mounting brackets with retaining bolts. Provisions must be made to allow the components to be assembled and maintained. The furnishings and equipment, like first aid kit, fire extinguisher, document holder, and seat adjust, must all be located and installed in the cockpit.

Round dials and display tapes, if used, must fit within the instrument panel space. If a dedicated caution/warning panel is used, the number of cautions/warnings must be determined, legends created, and colors assigned. If the cockpit is to have multifunction displays, then the displays must be specified as to pixel density, colors, update rates of display generators, formulas for movement of symbols on the screen generated, filter algorithms developed, bitmaps defining symbols provided, etc. The distance between switches must be large enough for fingers to use easily without becoming so great that the resulting panel does not fit into the mounting rails. Control grips for the collective and cyclic must be created and the functionality of switches determined.

The menu structure must be completed and the meaning of each pilot-selectable option defined. What had started out as a simple, clean menu usually must be modified to accommodate the emerging requirements to turn devices on and off, set up and adjust functions, enter local variations into the navigation system such as magnetic variations and coordinate system adjustments, and so forth. Allowances must be made for how the aircraft is started, checked for functionality, and shut down. Procedures for verification of functionality and means to reset/recycle balky systems must be considered. The menu now has a myriad of add-on steps and routines that obscure the simple and clear structure once envisioned. Display formats must be detailed out to the pixel level, identifying where each character or letter goes. Formats become burdened with additional information and festooned with buttons and options. Formats that once handled a few related functions now sport unrelated functions rather than add new branches to an already burdened tree. Controlling the structure of the menu as it grows from concept to practicality is a daunting task.

Timelines and throughput become major issues during detailed design. The display that was designed to have graphics updated at 30 Hz now might have to update at 15 Hz to help ease apparent processor overloads. This degrades the graphics and, coupled with lags in the speed at which the information is processed, might lead to control loop problems. Events now take seconds rather than the virtually instantaneous speed dreamed of during preliminary design, with obvious impacts on effectiveness, workload, and acceptability.

Detailed design can be summarized by stating that the devil is in the details. As hardware starts to mature, its functionality turns out to be less than envisioned, triggering a scramble to compensate. In other cases, capabilities are provided that one would really like to take advantage of but that had not previously been identified. The problem is that to meet software delivery schedules one has to stop making changes, and the earlier in the cycle if one stops making changes, the smoother is the development. Also, changes beget changes in that fixing a problem in one place in the system often forces rippling effects throughout the rest of the system. Finally, in this phase, the users of the helicopter start to see what had previously only been talked about. They may not like what is emerging and may start providing their own helpful hints about how things should be. All of this is during a time when the system is supposed to be finished and the development money has been spent. Every change during detailed design and subsequent phases is evaluated for cost and schedule impact by people not pleased by any change, good or bad.

17.3.4 Assembly Phase

During the assembly phase, the components of the aircraft come together and the helicopter is built. Final detailing of the design occurs in the cockpit. For instance, intercom cable lengths must be trimmed from the delivered length to that length that gives maximum freedom without excess cable draped over the crew. Decisions have to be made as to the ease with which adjustable items are allowed to move. These decisions and design adjustments come under the heading of "known unknowns," because it was known that the actual length, friction, setting, and so on were unknown until assembly. During assembly one may also encounter "unknown unknowns," problems that had not been anticipated. An example might be that a rotating control handle hits another installed component. Care had been taken to make sure the handle did not touch the airframe but another component had been moved or rotated itself, resulting in interference. These problems have to be dealt with as they arise.

An important aspect of the assembly phase is component testing. As hydraulic lines are completed, they are pressurized and checked for leaks. As electrical lines are laid in, they are tested and eventually power goes on the aircraft. The airframe side of the helicopter requires little human factors engineering support other than occasionally supplying information or details about how the crew or maintainer may use software-based interfaces to status and control systems. If the helicopter uses significant amounts of software to interface with the crew, and the human factors engineers have been involved in the design and development of the interface, then significant involvement in the testing phase can be expected. This involvement typically would be in the form of "Is this what was intended?" or, more to the point, "This can't be right!" If adequate documentation had been done during detailed design and person-in-the-loop simulation accomplished, then there should be no surprises. However, actual operation is always different from envisioned, and adjustments may be required.

17.3.5 Flight Test Phase

There are two major areas of involvement for the human factors engineer during the flight test phase. The first area is in obtaining the safety of flight releases for the aircraft's first flight. The second major area is in assessing workload and operational effectiveness. The specific nature and degree of effort are dependent on the kind of helicopter being built, whom it is being built for, and the aircraft's intended purpose.

The following human factors areas should be considered during a safety of flight evaluation:

- Ingress/egress—Can the crew enter and exit the cockpit both in normal and emergency situations?
- Visibility—Does adequate vision exist for flight test?
- Functional reaches—Are all flight-required controls within easy reach?
- Controls and display functional check—Do the controls and displays work as expected?
- Flight symbology—Is all the information necessary for flight available and are control lags and jitter acceptable?
- Emergency procedures—Have all conceivable failures of components in the aircraft been considered and emergency procedures created for the serious ones?

If the previous phases of development have included human factors engineers, addressing these questions should be merely a formality and a matter of providing documentation of work previously completed. If the program did not include significant human factors effort, then these questions may be difficult to answer or cause last-minute rework.

The other major area during flight test is workload and operational effectiveness. Depending on the customer and mission of the helicopter, testing may be required to demonstrate that workload levels, timelines, and situational awareness goals have been met. This may require dedicated flight time to conduct training and rehearsal of specific mission segments or tasks. It is cost-effective to conduct the majority of the workload and operational effectiveness studies in a high-fidelity full-mission simulator before flight test. During actual flight, selected tasks or segments can be evaluated to verify the simulation data. Full-mission simulation is highly recommended as a method of finding operational problems. If the full-mission simulation is done early in the development, then the cost to conduct operational effectiveness analysis in simulation during flight test phase is minimized and the impact of the discoveries to cost and schedule will also be minimal.

17.4 Workload in the Helicopter Cockpit

Workload in the helicopter cockpit is the result of the demands of the flight tasks, the tasks required to monitor, status, and control the helicopter systems, and the demands of the mission. This is no more than saying that flying results in workload. A more useful way of looking at the genesis of workload is to regroup the demands into those from outside the cockpit and those from within the cockpit. Demands from outside the cockpit are usually environmental, flight, and mission conduct related. The within cockpit demands are considered those that directly support flight demands (e.g., adjusting flight controls), mission goals, and those demands that are part of the routine housekeeping of the helicopter (startup, shutdown, navigation system alignment, etc.). This view of the sources of workload is useful because it allows the human factors engineer to recognize that the designer is trying to cope with the workload associated with an external task. The internal workload, especially the housekeeping workload, is the result of the cockpit design and is under the designer's control. It is always important when designing to recognize that whether the design is trying to cope with someone else's workload or is itself the source of workload. Although it is pleasant to think that the crewstation designer is always reducing the workload caused by others, it is more often the case that the crewstation designer is the source of workload.

17.4.1 Sources of Helicopter Cockpit Workload

The most common source of workload in the helicopter cockpit is tasks that require looking in two different places at the same time. That is, the pilot is usually trying to look out the window to attend to the demands of flight or mission tasks while simultaneously being required to look at displays within the cockpit. This results in having to switch visual attention constantly and increases workload while reducing performance. Head-up displays are recognized as effective workload reducers, because they reduce the amount of time spent switching from inside to outside tasks.

Another common workload problem has been to listen to multiple radios, a copilot, and the auditory cues of the aircraft simultaneously. Auditory workload can build to the point that the pilot has no choice but to stop listening or stop looking to be able to pay more attention to the audio. This may be why alerting systems that talk to the crew often receive low marks despite the apparently intuitive "fact" that a nonvisual interface is the solution to high workload. Visual overload may have been traded for auditory overload.

Another high workload problem is producing and maintaining situation awareness. The human factors engineer should recognize that situation awareness is a general term and that, in fact, many types of situation awareness have to exist simultaneously. For instance, pilots need to be situationally aware of the helicopter's movement through space and time. They must also be aware of where they are in the world and where they are trying to go. They must be aware of the health of the helicopter and its systems. They must be aware of the radio message traffic, from whom it is, what they want, and so on. Each of these types of awareness requires its own set of information that creates and maintains the awareness.

The pilot must continually be assimilating information as a result of the need for maintaining awareness. The information may be simply confirmation that nothing has changed or that a new element has been added for integration into awareness. This information demand will result in either a degradation of awareness without readily available information or increased workload as the pilot searches for the information needed. In a glass cockpit, the searching may require menu navigation to get information followed by menu navigation to return to the original display. This is the source of the constant button pushing seen in some glass-cockpit helicopters. Menu navigation formation to support situation awareness competes directly with situation awareness.

One part of the task of creating and maintaining situation awareness is not obvious. This is the problem that the information presented may be raw data that must be processed by the pilot. Processing raw data into that required for situation awareness results in more workload. For instance, the temperature of a critical section of the engine may be continuously displayed. A red range may be marked on the display to indicate excessive temperature. However, how long the temperature can stay in this range and how long it has been in this range are awareness issues that are typically considered as the responsibility of the pilot. The pilot must note when the red range is reached, track how long temperature stays red, and remember how long the engine can operate in the red. The desired awareness is built from processing the raw data of engine temperature, current time, and memorized information. A better situation would be one where these answers are displayed along with the raw data of temperature.

Another source of workload generated by cockpit design is the typical requirement for memorization of setup. This means that the pilots must know which switches, in which order, result in the desired effect. Many times an incorrectly set switch, or a switch thrown in the wrong order, precludes the desired effect but without clear indication of what the problem was. A typical result is to start the task over because the setup is remembered more as a sequence of actions than a table of positions.

This introduces another related source of workload, namely, error checking and error recovery. The pilot must recognize that goals are not being met because an action he or she thought had been initiated did not actually happen. Awareness of the lack of an action requires workload to monitor and compare the information over time to determine that the commanded action is not taking place. Determining why the action commanded is not taking place requires additional searching through information to compare the actual condition with the expected condition.

The ongoing external flight and mission demands continually conflict with internal flight, mission, and housekeeping, with the result that the pilot must constantly interrupt ongoing tasks to attend to a task that has become of higher priority. After the interruption, the pilot must return to the original task or, after assessing the situation, decide that another task has higher priority. The continued task interruptions result in workload to manage concurrent tasks and affects situation awareness.

Glass cockpits pose unique problems on their own. A multifunction display, by definition, presents different information at different times. The crew must control what is displayed by making selections from presented options. The options presented at any moment constitute a menu from which the pilot chooses. Each option chosen may change the information on the display or lead to a new menu of options. Moving through these menus to find information or controls is what is referred to as menu navigation. The structure of the menu determines how many selections must be made before accessing the information or control desired. Creating a mental map that guides the pilot to the proper display requires memorized knowledge of the shape of the menu trees and the contents of each associated display. Although the ability to customize a display with information or controls helps the problems of presentation format, allows each display to be presented in an easier-to-assimilate format, and allows squeezing all required functions into a small physical space, it brings the workload associated with menu navigation, the burden of memorizing the system structure, and the possibility of confusion.

As computers take over control of the helicopter and free the pilot from the routine tasks of status monitoring and function setup, and extend the number of things the crew can be expected to do, additional workload is created if something fails in the system. Should some part of the computer system or a device fail, there are many options to be examined to determine what exactly has failed and what can be done about it. As automation levels increase, the crew is less able to understand and control the automation, with the result that workload increases and awareness drops.

17.4.2 Engineering Solutions to Helicopter Workload

Once the sources of workload are understood, the human factors engineer can combine knowledge of required tasks (from mission and task analysis) with knowledge of sources of workload to create, during the detailed design phase of development, a cockpit that is low in workload and high in situation awareness. This is done by designing out the sources of workload and designing in the attributes of consistency, predictability, and simplicity. In early helicopters, the tasks were merely to fly and monitor the engine. Current and developing helicopters include a wide range of sensors and mission equipment and computerized control of the aircraft. Automation has been added to prevent workload overload. However, automation should not be added indiscriminately. Over automation can lead to the pilot not being aware of what the helicopter is doing, what the limitations are, and leave him or her helpless when the automation fails. The successful cockpit design will be a subtle blending of automation, information, and control options that allow the pilot to conduct flight and mission tasks with high situation awareness without the costs of being out of control. The difficulty is in translating the goals espoused here into solutions. The following are a number of guidelines for the cockpit designer taken from Hamilton (1993). The purpose is to help the designer understand what the attributes of a "good" design are and how to achieve those attributes

- Switching from one situation awareness to another should be recognized as an explicit task, and mechanisms should be provided to help the pilot switch and maintain situation awareness views.
- Displays should provide information in a format compatible with the information demands of the crew.
- Information needed to support decision making and situational awareness should be clustered together.
- All information required for tasks should be located together and controls for those tasks also should be located together.
- More options do not always make for a happier pilot. In fact, more choices increase pilot reaction times.
- Switch setups should be automated so that the pilot selects functions rather than settings.
- Incorrect switch choices (i.e., those whose order or setting does not apply) should not be presented.

- If the function is normally provided, but is not available because of faults, then the switch should indicate nonfunctionality and whether the function is broken or merely prevented from operation by an external state (e.g., data not available).
- Good menu structure organization is based on a human factors engineering task analysis at the mission segment level.
- Display pages should be based on a set of activities that must be performed by the pilot to complete a portion of the mission segment.
- Data that are commonly used, are frequently crosschecked, or may be time critical, should be readily available.
- Information, controls, and options should be segregated into groups that functionally represent maintenance, setup, normal operations, training, and tactical applications so that the menu structure is sensitive to mission phase.
- Consistency is the biggest ally of the pilot, whereas surprise is the biggest enemy. Pilots will avoid automation if it requires significant setup or option selection and will avoid systems that they are not completely familiar with.
- Bezel switch labels should indicate function, current status, and indicate impact on menu navigation task (do you go to another display page, just change information on the display, turn something on or off, etc.) to take the guesswork out of menu navigation.
- Switch labels and status indicators should be in the language of the pilot, not the engineer who designed the system.
- Data and tasks should be arranged so that the pilot does not have to go to the bottom of a menu tree to find high level or routine data and controls.
- All tasks will be interrupted before completion, so tasks should be designed so that interrupting tasks can be entered easily and quickly and it is simple to return to the interrupted task.
- Recognize the strengths and weaknesses of the various subsystems and integrate data from various subsystems to create better information than any one system can provide.

17.5 Requirements Documentation, Verification, and Flowdown

Computers have changed the way in which helicopters are designed and how they are operated. This change will be permanent in all but a few very restricted situations. This is because computers offer more capability, in less space, for less weight, and at cheaper costs than individual, purpose-built black boxes. To realize their potential, computers will have to have adequate throughput, buses connecting computers will have to have adequate bandwidth, and all functions of the aircraft will have to tie into the computer system. The problem is in defining what is to be built and what "adequate" means in the context of a specific program. The human factors engineer can provide significant insight into what needs to be done, what data at what rates are required, what kind of control lags can be tolerated, and how the crew interface displays information and controls the computer system that controls the aircraft. The system's designers benefit from the human factors engineer's knowledge, insight, and designs, but only if they are properly documented.

Software requirements are decomposed from general requirements to detailed, unique, and testable requirements suitable for software code development. This decomposition process results in the family of hierarchical documents as shown generically in Figure 17.4. These documents, as tailored to a program, describe what needs to be implemented, how to implement it, and how software interacts with software. Requirements start with a system specification that contains the top-level requirements. The next level of specification is the system/segment specification, depending on program scope. In a large system, there may be more than one segment, and system-level requirements may be allocated partially to one segment and partially to another. At this point, the goal is to separate requirements by type so that flight control requirements, for instance, are not mixed in with airframe requirements. Once segments have been created and segment level requirements derived (to do task X, functions A and B must be done),

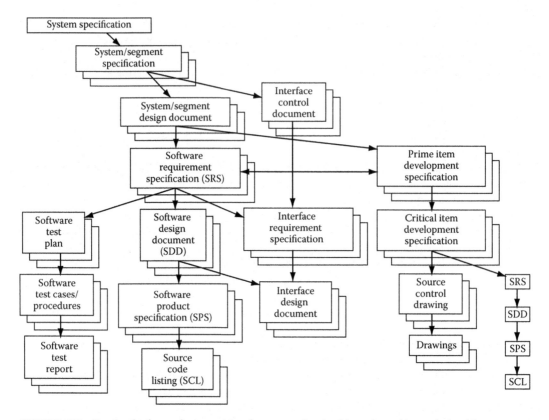

FIGURE 17.4   Family of software decomposition documents showing hierarchy and interrelationships.

then a system/segment design document is generated. The system/segment design document outlines how the segment is to be implemented, breaking all the requirements into functional groups (computer software configuration items) such as navigation, aircraft management, and crew interface. Software requirement specifications can then be created for these functional groups to describe the specific software routines to be developed and what the routines must do. The flow of information on the data buses must also be managed, so interface control documents are generated that define the messages and rate of transmission between the various computer software configuration items. Hardware requirements are specified in prime item development specifications and critical item development specifications. Finally, a host of test and qualification plans is generated to verify that the software does what the flowed down requirements dictate.

This decomposition process governs what gets implemented. It is this process that the human factors engineers need to influence. What is finally implemented in the cockpit and the pilot–vehicle interface are only those things called for by these documents. The decomposed requirements, however, are specific statements about what software does, and they do not address how the computer interacts with the pilot. For instance, a requirement in a software requirement specification might state that the routine is to present current location to the pilot and obtain a new position for purposes of updating the navigation system. The menu structure and display formats are not typically specified and are left to an informal process as to where in the system the update function is found, how the data is displayed, and how new data is entered. As a result, how the interface works is generally a fallout of design rather than a driver of the software decomposition process.

The only way to compete in the world of requirements decomposition and flowdown is for the human factors engineer to create his or her own set of requirements and flow them into the process.

How to do this is described in Hamilton and Metzler (1992). The pilot–vehicle requirements should include the mission and task analysis conducted to determine information requirements and general tasks. It should include specific design goals and implementation rules. The pilot–vehicle requirements should provide the specifics of each display format, each data entry or control method, and the specifics of the menu structure. The movement of symbols, definition of symbols (bitmaps), update rates, and smoothing and filtering algorithms should also be included. This document must be generated before the software requirements review but be iteratively updated before software preliminary design review as a function of person-in-the-loop testing and hardware/software development. After software preliminary design review, the requirements should be configuration managed (changed only by program directive) in response to known unknowns and unknown unknowns. Wherever possible, the user pilots of the helicopters should be consulted early in development and kept aware of the developing interface. A representative sample of the user community, environmental conditions, and missions should be included in a simulation program.

Creating a pilot–vehicle interface specification early in the program will address many of the questions and issues of cockpit development that have been raised in this chapter. Mission and task analysis are a requirement of most major development programs, as well as generating human engineering design approaches. The human engineering approach proposed here is cost-effective, because it centers on the continued iteration of analyses already conducted during the early stages of most programs. Eventually the display formats, control methods, and menu structure will have to be documented for testing, and for the development of training courses and devices. Although the effort shifts from an after-the-fact documentation to a design driver, it is not a new, unfunded activity. No new tasks are being defined by this approach, although some tasks are in more detail earlier in the program than previously. The human factors engineer must broaden his or her outlook and repertoire of skills, but the benefit is that the interface has the attributes and characteristics desired by design rather than by luck.

17.6 Summary

Helicopters present many of the same issues to the human factors engineer, as do fixed-wing aircraft, but helicopters do have unique challenges. These issues are related mostly to how helicopters generate and control lift and to what is done with the unique flight capabilities. Human factors engineers have always had an important role in designing a low-workload, high-situation-awareness cockpit, and that role will be more important in the computer age. Mission equipment development is now as expensive as airframe development, with a large portion of that cost being due to software. Human factors engineers must understand how computerized systems are developed and join in the process if acceptable cockpit workload and situation awareness are to be maintained in the face of ever increasing capabilities and expanding missions. Just as in airframe development, oversights in requirements and confusions in meaning can have very serious impacts on cost and schedule of software intensive systems. Like an airframe, software must have the inherent risks in the proposed design reduced by a systematic program of test and design maturation. This process of software requirement decomposition and verification will benefit from the participation of human factors engineers and will result in increased responsibilities for them. No new technological breakthroughs are required; the tools for design and test are available, but must be used in new ways.

Table 17.1 provides a list of references useful in the area of helicopter human factors. The list is composed primarily of military specifications (MIL-SPECs). MIL-SPECs have been condemned in recent years as the source of unnecessary cost and waste in defense procurement, and it may well be the case that elimination of chocolate chip cookie specifications may reduce the cost of federal cookies without impacting taste. However, not all MIL-SPECs are therefore bad. The ones listed here are generally very good in that they define a design space or processes rather than specify a solution. Most are applicable to either fixed- or rotary-wing aircraft.

TABLE 17.1 Recommended Readings

Reference	Title	Summary
ANSI/HFS 100	American National Standard for Human Factors Engineering of Visual Display Terminal Workstations	Specifies requirements for visual display terminals (VDTs), sometimes called visual display units, the associated furniture, and the office environment, where the use of a VDT is a required and substantial part of the work performed
ANSI-S1.11	Octave-Band and Fractional-Octave-Band Analog and Digital Filters, Specification for	Includes bandpass filter sets suitable for analyzing electrical signals as a function of frequency
ANSI-S1.13	Sound Pressure Levels, Methods for Measurement of	Provides performance requirements for fractional-octave-band bandpass filters, including, in particular, octave-band and one-third-octave-band filters. Basic requirements are given by equations with selected empirical constants to establish limits on the required performance
ANSI-S1.4	Meters, Sound Level, Specification for	Various degrees of accuracy are required for the practical measurement of sounds of various kinds for different purposes
ANSI-S1.40	Calibrators, Acoustical, Specification for	Provides specified performance requirements for coupler-type acoustical calibrators
ANSI-S3.2	Speech Over Communication Systems, Method for Measuring the Intelligibility of	Provides three alternative sets of lists of English words to be spoken by trained talkers over the speech communication system to be evaluated
ARI575	Method of Measuring Machinery Sound Within an Equipment Space	Establishes a uniform method of measuring and recording the sound pressure level of machinery installed in a mechanical equipment space
ASME-Y14.38 (supersedes MIL-STD-12)	Abbreviations and Acronyms	Establishes standard abbreviations and acronyms for documents and drawings submitted to government customer
ASTM-F1166	Marine Systems, Equipment and Facilities, Human Engineering Design for	Establishes general human engineering design criteria for marine vessels, and systems, subsystems, and equipment contained therein. It provides a useful tool for the designer to incorporate human capabilities into a design
ASTM-F1337	Standard Practice for Human Engineering Program Requirements for Ships and Marine Systems, Equipment, and Facilities	Establishes and defines the requirements for applying human engineering to the development and acquisition of ships and marine systems, equipment, and facilities
Boff and Lincoln (1988)	Engineering Data Compendium: Human Perception and Performance	Not a specification or standard. One of the best sources of human factors, human psychophysiology, and human performance available. Standard arrangement and user's guide make the data easily accessible. Highly recommended
Booher (2003)	Handbook of Human Systems Integration	Provides a comprehensive compilation of Human Systems Integration principles and methods
IEEE/EIA 12207 (supersedes MIL-STD-2167)	Software Life Cycle Processes	Provides guidelines for software development programs
Ketchel and Jenney (1968)	Electronic and Optically Generated Aircraft Displays: A Study of Standardization Requirements	Not a specification or standard. Provides excellent summary of information requirements, display principles, and human factors principles specifically oriented toward the aviation environment

Military Handbook MIL-HDBK-1473	Color and Marking of Army Materiel	This standard prescribes only general color and marking requirements—not finishes, surface preparations, related treatments for preservation and coating, or special requirements specified by Army design activities
Military Handbook MIL-HDBK-1908	Definitions of Human Factors Terms	This handbook defines terms frequently used in human factors standardization documents by providing common meanings of such terms to insure that they will be interpreted consistently and in the manner intended
Military Handbook MIL-HDBK-767	Design Guidance for Interior Noise Reduction in Light-Armored Tracked Vehicles	This handbook gives proven guidelines for designing quiet tracked vehicles and reducing interior tracked vehicle noise by redesigning vehicle components
Military Handbook MIL-HDBK-759	Human Engineering Design Guidelines	This handbook provides human engineering design guidelines and reference data for design of military systems, equipment, and facilities
Military Handbook MIL-HDBK-46855 (supersedes MIL-H-46855)	Human Engineering Program Process and Procedures	Establishes the requirements for applying human factors engineering to the development of all vehicles, equipment, and facilities for the U.S. military
Military Specification Data Item DI-HFAC-81204	Airborne Noise Analysis Report	Provides the format and content preparation instructions for a document that details the analysis to be performed to satisfy airborne noise requirements
Military Specification Data Item DI-HFAC-81278	Airborne Noise Control/Design History Booklet	Provides the format and content preparation instructions for a document that documents the efforts to meet the airborne noise requirements
Military Specification Data Item DI-HFAC-81399	Critical Task Analysis Report	Provides the format and content preparation instructions for a document that describes the results of analyses of critical tasks used to verify that human engineering technical risks have been minimized
Military Specification Data Item DI-HFAC-80270	Equipment Airborne Sound Measurement Plan	Provides a description of the airborne sound test procedure, facility and measurement system in advance of test to demonstrate planned compliance with requirements of MIL-STD-740-1
Military Specification Data Item DI-HFAC-80272	Equipment Airborne Sound Measurements Test Report	Provides the requirements for a report used to describe airborne sound tests and result that comply with the requirements of MIL-STD-740-1
Military Specification Data Item DI-HFAC-80274	Equipment Structureborne Vibration Acceleration Measurements Test Report	Provides a description of the structureborne sound tests and result to demonstrate compliance or noncompliance with MIL-STD-740-2
Military Specification Data Item DI-HFAC-80273	Equipment Structureborne Vibratory Acceleration Measurement Plan	Provides a description of the structureborne sound test procedure, facility and measurement system in advance of test
Military Specification Data Item DI-HFAC-80746	Human Engineering Design Approach Document—Operator	Provides the format and content preparation instructions for a document that provides a source of data to evaluate the extent to which equipment interfaces for operators having meets human performance requirements and criteria
Military Specification Data Item DI-HFAC-80747	Human Engineering Design Approach Document—Maintainer	Provides the format and content preparation instructions for HEDAD-M resulting from applicable tasks delineated by the statement of work
Military Specification Data Item DI-HFAC-80740	Human Engineering Program Plan	Establishes structure for the execution of formal human engineering program for development of vehicles, equipment, and facilities

(continued)

TABLE 17.1 (continued) Recommended Readings

Reference	Title	Summary
Military Specification Data Item DI-HFAC-80742	Human Engineering Simulation Concept	Provides the format and content preparation instructions for the Human Engineering Simulation concept to assess simulation approaches when there is a need to resolve potential critical human performance problems
Military Specification Data Item DI-HFAC-80745	Human Engineering System Analysis Report	Provides the format and content preparation instructions for the Human Engineering System Analysis Report used to evaluate the appropriateness and feasibility of system functions and roles allocated to operators and maintainers
Military Specification Data Item DI-HFAC-80743	Human Engineering Test Plan	Provides the format and content preparation instructions for a Human Engineering Test Plan
Military Specification Data Item DI-HFAC-80744	Human Engineering Test Report	Directs reporting structure and submittal of test results
Military Specification Data Item DI-HFAC-81202	Limited Noise Control Plan	Provides the format and content preparation instructions for a document that specifies and explains the approach, procedures and organization controls to insure compliance with noise control requirements
Military Specification Data Item DI-HFAC-80271	Sound Test Failure Notification And Recommendations Report	Provides the requirements for a report used to make the decision of whether to accept an equipment item, which fails to meet airborne sound criteria or structureborne vibratory acceleration criteria, or both
Military Specification MIL-STD-740	Airborne and Structureborne Noise Measurements and Acceptance Criteria of Shipboard Equipment	Describes instrumentation and procedures required for measurement of and analysis, and maximum acceptable sound level criteria for, airborne sound generated by shipboard equipment
Military Specification MIL-STD-1787	Aircraft Display Symbology	Controlled Distribution
Military Specification DOD-HDBK-743A	Anthropometry of U.S. Military Personnel (Metric)	Documents anthropometry of various military service populations
Military Specification MIL-C-81774	Control Panel, Aircraft General Requirement for	Delineates documents for design of control panels; establishes display and control selection, utilization, and arrangement; provides for verification data of these requirements
Military Specification MIL-STD-961 (supersedes MIL-STD-490)	Defense and Program-Unique Specifications Format and Content	Provides guidelines for types of specifications required, content, and layout of specifications
Military Specification MIL-D-23222	Demonstration Requirements	Defines demonstrations of the air vehicle. The only specific human engineering requirements are related to cockpit and escape system design
Military Specification MIL-STD-1472	Human Engineering	This standard establishes general human engineering criteria for design and development of military systems, equipment and facilities. Its purpose is to present human engineering design criteria, principles and practices to be applied in the design of systems, equipment and facilities so as to

Specification	Description
	(a) Achieve required performance by operator, control and maintenance personnel. (b) Minimize skill and personnel requirements and training time. (c) Achieve required reliability of personnel–equipment combinations. (d) Foster design standardization within and among systems.
Military Specification MIL-L-85762 — Lighting Aircraft, Interior, AN/AVS/6 Aviator's Night Vision Imaging System (ANVIS) Compatible	Provides performance requirements and testing methodology to insure effective and standardized aircraft interior lighting for ANVIS compatibility. This specification imposes very specific design and test requirements
Military Specification MIL-L-6503 (inactive except for replacements only) — Lighting Equipment, Aircraft, General Specification for Installation of	Requirements for installation of exterior and interior lighting, except instrument and aircrew station visual signals. Definition of emergency lighting system controls and fixtures, formation lights, position lights, landing and searchlights (and stick-mounted controls for same), etc. Specification used primarily by human factors engineering as information source on lighting intensities, cones of illumination, and orientation of lighting systems
Military Specification MIL-M-86506 — Mockups, Aircraft, General Specification for	General requirements for construction of aircraft and related systems mockups for formal evaluation and perpetration of mockup data. Provides mockup checklist for reviewers
Military Specification MIL-STD-1474 — Noise Limits	Describes three different types of "noise criteria" used to limit noise exposure have evolved (a) Hearing damage-risk criteria, (b) Hearing conservation criteria, and (c) Materiel design standards
Military Specification MIL-STD-1477 — Symbols for Army System Displays (Metric)	Prescribes the physical characteristics of ground and air track symbols, unit/installation symbols, control measures symbols, equipment symbols, and associated alphanumeric information for U.S. Army Combined Arms system displays, which are generated by electronic, optic, or infrared technology and presents information in real time or near-real time
SAE ARP 4033 — Pilot-System Integration	A recommended pilot-system integration (i.e., crew interface and system integration) approach for concept development is described
SAE-AS18012 (supersedes MIL-M-18012) — Markings for Aircrew Station Displays Design and Configuration of	Design requirements and configuration of letters, numerals, and identification for aircrew displays and control panels
SAE-AS18276 (supersedes MIL-L-18276) — Lighting, Aircraft Interior, Installation of	Requirements for primary, secondary, instrument, and emergency lighting systems and controls. Addresses visual signals for retractable gear warning
SAE-AS7788 (supersedes MIL-P-7788) — Panel, Information, Integrally Illuminated	Covers general requirements for integrally illuminated panels
Semple et al. (1971) — Analysis of Human Factors Data for Electronic Flight Display Systems	Not a specification or standard. Provides excellent background information related to human factors requirements for electronic displays. Provides historical information about a range of displays

References

Air-Conditioning and Refrigeration Institute, Inc. (1994). *Method of measuring machinery sound within an equipment space* (ARI 575-94). Arlington, VA: Air-Conditioning and Refrigeration Institute, Inc.

American National Standards Institute. (2001). *American National Standard Specification for sound level meters* (S1.4). New York: American National Standards Institute.

American National Standards Institute. (1998). *American National Standard Specification for sound level meters* (S1.11). New York: American National Standards Institute.

American National Standards Institute. (1998). *American National Standard Measurement of sound pressure levels in air* (S1.13). New York: American National Standards Institute.

American National Standards Institute. (1998). *American National Standard Measurement of sound pressure levels in air* (S1.40). New York: American National Standards Institute.

American National Standards Institute. (1999). *American National Standard Method for measuring the intelligibility of speech over communications systems* (S3.2). New York: American National Standards Institute.

American National Standards Institute/Human Factors Society. (1998). *ANSI/HFS American National Standard for human factors engineering of visual display terminal workstations.* Washington, DC/San Diego, CA: American National Standards Institute/Human Factors Society

American Society for Testing and Materials Standards. (2000). *Standard practice for human engineering design for marine systems, equipment and facilities* (ASTM-F1166). West Conshohocken, PA: American Society for Testing and Materials Standards.

American Society for Testing and Materials Standards. (2001). *Standard practice for human engineering program requirements for ships and marine systems, equipment, and facilities* (ASTM-F1337). West Conshohocken, PA: American Society for Testing and Materials Standards.

American Society of Mechanical Engineers. (1999). *Y14.38 Abbreviations and Acronyms.* American Society of Mechanical Engineers, NY.

Atkins, E. R., Dauber, R. L., Karas, J. N., & Pfaff, T. A. (1975). *Study to determine impact of aircrew anthropometry on airframe configuration* (Report No. TR 75-47). St. Louis, MO: U.S. Army Aviation System Command.

Boff, K. R., & Lincoln, J. E. (1988). *Engineering data compendium: Human perception and performance.* Wright-Patterson Air Force Base, OH: Harry G. Armstrong Aerospace Medical Research Laboratory.

Booher, H. R. (2003). *Handbook of human systems integration.* New York: John Wiley & Sons, Inc.

Hamilton, B. E. (1993, October). Expanding the pilot's envelope. Technologies for highly manoeuvrable aircraft. *North Atlantic Treaty Organization Advisory Group for Aerospace Research and Development, Conference Proceedings No. 548,* Annapolis, MD.

Hamilton, B. E., & Metzler, T. (1992, February 3–6). Comanche crew station design. In *Proceedings: 1992 Aerospace Design Conference, American Institute of Aeronautics and Astronautics* (AIAA-92-1049). Irvine, CA.

Institute of Electrical and Electronics Engineers/Electronic Industries Alliance. (1997). *12207 Software life cycle processes.* New York: Institute of Electrical and Electronics Engineers/Electronic Industries Alliance.

Ketchel, J. M., & Jenney, L. L. (1968). *Electronic and optically generated aircraft displays: A study of standardization requirements* (Report No. JANAIR 680505, AD684849). Washington, DC: Office of Naval Research.

Semple, C. A., Jr., Heapy, R. J., Conway, E. J., Jr., & Burnette, K. T. (1971). *Analysis of human factors data for electronic flight display systems* (Report No. A.FFDL-TR-70-174). Wright-Patterson Air Force Base, OH: Flight Dynamics Laboratory.

Society of Automotive Engineers, Inc. (1996). *Pilot-system integration* (ARP 4033). Washington, DC: Society of Automotive Engineers.

Society of Automotive Engineers, Inc. (1999). *AS 7788 Panel, Information, Integrally Illuminated.* Washington, DC: Society of Automotive Engineers.

Society of Automotive Engineers, Inc. (1998). *Markings for aircrew station displays design and configuration of* (AS 18012). Washington, DC: Society of Automotive Engineers.

Society of Automotive Engineers, Inc. (1999). *Lighting, aircraft interior, installation of* (AS 18276). Washington, DC: Society of Automotive Engineers.

U.S. Department of Defense (DoD). (1869). *Equipment airborne sound measurements test report* (DI-HFAC-80272). Washington, DC: Author.

U.S. Department of Defense (DoD). (1976). *Control panel, aircraft general requirement for* (MIL-C-81774A). Washington, DC: Author.

U.S. Department of Defense (DoD). (1983). *Anthropometry of military personnel (Handbook DOD-HDBK-743*). Washington, DC: Author.

U.S. Department of Defense (DoD). (1983). *Demonstration requirements* (MIL-D-23222A). Washington, DC: Author.

U.S. Department of Defense (DoD). (1986). *Equipment airborne sound measurement plan* (DI-HFAC-80270). Washington, DC: Author.

U.S. Department of Defense (DoD). (1986). *Equipment structureborne vibratory acceleration measurement plan* (DI-HFAC-80273). Washington, DC: Author.

U.S. Department of Defense (DoD). (1986). *Equipment structureborne vibration acceleration measurements test report* (DI-HFAC-80274). Washington, DC: Author.

U.S. Department of Defense (DoD). (1986). *Mockups, aircraft, general specification for* (MIL-M-8650C). Washington, DC: Author.

U.S. Department of Defense (DoD). (1986). *Sound test failure notification and recommendations report* (DI-HFAC-80271). Washington, DC: Author.

U.S. Department of Defense (DoD). (1986). *Airborne and structureborne noise measurements and acceptance criteria of shipboard equipment* (MIL-STD-740-1). Washington, DC: Author.

U.S. Department of Defense. (1988). *Job and task analysis handbook (*training and doctrine command pamphlet 351-4(T), TRADOC PAM 3514M). Washington, DC: Author.

U.S. Department of Defense (DoD). (1988). *Lighting aircraft, interior, AN/AVS/6 aviator's night vision imaging system (ANVIS) compatible* (MIL-L-85762A). Washington, DC: Author.

U.S. Department of Defense (DoD). (1991). *Airborne noise analysis report* (DI-HFAC-81203). Washington, DC: Author.

U.S. Department of Defense (DoD). (1991). *Limited noise control plan* (DI-HFAC-81202). Washington, DC: Author.

U.S. Department of Defense (DoD). (1992). *Airborne noise control/design history booklet* (DI-HFAC-81278). Washington, DC: Author.

U.S. Department of Defense (DoD). (1996). *Color and marking of army materiel* (MIL-HDBK-1473B). Washington, DC: Author.

U.S. Department of Defense (DoD). (1996). *Lighting equipment, aircraft, general specification for installation of* (MIL-L-6503H). Washington, DC: Author.

U.S. Department of Defense (DoD). (1997). *Noise limits* (MIL-STD-1474D). Washington, DC: Author.

U.S. Department of Defense (DoD). (1998). *Critical task analysis report* (DI-HFAC-81399A). Washington, DC: Author.

U.S. Department of Defense (DoD). (1998). *Human engineering design approach document–operator* (DI-HFAC-80746B). Washington, DC: Author.

U.S. Department of Defense (DoD). (1998). *Human engineering design approach document-maintainer* (DI-HFAC-80747B). Washington, DC: Author.

U.S. Department of Defense (DoD). (1998). *Human engineering simulation concept* (DI-HFAC-80742B). Washington, DC: Author.

U.S. Department of Defense (DoD). (1998). *Human engineering system analysis report* (DI-HFAC-80745B). Washington, DC: Author.

U.S. Department of Defense (DoD). (1998). *Human engineering test plan* (DI-HFAC-80743B). Washington, DC: Author.

U.S. Department of Defense (DoD). (1998). *Human engineering test report* (DI-HFAC-80744B). Washington, DC: Author.

U.S. Department of Defense (DoD). (2001). *Aircraft display symbology* (MIL-STD-1787C). Washington, DC: Author.

U.S. Department of Defense (DoD). (2002). *Symbols for army system displays (Metric)*, (MIL-STD-1477C). Washington, DC: Author.

U.S. Department of Defense (DoD). (2003). *Human engineering* (MIL-STD-1472F). Washington, DC: Author.

U.S. Department of Defense (DoD). (2003). *Defense and program-unique specifications format and content* (MIL-STD-961E). Washington, DC: Author.

U.S. Department of Defense (DoD). (2004). *Definitions of human factors terms* (MIL-HDBK-1908B). Washington, DC: Author.

U.S. Department of Defense (DoD). (2004). *Design guidance for interior noise reduction in light-armored tracked vehicles* (MIL-HDBK-767). Washington, DC: Author.

U.S. Department of Defense (DoD). (2004). *Human engineering program process and procedures* (MIL-HDBK-46855A). Washington, DC: Author.

Zimmermann, R. E., & Merrit, N. A. (1989). *Aircraft crash survival design guide Volume I-Design criteria and checklists* (Report No. TR 89-D-72). Fort Eustis, VA: Aviation Applied Technology Directorate.

18

Unmanned Aerial Vehicles

Nancy J. Cooke
Arizona State University

Harry K. Pedersen
L-3 Com

18.1 Benefits of the New Technology

The utilization of unmanned aerial vehicles (UAVs) has seen tremendous growth over the past few years. The military has employed UAVs for a number of years now in missions that are deemed too "dull, dirty, or dangerous" for manned aircraft. Systems such as the USAF Predator and U.S. Army Shadow are successfully deployed and have aided the U.S. Armed Forces in reconnaissance, surveillance, and even weapons deployment in theaters such as Afghanistan, Iraq, and Kosovo. This growth is expected to continue as civil, commercial, and private sectors begin to adopt UAVs for missions including, but not limited to, search and rescue, border patrol, homeland security, agricultural crop management, and communications relay. Such uses will benefit all by helping keep citizens safer, automating tedious jobs, and adding to the convenience of everyday life.

18.2 The Cost—Mishaps and Their Human Factor Causes

UAVs are commonly touted as a low-cost alternative to more expensive, less expendable manned aircraft. At the same time, UAV operations are becoming increasingly complex as trends such as reducing the number of operators per vehicle, weaponization, and operations in the National Airspace System (NAS) are considered. These factors have added to the operational cost of UAVs. In addition, UAVs have been subject to an unacceptable number of mishaps that also add to this cost. In fact, the number of UAV mishaps is, by some counts, 100 times higher than that of manned aircraft (Jackson, 2003). An examination by Schmidt and Parker (1995) of 107 mishaps that occurred between 1986 and 1993 revealed that 33% of all UAV mishaps were due to human errors such as crew selection and training, errors in teamwork and avionics control, aeromedical readiness, pilot proficiency, and operational tempo. Another analysis by Seagle (1997) of 203 mishaps from 1986 to 1997 revealed that 43% of all mishaps were due to human error (Ferguson, 1999).

For UAVs to be realized as the low-cost alternatives to manned aircraft they were meant to be, mishaps resulting in losses of the UAV (and potential injury to humans both in the air and on the ground) must be reduced to levels comparable with or below those of manned aircraft. This applies to all UAVs from the high altitude long endurance platforms to micro-UAVs, which can fit on the palm of a soldier's hand, for even small UAVs traveling at high speeds can cause large amounts of damage in a collision with a building or other aircraft.

18.3 A Misunderstood Technology

Despite the recent flurry of activity in UAV technologies and the gradual recognition of "hidden costs," there has been surprisingly minimal attention paid to human factors of these systems. Perhaps this is not at all surprising to human factor professionals who are commonly recruited at the end of system development to approve of or quickly fix the human interface to the system. However, for UAVs, the neglect seems to be at least partially due to some basic misconceptions about the role of the human in this system.

In particular, the term "unmanned" is an unfortunate choice not only because of the gender implications (i.e., there are no women in these vehicles), but also because of the implication that there are no humans involved in the system. This "feature" has been touted as an excuse for ignoring human factors. Of course, there *are* humans involved and for many platforms the persons involved outnumber those in manned systems. Humans are required to maintain, launch, control, operate, monitor, land, handoff, and coordinate UAV systems from the ground. There are also humans who are colocated in the same air or ground space as operators of manned vehicles or passersby.

Such "unmanned" notions are not only fueled by the unfortunate terminology, but also by over-confidence in the capabilities of automation. UAVs are highly automated. Platforms such as the Global Hawk are capable of taking off, flying missions, and landing autonomously. However, as we know from decades of research on automated systems (Parasuraman & Riley, 1997; Sheridan, 1987, 2002), automation does not relieve the humans from system responsibilities, it simply changes them. For instance, the human's task may change from one of control to one of oversight (Howell & Cooke, 1989). Furthermore, there are downsides of this changing role, including the loss of situation awareness that comes from being removed from the "raw data" (Endsley & Kiris, 1995). There are also some functions that are not easily automated (i.e., interpretation of target imagery, dynamic replanning in the face of change).

UAV systems also tend to evoke a number of analogies to air traffic control and manned aviation that, like all analogies, are imperfect. However, the mismatches between UAV systems and these other systems occur at critical human–systems integration junctures. For instance, the challenges of control of multiple UAVs by a single "pilot" on the ground are often underestimated because there are functions in UAV operation that do not exist in air traffic control (e.g., navigation, maneuvering, sensor operation, and coordination within the larger system). Further, cognitive workload associated with single vehicle operation can become quite intense when target areas are reached or when dynamic replanning is required (Wickens & Dixon, 2002; Wickens, Dixon, & Ambinder, 2006). Similarly, analogies to manned aviation overlook years of research demonstrating difficulties with remote operations such as time lag, loss of visual cues, and depth perception limits (Sheridan, 1992).

In sum, there are a number of human factor issues that are specific to this new technology, which should impact decisions about design, training, and certification. In the remainder of this chapter, we highlight some of these issues.

18.4 Some of the Same Human Factor Issues with a Twist

There are a number of human factor issues that are relevant to UAV operation and also shared with manned aviation. These include problems with spatial disorientation (SD), crew coordination, fatigue, and communication. Although the issues on the surface are similar, there are some unique twists that are peculiar to the remotely operated vehicle.

18.4.1 Spatial Disorientation

Spatial disorientation—the conscious interpretation of external cues and the subconscious addition of sensory cues (Young, 2003)—has always, and continues to be, a hazard of flying manned aircraft. SD affects all pilots, young and old, regardless of experience, and accounts for 15%–30% of fatalities in manned aviation (Young, 2003). SD for UAV operations, however, is quite furtive when compared with the more obvious cases in a manned cockpit. For example, UAV operators sitting in a stationary ground control station (GCS) are not subject to the peripheral visual stimuli, and proprioceptive inputs that a pilot experiences in the cockpit. Therefore, SD in UAV operations has been regarded with some skepticism. However, errors of misperception in UAV displays have been found to account for 10% of military UAV mishaps (Self, Ercoline, Olson, & Tvaryanas, 2006). The mechanisms of SD in UAVs include lack of visual flow due to poor displays, difficulty in discerning objects and judging distances between objects (especially at night), and visual–vestibular mismatches due to the operator not being physically located in the aircraft (Self et al., 2006). Possible solutions now being researched include improved display symbology, increasing sensor field-of-view, introducing higher levels of automation, and operator training.

18.4.2 Crew Coordination

The role of individual humans is often neglected in UAV system design and so too is the role of teams or crews of humans. Many UAV mishaps have been attributed to problems with crew coordination (Tvaryanas, Thompson, & Constable, 2005). Humans work in conjunction with other humans to control the UAV, to maintain it, to interpret imagery, and to coordinate activities with larger missions. The technology itself can also be conceptualized as playing the role of a team member in a larger mixed system of human and automation components. The ground operation of UAVs that requires multiple individuals (specific number depending on the platform) has been described as a command-and-control task (Cooke, Shope, & Rivera, 2000), which may present different crew requirements when compared with crews responsible for manned flight. Research on UAV command-and-control in a synthetic test bed (Cooke, DeJoode, Pedersen, Gorman, Connor, & Kiekel, 2004) has suggested that team interaction and process is central, team communication patterns change over time as the team gains experience, team skill is acquired in four 40 min missions and lost over periods of nonuse, distributed environments do not suffer large performance decrements when compared with colocated settings, and Internet video game teaming experience seems relevant to successful UAV team performance (Cooke, Pedersen, Gorman, & Connor, 2006). These are just some examples of crew coordination issues that may be unique to UAV operations.

18.4.3 Fatigue

UAV operators (i.e., USAF Predator) are called upon to work long shifts, often taking breaks only when they must visit the restroom (Goldfinger, 2004). These long shifts (due to the lack of operational standards), along with other environmental stressors, high workload, interruption of circadian rhythms, and lack of sleep lead to a state of fatigue in which operators must still function. Various studies have found that fatigue adversely affects the U.S. Armed Forces. For example, 25% of USAF

Class A mishaps from 1974 to 1992 and 12% of U.S. Naval Class A mishaps, and 4% of the U.S. Army's total number of mishaps from 1990 to 1999 were due to fatigue (Manning, Rash, LeDuc, Noback, & McKeon, 2004).

One way to mitigate the effects of fatigue is to design displays and systems by keeping in mind that the operator may be in a fatigued state. This could include the incorporation of auditory and tactile feedback to mitigate the workload from the visual modality, preprocessing information for the operator, but still engaging them by allowing them to make decisions, employing automation that monitors the operator, and employing basic ergonomics in the GCS to make the fatigued operator more comfortable (Tvaryanas Thompson, & Constable, 2005).

18.4.4 Communications

Communication is a large part of any command-and-control task. UAVs are no exception. Further, the communication is typically not face-to-face. UAVs being flown in the Middle East might be operated from a GCS in the Nevadan desert, in coordination with individuals in the eastern United States and on the ground in the Middle East to interpret data and take action. In heterogeneous teams like this with extreme division of labor, communication is paramount. However, it is not the case, as often is assumed, that effective communication means that every individual has access to all communications. This is not only impractical for larger systems, but is not even efficient for smaller teams. Unfortunately, advances in communication and information-sharing technologies enable widespread information access and the tendency for overinvolvement of leaders in lower level decision-making, moving away from the objective of decentralization. Thus, there are a number of important communication issues to be addressed in UAV operations. Toward this goal, researchers (e.g., Kiekel, Gorman, & Cooke, 2004) are examining communication patterns in a synthetic version of the task and are developing methods for rapid and online analysis of communications data.

18.5 Some New Issues

There are also a number of human factor issues that are unlike those found in other manned aviation systems. Some of these issues arise from the fact that this unmanned aviation system is operated remotely.

18.5.1 Remote Operation and Perceptual Issues

UAV designers are especially concerned about control latencies for sensor operators (SO) such that the SO cannot direct the UAV's sensor (i.e., camera) quickly. Considering that the majority of UAV sensors are controlled manually, this especially poses a problem for moving targets in busy urban environments. One possible solution is to increase the amounts of automation such that the SO can identify the target and allow a computer to "take over" based on target features. The automation could also track moving targets with more efficiency than a human—especially in crowded urban environments.

Soda straw views for operators are also a prevalent problem for UAVs that are controlled with traditional stick-and-rudder controls. These operators often report difficulty in flying. For example, Predator pilots have only a small, nose-mounted camera view by which to fly. Landings for Predators are especially difficult when compared with manned craft. In manned craft, the pilot has information about speed and position in his peripheral vision, whereas the Predator pilot does not. In fact, Predator pilots must land the UAV by first pointing it at the runway, and flaring just before touch down. Since the flight camera is not on a gimbal, the pilot loses sight of the runway until the UAV touches down. Possible remedies include synthetic vision overlays (Calhoun et al., 2005) and tools to help operators maintain a sense of scale within the environment.

18.5.2 Pilot Qualifications

Should UAV operators be experienced pilots of manned aircraft? Should training programs and selection criteria be the same as those for manned aircraft? If there are differences, what are they? This is a highly controversial topic in the UAV community and the current practices are as varied as the number of UAV platforms. For example, current Army systems (Hunter, Shadow UAVs) are piloted by individuals who are specifically trained to operate a UAV, but not a manned aircraft, whereas the Air Force's Predator operators are trained Air Force pilots of manned aircraft.

Training requirements and selection criteria boil down to the KSAs (knowledge, skills, and abilities) associated with the task of operating a UAV. Some believe that the KSAs are compatible with those of instrument flight conditions of manned flight (i.e., little or no visual feedback). Others believe that the task is so different that a completely new skill set is required. For instance, UAV operators need to be able to project themselves into the remote environment. Perhaps, this skill is more akin to experiences gained through Internet video gaming than through flight training. Research is being conducted (Schreiber, Lyon, Martin, & Confer, 2002), but the issue is complicated.

18.5.3 Presence

Views that UAV operation is like manned IFR flight neglect some subtle differences between the platforms. UAV operators are remote operators and lack presence. That is they have no tactile or motion feedback regarding the status of the aircraft. This precludes "seat of the pants" flying. Some might say that it also precludes adequate situation awareness. However, the need for the tactile and motion feedback in this very different environment raises a number of empirical questions that need to be addressed. Should we strive to make unmanned operations as similar as possible to the manned experience or is this new technology a completely different task environment with needs unlike that of manned flight?

18.5.4 Point-and-Click Control and Extensive Automation

A major area of discussion within the UAV community centers around what sort of control scheme is the most effective for the operation of UAVs. Many systems such as the Predator and Pioneer are operated with more traditional stick-and-rudder control schemes while others such as the Global Hawk and Shadow utilize a point-and-click interface. There are serious issues that surround each control scheme that requires research to determine which is optimal. For example, stick-and-rudder controls may require that the operator be a rated pilot, whereas point-and-click controls are more accessible to the nonpilots. However, point-and-click controls may also fail to provide the level of control needed in an emergency. Such was the case when a Global Hawk UAV, due to a failed actuator, entered a spin. The operator frantically clicked on the interface to regain control but could not and the UAV was lost. Although the spin was irrecoverable, the interface proved less than adequate for the task (Goldfinger, 2004).

18.5.5 UAV as a System

Issues also arise that make it clear that the target is not a single stand-alone vehicle, but rather a system that includes vehicle, sensor, and communication (and sometimes weapons) subsystems and which is itself embedded into a larger system such as the battlefield or the NAS. One could argue that most military vehicles are really complex systems, but with UAVs the system includes not only a vehicle system, but also sensor and communication systems. The vehicle merely supports the central tasks carried out by these central subsystems. Therefore, UAV operators often talk about "flying the camera," rather than flying the vehicle. Again, understanding differences in perspective or focus between manned flight and UAV operations may be critical in questions of design or training.

18.5.6 An Emotional Rollercoaster

As with manned flight, workload is not constant for UAV operations. There are long periods of vigilance and boredom interspersed with episodes of extreme terror and excitement. Much of the excitement occurs around planned targets and targets of opportunity with the boredom occurring when maneuvering between targets. There can also be situational changes (weather, enemy threat) that require vigilance and that will peak excitement even during routine maneuvering. Now with weaponized UAVs, the emotional variations are even more extreme. Pilots of manned aircraft experience these ups and downs too, but for the UAV operator these experiences occur during periods of remote operation (perhaps from a GCS in Nevada), which are further interspersed with periods off duty and at home. This integration of military mission and daily life has been described as going from the launching of hellfire missiles that morning to "soccer with the kids" that afternoon. Though the capabilities of remote operation enable the humans to be physically separate from the battlefield, the emotional separation between war and home has been increasingly blurred.

18.5.7 Lack of See-and-Avoid Capability

Another major issue with UAVs, particularly concerning operation in the NAS and the "swarming" concepts currently being researched by the U.S. Armed Forces, is the lack of see-and-avoid capabilities. This issue is of particular importance for UAV operation in the NAS in which operators will have to coordinate and interact with air traffic controllers, manned aircraft, and other UAVs. The small size of most UAVs prohibits the redundancy of avionics that other manned craft contain, let alone room for cameras or other such devices to be placed on the fuselage to act as "eyes" through which to see and avoid other aircraft. However, there are a number of projects that are currently exploring see-and-avoid technologies for future use (e.g., NASA's Environmental Research Aircraft and Sensor Technology (ERAST) program; Researchers Laud Collision-Avoidance Test Results, 2003).

18.5.8 Midair Handoffs

In UAV operations, there are shift changes in crews just as there are in manned aviation. However, for UAV operations these shift changes (i.e., handoffs) can occur in midair. For instance, it is often the case that one ground crew is tasked with launching the UAV and another with flight after take off. There have been a number of UAV mishaps attributed to coordination or communication difficulties associated with such handoffs. It seems clear that the task requires information sharing among crews. What are the human factor issues that are peculiar to these critical handoffs?

18.6 Conclusion

UAV operation is arguably a "different kind" of aviation. UAVs are different because they are remotely operated. In many ways, the task of "flying" a UAV has more in common with robotic operation than manned aviation.

Also, the UAV control interface seems to have more in common with video game interfaces (e.g., flight simulators), than they do with cockpit interfaces. Further, the communication and coordination required for UAV command-and-control is often exercised in Internet video games.

Although aviation human factors has much to contribute to the design of UAV technology, it is crucial that the unique nature of these unmanned aviation platforms be recognized. There are differences that result in new problems and likely novel solutions. Designing these new unmanned systems to replicate the old manned technology (e.g., including motion feedback in the GCS of a UAV) may needlessly constrain operations and fail to exploit the novel capabilities inherent in this exciting technology. Look up in the sky! Is it a plane? Is it a robot? Is it a video game? No, it is a UAV.

References

Calhoun, G. et al. (2005, May 25–26). *Synthetic Vision System for Improving UAV Operator Situation Awareness*. Paper presented at the CERI Second Annual Human Factors of UAVs Workshop, Mesa, AZ.

Cooke, N. J. et al. (2004). *The role of individual and team cognition in uninhabited air vehicle command-and-control* (Technical Report for AFOSR Grant Nos. F49620-01-1-0261 and F49620-03-1-0024). Mesa: Arizona State University.

Cooke, N. J., Pedersen, H. K., Gorman, J. C., & Connor, O. (2006). Acquiring team-level command and control skill for UAV operation. In N. J. Cooke, H. Pringle, H. Pedersen, & O. Connor (Eds.), *Human factors of remotely operated vehicles*. Volume in *Advances in human performance and cognitive engineering research Series*. Oxford, U.K.: Elsevier.

Cooke, N. J., Shope, S. M., & Rivera, K. (2000). Control of an uninhabited air vehicle: A synthetic task environment for teams. *Proceedings of the Human Factors and Ergonomics Society 44th Annual Meeting* (p. 389). San Diego, CA.

Endsley, M., & Kiris, E. (1995). (Should precede Ferguson in this list)The out-of-the-loop performance problem and level of control in automation. *Human Factors, 37*, 381–394.

Ferguson, M. G. (1999). *Stochastic modeling of naval unmanned aerial vehicle mishaps: Assessment of potential intervention strategies*. Unpublished master's thesis, Naval Post Graduate School, Monterey, CA.

Goldfinger, J. (2004, May 24–25). *Designing humans for UAVs: An Operator's perspective*. Paper presented at the CERI First Annual Human Factors of UAVs Workshop, Chandler, AZ.

Howell, W. C., & Cooke, N. J. (1989). Training the human information processor: A look at cognitive models. In I. L. Goldstein, & Associates (Eds.), *Training and development in organizations* (pp. 121–182). New York: Jossey Bass.

Jackson, P. (Ed.). (2003). *Jane's all the world's aircraft 2003–2004*. Alexandria, VA: Janes Information Group.

Kiekel, P. A., Gorman, J. C., & Cooke, N. J. (2004). Measuring speech flow of co-located and distributed command and control teams during a communication channel glitch. *Proceedings of the Human Factors and Ergonomics Society 48th Annual Meeting*. New Orleans, LA.

Manning, S. D., Rash, C. E., LeDuc, P. A., Noback, R. K., & McKeon, J. (2004). *The role of human causal factors in U.S. Army unmanned aerial vehicle accidents* (USAARL Report No. 2004-11). Washington, DC: United States Army Aeromedical Research Laboratory, Aircrew Health and Performance Division.

Parasuraman, R., & Riley, V. (1997). Humans and automation: Use, misuse, disuse, abuse. *Human Factors, 39*, 230–253.

Researchers laud collision-avoidance test results. (2003, Summer). Retrieved June 29, 2005, from the National Aeronautics and Space Administration Web site: http://ipp.nasa.gov/innovation/innovation112/5-aerotech3.html.

Schmidt, J., & Parker, R. (1995). Development of a UAV mishap factors database. *Proceedings of the 1995 Association for Autonomous Vehicle Systems International Symposium* (pp. 310–315). Washington, DC.

Schreiber, B. T., Lyon, D. R., Martin, E. L., & Confer, H. A. (2002). *Impact of prior flight experience on learning predator UAV operator skills* (AFRL-HE-AZ-TR-2002-0026). Mesa, AZ: United States Air Force Research Laboratory, Warfighter Training Research Division.

Seagle, Jr., J. D. (1997). *Unmanned aerial vehicle mishaps: A human factors approach*. Unpublished master's thesis, Embry-Riddle Aeronautical University, Norfolk, VA.

Self, B. P., Ercoline, W. R., Olson, W. A., & Tvaryanas, A. P. Spatial disorientation in uninhabited aerial vehicles. In N. J. Cooke, H. Pringle, H. Pedersen, & O. Connor (Eds.), *Human factors of remotely operated vehicles*. Oxford, U.K.: Elsevier.

Sheridan, T. (1987). Supervisory control. In G. Salvendy (Ed.), *Handbook of human factors* (pp. 1244–1268). New York: John Wiley & Sons.

Sheridan, T. (1992). *Telerobotics, automation, and supervisory control*. Cambridge, MA: MIT Press.

Sheridan, T. (2002). *Humans and automation*. Santa Monica, CA: Human Factors and Ergonomics Society and New York: Wiley.

Tvaryanas, A. P., Thompson, B. T., & Constable, S. H. (2005, May 25–26). *U.S. military UAV mishaps: Assessment of the role of human factors using HFACS*. Paper presented at the CERI Second Annual Human Factors of UAVs Workshop, Mesa, AZ.

Wickens, C. D., & Dixon, S. (2002). *Workload demands of remotely piloted vehicle supervision and control: (I) Single vehicle performance* (AHFD-02-10/MAAD-02-1). Savoy, IL: University of Illinois, Aviation Human Factors Division.

Wickens, C. D., Dixon, S. R., & Ambinder, M. Workload and automation reliability in unmanned aerial vehicles. In N. J. Cooke, H. Pringle, H. Pedersen, & O. Connor (Eds.), *Human factors of remotely operated vehicles*. Amsterdam, the Netherlands: Elsevier.

Young, L. R. (2003). Spatial orientation. In P. Tsang, & M. Vidulich (Eds.), *Principles and practice of aviation psychology* (pp. 69–113). Mahwah, NJ: Lawrence Erlbaum Associates.

IV

Air-Traffic Control

19

Flight Simulation

William F. Moroney
University of Dayton

Brian W. Moroney
InfoPrint Solutions Company

The U.S. Army Signal Corps' Specification Number 486 (1907) for the first "air flying machine" has a very straightforward "human factor" requirement:

> It should be sufficiently simple in its construction and operation to permit an intelligent man to become proficient in its use within a reasonable period of time.

Less than 3 years later, Haward (1910, as quoted in Rolfe & Staples, 1986) described an early flight simulator as

> a device which will enable the novice to obtain a clear conception of the workings of the control of an aeroplane, and of the conditions existent in the air, without any risk personally or otherwise. (p. 15)

The capabilities of both aircraft and flight simulators have evolved considerably since that time. Modern flight simulators have the same purpose except that they are used not only by novices but also by fully qualified aviators seeking a proficiency rating in a particular type of aircraft. After qualifying on a simulator, commercial pilots may proceed directly from a simulator to a revenue-producing flight. Similarly, if a two-seat training version is not available, pilots of single seat military aircraft proceed directly from the simulator to the aircraft.

In 1994, flight simulation was a worldwide industry with many competitors (Sparaco), and sales of $3 billion per year for commercial airlines and $2.15 billion per year for the U.S. Department of Defense. Individual simulators ranged in price from $3000 for a basic personal computer (PC)-based flight simulator with joystick controls up to $10–$13 million for a motion-based simulator (down from $15–$17 million in the early 1990s). In 2006, Frost and Sullivan, an organization that studies markets, reported that within North America, revenues for commercial and military ground based flight simulation (GBFS) totaled $2.01B in 2005 and were expected to reach $2.78B in 2012. Their August 2006 report notes that, within North America, in 2005 the commercial GBFS segment accounted for 36.3% of the total revenues, while the military segment accounted for 63.7%. They predict growth to the GBFS market based on the introduction of new aircraft (such as the B787 and the A380), high fuel and maintenance costs. The military sector has the additional pressure of aircraft and instructor unavailability owing to operational commitments. Frost and Sullivan also predict simulator growth in the very light jet air taxi and business jet markets, finally, there will be the need to train individuals with no or minimal flight experience as operators of unmanned air vehicles (UAVs), which will be sharing the airspace with manned vehicles carrying passengers.

Consolidation within the aviation industry has reduced the number of competitors in the GBFS market since the 1990s. This consolidation has contributed to advances in technology and the emphasis on leaner more efficient core-business focused organizations (Wilson, 2000). According to L-3 CEO Frank Lanza, "Simulators that used to cost $40 million per copy now cost $7–$10 million each. And we are driven by commercial technology because the explosion of precision graphics and visuals for the multimedia industry is directly transferable to the military" (Wilson, 2000, p. 19). This reduction in hardware cost for the display technologies has changed the traditional focus from hardware-driven to software-driven. Major suppliers in the simulation industry are now focusing on incorporation of the Internet and virtual-reality into training systems.

Flight simulation is essentially the representation of aircraft flight and system characteristics with varying degrees of realism for research, design, or training purposes (Cardullo, 1994a). Cardullo listed three categories of training simulators: (a) the operational flight trainer (OFT), used to train individual pilots or crews in all aspects of flight and the use of flight, navigation, and communication systems; (b) the weapons systems trainer, used to train in the use of offensive and defensive systems; (c) the part task trainer, used to train flight crews for specific tasks (e.g., in-flight refueling).

Most flight simulators have the following features:

1. Visual displays: Most simulators provide an external view of the world along with cockpit flight, navigation, and communication instruments. In addition, depending on its mission, some simulators display radar and infrared data.
2. Control/input devices: Usually, a yoke or a stick combined with a control loader is used to mimic the "feel" of the real aircraft. The original control loaders were mechanical devices, that used weights, cables, and springs to mimic the response of a control stick to aerodynamic forces. These were replaced by hydraulic and electro-hydraulic control loaders, which improved fidelity and reliability but were bulky and required considerable maintenance. Modern control loaders are much smaller and utilize computer-controlled electric motors to mimic the feel of control device. In high-fidelity simulators, switches and knobs identical to those in the aircraft are used, whereas in lower fidelity devices, a mouse or a keyboard may be used to input changes to a switch position.
3. An auditory display: These may include a synthetically generated voice, warning and advisory tones, and/or intercommunication systems.
4. Computational systems: These units may include the flight dynamics model, image generation, control, and data collection software.

In addition, some simulators (usually OFTs) have a motion base that provides rotation and translation motion cues to the crewmember(s). While others may use G-seats or anti-G suits to simulate motion and G cues.

Typically, the more sophisticated simulators are used in the commercial and military aviation communities, whereas less sophisticated simulators are used by general aviation communities. Some military simulators are full mission simulators and may include enemy threats (e.g., surface-to-air missiles, communications jamming, etc.) as well as other simulated aircraft with aircrew, simulating wingmen or enemy aircraft, and other "players" simulating air traffic controllers, airborne command posts, and so on.

This chapter is intended to provide a broad overview of flight simulation with an emphasis on emerging areas. It begins with a brief history of flight simulators and a discussion of the advantages and disadvantages of flight simulators. Following this, the topics of simulator effectiveness, including cost and transfer measurement strategies, and the issue of fidelity are examined. Next is a description of different types of visual and motion systems, as well as a discussion of the debate surrounding the use of motion in flight simulators. Considerable discussion is then devoted to the issue of simulator sickness and strategies to minimize its deleterious effects. Implications of using virtual reality/virtual environment technology are discussed, with an emphasis on cybersickness. To broaden the reader's appreciation of the wide variety of simulators, a brief overview of five unique simulators is presented. Next, the often ignored, but critical area of instructional features is explored, followed by an overview of an area of tremendous potential growth—PC-based simulation. Finally, the technical differences between simulators and training devices (TDs) are delineated. The chapter ends with a listing of the authors' perceptions of the future and opportunities.

19.1 History of Flight Simulators

Adorian, Staynes, and Bolton (1979) described one of the earliest simulators, an Antoinette trainer (circa 1910), in which a student was expected to maintain balanced flight while being seated in a "barrel" (split the long way) equipped with short "wings." The barrel, with a universal joint at its base, was mounted on a platform slightly above shoulder height so that instructors could push or pull on these "wings" to simulate "disturbance" forces. The student's task was to counter the instructors' inputs and align a reference bar with the horizon by applying appropriate control inputs through a series of pulleys.

In an attempt to introduce student pilots to the world of flying prior to actual liftoff, the French Foreign Legionnaires realized that an airframe with minimal fabric on its wings would provide trainees with insight into the flight characteristics of the aircraft while limiting damage to the real aircraft and the student (Caro, 1988). Winslow (1917), as reported in Rolfe and Staples (1986), described this device as a "penguin" capable of hopping at about 40 miles per hour. Although this may seem to be of limited use, it was a considerable improvement from the earlier flight training method of self-instruction in which trainees practiced solo until basic flight maneuvers had been learned. Instructors would participate in the in-flight training only after the trainees had, through trial and error, learned the relationship between input and system response (Caro, 1988). Apparently, the Legionnaires understood the value of a skilled flight instructor.

In 1917, Lender and Heidelberger developed a rudimentary simulator that utilized compressed air to induce deflections (Kuntz Rangal, Guimaraes, & De Assis Correa, 2002). In an effort to increase fidelity, they provided noise and used visual imagery to simulate speed. In 1929, Buckley patented a moving-based simulator driven by an electric motor. In-flight disturbances were introduced by the use of perforated tapes. The flight simulator industry began in 1929, when Edward A. Link received a patent for his generic ground-based flight simulator (Fischetti & Truxal, 1985). His initial trainer was designed to demonstrate simple control surface movements and was later upgraded for instrument flight instruction. Link based his design on the belief that the trainer should be as analogous to the operational setting as possible. Through the use of compressed air, which actuated bellows (adapted by Link from his father's pipe organ factory), the trainer had motion capabilities of pitch, yaw, and roll that enabled student pilots to gain insight into the relationship between stick inputs and movement in three flight dimensions. Originally, marketed as a coin-operated amusement device (Fischetti & Truxal, 1985), the value of Link's simulator was recognized when the Navy and Army Air Corps began purchasing trainers in 1934. Flight instructors, watching from outside the "Blue Box," would monitor the movements of the ailerons, elevator, and rudder to assess the student's ability to make the correct stick movements necessary for various flight maneuvers.

When the United States entered World War II, there were over 1600 trainers in use throughout the world. The necessity for the trainers increased as the Allied forces rushed to recruit and train pilots. As part of this massive training effort, 10,000 Link trainers were used by the United States military during the war years (Caro, 1988; Stark, 1994). In 1944, the U.S. Navy funded the Massachusetts Institute of Technology to develop Whirlwind, an experimental computer designed as part of a flight simulator (Waldrop, 2001). This unique computer was essentially a calculator/batch processor but worked in real time with an interactive capability. By 1948, this billion dollar a year (in 1948 dollars) project had evolved into the first general-purpose real-time computer. This interactive capability laid the groundwork for today's PCs. Although this computer occupied the space of a small house, its computing power was equivalent to that of an early 1980 TRS-80 (1.774 MHz, 12K ROM, 4–48K RAM).

After the war, simulations developed for military use were adapted by commercial aviation. Loesch and Waddell (1979) reported that by 1949, the use of simulation had reduced airline transition flight training time by half. Readers interested in details on the intriguing history of simulation would do well to consult the excellent three-volume history entitled *50 Years of Flight Simulation* (Royal Aeronautical Society, 1979). Also, Jones, Hennessy, and Deutsch (1985), in *Human Factors Aspects of Simulation*, provided an excellent overview of the state of the art in simulation and training through the early 1980s.

Following the war and throughout the 1950s, increases in aircraft diversity and complexity resulted in the need for aircraft-specific simulators, that is, simulators that represent a specific aircraft in instrument layout, performance characteristics, and flight-handling qualities. Successful representation of instrument layout and performance characteristics was readily accomplished; however, the accurate reproduction of flight-handling qualities was a more challenging task (Loesch & Waddell, 1979). Precise replication of the control, display, and environmental dynamics is based on the unsupported belief that higher fidelity simulation results in greater transfer of training from the simulator to the actual aircraft. This belief has prevailed for many years and continues today. However, even 55 years ago, researchers were questioning the need for duplicating every aspect of flight in the simulator (Miller, 1954; Stark, 1994).

Caro (1979) described the purpose of a flight-training simulator as "to permit required instructional activities to take place" (p. 84). However, from his examination of the existing simulators, simulator design procedures, and the relevant literature, Caro concluded that "designers typically are given little information about the instructional activities intended to be used with the device they are to design and the functional purpose of those activities" (p. 84). Fortunately, some progress has been made in this area. Today, as part of the system development process, designers (knowledgeable about hardware and software), users/instructors (knowledgeable about the tasks to be learned), and trainers/psychologists (knowledgeable about skill acquisition and evaluation) interact as a team in the development of training systems (Stark, 1994). The objective of this development process is to maximize the training effectiveness while minimizing the cost and time required to reach the training objective (Stark, 1994).

19.2 Why Use Simulation?

Simulation is both effective and efficient. As a tool within a broader training program, it provides an excellent training environment that is well accepted by the aviation community.* It provides an opportunity for initial qualification or requalification in type and is a means for experiencing critical conditions

---

* Simulator effectiveness is discussed in a separate section of this chapter.

that may never be encountered in flight. However, like all attempts at education and training, it has both advantages and disadvantages, which are discussed next.

19.2.1 Advantages

Part of the efficiency of simulators may be attributed to their almost 24 h a day availability, and their ability to provide immediate access to the operating area. For example, simulators allow a student to complete an instrument landing system (ILS) approach and return immediately to the final approach fix for the next ILS approach, without consuming time and fuel. Indeed, because simulators are not realistic, conflicting traffic in the landing approach can be eliminated to further increase the number of approaches flown per training session. In short, simulators provide more training opportunities than could be provided by an actual aircraft in the same time. As noted by Jones (1967), simulators can provide training time in nonexistent aircraft or in aircraft where an individual's first performance in a new system is critical (consider the first space shuttle landings or single seat aircraft).

Because of safety concerns, simulators may be the only way to teach some flight maneuvers or to expose aircrew to conditions that they are unlikely to experience under actual flight conditions (e.g., wind sheer, loss of hydraulic systems, engine loss, engine fire, exposure to wake turbulence, and clear air turbulence). Additionally, automation has increased the need for simulators, as Wiener and Nagel (1988, p. 453) commented "It appears that automation tunes out small errors and creates the opportunities for larger ones." In automated glass (cathode ray tube [CRT] or liquid crystal equipped) cockpits, improvements in system reliability have reduced the probability and frequency of system problems, thus inducing a sense of complacency among the aircrew. However, when an unanticipated event occurs, the crew must be trained to respond rapidly and correctly. Simulators provide an opportunity for training under these conditions.

Simulator usage also reduces the number of flight hours on the actual aircraft, which in turn reduces mechanical wear and tear, associated maintenance costs, and the load on the national airspace system. Additionally, airlines do not incur the loss of revenue associated with using an aircraft for in-flight training. Simulator usage also reduces environmental problems, not only air and noise pollution but, in the case of military training, damage to land and property.

Simulators also provide an improved training environment by incorporating instructional features that enhance student learning, and facilitate instructor intervention. Such features are described later in this chapter. Additionally, simulators provide standardized training environments with identical flight dynamics and environmental conditions. Thus, the same task can be repeated until the required criteria are attained, and, indeed, until the task is overlearned (i.e., automated). Unlike the airborne instructor, the simulator instructor (SI) can focus on the teaching task without safety of flight responsibilities, or concerns about violations of regulations. Thus, he or she may deliberately allow a student to make mistakes such as illegally entering a terminal control area or exceeding the aircraft's aerodynamic capability.

Simulators allow performance data to be collected, which according to Stark (1994) permits:

1. Performance comparison: As part of the diagnosis process, the instructor pilot (IP) can compare the student's performance with the performance criteria, and the performance of students at the same stage of training.
2. Performance and learning diagnosis: Having evaluated the student's performance, the IP can gain some insight into the student's learning process and suggest new approaches in problem areas.
3. Performance evaluation: Performance measurement can be used to evaluate the efficacy of different approaches to training a particular task.

Despite the emphasis on high fidelity and "realism," *simulators are not realistic*. In a sense, the lack of realism may contribute to their effectiveness. Indeed, Lintern (1991) believed that transfer can be enhanced by "carefully planned distortions of the criterion task" (p. 251). Additionally, most instructional features found in simulators do not exist in the cockpit being simulated. Indeed, if real cockpits had the same features as simulators, the "PAUSE" button would be used routinely.

19.2.2 Disadvantages

Let us now examine some of the alleged "disadvantages" of simulators. We must recognize that performance in a simulator does not necessarily reflect how an individual will react in flight. Because there is no potential for an actual accident, the trainee's stress level may be lower in a simulator. However, the stress level can be high when an individual's performance is being evaluated or when he or she is competing for a position or a promotion.

To the extent that aircrew being evaluated or seeking qualification-in-type expects an emergency or unscheduled event to occur during their time in the simulator, their performance in a simulator may not reflect in-flight performance, since the aircrew would, in all probability, have reviewed operating procedures prior to being evaluated in the simulator. Nonetheless, it should be recognized that a review of procedures even in preparation for a check ride is of value. Performance in simulators rarely reflects the fatigue and/or boredom common to many cockpits. Therefore, performance in a simulator may be better than that actually expected in flight.

In addition, simulators reduce the utilization of actual aircraft, which leads to fewer maintenance personnel and reduced supply requirements. These apparent savings may create personnel shortages and logistic problems when the operational tempo rises beyond the training level.

Simulators, particularly dome and motion-based simulators, usually require unique air-conditioned facilities, and maintenance personnel, which reduces the assets available to operational personnel. When used excessively, simulators may have a negative effect on morale and retention. This attitude is usually reflected as, "I joined to fly airplanes, not simulators." Finally, the acceptance and use of simulators is subject to the attitudes of simulator operators, instructors, aircrew, and corporate management and the evaluating agency.

Overall, the advantages significantly outweigh any real or perceived disadvantages as evidenced by the general acceptance of simulators by the aviation community and regulatory agencies.

19.3 Simulator Effectiveness

Simulation is a means for providing the required training at the lowest possible cost. Baudhuin (1987) stated, "the degree of transfer from the simulator to the system often equates to dollars saved in the operation of the real system and in material and lives saved" (p. 217). The aviation industry could not function without simulators and flight training devices (FTDs), whose existence is mandated by Federal Aviation Administration (FAA) regulations (1991, 1992).

In a very detailed analysis of cost-effectiveness, Orlansky and String (1977) reported that flight simulators for military training can be operated at between 5% and 20% of the cost of operating the aircraft being simulated; median savings is approximately 12%. They also reported that commercial airlines can amortize the cost of a simulator in less than 9 months and the cost of an entire training facility in less than 2 years.

Roscoe (1980) has provided sufficient data illustrating the effectiveness of fixed-base simulators for teaching the skills needed in benign flight environments. Spears, Sheppard, Roush, and Richetti (1981a, 1981b) provided detailed summaries and evaluations of 196 research and development reports related to simulator requirements and effectiveness. Pfeiffer, Horey, and Butrimas (1991) supplied additional support in their report of positive transfer of instrument training to instrument and contact flight in an operational flight training aircraft (U.S. Navy T-2C).

Often, because of the high cost of true transfer of training experiments, quasi-experiments are performed to determine the transfer between an FTD or part-task trainer and a representative high-fidelity simulator that serves as a surrogate for the real aircraft. However, Jacobs, Prince, Hays, and Salas (1990) in a meta-analysis of data culled from 247 sources identified 19 experiments in which training transfer between the simulator and the actual aircraft was evaluated. They concluded that simulators reliably produced superior training relative to aircraft only training. They also reported that for jet aircraft, take-offs, landings, and approaches benefited from the use of a simulator, with the landing approach showing the greatest benefit. However, similar conclusions regarding the effectiveness of helicopter simulators could not be drawn because only seven experiments involving helicopters meet the criterion for inclusion in the meta-analysis.

Today's effectiveness questions are focused on how the required skills can be taught rapidly and inexpensively. Thus, we have seen an emphasis on the systems approach to training (systems approach to timing, Department of the Army, 1990), similar to the instructional systems development approach, which emphasizes requirement definition and front-end analysis early in the system development process and an evaluation at the end. Roscoe and Williges (1980), Roscoe (1980), and Baudhuin (1987) provided excellent descriptions of strategies for evaluating transfer of training, including the development of transfer effectiveness ratios (TERs), and of incremental transfer effectiveness functions, and cumulative transfer effectiveness functions. All of these approaches attempt to measure the degree to which performing the desired task in the actual aircraft is facilitated by learning an intervening task on a TD or simulator. The resulting measure is usually expressed in terms of time saved. The critical concern, as emphasized by Roscoe (1980), was not simply measuring training effectiveness but determining cost-effectiveness. Specifically, Roscoe was concerned with identifying the region in which increasing the investment in the TD (by improving fidelity, adding additional instructional features, etc.) did not result in a significant increase in transfer. However, as noted by Beringer (1994), because the cost of simulation has decreased as the capabilities of simulators have increased, today's question is more often phrased as, "If we can get more simulation for the same investment, what is the 'more' that we should ask for?" Thus, according to Beringer, cost is seen as a facilitating, rather than a prohibitive factor.

Measuring effectiveness is a fairly complicated process that has performance measurement at its core. Lane's (1986) report is "must reading" for individuals interested in measuring performance in both simulators and the real world. Mixon and Moroney (1982) provided an annotated bibliography of objective pilot performance measures in both aircraft and simulators. Readers interested in measuring transfer effectiveness are referred to Boldovici's (1987) chapter on sources of error and inappropriate analysis for estimating transfer effectiveness.

19.4 Fidelity

Hays and Singer, in their book *Simulation Fidelity in Training System Design* (1989), provided an excellent, comprehensive examination of the complex issue of fidelity. They defined simulation fidelity as

> the degree of similarity between the training situation and the operational situation which is simulated. It is a two dimensional measurement of this similarity in terms of (1) the physical characteristics, for example, visual, spatial, kinesthetic, etc.; and (2) the functional characteristics (for example, the informational, and stimulus response options of the training situation). (p. 50)

The simulation community appears to be divided into two camps on the issue of fidelity. One group (usually the simulator developers and regulatory agencies) believes that the simulation should be as realistic as technically possible. Thus, they emphasize high fidelity of both the simulator cockpit and the environment. They are concerned that failure to properly represent the cockpit of the environment may increase the probability of a critical error, which could result in the loss of life. The other group (behavioral scientists and trainers) emphasizes the functional characteristics. They contend, as Bunker (1978) stated, that "instead of pondering on how to achieve realism, we should ask how to achieve training" (p. 291). Lintern (1991) notes that "similarity, as it is normally viewed is not a sufficient element of a conceptual approach to skill transfer" (p. 253). He argued further that simulator designers must distinguish between the informational "invariants" critical for skill acquisition and the irrelevant elements (i.e., the "extras" often included in the name of fidelity). The "invariants" according to Lintern are the properties of the events that remain unchanged as other properties change. Such a property remains constant across events that are perceived as similar but differs between events that are perceived as different.

Owing, at least in part, to economic considerations, the two camps now interacting more often should lead to improved training systems. Increased computing capability, increased memory capacity and rapid access, multiple-processor architecture, improved image generation capability, and so on have led to new simulation technologies ranging from PC-based flight simulation to VEs (Garcia, Gocke, & Johnson, 1994), to distributed simulations such as SIMNET (Alluisi, 1991), to even more complex real-time virtual world interactions (Seidensticker, 1994). Because there are such widely varied technologies available in very disparate cost ranges, the focus appears to be gradually evolving from developing new technology to deciding what needs to be simulated. The Department of the Army's (1990) systems approach to training, emphasizes the use of task analysis to identify training requirements and has tasked the training community with defining the simulator requirements for the developers. Thus, today's simulators are integrated into a training system, which may include a variety of TDs, media, and educational strategies to achieve the desired outcome. While the systems approach to training is a good start, much progress remains to be made in achieving a trainee-centered approach to training. Perhaps DoD's human system's integration efforts will accelerate the process.

Hays and Singer (1989) advised that the effectiveness of a simulator is not only a function of the characteristics and capabilities of the simulator, but how those features support the total training system. They indicate that simulator fidelity should vary as a function of stage of learning, type of task, and type of task analysis. Each of these factors is described independently.

19.4.1 Stage of Learning

Fitts (1962) tripartite model of skill development consisting of a cognitive phase, an associative phase, and an autonomous phase has served the aviation community well. Although the boundaries between the phases are not clearly delineated, the skills needed in aviation progress in this sequence. During the cognitive phase, the novice attempts to understand the task, the expected behavior, the sequence of required procedures, and the identification of relevant cues. Instructions and demonstrations are most effective during this phase. During the associative phase, the student integrates skills learned during the cognitive phase and new patterns emerge, errors are gradually eliminated, and common features among different situations begin to be recognized. Hands-on practice is most appropriate during this phase. Finally, during the autonomous phase, the learner's performance becomes more automatic, integrated, and efficient, thus requiring less effort. Individuals at this level of skill development are more resistant to the effects of increased workload because they have well-developed "subroutines." At this stage, learners can perform previously learned tasks while a new skill is being acquired. Whole task and mission simulations are most appropriate for individuals at this skill level.

During the early phase of learning, less expensive, lower fidelity simulators will suffice. Caro (1988) provided an interesting case study in which wooden mockups with fairly simple displays were as effective as a much more expensive cockpit procedures trainer. Warren and Riccio (1985) noted that simulations providing stimuli that experienced pilots tend to ignore make learning more difficult, because the trainee has to learn how to ignore those irrelevant stimuli. More recently, Kass, Herscheler, and Campanion (1991) demonstrated that students trained in a "reduced stimulus environment that presented only task-relevant cues performed better in a realistic battle field test condition than did those who were trained in the battle field test condition" (p. 105). Similarly, Lintern, Roscoe, and Sivier (1990) trained two groups of flight-naive subjects in landing procedures: one group trained with crosswinds and the other group trained without crosswinds. When the performance of both groups was evaluated on a 5 knot crosswind landing task, the group trained without the crosswinds performed better. Apparently, training with the crosswinds confounded the students' understanding of the relationship of control action and system response, whereas training without the crosswinds did not interfere with the students' learning. Thus, it has been demonstrated that higher fidelity does not necessarily lead to more efficient transfer of training.

19.4.2 Type of Task

The fidelity required to teach a cognitive task (information processing) is very different from the fidelity required to learn a psychomotor (tracking) task. For example, the type of simulation required to facilitate the development of a cognitive map of a fuel flow system is very different from the type of simulation required to demonstrate an individual's knowledge of the same fuel flow system under emergency conditions. In the former case, a model board indicating valve positions, fuel tank locations and quantities, and so forth would be appropriate, whereas in the latter, a full-cockpit simulation is more appropriate. However, even in very complex flight conditions such as Red-Flag simulated air combat exercises, it has been demonstrated that individuals trained on lower fidelity simulators showed a higher level of ground attack skills than individuals who did not receive any simulator training (Hughes, Brooks, Graham, Sheen, & Dickens, 1982).

19.4.3 Type of Task Analysis

The type of task analysis performed will significantly influence the level of fidelity incorporated into the simulator. Baudhuin (1987) emphasized the need for comprehensive front-end analysis in developing meaningful task modules that ultimately are incorporated into the specifications to which the system is developed. Warren and Riccio (1985) and Lintern (1991) argued that appropriate task analysis would help distinguish between the necessary and the irrelevant cues. Inappropriate task analysis will lead to inadequate, perhaps even inappropriate, training and low transfer to the operational setting.

Thus, decisions regarding the required level of fidelity are multifaceted. Alessi (1988) provides a taxonomy of fidelity considerations to be addressed in examining the relationship between learning and simulation. Cormier (1987) and Lintern (1991) provided insights into the role of appropriate cues in the transfer of training, whereas Baudhuin (1987) provided guidance on simulator design. In 1998, Salas, Bowers, and Rhodenizer provided an insightful perspective on the overreliance on high-fidelity simulators and the misuse of simulation in enhancing the learning of flying skills. They emphasize that this area is approached from a variety of perspectives. On the one hand, engineers, computer scientists, and simulation designers focus on technology and are driven by requirements and specifications. On the other hand, human factors personnel, reflecting their psychology training, focus on understanding the processes involved in the acquisition of knowledge skills and attitudes. Despite a wealth of knowledge about training and learning, little of that knowledge has been put into practice in flight simulation. The authors believe that

> …the solution to this problem lies in bridging the gap between training research findings and the current capabilities that simulations offer for the aviation domain. This will require, we suggest a rather drastic paradigm shift. Specifically, scientists and engineers must identify, confront, and engage in a dialog about the assumptions in the use and applications of simulations in aviation training; assumptions that as we demonstrate are probably not valid, appropriate, correct, or useful for the advancement of aviation training. (p. 199)

Salas, Bowers, and Rhodenizer (1998) describe three problematic assumptions:

1. Simulation is all you need.
   Based on this assumption, large amounts of money have been expended on the development of training/simulation devices. This has occurred without appropriate training needs analysis and

the development of testable criteria. The emphasis has been on increasing realism rather than improving a crewmember's proficiency at a lower cost. They note that "There are strict guidelines and specifications for the development of simulators but not for the training conducted in them p. 201." They argue that appropriate use of instructional features (see Table 19.1) determine the success of training more than the fidelity of the simulation.

2. More is better.

   It is commonly believed that higher fidelity leads to greater transfer of training and improved learning. The authors cite research (as far back as 1987), demonstrating that this assumption is incorrect. They also provide current research documenting the effectiveness of low fidelity simulations in achieving cost-effective transfer of training.

3. If the aviators like it, it is good.

   The initial evaluation of simulator effectiveness is usually performed by subject matter experts (SMEs) whose primary focus is face validity. Subsequently, the trainee's opinion of the effectiveness of the simulation is solicited. The emphasis is on how well the simulator performs as opposed to how well the training transfers. Data are collected at what Kirkpatrick (1998) describes as the lowest level of system evaluation: reaction (did the trainee like it). Rarely are data gathered at the learning (measurement of what was learned), behavior (transfer of training), and results (how did the training impact the organization) levels.

The authors conclude with the following recommendations for improving the use of simulators in the learning environment:

1. Acquisition managers in organizations acquiring simulators should focus on processes that achieve the desired learning and not on technology development.
2. More sophisticated measures of effectiveness must be developed and used in evaluating simulations. A good example of this is the work of Taylor, Lintern, and Koonce (2001), which provides an approach to predicting transfer from a simulator to an aircraft in a quasitransfer study.
3. The assumptions listed above must be abandoned.
4. Engineers, system designers, and behavioral scientists must work as partners in developing training systems.
5. Behavioral scientists must translate their knowledge about learning, instructional design, and human performance into guidelines that can be used by simulation developers.

Salas et al. (1998) have described the need for a paradigm shift from technology design to trainee-centered design. The critical element to remember with respect to fidelity is that simulator fidelity is not the end, but rather it is a means to the end—effective, efficient training.

19.5 Visual and Motion Systems

This section introduces the reader to visual- and motion-based systems, describes display strategies, discusses the motion versus no-motion controversy, and force cueing devices.

19.5.1 Visual Systems

Early simulators, such as Link's "Blue Box," served primarily as instrument flight rules (IFR) trainers and thus provided no information about the correspondence between control inputs (pushing the stick) and changes in the external visual scene. Gradually, simple but effective visual displays such as a representation of the horizon on a tiltable blackboard to represent varying glide slopes evolved (Flexman, Matheny, & Brown, 1950). During the 1950s, simulator designers, in their quest for realism, developed additional methods to present external visual information. For example, model boards using closed-circuit television in which a gantry mounted video camera (steered by the pilot) moved over a

TABLE 19.1 Advanced Instructional Features

Simulator instructor (SI) options	
Preset/reset	Starts/restarts the task at preset coordinates, with a predetermined aircraft configuration and environmental conditions
Demonstration	Demonstrates desired performance to the student
Briefing	Provides student with an overview of the planned training
Slew/reposition[a]	Moves the aircraft to a particular latitude, longitude, and altitude
Repeat/fly out	Allows the student to return to an earlier point, usually to where a problem has occurred in the training exercise, and "fly the aircraft out" from those conditions
Crash/kill override	Allows simulated flight to continue after a crash or kill
Playback, replay	Replays a selected portion of the flight. The playback may be time in real time, compressed time (higher rate) or expanded time (slower rate)
System freeze[a]	Temporarily stops the simulation while maintaining the visual scene and other data
Automated/adaptive training	Computer algorithms vary level of task difficulty based on student performance. Under predetermined conditions, augmented feedback may be provided
Record	Records student performance usually for either a set period of time or a portion of the training
Motion[a]	Turns motion parameters on or off
Sound	Turns sound on or off
Partial panel	Selects which instrument to blank, thus simulating instrument failure
Reliability	Assigns a probability of failure to a particular system or display
Scenery select	Selects terrain over which aircraft is to travel, defines level of detail, and amount/type of airborne traffic
Task features	
Malfunction	Simulates sensor or instrument malfunction or failure
Time compression	Reduces time available to perform the required tasks
Time expansion	Increases the time available to accomplish the required tasks
Scene magnification	Increases/decreases the magnification of the visual scene
Environmental	Manipulates time of day, seasons, weather, visibility, wind direction, and velocity, etc.
Flight dynamics	Manipulates flight dynamic characteristics such as stability, realism, gain, etc.
Parameter freeze	"Locks in" a parameter such as altitude or heading; used to reduce task difficulty
Performance analysis/monitoring features	
Automated performance measurement and storage	Collects data on student's performance during training and is used by the SI to evaluate student performance. On some systems, data on the student's prior performance may be recovered and used for comparison purposes. These data can also become part of the normative database for the system
Repeaters	Displays cockpit instruments and switch status at the SI's console
Closed circuit	Allows SI to visually monitor student's performance
SI displays	Presents student performance in an integrated or pictorial format such as a map overlay or sideview of an instrument approach
Warnings	Advises SI that student has exceeded a preset parameter (e.g., lowering gear above approved airspeed). Sometimes alerts, advising the student that a performance parameter is about to be exceeded or has been exceeded, are also presented in the cockpit during the training
Debriefing aids	Presents student performance in a pictorial format; on some simulators, selected parameters (airspeed, range to target) are also displayed or may be called up. Measures of effectiveness may also be provided
Automated checkride	An evaluation on a predetermined series of maneuvers for which PTSs have been specified

Note: Not all advanced instructional features (AIFs) will be utilized during a given training period and not all simulators have all features. Material integrated from Caro (1979), Polzella, Hubbard, Brown and McLean (1987), Hays and Singer (1989), and Sticha, Singer, Blackensten, Morrison, and Cross (1990).

[a] See Section 19.6 for cautions regarding use of these features.

prefabricated terrain model were developed. Although model board technology systems were used successfully in developing procedural skills requiring vision, the resolution and depth-of-field constraints imposed by video camera limitations reduced their ability to help develop complex perceptual and psychomotor skills (Stark, 1994). Additionally, model boards were expensive to construct and modify, and due to physical limits in the area that they could represent, aircrew quickly learned the terrain.

The development of the digital computer in the 1960s and improved mathematical models afforded the creation of complex external visual scenes. The use of computer-generated imagery allows for the dynamic presentation of an enormous amount of visual input.* However, simulator designers must distinguish between the required information (cues) and the noise content of the visually presented material to define the necessary image fidelity (Chambers, 1994). Armed with knowledge of the system requirements, the available technology, and the associated life-cycle cost, designers must then make trade-offs to determine whether the external view from the cockpit should be displayed as either a real or virtual image. Real image displays project an image onto a surface 10 or 20 ft away from the pilot's eye, whereas virtual image displays project an image at or near optical infinity.

19.5.1.1 Real Image Displays

Real images are usually projected onto flat screens. However, to provide a large field of view (FOV), dome-shaped screens are often used to ensure that the entire image is presented at a constant distance from the observer. Large FOV images, greater than 40°–50° horizontally and 30°–40° vertically, are generally achieved by coordinating a number of projectors (Stark, 1994). Currently, however, systems that provide a large FOV with very accurate scene detail are technically difficult to build and maintain, and extremely expensive to develop and operate.

Therefore, designers developed systems that maintain a high degree of detail within the pilot's area of interest (see Figure 19.1). These area-of-interest systems operate in a number of ways.

FIGURE 19.1 Generic tactical aircraft simulator installed in a dome. Note high-quality imagery immediately forward of the "nose" of the simulator. (Courtesy of McDonnell Douglas Corporation.)

---

* Fortin (1994) provides an excellent technical presentation on computer image generation.

Many simply provide the greatest detail off the nose of the aircraft, with less scene detail in the periphery. An alternate strategy utilizes a head-slaved area-of-interest display that creates a highly detailed scene based on the head movements of the pilot. ESPRIT (eye-slaved projected raster inset) uses two projectors to present the image to the pilot. The first projector displays a low-resolution background scene with information to be processed by the peripheral visual system. The second projector positions a very detailed scene along the pilot's line of sight. The positioning of the second image is controlled through the use of a servo system, which utilizes an oculometer to monitor the movements of the pilot's eyes. The interval between visual fixations is sufficiently long to allow the servo system to position the highly detailed image in the subject's line of sight (Haber, 1986; Stark, 1994). While the pilot moves his or her eyes to another fixation point, no visual information is processed, and thus the pilot does not see the image as it is moved to the new fixation point. ESPIRIT has been used in some Royal Air Force Simulators.

The high cost of dome systems and their support requirements has led to the development of a relatively low-cost (approximately $1.0 million) air combat simulator (Mosher, Farmer, Cobasko, Stassen, & Rosenshein, 1992). Originally tested on a 19 in. CRT with a 30° vertical and 35° horizontal FOV, the innovative approach uses aircraft icons. When an aircraft is outside the pilot's FOV, an aircraft icon is presented at the appropriate location at the edge of the display. This icon provides information to the pilot about the relative position, orientation, and closure rate of aircraft outside the displayed FOV. This approach has led to the development of the F-16 unit TD, which utilizes a rear projection display, to create a 60° vertical and 78° horizontal FOV out-of-the-cockpit "virtual dome system."

19.5.1.2 Virtual Image Displays

Virtual image displays present collimated images (i.e., images at or near optical infinity) to the pilot, who must be positioned at the correct focal plane to observe the image. Collimation is sometimes achieved by projecting a CRT image through a beamsplitter onto an appropriately designed spherical mirror. Head-up displays installed in aircraft also use collimating optics to project CRT generated images onto a combiner (beamsplitter) surface mounted in the front of the cockpit. As noted by Randle and Sinacori (1994), when collimated systems are used, the optical message is that "all elements in the scene are equally distant and far away." However, the absence of parallax cues (since the entire image is at the same distance), makes it difficult for pilots to discriminate objects in the foreground from objects in the background. Distance must be inferred from perspective, occlusion, and texture cues, without the support of stereopsis, vergence, and accommodation (Randle & Sinacori, 1994; Stark, 1994). Despite the loss in perceptual fidelity, the illusion is compelling and becomes "virtual reality" when the pilot becomes involved in his or her flying tasks.

Collimation technology is also employed in helmet-mounted displays (HMDs) used for simulation. Training HMDs usually consist of a helmet with two half-silvered mirrors mounted on the helmet and positioned in front of the eyes. The presentation of the optical infinity image directly in front of the subject's eye(s) eliminates many of the problems associated with domes. However, HMDs require precise alignment of the two images. While the weight of early versions of HMDs limited user acceptance, recent advances in image projection (lighter weight CRTs, liquid crystals, fiber optics, polycarbonate optics, etc.) have now decreased the weight of these systems to more acceptable levels.

Both real and virtual imagery systems work well under conditions (landing, takeoff, air combat) in which most of the elements in the real-world scene are at a considerable distance from the pilot's eye position. Nevertheless, there are still many unknowns regarding visual cues and the appropriate dynamics for low-level (nap-of-the-earth) rotary-wing simulations. As will be discussed later, the highest rates of simulator sickness are reported in helicopter simulators. Part of this may perhaps be attributed to the nature of the helicopter, which Bray (1994) describes as a small, very agile, low stability, highly responsive aircraft capable of motions that are difficult to simulate. In addition, at the nap-of-the-earth levels, the visual display requirements of a helicopter simulator are demanding. Randle and Sinacori (1994) described the pilots gaze-points as being distributed primarily in the "immediate impact field from 3 to 5 s ahead." The requirements for "in-close" viewing need to be specified carefully and they will vary as a function of altitude and airspeed. Much work remains to be done in defining and justifying the requirements for rotary wing simulators.

19.5.2 Motion Systems

This section describes degrees of freedom (DOFs) and synergistic platform motion simulators; the current debate over the use of motion platforms and a brief description of unique simulators follow.

Simulators have from zero (no movement) to six DOF (yaw, pitch, and roll; heave, surge, and sway). The first three—yaw, pitch, and roll—require rotation about an axis. Yaw is rotation about the aircraft's vertical axis, pitch is rotation about the lateral axis, and roll is rotation about the longitudinal axis of the aircraft. The latter three—heave, surge, and sway—require displacement. Heave refers to up and down displacement, surge refers to forward and backward displacement, and sway refers to lateral displacement.

Synergistic platform motion simulators are the most common type of motion simulators. The hexapod platform uses six hydraulic posts, whereas other platforms use combinations of lifting devices and posts (see Figure 19.2). Although hexapod platforms have several DOFs, their nominal excursion ranges are perhaps 40°–50° for yaw, pitch, and roll with up to 6 ft in heave, surge, and sway. Typical yaw, pitch, and roll velocities may range from 5°/s to 20°/s with displacement rates of 1–2 ft/s. The motion capabilities of a system are based on both the characteristics of the simulated aircraft and the physical limitations of the individual components of the motion platform. In commercial and transport systems, a "cockpit" (see Figure 19.3) with stations for the aircrew and sometimes a checkride pilot or instructor are mounted atop the platform. The movements of the posts are coordinated to produce the motion required for vestibular and kinesthetic input "similar" to the movements of the actual vehicle. Acceleration and

FIGURE 19.2   Motion-based platform with Boeing 737 simulator mounted on platform. (Courtesy of Frasca Corporation.)

FIGURE 19.3 Boeing 737 simulator cockpit installed on motion-based platform shown in Figure 19.2. (Courtesy of Frasca Corporation.)

displacement provide the initial sensory input, and washout techniques are then used to return the platform to its initial position. Because washout of the movement theoretically occurs below the pilot's motion detection threshold, the pilot's perception is that his or her vehicle is still moving in the direction of the initial motion (Rolf & Staples, 1986).

An additional concern with motion systems is the magnitude of the lag between an airborne pilot's input and the corresponding movement of the platform. Delays of approximately 150 ms between the aircraft and the simulator have minimal effect; however, delays of greater than 250 ms significantly reduce the quality and transferability of the simulation (McMillan, 1994; Stark, 1994).

Although synergistic platform motion simulators do induce the feeling of motion, it is difficult to coordinate the smooth movement of the hydraulic components due to interactions between the various DOFs. Finally, these platforms are expensive to operate and maintain, and require special facilities.

19.5.3 Motion versus No Motion Controversy

The focus of most simulator research has been on measuring the training provided by a specific simulator in a particular training program. As discussed previously, there has been an untested belief, based primarily on face validity, that the more closely the simulation duplicates the aircraft and the flight environment, the better the transfer. In part because of the costs associated with motion systems, there has been considerable controversy about the contribution of platform motion to training. To examine the contribution of platform motion to simulator training effectiveness for basic contact (non-IFR) flight, Martin and Wagg (1978a, 1978b) performed a series of studies. They reported (1978a) no differences between the groups trained in either the fixed or motion-based simulators. However, students in both simulator groups performed better than students in the control group who received all their training in the T-37 aircraft. Later, they extended the study (1978b) to include aerobatic tasks and found that platform motion did not enhance performance in the simulator or in the aircraft. They concluded that aerobatic skills may be more cost-effectively trained in the aircraft. In her review of the six studies in the series, Martin (1981) concluded that the procurement of six postsynergistic platform motion systems was not necessary for teaching pilot contact skills.

More recently, two different groups have examined the issue of simulator motion from different perspectives. Jacobs et al. (1990) performed a meta-analysis of flight simulator training research, whereas Boldovici (1992) performed a qualitative analysis based on the opinions of 24 well-known authorities in the field of simulator motion. Jacobs et al. concluded that, for jet aircraft, motion cueing did not add to simulator effectiveness and in some cases may have provided cues that reduced the effectiveness of the simulator. However, they advised that this conclusion be accepted with caution because (a) the calibration of the motion cueing systems may not have been performed as frequently as necessary and (b) the conclusion is based on all tasks combined not on specific tasks (thus, any gain on a task that could have been attributed to motion may have been canceled by a decrement on another task). No conclusion was possible for helicopter simulators because only one study compared the transfer between the simulators and the actual aircraft. However, one study by McDaniel, Scott, and Browning (1983) reported that certain tasks (aircraft stabilization equipment off, free-stream recovery, and coupled hover) benefited from the presence of motion, whereas takeoffs, approaches, and landings did not. Bray (1994) believes that platform motion "might offer a bit more in the helicopter simulator than it does in the transport aircraft simulator, because control sensitivities are higher and stability levels are lower in helicopters." Regarding motion platforms for helicopters, he comments that "if the benefits of six-DOF cockpit motion are vague, its cost is not."

With respect to the motion simulation literature in general, Boldovici (1992) argued that finding no differences (null hypothesis) between the effect of motion and no-motion conditions does not prove that an effect does not exist, only that no effect was obtained. He also noted that the statistical power of some of the literature examined may be inadequate to detect the existing differences and that most of the literature failed to adequately describe the characteristics of the motion platform. Sticha, Singer, Blacksten, Morrison, and Cross (1990) suggested that perhaps there are real differences between the effectiveness of fixed and motion-based systems but inappropriate lags in the motion systems, problems in drive algorithms, lack of synchronization of the visual and motion systems, and so on may preclude the advantage of motion-based simulation from being noted. They propose that the results "may simply show that no motion is better than bad motion" (p. 60). Lintern and McMillan (1993) support their position and suggest that motion provides neither an advantage nor a disadvantage, since most flight transfer studies show no transfer effects attributable to motion.

Boldovici (1992), on the other hand, asked 24 well-known authorities in the field of simulator motion to provide arguments both for and against the use of motion platforms.

Their arguments for using motion platforms included reducing the incidence of motion sickness, low cost when compared with aircraft use, user's and buyer's acceptance, trainee motivation, learning to perform time constrained, dangerous tasks, motion as a distraction to be overcome by practice, application of adaptive or augmenting techniques, and finally the inability to practice some tasks without motion. Their arguments against the use of motion platforms included absence of supporting research results, possible learning of unsafe behavior, possible achievement of greater transfer by means other than motion cueing, undesirable effects of poor synchronization of the motion cues, direct, indirect, and hidden costs, existing alternatives to motion bases for producing motion cueing, and finally, the relatively benign force environments encountered under most flight conditions. Boldovici examined each of the sometimes conflicting positions previously listed above and concluded:

1. Results of transfer-of-training studies are insufficient to support the decisions about the need for motion systems.
2. Greater transfer can be achieved by less expensive means than using motion platforms. Therefore, if cost-effectiveness is used as a metric, motion platforms will never demonstrate an advantage.

3. From a statistical viewpoint, the research results concluding no differences in transfer to parent vehicles do not prove that no differences exist. Boldovici recommended that researchers report the results of power tests to determine the number of subjects required to detect treatment effects.

4. Because much of the transfer-of-training literature does not adequately address test reliability, we cannot adequately assess the validity of our inferences.

5. Because some of the conditions under which a simulator is "flown" cannot be repeated safely in the aircraft, some transfer of training cannot be evaluated. On the other hand, adequate training for flying an aircraft in benign environments can be provided by a fixed-base simulator.

6. Training in either motion-based or fixed-base simulators can promote learning unsafe or counterproductive behavior.

7. No evidence exists regarding the effect of motion on trainee motivation.

8. The use of motion-based platforms to reduce simulator sickness is inappropriate (see also Sharkey & McCauley, 1992).

9. User's and buyer's acceptance is not an appropriate reason for the use of motion platforms.

10. Incentives (such as job advancement for working in high-tech projects) for purchasing expensive simulators may be greater than incentives for purchasing less expensive simulators.

11. Some tasks may require force motion cueing, which can be provided by seat shakers, G-seats, and motion bases. Sticha et al. (1990) developed a rule-based model for determining which, if any, of these force cueing strategies is necessary. Their model for the optimization of simulation-based training systems requires the developer of the training system to develop specifications, which identify the cues required for proper learning.

While the controversy continues, Caro (quoted in Boldovici, 1992) asked the incisive question: "Does the motion permit the operator to discriminate between conditions that otherwise could not be distinguished?" (p. 20). Although it appears that the answer to this question will be more often negative, if the discrimination is essential and cannot be induced visually, then perhaps the use of motion should be considered seriously. This position is reflected in the work of Berki-Cohen, Soja, and Longridge (1998), who reviewed 20 years of literature in this area. The current FAA perspective is that simulators used to determine the ability of a pilot to immediately perform the required in-flight actions (e.g., recovery from a sudden engine failure) are very different from simulators used to provide transfer of training. "Consequently, the simulator must be capable of supporting a 100% transfer of performance to the aircraft. Anything less would compromise safety. The existing standards for full flight simulator qualification, all of which entail a requirement for platform-motion cueing, have a 20 year record of meeting the requisite criterion for transfer of performance. In the absence of compelling evidence to the contrary, it is, therefore prudent to maintain the standards in the interest of public safety" (p. 296). The alternate perspective is that the existing requirements for simulators are based primarily on the SME opinions. The reliability and validity of this SME evaluation strategy has never been systematically quantified. In an interesting exchange, the article provides arguments both for and against the requirements for motion-based simulation and concludes with the position that the requirement for motion will remain in place until there is definitive research to the contrary. The chapter provides guidelines for this additional research.

19.5.4 Force Cueing Devices

Force cueing devices have been used to simulate motion in fixed-base platforms. Two devices, the G-suit and G-seat, have been used to simulate the effects of motion on the pilot's body during high G-load situations (Cardullo, 1994b). The G-suit (more properly the anti-G suit) is used in aircraft to maintain the blood level in the brain by preventing the blood from pooling in the pilot's lower extremities. The G-suit used in simulators consists of a series of air bladders, imbedded in a trouser-like assembly, which inflate as a function of the simulated G-load. Thus, the pilot has some of the sensations of being exposed to G-forces. On the other hand, the G-seat consists of independently operating seat and backrest panels, and mechanisms that vary the pressure exerted on the restrained pilot. The properties of the seat (shape, angle, and hardness) are manipulated to correspond with changes in the G-load imposed by specific maneuvers. The use of a G-suit or g-seat during simulation provides the pilot with additional cues regarding the correct G-load needed for certain flight maneuvers (Stark, 1994). Some helicopter simulators use seat shakers to simulate the vibratory environment unique to rotary wing aircraft.

In addition to tactile cues, dimming the image intensity has been used to mimic the "graying" of the visual field, which occurs under high G-loads (Cardullo, 1994b). Cardullo also describes other strategies such as variable transparency visors, which mimic the graying of the visual field by varying the amount of light transmitted through the visor as a function of G-load. Harness loading devices, usually used in conjunction with G-seats, simulate the G-load by tightening and releasing the crewmember's restraint system as a function of G-load.

19.6 Simulator Sickness

People and other animals show symptoms of motion sickness in land vehicles, ships, aircraft, and spacecraft (Money, 1970). Consequently, while attempting to simulate the motion and the external visual environment of these vehicles, it was reasonable to expect a form of motion sickness to occur. This form of motion sickness is referred to as simulator sickness. As noted by Kennedy and Fowlkes (1992), simulator sickness is polygenic and polysymptomatic. It is polygenic, since it may be induced by the severity, frequency and/or duration of certain physical motions, the lack of appropriate motion cues, the apparent motion in visual displays with varying FOV and levels of detail, or some interaction of these variables. The multiple symptoms of motion sickness (cold sweats, stomach awareness, emesis, etc.) are, at the very least, disruptive in the operational environment. Simulator sickness threatens and perhaps destroys the efficacy of the training session and may decrease simulator usage (Frank, Kennedy, Kellog, & McCauley, 1983; Kennedy, Hettinger, & Lilienthal, 1990; McCauley, 1984).

During and after a simulator session, the foremost concern is the safety and health of the trainee. Secondary to safety is the value of the training session. Trainees more concerned about avoiding simulator sickness than learning the assigned task are unlikely to benefit from simulator training. Additionally, if simulators produce effects that differ from the real-world situation, then the skills learned in the simulator may be of limited value in the operational setting. Furthermore, the perceptual after-effects of a simulator session may interfere with the pilot's flight readiness, that is, the ability to fly an aircraft safely or operate a vehicle immediately or shortly after a simulator training session (Kennedy et al., 1990; McCauley, 1984).

To determine the incidence rate of simulator sickness, Kennedy, Lilienthal, Berbaum, Baltzley, and McCauley (1989) surveyed 1186 "flights," conducted in 10 different US Navy simulators during a 30 month period. All the simulators had a wide field-of-view visual system. The reported incidence rate, based on the Motion Sickness Symptom Questionnaire, ranged from 10% to an astonishing 60%. The lowest incidence rates occurred in fixed-wing, fixed-base, dome-display simulators, whereas the highest reported sickness rate occurred in rotary wing (helicopter) simulators that employed six-DOF motion systems. It should be noted that in many instances, simulator sickness was induced even in stationary simulators. The latter case may be explained by the strong correlation between simulator sickness and the perception of vection, that is, the sensation of self-motion (Hettinger, Berbaum, Kennedy, Dunlap, & Nolan, 1990). A major contributor to the perception of vection is visual flow (i.e., the movement of the surround as the observer moves past it). Sharkey and McCauley (1991) have reported that increased levels of global visual flow are associated with an increased incidence of simulator sickness. McCauley (personal communication, October 1994) believes that, "In fixed-base and typical hexapod motion bases, sickness occurs only with a wide field-of-view representation of the outside world, which leads to vection." These higher levels of visual flow are more common in aircraft (or simulators) flying at lower altitudes than in aircraft flying at higher altitudes. Thus, the higher incidence of simulator sickness in rotary wing simulators may be attributed in part to the higher visual flow rates common at lower altitudes. More specifically, as reported by Sharkey and McCauley (1991), the increased incidence may be associated with changes in that visual flow.

Sensory conflict theory, the more commonly accepted explanation for simulator sickness, states that motion sickness occurs when current visual, vestibular, and other sensory inputs are discordant with expectations based on prior experience (Reason, 1978). Support for this theory is found in studies that indicate that individuals with more experience (higher flight hours) in the operational vehicle report a higher incidence of simulator sickness than less experienced individuals (Kennedy et al., 1989). The authors attributed this finding to a greater sensitivity to the disparity between the operational system and the simulator among experienced individuals (Kennedy et al., 1990).

Stoffregen and Riccio (1991) noted that the disparity between actual and expected sensory input may be impossible to measure because the baseline cannot be determined. They proposed an alternate theory, which contends that simulator sickness is produced by prolonged postural instability. This theory predicts that individuals who become sick in a simulator have not identified the appropriate constraints on bodily motion imposed by the simulator and thus have failed to implement the correct postural control strategies necessary for that situation. Irrespective of which theory is correct, the presence of simulator sickness may be detrimental to learning and performance.

If we are to improve the efficacy of simulators and TDs, we must identify the possible causal factors contributing to simulator sickness. Factors proposed include

1. Mismatch between visual and vestibular cueing (Kennedy et al., 1990).
2. Visual and inertial lag discrepancies produced by the computational limitations of the simulator computer system (Kennedy et al., 1990).
3. Motion systems with resonant frequencies in the nausoegenic region (Frank et al., 1983).
4. Geometric distortions of the visual field that occur when the crewmember moves his or her head outside the center of projection (Rosinski, 1982).

Although this may seem to be an area amenable to additional research, both Guedry (1987) and Boldovici (1992) noted that, without incidence data obtained in the actual aircraft, objective assessments of the contribution of platform motion to simulator sickness will be difficult to obtain. Indeed, would the elimination of all simulator sickness be desirable, as that would change trainees' expectancies when they start flight training? Nonetheless, simulator sickness is a problem that interferes with learning and even leads individuals to avoid using some simulators. Therefore, the following preventative strategies, proposed by McMillan (1994), McCauley and Sharkey (1991), Kennedy et al. (1990), and/or are contained in the *Simulator Sickness Field Manual* (Naval Training Systems Center, 1989) should be applied:

1. Monitor trainees new to the simulator more closely. Trainees with considerable flight time are especially vulnerable to simulator sickness.
2. Only use trainees, who are in their usual state of fitness. Avoid subjects with symptoms of fatigue, flu, ear infections, hangover, emotional stress, upset stomach, and so on.
3. For optimal adaptation, there should be a minimum of 1 day and a maximum of 7 days between simulator sessions.
4. Simulator sessions should not exceed 2 h; indeed, shorter sessions are more desirable.
5. Minimize changes in orientation, especially when simulating low-level flights.
6. Take steps to minimize abrupt changes in direction (e.g., altitude, roll, porpoising).
7. Use the freeze option only during straight and level flight.
8. Do not slew the projected image while the visual scene is visible to the trainee.
9. Use a reduced FOV in nauseogenic situations.
10. If the trainee shows initial signs of sickness have the trainee use flight instruments. If the symptoms increase, the trainee should not return to the simulator until all symptoms have subsided (10–12 h).

11. Advise the trainee to minimize head movements during new situations.
12. When the trainee enters and exits the simulator, the visual display should be off and the simulation should be at 0° of pitch, yaw, and roll.
13. Maintain proper calibration of the visual and motion systems.

The undesirable side-effects of simulation will become more apparent as ordinary citizens utilize simulators in places like amusement parks and in other recreational activities. As will be seen in Section 19.7, the occurrence of simulator sickness also has significant implications for virtual reality/virtual environments.

19.7 Virtual-Reality/Virtual Environments

Having examined the issue of simulator sickness, it is now appropriate to discuss the implications of using virtual reality/virtual environment in teaching flying skills. virtual reality/virtual environment, which are sometimes referred to as artificial reality/virtual worlds, has been described as

1. An artificial environment, created with computer hardware and software, presented to the user in such a way that it appears and feels like a real environment (Webopedia, 2004).
2. The simulation of a real or imagined environment that can be experienced visually in the three dimensions of width, height, and depth and that may additionally provide an interactive experience visually in full real-time motion with sound and possibly with tactile and other forms of feedback. The simplest form of virtual reality is a 3-D image that can be explored interactively at a PC, usually by manipulating keys or the mouse so that the content of the image moves in some direction or zooms in or out.... More sophisticated efforts involve such approaches as wrap-around display screens, actual rooms augmented with wearable computers, and haptics joystick devices that let you feel the display images (Searchsmallbizit, 2004, searchsmallbizit.techtarget.com).

Virtual-reality/virtual environment exist in one of the three possible contexts: as a simulation of an *existing* environment such as the interior of a building; as a *proposed* environment such as a Mars-bound space station; or as an *imaginary* environment such as that found in PC-based adventure games. These environments are designed to achieve an educational or entertainment goal.

There are two virtual reality/virtual environment levels: nonimmersion and immersion (Kuntz Rangal et al., 2002). At the nonimmersion level, images are presented on a computer display and the user is aware of his or her real-world surroundings. While at the immersive level, efforts are made to convince the individual that he/she is actually present in the environment by the use of devices such as HMDs. HMDs project computer-generated images onto the inside of a visor, while preventing the individual from seeing the real world. Controlling the auditory input through earpieces/surround sound often increases the depth of immersion. Haptic (tactile) information can be provided by the use of body gloves. When locomotion is being simulated, the virtual environment may include a small treadmill-like platform on the individual walks with minimal real displacement. At a higher level of immersion, the trainee wearing an appropriate virtual reality projections system is enclosed in an 8.5 ft diameter sphere, which rotates as she/he walks, runs, crawls, etc., in any direction (VirtuSphere, 2006). Readers desiring additional information on virtual environments should consult the *Handbook of Virtual Environments: Design, Implementation, and Applications*, edited by Stanney (2002).

Since virtual reality technology (particularly helmet-mounted technology) requires less space than the traditional simulators, the US Navy is considering using it on ships while they are underway. In that environment, the trainee is simultaneously subjected to both the ship's movement and the moving visual imagery seen on the virtual reality visor; clearly sensory conflict is a concern. During a demonstration, in 1998, an F/A-18 weapons system trainer was deployed aboard the carrier USS Independence (Muth & Lawson, 2003). While this was a demonstration and not a controlled experiment, there were no reports of sickness by the participants. However, it should be noted that there were no major storms at sea during this demonstration. Subsequently, Muth and Lawson (2003) demonstrated that test participants showed minimal symptoms of nausea and simulator sickness after completing a 1 h simulated flight while riding aboard a 108 ft coastal patrol boat. The authors note that their study examined a best-case scenario " in which a minimal provocative ship motion stimulus was combined with a minimally provocative flight simulator (p. 504)." Additional testing in more provocative environments will be required. The authors note that to minimize the interaction between ships motion and the apparent motion of the aircraft/earth in a shipboard flight simulator, the flight simulator should be located near the ship's center of rotation, where ship motion is less provocative. Since virtual reality systems require considerably less space than the current weapon systems trainers, it is reasonable to assume that additional efforts will be made to use virtual reality/virtual environment at sea and in flight (consider an airborne controller of UAVs). Research on the use of virtual reality/virtual environment reality in dynamic environments can be expected to increase. This research could build on basic etiological research on the undesirable side-effects of virtual environments in static environments and the development of appropriate countermeasures.

19.8 Instructional Features of Simulators

Simulators incorporate many advanced instructional features, designed to enhance training. Although the list of AIFs presented in Table 19.1 is impressive, Polzella and Hubbard in 1986 reported that most AIFs are underutilized because of the minimal training provided to SIs. Apparently, the situation has not changed, for in 1992, Madden reported that most SI training was on-the-job, and indeed only 10% of training was classroom training or involved the use of an instructor's manual. Many manuals were described as "written for engineers," "user unfriendly," and "too technical." Six years later, Salas et al. (1998) repeat the plea for a paradigm shift in which the knowledge gathered by psychologists and cognitive engineers be applied to aviation training (see the earlier discussion on fidelity in this chapter).

Polzella and Hubbard (1986) reported that some AIFs may be more appropriate for initial-level training than for more advanced training. For example, the use of AIFs during initial level-training affords an opportunity for immediate feedback, whereas during advanced training, the use of AIFs would disrupt the continuity of a realistic scenario. Jacobs et al. (1990) in their meta-analysis noted that the use of AIFs was rarely reported in the literature that they examined.

Little research has been performed on the training efficacy of AIFs in flight simulation, although most of the AIF strategies are based on the training and education literature. Hughes, Hannon, and Jones (1979) reported that playback was more effective in reducing errors during subsequent performance than demonstration. However, record/playback was no more effective than simple practice. Moreover, inappropriate use of the AIFs can contribute to problems. For example, use of the rewind and slew features while the scene is being observed by the trainee, or freezing the simulator in an unusual attitude, can contribute to simulator sickness (Kennedy et al., 1990).

The research specific to the use of AIFs in flight simulation indicates that appropriate use of AIFs can greatly facilitate learning. Backward chaining, a teaching strategy in which a procedure is decomposed into a chain of smaller elements and the student's training starts at the endpoint and proceeds back along the chain, appears to have considerable promise. For example, using backward chaining, a student would learn touchdown procedures first and gradually be repositioned further back on the glideslope. Backward chaining has been utilized successfully to train 30° dive-bombing maneuvers (Bailey, Hughes, & Jones, 1980) and simulated carrier landings (Wightman & Sistruck, 1987).

Recently, under laboratory conditions, the time manipulation capability of simulators has produced some promising results. Using the above real-time training (ARTT), in an F-16 part-task flight simulator, Guckenberger, Uliano, and Lane (1993) evaluated performance by F-16 pilots trained under varying rates of time compression (1.0x, 1.5x, 2.0x, and random order of time compression). When tested under real-time conditions and required to perform an emergency procedure in a simulated air combat task, the following differences were noted. Groups trained under ARTT conditions performed the emergency procedures tasks significantly more accurately than the group trained under the real-time condition. In addition, the ARTT groups "killed" six times more MIGs than the 1.0x group. Thus, it appears that ARTT can be used to train individuals to perform procedural tasks more accurately and in less time than in traditional techniques. Although advanced instructional features have considerable promise, their use must be justified in terms of savings and transfer to the real world. A theory or model that estimates the amount of transfer and savings resulting from the use of particular advanced instructional features is needed.

The authors believe that, as in most training, the skills, knowledge, and enthusiasm of the instructor as well as the management policy (and level of enforcement) greatly determine how the simulator is used and its ultimate effectiveness. Unfortunately, the SI is the forgotten component in the simulator system. As Hays, Jacobs, Prince, and Salas (1992) note, much simulator research is dependent on the subjective judgment of the SI. This is also true for pilot performance evaluations. In both the research and the operational world, strategies for improving the reliability and validity of subjective ratings need to be developed and evaluated. Greater emphasis on instructor training in the proper use of advanced instructional features, and improved evaluation procedures, possibly combined with the development of expert system "trainers" as part of the software package promises considerable payoff.

19.9 PC-Based Flight Simulations

The increased capability of PC-based flight simulation has benefited from advances in computer technology (increased memory capability and processing speed) and reducing hardware and software costs (Sinnett, Oetting, & Selberg, 1989). The increased use of PC-based flight simulation had been documented at the American Society of Mechanical Engineers' Symposium (Sadlowe, 1991), and by Peterson (1992) and Williams (1994). In 2005, Grupping documented the development of the PC-based software program known as "Flight Simulator." It was originally developed by Bruce Artwick, as part of his 1975 Masters thesis at the University of Illinois. Grupping's timeline documents its evolution. It was originally released by subLOGIC as FS1 for the Apple II in January 1980. In November 1982, it began to be distributed by Microsoft as Microsoft Flight Simulator 1.01, and there have been multiple upgrades since then. In 2006, Microsoft released the newest version as Microsoft Flight Simulator X, which has simulations of 21 new or current aircraft (e.g., Airbus 321, Boeing 737-800) and legacy aircraft (e.g., Piper Cub and Cessna 172). These simulated aircraft can be flown between thousands of airports (with dynamic airborne and ground traffic and ground support equipment). It also includes an air traffic control simulation.

Of particular significance are the capabilities provided by today's software: (a) more realistic characterization of instrument navigation aids, (b) more realistic presentations of aircraft handling characteristics and instrumentation, and (c) a wide range of instructional features. Not only have computer hardware/software improvements resulted in near real-time flight simulation characteristics, but sophisticated interface media to represent yoke, throttle, and cockpit controls have been developed to better emulate the psychomotor aspects of aircraft control. As would be expected, most PC-based simulations have considerably less fidelity and lower cost than full-scale simulations. However, the lower levels of fidelity may be adequate in many research and training situations. Lower fidelity simulations have proven effective in evaluating the effects of (a) automation (Bowers, Deaton, Oser, Prince, & Kolb, 1993; Thornton, Braun, Bowers, & Morgan, 1992), (b) scene detail and FOV during the introductory phases of flight training (Lintern, Taylor, Koonce, & Talleur, 1993), and (c) the development of aircrew coordination behavior (Bowers, Braun, Holmes, Morgan, & Salas, 1993). Beringer (1994) networked five PCs and combined two commercially available flight simulation packages to develop the simulator presented in Figure 19.4. This approximately $25,000 apparatus was used at the FAA's Civil Aeromedical Institute to compare two levels of navigational displays.

Because of the availability, flexibility, and low costs of PC-based simulations, efforts to determine the effectiveness of their transfer of training for general aviation (private, noncommercial) have been undertaken. Taylor (1991) described a series of studies, which utilized the ILLIMAC (University of Illinois Micro Aviation Computer) flight simulation system. The system utilizes an 8086, 16 bit microprocessor to control a fixed base, general aviation trainer with the flight characteristics of the Piper Lance.

FIGURE 19.4 FAA's PC-based simulation facility. (Courtesy of FAA Civil Aeromedical Institute.)

Taylor reported that providing students, who have completed their private pilot certification program, with a concentration of instrument procedures on the ILLIMAC prepared them well for their commercial training. Based on the findings of these studies, an accelerated training program was developed and approved by the FAA. Under this program, students saved a full semester of flight training.

In a study at Embry-Riddle Aeronautical University, Moroney, Hampton, Beirs, and Kirton (1994) compared the in-flight performance of 79 aviation students trained on one of two PC-based aircraft training devices (PCATDs) or an FAA approved generic TD. Student performance on six maneuvers and two categories of general flight skills was evaluated, based on the criteria specified in the FAA's performance test standards (PTSs) for an instrument rating (FAA, 1989). For those factors evaluated, no significant difference in either the number of trials or hours to instrument flight proficiency in the aircraft was noted among those students taught in any of the three training devices. However, differences in student performance were noted in the number of trials/hours to proficiency in the TDs. When compared with students trained in the approved generic training device, students trained in the PCATDs required (a) significantly fewer total trials, trials per task, and hours to reach the overall PTS, and (b) significantly fewer trials to reach proficiency in the following maneuvers: precision approach, nonprecision approach, timed turn to magnetic compass heading, and general flight skills (partial panel). Relative to cost, the training received in the PCATDs cost 46% less than the training received in the approved generic training device (mean savings of $463). Finally, the initial cost of the PCATDs, associated hardware, and software was approximately 8% of that of the approved TD ($4,600 and $60,000 respectively). Based on these findings, the authors recommended (a) the use of PCATDs by general aviation and (b) that steps be initiated to PCATDs as FTDs, which could be used to accrue instrument rating credit.

In 1997, Taylor et al. compared the performance of a group trained on a PCATD with the performance of a group trained entirely in an aircraft. Students trained on the PCATD completed the course in significantly less time than students trained in the aircraft. Taylor et al. (1997) reported substantial transfer from the PCATD to the aircraft for tasks such as ILS approaches, localizer back-course, and nondirectional beacon approaches. However, they reported lower transfer when the PCATD was used for reviewing tasks learned earlier in the course. They recommended that PCATD training be focused on those areas in which substantial transfer to the aircraft has been documented. The studies cited above contributed to the FAA's release of Advisory Circular 61-126 (1997), *Qualification and approval of personal computer-based aviation training devices*). Advisory Circular 61-126 permits approved PCATDs, meeting the qualification criteria, to be used in lieu of up to 10 h of time that ordinarily may be acquired in a flight simulator or flight training device authorized for use under Part 61 or Part 141 regulations.

In a related effort, Moroney, Hampton, and Beirs (1997) surveyed flight instructors, and used instructor focus groups to describe how the instructional features of PCATDs could be best utilized, how the instructor–software interface could be improved, and strategies for presenting and evaluating student performance. In 1999, Taylor et al. evaluated the transfer of transfer effectiveness of a PCATD utilized by students in beginning and advanced instrument courses. Their performance was compared with that of students who received all training in an aircraft. TERs were determined for various flight lessons. In general, transfer savings were positive for new tasks but much lower when previously learned tasks were reviewed. Students in the PCATD group completed training in an average of 3.9 h less than students in the airplane control group. Most PCATD studies have examined or been related to the transfer of flying skills from the PCATD to the aircraft. However, PCATDs have been used for other purposes. Jentsch and Bowers (1998) examined the application of PCATDs to teaching and evaluating aircrew coordination training. Their review of more than 10 years of research documents the validity of using PCATDs for this purpose. They also provide guidelines, which can improve the validity of PCATD simulations.

19.10 Simulator or Training Device?

Throughout this chapter, we have referred to all devices that use simulation as simulators. However, the FAA does not classify all such devices as "simulators." The FAA's Advisory Circular, *Airplane Simulator Qualification* (AC120-40B; FAA, 1991), defines an airplane simulator as:

> a full size replica of a specific type or make, model and series airplane cockpit, including the assemblage of equipment and computer programs necessary to represent the airplane in ground and flight operations, a visual system providing an out-of-the-cockpit view, and a force cueing system. (p. 3, 4).

The FAA specifies four levels of simulators, ordered in increasing complexity from Level A through Level D. For example the optical systems for levels A and B must have a minimum field of view (FOV) of 45 degrees horizontal and 30 degrees vertical, and level C and D must provide a field of view of the least a 150 degrees horizontal and 75 degrees vertical. Thus, many of the simulators discussed in this chapter do not meet the FAA's definition of a simulator but rather are classified as airplane training devices, which are defined in the Airplane Flight Training Device Circular (AC 120-45A; FAA, 1992) as

> a full scale replica of an airplane's instruments, equipment, panels, and controls in an open flight deck area or an enclosed aircraft cockpit, including the assemblage of equipment and computer software programs necessary to represent the airplane in ground and flight conditions to the extent of the systems installed in the device; does not require force (motion) cueing or visual system. (p. 3)

There are seven levels of FTDs. Level 1, the lowest level, is deliberately ambiguous and perhaps PC-based systems may qualify for this level. However, Level 7 FTDs must have the same lighting as the aircraft; use aircraft seats that can be positioned at the design-eye position; simulate all applicable flight, navigation, and systems operation; and provide significant aircraft noises (precipitation, windshield wipers); and so on.

Boothe (1994) commented that in simulators and FTDs, the emphasis is not just on accomplishing the required task, but on obtaining maximum "transfer of behavior" the task must be performed exactly as it would be in the aircraft. Thus, the same control strategies and control inputs must be made in both the aircraft and the simulator. He believed that the emphasis should be on appropriate cues, as identified by pilots, who are the subject-matter experts. To achieve this end, Boothe argued for replication of form and function, flight and operational performance, and perceived flying (handling) qualities. He noted that these advisory circulars are developed by government and industry working groups, which utilize realism as their reference and safety as their justification.

Roscoe (1991) offered a counterposition. He argued that "qualification of ground-based training devices for training needs to be based on their effectiveness for that purpose and not solely on their verisimilitude to an airplane" (p. 870). Roscoe concluded that pilot certification should be based on demonstrated competence, not hours of flight experience. Lintern et al. (1990) argued further that for "effective and economical training, absolute fidelity is not needed nor always desirable, and some unreal-worldly training features can produce higher transfer than literal fidelity can" (p. 870). Caro (1988) added: "The cue information available in a particular simulator, rather than stimulus realism per se, should be the criterion for deciding what skills are to be taught in that simulator" (p. 239). Thus, there are significant differences of opinion regarding both the definition and the requirements for the qualification of simulators.*

19.11 Unique Simulators

Since the beginning of manned flight, a variety of unique simulators have been developed. This section describes five different simulators. While some of these simulators are no longer in use, this material will provide the reader with a historical perspective on simulators and the efforts of their designers. For a historical review of other unique simulators, the reader should consult Martin (1994). The first unique simulator to consider is the LAMARS (large amplitude multimode aerospace research simulator) located at Wright–Patterson Air Force Base. The LAMARS (Martin, 1994) has a flight cab located at the end of a 20 ft movable arm. The cab can heave, sway, yaw, pitch, and roll, but cannot surge.

Second, the dynamic flight simulator (DFS, see Figure 19.5), located at the Naval Air Warfare Center (NAWC), has a cockpit in a two-axis gimbaled gondola at the end of a 50 ft arm in a centrifuge. The DFS

FIGURE 19.5 Dynamic flight simulator and centrifuge arm. (Courtesy of Naval Air Warfare Center, Aircraft Division.)

---

* Readers interested in U.S. Air Force requirements for flight simulators are referred to AFGS-87241—Guide Specification Simulators, Flight (U.S. Air Force, 1990).

FIGURE 19.6 Cockpit installed in gondola of DFS (the top and bottom portions of the gondola have been removed). (Courtesy of Naval Air Warfare Center, Aircraft Division.)

can generate 40 g and has an onset rate of 13 g/s (Eyth & Heffner, 1992; Kiefer & Calvert, 1992). The pilot views three CRTs (see Figure 19.6), which present the outside scene while the enclosed gondola responds with yaw, pitch, and roll appropriate to the pilot's input (Cammarota, 1990). The DFS has been used to simulate G-forces sustained in air combat maneuvering, recoveries from flat spins, and high angle of attack flight. Wolverton (2007) provides a history of the DFS and the centrifuge from which it evolved.

A third type of motion simulator uses a cascading motion platform. Cascading refers to the approach of stacking one moveable platform (or DOF) on another so that although each platform in the stack has only one DOF, because it is mounted on other platforms additional DOF can be achieved without interactions between the platforms. The vertical motion simulator (VMS) located at NASA Ames is used to simulate handling qualities of vertical takeoff and landing (VTOL) aircraft. The VMS (Martin, 1994) has a 50 ft heave capability with a 16 ft/s velocity. Limitations of cascading platforms include the size and cost of the facility.

The fourth simulator, TIFS (total-in-flight simulation) is owned by the US Air Force but operated by Calspan (see Figure 19.7). TIFS is a simulator installed in a turboprop aircraft, and can be adapted to provide a variety of handling characteristics. The aircraft being "simulated" is flown from the simulator located in the nose of the aircraft, while a "safety" crew located in the aft cockpit is ready to take control of the aircraft if a hazardous condition arose. The TIFS has been used to simulate the handling qualities of aircraft as diverse as the Concorde, C-5, B-2, X-29, YF-23, and the space shuttle (V. J. Gawron, personal communication, September 1994).

The final simulator is the SIRE (synthesized immersion research environment) facility located within the Air Force Research Laboratory at Wright–Patterson Air Force Base. SIRE (see Figure 19.8) is a VE research facility designed to develop and evaluate advanced, multi-sensory virtual interfaces for future US Air Force crewstations. The main station of the SIRE facility is a 40 ft diameter dome, with a high resolution, large FOV (70° vertical by 150° horizontal) visual imaging system. The station can be used to present 3-D sound information and has an electro-hydraulic control loader system. Several smaller independent cockpit simulators are tied into the main station, thus providing the capability for wingmen and adversary aircraft.

FIGURE 19.7   Total-in-flight simulation aircraft. (Courtesy of Calspan Corporation.)

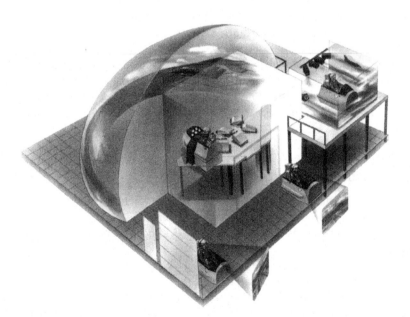

FIGURE 19.8   The SIRE facility. (Courtesy of the Crew Systems Interface Division of the Air Force Research Laboratory.)

19.12 The Future of Flight Simulation

Prior to the traditional chapter summary, the authors feel that, at a minimum, a brief listing of their expectancies and research opportunities regarding flight simulation would be appropriate.

19.12.1 Expectancies

1. The use of flight simulation as a cost-effective alternate to in-flight training will increase. Although face validity (the look and feel of the vehicle being simulated) will remain a factor in system design, the advantages of effective, less costly, lower fidelity simulation will reduce the emphasis on face validity.
2. Knowledge from the field of flight simulation will be transferred to fields as diverse as elementary schools, entertainment, and nuclear power plants. Advance technologies will trickle down and become more commonplace.
3. Simulators and simulation will be used for distance learning/training and will be linked into even larger, real-time interactive networks, as network speed and bandwidth increase.
4. Simulators will continue to be used as procedure trainers and their role in evaluating and providing training in decision-making skills and cockpit/crew resource management will increase.
5. Most large-dome type simulators will be "retired" because of high operating and support costs and changes in technology. However, some dome type simulators will be retained to serve as research tools.
6. The use of PC-based simulations and VR helmet-mounted display simulations will increase, as will the environments in which they are used.
7. The systems approach to training or its analogs will become even more trainee-centered. Customers will encourage/mandate the use of (a) front-end analysis, (b) lower fidelity simulators, (c) virtual and augmented reality, and (d) artificial intelligence and expert systems in training.
8. With the increasing use of simulators, greater emphasis should be placed on the role of the instructor. In time, training systems will incorporate instructor associates (i.e., interactive expert systems that will describe the goal of the training, demonstrate what is expected, and provide diagnostic feedback). The development of expert-system-based instructor associates promises considerable payoff.

19.12.2 Research Opportunities

1. Studies are needed that differentiate between tasks that can be learned most effectively and efficiently in training devices, simulators (fixed/motion-based), and aircraft. Once the common characteristics of selected tasks have been identified, it should be possible to generalize these findings to other tasks.
2. Studies are needed that identify the cues necessary for effective and efficient transfer of training. In addition to these studies, task analytical techniques that can identify the essential cues need to be developed and validated. To maximize the return on investment, we need to identify the critical visual and motion cues and communicate that knowledge, in an appropriate form, to the system developer.
3. The role of the instructor is pivotal to flight simulation. The evaluation function performed by the instructor is primarily subjective, and the reliability and validity of instructor ratings could be improved. Additionally, the use of objective performance measures, and more formal strategies for displaying and evaluating student/team performance would greatly improve the contribution of the instructors.

4. Presently, a variety of advanced instructional features are available. To learn when and how to use advanced instructional features, we need a theoretical base and a model that would provide valid estimates of the amount of transfer and savings that would result from the use of a particular advanced instructional features.

5. Developments in the area of VR/VE will require increased knowledge regarding cue presentation and human perception. Higher fidelity is not necessarily better; indeed it may be more costly and result in increased cybersickness.

6. As simulation expands into the vertical flight environment, we need to increase our knowledge of the control, display, and cue requirements unique to that environment.

7. Studies documenting the cost-effectiveness of lower fidelity simulations are needed. These simulations could then be utilized by general, military, and commercial aviation.

19.13 Conclusion

This chapter began with the specification of the "human factor requirements" for the US Army's first heavier than air flying machine. Today, a revised version of this specification might read:

> The flight simulator's cost effective design should incorporate only those cues, (at the appropriate level of fidelity) and instructional features necessary to permit an intelligent person to effectively learn and demonstrate the required skills at an appropriate level of proficiency within a reasonable period of time.

Wilbur and Orville Wright delivered their heavier than air flying machine within 7 months after contract award. However, responding to the revised specification will take considerably longer and require more assets. Nonetheless, this specification is presented as a challenge to individuals involved in flight simulation. Indeed, if flight simulation is to advance, we must respond to the elements of this revised specification.

Acknowledgments

The authors acknowledge the assistance of their colleagues and students at the University of Dayton and the University of Cincinnati. The critical reading by Shilo Anders deserves particular recognition. The assistance of friends in the aviation community, who not only shared material from their archives with us but reviewed portions of the chapter is also recognized. Finally, the authors thank their wives, Kathy and Hope, for their patience and support.

References

Adorian, P., Staynes, W. N., & Bolton, M. (1979). The evolution of the flight simulator. *Proceedings of Conference, 50 Years of Flight Simulation* (Vol.1, pp. 1–23). London: Royal Aeronautical Society.

Alessi, S. M. (1988). Fidelity in the design of instructional simulations. *Journal of Computer-Based Instruction, 15*(2), 40–47.

Alluisi, E. A. (1991). The development of technology for collective training: SIMNET, a case history. *Human Factors, 33,* 343–362.

Bailey, J., Hughes, R., & Jones, W. (1980). *Application of backward chaining to air to surface weapons delivery training* (AFHRL Tech. Rep. No. 79-63). Brooks Air Force Base, TX: Air Force Human Resources Laboratory.

Baudhuin, E. S. (1987). The design of industrial and flight simulators. In S. M. Cormier, & J. D. Hagman (Eds.), *Transfer of learning* (pp. 217–237). San Diego, CA: Academic Press.

Beringer, D. B. (1994). Issues in using off-the-shelf PC-based flight simulation for research and training: Historical perspective, current solutions and emerging technologies. *Proceedings of the Human Factors and Ergonomics Society 38th Annual Meeting-1994* (pp. 90–94). Santa Monica, CA: Human Factors and Ergonomics Society.

Berki-Cohen, J., Soja, N. N., & Longridge, T. (1998). Simulator platform motion—the need revisited. *The International Journal of the Aviation Psychology, 8*(3), 293–317.

Boldovici, J. A. (1987). Measuring transfer in military settings. In S. M. Cormier, & J. D. Hagman (Eds.), *Transfer of learning* (pp. 239–260). San Diego, CA: Academic Press.

Boldovici, J. A. (1992). *Simulator motion* (TR 961, AD A257 683). Alexandria, VA: U.S. Army Research Institute.

Boothe, E. M. (1994). A regulatory view of flight simulator qualification. *Flight simulation update-1994* (10th ed.). Binghamton, NY: SUNY Watson School of Engineering.

Bowers, C. A., Braun, C. C., Holmes, B. E., Morgan, B. B. J., & Salas, E. (1993). The development of aircrew coordination behaviors. In R. S. Jensen, & D. Neumeister (Eds.), *Proceedings of the Seventh International Symposium on Aviation Psychology* (pp. 758–761). Columbus: Aviation Psychology Laboratory, Ohio State University.

Bowers, C., Deaton, J., Oser, R., Prince, C., & Kolb, M. (1993). The impact of automation on crew communication and performance. In R. S. Jensen, & D. Neumeister (Eds.), *Proceedings of the Seventh International Symposium on Aviation Psychology* (pp. 573–577). Columbus: Aviation Psychology Laboratory, Ohio State University.

Bray, R. S. (1994). Cockpit motion in helicopter simulation. In W. E. Larsen, R. J. Randle, & Popiah, L. N. (Eds.), *Vertical flight training: An overview of training and flight simulator technology with reference to rotary wing requirements* (NASA 1373, DOT/FAA/CT-94/83). Moffett Field, CA: NASA Ames.

Bunker, W. M. (1978). Training effectiveness versus simulation realism. *The Eleventh NTEC/Industry Conference Proceedings.* Orlando, FL: Naval Training Equipment Center. As cited on p. 53 of Hays, R. T., & Singer, M. J. (1989). *Simulation fidelity in training system design.* New York: Springer-Verlag.

Cammarota, J. P. (1990). Evaluation of full-sortie closed-loop simulated aerial combat maneuvering on the human centrifuge. *Proceedings on the National Aerospace and Electronics Conference* (pp. 838–842). Piscataway, NJ: IEEE.

Cardullo, F. M. (1994a). Motion and force cueing. *Flight simulation update-1994* (10th ed.). Binghamton, NY: SUNY Watson School of Engineering.

Cardullo, F. (1994b). Simulation purpose and architecture. *Flight simulation update-1994* (10th ed.). Binghamton, NY: SUNY Watson School of Engineering.

Caro, P. W. (1979). Development of simulator instructional feature design guides. *Proceedings of Conference: 50 Years of Flight Simulation* (pp. 75–89). London: Royal Aeronautical Society.

Caro, P. W. (1988). Flight training and simulation. In E. L. Wiener, & D. C. Nagel (Eds.), *Human factors in aviation* (pp. 229–261). New York: Academic Press.

Chambers, W. (1994). Visual simulation overview. *Flight simulation update-1994* (10th ed.). Binghamton, NY: SUNY Watson School of Engineering.

Cormier, S. M. (1987). The structural process underlying transfer of training. In S. M. Cormier, & J. D. Hagman (Eds.), *Transfer of learning* (pp. 151–181). San Diego, CA: Academic Press.

Department of the Army. (1990). *Systems approach to training analysis* (TRADOC Pamphlet 351-4). Fort Monroe, VA: U.S. Army Training and Doctrine Command.

Eyth, J., & Heffner, P. (1992). *Design and performance of the centrifuge-based dynamic flight simulator* (AIAA Paper No. 92-4156, Flight Simulation Technologies Conference, Hilton Head, SC). Washington, DC: American Institute of Aeronautics and Astronautics.

Federal Aviation Administration. (1989). *Instrument rating: Practical test standards.* Washington, DC: Author.

Federal Aviation Administration. (1991). *AC120-40C: Airplane simulator qualification.* Washington, DC: Author.

Federal Aviation Administration. (1992). *AC 120-45A: Airplane flight training device qualification.* Washington, DC: Author.

Federal Aviation Administration. (1997). *AC No: 61-126: Qualification and approval of personal computer-based aviation training devices.* Washington, DC: Author.

Fischetti, M. A., & Truxal, C. (1985, March). Simulating "The right stuff." *IEEE Spectrum, 22,* 38–47.

Fitts, P. M. (1962). Factors in complex skill training. In R. Glaser (Ed.), *Training research and education* (pp. 77–197). Pittsburgh, PA: University of Pittsburgh Press.

Flexman, R. E., Mathney, W. G., & Brown, E. L. (1950). *Evaluation of the school link and special method of instruction in a 10 h private pilot flight training program* (Bull. 47, No. 80). Champaign-Urbana: University of Illinois.

Frank, L. H., Kennedy, R. S., Kellogg, R. S., & McCauley, M. E. (1983). *Simulator sickness: Reaction to a transformed perceptual world. I. Scope of the problem* (Rep. No. NAVTRAEQUIPCEN TN-65). Orlando, FL: Naval Training Equipment Center.

Frost & Sullivan. (December 2006). Frost: New aircraft spur N.American ground-based flight simulation solutions market. Retrieved June 18, 2007 from http://aero-defense.ihs.com/news/2006/frost-flight-simulation.htm

Frost & Sullivan. Research Services. (August 2006). North American commercial and military ground-based flight simulation market. Retrieved June 18, 2007 from http://www.frost.com/prod/servlet/report-brochure.pag?id = F898-01-00-00-00

Garcia, A. B., Gocke, R. P. J., & Johnson, N. P. (1994). *Virtual prototyping: Concept to production* (Rep. No. DSMC 1992–93). Fort Belvoir, VA: Defense Systems Management College Press.

Grupping, J. (2005). The story of Flight Simulator. Retrieved June 26, 2007 from http://fshistory.simflight.com/fsh/versions.htm

Guckenberger, D., Uliano, K. C., & Lane, N. E. (1993). *Teaching high-performance skills using above-real-time training* (NASA Contractor Rep. No. 4528). Edwards AFB, CA: NASA Dryden.

Guedry, F. E. (1987). *Motion cues in flight simulation and simulator induced sickness* (Advisory Group for Aerospace Research-CP-433). Neuilly Sur Seine, France: NATO.

Haber, R. N. (1986). Flight simulation. *Scientific American, 255*(1), 96–103.

Hays, R. T., & Singer, M. J. (1989). *Simulation fidelity in training system design.* New York: Springer-Verlag.

Hays, R. T., Jacobs, J. W., Prince, C., & Salas, E. (1992). Requirements for future research in flight simulation training: Guidance based on a meta-analytic review. *International Journal of Aviation Psychology, 2*(2), 143–158.

Hettinger, L. J., Berbaum, K. S., Kennedy, R. S., Dunlap, W. P., & Nolan, M. D. (1990). Vection and simulator sickness. *Military Psychology, 2,* 171–181.

Hughes, R. G., Hannon, S. T., & Jones.W. E. (1979). *Application of flight simulator record/playback feature* (Rep. No. AFHRL-TR-79-52, AD A081 752). Williams AFB, AZ: Air Force Human Resources Laboratory.

Hughes, R., Brooks, R., Graham, D., Sheen, R., & Dickens, T. (1982). Tactical ground attack: On the transfer of training from flight simulator to operational Red Flag range exercise. *Proceedings of the Human Factors and Ergonomics Society 26th Annual Meetin -1982* (pp. 596–600). Santa Monica, CA: Human Factors and Ergonomics Society.

Jacobs, J. W., Prince, C., Hays, R. T., & Salas, E. (1990). *A meta-analysis of the flight simulator research* (Tech. Rep. No. 89-006). Orlando, FL: Naval Training Systems Center.

Jentsch, F., & Bowers, C. A. (1998). Evidence for the validity of PC based simulations in studying aircrew coordination. *The International Journal of Aviation Psychology, 8*(3), 243–260.

Jones, E. R. (1967). *Simulation applied to education.* McDonnell-Douglas Corporation Paper. St. Louis, MO: McDonnell-Douglas Corporation.

Jones, E., Hennessy, R., & Deutsch, S. (Eds.). (1985). *Human factors aspects of simulation.* Washington, DC: National Academy Press.

Kass, S. J., Herschler, D. A., & Companion, M. A. (1991). Training situational awareness through pattern recognition in a battlefield environment. *Military Psychology, 3*(2), 105–112.

Kennedy, R. S., & Fowlkes, J. E. (1992). Simulator sickness is polygenic and polysymptomatic: Implications for research. *International Journal of Aviation Psychology, 2*(1), 23–38.

Kennedy, R. S., Hettinger, L. J., & Lilienthal, M. G. (1990). Simulator sickness. In G. H. Crampton (Ed.), *Motion and space sickness* (pp. 317–341). Boca Raton, FL: CRC Press.

Kennedy, R. S., Lilienthal, M. G., Berbaum, K. S., Baltzley, D. R., & McCauley, M. E. (1989). Simulator sickness in U.S. Navy flight simulators. *Aviation, Space and Environmental Medicine, 60,* 10–16.

Kiefer, D. A., & Calvert, J. F. (1992). *Developmental evaluation of a centrifuge flight simulator as an enhanced maneuverability flying qualities tool* (Paper No. 92-4157). Washington, DC: American Institute of Aeronautics and Astronautics.

Kirkpatrick, D. D. (1998). *Evaluating training programs. The four levels* (2nd ed.). San Francisco, CA: Berrett-Koehler Publishers.

Kuntz Rangal, R., Guimaraes, L. N. F., & De Assis Correa, F. (2002). Development of a virtual flight simulator. *CyberPsychology and Behavior, 5*(5), 461–470.

Lane, N. E. (1986). *Issues in performance measurement for military aviation with applications to air combat maneuvering.* Orlando, FL: Naval Training Systems Center.

Lintern, G., & McMillan, G. (1993). Transfer for flight simulation. In R. A. Telfer (Ed.), *Aviation instruction and training* (pp. 130–162). Brookfield, VT: Ashgate Publishing Co.

Lintern, G. (1991). An informational perspective on skill transfer in human-machine systems. *Human Factors, 33*(3), 251–266.

Lintern, G., Roscoe, S. N., & Sivier, J. E. (1990). Display principles, control dynamics, and environmental factors in augmentation of simulated visual scenes for teaching air-to-ground attack. *Human Factors, 32,* 299–371.

Lintern, G., Taylor, H. L., Koonce, J. M., & Talleur, D. A. (1993). An incremental transfer study if scene detail and field of view effects on beginning flight training. *Proceedings of Seventh International Symposium on Aviation Psychology* (pp. 737–742). Columbus: Ohio State University.

Loesch, R. L., & Waddell, J. (1979). The importance of stability & control fidelity in simulation. *Proceedings of Conference: 50 Years of Flight Simulation* (pp. 90–94). London: Royal Aeronautical Society.

Madden, J. J. (1992). Improved instructor station design. *Proceedings of the 15th Interservice/Industry Training Systems Conference* (pp. 72–79). Orlando, FL: Naval Training Systems Center.

Martin, E. A. (1994). Motion and force simulation systems I. *Flight simulation update-1994* (10th ed.). Binghamton, NY: SUNY Watson School of Engineering.

Martin, E. L. (1981). *Training effectiveness of platform motion: Review of motion research involving the advanced simulator for pilot training and the simulator for air-to-air combat* (Rep. No. AFHRL-TR-79-51). Williams Air Force Base, AZ: Air Force Human Resources Laboratory.

Martin, E. L., & Wagg, W. L. (1978a). *Contributions of platform motion to simulator training effectiveness: Study 1-Basic Contact* (Tech. Rep. No. AFHRL TR-78-15, AD A058 416). Williams Air Force Base, AZ: Air Force Human Resources Laboratory.

Martin, E. L., & Waag, W. L. (1978b). *Contributions of platform motion to simulator training effectiveness: Study II-aerobatics* (Tech. Rep. No. AFHRL TR-78-52. AD A064 305). Williams Air Force Base, AZ: Air Force Human Resources Laboratory.

McCauley, M. E. (Ed.). (1984). *Research issues in simulator sickness.* Washington, DC: National Academy Press.

McCauley, M. E., & Sharkey, T. J. (1991). Spatial orientation and dynamics in virtual reality systems: Lessons from flight simulation. *Proceedings of the Human Factors and Ergonomics Society 35th Annual Meeting-1991* (pp. 1348–1352). Santa Monica, CA: Human Factors and Ergonomics Society.

McCauley, M. E., & Sharkey, T. J. (1992). Cyberspace: Perception of self-motion in virtual environments. *Presence, 1*(2), 311–318.

McDaniel, W. C., Scott, P. G., & Browning, R. F. (1983). Contribution of platform motion simulation in SH-3 helicopter pilot training (TAEG Rep 153). Orlando, FL: Training Analysis and Evaluation Group, Naval Training Systems Center.

McMillan, G. (1994). System integration. *Flight simulation update-1994* (10th ed.). Binghamton, NY: SUNY Watson School of Engineering.

Miller, R. B. (1954). *Psychological considerations in the design of training equipment* (Tech. Rep. No. 54-563). Wright Patterson AFB, OH: Wright Air Development Center.

Mixon, T. R. & Moroney, W. F. (1982). *An annotated bibliography of objective pilot performance measures* (Rep. No. NAVTRAEQUIPCEN IH 330). Orlando, FL: Naval Training Equipment Center.

Money, K. E. (1970). Motion sickness. *Physiological Reviews, 50,* 1–39.

Moroney, W. F., Hampton, S., & Beirs, D. W. (1997). Considerations in the design and use of personal computer-based aircraft training devices (PCATDs) for instrument flight training: A survey of instructors. *Proceedings of Ninth International Symposium on Aviation Psychology.* Columbus: Ohio State University.

Moroney, W. F., Hampton, S., Beirs, D. W., & Kirton, T. (1994). The use of personal computer-based training devices in teaching instrument flying: A comparative study. *Proceedings of the Human Factors and Ergonomics Society 38th Annual Meeting-1994* (pp. 95–99). Santa Monica, CA: Human Factors and Ergonomics Society.

Mosher, S., Farmer, D., Cobasko, J., Stassen, M., & Rosenshein, L. (1992). *Innovative display concepts for field-of-view expansion in air combat simulation* (WLTR-92-3091). Wright Patterson AFB, OH: Flight Dynamics Directorate, Wright Laboratory.

Muth, E. R., & Lawson, B. (2003). Using flight simulators aboard ships: Human side effects, optimal scenario with smooth seas. *Aviation, Space and Environmental Medicine, 77*(5), 497–505.

Naval Training Systems Center. (1989). *Simulator sickness.* Orlando, FL: Naval Training Systems Center.

Orlansky, J., & String, J. (1977). *Cost-effectiveness of flight simulator for military training* (Rep. No. IDA NO. HQ 77-19470). Arlington, VA: Institute for Defense Analysis.

Peterson, C. (1992, June). Simulation symposium: Are PC-based flight simulators coming of age? *Private Pilot,* pp. 75–79.

Pfeiffer, M. G., Horey, J. D., & Butrimas, S. K. (1991). Transfer of simulated instrument training to instrument and contact flight. *International Journal of Aviation Psychology, 1*(3), 219–229.

Polzella, D. J., & Hubbard, D. C. (1986). Utility and utilization of aircrew training device advanced instructional features. *Proceedings of the Human Factors Society-30th Annual Meeting-1986* (pp. 139–143). Santa Monica, CA: Human Factors Society.

Polzella, D. J., Hubbard, D. C., Brown, J. E., & McLean, H. C. (1987). *Aircrew training devices: Utility and utilization of advanced instructional features* (Tech. Rep. No. AFHRL-TR-87-21). Williams Air Force Base, AZ: Operation Training Division.

Randle, R. J., & Sinacori, J. (1994). Visual space perception in flight simulators. In W. E. Larsen, R. J. Randle, & Popiah, L. N. (Eds.), *Vertical flight training: An overview of training and flight simulator technology with reference to rotary wing requirements* (NASA 1373, DOT/FAA/CT-94/83). Moffett Field, CA: NASA Ames.

Reason, J. T. (1978). Motion sickness adaptation: A neural mismatch model. *Journal of the Royal Society of Medicine, 71,* 819–829.

Rolfe, J. M., & Staples, K. J. (Eds.). (1986). *Flight simulation.* Cambridge, U.K.: Cambridge University Press.

Roscoe, S. N. (1991). Simulator qualification: Just as phony as it can be. *International Journal of Aviation Psychology, 1*(4), 335–339.

Roscoe, S. N. (1980). Transfer and cost effectiveness of ground-based flight trainers. In S. N. Roscoe (Ed.), *Aviation psychology* (pp. 194–203). Ames: Iowa State University Press.

Roscoe, S. N., & Williges, B. H. (1980). Measurement of transfer of training. In S. N. Roscoe (Ed.), *Aviation psychology* (pp. 182–193). Ames: Iowa State University Press.

Rosinski, R. R. (1982). *Effect of projective distortions on perception of graphic displays* (Report No. 82-1). Washington, DC: Office of Naval Research.

Royal Aeronautical Society. (1979). *50 Years of flight simulation* (Vols. I, II, and III). London, England: Royal Aeronautical Society.

Sadlowe, A. R. (Ed.). (1991). *PC-based instrument flight simulation- A first collection of papers*. New York: The American Society of Mechanical Engineers.

Salas, E., Bowers, C. A., & Rhodenizer, L. (1998). It is not how much you have but how you use it: Toward a rational use of simulation to support aviation training. *International Journal of Aviation Psychology*, 8(3), 197–208.

Searchsmallbizit (2004). Homepage, Search SMB.com Retrieved on July 24, 2004, from http://searchsmall-bizit.techtarget.com/sDefinition/0,,sid44_gci213303,00.html

Seidensticker, S. (1994). Distributed interactive simulation (DIS). *Flight simulation update-1994* (10th ed.). Binghamton, NY: SUNY Watson School of Engineering.

Sharkey, T. J., & McCauley, M. E. (1991). The effect of global visual flow on simulator sickness. *Proceedings of the AIAA Flight Simulation Technologies Conference* (pp. 496–504) (Report No. AIAA-91-2975-CP.). Washington, DC: American Institute of Aeronautics and Astronautics.

Sharkey, T. J., & McCauley, M. E. (1992). Does a motion base prevent simulator sickness?. *Proceedings of the AIAA Flight Simulation Technologies Conference* (pp. 21–28). Washington, DC: American Institute of Aeronautics and Astronautics.

Sinnett, M. K., Oetting, R. B., & Selberg, B. P. (1989). Improving computer technologies for real-time digital flight simulation. *SAE Aerospace Technology Conference and Exposition* (pp. 1826–1829). Long Beach, CA: SAE.

Sparaco, P. (1994, June). Simulation acquisition nears completion. *Aviation Week & Space Technology*, pp. 71–72.

Spears, W. D., Sheppard, H. J., Roush, M. D., & Richetti, C. L. (1981a). *Simulator training requirements and effectiveness study (STRES)* (Tech. Rep. No. AFHRL-TR-80-38, Part I). Dayton, OH: Logistics and Technical Training Division.

Spears, W. D., Sheppard, H. J., Roush, M. D. I., & Richetti, C. L. (1981b). *Simulator training requirements and effectiveness study (STRES)* (Tech. Rep. No. AFHRL-TR-80-38, Part II). Dayton, OH: Logistics and Technical Training Division.

Stanney, K. M. (2002). *Handbook of virtual environments: Design, implementation, and applications*. Mahwah, NJ: Lawrence Erlbaum Associates.

Stark, E. A. (1994). Training and human factors in flight simulation. *Flight simulation update* (10th ed.). Binghamton, NY: SUNY Watson School of Engineering.

Sticha, P. J., Singer, M. J., Blacksten, H. R., Morrison, J. E., & Cross, K. D. (1990). *Research and methods for simulation design: State of the art* (Rep. No. ARI-TR-914). Alexandria, VA: U.S. Army Research Institute for the Behavioral and Social Sciences.

Sticha, P. J., Buede, D. M., Singer, M. J., Gilligan, E. L., Mumaw, R. J., & Morrison, J. E. (1990). *Optimization of simulation-based training systems: Model description, implementation, and evaluation* (Tech. Rep. No. 896). Alexandria, VA: U.S. Army Research Institute.

Stoffregen, T. A., & Riccio, G. E. (1991). An ecological critique of the sensory conflict theory of motion sickness. *Ecological Psychology*, 3(3), 159–194.

Stone, R. (2001). Virtual-reality for interactive training: An industrial practitioner's viewpoint. *International Journal of Human Computer Studies*, 55, 699–711.

Taylor, H. L. (1991). ILLIMAC: A microprocessor based instrument flight trainer. In A. R. Sadlowe (Ed.), *PC-based instrument flight simulation- A first collection of papers* (pp. 1–11). New York: The American Society of Mechanical Engineers.

Taylor, H. L., Lintern, G., Hulin, C. L., Talleur, D. A., Emanuel, T. W., & Phillips, S. I. (1997). Effectiveness of personal computers for instrument training. *Proceedings of Ninth International Symposium on Aviation Psychology*. Columbus: Ohio State University.

Taylor, H. L., Lintern, G., Hulin, C. L., Talleur, D. A., Emanuel, T. W., & Phillips, S. I. (1999). Transfer of training effectiveness of a personal computer aviation training device. *International Journal of Aviation Psychology, 9*(4), 319–335.

Taylor, H. L., Lintern, G., & Koonce, J. M. (2001). Quasi-transfer as a predictor of transfer from simulator to aircraft. *The Journal of General Psychology, 120*(3), 257–276.

Thornton, C., Braun, C., Bowers, C., & Morgan, B. B. J. (1992). Automation effects in the cockpit: A low fidelity investigation. *Proceedings of the Human Factors Society 36th Annual Meeting-1992* (pp. 30–34). Santa Monica, CA: Human Factors and Ergonomics Society.

U.S. Air Force (1990). *USAF Guide Specification: Simulators* (AFSGS-8724A ed.). Wright Patterson AFB, OH: Aeronautical Systems Directorate (Code ENES).

U.S. Army Signal Corps Specification No. 486. (1907). *Advertisement and specification for a heavier than air flying machine.* Washington, DC: Department of the Army. (On display at USAF Museum, Wright Patterson AFB, Dayton, OH.)

VirtuSphere. (2006) VirtuSphere Home Page, Retrieved June 29, 2007, from http://www.virtusphere.com/

Warren, R., & Riccio, G. E. (1985). Visual cue dominance hierarchies: Implications for simulator design. *Proceedings of the Aerospace Technology Conference* (pp. 61–74). Washington, DC: AIAA.

Webopedia. (2004). Virtual Reality. Retrieved on July 24, 2004 from http://webopedia.internet.com/TERM/V/virtual_reality.html

Waldrop, M. M. (2001). The origins of personal computing. *Scientific American, 285*(6), 84–91.

Wiener E. L., & Nagel, D. C. (1988). *Human Factors in Aviation.* New York: Academic Press.

Williams, K. W. (1994). *Proceedings of the joint industry-FAA conference on the development and use of PC-Based aviation training devices* (Rep. No. DOT/FAA/AM-94.25). Oklahoma City, OK: Civil Aeromedical Institute.

Wilson, J. R. (2000, June). Technology brings change to simulation industry. *Interavia Business and Technology, 55*(643), 19–21.

Wightman, D. C., & Sistrunk, F. (1987). Part-task training strategies in simulated carrier landing final approach training. *Human Factors, 29*(3), 245–254.

Wolverton, M. (April/May 2007). The G machine. *Air and Space,* 68–71.

20

Air-Traffic Control

Michael S. Nolan
Purdue University

The primary function of an air-traffic control (ATC) system is to keep aircraft participating in the system separated from one another. Secondary reasons for the operation of an ATC system are to make more efficient use of airspace, and to provide additional service to pilots such as traffic information, weather avoidance, and navigational assistance.

Not every aircraft may be required to participate in an ATC system, however. Each nation's regulations only obligate certain aircraft to participate in the ATC system. ATC participation in each country may range from mandatory participation of all aircraft, to no ATC services offered at all.

The level of ATC services provided is usually based on each nation's priorities, technical abilities, weather conditions, and traffic complexity. To more specifically define and describe the services that can be offered by an ATC system, the International Civil Aviation Organization (ICAO) has defined different aircraft operations and classes of airspace within which aircraft may operate. Different rules and regulations apply to each type of aircraft operation, and these rules vary depending on the type of airspace within which the flight is conducted. Although ICAO publishes very specific guidelines for the classification of airspace, it is the responsibility of each country's aviation regulatory agency to categorize its national airspace.

20.1 Aircraft Operations

Visual meteorological conditions (VMC) are defined as weather conditions where pilots are able to see and avoid other aircraft. In general, pilots flying in VMC conditions comply with *visual flight rules* (VFR). VFR generally require that 3–5 miles of flight visibility be maintained at all times, that the aircraft remain clear of clouds, and that pilots have the responsibility to see and avoid other aircraft. Pilots provide their own air-traffic separation. The ATC system may assist the pilots, and may offer additional services, but the pilot has the ultimate responsibility to avoid other air-traffic.

Instrument meteorological conditions (IMC) are generally defined as weather conditions where the visibility is below that required for VMC or whenever the pilot cannot remain clear of clouds. Pilots operating in IMC must comply with *instrument flight rules* (IFR), which require the filing of a flight plan, and ATC normally provides air-traffic separation. Pilots may operate under IFR when flying in VMC conditions. Under these circumstances, ATC will separate only those aircraft complying with IFR. VFR aircraft provide their own separation, and IFR aircraft have the responsibility to see and avoid VFR aircraft.

20.2 Airspace Classes

National governments define the extent to which they wish to offer ATC services to pilots. In general, ICAO recommendations suggest three general classes of airspace within which different services are provided to VFR and IFR pilots. These three general classes are uncontrolled, controlled, and positive-controlled airspace.

Uncontrolled airspace is that within which absolutely no aircraft separation is provided by ATC, regardless of weather conditions. Uncontrolled airspace is normally that airspace with little commercial aviation activity.

Controlled airspace is that within which ATC separation services may be provided to certain select categories of aircraft (usually those complying with IFR). In controlled airspace, pilots flying VFR must remain in VMC, and are not normally provided ATC separation, and therefore, must see and avoid all other aircraft. Aircraft who wish to utilize ATC services in controlled airspace must file a flight plan and comply with IFR. IFR aircraft are permitted to operate in VMC and IMC. When operating within controlled airspace, IFR aircraft are separated by ATC from other aircraft operating under IFR. When operating in VMC in controlled airspace, IFR pilots must see and avoid aircraft operating under VFR.

In positive-controlled airspace, all aircraft, whether IFR or VFR, are separated by ATC. All aircraft operations require an ATC clearance. VFR pilots must remain in VMC conditions, but are separated by ATC from both VFR and IFR aircraft. IFR aircraft are also separated from both IFR and VFR aircraft.

TABLE 20.1 Requirements for Operation and ATC Services Provided to Flight Operations within General Airspace Categories

	Uncontrolled Airspace (Class G Airspace)	Controlled Airspace (Classes C, D, and E Airspace)	Positive-Controlled Airspace (Classes A and B Airspace)
VFR Flight Operations	Must remain in VMC (VMC minima are fairly low, typically clear of clouds and 1 mile visibility) If VMC conditions exist, VFR operations are permitted but an ATC clearance may be required to operate in certain areas If VMC conditions exist, VFR operations are permitted and no ATC clearance is required If IMC conditions exist, VFR operations are not authorized No ATC separation services are provided Pilot's responsibility to see and avoid both IFR and VFR aircraft	Must remain in VMC (VMC minima are higher, typically a specified distance from clouds and 3–5 miles visibility) If permitted, VFR aircraft must remain in VMC (VMC minima are higher, typically a specified distance from clouds and 3–5 miles visibility) If IMC conditions exist, VFR operations are not authorized No ATC separation services are provided to VFR aircraft Pilot's responsibility to see and avoid both IFR and other VFR aircraft VFR aircraft operating in controlled airspace may be required to meet additional, class-specific operating rules and procedures	VFR flight operations might not be permitted If VMC conditions exist, VFR operations may be permitted but an ATC clearance would be required If IMC conditions exist, VFR operations are not authorized Separation services are provided to all aircraft All aircraft will be separated by ATC VFR aircraft operating in positive-controlled airspace may be required to meet additional, class-specific operating rules and procedures
IFR Flight Operations	IFR operations permitted without ATC clearance, nor will it be issued ATC separation services not provided Pilot's responsibility to see and avoid both IFR and VFR aircraft	ATC clearance required ATC separation will be provided between IFR aircraft; IFR pilots must see and avoid VFR aircraft while in VMC IFR aircraft operating in controlled airspace may be required to meet additional, class-specific operating rules and procedures	ATC clearance required ATC separation will be provided between all aircraft All aircraft operating in positive-controlled airspace may be required to meet additional, class-specific operating rules and procedures

Table 20.1 describes the general rules that both IFR and VFR pilots must comply with when operating in these three classes of airspace.

20.3 Air-Traffic Control Providers

In most countries, a branch of the national government normally provides ATC services. The ATC provider may be civilian, military, or a combination of both. Some national ATC services are now being operated by private corporations funded primarily by user fees. Other governments are experimenting with ATC-system privatization. Some of these initiatives propose to transfer all ATC responsibility to private agencies, whereas others propose to transfer only certain functions, such as weather dissemination and the operation of low-activity control towers, to private or semipublic entities.

Privatized ATC is a fairly recent historical development with roots tracing back to the 1930s. When an ATC system was first started in the United States, control towers were operated by the municipalities that owned the airports. En route ATC was provided through a consortium of airlines. Only in the 1940s was ATC taken over and operated by the national government.

The concept behind privatized ATC is that if freed from cumbersome government procurement requirements, employment regulations, and legislative pressures, private corporations might provide service at less cost, be more efficient, and be more responsive to users' needs because they would be funded and controlled by the users. Possible disadvantages of such a system include lack of governmental oversight and responsibility, possible conflict of interest between system users and operators, little incentive to assist military aviation activities, and restricted access to the capital funding needed to upgrade and operate such a complex system.

20.4 Air-Traffic Control Assignments

Every nation is responsible for providing ATC services within its national borders. In order to provide for a common method of ATC, ICAO promulgates standardized procedures that most countries generally adhere to. These standards include universally accepted navigation systems, a common ATC language (English), and general ATC separation standards. ICAO is a voluntary organization of which most countries are members. Every ICAO signatory nation agrees to provide ATC services to all aircraft operating within its boundaries and agrees to require that their pilots abide by other national ATC systems when operating within foreign countries.

Every nation's ATC procedures can and do occasionally deviate from ICAO recommended practices. Each operational procedure that deviates from ICAO standards is published by the national ATC service provider in the Aeronautical Information Publication.

ICAO has been granted the responsibility for providing ATC services in international airspace, which is comprised mostly of oceanic and polar airspace. ICAO has assigned separation responsibility in those areas to individual states both willing and able to accept that responsibility. Some countries that have accepted this responsibility include the United States, United Kingdom, Canada, Australia, Japan, Portugal, and the Philippines.

20.5 Air-Traffic Control Services

Airspace with little or no potential traffic conflicts requires little in the way of sophisticated ATC systems. If air-traffic density increases, if aircraft operations increase in complexity, or if special, more hazardous operations are routinely conducted, additional control of aircraft is usually required to maintain an acceptable level of safety. The easiest method of defining these increasing ATC-system requirements and their associated operating rules is to define different classes of airspace within which different ATC services and requirements exist.

Standard ICAO airspace classifications include classes labeled A, B, C, D, E, F, and G. In general, Class A airspace is positive controlled, where ATC services are mandatory for all aircraft. Class G is uncontrolled airspace where no ATC services are provided to either IFR or VFR aircraft. Classes B, C, D, E, and F provide declining levels of ATC services and requirements.

It is each nation's responsibility to describe, define, explain, and chart the various areas of airspace within its respective boundaries. In general, areas with either high-density traffic or a mix of different aircraft operations are classified as class A, B, or C airspace. Areas of low-density traffic are usually designated as class D, E, F, or G.

20.6 Air-Traffic Control Services Offered within Each Type of Airspace

The requirements to enter each airspace classification and the level of ATC services offered within each area are listed here.

Class A Airspace: All operations must be conducted under IFR and are subject to ATC clearances and instructions. ATC separation is provided to all aircraft. Radar surveillance of aircraft is usually provided.

Class B Airspace: Operations may be conducted under IFR or VFR. However, all aircraft are subject to ATC clearances and instructions. ATC separation is provided to all aircraft. Radar surveillance of aircraft is usually provided.

Class C Airspace: Operations may be conducted under IFR or VFR; however, all aircraft are subject to ATC clearances and instructions. ATC separation is provided to all aircraft operating under IFR and, as necessary, to any aircraft operating under VFR when any aircraft operating under IFR is involved. All VFR operations will be provided with safety alerts and, on request, conflict-resolution instructions. Radar surveillance of aircraft is usually provided.

Class D Airspace: Operations may be conducted under IFR or VFR; however, all aircraft are subject to ATC clearances and instructions. ATC separation is provided to aircraft operating under IFR. All aircraft receive safety alerts and, on pilot request, conflict-resolution instructions. Radar surveillance of aircraft is not normally provided.

Class E Airspace: Operations may be conducted under IFR or VFR. ATC separation is provided only to aircraft operating under IFR within a surface area. As far as practical, ATC may provide safety alerts to aircraft operating under VFR. Radar surveillance of aircraft may be provided if available.

Class F Airspace (United States does not utilize this class): Operations may be conducted under IFR or VFR. ATC separation will be provided, so far as practical, to aircraft operating under IFR. Radar surveillance of aircraft is not normally provided.

Class G Airspace: Operations may be conducted under IFR or VFR. Radar surveillance of aircraft is not normally provided.

20.7 Aeronautical Navigation Aids

Air-traffic separation can only be accomplished if the location of an aircraft can be accurately determined. Therefore, an ATC system is only as accurate as its ability to determine an aircraft's position. The navigation systems currently in use were developed in the 1950s, but are undergoing a rapid change in both technology and cost. As integrated circuitry and computer technology continue to become more robust and inexpensive, the *global navigation satellite system* (GNSS) global-positioning system promises unprecedented navigational performance at a relatively low cost. ICAO has affirmed its preference for GNSS as the future primary international navigation standard. Various experts predict that existing navigation systems will be either decommissioned or relegated to a GNSS backup system within the decade.

In general, the accuracy of existing navigation aids is a function of system cost and/or aircraft distance from the transmitter. Relatively inexpensive navigation systems are generally fairly inaccurate. The most accurate systems tend to be the most expensive. Table 20.2 describes the type, general cost, advantages, and disadvantages of many common aeronautical navigation systems.

20.8 Global Navigation Satellite System

GNSSs have just recently been adopted as the future navigation standard by ICAO. Currently, GNSS systems are as accurate as most current en route navigation systems. Inherent inaccuracies (and some intentional signal degradation) require that GNSS be augmented if it is to replace the instrument-landing system (ILS) as a precision navigation system. Satellite accuracy augmentation (*wide-area augmentation system*, WAAS) has been proposed as one method to provide general improvements to accuracy that may permit GNSS to replace ILS as the precision approach standard. Ground-based augmentation (*local-area augmentation system*, LAAS) may be required before GNSS will be sufficiently accurate for all-weather automatic landings. Which system or combination of systems will be eventually used is still undermined.

TABLE 20.2   Navigation System Capabilities, Advantages, and Disadvantages

System	General Cost	Effective Range	Accuracy	Ease of Use	Advantages	Disadvantages
Nondirectional Beacon	Inexpensive transmitter Inexpensive receiver	50–1000 nautical miles	Fairly inaccurate	Somewhat difficult to use	Inexpensive and easy transmitter installation	Susceptible to atmospheric interference No data transmission capability
VORTAC	Moderately expensive transmitter Fairly inexpensive receiver	25–200 nautical miles	Fairly accurate	Fairly easy to use	Current en route international standard Can provide distance information if aircraft suitably equipped	Large number of transmitters required to provide adequate coverage Primarily a point-to-point navigation system
LORAN-C	Expensive transmitter Inexpensive receiver	Up to 1000 nautical miles	Fairly accurate	Fairly easy to use	Only a limited number of transmitters needed to proved adequate signal coverage Provides direct routing between pilot-selected locations	Originally a marine navigation system, therefore systems primarily located in marine areas Somewhat obsolete but considered as a backup system
Inertial Navigation Systems (INS)	No transmitters required Very expensive receivers	Unlimited	Fairly accurate	Fairly easy to use	Very independent operation as no ground transmitters required	Expensive Needs to be programmed. Accuracy deteriorates over time without external input
GNSS	Extremely expensive, space-based transmitters Inexpensive receivers	Worldwide	Very accurate; can be made more accurate with augmentation	Fairly easy to use	Inexpensive, world-wide coverage with point-to-point navigation capability	Only one or two independent systems currently available; other countries considering systems but very expensive to create, operate, and maintain
ILS	Expensive transmitter Fairly inexpensive receiver	10–30 nautical miles	Fairly accurate	Fairly easy to use	Current worldwide standard for precision approaches	Limited frequencies available Only one approach path provided by each transmitter. Extensive and expensive site preparation might be required

TABLE 20.3   Air-Traffic Control Radar Systems

Radar System	Operational Theory	Information Provided	Advantages	Disadvantages
Primary surveillance radar	Very powerful electrical transmission is reflected by aircraft back to radar receiver which is then displayed to ATC personnel	Range and azimuth	Detects all aircraft within range regardless of aircraft equipment	Also detects unwanted objects. Weather and terrain can reflect and block signal. System prone to numerous false targets
Secondary surveillance radar (also known as the Air-Traffic Control Radar Beacon System or ATCRBS)	Low-powered electrical signal transmitted from ground station triggers response from airborne equipment	Range and azimuth, assigned aircraft code and altitude	Detects only aircraft. If ground system is properly equipped, aircraft identity and altitude can be displayed to ATC	System requires aircraft to be equipped with operable transponder. Operation restricted to common frequency that can be overwhelmed if too many aircraft respond
Mode-S	Selective low-powered signal transmitted from ground triggers response from individual aircraft	Range, azimuth, aircraft identity and altitude. Capability exists to transmit additional data both to and from aircraft	Detects only those aircraft specifically interrogated by the ground equipment	Requires all aircraft to be reequipped with Mode-S-capable transponder

20.9 Radar Surveillance in Air-Traffic Control

Radar is used by air-traffic controllers to monitor aircraft position, detect navigational blunders, reduce separation if possible, and make more efficient use of airspace. Controllers can utilize radar to provide aircraft navigational assistance during both the en route and approach phases of flight. If radar is able to provide more accurate aircraft-positional information than existing navigation systems can provide, it may be possible to reduce the required separation between aircraft.

Three different types of radar are used in ATC systems. Primary surveillance radar was first developed during World War II, and can detect aircraft without requiring onboard aircraft equipment. Secondary surveillance radar (SSR) requires an interrogator on the ground and an airborne transponder in each aircraft. SSR provides more accurate aircraft identification and position, and can transmit aircraft altitude to the controller. Mode-S secondary radar is a recent improvement to secondary radar systems that will provide unique aircraft identification and the ability to transmit flight information to the controller, and ATC instructions and other information directly to the aircraft. Table 20.3 lists the functional advantages and disadvantages of each radar surveillance system.

20.10 Aircraft Separation in an Air-Traffic Control System

The airspace within which ATC services are provided is normally divided into three-dimensional blocks of airspace known as sectors. Sectors have well-defined lateral and vertical limits, and normally are shaped according to traffic flow and airspace structure. Only one controller has ultimate responsibility for the separation of aircraft within a particular sector. The controller may be assisted by other controllers, but is the one person who makes the decisions (in accordance with approved procedures), concerning the separation of aircraft within that particular sector.

If pilots of participating aircraft within the sector can see other nearby aircraft, the pilots can simply "see and avoid" nearby aircraft. Or if a controller can see one or both aircraft, the controller may issue heading and/or altitude instructions that will keep the aircraft separated. This informal but effective method of aircraft separation is known as visual separation. Although a simple concept, it is very effective and efficient when properly used. As long as aircraft can be spotted and remain identified, the use of visual separation permits aircraft to operate in much closer proximity than if the aircraft cannot be seen. Most airports utilize visual separation and visual approaches during busy traffic periods. If weather conditions permit visual separation to be applied, the capacity of most major airports can be significantly increased.

Visual separation can only be employed if one pilot sees the other aircraft, of if the controller can see both aircraft. The primary disadvantage of visual separation is that it can only be employed when aircraft are flying fairly slowly. It would be next to impossible to utilize visual separation during high-altitude, high-speed cruising conditions common to modern aircraft. Visual separation can therefore only be effectively employed within the immediate vicinity of airports. The use of visual separation near airports requires that aircraft remain continuously in sight of one another. This is a difficult proposition at best during the approach to landing or departure phase of flight because these are two of the busiest times for pilots.

20.11 Nonradar Separation

When visual separation cannot be employed, controllers must use either radar or nonradar separation techniques. Due to range and curvature-of-the-earth limitations inherent to radar, there are many situations where radar cannot be used to identify and separate aircraft. Radar coverage exists near most medium- and high-density airports, and at altitudes of 5000 ft or above in the continental United States and Europe. Outside of these areas, and over the ocean, radar surveillance may not exist and the controller must employ some form of nonradar separation to provide ATC.

Nonradar separation depends on accurate position determination and the transmittal of that information to the controller. Due to navigation and communication-system limitations, ATC is unable to precisely plot the position of each aircraft in real time. Because navigation systems have inherent inaccuracies, it is impossible to know exactly where each aircraft is at any given time. Nonradar separation therefore assumes that every aircraft is located within a three-dimensional block of airspace. The dimensions of the airspace are predicated on the speed of the aircraft and the accuracy of the navigation system being used. In general, if VORs [Very-high-frequency (VHF) Omnidirectional Ranges] are being utilized for aircraft navigation, the airspace assigned to each aircraft may have a lateral width of about 8 nautical miles, a vertical height of 1000 ft, and a longitudinal length that varies depending upon the speed of the aircraft. In general, the longitudinal extent of the airspace box extends about 10 min of flight time in front of the aircraft. Depending on the speed of the aircraft, this longitudinal dimension could extend from 10 to 100 miles in front of the aircraft.

Because neither the controller nor the pilot knows exactly where within the assigned airspace box each aircraft is actually located, the controller must assume that aircraft might be located anywhere within the box. The only way to insure that aircraft do not collide is to insure that airspace boxes assigned to different aircraft never overlap. Airspace boxes are permitted to get close to one another, but as long as they never overlap, aircraft separation is assured.

Nonradar separation is accomplished by assigning aircraft either different altitudes or nonoverlapping routes. If aircraft need to operate on the same route at the same altitude, they must be spaced accordingly to prevent longitudinal overlap. Controllers may separate potentially conflicting aircraft either through the use of nonoverlapping holding patterns, or by delaying departing aircraft on the ground. If there is a sufficient speed differential between two conflicting aircraft, the controller can normally permit the faster aircraft to lead the slower aircraft using the same route and the same altitude. Depending on the speed difference between the aircraft, the longitudinal separation criteria can normally be reduced.

The controller uses flight progress strips to visualize the aircraft's position and therefore effect nonradar separation. Pertinent data are written on a flight strip as the aircraft progresses through each controller's sector. The controller may request that the pilot make various position and altitude reports, and these reports are written on the flight strip.

The primary disadvantage of nonradar separation is that its application depends on the pilot's ability to accurately determine and promptly report the aircraft's position, and the controller's ability to accurately visualize each aircraft's position. To reduce the probability of an in-flight collision occurring to an acceptably low level, the separation criteria must take into account these inherent inaccuracies and built-in communications delays. This requires that fairly large areas of airspace be assigned to each aircraft. An aircraft traveling at 500 knots might be assigned a block of airspace 1000 ft in height, covering close to 400 square miles! This is hardly an efficient use of airspace.

20.12 Radar Separation

Radar can be utilized in ATC to augment nonradar separation, possibly reducing the expanse of airspace assigned to each aircraft. Radar's design history causes it to operate in ways that are not always advantageous to ATC, however. Primary radar was developed in World War II as a defensive, anti-aerial invasion system. It was also used to locate enemy aircraft and direct friendly aircraft on an intercept course. It was essentially designed to bring aircraft together, not keep them apart.

Primary radar is a system that transmits high-intensity electromagnetic pulses focused along a narrow path. If the pulse is reflected off of an aircraft, the position of the aircraft is displayed as a bright blip, or target, on a display screen known as a plan position indicator (PPI). This system is known as primary surveillance radar.

The radar antenna rotates slowly to scan in all directions around the radar site. Most radars require 5–15 s to make one revolution. This means that once an aircraft's position has been plotted by radar, it will not be updated until the radar completes another revolution. If an aircraft is moving at 600 knots, it might move 2–3 miles before it is replotted on the radar display.

Primary radar is limited in range based on the curvature of the earth, the antenna rotational speed, and the power level of the radar pulse. Radars used by approach-control facilities have an effective range of about 75 nautical miles. Radars utilized to separate en route aircraft have a range of about 300 nautical miles.

SSR is a direct descendent of a system also developed in World War II known as *identification friend or foe* (IFF). Secondary radar enhances the radar target and can be integrated with a ground-based computer to display the aircraft's identity, altitude, and ground speed. This alleviates the need for the controller to constantly refer to flight progress strips to correlate this information. However, flight progress strips are still used by radar controllers to maintain other information, and as a backup system utilized in case of radar-system failure.

Although one might think that radar dramatically reduces aircraft separation, in fact, it only normally significantly reduces the longitudinal size of the airspace box assigned to each aircraft. The vertical dimension of the airspace box remains 1000 ft, the lateral dimension may be reduced from 8 to 5 nautical miles (sometimes 3 miles), but longitudinal separation is reduced from 10 flying minutes to 3–5 nautical miles.

20.13 Radar-System Limitations

There are various physical phenomena that hamper primary radar effectiveness. Weather and terrain can block radar waves, and natural weather conditions such as temperature inversions can cause fake or false targets to be displayed by the system. Radar also tracks all moving targets near the airport, which may include highway, train, and in some cases ship traffic. While controlling air-traffic, the controller can be distracted and even momentarily confused when nonaircraft targets such as these are displayed on the radar. It is difficult for the controller to quickly determine whether a displayed target is a "false target" or an actual aircraft.

Another major limitation of radar is its positional accuracy. Because the radar beam is angular in nature (usually about half a degree wide), the beam widens as it travels away from the transmitter.

At extreme ranges, the radar beam can be miles wide. This makes it difficult to accurately position aircraft located far from the antenna, and makes it impossible to differentiate between two aircraft operating close to one another. Because radar-system accuracy decreases as the aircraft distance from the radar antenna increases, aircraft close to the radar antenna (less than about 40 miles) can be laterally or longitudinally separated by 3 miles. Once the aircraft is greater than 40 miles from the radar antenna, 5 nautical miles of separation must be used. The size of the airspace box using radar is still not reduced vertically, but can now be as little as 9 square miles (compared to 600 when using nonradar separations).

20.14 Additional Radar Services

Radar can also be used by the controller to navigate aircraft to provide a more efficient flow of traffic. During the terminal phase of flight, as the aircraft align themselves with the runway for landing, radar can be used by the controller to provide navigational commands (vectors) that position each aircraft at the optimal distance from one another, something impossible to do if radar surveillance is not available. This capability of radar is at least as important as the ability to reduce the airspace box assigned to each aircraft.

Air-traffic controllers can also utilize radar to assist the pilot to avoid severe weather, although the radar used in ATC does not optimally display weather. The controller can also advise the pilot of nearby aircraft or terrain. In an emergency, the controller can guide an aircraft to the nearest airport, and can guide the pilot through an instrument approach. All of these services are secondary to the primary purpose of radar, which is to safely separate aircraft participating in the ATC system.

20.15 Radar Identification of Aircraft

Before controllers can utilize radar for ATC separation, they must positively identify the target on the radar. Due to possible false target generation, unknown aircraft in the vicinity, and weather-induced false targets, it is possible for a controller to be unsure of the identity of any particular radar target. Therefore, the controller must use one or more techniques to positively verify the identity of any target before radar separation criteria can be utilized. If positive identity cannot be ascertained, nonradar separation techniques must be utilized.

Controllers can verify the identity of a particular target using either primary or secondary radar. Primary methods require that the controller correlate the pilot's reported position with a target on the radar, or by asking the pilot to make a series of turns and watching for a target to make similar turns. Secondary radar identification can be established by asking the pilot to transmit an IDENT signal (which causes a distinct blossoming of the radar target), or, if the radar equipment is so equipped, asking the pilot to set the transponder to a particular code, and verifying that the radar displays that code (or the aircraft identification) next to the target symbol on the radar.

None of these methods are foolproof, and all have the potential for aircraft misidentification. During the identification process, the wrong pilot may respond to a controller's request, equipment may malfunction, or multiple aircraft may follow the controller's instruction. If an aircraft is flying too low or is outside the limits of the radar display, the target may not even show up on the radar scope. Once identified, the controller may rely completely on radar-positioning information when applying separation, so multiple methods of radar identification are usually utilized to insure that a potentially disastrous misidentification does not occur and that the aircraft remains identified. If positive radar identification or detection is lost at any time, the controller must immediately revert to nonradar separation rules and procedures until aircraft identity can be reestablished.

TABLE 20.4   Radar Separation Criteria

Aircraft Distance for Radar Antenna	Vertical Separation	Lateral Separation	Longitudinal Separation
Less than 40 nautical miles	1000 ft	3 nautical miles	3 nautical miles. Additional separation may be required for wake turbulence avoidance
40 nautical miles or greater	1000 ft	5 nautical miles	5 nautical miles. Additional separation may be required for wake turbulence avoidance

20.16 Radar Separation Criteria

Radar accuracy is inversely proportional to the aircraft's distance from the radar antenna. The further away an aircraft, the less accurate is the radar positioning of that aircraft. Radar separation criteria have been developed with this limitation in mind. One set of criteria has been developed for aircraft that are less than 40 nautical miles from the radar site. An additional set of criteria has been developed for aircraft 40 or more nautical miles from the antenna. Because the display system used in air route traffic control centers uses multiple radar sites, controllers using this equipment must always assume that aircraft might be 40 miles or farther from the radar site when applying separation criteria. Table 20.4 describes the separation criteria utilized by air-traffic controllers when using radar. The controller must utilize at least one form of separation.

As stated previously, radar serves only to reduce the nonradar separation criteria previously described. It does nothing to reduce the vertical separation between aircraft. Radar primarily serves to reduce lateral and longitudinal separation. Nonradar lateral separation is normally 8 nautical miles, but the use of radar permits lateral separation to be reduced to 3–5 nautical miles. Radar is especially effective when reducing longitudinal separation, however. Nonradar longitudinal separation requires 5–100 nautical miles, whereas radar longitudinal separation is 3–5 nautical miles. It is this separation reduction that is most effective in maximizing the efficiency of the ATC system. Instead of lining up aircraft on airways 10–50 miles in trail, controllers using radar can reduce the separation to 3–5 miles, therefore increasing the airway capacity 200%–500%. While under radar surveillance, pilots are relieved of the responsibility of making routine position and altitude reports. This dramatically reduces frequency congestion and pilot/controller miscommunications.

Another advantage of radar is that controllers are no longer restricted to assigning fixed, inflexible routes to aircraft. Because aircraft position can be accurately determined in near real time, controllers can assign new routes to aircraft that may shorten the pilot's flight, using the surrounding airspace more efficiently.

Radar vectors such as these are most effective in a terminal environment where aircraft are converging on one or more major airports, and are in a flight transitional mode where they are constantly changing altitude and airspeed. A controller using radar is in a position to monitor the aircraft in the terminal airspace, and can make overall adjustments to traffic flow by vectoring aircraft for better spacing, or by issuing speed instructions to pilots to close or widen gaps between aircraft. It is because of these advantages that most national ATC organizations first install radar in the vicinity of busy terminals. Only later (if at all) are en route navigation routes provided radar monitoring.

20.17 Current Trends in Automation

Early forms of radar provided for the display of all moving targets within the radar's area of coverage. This included not only aircraft, but weather, birds, vehicular traffic, and other atmospheric anomalies. Using technology developed in World War II, air-traffic controllers have been able to track and identify aircraft using the *air-traffic control radar beacon system* (ATCRBS). ATCRBS, sometimes known as secondary surveillance radar, or simply secondary radar, requires a ground-based interrogator and an airborne transponder installed in each aircraft. When interrogated by the ground station, the transponder replies with a unique code that can be used to identify the aircraft, and if so equipped can also transmit the aircraft's altitude to the controller.

This system is tremendously beneficial to the controller because all aircraft can easily be identified. Nonpertinent aircraft and other phenomena observed by the radar can be ignored by the controller. If the ground-based radar is properly equipped, aircraft identity and altitude can also be constantly displayed on the radar screen, relieving the controller of mentally trying to keep each radar target properly identified.

The ground-based component of the secondary radar system has since been modified to perform additional tasks that benefit the air-traffic controller. If the ground radar is properly equipped, and the computer knows the transponder code a particular aircraft is using, the aircraft can be tracked and flight information can be computer processed and disseminated. As the radar system tracks each aircraft, basic flight information can be transmitted to subsequent controllers automatically as the aircraft nears each controller's airspace boundary. Future aircraft position can also be projected based on past performance, and possible conflicts with other aircraft and with the ground can be predicted and prevented. These last two systems (known as *conflict alert* for aircraft–aircraft conflicts, and *minimum safe-altitude warning* for aircraft–terrain conflicts) only provide the controller with a warning when aircraft are projected to be in danger. The system does not provide the controller with any possible remediation of the impending problem. Future enhancements to the computer system should provide the controller with options that can be selected to resolve the problem. This future system is to be known as *conflict-resolution advisories*.

20.18 Airborne Systems

Engineers and researchers have experimented with aircraft-based traffic-avoidance systems since the 1960s. These prototype systems were not designed to replace but rather to augment and back up the current ground-based ATC system. The Federal Aviation Administration (FAA) has approved and users have begun installing an airborne traffic-avoidance system. This device is known as *traffic-alert/collision-avoidance system* (TCAS). TCAS was developed with three different levels of services and capabilities.

TCAS is an aircraft-based system that monitors and tracks nearby transponder-equipped aircraft. This position and relative altitude of nearby aircraft are constantly displayed on a TCAS display located in the cockpit of each aircraft. TCAS I provides proximity warning only, to assist the pilot in the visual acquisition of intruder aircraft. No recommended avoidance maneuvers are provided nor authorized as a direct result of a TCAS I warning. It is intended for use by smaller commuter aircraft holding 10–30 passenger seats, and general aviation aircraft. TCAS II provides traffic advisories and resolution advisories. Resolution advisories provide recommended maneuvers in a vertical direction (climb or descent only) to avoid conflicting traffic. Airline aircraft, and larger commuter and business aircraft holding 31 passenger seats or more, use TCAS II equipment. TCAS III provides all the capabilities of TCAS II but adds the capability to provide horizontal maneuver commands. All three versions of TCAS monitor the location of nearby transponder-equipped aircraft. Current technology does not permit TCAS to monitor aircraft not transponder equipped.

20.19 Conflict-Alert/Visual-Flight-Rule Intruder

ATCRBS has been enhanced with a conflict-alert program known as *conflict-alert/VFR intruder*. The old conflict-alert program only advised the controller of impending collisions between participating IFR aircraft. It did not track nonparticipating aircraft such as those operating under VFR. Conflict-alert/VFR intruder tracks all IFR and VFR aircraft equipped with transponders, and alerts the controller if a separation error between the VFR and a participating IFR aircraft is predicted. The controller can then advise the pilot of the IFR aircraft and suggest alternatives to reduce the risk of collision.

20.20 Traffic Management Systems

It has become apparent that the current ATC system may not be able to handle peak traffic created in a hub-and-spoke airline system. Much of this is due to inherent limitations of the ATC system. ATC-system expansion is planned in many countries, but until it is completed other methods of ensuring aircraft safety have been developed. To preserve an acceptable level of safety, special traffic management programs have been developed to assist the controllers in their primary function, the safe separation of aircraft.

20.20.1 Airport-Capacity Restrictions

During hub-and-spoke airport operations, traffic can become intense for fairly short periods of time. During these intense traffic periods, if optimal weather and/or airport conditions do not exist, more aircraft may be scheduled to arrive than the airport and airspace can safely handle. In the past, this traffic overload would be handled through the use of airborne holding of aircraft. Controllers would try to land as many aircraft as possible, with all excess aircraft assigned to nearby holding patterns until space became available.

This method of smoothing out the traffic flow has many disadvantages. The primary disadvantage is that while holding, aircraft consume airspace and fuel. In today's highly competitive marketplace, airlines can ill afford to have aircraft circle an airport for an extended period of time.

In an attempt to reduce the amount of airborne holding, the FAA has instituted a number of new traffic management programs. One program seeks to predict near-term airport-acceptance rates (AAR), and match arriving aircraft to that number. One program in use is the controlled-departure program. This program predicts an airport's acceptance rate over the next 6–12 h and matches the inbound flow of aircraft to that rate. Aircraft flow is adjusted through the delaying of departures at remote airports. Overall delay factors are calculated, and every affected aircraft is issued a delayed departure time that will coordinate its arrival to the airport's acceptance rate.

The primary disadvantage of such a system is twofold. First, it is very difficult to predict 6–12 h in advance conditions that will affect a particular airport's acceptance rate. These conditions include runway closures, adverse weather, and so on. As unforeseen events occur that require short-term traffic adjustments, many inbound aircraft are already airborne, and therefore cannot be delayed on the ground. This means that the only aircraft that can be delayed are those that have not yet departed and are still on the ground at nearby airports. This system inadvertently penalizes airports located close to hub airports because they absorb the brunt of these unpredictable delays. In other situations, traffic managers may delay aircraft due to forecasted circumstances that do not develop. In these situations, aircraft end up being delayed unnecessarily. Unfortunately, once an aircraft has been delayed, that time can never be made up.

Once aircraft are airborne, newer traffic flow management programs attempt to match real-time airport arrivals to the AAR. These programs are known as aircraft metering. Metering is a dynamic attempt to make short-term adjustments to the inbound traffic flow to match the AAR. In general terms, a metering program determines the number of aircraft that can land at an airport during a 5–10 min period, and then applies a delay factor to each inbound aircraft so that they land in sequence with proper spacing. The metering program dynamically calculates the appropriate delay factor, and reports this to the controller as a specific time at which each aircraft should cross a specific airway intersection. The controller monitors the progress of each flight, and issues speed restrictions to ensure that every aircraft crosses the appropriate metering fix at the computer-specified time. This should, in theory, ensure that aircraft arrive at the arrival airport in proper order and sequence.

20.21 Air-Traffic Control System Overloads

Due to the procedural limitations placed upon aircraft participating in the ATC system, many ATC sectors far away from major airports can become temporarily overloaded with aircraft. In these situations, controllers would be required to separate more aircraft than they could mentally handle. This is one major limitation to the expansion of many ATC systems.

Various programs are being researched to counteract this problem. A prototype system has been developed in the United States known as *en route sector loading* (ELOD). The ELOD computer program calculates every sector's current and predicted traffic load and alerts ATC personnel whenever it predicts that a particular sector may become overloaded. When this occurs, management personnel determine whether traffic should be rerouted around the affected sector. This particular program is successful at predicting both systemic overloads and transient overloads due to adverse weather and traffic conditions.

20.22 Pilot/Controller Communications-Radio Systems

Most ATC instructions, pilot acknowledgments, and requests are transmitted via voice radio communications. By international agreement, voice communication in ATC is usually conducted in the English language using standardized phraseology. This phraseology is specified in ICAO documents and is designed to formalize phrases used by all pilots and controllers, regardless of their native language. This agreement permits pilots from the international community to be able to fly to and from virtually any airport in the world with few communication problems.

Voice communications between pilots and controllers are accomplished using two different formats and multiple frequency bands. The most common form of voice communication in ATC is simplex communications, where the controller talks to the pilot and vice versa utilizing a single radio frequency. This method makes more efficient use of the narrow radio-frequency bands assigned to aviation, but has many inherent disadvantages. Because one frequency is used for both sides of the conversation, when one person is transmitting, the frequency is unavailable to others for use. To prevent radio-system overload, simplex radios are designed to turn off their receiver whenever transmitting.

These conditions make it difficult for a controller to issue instructions in a timely manner when using simplex communications. If the frequency is in use, the controller must wait until a break in communications occurs. More problematic is the occasion when two or more people transmit at the same time or if someone's transmitter is inadvertently stuck on. Due to the way radios operate, if two people try to transmit at the same time, no one will be able to understand the transmission, and neither of the individuals transmitting would be aware of the problem, because their receivers are turned off when transmitting.

Duplex transmission utilizes two frequencies, one for controller-to-pilot communications, and another for pilot-to-controller communications. This communication method is similar to that utilized during telephone conversations. Both individuals can communicate simultaneously and independently, are able to interrupt one another, and can listen while talking. Duplex-transmission schemes have one major disadvantage, however. To prevent signal overlap, two discrete frequencies must be assigned to every controller-pilot communication. This essentially requires that double the number of communications frequencies be made available for ATC. Due to the limited frequencies available for aeronautical communications, duplex transmissions can seldom be used in ATC.

Most short-range communications in ATC utilize the VHF radio band located just above those used by commercial FM radio stations. Just as FM radio stations, aeronautical VHF is not affected by lightning and other electrical distortion, but is known as a line-of-sight frequency band, which means that the radio signal travels in a straight line and does not follow the curvature of the earth. Airborne VHF radios must be above the horizon line if they are to receive any ground-based transmissions. If an aircraft is below the horizon, it will be unable to receive transmissions from the controller and vice versa.

This problem is solved in the ATC system through the use of *remote-communications outlets* (RCO). RCOs are transmitters/receivers located some distance from the ATC facility. Whenever a controller transmits, the transmission is first sent to the RCO using land-based telephone lines, and then is transmitted to the aircraft. Aircraft transmissions are relayed from the RCO to the controller in the same manner. Each RCO is assigned a separate frequency to prevent signal interference. This system permits a single controller to communicate with aircraft over a wide area, but requires the controller to monitor and operate multiple radio frequencies. The use of RCOs extends the controller's communications range, but also makes the ATC communications system vulnerable to ground-based telephone systems that may malfunction or be damaged, thereby causing serious ATC communication problems.

Most civil aircraft utilize VHF communications equipment. Military aircraft utilize ultra-high-frequency (UHF) band transmitters. UHF is located above the VHF band. UHF communications systems are preferred by most military organizations because UHF antennas and radios can be made smaller and more compact than those utilized for VHF. UHF is also a line-of-sight communications system. Most ATC facilities are equipped with both VHF and UHF radio-communications systems.

Extended-range communication is not possible with VHF/UHF transmitters. RCOs can help extend the range of the controller, but need solid ground on which to be installed. VHF/UHF radios are unusable over the ocean, the poles, or in sparsely populated areas. For long-range, over-ocean radio communications, high-frequency (HF) radios are used. HF uses radio frequencies just above the medium-wave or AM radio band. HF radios can communicate with line-of-sight limitations, as far as 3000 miles in some instances, but can be greatly affected by sunspots, atmospheric conditions, and thunderstorm activities. This interference is hard to predict and depends on the time of day, season, sunspot activity, local and distant weather, and the specific frequency in use. HF radio communication requires the use of multiple frequencies, with the hope that at least one interference-free frequency can be found for communications at any particular time. If controllers cannot directly communicate with aircraft, they may be required to use alternate means of communications, such as using the airline operations offices to act as communication intermediaries. This limitation requires that controllers who rely on HF communications not place the aircraft in a position where immediate communications may be required.

Experiments have been conducted using satellite transmitters and receivers to try to overcome the limitations of HF/VHF/UHF transmission systems. Satellite transmitters utilize frequencies located well above UHF and are also line-of-sight. But if sufficient satellites can be placed in orbit, communications anywhere in the world will be virtually assured. Satellite communications have already been successfully tested on overseas flights and should become commonplace within a few years.

20.23 Voice Communications Procedures

As previously stated, virtually, every ATC communication is currently conducted by voice. Initial clearances, taxi and runway instructions, pilot requests, and controller instructions are all primarily conducted utilizing voice. This type of communication is fairly unreliable due to both the previously mentioned technical complications and communications problems inherent in the use of one common language in ATC. Although all air-traffic controllers utilize English, they may not be conversationally fluent in the language. In addition, different cultures pronounce words and letters in different ways. Many languages do not even use the English alphabet. And every controller has idioms and accents peculiar to their own language and culture. All these factors inhibit communications and add uncertainty to ATC communications.

When using voice radio communications, it can be very difficult for a controller to insure that correct and accurate communication with the pilot has occurred. Pilots normally read back all instructions, but this does not solve the miscommunication problem. Informal and formal surveys lead experts to believe that there are literally millions of miscommunications worldwide in ATC every year. Obviously, most of these are immediately identified and corrected, but some are not, leading to potential problems in the ATC system.

20.24 Electronic Data Communications

In an attempt to minimize many of these communications problems, various schemes of nonvoice data transmission have been tried in ATC. The most rudimentary method still in use is the ATCRBS transponder. If the aircraft is properly equipped, its identity and altitude will be transmitted to the ground station. Existing ATCRBS equipment is currently incapable of transmitting information from the controller to the aircraft. The new Mode-S transponder system will be able to transmit more information in both directions. This information might include aircraft heading, rate of climb/descent, airspeed, and rate of turn, for example. Mode-S should also be able to transmit pilot requests and controller instructions. Mode-S is slowly being installed on the ground and airborne equipment is gradually being upgraded. Until a sufficient number of aircraft have Mode-S capability, the ATCRBS system will still be utilized.

An intra-airline data communications system known as the *aircraft communications addressing and reporting system* (ACARS) has been utilized by the airlines for years to send information to and from properly equipped aircraft. ACARS essentially consists of a keyboard and printer located on the aircraft, and corresponding equipment in the airline's flight operations center. ACARS is currently used by the airlines to transmit flight planning and load information. A few ATC facilities are now equipped to transmit initial ATC clearances to aircraft using ACARS. This limited service will probably be expanded until Mode-S becomes widespread.

20.25 Controller Coordination

Because controllers are responsible for the separation of aircraft within their own sector, they must coordinate the transfer of aircraft as they pass from one sector to another. In most situations, this coordination is accomplished using voice communications between controllers. In most cases, unless the controllers are sitting next to each other within the same facility, coordination is accomplished using the telephone.

Hand-offs are one form of coordination and consist of the transfer of identification, communications, and control from one controller to the next. During a hand-off, the controller with responsibility for the aircraft contracts the next controller, identifies the aircraft, and negotiates permission for the aircraft to cross the sector boundary at a specific location and altitude. This is known as the transfer of identification. Once this has been accomplished, and all traffic conflicts are resolved, the first controller advises the pilot to contact the receiving controller on a specific radio frequency. This is known as the transfer of communication. Separation responsibility still remains with the first controller until the aircraft crosses the sector boundary. Once the aircraft crosses the boundary, separation becomes the responsibility of the receiving controller. This is known as the transfer of control.

To simplify hand-offs, standardized procedures and predefined altitudes and routes are published in a document known as *letter of agreement* (LOA). LOAs simplify the coordination process because both controllers already know what altitude and route the aircraft will be utilizing. If the controllers wish to deviate from these procedures, they must agree to an *approval request (appreq)*.

The transferring controller usually initiates an appreq verbally, requesting a different route and/or altitude for the aircraft to cross the boundary. If the receiving controller approves the appreq, the transferring controller may deviate from the procedures outlined in the LOA. If the receiving controller does not approve the deviation, the transferring controller must amend the aircraft's route/altitude to conform to those specified in the LOA.

There are many problems inherent in this system of verbal communication/coordination. When both controllers are busy, it is very difficult to find a time when both are not communicating with aircraft. Controllers are also creatures of habit, and may sometimes "hear" things that were not said. There are many situations in ATC where aircraft are delayed or rerouted, not due to conflicting traffic, but because required coordination could not be accomplished in a timely manner.

Automated hand-offs have been developed in an attempt to reduce these communication/coordination problems. An automated hand-off can be accomplished if the two sectors are connected by a computer, and the routes, altitudes, and procedures specified in the LOA can be complied with. During an automated hand-off, as the aircraft nears the sector boundary, the transferring controller initiates a computer program that causes the aircraft information to be transferred and start to flash on the receiving controller's radar display. This is a request for a hand-off and implies that all LOA procedures will be complied with. If the receiving controller determines that the hand-off can be accepted, computer commands are entered that cause the radar target to flash on the transferring controller's display.

This implies that the hand-off has been accepted, and the first controller then advises the pilot to contact the next controller on the appropriate frequency. Although this procedure may seem quite complex, in reality it is very simple and efficient, and reduces voice coordination between controllers significantly. Its primary disadvantage is that the route and altitudes permissible are reduced and the ATC system becomes less flexible overall.

20.26 Flight Progress Strips

Virtually all verbal communications are written down for reference on paper flight progress strips. Flight strips contain most of the pertinent information concerning each aircraft. When a controller verbally issues or amends a clearance or appreqs a procedural change with another controller, this information is handwritten on the appropriate flight progress strip. Flight progress strips are utilized so that controllers do not need to rely on their own memory for critical information. Flight strips also make it easier for other controllers to ascertain aircraft information if the working controller needs assistance or when a new controller comes on duty. Due to differences in each controller's handwriting, very specific symbology is used to delineate this information. Figure 20.1 contains examples of some common flight strip symbology.

Symbol	Meaning
↑	Climb and maintain
↓	Descend and maintain
⟶M⟶	Maintain
RR	Report reaching
RL	Report leaving
Rx	Report crossing
⤊	Cross at or above
⤋	Cross at or below
X	Cross
@	At
C	Contact
≢	Join an airway
>	Before
<	After
¢	Cancel flight plan
RV	Radar vectors

FIGURE 20.1 Sample flight progress strip symbology.

20.27 Flight Information Automation

The constant updating of flight progress strips and the manual transferring of information consume much of a controller's time, and may necessitate the addition of another controller to the sector to keep up with this essential paperwork. This process is forecast to become somewhat more auto-mated in the future. Future ATC systems have been designed with flight strips displayed on video screens. It is theoretically possible that as controllers issue verbal commands, these commands will be automatically interpreted and the electronic flight strips will be updated. Future enhancements may make it possible for the controller to update an electronic flight strip, and that information might be automatically and electronically transmitted to the pilot or even to the aircraft's flight control system.

20.28 Controller Responsibilities in the Air-Traffic Control System

Controllers are responsible for the separation of participating aircraft within their own sector. They also provide additional services to aircraft, such as navigational assistance and providing weather advisories. Additional responsibilities placed on the controller include maximizing the use of the airspace and complying with air-traffic management (ATM) procedures.

To accomplish these tasks, the controller must constantly monitor both actual and predicted aircraft positions. Due to rapidly changing conditions, a controller's plan of action must remain flexible and subject to constant change. The controller must continuously evaluate traffic flow, plan for the future, evaluate the problems that may occur, determine appropriate corrective action, and implement this plan of action. In the recent past, when traffic moved relatively slowly and the airspace was not quite as crowded, a controller might have minutes to evaluate situations and decide on a plan of action. As aircraft speeds have increased, and the airspace has become more congested, controllers must now make these decisions in seconds. As in many other career fields, experts feel that the current system may have reached its effective limit, and increased ATC-system expansion will not be possible until many of the previously mentioned tasks become automated.

20.29 Future Enhancements to Air-Traffic Control Systems

ICAO has recently agreed that GNSS should become the primary aircraft-positioning system. It appears at this time that uncorrected GNSS systems should supplant VORTAC as both an en route and a nonprecision instrument approach aid. WAAS should permit GNSS to be used as a CAT I preci-sion approach replacement for ILS. LAAS should correct GNSS to meet CAT II and possibly CAT III ILS standards.

The GNSS system can be modified to permit the retransmission of aircraft position back to ATC facilities. This system, known as *automatic-dependent surveillance* (ADS), should supplant radar as a primary aircraft-surveillance tool. Not only should this system be more accurate than radar surveil-lance, but also it will not have the range and altitude limitations of radar and will be able to transmit additional data both to and from the controller. This might include pilot requests, weather information, traffic information, and more. ADS has already been demonstrated experimentally and is being tested for aircraft separation over oceanic airspace.

Many other changes are planned. ICAO has completed a future air navigation system (FANS) that defines changes to navigation, communication, and surveillance systems. FANS is a blueprint for the future of international aviation and ATC. Table 20.5 summarizes FANS.

TABLE 20.5 Future ATC and Navigation System Improvements

Function	Type	Current Standard	Future Standard
Navigation	En route	VORTAC	GNSS
	Approach	ILS	Augmented GNSS
Communication	Short range	VHF and UHF	VHF and UHF
	Long range	HF	Satellite
Surveillance		Radar	Radar and automatic-dependent surveillance
Data link		ATCRBS	Mode-S

Once these improvements have taken place, automated ATC systems can be introduced. Various research programs into automation have been initiated by many ATC organizations, but it is highly likely that it will be well past the year 2010 before automated systems such as Automated Enroute Air-Traffic Control (AERA) can be designed, constructed, installed, and made operational. In the meantime, the FAA has begun to study an ATM system called "free flight."

The concept of free flight has been discussed since the early 1980s. Only since the demise of the FAA's planned advanced automation system (AAS) has it come into favor in the United States. Free flight proposes to change ATC separation standards from a static, fixed set of standards to dynamic separation that takes into account aircraft speed, navigational capability, and nearby traffic. Based on these parameters, each aircraft will be assigned a "protected" zone that will extend ahead, to the sides, above, and below the aircraft. This zone will be the only separation area protected for each aircraft. This differs from the current system that assigns fixed airway dimensions and routes for separation.

Assuming that each aircraft is equipped with an accurate flight management system (FMS), free flight proposes that each aircraft transmit to ground controllers its FMS-derived position. On the ground, computer workstations will evaluate the positional data to determine whether any aircraft conflicts are predicted to exist, and if so, offer a resolution instruction to the air-traffic controller. The controller may then evaluate this information and pass along appropriate separation instructions to the aircraft involved.

The free flight concept is still being developed, but if found feasible will soon be implemented at high altitudes within the U.S. airspace structure. As confidence in the system is gained, it will likely be extended overseas and into the low-altitude flight structure.

Suggested Readings

Federal Aviation Administration. (1976). *Takeoff at mid-century.* Washington, DC: Department of Transportation.

Federal Aviation Administration. (1978). *Bonfires to beacons.* Washington, DC: Department of Transportation.

Federal Aviation Administration. (1979). *Turbulence aloft.* Washington, DC: Department of Transportation.

Federal Aviation Administration. (1980). *Safe, separated and soaring.* Washington, DC: Department of Transportation.

Federal Aviation Administration. (1987). *Troubled passage.* Washington, DC: Department of Transportation.

International Civil Aviation Organization. (various). *ICAO Bulletin*, Montreal, Canada.

International Civil Aviation Organization. (various). *Annexes to the Convention of International*.

Jackson, W. E. (1970). *Civil Aviation. The federal airways system* (Institute of Electrical and Electronic Engineers). Montreal, Canada: Author.

Nolan, M. S. (2003). *Fundamentals of air-traffic control*. Belmont, CA: Brooks-Cole.

Other FAA Publications

Aeronautical Information Publication (1988)

Air-Traffic Handbook (1995)

National Airspace System Plan (1988)

21

Air-Traffic Controller Memory

Earl S. Stein
Federal Aviation Administration

Daniel J. Garland
The Art Institute of Atlanta

John K. Muller
IBM Corporation

Human beings are fallable elements in any system. They are also a resource on which we continue to depend. System designers cannot seem to replace them with automation. Forgetting is part of what people do. Forgetting key elements of information in a dynamic, information-rich environment is all too common. Short-term memory, often referred today as working memory, is highly vulnerable when intervening events disrupt it. Add a distraction here and a little time pressure there, and presto, people forget even very important things (Lauber, 1993, pp. 24–25).

The authors originally wrote this chapter and published it years ago to raise awareness. The current document examines the memory literature in light of real-world air-traffic controller (ATC) memory requirements in tactical operations as they exist today and are anticipated for the future. The chapter presents information on working memory processes in ATC tasks and shows the vulnerability of these processes to disruption. This chapter focuses on the role that working memory plays in ATC performance and emphasizes on the mechanisms of working memory, with its limitations and constraints. It also examines how controllers might overcome or minimize memory loss of critical ATC information. Awareness of the limitations and constraints of working memory and the conditions under which they occur is critically necessary to avoid situations that can result in airspace incidents and accidents. However, controllers should not be the last defense against predictable memory errors. If we design systems correctly, then the operators' need for memory may be realistic rather than optimistic.

Current predictions suggest that more traffic will occur and there may or may not be more controllers to work on it. Planners both within and beyond the Federal Aviation Administration (FAA) believe and assume that technology is the solution so that controllers can do more with less. Perhaps, we may offload the separation responsibilities on aircrews, who may then self-separate using airborne-based technology. This is currently anticipated by an intergovernmental organization known as the Joint Planning Development Office (JPDO, 2005).

According to the JPDO, "In designated airspace, flight crews can cooperatively separate aircraft from each other using on-board automation capabilities. Ground automation may facilitate the interaction of aircraft, but pilots will retain managerial control over the flight P.12, JPDO." This approach and philosophy represents a potential meteoric change in control responsibilities. It implies some role for ATCs in a passive monitoring capacity. Human operators generally are not at their best in a monitoring role, and lessened active involvement may impact both situational awareness and memory. Metzger and Parasuraman (2001) demonstrated that controllers placed in a passive monitoring task had poorer recall of aircraft altitudes, than when actively controlling traffic. According to Charlie Keegan (2005), the JPDO is already planning for the resolution of all of these issues and has established an integrated product team to deal with them.

Unless the issues are resolved, they will have an impact on operator workload and memory, as well as raise vigilance issues that are yet to be fully resolved or researched. The final section of this chapter briefly deals with some of the potential human-factors consequences of new automated technologies on ATC working memory.

During the last several decades with some dip in demand after the events surrounding September 11, 2001, the ATC system has been strained owing to increases in the amount and changes in the distribution of air traffic in the United States. The FAA continues to anticipate that we will see increases well through the first quarter of the 21st century.

The current system has evolved from the nonradar days of the late 1940s to the present time. The infrastructure of the current system was not designed to handle the predicted traffic loads. Support technology for maintaining the National Airspace system has grown by fits and starts over that time period, where today, what is actually in the field and working, varies considerably from facility to facility across the country.

The safe, orderly, and expeditious flow of air traffic is traditionally the fundamental objective of air-traffic control (Federal Aviation Administration, 1989). There were 63.1 million instrument operations logged by the FAA's airport control towers in 2004 (FAA, 2006) with forecasts for 78.9 million Ops in 2016. The FAA forecasts 1.6 trillion available seat miles in 2016 when compared with 953.6 billion in 2004, with a growth of 4.9% a year. This assumes the validity of using past data to predict future activity. Further, the FAA estimates that, in the United States alone, delays resulting in air-traffic problems result in economic losses of over 5 billion per year.

Human ATCs are the backbone of the system. It is their strengths that keep the system going as well as their qualities that may cause it to break down if they do not get the appropriate tools to keep up. One of these human qualities is the fact that we have limited working memory capacity, which is further constrained by the dynamic nature of the control process. Things keep happening, often causing interference with the coding and storage process (Stein & Bailey, 1994). Working memory is all about the here and now. While it can be influenced by what you already know, it is most characterized by events surrounding the individual at the moment. We know that among pilots it is likely that as many as 6% of errors that they report through the aviation-safety reporting system are owing to memory failures (Nowinsky, Holbrook, & Dismukes, 1987). Unfortunately, there is no data available like this based on self-reporting for ATCs, because pilots are given incentives to report their mistakes, but controllers at the time when that paper was written, were not.

Working memory allows the controller to retain intermediate (i.e., transient) products of thinking and the representations generated by the perceptual system. Functionally, working memory is where all cognitive operations obtain their information and produce their outputs (i.e., responses). It allows the controller to retain relevant information for tactical operations. Such tactically relevant information may include altitude, airspeed, heading, call sign, type of aircraft, communications, weather data, runway conditions, current traffic "picture," projected traffic "picture," immediate and projected conflicts, and so forth. Working memory is dependent on long-term memory for cognitive tasks such as information organization, decision making, and problem solving. It is also heavily constrained and limited by time-dependent processes such as attention, capacity, and forgetting. It can be influenced by how material is presented and how that presentation influences the attention that operator focuses on the display (Foos & Goolkasian, 2005). It may even be influenced by what you brought with you to work that day, such as personal problems or issues. Essentially, working memory permeates every aspect of the controller's ability to process air-traffic information and control live traffic.

The air-traffic control environment is characterized by a continuous sequence of ever-changing, transient information (e.g., series of aircraft being handled by an ATC), which must be encoded, retained primarily for tactical use (3–5 min), and secondarily for strategic planning and subsequently discarded. The ability to manage flight information is complicated by limitations and constraints of human memory, in particular working memory (Finkelman & Kirschner, 1980; Kirchner & Laurig, 1971; Wickens, 1992). Working memory limitations and constraints are routinely severe enough to significantly degrade performance. Degraded performance can lead to operational errors in the FAA's ATC system.

The FAA invests considerable energy in attempting to discover the causes and methods for preventing actual and potential operational errors of air-traffic control (Operational Error Analysis Work Group, 1987). There will be a need for effective transition training for controllers who must be able to use new technologies and procedures to control live traffic. New technology by whatever name it is given will change the way ATCs do their job. New technologies and procedures may impose requirements on the controller which are incompatible with the way he or she processes information and the way a controller attends, perceives, remembers, thinks, decides, and responds. We are already witnessing this as new systems are added in a piecemeal fashion to the legacy systems in place. This reached a point in tower cabs years ago where executives have mandated that there shall be no new glass (meaning additional displays) until system developers can somehow integrate the development and establish the requirements for the information controllers which they really need.

The cognitive requirements of air-traffic control as it currently exists have involved the processing of a great volume of dynamically changing information (Kirchner & Laurig, 1971; Means et al., 1988). Cognitive processing of flight data (i.e., call/sign, aircraft type, sector number, planned route, assigned speed, heading, altitude, time over posted fix, etc.) is crucial to virtually every aspect of a controller's performance. It is essential for the controller to be able to manage available information resources in such a way that accurate information is available when needed. The ease with which information (e.g., flight data) is processed and remembered depends on how it is displayed and how the operator interacts with it. As information displays change with evolving technology, controllers may process flight information in different ways, potentially affecting ATC performance and possibly influencing flight safety and efficiency. It is important to understand these cognitive processes.

21.1 Human Information-Processing System

Researchers have studied memory issues for a considerable period of time, as shown by a three-volume work providing an annotated compilation of 798 references dealing with short-term memory, covering the time period from 1959 to 1970 (Fisher, 1969, 1971; Fisher & Wiggins, 1968). Unfortunately, many of the early memory studies had nothing to do with understanding. In fact, early studies often deliberately employed nonsense syllables (also known as CVC Trigrams such as "PAG") because they were incomprehensible (Hopkin, 1982). Studies of this type did not require the participants to incorporate new material with existing knowledge, and therefore, have no direct relevance to memory for complex material in applied operational settings. Such studies gained popularity in academia where research in memory is viewed as pure rather than applied science. Memory as a research topic has not faded with time, as shown by the frequencies of articles, chapters, and books on memory (e.g., Baddeley, 1996; Cowan, 1995; Healy & McNamara, 1996; Jonides, 1995; Logie, 1995; Lyon & Krasnegor, 1996; Shaffer, 1993; Squire, Knowlton, & Musen, 1993).

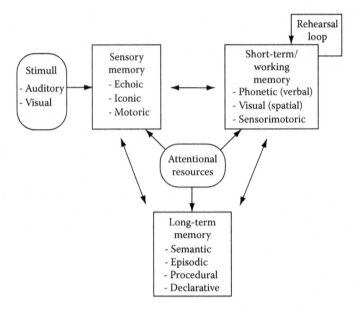

FIGURE 21.1 Memory model.

Human information processing may provide clues to how working memory influences controller behavior. Several information-processing models have been developed (e.g., Broadbent, 1958; Card, Moran, & Newell, 1986; Smith, 1968; Sternberg, 1969; Welford, 1976; Wickens, 1984). Each assumes various stages of information processing, characterized by stage-specific transformations on the data. The present approach follows a simplified description of human information processing, consisting of three interacting subsystems, similar to the Card et al. (1986) model. This model was originally proposed by Atkinson and Shiffrin (1968) (Figure 21.1).

The model is based on the premise that the human operator is actively reaching out and taking in necessary information. This information comes through the senses and if the individual attends to it for whatever reason, it will make it into working memory. In working memory, it is fragile and is either used and discarded or processed to the point that it makes it into long-term memory.

The subsystems dynamically interact. The three subsystems may interact in series or in parallel. For example, some tasks (e.g., marking the flight-strip in response to an altitude change) require serial processing. Other tasks (e.g., radar/flight-strip scanning, flight-strip marking, and ground-air-ground communications) may require integrated, parallel operation of the three subsystems.

The following brief description of information processing in the ATC system demonstrates the interaction of the three information-processing subsystems. Human information processing is a necessary component of all ATC operations as they currently exist. Although technical support is necessary for communication between the ATC system and the aircraft, the controller is the primary information processor. Technical equipment supports the guidance of aircraft from the ground. It provides feedback that serves to guide the execution of controller instructions and provides new formation about the changed situation for guidance of future controller actions.

After receiving information about the present condition of the traffic, the controller evaluates the situation based on safety and efficiency criteria as well as factoring in his or her comfort zone for how tightly he or she wants to press the edges of the safety envelope. If a potential conflict arises, which demand intervention, the controller takes the necessary control actions. The control actions, once implemented, change the situation, providing new information to the controller. The control actions require two basic information-processing steps. First, the present situational information is received, analyzed, and evaluated. The operator must have an adequate knowledge base, training, and experience.

Second, the controller responds based on the available data, training, and experience. In addition to the immediate demands on information processing, the controller must process additional system information derived from coordination between different controllers.

This coordination is essential for traffic planning and keeping the "picture" of the traffic under control (for more detailed information, see Ammerman et al., 1983; Ammerman & Jones, 1988; Bisseret, 1971; Kinney, Spahn, & Amato, 1977; Kirchner & Laurig, 1971; Means et al., 1988). For example, the controller detects a potential conflict between TCA483 and TWA358. If the position and the facility employ flight data strips, then the controller may place these flight strips next to each other to call attention to them. This is a memory cue that the controller may use. The TCA483 is contacted and instructed to climb to altitude 320. The controller crosses out 300 and writes 320, the new proposed altitude. Concurrently, TWA358 informs the controller that it has reached the assigned cruising altitude of 300 and the controller makes a notation next to the altitude.

This illustration is an obvious simplification of the ATC system. In practice, there would be a far greater number of aircraft in the traffic pattern, and the controller would potentially have to resolve a number of conflicts simultaneously. However, this illustration provides a characterization of the information-processing components in the ATC system and the basis for a closer examination of the mechanisms underlying information processing, with particular attention to cognitive research on memory and its application to the ATC system. As we choose to change the technology, moving to either a stripless or electronic strip-based system, the cues available for short-term memory support will change as well. This is neither inherently bad nor good. It will depend on what demands are made on the controller and what tools he or she has available to support performance against those demands. Controllers now using the User Request Enroute Tool (URET) that has electronic strips find that they can work enroute traffic without the paper strips.

21.2 Air-Traffic Controller Memory

Controllers are human, and human memory can be viewed as a continuously active system that receives, retrieves, modifies, stores, and acts on information (Baddeley, 1976, 1986; Klatzky, 1980). Researchers have referred to working memory as the "site of ongoing cognitive activities. These include the meaningful elaboration of words, symbol manipulation such as that involved in mental arithmetic and reasoning" (Klatzky, 1980, p. 87). The discussion here focuses more on the transient characteristics of working memory than on long-term memory. This emphasis is based on the psychological knowledge that long-term memory storage and retrieval are relatively automatic processes. They present fewer formidable disruptions to performance (Baddeley, 1976, 1986; Klatzky, 1980; Wickens, 1992). While memory lapses are a common cause underlying controller systems errors. The majority of these are failures in working memory and not long-term memory.

Working memory is severely affected by the limitations and constraints of limited processing resources. Wickens (1992) emphasized that occasional limitations of, and constraints on working memory are often responsible for degraded decision making. Working memory allows the controller to retain intermediate (i.e., transient) products of thinking and the representations generated by the perceptual system. The mechanisms of working memory and the nature of its limitations and constraints that directly and/or indirectly influence ATC are the focus of this chapter and are presented in the following sections covering memory codes, code interference, attention, capacity, chunking, organizing, and forgetting.

21.2.1 Memory Codes

Immediately after the presentation of an external visual stimulus such as an aircraft target with accompanying data tag on the radar display, a representation of the stimulus appears in the visual image store (i.e., iconic memory) of the perceptual system. There is also a corresponding auditory image store (i.e., echoic memory) for external auditory stimulus (e.g., ground-air-ground communications). These sensory codes or memories are representations of external physical stimuli. They are vital for working memory, in that they sufficiently prolong the external stimulus representations (usually measured in milliseconds) for relevant processing of the stimulus representations to take place in working memory (Card et al., 1986). The sensory memories, although not demanding the operator's limited attentional resources, are important for partial activation of the visual (i.e., iconic) and phonetic (i.e., echoic) primary codes in working memory (Baddeley, 1986; Wickens, 1984, 1992).

Although the sensory codes are generated exclusively by external physical stimuli, primary visual and phonetic codes may be activated by external stimuli via the perceptual system (i.e., sensory codes) or from inputs into working memory from long-term memory. The primary visual and phonetic codes, along with semantic and motoric codes, form the foundation of our attention demanding working memory, which is necessary for all ATC tactical operations (Baddeley & Hitch, 1974).

Semantic codes take the process one step further. Semantic codes are abstract representations based on the meaningfulness of the stimuli (e.g., the controller's knowledge of specifics of the sector map, the location of data on the flight strip, aircraft characteristics, etc.). They are constructed using retrieval from long-term memory to attach meaning to new stimuli.

Motoric codes are sensory and motor, usually involving some sort of movement or anticipated movement, representations of actions, which are involved in the encoding of past and future activities (Koriat, Ben-Zur, & Nussbaum, 1990). The encoding of future actions, which has been a neglected issue in memory research, is important for air-traffic control operations. It is gaining more acceptances with the research being conducted under the heading of situational awareness. This work in the cockpit environment has made inroads into ATC research. Endsley (1990) coined a new term for this: "Prospective Memory" (Sollenberger & Stein, 1995). For example, a controller instructs TWA348 to climb to a new cruising altitude of 290, having forgotten to previously instruct AAL584 to descend from 290 to 270 for eventual hand off. This forgotten-to-be-performed action, located in prospective memory (to use the language of situational awareness), may subsequently result in an airspace conflict. In the subsequent paragraphs, the visual, phonetic, semantic, and motoric codes have been linked more closely to what the ATCs actually do. These memory codes play a significant role in the ATC process. Information is provided about the characteristics of these codes and their susceptibility for disruption and enhancement.

21.2.2 Visual Codes

Visual representations or images of spatial information (e.g., a controller's "pictorial" mental representation of an aircraft's location, orientation,) are based on scanning the available information) and are normally maintained in working memory using visual codes (Wickens, 1992). However, visual input is not necessary or sufficient for the generation of visual representations. External visual stimuli do not automatically produce visual images. That is, simply looking at something will not ensure its processing working memory. Kosslyn (1981) reported evidence indicating visual images can be generated by nonvisual sources, such as information that has been experienced and subsequently stored in long-term memory (e.g., sector map, previous conflict situations), and by verbal (i.e., phonetic) stimulus material.

Primary visual codes are highly transient in nature, requiring a great deal of effortful attention. They demand processing (i.e., maintenance rehearsal) to persist in working memory (Goettl, 1985; Posner, 1973, 1978; Posner & Mitchell, 1967). Research conducted by Bencomo and Daniel (1975), using a same–different recognition task, suggests that visual codes (i.e., visual representations or images) are more likely to persist when processing involves more natural visual/spatial materials (e.g., sector map, radar display), than verbal or auditory materials. For example, it is easier for a controller to visualize incoming weather if he or she has a weather display, when compared with a textual description of the weather forecast for the sector, and this was recently demonstrated. Research conducted at the FAA Human Factors Laboratory in Atlantic city showed that controllers with current weather graphically displayed, when compared with a text-based weather update, could move between 6% and 10% more traffic safely and expeditiously (Ahlstrom & Friedman-Berg, 2005; Ahlstrom, Keen, & Mieskolainen, 2004).

21.2.3 Phonetic Codes

Verbal information (e.g., the controller at Chicago Center instructs TWA484 to "descend and maintain one thousand, report leaving one two thousand") is normally maintained in working memory by phonetic or auditory rehearsal (Wickens & Flach, 1988). This process in working memory is known as "maintenance rehearsal" (also called Type I, primary, or rote rehearsal) and is used only to maintain information in working memory, presumably by renewing the information before it is subject to time-dependent loss (Bjork, 1972; Craik & Watkins, 1973; Klatzky, 1980). The phonetic primary code is automatically generated from an echoic sensory code and represents continued processing at a shallow, acoustic level (Wickens, 1992). In addition, Conrad (1964) demonstrated that phonetic codes can be automatically generated from visual stimuli (i.e., iconic codes). Conrad's (1964) results indicated that when subjects were to recall visually presented letters, the recall intrusion errors tended to be acoustic rather than visual. For example, an ATC may have a tendency to write, by mistake, letters such as Z instead of T. There is more potential intrusion or opportunity for error based on the associated sounds rather than on visual appearance. Further, Conrad and Hull (1964) demonstrated that recall information that was phonetically similar created greater recall confusion than information that was phonetically dissimilar.

A series of laboratory studies on phonetic codes and information presentation concluded that verbal working memory can be enhanced by employing speech (i.e., verbal information) as an information display mode (Murdock, 1968; Nilsson, Ohlsson, & Ronnberg, 1977; Wickens, Sandry, & Vidulich, 1983). This conclusion is based on the facts that echoic (i.e., auditory) memory is retained longer that iconic (i.e., visual) memory, and that auditory displays are more compatible with the auditory nature of maintenance rehearsal in working memory (Wickens, 1992).

There are also significant human-factors implications of using an auditory information display for the presentation of transient information to be used in working memory. Such information will be less susceptible to loss when presented via auditory channels, such as natural or synthetic speech. For example, Wickens et al. (1983) demonstrated that pilots can retain navigational information better with auditory display when compared with visual display, and this finding was enhanced under high-workload conditions. These findings suggest that auditory display of information may be advantageous when rapid information presentation is necessary; the information is of a transient nature, is not overly complex, and visual display space cannot afford further cluttering (Wickens, 1992). However, auditory displays present formidable challenges to the human-factors specialist. They cannot be easily monitored on a time-sharing basis, and once the information is gone from the working memory, it cannot be returned similar to the visual displays.

21.2.4 Semantic Codes

Semantic codes are responsible for representing information in working memory in terms of meaning rather than physical (i.e., auditory, visual) attributes. They provide the critical link between working memory and the permanent long-term memories. Card et al. (1986), when noting the intimate association between working memory and long-term memory, suggested that "structurally, working memory consists of a subset of the elements in long-term memory that has become activated" (pp. 45–47). Semantic codes are primarily responsible for information storage and organization in working memory, and subsequently in long-term memory. The creation and use of semantic codes involves a process that is substantively different from maintenance rehearsal. This is elaborative rehearsal.

Elaborative rehearsal involves deep, meaningful processing in which new information is associated with existing meaningful knowledge in long-term memory. This processing, in contrast to the previously cited maintenance rehearsal, facilitates the retention of information in working memory and enhances information transfer to long-term memory by way of semantic codes. Elaborative rehearsal in working memory requires thinking about information, interpreting the information, and relating the information to other information in long-term memory. These processes enhance the retrieval of information from long-term memory and facilitate planning future actions (Klatzky, 1980).

Semantic codes afford the working memory the ability to actively retain and analyze information. Wingfield and Butterworth (1984) suggested that, rather than passively retaining auditory and visual information in working memory, we are "continuously forming hypotheses about the structure of what they are hearing and forming predictions about what they have yet to hear. These are working hypotheses, either confirmed or modified with the arrival of new information" (p. 352). Klapp (1987) noted that working memory actively formulates and stores hypotheses, resulting in abstract representations (i.e., semantic codes) in working memory, in addition to auditory or visual codes.

Semantic codes are vital for the organization, analyses, and storage of ATC tactical information in working memory and long-term memory. They are the invaluable link between working memory and long-term memory, providing and facilitating the ability to actively manipulate and analyze data, and to generate decision-making and problem resolution alternatives. For example, for an ATC to make an informed and accurate assessment of a potential conflict between two aircraft, a great deal of flight information is required about the two aircraft (e.g., altitude, heading, airspeed, type of aircraft, current traffic "picture," projected traffic "picture," etc.).

These flight data must in turn be analyzed and interpreted against a knowledge and experience database in long-term memory to accurately construct and assess a "pictorial" mental representation of the current and projected airspace. Alternative hypotheses about the traffic situation can be generated from long-term memory and retained in working memory to be analytically integrated with the flight data. This process of hypothesis formulation and evaluation is complicated by the limitations and constraints of working memory and long-term memory decision biases (Wickens & Flach, 1988).

Semantic coding is both a strength and weakness of the human operator. Under time and space pressure, he or she may not have the resources to create new codes and must work with what is available. Semantic codes can be developed over time and retrieved from long-term memory as powerful tools. The name of the game in ATC is decisiveness and accuracy. Controller training essentially emphasizes the development of semantic codes that some call templates or schemas that controllers can use rapidly to resolve their more common tactical issues. Loft, Humphreys, and Neal (2004) studied the controllers' conflict detection after participants were exposed to scenarios in which aircraft pairs established patterns where they repeatedly conflicted in the airspace or remained safely separated. The researchers found that detection of a conflict was faster when a pair of aircraft resembled a pair that had conflicted in the past. Their response was slower when a new conflict happened between a pair of aircraft that had safely separated in the past. This suggested that there was a process that invoked memory patterns for controllers' prior experiences. Such cognitive tools can be both empowering as well as a problem if the controller depends too much on memories of past events.

Researchers have been searching for new tools to help controllers use their memory more effectively. Stein and Sollenberger (1995) summarized the FAA's program on controller memory at the eighth Biannual Symposium on Aviation Psychology. The program has focused on the system as it exists today and how controllers can use tools available to them to avoid systems errors. Most complex command and control systems involve memory processing using multiple levels of coding. In addition, semantic coding, motoric codes also play an important role.

Recent research on memory for action events has focused on memory for past activities (e.g., Anderson, 1984; Backman, Nilsson, & Chalom, 1986; Cohen, 1981, 1983; Johnson, 1988; Kausler & Hakami, 1983; Koriat & Ben-Zur, 1988; Koriat, Ben-Zur, & Sheffer, 1988). A consistent and general finding of these studies is that memory for performing a task is superior to that for verbal materials, owing to the beneficial effects of motoric enactment. That is, the process of physically performing a task seems to enhance the encoding of and subsequent memory for the task. You learn by doing. The superior memory for performing tasks "has been generally attributed to their multimodal, rich properties, assumed to result in richer memorial representations than those formed for the verbal instructions alone" (Koriat et al., 1990, p. 570). The more active a controller is in the process, both psychologically and physically, the more likely it is that memories will be sufficiently coded and stored so that they can be effectively used.

These results have direct human-factors implications for the use of automation in ATC. Several researchers (e.g., Hopkin, 1988, 1989, 1991b; Narborough-Hall, 1987; Wise & Debons, 1987; Wise et al., 1991) suggested that routine task performance facilitates controller tactical operations (e.g., the understanding of and the memory for traffic situations). Hopkin (1991b) asserted that physical interaction with the flight progress strip was essential to support a controller's memory for immediate and future traffic situations.

However, a series of studies conducted by the University of Oklahoma and the Civil Aero medical Institute have called this belief into question. They have demonstrated that under very limited simulated conditions the controllers can work without strips without significant loss to their working memory (Vortac, Edwards, Fuller, & Manning, 1993, 1994, 1995; Vortac, Edwards, & Manning, 1995). These studies were conducted primarily with FAA academy instructors and using a low to moderate fidelity simulator. Zingale, Gromelski, and Stein (1992) attempted to study the use of flight strips using aviation students at a community college. The study demonstrated the importance of using actual controllers, because the students could not keep up with or without strips. Zingale, Gromelski, Ahmed, and Stein (1993), in a follow-on study using controllers and the same low-fidelity simulator, found that controllers did find the strips useful but were uncomfortable using a PC-based simulation that required them to key in their responses. The FAA decided to maintain paper flight strips as operational tools in the early 1990s. This decision has gradually changed with the advent of systems such as the URET, which provides electronic strips in the enroute environment. Also, a number of air-traffic control towers received FAA authorization to experiment with a stripless environment where controllers could use a notepad, if they wished, as a memory tool. Contrary to Hopkin's assertions, controllers have managed to separate aircraft under these conditions. Strips or some other cue can stimulate both memories for past actions as well as serve as a tool for actions that controllers intend to take later.

The flight-strip studies do not resolve other automation issues that could reduce the controllers' physical and cognitive activity level. People learn by doing and they retain knowledge and skill bases by using what they know. Systems which move a controller toward a more passive role will possibly threaten both current memory and memory for prospective actions (that the machine will have).

Memory for future activities is known as prospective memory (Harris, 1984; Wilkins & Baddeley, 1978; Winograd, 1988). In some cases, information for future control actions need only be retained for a short period of time. A recent study investigating the nature of the representations underlying the memory for future actions (i.e., prospective memory) found a significant beneficial effect of imaginal-motoric enactment of the future activity (Koriat et al., 1990). This involves the controller's thinking through the steps he or she will take to include both the cognitive and physical aspects of the task. This imaginal enactment of the future activity is consistent with the research on memory for past activities. This beneficial effect can also be attributed to the multimodal and contextual properties of having actually performed the task. It is also seen with the intentional (or unintentional) visualization of the task, which promotes visual and motor encoding (Backman & Nilsson, 1984; Koriat et al., 1990).

Koriat et al. (1990) suggested that the process of encoding future activities involves an internal, symbolic enactment of the tasks, which enhances the memory. This implies that rehearsal (i.e., maintenance and/or elaborative) or repeated internal simulation of the procedure to be performed will enhance the memory at the time of testing, in greatly the same manner that maintenance rehearsal retains the verbal material in working memory. Koriat et al. (1990) also suggested that if rehearsal takes the advantage of the modality-specific properties of the future task, not only will the memory for the content be enhanced, but the memory retrieval cues will be enhanced under proper conditions. The question will remain as to how to motivate operators to do this to remain engaged.

Given the previous example of a potential conflict between TCA483, AAL284, and TWA343 before TCA483 is displayed on the radar display, the controller is responsible for retaining and eventually conveying this information to the relief controller on the next shift, along with additional information concerning the status of other aircraft under control. To remember this potential crisis situation, the controller encodes the future task (i.e., briefing or execution of control actions needed to avoid a pending crisis situation) in terms of the sensorimotor properties (e.g., internal visual representation of the projected traffic picture and/or physical performance requirements of the control action) of the task which will enhance the actual performance at the time of task. This type of encoding will facilitate the activation of memory with the appropriate external retrieval cues, for example, the flight strips for TCA483, AAL284, and TWA343 being placed adjacent to each other, with the TCA483 flight strip being cocked (Koriat et al., 1990). The previous example is indicative of the significant role that flight strips can play in facilitating motoric encoding and planning future actions.

Several researchers have identified the significant cognitive value of flight strips in preparing for future actions (Hopkin, 1989, 1991b; Vortac, 1991). One reason for the cognitive value of flight strips is that they represent the history of actions, goals, intentions, and plans of pilots and controllers. These functions are elaborated in the following controller interview extract (Harper, Hughes, & Shapiro, 1989): "It's a question of how you read those strips. An aircraft has called and wants to descend, now what the hell has he got in his way? And you've got ping, ping, ping, those three, where are those three, there they are on the radar. Rather than looking at the radar, one of the aircraft on there has called, now what has he got in his way? Well, there's aircraft going all over the place, now some of them may not be anything to do with you, your strips will show you whether the aircraft are above or below them, or what aircraft are below you if you want to descend an aircraft, and which will become a confliction. You go to those strips and you pick out the ones that are going to be in conflict if you descend an aircraft, and you look for those on the radar and you put them on headings of whatever, you find out whether those, what those two are which conflict with your third one. It might be all sorts of conflicts all over the place on the radar, but only two of them are going to be a problem, and they should show up on my strips" (p. 9).

This interview extract provides a good example of the role that flight strips may play in assisting information processing and its significance in planning future actions. Harper et al. (1989) pointed out that "paradoxically, the "moving" radar screen is from an interpretative point of view relatively static, while the "fixed," "hard copy" strip is interpretatively relatively dynamic" (p. 5). For ATC tactical operations, planned actions are the purview of flight progress strips, and past actions are reflected in feedback on the radar and flight-strip markings (Vortac, 1991). This suggests that controllers using strips have found them useful for a number of reasons. However, it does not demonstrate that some other cuing system may not work just as well if it is designed as technology evolves. Whatever the controllers use, they have to be actively involved in it. Controllers now using URET with electronic strips can annotate these electronically without using a pencil. This type of annotation may be sufficient to generate memory encoding.

The "generation effect" is directly related to memory codes, particularly motoric encoding (Dosher & Russo, 1976; Erdelyi, Buschke, & Finkelstein, 1977; Johnson, Taylor, & Raye, 1977; Slamecka & Graf, 1978). Simply stated, the generation effect refers to the fact that information actively and effortfully generated (or information that you are actively involved) are more memorable than passively perceived information. The essence of this memory phenomenon is expressed in the "sentiment that there is a special advantage to learning by doing, or that some kind of active or effortful involvement of the person in the learning process is more beneficial than merely passive reception of the same information" (Slamecka & Graf, 1978, p. 592).

The generation effect has direct relevance to ATC tactical operations, where the active integration of the controller's information-processing capabilities with the relevant support systems (e.g., flight progress strips, radar, etc.) is a critical component of how controllers work with traffic. Means et al. (1988), using a "blank flight strip recall task," demonstrated that controllers' memory for flight data is a function of the level of control exercised. Their data indicated that memory for flight information of "hot" aircraft, which required extensive control instructions, was significantly better than that for flight information for "cold" aircraft, which required little controller intervention (e.g., overflight).

The foregoing discussion suggests the importance of a direct manipulation environment (Hutchins, 1986; Jackson, 1989; Jacob, 1986; Schneiderman, 1983) for ATC. Such an environment seems essential to maintain and potentially enhance the integrity of ATC tactical operations. David Hopkin (1991b) commented: "Whatever form electronic flight strips take, it is essential to define beforehand all the functions of paper flight strips, in order to discard any unneeded functions deliberately and not inadvertently, to confirm that familiar essential functions can still be fulfilled electronically, and to appreciate the functional and cognitive complexity of paper flight strips. Electronic flight strips have major advantages in compatibility with computer-based air traffic control systems, but their compatibility with human roles is less obvious, requires positive planning, and depends on matching functions correctly with human capabilities" (p. 64).

Manipulative control actions, both routine and strategic, required by the controller are important. Although not everyone in or working for the FAA agree with this and although some TRACON facilities have actually gone to a stripless environment, the controversy about flight strips as memory tool has continued for many years. Technology appears to be winning this conflict and electronic strips are the future. This is more of a truth in the enroute control environment of today, but will undoubtedly become the situation in terminals as well. Some approach controls have gone without strips for some time, with the controllers using the data block and a notepad if desired or required. How controllers code and store information is a very important aspect of their ability to retrieve what they need when they need it.

21.2.5 Code Interference in Working Memory

The primary phonetic (i.e., acoustic, verbal) and visual (i.e., spatial) codes essentially form two independent systems in working memory (Baddeley & Hitch, 1974; Baddeley & Lieberman, 1980; Baddeley, Grant, Wight, & Thompson, 1975; Brooks, 1968; Crowder, 1978; Healy, 1975). Different concurrent tasks can cause interference in these two systems (Baddeley, Grant, Wight, & Thompson, 1975). Essentially, recall declines as the items become more similar in memory. This similarity refers to the mental representation (e.g., phonetic, visual) of the item retained in the working memory (Card et al., 1986). Given the phonetic or verbal rehearsal (i.e., maintenance rehearsal) as the primary maintenance technique for retaining information in the working memory, items in working memory will be more susceptible to phonetic interference. For example, intrusion errors are more likely to occur between items that sound similar (e.g., B for P, K for J).

We should design tasks to minimize code interference and take the advantage of the cooperative nature of the two primary codes (Posner, 1978). For example, ATCs must create and maintain a transient, dynamic "pictorial" representation or mental model of the airspace traffic under control (Schlager, Means, & Roth, 1990; Sperandio, 1974; Whitfield, 1979; Whitfield & Jackson, 1982). The construction (and/or reconstruction) of this airspace traffic "picture" requires a great deal of spatial working memory. To minimize visual code interference and maintain the integrity of spatial working memory, this primary task should not be performed concurrently with tasks that require similar spatial demands in working memory. Rather, concurrent tasks will be better served if they take advantage of phonetic (i.e., verbal, acoustic) representations in working memory (Wickens, 1992).

Questions still remain as to whether the codes just described are an exhaustive representation of those present in the working memory. For example, if there are auditory–verbal and visual–spatial codes or systems, perhaps, there are also olfactory or kinesthetic codes (Klapp, 1987). It is also not clear whether separate systems exist within each working memory with specific processing codes, or different codes within the same working memory system (Phillips & Christie, 1977; Klapp, 1987). Several memory-loading studies have concluded that a single-system view of working memory is tenuous at best and largely unsubstantiated (e.g., Roediger, Knight, & Kantowitz, 1977; Hellige & Wong, 1983; Klapp & Philipoff, 1983; Klapp, Marshburn, & Lester, 1983). A general implication of these studies is that tasks using systems with different codes (e.g., visual vs. auditory) will not result in performance degradation owing to interference as readily as tasks using similar system codes. These studies are consistent with the multiple-resource view of information processing (Monsell, 1984; Navon & Gopher, 1979; Wickens et al., 1983), which essentially predicts that if two tasks use the same resources (e.g., auditory–verbal), interference will be reliably greater than when the two tasks use different resources (e.g., auditory–verbal vs. visual–spatial).

This means that better system designs take the advantage of the operators' abilities to parallel process more effectively if the demands made on them use more than one processing modality. If there is too much in the visual or acoustic store, then the system gets overloaded, resulting in coding and/or retrieval errors.

21.2.6 Attention and Working Memory

The volatility of information in working memory is potentially the greatest contributor to operational errors in ATC tactical operations. A series of experiments in the late 1950s demonstrated that in the absence of sustained attention, information is forgotten from working memory in approximately 15 s (Brown, 1958; Peterson & Peterson, 1959). Over the past 30 years, hundreds of experiments have confirmed this finding. Working-memory information loss is particularly profound when distracting or concurrent events demand an attention shift. Controllers, for example, frequently find themselves in situations where they must perform some kind of distracting activity (e.g., notations on flight strips, cocking a flight strip, consulting a chart, adjusting their eyeglasses between the time when primary information is received and the time this information must be acted on). These concurrent activities diminish information retention. Further, while ATCs usually have the status of relevant information (e.g., aircraft, flight data) continuously available on the radar display or in the flight-strip bay, allowing responses based on perceptual data rather than memory data, there are often occasions when attention is directed away from the displays.

In a memory study of simulated communications, Loftus, Dark, and Williams (1979) obtained results similar to hundreds of studies on retention in working memory when rehearsal was prevented. They found that performance was very high at a retention interval of 0 and then declined to a stable level by about 15 s, with minimal information being retained after 15 s. The authors concluded that because "forgetting occurs over an interval of 15 (s) following the initial reception of a message, a message should be responded to as soon as possible after it is received" (p. 179). In addition, the authors replicated the research findings (e.g., Murdock, 1961) indicating that as working-memory load increases, the probability of correctly recalling the information from the working memory decreases. The practical implication of this finding is that "whenever possible, as little information as is feasible should be conveyed at any one time. In particular, no instruction should be conveyed until 10 (s) or so after the previous instruction has been acted upon" (p. 179).

Based on the foregoing discussion of the fragile nature of information in working memory, one might conclude that sustained attention (e.g., maintenance rehearsal) to one item of information is necessary to maintain the information in working memory. In addition to this intuitive conclusion, several studies have demonstrated that information is more volatile early in the retention interval (e.g., Dillon & Reid, 1969; Kroll, Kellicut, & Parks, 1975; Peterson & Peterson, 1959; Stanners, Meunier, & Headley, 1969). These studies generally concluded that early rehearsal of information reduced the amount lost during a retention period. Klapp (1987) further elaborated that: A few seconds of rehearsal can largely protect (working memory) from the usual loss attributed to distraction. The potential human-factors implications of this finding appear to have been overlooked. One would suppose that retention of information, such as directives from air-traffic control, would be improved by brief rehearsal when that information cannot be used immediately. Therefore, the practical implication of these studies is that if information is rehearsed immediately after it is received (early rehearsal), the process will enhance the information retention in working memory (Klapp, 1987).

21.2.7 Automated versus Controlled Human Information Processing

Considerable researches have identified two qualitatively distinct ways in which we process and/or respond to information. These are automatic and controlled processing (e.g., Fisk, Ackerman, & Schneider, 1987; James, 1890; Kahneman & Treisman, 1984; LaBerge, 1973, 1975, 1976, 1981; Logan, 1978, 1979, 1985a, 1985b; Norman, 1986; Posner & Snyder, 1975; Schneider, Dumais, & Shiffrin, 1984; Schneider & Shiffrin, 1977; Shiffrin & Schneider, 1977). Experts and novices in any domain may well process information differently. Automatic and controlled processing can serve as a means for explaining how experienced and new controllers think and solve problems in different ways.

A well-formed representation of the stimuli in memory, as a result of extensive practice, is a component of automatic processing or automaticity (Schneider & Shiffrin, 1977). This extensive practice affords the development of automatic links or associations between replicated research findings (e.g., Murdock, 1961) indicating that as working-memory load increases, the probability of correctly recalling information from working memory decreases.

The influence of practice on performance is another important aspect of the dramatic attentional demands on the working memory. It is well known that practice is the single most powerful factor for improving the controller's ability to perform ATC tasks. Nothing is as likely to offset the frailties of working memory. The framework of automatically and controlled processing serves to help explain the influence of practice on the attentional demands of working memory (Schneider & Shiffrin, 1977; Shiffrin & Schneider, 1977).

A well-formed representation of the stimuli in memory as a result of extensive practice is a component of automatic processing or automaticity (Schneider and Shiffrin, 1977). Extensive practice affords the development of automatic links or associations between stimulus and response that can be operated with minimal processing effort (Gopher & Donchm, 1986). The defining characteristics of automaticity are empirically well understood and documented. Automatic processing is fast, parallel (Logan, 1988a; Neely, 1977; Posner & Snyder, 1975), effortless (Logan, 1978, 1979; Schneider & Shiffrin, 1977), autonomous (Logan, 1980; Posner & Snyder, 1975; Shiffrin & Schneider, 1977; Zbrodoff & Logan, 1986), consistent (Logan, 1988a; McLeod, McLaughlin, & Nimmo-Smith, 1985; Naveh-Benjamim & Jonides, 1984), and not limited by working-memory capacity (Fisk et al., 1987). It also requires no conscious awareness of the stimulus input (Carr, McCauley, Sperber, & Parmalee, 1982; Marcel, 1983), and it can be learned with extensive practice in consistent environments (Durso, Cooke, Breen, & Schvaneveldt, 1987; Fisk, Oransky, & Skedsvold, 1988; Logan, 1979; Schnedider & Fisk, 1982; Schneider & Shiffrin, 1977; Shiffrin & Schneider, 1977). On the other hand, controlled processing is relatively slow, serial, mentally demanding, dependent on working memory capacity, and requires less practice to develop. Controlled processing is also used to process novel or inconsistent information, and essentially characterizes novice performance where the operator consciously applies rules and templates to the situation.

Although initial theoretical treatments viewed automaticity in terms of little or no attentional resource demands (Hasher & Zacks, 1979; Logan, 1979, 1980; Posner & Snyder, 1975; Shiffrin & Schneider, 1977), new theoretical treatments of automaticity as a memory phenomenon appear to be the most viable, particularly, in terms of skill acquisition and training applications. According to the memory view, automaticity is achieved when performance is dependent on "single-step, direct-access retrieval of solutions from memory" (Logan, 1988b, p. 586). For example, an experienced controller who is familiar with the spatial layout of the ATC console visually searches for information automatically. The search goal, along with current display features, allows retrieval of prescriptive search strategies from the memory. An inexperienced controller might not search automatically, because the necessary visual search strategies would not be present in the memory, requiring reliance on the general search skills and deliberate attention to all the potentially relevant information. These conclusions are valid for any environment in which complex skill grows with experience.

The training of automatic processing could have tremendous implications for ATC and the integrity of the controller's working memory. We have seen that the volatility of information in working memory places a tremendous burden on a controller's flight information management performance. Automaticity allows increased information processing (e.g., parallel processing) without decrements in working-memory performance. The viability of automaticity training for complex tasks, such as ATC, has been questioned by several researchers, who suggested that only "simple" tasks can be automated (e.g., Hirst, Spelke, Reaves, Caharack, & Neisser, 1980). However, Fisk et al. (1987) questioned this suggested limitation of automaticity, noting that researchers do not clearly define what makes a task simple or complex. Complex tasks can be performed via automatic processing, via controlled processing, or most likely, through a combination of both processes. Simple tasks can also be performed by either automatic or controlled processing. The type of processing is not determined by the complexity (or simplicity) of a task, but rather by the consistency, and if the task is consistent, that is, the amount of practice (p. 191) (see Fisk et al., 1987; Fisk & Schneider, 1981, 1983, and Logan, 1988b, for a discussion of automaticity training principles and guidelines). However, the extent to which automaticity can lead to profitable training guidelines and recommendations that can be implemented in the complex and dynamic ATC environment is not clear and needs further investigation. The identification of the ATC tasks and subtasks that would afford automatic processing, and those that would afford controlled processing is a fundamental part of such an investigation.

Further research is also needed to investigate the influence of ATC automation on automatic processing. Specifically, what influence will ATC automation have on the development of overlearned (i.e., automatized) patterns of behavior, which are important for reducing the attentional demands of a controller's working memory? This will undoubtedly be an ongoing issue as automation increases, but may leave the controller still primarily responsible for separation.

In addition, another issue must be addressed. This is the concern that the cognitive structures (e.g., memory processes, conceptual knowledge) associated with overlearned patterns of behavior, which work to reduce the current load on working memory, may not be available to those controllers who "grow up" in a more automated ATC environment. The cognitive requirements of ATC will be ever changing with continued increases in ATC automation, making it difficult to reliably appreciate the nature of automatic processing in future ATC systems. How will future ATC systems afford automatic processing for the controller? One can safely conclude that the development of automaticity in future systems will be different than automaticity development in the current system. Although there is an extensive literature on the psychology of memory and its influences on automaticity and the allocation of attention, questions still remain as to whether increased attention facilitates improved memory (Vortac, 1991). In particular, is additional attention beyond the minimum attentional threshold for a stimulus (i.e., the minimum amount of attention needed to activate a memory representation), necessary or sufficient for memory improvement? Several empirical studies (e.g., Mitchell, 1989) demonstrated that if sufficient attentional resources are available to allow the activation of a memorial process, additional attentional resources will not strengthen the activation nor improve the memory operation. Rather, the additional, unnecessary attentional resources will result in unnecessary memory loading and decreased working-memory efficiency.

The previous brief discussion of attention and memory suggests that depending on the memory processes required for a task, deliberate attention may or may not be necessary or sufficient for activation. For example, automatic processes will be activated regardless of the attentional resources available or expended. However, controlled or nonautomatic processes will not operate without the attentional resources necessary to exceed the minimum attentional threshold.

21.2.8 Working-Memory Capacity

A number of textbooks in cognitive psychology (see Klapp et al., 1983, for a review) and human factors (e.g., Adams, 1989; Kantowitz & Sorkir, 1983; Sanders & McCormick, 1993) have proposed a single, limited-capacity system theory of working memory. This is based primarily on laboratory methods designed to measure static memory (e.g., recall of randomly presented alphanumerics or words). Much of the ground-breaking original memory research was built around paradigms like this. The standard claim was that the maximum capacity of working memory is limited to "seven plus or minus two chunks" (Miller, 1956). This one paper has had a tremendous impact on theory and practice in memory research. A "chunk" is a single unit of information temporarily stored in working memory. This view of memory assumes that it is a single limited-capacity system and that it serves as the foundation of working memory. This standard single-system theory suggests that once working memory is filled to its five to nine chunks, maximum capacity is reached, full attention is deployed, and no further memory-involved tasks can be processed without degrading the performance on the current task performance in situations, such as strategic planning, decision making, and the processing of visual–spatial material, where extensive amounts of information are processed and retained (Chase & Ericsson, 1982; Klapp & Netick, 1988). However, it may be unreasonably optimistic in dynamic-memory situations, "in which an observer must keep track of as much information as possible, when signals arrive in a continuous stream with no well-defined interval for recall" (Moray, 1986, p. 40-27). It is also possible that the number of features or attributes associated with each chunk or object may further complicate an individual's ability to recall the needed information when required (Davis & Holmes, 2005).

Recalling what the controller intended to do as part of his or her current plan is particularly challenging, given the dynamic nature of air-traffic control. In a series of laboratory studies using high workload and interruptions, researchers have found rapid forgetting of the intentions that participants set for themselves (Einstein, Williford, Pagan, McDaniel, & Dismukes, 2003). The authors commented that: "The results suggest that maintaining intentions over brief delays is not a trivial task for the human cognitive system (p. 147)."

Several authors have presented data to support a multi-component working-memory system, which includes, but is not limited to, static memory (e.g., Baddeley, 1986; Brainerd, 1981; Chase & Ericsson, 1982; Hitch, 1978; Klapp et al., 1983; Klapp & Netick, 1988; Moray, 1980). For example, Baddeley (1986) described a working-memory system that consists of a "central executive" that coordinates and directs the operations of two "slave" systems, the articulatory loop, and the visual–spatial "scratchpad." Essentially, these two slave systems are responsible for processing verbal and nonverbal information, respectively. Baddeley's model is very much a multiple-resource model like Wickens's (1984) model. Information on the three lines of research, multiple resources, dynamic memory, and the skilled memory effect, is briefly presented subsequently, documenting the alternative approaches to norming memory dynamics.

21.2.9 Multiple Resources

The literature on working-memory capacity suggests that rather than a single working-memory system, capable of being easily overloaded, there appear to be several systems with multiple resources, each system capable of being overloaded without interference from the other (Klapp, 1987). Multiple-resource theory has been successful in describing performance in dual-task situations (Navon & Gopher, 1979; Wickens et al., 1983). For example, Klapp and Netick (1988), in examining dual-task performance in working memory, reported data suggesting that there are at least two working-memory systems (i.e., auditory–verbal and visual–spatial) that differ in resources (i.e., composition). Essentially, the data demonstrated that if two tasks use the same resources (e.g., auditory–verbal), interference will be reliably greater than if the two tasks use different resources (e.g., auditory–verbal vs. visual–spatial). There are additional advantages of multiple resources theory that have the potential for improving the use of memory aids, so that we can recall more information. Wickens et al. (1983) developed the principle of "stimulus/central processing/response compatibility." It described the optimal relationship between how information is displayed and human resources are effectively used in the form of memory codes. Displays should be designed in a format that actively helps the individual encode information into working memory. Essentially, the presentation display format should be compatible with the code used in working memory for the particular task. For example, the encoding and storage of air-traffic control information is better served if it is presented in a visual–spatial format. The authors also suggested that retrieval of material from memory aids (e.g., computerized menu systems, spatially organized aids such as a "mouse") would be more effective if the resource modality needed to operate the memory aid does not look like or sound like modality in working memory. This would reduce the retrieval interference. For example, air-traffic control tasks, which are heavily dependent on visual–spatial resources, may be better served by semantic-based computer menu systems or auditory–verbal systems for memory aiding.

Multiple-resource theory has the potential for new approaches for improving complex and dynamic tasks, such as ATC. Klapp and Netick's (1988) data suggested that to optimize working memory resources, tasks and subtasks need to be appropriately allocated across independent subsystems of working memory. The data also indicated that training to make the most out of task configuration may also help in the management of working memory. The general guidelines offered by multiple resource theory need to be extensively investigated to determine their profitability in improving ATC tactical operations.

21.2.10 Dynamic Memory

Dynamic-memory tasks that require operators to keep track of a continuous sequence of information with no well-defined recall intervals are more analogous to the complex and multidimensional nature of "real-life" tasks. For example, ATCs must be competent in responding to the nature of an individual aircraft under control, while concurrently "handling" the entire series of the aircraft. The multidimensional nature of this task requires the controller to keep track of a large number of identified aircraft, each varying in flight data (e.g., altitude, heading, location, type), with flight data further varying along a number of values (e.g., 12,000 ft, 45° north, 350 mph). Furthermore, the number of aircraft and associated flight data are periodically updated, requiring the controller to continually acquire and forget the no longer needed flight information. This is done to revise the memory representation of the airspace traffic. The existing researches overwhelmingly suggests that dynamic-memory capacity is only about three items much less than the traditional memory capacity of seven items, using a static memory paradigm (Baker, 1963; Kvalseth, 1978; Mackworth, 1959; Moray, 1980; Rouse, 1973a, 1973b; Yntema, 1963; Zeitlin & Finkleman, 1975). Based on a dynamic-memory task, analogous to that of an ATC, Yntema (1963) suggested three corrective solutions to reduce the severe limitations of dynamic-memory capacity. First, recall performance is much better in a monitoring situation when the operator is responsible for only a few objects (e.g., aircraft) that vary on a large number of attributes (e.g., flight data), than for a large number of objects with few attributes.

This recommendation is consistent with work on "conceptual chunking," which indicates that recall of a primary object or concept (e.g., aircraft) precipitates recall of associative elements or attributes (e.g., flight data) from long-term memory (Egan & Schwartz, 1979; Garland & Barry, 1990a, 1990b, 1991, 1992). Additional information on conceptual chunking is presented in the subsequent section. Also, the amount of information about each attribute (e.g., altitude, heading) has relatively little influence on the dynamic-memory integrity. This result is also consistent with conceptual chunking. Therefore, information precision can be increased without degrading the dynamic-memory performance. Dynamic-memory performance is enhanced when each attribute value has its own unique scale. Such attribute-value discriminability reduces the influence of interference owing to item similarity.

Yntema's (1963) suggestions for dynamic-memory enhancement warrant a note of qualification, particularly if applying them to an ATC environment. The conclusions were based on sound-controlled laboratory experimentation. There are no data currently available that links these conclusions specifically with air-traffic control. Yntema's (1963) subjects were not controllers and the task stimuli were "meaningless" to the subjects. However, an investigation of the applicability of these suggestions to an ATC setting is needed.

The nature of the dynamic-memory tasks present in the available literature invariably involve the presentation of a time series of "random" and "meaningless" information observed (i.e., monitors)

(Baker, 1963; Kvalseth, 1978; Mackworth, 1959; Moray, 1980; Rouse, 1973a, 1973b; Yntema, 1963). The general finding of a limited dynamic-memory capacity of approximately three items may simply be a product of these task characteristics (Moray, 1986). For example, skilled operators (e.g., ATCs) who have to deal with complex and multidimensional information often exceed the three-item capacity that has been proposed. These operators process the heavy information loads and are competent recalling a considerable amount of information from their dynamic displays on demand. This superior ability may be a result of meaningful information processing as a result of dynamic interaction and direct manipulation of the displays. This direct manipulation (vs. monitoring) may allow the operator more meaningful encoding and retrieval strategies, which facilitate recall of the information. This explanation has definite ATC automation implications. More specifically, direct manipulation environments with motoric enactment may facilitate dynamic-memory performance, while monitoring may degrade or unnecessarily restrict dynamic memory to the three-item limit. This was a specific concern for the future with the advent of free-flight concepts as described in the RTCA concept document (Radio Technical Commission for Aeronautics, 1995).

It is tenuous at best to generalize the available dynamic-memory results found in the laboratory (using meaningless material) to "real-life" dynamic environments, where operators skillfully construct the form and content of the information that they need to remember (Moray, 1986). Extensive research is needed to identify the features of controllers' dynamic memory that will contribute to the development of corrective solutions and training guidelines to reduce the effects of severe memory constraints of an ATC setting. Such research is especially important with the growth of ATC automation, where the integrity of system decision-making (which is based on information monitoring) is highly dependent on the dynamic-memory capacity. Based on the work by Megan and Richardson (1979), Moray (1986) suggested that dynamic-memory research may be better served if the research objectives view "the gathering of information as a cumulative process, one whose outcome (is) the convolution of a data acquisition function and a forgetting function" (p. 40). Experts in many fields appear to use memory more effectively than that would have been anticipated, based on either the static or dynamic-memory research. This may be partly owing to the skilled memory effect.

21.2.11 Skilled Memory Effect

The intimate relationship between working memory and long-term memory provides the means to substantially increase the working-memory capacity beyond the traditional limits. Baddeley et al. (1976, 1981, 1986) functionally described the working memory as a product of several memory system (i.e., components), which in combination, allow skilled tasks (e.g., reading) to exceed the traditional working-memory capacity limits. Further, in a sense of experiments examining memory performance as a function of practice, Chase and Ericsson (1982) demonstrated that individuals can substantially increase their working memory (i.e., capacity). The authors suggested that with increased practice, working memory develops rapid-access mechanisms in long-term memory.

Researchers have built a solid base of empirical evidence for the "skilled memory effect" (see Chase, 1986, for a review). The literature that covers research on a wide range of perceptual–motor and cognitive skills, generally concludes that experts in their area of expertise are able to retain information far in excess of the traditional limits of working memory (Chase, 1986; Chase & Ericsson, 1982). Based on the now-classic studies with the game of chess, Chase and Simon (1973a, 1973b) theorized that for search-dependent domains like chess, domain-specific expertise can be differentiated based on how memory is organized. They suggested "that the chess master has acquired a very large repertoire of chess patterns in long-term memory that he or she can quickly recognize, although both masters and weaker players have the same (working memory) capacity" (Chase, 1986, pp. 28–55).

The skilled memory effect has been replicated many times in various search-dependent domains, such as chess (Charness, 1976; Chi, 1978; Frey & Adesman, 1976; Goldin, 1978, 1979; Lane & Robertson, 1979), Go (Reitman, 1976), Gomoku (Rayner, 1958; Eisenstadt & Kareev, 1975), bridge (Charness, 1979;

Engle & Bukstel, 1978), and in nonsearch domains such as music (Slaboda, 1976), computer programming (McKeithen, Reitman, Rueter, & Hirtle, 1981; Schneiderman, 1976), baseball events (Chiesi, Spilich, & Voss, 1979), electronics (Egan & Schwartz, 1979), architecture (Akin, 1982), and sport (see Garland & Barry, 1990a, and Starkes & Deakin, 1984, for reviews). Research in nonsearch domains has identified "hierarchical knowledge structures" as a fundamental property of the skilled memory effect (e.g., Akin, 1982; Egan & Schwartz, 1979; Garland & Barry, 1990a, 1991). Specifically, these studies suggest that experts use domain-specific conceptual knowledge from long-term memory to organize information, and this organization serves to facilitate storage and retrieval.

Based on the accumulated knowledge, Chase (1986) concluded that the skilled memory effect is owing to the existence of a vast domain-specific, long-term memory knowledge base built up by the expert with years of practice. This knowledge base can be used to serve two important mnemonic functions: (1) patterns can be used to recognize familiar situations, and (2) conceptual knowledge can be used to organize new information (pp. 28–61).

The research literature suggests that the traditional view of working memory as a single, limited-capacity system is not viable. Working-memory capacity appears to be directly or indirectly related to several factors, such as the nature of the multiple working-memory components (e.g., resources, conceptual organization), task parameters, meaningfulness of materials, and operator skill and experience. Despite the incredibly vast research literature on memory, Klapp (1987) asserted that a "detailed breakdown and mapping of (working) memory systems onto tasks is not yet understood" (p. 6), "largely because of our ignorance concerning the nature of the memory systems" (p. 17).

21.2.12 Chunking and Organization

Researchers have long recognized the principle of "chunking" as a means to expand the limits of working memory (Miller, 1956). Essentially, chunking is any operation (or operations) that can combine two or more items of information into one. The resulting one item or "chunk" can then be stored as a single information unit in working memory, making available the additional working-memory capacity to allocate elsewhere. For example, a controller may become familiar with the aircraft call sign TWA354 and process it as a single chunk, requiring only one space in working memory, rather than a series of six alphanumeric, requiring six spaces in the working memory. Further, a potential conflict between three aircraft—AAL348, TWA548, DAL35—may probably be organized as one chunk, rather than three, because the controller might not think of one without recalling the others.

Before addressing this topic, a qualification is necessary to clarify the relationship between "chunking" and "organization." It is suggested that these terms refer to essentially the same processes; however, their applications are traditionally different (Klatzky, 1980). Chunking is generally associated with recent working-memory storage of a relatively small number of items that will be available for immediate recall. Organization, on the other hand, is generally associated with long-term storage of a considerable amount of information. Although the terms traditionally apply to different situations, they share the underlying process of combining (organizing/chunking) two or more items of information into one. Further, as chunking is recognized as a process for the initial organization and encoding of information into long-term memory (i.e., elaborative rehearsal), it is reasonable to conclude that organization also occurs in working memory (Klatzky, 1980).

In general, chunking operations can be divided into two related forms. First, chunking may be facilitated by combining items based on temporal or spatial properties, that is, combining items that occur closely in time or space. In this manner, chunking occurs without the items necessarily forming a meaningful unit (Bower & Winzenz, 1969; Huttenlocher & Burke, 1976). This sort of chunking is often referred to as "grouping" (Klatzky, 1980). Parsing is closely related to grouping. Parsing is the process of "placing physical discontinuities between subsets that are likely to reflect chunks" (Wickens, 1984, p. 222). You can improve retention of relatively meaningless information by putting gaps or breaks within the information sequence. For example, someone could recall the telephone number 516 347 0364, better than 5163470364 (Wickelgren, 1964). Loftus et al. (1979), in their study of working-memory retention of air-traffic control communications, reported that in certain circumstances, four-digit items (e.g., 7382) were better retained when parsed into two pairs of double digits (e.g., "seventy-three, eighty-two").

Second, chunking may be facilitated if it "utilizes information from (long-term memory) to meaningfully relate many incomplete items to a single known item" (Klatzky, 1980, p. 92). This would be chunking with a plan. The degree of the inherent meaningful relationship between the separate items is also important and can help or hinder chunking. For example, the potential conflict between AAL348, TWA548, and DAL35 allows these three aircraft to be chunked as one item (i.e., potential conflict), owing to the shared meaningfulness of each being a contributor to a potential conflict. Chunking essentially benefits two qualitatively distinct processes in working memory (Wickens, 1992). First, chunking helps the retention (i.e., maintenance) of information in working memory for a brief period of time, after which the information is directly or indirectly "dumped." For example, controllers typically deal with a continuous flow of aircraft through their sector of responsibility. When aircraft are handed off, the associative information for that aircraft is no longer needed. Therefore, it is beneficially dumped from the memory. Second, chunking facilitates the transfer of information into long-term memory. Controllers must process a considerable amount of information concerning the status of several aircraft, which must be integrated and stored in long-term memory to initially create and subsequently revise the memorial representation (i.e., "picture") of the airspace traffic. The psychological literature has clearly documented the contribution of organizational processes (e.g., chunking) to good memory (e.g., Ellis & Hunt, 1989). How well someone organizes the material is often a clear indication of their level of expertise in any given area. Experts can take in a large quantity of task-specific information in a brief period of time and subsequently recall the information in meaningful units or chunks. Chase and Simon's (1973a, 1973b) study of chunking of stimulus information by chess experts demonstrated that experts are able to encode more information in a limited time when compared with nonexperts.

Chase and Simon (1973a, 1973b; Simon & Chase, 1973; Simon & Gilmartin, 1973) proposed a perceptual chunking hypothesis. Chunking behavior in the recall of task-specific stimulus information can be explained using "Perceptual chunking" that involves perception by coding the position of the entire chunks or several items, storing chunk labels in working memory, and subsequently decoding at the time of recall. Two critical features of the perceptual chunking hypothesis are that chunks are independently perceived and that recall requires decoding chunk labels in working memory. This means that heavy processing demands are placed on working memory. However, Egan and Schwartz (1979) pointed out several problems with these critical features. First, chunk independence does not allow for global processing. For example, an air-traffic control specialist can perceive the global characteristics (e.g., "a developing conflict situation") of a traffic pattern on the radar display in addition to the individual features (e.g., individual aircraft). Second, a group of display features (e.g., aircraft) may not form a functional unit or chunk, independent of other functional units. The functional units (chunks) must be context-dependent. As another example, the controller in identifying and processing two concurrent potential conflict situations will form two chunks, for example, "conflict A" and "conflict B." These chunks are not independent of each other, in that the resolution of conflict A will have an influence on the resolution of conflict B and vice versa. This is owing to shared airspace. In addition, the two conflict resolutions will influence and be influenced by the surrounding noninvolved air traffic. Third, some studies have shown that various interpolated tasks have no influence on recall performance of skilled chess players (Charness, 1976; Frey & Adesman, 1976). These studies strongly question Chase and Simon's position that task-specific information places substantial demands on working-memory capacity.

As an alternative to perceptual chunking, Egan and Schwartz (1979; also see Garland & Barry, 1990a, 1991) proposed a conceptual chunking hypothesis that links chunking (and skilled memory) to the organization of concepts in long-term memory. Conceptual chunking consists of a few primary features. First, skilled operators rapidly identify a concept(s) for the entire display or segments of the display (e.g., overflights, climbing aircraft, descending aircraft, and military aircraft). Second, skilled operators may systematically retrieve functional units and their elements that are related to the identified conceptual category stored in long-term memory (e.g., flights DAL1134, TWA45, UAL390, and TCA224 are elements identified as part of the conceptual category "overflights"). Third, conceptual knowledge of the display enables skilled operators to systematically search displays to verify details suggested by the conceptual category. For example, a controller is able to systematically search and detect aircraft that possess identifying flight characteristics that are consistent with the defining characteristics of the conceptual category "overflights."

Based on the available research, the conceptual chunking hypothesis appears to overcome the problems of the perceptual chunking hypothesis, by associating skilled memory and chunking to the organization of concepts in long-term memory (Egan & Schwartz, 1979; Garland & Barry, 1990a, 1991). The available data indicate that skilled operators are reliably better at recalling display features even after a brief exposure time. This superior recall performance may be based on the use of a "generate and test" process (Egan & Schwartz, 1979). This means that emphasis on processing information related to a conceptual category (e.g., potential air-traffic conflict) allows skilled operators to systematically retrieve elements (e.g., the defining features of the potential conflict and the involved aircraft) that are meaningfully associated with the conceptual category. The readers may recall Yntema's (1963) research on dynamic memory, which indicated that recall performance was better in a monitoring situation when the subject was responsible for a few objects (e.g., aircraft) that vary on a number of attributes (e.g., flight data) rather than when subjects were responsible for a large number of objects with few attributes. These findings are consistent with conceptual chunking, in that recall of the primary object or concept (e.g., aircraft) facilitated recall of the associative elements or attributes (e.g., flight data) from long-term memory. Tulving (1962) suggested that the ability to access the whole functional unit allows for systematic retrieval of all the information within a unit or chunk. He stressed that this ability is contingent on a good organizational structure of the task-specific knowledge in long-term memory. Ellis and Hunt (1989) noted that the question of how organization affects the memory is very important and equally complex. Although memory and organization are two different processes, Ellis and Hunt suggested that the two processes are positively correlated, resulting in the assumption that "organization processes contribute to good memory." Mandler (1967) provided support for this assumption, suggesting that organization is effective because of "economy of storage." Simply, organization is similar to chunking, in that individual units are grouped into large functional units, reducing the number of items to be stored in working memory and/or long-term memory. Mandler's approach assumes that organization occurs during encoding.

In a supportive yet alternative approach, Tulving (1962) suggested that organization benefits memory because of its "effects at retrieval." Tulving agreed that the organization of information occurs at encoding. However, he stressed that the ability to access the functional units or the whole entity at retrieval facilitates memory. This ability to access the whole functional unit allows for systematic retrieval of all the information within a unit. Tulving's arguments are consistent with conceptual chunking, in that knowledge of a conceptual display would allow subjects to systematically retrieve functional units that are related to the previously identified conceptual category that has been accessed in long-term memory. In addition, conceptual knowledge of the display would enable skilled operators to systematically search the conceptual category in long-term memory to verify the details suggested by the initial conceptual category. Ericsson (1985) pointed out apparent parallel between experts' superior memory performance in their domain of expertise and normal memory for meaningful materials, such as texts and pictures. Kintsch (1974) demonstrated that a competent reader can form a long-term representation for the text's meaning very rapidly and extensively, without deliberate effort (automatic processing). In addition, pictures (e.g., spatial information) appear to be fixated in long-term memory in less than 1 s (Potter & Levy, 1969). Those results appear consistent with the process of conceptually driven pattern recognition, which involves recognition decisions being guided by long-term memory rather than by sensory information (Ellis & Hunt, 1989).

The superior perceptual skill of experts in a variety of skill domains may not involve rapidly decoding independent chunk labels from a limited-capacity working memory; rather, as Egan and Schwartz

(1979) proposed, perceptual skill may be linked to the organization of task-specific concepts in long-term memory. It is suggested that expert memory performance may be more conceptual in nature, enabling skilled operators to (a) rapidly identify a concept for an entire stimulus display, (b) systematically retrieve functional units (chunks) that are related to the conceptual category stored in long-term memory through a "generate and test" process, and (c) systematically search displays to verify details suggested by the activated conceptual category.

These findings and the theoretical foundations behind them re-emphasize the importance of both initial and recurrent training in any command and control environment, where the situation is fluid and memory resources are in demand. Working memory might probably be used more effectively when the operator is completely up to the speed in the higher-order tasks and concepts. This will lead to more effective and less effortful organization in working memory. The compatibility of encoding processes with those of retrieval can have a major impact on memory organization and subsequent success or failure. Essentially, information retrieval is enhanced when the meaningful cues used at encoding are also present at retrieval. If the encoding and retrieval cues are not compatible, then memory will fail (e.g., Godden & Baddeley, 1980). For example, in the ATC setting, the flight progress strips and their manipulation served as significant retrieval cues, because they essentially contained the same information present during initial encoding. Although research on air-traffic control memory, specifically controller memory organization and chunking behavior has been limited, a study by Means et al. (1988) of controller memory provided some interesting data. In an airspace traffic drawing task, controllers were presented a sector map at the end of a 30–45 min ATC simulation, and subsequently were instructed to group the associated aircraft in the sector by drawing a circle around them. It was assumed that the aircraft groupings reflect the manner in which controllers organize the airspace traffic. The findings indicated that aircraft groupings could be characterized by various kinds of traffic properties or concepts (e.g., landing aircraft, overflights, climbing aircraft, traffic crossing over a fix, etc.). In addition, the researchers gathered data indicating that controllers who performed in a radar scenario condition (control traffic with radar and flight progress strips) tended to group aircraft based on the potential to "conflict," whereas those in a manual scenario condition (control traffic with flight progress strips only) tended to group aircraft based on geographical proximity.

Controllers in the manual scenario condition failed to control traffic without radar for a number of years, and therefore, were less up-to-date in controlling traffic under the experimental conditions, than the radar scenario controllers, who had the necessary displays available. These data suggest that the more current controllers tended to use higher-order grouping criteria (e.g., potential conflict) than the "handicapped" controllers, who tended to use simpler grouping criteria (e.g., geographical proximity). These data are consistent with conceptual chunking, in that the controllers tended to group (organize) the airspace around a number of ATC concepts and potential control problems. Further, the radar scenario controllers appeared to use more discriminating grouping criteria based on the strategic dynamics (e.g., conceptual nature) of the airspace, unlike the manual controllers, who appeared to use the criteria based on it simpler airspace spatial properties (e.g., aircraft are close to one another). These results suggest that the more experienced and skilled controller uses a larger, more discriminating conceptual knowledge base to control traffic. These results were consistent with the findings of Sollenberger and Stein (1995). Controllers were generally more successful in recalling aircraft in a simulation scenario based on the concept of what role they played, than on what the call signs were. Aircraft were chunked around spatio temporal concepts. Although controllers could only recall a small percentage of the call signs, they had little difficulty in determining what had been occurring in the airspace that they had under control.

Several times throughout this chapter, the rather common ATC phrase, the "controller's picture," appears referring to the controller's mental representation of the airspace. This is a key concept in the controllers' work and how they deal with their situational awareness. This mental modeling of the airspace and what is occurring in it plays an important role in ATC memory and tactical operations.

A mental model is a theoretical construct that provides the user with a framework for thinking about a complex domain of which they are a part. Mental models may be specific to a situation (e.g., VFR traffic) or more global to the entire task domain (e.g., the entire flight sector). They may, or may not, include abstractions concerning, functional relationships, operating guidelines, and systems goals and objectives (Mogford, 1991; Norman, 1986; Rasmussen, 1979; Wilson & Rutherford, 1989; Wickens, 1992). Theoretical descriptions of mental models are varied (Mogford, 1991). For example, Rouse and Morris (1986) suggested: "Mental models are the mechanisms whereby humans are able to generate descriptions of system purpose and form, explanations of system functioning and observed system states, and predictions of future system states" (p. 351). Further, Norman (1986) stated: "Mental models seem a pervasive property of humans. I believe that people form internal mental models of themselves and of the things and people with whom they interact. These models provide predictive and explanatory power for understanding the interaction. Mental models evolve naturally through interaction with the world and with the particular system under consideration. These models are highly affected by the nature of the interaction, coupled with the person's prior knowledge and understanding. The models are neither complete nor accurate, but nonetheless they function to guide much human behavior (p. 46)."

Research on mental models and conceptual structures in the air-traffic control environment is disappointingly limited (see Mogford, 1991, for a review). However, the research that is available does suggest a connection between a controller's "picture" and the controller's understanding and memory for the traffic situation (e.g., Bisseret, 1970; Means et al., 1988; Moray, 1980; Landis, Silver, Jones, & Messick, 1967; Whitfield, 1979). A general conclusion of these studies is that skilled controllers, when compared with the less-skilled controllers, use their picture as a supplementary display to enhance the memory for aircraft. In addition, it is generally concluded that the quality and functionality of the controller's picture is directly related to ATC expertise.

According to Whitfield (1979), who was one of the first to study the picture systematically, the skilled controller's picture seems to use three kinds of memory: (a) static memory (e.g., sector characteristics, separation standards), (b) dynamic memory (e.g., continual updating of aircraft flight data), and working memory (e.g., current status of aircraft). Further, Mogford (1991) suggested that the controller's "picture" is probably maintained in working memory, with substantial influences from "unconscious rules" stored in long-term memory. He stated that "it appears that the controller's mental model possesses various kinds of information which are reliant on different types of memory. Maps, flight plans, aircraft performance information, separation standards, and procedures are learned through training and experience and stored in memory." The extent to which mental models can provide assistance with the practical problems of ATC memory enhancement remains unclear. However, the available research has not yet revealed empirical evidence suggesting how the controller's picture may assist in enhancing the controller's working memory and improving ATC tactical operations. Research on mental models in air-traffic control is needed as ATC systems become more automated, forcing the controller into ever increasing levels of supervisory control. The dramatic changes with future automation will not only replace ATC technology and equipment, but will also change the way in which controllers conduct their job. Research is needed to investigate how increased computerization of ATC tasks influences the development of the controller's picture and its potential supporting influence on controller's working memory.

Hopkin (1980) addressed this problem, and noted that controllers frequently report that computer aids seem to increase the probability that they will lose the picture, their mental model of traffic. This is a working memory issue. If, as Neisser (1976) claimed, images are anticipatory phases of perceptual activity and are plans for obtaining information from potential environments, then this may provide a theoretical framework and suggest appropriate measures for evaluating the efficacy of various forms of computer assistance, particularly predictions, as aids to imagining. It could also provide hypotheses for specifying conditions when forgetting is most likely to occur (p. 558). An understanding of ATC mental models may prove beneficial for understanding the impact of automation on designing controller's training and memory aids. To be effective, such aids must interact with the cognitive processes of the controller (e.g., Hollnagel & Woods, 1983; Moray, 1988). It is important that data and job aids be designed and presented in such a way as to work with the controllers' internal representation of the airspace, rather than against it. Wickens (1992) stated: Within the last decade, designers of computer systems are beginning to capitalize on the fact that people have a lifetime's worth of experience in negotiating in a three-dimensional environment and manipulating three-dimensional objects (Hutchins, Hollan, & Norman, 1985). Therefore, the spatial metaphor is an important emerging concept in human–computer interaction (p. 154).

Further, Wickens (1984) commented on advanced automation and the design implications for computer-based data entry and retrieval systems: How does the computer model or understand the user's conception of the data and logic within the computer itself? Clearly, the computer should organize data in a form compatible with the user's mental model. But, what if different individuals possess different styles of organization? Are different organizational formats appropriate for spatial versus verbal modes of thinking, as suggested by Schneiderman (1980)? A related question concerns the assumptions that the computer should have the level of knowledge of the user. For the same program, a computer's interaction with a novice should probably be different from the interaction with an expert user. A novice, for example, would benefit from a menu selection program in which all the options are offered, as many of them are not likely to be stored in long-term memory. For the expert, this format will probably give unnecessary clutter, as the alternatives are stored and available in long-term memory in any case. An intriguing question from the viewpoint of systems designs is how the computer can either explicitly assess or implicitly deduce the level of knowledge or the format of organization employed by the user (p. 237).

Although several researchers have suggested potential implications of the mental models for both training and display design (e.g., Mogford, 1991; Wickens, 1992), Wilson and Rutherford (1989) asserted that "We have shown the several different interpretations of the concept (mental models) and its utility to be a weakness, which militates against the widespread use of mental models in system design" (p. 629). Obviously, further work is needed on the ATC's picture. This brief overview of the work on chunking and organization, and its relevance to ATC tactical operations leads to the primary conclusion that more research is needed. In particular, research is needed in an ATC setting to better understand the conceptual structures that guide the synthesis and organization of present and future traffic situations. In support of this line of research, Whitfield (1979), many years ago, suggested that a controller's mental model is required for current and future planning of the traffic situation. A further line of research is suggested by the available work on dynamic memory and conceptual organization (e.g., mental model). Perhaps, the ability of controllers to exceed the traditional limits of dynamic memory (i.e., three items) is associated with the controller's conceptualization (e.g., mental model) of the ATC domain. If so, what are the features of the controller's conceptualization that may contribute to dynamic-memory enhancement? How can this be used to help train and maintain controller skill? Do ATC conceptual structures fundamentally change with experience and expertise, thus, facilitating the enhancement of dynamic memory and skilled memory? There are obviously more questions than answers at this point; however, with increased ATC automation, the time (although limited) is ripe for extensive investigations to address these crucial questions. Harwood, Murphy, and Roske-Hofstrand (1991) pointed out that the complexity of ATC must be recognized; otherwise, research and applications will not be useful or meaningful. Even as we witness the evolution of ATC technology in the 21st century, we can observe systems designed that do not fully consider what the controllers need to avoid forgetting critical information.

21.2.13 Forgetting in Working Memory

The primary objective of this chapter is to examine the relationship of working memory with controller operational errors. An FAA Administrators Task Force identified controller memory lapses (i.e., forgetting) as a significant issue related to revising and retrieving critical operational information

(Operational Error Analysis Work Group, 1987). Although considerable information on forgetting is available in the psychological literature (see Klatzky, 1980, pp. 124–150 for a review), the profitable application of this material to the real-life setting of ATC is unclear. In contrast to the unforgiving nature of unintended memory failure, Hopkin (1995) noted: Forgetting as a boon rather than a bane has scarcely been studied at all. Yet, it is not always an advantage in air-traffic control to be able to recall all the details of what happened previously, as this could encourage unwarranted presumptions that any intervening changes of circumstance are trivial and that previous solutions can be adopted again, whereas, it might be better to work out fresh solutions without such remembered preconceptions. Some limited guidance on how to code air-traffic control information to make it more memorable can be offered, but there is no comparable practical guidance on how to code air-traffic control information so that it is easy and efficient to use while it is needed, but is readily forgotten after it has served its purpose and there is no benefit in remembering it, given the perennial problem of too much information in air-traffic control recommendations on how to render the useless forgettable would have real practical value (pp. 55–56).

Forgetting is also desirable, because it provides storage space for incoming new information in working memory. The level at which information is processed plays a large role in determining how difficult it will be to remember or forget that information (Murphy & Cardosi, 1995, pp. 179–191). Thus, the nature of forgetting information in the ATC setting is paradoxical, in that it has both desirable and undesirable implications. Essentially, research on both unintentional and intentional forgetting is necessary to develop aids to eliminate and/or enhance forgetting depending on the situation. The following discussion presents some of the available information on forgetting that may be applicable to the ATC setting.

Information processing models generally incorporate two mechanisms that produce memory retrieval failures. These are (a) spontaneous decay, which refers to a time-dependent process of information becoming less available over time, and (b) interference, which refers to the disruption of the memory trace owing to competing activity. Considerable research effort has gone into trying to determine which of these mechanisms really drives forgetting (Card et al., 1986).

Decay Research by Reitman (1974) initially demonstrated the separate roles of decay and interference in working memory. This research, along with others, has generally implicated time-dependent processes as being attributable to the rapid rate of decay or complete loss of information availability, if the individual takes no or inefficient action to process the information for temporary short-term or permanent long-term memory. In addition to the rapid decay of information that has been actively attended to and encoded, forgetting as a result of decay is also, in part, a function of the initial level to which the material is processed. Preattentive processing of information, without higher-order encoding, will inevitably result in decay. In addressing research on the decay mechanism as a means of forgetting, Wickens (1984) stated: When verbal information is presented by sound, the decay may be slightly postponed because of the transient benefits of the echoic (auditory) code. When information is presented visually, the decay will be more rapid. The consequence of decay before the material is used is the increased likelihood of error. The pilot may forget navigational instructions delivered by the ATC before they are implemented. In fact, Moray (1980) concluded that "the task of monitoring a large display with many instruments is one for which human memory is ill suited, especially when it is necessary to combine information from different parts of the display and the information is dynamic" (p. 216). The time-dependent decay process operates to significantly attenuate the fidelity of the memory trace (Klatzky, 1980; Wickens, 1992). The extent to which the decay process is disruptive or beneficial to the controller is situation-specific. The development of techniques to optimize the decay of information seems to be a viable line of research. If the controller was able to reliably control the decay of information, then the information management would be facilitated. This is clearly an area in which good display design can be beneficial. Controllers and other operators should never be forced to depend on memory with all its foibles if there is a viable way of organizing information so that it is present and available when needed.

21.2.14 Interference

Considerable research has demonstrated that it is more difficult to retrieve an item from working memory and long-term memory if there are other similar materials in the respective memory system (e.g., Conrad, 1964; Underwood, 1957). The similarity of items in memory is contingent on the memory representation of each item. For example, interference in working memory is more likely for items that sound alike (i.e., acoustic/phonetic interference). Long-term memory is more susceptible to semantic interference. That is, items (e.g., chunks) that share similar meanings are likely to share the same retrieval cues, which in turn disrupt information retrieval. Research on the interference effect has demonstrated that much of what is commonly referred to as forgetting is simply failure to retrieve, not actual loss (e.g., decay) from the memory (Card et al., 1986). However, in a dynamic environment, retrieval cues can be lost, and under time pressure, the operator makes decisions with what he has.

Generally, the literature recognizes three sources that may contribute to the interference effect: within-list (or information redundancy) interference, retroactive interference, and proactive interference (Wickens, 1984). Within-list interference is attributable to the increased similarity of items within a group that must be processed in the working memory. For example, Wickens (1984) illustrated that "when an ATC must deal with a number of aircraft from one fleet, all possessing similar identifier codes (AI3404, AI3402, and AI3401), the interference due to the similarity between items makes it difficult for the controller to maintain their separate identity in working memory" (p. 224). Obviously, to alleviate within interference, information must be presented in a manner that reduces the information redundancy interference. Also, there can be a detrimental effect of recently acquired information (retroactively) interfering with previously learned material (Underwood, 1957). This is retroactive interference. For example, a controller may forget a newly assigned altitude of an aircraft, because an additional item of information intervened and prevented sufficient maintenance rehearsal of the new altitude and/or notation on the flight progress strip. Further, increased similarity between the item to be retained and the intervening item will increase the probability of interference.

Proactive interference is the detrimental effect of the previously acquired information (proactively) interfering with recently learned material (Keppel & Underwood, 1962; Underwood, 1957). This effect may be especially profound during labor and time-intensive situations, where there is a tendency of cognitively regress back to former firmly established ways of thinking. This is a situation where the power of long-term memory to help in organize information in working memory can work against you. This creates challenges for training managers who are planning for transitions to new equipment and/or systems. Proactive interference must be considered or could diminish the potential benefits of new technology. A considerable amount of research has been conducted to examine the processes that reduce the effects of proactive interference, or as the literature commonly refers to it, a release from proactive interference (e.g., Keppel & Underwood, 1962). This phenomenon refers to the fact that if the type of stimulus material (e.g., letters, numbers) is changed from trial to nontrial (e.g., from numbers on trial to letters on trial), then proactive interference will be reduced, resulting in a substantial decrease in forgetting of the recently acquired material (e.g., the stimulus material on trial) (Loftus et al., 1979). Explanations for this phenomenon are generally consistent with the following example provided by Loftus et al. (1979): Consider that a subject must remember two pieces of information, A and B, which are presented in close temporal proximity. To the extent that A and B may be differently encoded, they will be less confusable, and hence, easier to recall. Carrying this notion over to the controller/pilot situation, it seems reasonable to expect that two pieces of numerical information will be easier to remember to the extent that they are uniquely encoded (p. 172).

In a study of simulated communications between controllers and pilots, Loftus et al. (1979) found evidence to indicate that a "unique-encoding system" as compared to a "same-encoding system" of ATC communications led to superior memory performance. The same-encoding system refers to the current relatively standard ATC practice of transmitting virtually all numerical data in a digit-by-digit manner (e.g., the radio frequency 112.1 would be transmitted as "one, one, two, point, one").

In contrast, "an example of the unique-encoding system, would be to encode radio frequencies in the digit-by-digit manner described above but to encode transponder codes as two pairs of double digits (e.g., '7227' would be encoded as 'seventy-two, twenty-seven')" (p. 171). This finding has definite memory implications for recall of multidimensional flight data. Loftus et al. (1979) concluded that: Attention is traditionally paid to the question of how transmitted information should be encoded, so as to minimize errors in perception (e.g., by use of the phonemic alphabet). However, virtually no attention has been paid to the question of how information may be encoded, so as to minimize errors in memory. The unique-encoding system represents only one possible improvement in encoding of transmitted information. Potentially, there are many others (p. 180).

Information on dynamic memory is also available in support of the utility of the unique-encoding system. In particular, dynamic-memory studies by Yntema (1963) and Yntema and Mueser (1960) provided the most applicable evidence. In these studies, subjects were required to keep track of a large number of objects, which varied on a number of attribute, which in turn varied on a number of unique values. These studies indicated that memory fidelity was enhanced when the attribute values each had their own unique codes (e.g., feet, speed, miles), when compared with those sharing common codes. For example, consider that a controller must identify and then enter the status of several aircraft along several flight data dimensions. As the flight data are coded differently (e.g., altitude/feet, distance/ nautical miles), performance will be superior if the controller deals in turn with all the relevant flight data of one aircraft before progressing to the next aircraft, rather than dealing with all the aircraft on only one flight data dimension (e.g., altitude/feet), before progressing to the next flight data dimension. The unique-encoding system appears to be a profitable means by which information can be optimally encoded, thus, enhancing the working-memory retention and minimizing retrieval failures of critical information. Research is needed to examine the viability of such an information-encoding system in an ATC environment. Based on the available research on interference effects, Wickens and Flach (1988) suggested four ways to reduce the effects of interference on forgetting in working memory. They are:

1. "Distribute the material to be held in (working) memory over time." This will allow proactive interference from previously acquired information to be reduced.
2. "Reduce similarity between items." This is suggested as similar-looking or similar-sounding (Conrad, 1964) items lead to greater interference.
3. "Eliminate unnecessary redundancy." This suggestion is intended to reduce the effects of within-list interference.
4. "Minimize within-code interference." This suggestion is consistent with the previously presented information on code interference in working memory. For example, in the predominantly visual/ spatial ATC environment, concurrent secondary tasks should minimize the use of visual/spatial codes, and instead, should utilize auditory/speech encoding (e.g., voice recognition technology) (pp. 124–126).

21.2.15 Directed Forgetting

As mentioned earlier, in addition to enhancing the integrity of working-memory performance through the reduction of memory lapses, there are also times when the intentional "purging" of information from working memory will work to enhance memory. Hopkin (1988) asserted that intentional forgetting may be beneficial in that the "controller dealing with an immediate problem is not overburdened by recalling other problems not sufficiently similar to be helpful in solving the present one" (p. 12). Further, Hopkin (1980) noted the importance of identifying and developing ATC techniques intended to aid the controller in the forgetting of "unwanted baggage," which may prove to proactively interfere with the current information processing. Such "directed forgetting" (also referred to as "motivated" or "intentional" forgetting in the cognitive literature) of information that is no longer useful would seem to be a necessary skill in a dynamic-memory setting, such as ATC flight management, in which the ability to process incoming sequential information is contingent upon the availability of processing space in working memory. The available research indicates that when subjects are instructed to intentionally forget unwanted information, there are additional attention resources for dealing with concurrent tasks (e.g., Bjork, 1972; Bjork, Bjork, & Kilpatrick, 1990; Martin & Kelly, 1974). In addition, Bjork (1972) suggested that directed forgetting can be trained.

The information presented earlier on the effects of decay on forgetting is relevant to the present discussion of directed forgetting. If techniques can be identified to assist the controller in reliably controlling the decay of information, directed forgetting would be a valuable product. As mentioned previously, two qualitatively different types of rehearsal strategies are involved in working-memory maintenance and elaborative rehearsal. Short-term retention of information in working memory is achieved through maintenance rehearsal, which emphasizes the phonetic aspects (i.e., auditory, speech) of the stimuli, whereas elaborative rehearsal is important for transfer of information into long-term memory by emphasizing the semantic aspects (i.e., meaningfulness) of the stimuli and their association with the conceptual information of the controller's mental model stored in the long-term memory. As information transfer to long-term memory facilitates the undesirable effects of proactive interference (see Craik & Watkins, 1973; Glenberg, Smith, & Green, 1977), information to be retained for only a short period of time should only use phonetic maintenance rehearsal as opposed to the semantic elaborative rehearsal (Wickens, 1992). This strategy, along with directed forgetting strategies, may prove useful in enhancing memory availability (Bjork, 1972; Bjork et al., 1990). Based on the available data from laboratory studies, Wickens (1984) suggested "that this technique (directed forgetting), like chunking, is a potentially valuable strategy that can be learned and subsequently employed for more efficient storage and retrieval of subsequent memory items" (p. 226). However, a note of qualification is warranted. Specifically, research is needed to determine the applicability of the laboratory findings to the ATC setting. The preceding suggestion was based on data gathered in a laboratory setting with college students (e.g., sophomores) who were required to forget meaningless information, which they had no experience in actively using and/or processing. Information is needed to determine the utility of purposefully forgetting meaningful information in a real-life, labor-intensive, time-intensive environment such as ATC. Until such data is available, instructional guidelines for the training of directed forgetting in an ATC setting will not be useful. In the near term, we are more concerned on how to help controllers retain what they need in some memory stage. Helping them forget is a much lower priority.

21.3 What Does the Future Hold for Working Memory in ATC?

The preceding discussion of working memory and its implications for air-traffic control is by no means an exhaustive, definitive treatment of the working-memory requirements for air-traffic control tactical operations. Although considerable information on working memory is available (e.g., Baddeley, 1986; Klatzky, 1980), there remain more questions than answers. Working memory permeates every aspect of the human information-processing system, making it difficult to get a "handle" on all the parameters that define its functionality. This chapter has attempted to raise an awareness of a few of the most salient and transient characteristics of working memory and their implications for ATC. Additional areas of research that directly or indirectly influence working memory are beyond the scope of this chapter. These include, but are not limited to, long-term memory, stress, decision making, and workload. The limiting factor in gaining a more comprehensive understanding of the working-memory requirements for ATC tactical operations is the simple fact that there is not a great deal of human-factors research on the cognitive aspects of ATC, especially on working memory. In 1980, Hopkin, in noting the importance of memory research in the ATC environment, concluded that "the application of theories of memory to practical air traffic control problems must be developed more in the future" (p. 558). In calling attention to the necessity to reinterpret the ATC's tasks in relation to cognitive psychology constructs, Hopkin (1995) stated: Some of the roles of memory in air-traffic control do not fit the main theories of memory very well. Theories tend to emphasize timescales of a few seconds for short-term memory or relative permanence for long-term memory, or refer to active task performance for working memory (Baddeley, 1990; Logie, 1993; Stein & Garland, 1993).

The controller relies on a mental picture of the traffic that is based on a synthesized integration of radar, tabular, and communicated information, interpreted according to professional knowledge and experience (Rantanen, 1994; Whitfield & Jackson, 1982). Although a simplified form of the picture can be built in a few seconds, as is routinely done at watch handover, building the complete picture requires more processing (Craik & Lockhart, 1972) and typically takes about 15–20 min, by which time the controller knows the full history and intentions of all current and pending traffic and can plan accordingly (pp. 54–55).

The application of psychological theories (i.e., memory theories) to practical air-traffic control problems is challenging. This is unsettling with the onset of the progression of automated systems, which will substantially alter the way in which controllers manage live traffic (Wise et al., 1991). The implications of increased ATC automation on the controller's cognitive processes are unknown. How can we gain an adequate understanding of the cognitive (e.g., working memory) requirements of advanced automation when the cognitive requirements of the current system remain elusive? We do know that the controller systems error rate has not changed much over the last 10–15 years, with the introduction of some automation such as the URET. Comprehensive task analyses of controllers have evolved over the years to a point till today, where scientists begin to understand the scope and complexity of the controller's job (Nickels, Bobko, Blair, Sands, & Tartak, 1995). There are potential human-factors consequences of increasing ATC automation. These include the impact of memory aids on ATC working memory. After considering the cognitive psychology research on the working memory system, one can safely conclude that the ATC system, given the current structure and technology, will only be as efficient and reliable as the controller's working-memory system as we currently understand the system. A controller's working memory directly or indirectly influences every aspect of his or her ability to control traffic. With ever-increasing amounts of ATC automation, human problem-solving and other cognitive processes will change or become additional complicating factors. Researchers need a new set of cognitive performance measures to fully appreciate the consequences of automation on controller performance. In the future, better measurement tools will be needed to show the consequences of automation not only in terms of performance, but also in terms of associated cognitive skills. Some cognitive effects are not currently measured at all, for example, on understanding or memory, but they may be more significant than the routinely measured effects on performance (p. 558).

Grolund, Ohrt, Dougherty, Perry, and Manning (1998) proposed a novel new variable that they suggest may intervene in how controllers use memory and how well they recall the needed information. According to the authors, controllers are more likely to recall information, especially altitude and relative position of the aircraft, if they have classified those aircraft as "Important." The definition of an important aircraft is one that was viewed as current or potential traffic for any other aircraft under control. Grolund et al. observed memory as a foundation for situation awareness, and the situation awareness as a critical element in safely managing the traffic. One might speculate that knowing or determining the real-world role of the controller's classification process may have training implications for more effective use of working memory.

Throughout this chapter, considerable information has been presented emphasizing the critical importance of working memory in ATC tactical operations. Unfortunately, the available research on working memory in ATC and nonATC settings has largely gone unnoticed in current and future ATC system design. As Hopkin noted in 1980, it is definitely the time to apply the existing (and new) memory research to the practical problems of air-traffic control. Although there are considerable researches on the frailties of working memory and ways to overcome them, there also exists a fundamental problem in making the appropriate knowledge influence the ATC-system design process. It is easier for designers to ignore memory issues than to integrate them into the designs. Hopkin (1991c) commented: It is not sufficient to plan and conduct research if the only products are journal articles, standards, or handbooks, though these are essential. The research evidence has to be applied and integrated into the design. Nowhere does this seem to be done satisfactorily. Lack of appropriate mechanisms to apply research findings to design processes appears to be the main difficulty. This problem is linked to some uncertainty about how valid and general some of the existing data and findings are. Is all the existing evidence actually worth applying to the design process? If not, how do we determine which should be applied and which should not? What criteria could serve such a purpose? What should the balance of evidence be between previous research and current and future research? What are the best measurements to gain evidence in the form of practical advice at the design stages? How can research findings be made more acceptable to designers, so that they are more willing to adapt design processes of future air-traffic control systems to incorporate evidence from research?

For several decades, an implicit philosophy of automation has existed that adopted the assumption that maximum available automation is always appropriate (invest in hardware, not people). This philosophy has been based, in part, on the availability of increasingly sophisticated and advanced technological innovations, the assumed need to reduce human workload, the need for increased safety of flight, and perhaps, primarily, on the assumption that the human "mind" (especially human memory) is similar to a silicon-based system that cannot be easily overloaded. Although automated systems have provided substantial benefits, several human-factors consequences have arisen and incidents/accidents have occurred. These problems often end up by calling for the human-factors professionals and the aviation community to reexamine automation practices. We are continuing to automate without building the human factors in the design process. This is not limited to air-traffic control. It is a recurrent issue for systems evolution, especially when operational demands continue during modernization.

There is an increasing awareness of the lack of a scientifically based philosophy of automation. This philosophy must be based on an understanding of the relative capabilities (e.g., frailties of working memory) of the controller in the system, and the circumstances under which automation should assist and augment the capabilities of the controller. What is needed is an approach that has a better philosophical base for what automation seeks to achieve and a more human-centered approach, to avoid the most adverse human-factors consequences of automated systems and provide a better planned progressive introduction of automated aids in step with user needs (e.g., Garland, 1991).

Such a comprehensive, scientifically based design philosophy for human-centered automation must be developed to avoid inevitable one step forward and two steps backward progression. For the timebeing, the human controller, despite the limitations and constraints of the working-memory system, will remain an essential part of the ATC system. Furthermore, it is suggested that with ever increasing levels of ATC automation, the significance of the human controller in the system and the controller's working-memory system should no longer be taken for granted. The purpose, intent, and nature of this chapter are perhaps best reflected in ideas that Levesley (1991) put forth about the way he saw the ATC system in 50 years. Levesley commented: "What I actually predict will happen is that the lessons of the last 50 years will be repeated in the next 50. Airlines will still prefer to spend $500 on aircraft for every $1 spent on ATC. Will the cost of potential super-systems actually prohibit their introduction, as they prove totally cost-ineffective? If I survive to the age of 93 and I fly somewhere in 2040, I suspect that there will still be a human problem solver on the ground in control of my flight, who will rejoice in the title of 'the controller.' And I don't think that controllers will be there because they are irreplaceable, or because the public wants someone there. I think that, with the right tools to help, the controller will still be there as the most cost effective, flexible system solution to the problem of safely guiding pilots and passengers to their destination. And that is what air traffic control is really all about (p. 539)."

References

Adams, J. A. (1989). *Human factors engineering*. New York: Macmillan.

Akin, O. (1982). *The psychology of architectural design*. London: Pion.

Ahlstrom, U., Keen, J., & Mieskolainen, A. J. (2004). Weather information display system (WIDS). *Journal of Air Traffic Control, 46*(3), 7–14.

Ahlstrom, U., & Friedman-Berg, F. (2005). *Subjective workload ratings and eye movement activity measures* (DOT/FAA/CT-05/32). Atlantic City International Airport, NJ: FAA William Hughes Technical Center.

Ammerman, H., Fligg, C., Pieser, W., Jones, G., Tischer, K., & Kloster, G. (1983). *Enroute/terminal ATC operations concept* (Report No. DOT /FAA/ AP-83-16). Washington, DC: Federal Aviation Administration.

Ammerman, H., & Jones, G. (1988). *ISSS impact on ATC procedures and training* (Report No. CDRL C108). Washington, DC: Federal Aviation Administration.

Anderson, R. E. (1984). Did I do it or did I only imagine doing it? *Journal of Experimental Psychology: General, 113*, 594–613.

Atkinson, R. C., & Shiffrin, R. M. (1968). Human memory: A proposed system and its controlled processes. In K. W. Spence, & J. T. Spence (Eds.), *The psychology of learning and motivation* (Vol. 2, pp.89–195). New York: Academic.

Backman, L., & Nilsson, L. G. (1984). Aging effects in free recall: An exception to the rule. *Human Learning, 3*, 53–69.

Backman, L., Nilsson, L. G., & Chalom, D. (1986). New evidence on the nature of the encoding of action events. *Memory & Cognition, 14*, 339–346.

Baddeley, A. D. (1976). *The psychology of memory*. New York: Basic Books.

Baddeley, A. D. (1981). The concept of working memory: A view of its current state and probable future development. *Cognition, 10*, 17–23.

Baddeley, A. D. (1986). *Working memory*. Oxford: Clarendon Press.

Baddeley, A. D. (1990). *Human memory: Theory and practice*. Boston: Allyn & Bacon.

Baddeley, A. D (1996). Exploring the central executive. *Quarterly Journal of Experimental Psychology: Human Experimental Psychology, 49A*(1), 5–28.

Baddeley, A. D., Grant, S., Wight, E., & Thompson, N. (1975). Imagery and visual working memory. In P. M. Rabbitt, & S. Domic (Eds.), *Attention and performance V*. New York: Academic Press.

Baddeley, A. D., & Hitch, G. (1974). Working memory. In G. Bower (Ed.), *The psychology of learning and motivation: Advances in research and theory*. New York: Academic Press.

Baddeley, A. D., & Lieberman, K. (1980). Spatial working memory. In R. S. Nickerson (Ed.), *Attention and performance VIII*. Hillsdale, NJ: Lawrence Erlbaum Associates.

Baker, C. (1963). Further towards a theory of vigilance. In D. Buckner, & J. McGrath (Eds.), *Vigilance: A symposium*. New York: McGraw-Hill.

Bencomo, A. A., & Daniel, T. C. (1975). Recognition latency for pictures and words as a function of encoded-feature similarity. *Journal of Experimental Psychology: Human Learning and Memory, 1*, 119–125.

Bisseret, A. (1970). Memoire operationelle et structure du travail. *Bulletin de Psychologie, 24*, 280–294.

Bisseret, A. (1971). Analysis of mental processes involved in air traffic control. *Ergonomics, 14*, 565–570.

Bjork, E. L., Bjork, R. A., & Kilpatrick, H. A. (1990, November). *Direct and indirect measures of inhibition in directed forgetting*. Paper presented at the 31st annual meeting of the Psychonomic Society, New Orleans, LA.

Bjork, R. A. (1972). Theoretical implications of directed forgetting. In A. W. Melton, & E. Martin (Eds.), *Coding processes in human memory*. Washington, DC: Winston.

Bower, G. H., & Winzenz, D. (1969). Group structure, coding, and memory for digit series. *Journal of Experimental Psychology Monograph Supplement, 80*, 1–17.

Brainerd, C. J. (1981). Working memory and the developmental analysis of probability judgement. *Psychological Review, 88*, 463–502.

Broadbent, D. E. (1958). *Perception and communications*. New York: Pergamon Press.

Brooks, L. R. (1968). Spatial and verbal components in the act of recall. *Canadian Journal of Psychology, 22*, 349–368.

Brown, J. (1958). Some tests of the decay theory of immediate memory. *Quarterly Journal of Experimental Psychology, 10*, 12–21.

Card, S. K., Moran, T. P., & Newell, A. (1986). The model of human processor: An engineering model of human performance. In K. R. Boff, L. Kaufman, & J. P. Thomas (Eds.), *Handbook of perception and human performance: Volume II, Cognitive processes and performance* (pp. 45-1–45-35). New York: Wiley-Interscience.

Carr, T. H., McCauley, C., Sperber, R. D., & Parmalee, C. M. (1982). Words, pictures, and priming: On semantic activation, conscious identification, and the automaticity of information processing. *Journal of Experimental Psychology: Human Perception and Performance, 8,* 757–777.

Charness, N. (1976). Memory for chess positions: Resistance to interference. *Journal of Experimental Psychology: Human Learning and Memory, 2,* 641–653.

Charness, N. (1979). Components of skill in bridge. *Canadian Journal of Psychology, 33,* 1–50.

Chase, W. G. (1986). Visual information processing. In K. R. Boff, L. Kaufman, & J. P. Thomas (Eds.), *Handbook of perception and human performance: Volume II, Cognitive processes and performance* (pp. 28-1–28-71). New York: Wiley-Interscience.

Chase, W. G., & Ericsson, K. A. (1982). Skill and working memory. In G. Bower (Ed.), *The psychology of learning and motivation.* New York: Academic Press.

Chase, W. G., & Simon, H. A. (1973a). The mind's eye in chess. In W. G. Chase (Ed.), *Visual information processing* (pp. 215–272). New York: Academic Press.

Chase, W. G., & Simon, H. A. (1973b). Perception in chess. *Cognitive Psychology, 4,* 55–81.

Chi, M. T. H. (1978). Knowledge structures and memory development. In R. S. Siegler (Ed.), *Children's thinking: What develops?* (pp. 144–168). Hillsdale, NJ: Lawrence Erlbaum Associates.

Chiesi, H. L., Spilich, G. I., & Voss, I. F. (1979). Acquisition of domain-related information in relation to high and low domain knowledge. *Journal of Verbal Learning and Verbal Behavior, 18,* 257–273.

Cohen, S. (1981). On the generality of some memory laws. *Scandinavian Journal of Psychology, 22,* 267–281.

Cohen, S. (1983). The effect of encoding variables on the free recall of words and action events. *Memory & Cognition, 11,* 575–582.

Conrad, R. (1964). Acoustic comparisons in immediate memory. *British Journal of Psychology, 55,* 75–84.

Conrad, R., & Hull, A. I. (1964). Information, acoustic confusions, and memory span. *British Journal of Psychology, 55,* 429–432.

Cowan, N. (1995). *Attention and memory: An integrated framework* (Oxford psychology series, no. 26). New York: Oxford University Press.

Craik, F. I. M., & Lockhart, R. S. (1972). Levels of processing: A framework for memory research. *Journal of Verbal Learning and Verbal Behaviour, 11,* 671–684.

Craik, F. I. M., & Watkins, M. I. (1973). The role of rehearsal in short-term memory. *Journal of Verbal Learning and Verbal Behavior, 12,* 599–607.

Crowder, R. (1978). Audition and speech coding in short-term memory. In I. Requin (Ed.), *Attention and performance VII* (pp. 248–272). Hillsdale, NJ: Lawrence Erlbaum Associates.

Davis, G., & Holmes, A. (2005). The capacity of visual short-term memory is not a fixed number of objects. *Memory & Cognition, 33*(2), 185–195.

Dillon, R. F., & Reid, L. S. (1969). Short-term memory as a function of information processing during the retention interval. *Journal of Experimental Psychology, 81,* 261–269.

Dosher, B. A., & Russo, I. E. (1976). Memory for internally generated stimuli. *Journal of Experimental Psychology: Human Learning and Memory, 2,* 633–640.

Durso, F. T., Cooke, N. M., Breen, T. I., & Schvaneveldt, R. W. (1987). Is consistent mapping necessary for high-speed scanning? *Journal of Experimental Psychology: Learning, Memory, and Cognition, 13,* 223–229.

Egan, D. E., & Schwartz, B. I. (1979). Chunking in recall of symbolic drawings. *Memory & Cognition, 7,* 149–158.

Einstein, G. O., Williford, C. L., Pagan, J. L., McDaniel, M. A., & Dismukes, R. K. (2003). Forgetting intentions in demanding situations is rapid. *Journal of Experimental Psychology, 9*(3), 147–162.

Eisenstadt, M., & Kareev, Y. (1975). Aspects of human problem solving: The use of internal representations. In D. A. Norman, & D. E. Rumelhart (Eds.), *Exploration in cognition* (pp. 87–112). San Francisco, CA: Freeman.

Ellis, H. C., & Hunt, R. R. (1989). *Fundamentals of human memory and cognition*. Dubuque, IA: Brown.

Endsley, M. R. (1990, June). *Situational awareness global assessment technique (SAGAT)-Air to air tactical version* (Rep. No. NOR DOC 89-58). Hawthorne, CA: Northrop Corporation.

Engle, R. W., & Bukstel, L. (1978). Memory processes among bridge players of differing expertise. *American Journal of Psychology*, 91, 673–690.

Erdelyi, M., Buschke, H., & Finkelstein, S. (1977). Hypermnesia for socratic stimuli: The growth of recall for an internally generated memory list abstracted from a series of riddles. *Memory & Cognition*, 5, 283–286.

Ericsson, K. A. (1985). Memory skill. *Canadian Journal of Psychology*, 39, 188–231.

Federal Aviation Administration. (1989). Air traffic control: Order No. 7110.65F. Washington, DC: Air Traffic Operations Service, U.S. Department of Transportation.

Federal Aviation Administration. (2006). Aviation forecasts-fiscal years 2005–2016. Retrieved January 26, 2006, from http://www.faa.gov/data_statistics/aviation/aerospace_forecasts/2005-2016/.

Finkelman, I. M., & Kirschner, C. (1980). An information-processing interpretation of air traffic control stress. *Human Factors*, 22(5), 561–567.

Fisher, D. F. (1969). *Short-term memory: An annotated bibliography supplement I*. Aberdeen Proving Ground, MD: Human Engineering Laboratories, Aberdeen Research & Development Center.

Fisher, D. F. (1971). *Short-term memory: An annotated bibliography supplement II*. Aberdeen Proving Ground, MD: Human Engineering Laboratories, Aberdeen Research & Development Center.

Fisher, D. F., & Wiggins, H. F. (1968). *Short-term memory: Annotated bibliography*. Aberdeen Proving Ground, MD: Human Engineering Laboratories.

Fisk, A. D., Ackerman, P. L., & Schneider, W. (1987). Automatic and controlled processing theory and its applications to human factors problems. In P. A. Hancock (Ed.), *Human factors psychology* (pp. 159–197). New York: North Holland.

Fisk, A. D., Oransky, N., & Skedsvold, P. (1988). Examination of the role of "higher-order" consistency in skill development. *Human Factors*, 30, 567–581.

Fisk, A. D., & Schneider, W. (1981). Controlled and automatic processing during tasks requiring sustained attention: A new approach to vigilance. *Human Factors*, 23, 737–750.

Fisk, A. D., & Schneider, W. (1983). Category and work search: Generalizing search principles to complex processing. *Journal of Experimental Psychology: Learning, Memory, and Cognition*, 9, 117–195.

Foos, P. W., & Goolkasian, P. (2005). Presentation format effects in working memory: The role of attention. *Memory & Cognition*, 33(3), 499–513.

Frey, P. W., & Adesman, P. (1976). Recall memory for visually presented chess positions. *Memory & Cognition*, 4, 541–547.

Garland, D. J. (1991). Automated systems: The human factor. In J. A. Wise, V. D. Hopkin, & M. L. Smith (Eds.), *Automation and systems issues in air traffic control* (pp. 209–215). Berlin: Springer-Verlag.

Garland, D. J., & Barry, J. R. (1990a). Sport expertise: The cognitive advantage. *Perceptual and Motor Skills*, 70, 1299–1314.

Garland, D. J., & Barry, J. R. (1990b, August). *An examination of chunking indices in recall of schematic information*. Paper presented at the 98th Annual Convention of the American Psychological Association, Boston, MA.

Garland, D. J., & Barry, J. R. (1991). Cognitive advantage in sport: The nature of perceptual structures. *American Journal of Psychology*, 104, 211–228.

Garland, D. J., & Barry, J. R. (1992). Effects of interpolated processing on expert's recall of schematic information. *Current Psychology: Research & Reviews*, 4, 273–280.

Glenberg, A., Smith, S. M., & Green, C. (1977). Type 1 rehearsal: Maintenance and more. *Journal of Verbal Learning and Verbal Behavior*, 16, 339–352.

Godden, D., & Baddeley, A. D. (1980). When does context influence recognition memory? *British Journal of Psychology*, 71, 99–104.

Goettl, B. P. (1985). The interfering effects of processing code on visual memory. *Proceedings of the 29th Annual Meeting of the Human Factors Society* (pp. 66–70). Santa Monaca, CA: Human Factors Society.

Goldin, S. E. (1978). Effects of orienting tasks on recognition of chess positions. *American Journal of Psychology, 91* 659–672.

Goldin, S. E. (1979). Recognition memory for chess position. *American Journal of Psychology, 92*, 19–31.

Gopher, D., & Donchm, E. (1986). Workload-an examination of the concept. In K. R. Boff, L. Kaufman, & J. P. Thomas (Eds.), *Handbook of perception and human performance: Volume II, Cognitive processes and performance* (pp. 41-1–41-49). New York: Wiley-Interscience.

Grolund, S. D., Ohrt, D. D., Dougherty, R. P., Perry, J. L., & Manning, C. A. (1998). Role of memory in air traffic control. *Journal of Experimental Psychology: Applied, 4*(3), 263–280.

Harper, R. R., Hughes, A., & Shapiro, D. Z. (1989). *The functionality of flight strips in ATC work* (Report to the U.K. Civil Aviation Authority). Lancaster, U.K.: Lancaster Sociotechnics Group.

Harris, J. E. (1984). Remembering to do things: A forgotten topic. In J. E. Harris, & P. E. Morris (Eds.), *Everyday memory actions and absent mindedness* (pp. 71–92). London: Academic Press.

Harwood, K., Murphy, E. D., & Roske-Hofstrand, R. J. (1991). Selection and refinement of methods and measures for documenting ATC cognitive processes. Unpublished Draft Technical Note.

Hasher, L., & Zacks, R. T. (1979). Automatic and effortful processing in memory. *Journal of Experimental Psychology: General, 108*, 356–388.

Healy, A. F. (1975). Temporal-spatial patterns in short-term memory. *Journal of Verbal Learning and Verbal Behavior, 14*, 481–495.

Healy, A. F., & McNamara, D. S. (1996). Verbal learning and memory: Does the modal model still work? *Annual Review of Psychology, 47*, 143–72.

Hellige, J. B., & Wong, T. M. (1983). Hemisphere specific interference in dichotic listening: Task variables and individual differences. *Journal of Experimental Psychology: General, 112*, 218–239.

Hirst, W., Spelke, E. S., Reaves, C. C., Caharack, G., & Neisser, U. (1980). Dividing attention without alternation or automaticity. *Journal of Experimental Psychology, 109*, 98–117.

Hitch, G. J. (1978). The role of short-term memory in mental arithmetic. *Cognitive Psychology, 10*, 302–323.

Hollnagel, E., & Woods, D. D. (1983). Cognitive systems engineering: New wine in new bottles. *International Journal of Man-Machine Studies, 18*, 583–600.

Hopkin, V. D. (1980). The measurement of the air traffic controller. *Human Factors, 22*(5), 547–560.

Hopkin, V. D. (1982). *Human factors in air traffic control* (Gradiograph No. 275). Neuilly-sur-Seine, France: NATO.

Hopkin, V. D. (1988). *Human factors aspects of the AERA 2 program.* Famborough, Hampshire, U.K.: Royal Air Force Institute of Aviation Medicine.

Hopkin, V. D. (1989). Man-machine interface problems in designing air traffic control systems. *Proceedings IEEE, 77*, 1634–1642.

Hopkin, V. D. (1991a, January). *Automated flight strip usage: Lessons from the functions of paper strips.* Paper presented at the conference on Challenges in Aviation Human Factors: The National Plan, Washington, DC.

Hopkin, V. D. (1991b). Closing remarks. In J. A. Wise, V. D. Hopkin, & M. L. Smith (Eds.), *Automation and systems issues in air traffic control* (pp. 553–559). Berlin: Springer-Verlag.

Hopkin, V. D. (1995). *Human factors in air traffic control.* Bristol, PA: Taylor & Francis.

Hutchins, E. (1986). Direct manipulation interface. In D. Norman, & S. Draper (Eds.), *User centered system design: New perspectives in human-computer interaction.* Hillsdale, NJ: Lawrence Erlbaum Associates.

Hutchins, E. L., Hollan, J. D., & Norman, D. A. (1985). Direct manipulation interfaces. *Human-Computer Interaction, 1*(4), 311–338.

Huttenlocher, D., & Burke, D. (1976). Why does memory span increase with age? *Cognitive Psychology, 8*, 1–31.

Jackson, A. (1989). *The functionality of flight strips* (Report to the U.K. Civil Aviation Authority). Farnborough, U.K.: Royal Signals and Radar Establishment.

Jacob, R. (1986). Direct manipulation. *Proceedings of the IEEE International Conference on Systems, Man, and Cybernetics* (pp. 348–359). Atlanta, GA.

James, W. (1890). *Principles of psychology*. New York: Holt.

Johnson, M. K. (1988). Reality monitoring: An experimental phenomenological approach. *Journal of Experimental Psychology: General, 117*, 390–394.

Johnson, M. K., Taylor, T. H., & Raye, C. L. (1977). Fact and fantasy: The effects of internally generated events on the apparent frequency of externally generated events. *Memory & Cognition, 5*, 116–122.

Joint Planning and Development Office (JPDO). (2005). Next generation air transportation system integrated plan. Retrieved 12/2/2005 from http://www.jpdo.aero/site_content/index.html

Jonides, J. (1995). Working memory and thinking. In E. E. Smith, & D. N. Osherson (Eds.), *Thinking: An invitation to cognitive science* (Vol. 3, 2nd ed., pp. 215–265). Cambridge, MA: MIT Press.

Kahneman, D., & Treisman, A. M. (1984). Changing views of attention and automaticity. In R. Parasuraman, & R. Davies (Eds.), *Varieties of attention* (pp. 29–61). New York: Academic Press.

Kantowitz, B. H., & Sorkir, R. D. (1983). Human factors: Understanding people-system relationships. New York: Wiley.

Kausler, D. H., & Hakami, M. K. (1983). Memory for activities: Adult age differences and intentionality. *Developmental Psychology, 19*, 889–894.

Keegan, C. (2005). *The next generation air transportation system*. Paper presented at the annual meeting of the air traffic controllers association, Garden Spring, TX.

Keppel, G., & Underwood, B. J. (1962). Proactive inhibition in short-term retention of single items. *Journal of Verbal Learning and Verbal Behavior, 1*, 153–161.

Kinney, G. C., Spahn, M. J., & Amato, R. A. (1977). *The human element in air traffic control: Observations and analysis of the performance of controllers and supervisors in providing ATC separation services* (Rep. No. MTR-7655). McLean, VA: MITRE Corporation.

Kintsch, W. (1974). *The representation of meaning in memory*. Hillsdale, NJ: Lawrence Erlbaum Associates.

Kirchner, J. H., & Laurig, W. (1971). The human operator in air traffic control systems. *Ergonomics, 14*(5), 549–556.

Klapp, S. T. (1987). Short-term memory limits in human performance. In P. Hancock (Ed.), *Human factors psychology* (pp. 1–27). Amsterdam, the Netherlands: North-Holland.

Klapp, S. T., Marshburn, E. A., & Lester, P. T. (1983). Snort-term memory does not involve the "working memory" of information processing: The demise of a common assumption. *Journal of Experimental Psychology: General, 112*, 240–264.

Klapp, S. T., & Netick, A. (1988). Multiple resources for processing and storage in short-term working memory. *Human Factors, 30*(5), 617–632.

Klapp, S. T., & Philipoff, A. (1983). Short-term memory limits in performance. *Proceedings of the Human Factors Society* (Vol. 27, pp. 452–454). Norfolk, VA.

Klatzky, R. L. (1980). *Human memory: Structures and processes*. San Francisco, CA: Freeman.

Koriat, A., & Ben-Zur, H. (1988). Remembering that I did it: Process and deficits in output monitoring. In M. Grunegerg, P. Morris, & R. Sykes (Eds.), *Practical aspects of memory: Current research and issues* (Vol. I, pp. 203–208). Chichester: Wiley.

Koriat, A., Ben-Zur, H., & Nussbaum, A. (1990). Encoding information for future action: Memory for to-be-performed versus memory for to-be-recalled tasks. *Memory & Cognition, 18*, 568–583.

Koriat, A., Ben-Zur, H., & Sheffer, D. (1988). Telling the same story twice: Output monitoring and age. *Journal of Memory & Language, 27*, 23–39.

Kosslyn, S. (1981). The medium and the message in mental imagery: A theory. *Psychological Review, 88*, 46–66.

Kroll, N. E. A., Kellicut, M. H., & Parks, T. E. (1975). Rehearsal of visual and auditory stimuli while shadowing. *Journal of Experimental Psychology: Human Learning and Memory, 1*, 215–222.

Kvalseth, T. (1978). Human and Baysian information processing during probabilistic inference tasks. *IEEE Transactions on Systems, Man, and Cybernetics, 8*, 224–229.

LaBerge, D. (1973). Attention and the measurement of perceptual learning. *Memory & Cognition,* *1,* 268–276.

LaBerge, D. (1975). Acquisition of automatic processing in perceptual and associative learning. In P. M. A. Rabbit, & S. Dornic (Eds.), *Attention and performance V* (pp. 78–92). New York: Academic Press.

LaBerge, D. (1976). Perceptual learning and attention. In W. K. Estes (Ed.), *Handbook of learning and cognitive processes.* Hillsdale, NJ: Lawrence Erlbaum Associates.

LaBerge, D. (1981). Automatic information processing: A review. In J. Long, & A. D. Baddeley (Eds.), *Attention and performance IX* (pp. 173–186). Hillsdale, NJ: Lawrence Erlbaum Associates.

Landis, D., Silver, C. A., Jones, J. M., & Messick, S. (1967). Level of proficiency and multidimensional viewpoints about problem similarity. *Journal of Applied Psychology, 51,* 216–222.

Lane, D. M., & Robertson, L. (1979). The generality of levels of processing hypothesis: An application to memory for chess positions. *Memory & Cognition, 7,* 253–256.

Lauber, J. K. (1993, July). Human performance issues in air traffic control. *Air Line Pilot,* pp. 23–25.

Levesley, J. (1991). The blue sky challenge: A personal view. In J. A. Wise, V. D. Hopkin, & M. L. Smith (Eds.), *Automation and systems issues in air traffic control* (pp. 535–539). Berlin: Springer-Verlag.

Loftus, G. R., Dark, V. J., & Williams, D. (1979). Short-term memory factors in ground controller/pilot communications. *Human Factors, 21,* 169–181.

Loft, S., Humphreys, M., & Neal, A. (2004). The influence of memory for prior instances and performance on a confllict detection task. *Journal of Experimental Psychology, 10*(3), 173–187.

Logan, G. D. (1978). Attention in character-classification tasks: Evidence for the automaticity of component stages. *Journal of Experimental Psychology: General, 107,* 32–63.

Logan, G. D. (1979). On the use of a concurrent memory load to measure attention and automaticity. *Journal of Experimental Psychology: Human Perception and Performance, 5,* 189–207.

Logan, G. D. (1980). Attention and automaticity in stroop and priming tasks: Theory and data. *Cognitive Psychology, 12,* 523–553.

Logan, G. D. (1985a). Skill and automaticity: Relations, implications, and future directions. *Canadian Journal of Psychology, 39,* 367–386.

Logan, G. D. (1985b). On the ability to inhibit simple thoughts and actions: ll. Stop signal studies of repetition priming. *Journal of Experimental Psychology: Learning, Memory, and Cognition, 11,* 65–69.

Logan, G. D. (1988a). Toward an instance theory of automatization. *Psychological Review, 95,* 95–112.

Logan, G. D. (1988b). Automaticity, resources, and memory: Theoretical controversies and practical implications. *Human Factors, 30*(5), 583–598.

Logie, R. H. (1993). Working memory and human-machine systems. In J. A. Wise, V. D. Hopkin, & P. Stager (Eds.), *Verification and validation of complex systems: Human factors issues* (NATO ASI Series Vol. F110, pp. 341–353). Berlin: Springer-Verlag.

Logie, R. H. (1995). *Visuo-spatial working memory.* Series: *Essays in cognitive psychology.* Hove, U.K.: Lawrence Erlbaum Associates.

Lyon, G. R., & Krasnegor, N. A. (Eds.). (1996). *Attention, memory, and executive function.* Baltimore, MD: P. H. Brookes.

Mackworth, J. (1959). Paced memorizing in a continuous task. *Journal of Experimental Psychology, 58,* 206–211.

Marcel, A. T. (1983). Conscious and unconscious perception: An approach to the relations between phenomenal experience and perceptual processes. *Cognitive Psychology, 15,* 238–300.

Martin, P. W., & Kelly, R. T. (1974). Secondary task performance during directed forgetting. *Journal of Experimental Psychology, 103,* 1074–1079.

McKeithen, K. B., Reitman, J. S., Rueter, H. H., & Hirtle, S. C. (1981). Knowledge organization and skill differences in computer programmers. *Cognitive Psychology, 13,* 307–325.

McLeod, P., McLaughlin, C., & Nimmo-Smith, I. (1985). Information encapsulation and automaticity: Evidence from the visual control of finely timed action. In M. I. Posner, & O. S. Marin (Eds.), *Attention and performance XI* (pp. 391–406). Hillsdale, NJ: Lawrence Erlbaum Associates.

Means, B., Mumaw, R., Roth, C., Schlager, M., McWilliams, E., Gagne, V. R., et al. (1988). *ATC training analysis study: Design of the next-generation ATC training system*. Washington, DC: Federal Aviation Administration.

Megan, E., & Richardson, J. (1979). Target uncertainty and visual scanning behavior. *Human Factors, 21*, 303–316.

Metzger, U., & Parasuraman, R. (2001). The role of air traffic controller in future air traffic management: An empirical study of active control versus passive monitoring. *Human Factors, 43*(4), 519–528.

Miller, G. A. (1956). The magical number seven plus or minus two: Some limits on our capacity for processing information. *Psychological Review, 63*, 81–97.

Mitchell, D. B. (1989). How many memory systems? Evidence from aging. *Journal of Experimental Psychology: Learning, Memory, and Cognition, 15*, 31–49.

Mogford, R. H. (1991). Mental model in air traffic control. In J. A. Wise, V. D. Hopkin, & M. L. Smith (Eds.), *Automation and systems issues in air traffic control* (pp. 235–242). Berlin: Springer-Verlag.

Monsell, S. (1984). Components? If working memory underlying verbal skills: A "distributed capacities" view. 11. In H. Bouma, & D. BouwhulS (Eds.), *Attention and performance X* (pp. 142–164). Hillsdale, NJ: Lawrence Erlbaum Associates.

Moray, N. (1980). *Human information processing and supervisory control* (Tech. Rep.). Cambridge, MA: MIT Man-Machine Systems Laboratory.

Moray, N. (1986). Monitoring behavior and supervisory control. In K. R. Boff, L. Kaufman, & J. P. Thomas (Eds.), *Handbook of perception and human performance: Volume II, Cognitive processes and performance* (pp. 40-1–40-51). New York: Wiley-Interscience.

Moray, N. (1988). Intelligent aids, mental models, and the theory of machines. In E. Hollnagel, G. Mancini, & D. D. Woods (Eds.), *Cognitive engineering in complex dynamic worlds* (pp. 165–175). London: Academic Press.

Murdock, B. B. (1961). The retention of individual items. *Journal of Experimental Psychology, 62*, 618–625.

Murdock, B. B. (1968). Modality effects in short-term memory: Storage or retrieval? *Journal of Experimental Psychology, 77*, 79–86.

Murphy, E. D., & Cardosi, K. M. (1995). Human information processing. In K. M. Cardosi, & E. D. Murphy (Eds.), *Human factors in the design and evaluation of air traffic control systems* (Report No. DOT-FAARD-95-3, pp. 135–218). Washington, DC: FAA.

Narborough-Hall, C. S. (1987). Automation: Implications for knowledge retention as a function of operator control responsibility. *Proceedings of the Third Conference of the British Computer Society* (pp. 269–282). Cambridge, U.K.: Cambridge University Press.

Naveh-Benjamin, M., & Jonides, J. (1984). Maintenance rehearsal: A two-component analysis. *Journal of Experimental Psychology: Learning, Memory, and Cognition, 10*, 369–385.

Navon, D., & Gopher, D. (1979). On the economy of the human processing system. *Psychological Review, 86*, 214–255.

Neely, J. H. (1977). Semantic priming and retrieval from lexical memory: Roles of inhibitionless spreading activation and limited-capacity attention. *Journal of Experimental Psychology: General, 106*, 226–254.

Neisser, U. (1976). *Cognition and reality*. San Francisco, CA: Freeman.

Nickels, B. J., Bobko, P., Blair, M. D., Sands, W. A., & Tartak, E. L. (1995). *Separation and control hiring assessment (SACHA): Final job analysis report*. Bethesda, MD: University Research Corporation.

Nilsson, L. G., Ohlsson, K., & Ronnberg, J. (1977). Capacity differences in processing and storage of auditory and visual input. In S. Domick (Ed.), *Attention and performance VI*. Hillsdale, NJ: Lawrence Erlbaum Associates.

Norman, D. A. (1986). Cognitive engineering. In D. A. Norman, & S. W. Draper (Eds.), *User centered system design* (pp. 31–61). Hillsdale, NJ: Lawrence Erlbaum Associates.

Operational Error Analysis Work Group. (1987, August). *Actions to implement recommendations of April 17, 1987*. Unpublished manuscript, Federal Aviation Administration, Washington, DC.

Nowinsky, J. L., Holbrook, J. B., & Dismukes, R. K. (1987). Human memory and cockpit operations: An ASRS study. In *Proceedings of the International Symposium on Aviation Psychology*. Columbus: Ohio State University.

Peterson, L. R., & Peterson, M. J. (1959). Short-term retention of individual verbal items. *Journal of Experimental Psychology, 58*, 193–198.

Phillips, W. A., & Christie, F. M. (1977). Interference with visualization. *Quarterly Journal of Experimental Psychology, 29*, 637–650.

Posner, M. I. (1973). *Cognition: An introduction*. Glenview, IL: Scott, Foresman.

Posner, M. I. (1978). *Chronometric explorations of the mind*. Hillsdale, NJ: Lawrence Erlbaum Associates.

Posner, M. I., & Mitchell, R. F. (1967). Chronometric analysis of classification. *Psychological Review, 74*, 392–409.

Posner, M. I., & Snyder, C. R. R. (1975). Attention and cognitive control. In R. L. Solso (Ed.), *Information processing and cognition: The Loyola symposium* (pp. 212–224). Hillsdale, NJ: Lawrence Erlbaum Associates.

Potter, M. C., & Levy, E. I. (1969). Recognition memory for a rapid sequence of pictures. *Journal of Experimental Psychology, 81*, 10–15.

Rantanen, E. (1994). The role of dynamic memory in air traffic controllers' situational awareness. In R. D. Gilson, D. J. Garland, & J. M. Koonce (Eds.), *Situational awareness in complex systems* (pp. 209–215). Daytona Beach, FL: Embry-Riddle Aeronautical University Press.

Rasmussen, J. (1979). *On the structure of knowledge: A morphology of mental models in a man-machine systems context* (Rep. No. RISO-M-2192). Roskilde, Denmark: Riso National Laboratory.

Rayner, E. H. (1958). A study of evaluative problem solving. Part I: Observations on adults. *Quarterly Journal of Experimental Psychology, 10*, 155–165.

Reitman, J. S. (1974). Without surreptitious rehearsal: Information and short-term memory decays. *Journal of Verbal Learning and Verbal Behavior, 13*, 365–377.

Reitman, J. S. (1976). Skilled performance in go: Deducing memory structures from inter-response times. *Cognitive Psychology, 8*, 336–356.

Roediger, H. L., III, Knight, J. L., & Kantowitz, B. H. (1977). Inferring decay in short-term memory: The issue of capacity. *Memory & Cognition, 5*, 167–176.

Rouse, W. B. (1973a). Model of the human in a cognitive prediction task. *IEEE Transactions on Systems, Man, and Cybernetics, 3*, 473–478.

Rouse, W. B. (1973b). *Models of man as a suboptimal controller* (NTIS Rep. No. N75-19126/2). Cambridge, MA: MIT, *Ninth Annual NASA Conference on Manual Control*, Cambridge, MA.

Rouse, W. B., & Morris, N. M. (1986). On looking into the black box: Prospects and limits in the search for mental models. *Psychological Bulletin, 100*, 349–363.

Radio Technical Commission for Aeronautics. (1995). *Report of the RTCA Board of Directors select committee on free flight*. Washington, DC.

Sanders, M. S., & McCormick, E. J. (1993). *Human factors in engineering and design*. New York: McGraw-Hill.

Schlager, M. S., Means, B., & Roth, C. (1990). Cognitive task analysis for the real-time world. In *Proceedings of the Human Factors Society 34th Annual Meeting* (pp.1309–1313). Santa Monica, CA: Human Factors Society.

Schneider, W., Dumais, S. T., & Shiffrin, R. M. (1984). Automatic and control processing and attention. In R. Parasuraman, & R. Davies (Eds.), *Varieties of attention* (pp. 1–27). New York: Academic Press.

Schneider, W., & Fisk, A. D. (1982). Degree of consistent training: Improvements in search performance and automatic process development. *Perception and Psychophysics, 31*, 160–168.

Schneider, W., & Shiffrin, R. M. (1977). Controlled and automatic human processing: I. Detection, search, and attention. *Psychological Review, 84*, 1–66.

Schneiderman, B. (1976). Exploratory experiments in programmer behavior. *Journal of Computer and Information Sciences, 5*, 123–143.

Schneiderman, B. (1980). *Software psychology*. Cambridge, MA: Winthrop.

Schneiderman, B. (1983). Direct manipulation: A step beyond programming languages. *IEEE Computer*, *16*(8), 57–69.

Shaffer, L. H. (1993). Working memory or working attention? In A. Baddeley, & L. Weiskrantz (Eds.), *Attention: Selection, awareness, and control: A tribute to Donald Broadbent*. New York: Clarendon Press.

Shiffrin, R. M., & Schneider, W. (1977). Controlled and automatic human information processing: II. Perceptual learning, automatic attending, and a general theory. *Psychological Review, 84*, 127–190.

Simon, H. A., & Chase, W. G. (1973). Skill in chess. *American Scientist, 61*, 394–403.

Simon, H. A., & Gilmartin, K. A. (1973). Simulation of memory for chess positions. *Cognitive Psychology, 5*, 29–46.

Slaboda, J. (1976). Visual perception of musical notation: Registering pitch symbols in memory. *Quarterly Journal of Experimental Psychology, 28*, 1–16.

Slamecka, N. J., & Graf, P. (1978). The generation effect: Delineation of a phenomenon. *Journal of Experimental Psychology: Learning, Memory, and Cognition, 4*, 592–604.

Smith, E. (1968). Choice reaction time: An analysis of the major theoretical positions. *Psychological Bulleting, 69*, 77–110.

Sollenberger, R. L., & Stein, E. S. (1995). A simulation study of air traffic controller situational awareness. *Proceedings of the International Conference on Experimental Analysis and Measurement of Situational Awareness* (pp. 211–217). Daytona Beach, FL: Embry Riddle Aeronautical University.

Sperandio, J. C. (1974). Extension to the study of the operational memory of air traffic controllers (RSRE Translation No. 518). Unpublished manuscript.

Squire, L. R., Knowlton, B., & Musen, G. (1993). The structure and organization of memory. *Annual Review of Psychology, 44*, 453–495.

Stanners, R. F., Meunier, G. F., & Headley, D. B. (1969). Reaction time as an index of rehearsal in short-term memory. *Journal of Experimental Psychology, 82*, 566–570.

Starkes, J. L., & Deakin, J. (1984). Perception in sport: A cognitive approach to skilled performance. In W. F. Straub, & J. M. Williams (Eds.), *Cognitive sport psychology* (pp. 115–128). Lansing, NY: Sport Science Associates.

Stein, E. S., & Bailey, J. (1994). *The controller memory guide. Concepts from the field* (Rep. No. AD-A289263; DOT/FAA/CT-TN94/28). Atlantic City, NJ: Federal Aviation Administration.

Stein, E. S., & Garland, D. (1993). *Air traffic controller working memory: Considerations in air traffic control tactical operations* (DOT/FAA/CT-TN93/37). Atlantic City International Airport: Federal Aviation Administration Technical Center.

Stein, E. S., & Sollenberger, R. (1995, April 24–27). The search for air traffic controller memory aids. *Proceedings Eighth International Symposium on Aviation Psychology*, Columbus, OH, 1(A96-45198 12-53), 360–363.

Sternberg, S. (1969). The discovery of processing stages: Extension of Donder's method. *Acta Psychological, 30*, 276–315.

Tulving, E. (1962). Subjective organization in free recall of "unrelated" words. *Psychological Review, 69*, 344–354.

Underwood, B. J. (1957). Interference and forgetting. *Psychological Review, 64*, 49–60.

Vortac, O. U. (1991). *Cognitive functions in the use of flight progress strips: Implications for automation*. Unpublished manuscript, University of Oklahoma, Norman, OK.

Vortac, O. U., Edwards, M. B., Fuller, D. K., & Manning, C. A. (1993). Automation and cognition in air traffic control: An empirical investigation [Special issue]. Practical aspects of memory: The 1994 conference and beyond. *Applied Cognitive Psychology, 7*(7), 631–651.

Vortac, O. U., Edwards, M. B., Fuller, D. K., & Manning, C. A. (1994). *Automation and cognition in air traffic control: An empirical investigation* (Report No. FAA-AM-94-3). Oklahoma City, OK: FAA Office of Aviation Medicine Reports.

Vortac, O. U., Edwards, M. B., Fuller, D. K., & Manning, C. A. (1995). *Automation and cognition in air traffic control: An empirical investigation* (Final report, Rep. No. AD-A291932; DOT/FAA/AM-95/9). Oklahoma, City, OK: Federal Aviation Administration.

Vortac, O. U., Edwards, M. B., & Manning, C. A. (1995). *Functions of external cues in prospective memory* (Final report, Report No. DOT/F AA/AM-95/9). Oklahoma City, OK: Civil Aeromedical Institute.

Welford, A. T. (1976). *Skilled performance.* Glenview, IL: Scott, Foresman.

Whitfield, D. (1979). A preliminary study of air traffic controller's picture. *Journal of the Canadian Air Traffic Controller's Association, 11,* 19–28.

Whitfield, D., & Jackson, A. (1982). The air traffic controller's "picture" as an example of a mental model. In G. Johannsen, & J. E. Rijnsdorp (Eds.), *Analysis, design, and evaluation of man-machine systems* (pp. 45–52).

Dusseldorf, West Germany: International Federation of Automatic Control.

Wickelgren, W. A. (1964). Size of rehearsal group in short-term memory. *Journal of Experimental Psychology, 68,* 413–419.

Wickens, C. D. (1984). *Engineering psychology and human performance.* Columbus, OH: Charles E. Merrill.

Wickens, C. D. (1992). *Engineering psychology and human performance* (2nd ed.). New York: HarperCollins.

Wickens, C. D., & Flach, D. M. (1988). Information processing. In E. L. Wiener, & D. C. Nagel (Eds.), *Human factors in aviation* (pp. 111–155). San Diego, CA: Academic Press.

Wickens, C. D., Sandry, D., & Vidulich, M. (1983). Compatibility and resource competition between modalities of input, central processing, and output: Testing a model of complex task performance. *Human Factors, 25,* 227–248.

Wilkins, A. J., & Baddeley, A. D. (1978). Remembering to recall in everyday life: An approach to absent-mindedness. In M. Gruneberg, P. Morris, & R. Sykes (Eds.), *Practical aspects of memory.* London: Academic Press.

Wilson, J. R., & Rutherford, A. (1989). Mental models: Theory and application in human factors. *Human Factors, 31,* 617–634.

Wingfield, A., & Butterworth, B. (1984). Running memory for sentences and parts of sentences: Syntactic parsing as a control function in working memory. In H. Bouma, & D. Bouwhuis (Eds.), *Attention and performance X.* Hillsdale, NJ: Lawrence Erlbaum Associates.

Winograd, E. (1988). Some observations on prospective remembering. In M. Gruneberg, P. Morris, & R Sykes (Eds.), *Practical aspects of memory: Current research and issues* (Vol. 1, pp. 348–353). Chichester: Wiley.

Wise, J. A., & Debons, A. (1987). *Information systems: Failure analysis.* Berlin: Springer-Verlag.

Wise, J. A., Hopkin, V. D., & Smith, M. L. (1991). Automation and System Issues in Air Traffic Control, (Eds.), Berlin: Springer-verlag.

Yntema, D. B. (1963). Keeping track of several things at once. *Human Factors, 6,* 7–17.

Yntema, D. B., & Mueser, G. B. (1960). Remembering the states of a number of variables. *Journal of Experimental Psychology, 60,* 18–22.

Zbrodoff, N. J., & Logan, G. D. (1986). On the autonomy of mental processes: A case study of arithmetic. *Journal of Experimental Psychology: General, 115,* 118–130.

Zeitlin, L. R., & Finkleman, J. M. (1975). Research note: Subsidiary task techniques of digit generation and digit recall as indirect measures of operator loading. *Human Factors, 17,* 218–220.

Zingale, C., Gromelski, S., Ahmed, S. B., & Stein, E. S. (1993). *Influence of individual experience and flight strips on air traffic controller memory/situational awareness* (Rep. No. DOT /F AA/ CT-TN93/31). Princeton, NJ: PERI, Inc.

Zingale, C., Gromelski, S., & Stein, E. S. (1992). *Preliminary studies of planning and flight strip use as air traffic controller memory aids* (Rep. No. DOT/FAA/CT-TN92/22). Princeton, NJ: PERI, Inc.

22

Air-Traffic Control Automation

V. David Hopkin
Human Factors Consultant

22.1 The Need for Automation

Throughout most parts of the world, aviation as an industry is expanding. Though air traffic demands are notoriously difficult to predict, being vulnerable to powerful and unforeseeable extraneous influences beyond the aviation community, all current forecasts concur about substantial future increases in aircraft numbers. As a consequence, air-traffic control must seek to accommodate increasing demands for its services. Even the most efficient current air-traffic control systems cannot remain as they are, because they were never designed to handle the quantities of air traffic now expected in the longer term, and they could not readily be adapted to cope with such increases in traffic volume. The combined sequential processes of devising, proving, and introducing major changes in an air-traffic control system may take many years to implement, but to make no changes is not a practical option. Hence, air-traffic control must evolve (Wise, Hopkin, & Smith, 1991).

A relevant parallel development is the major expansion in the quantity and quality of the information available about each flight, which has brought significant changes and benefits in the past and will bring further ones in the future, applicable to the planning and conduct of each flight. The quality and frequency of updating of the information about the position and progress of each aircraft were enhanced when radar was introduced, and with each refinement of it further enhancements accrued as information became available from satellites, data links, and other innovations (Hopkin, 1989). In principle, practical technical limitations on data gathering might disappear altogether, because whatever information was deemed to be essential for safety or efficiency could be provided.

With limited and finite airspace, the only way to handle more air traffic in regions that are already congested is to allow aircraft to approach each other more closely in safety. Flight plans, navigational data, onboard sensors, prediction aids, and computations can collectively provide very complete and frequently updated details about the current state and future progress of each flight, in relation to other flights, hazards nearby, and the flight objectives. The provision of high-quality information about where each aircraft is and where it is going could allow the minimum separation standards between aircraft to be reduced safely. However, the closer that aircraft are permitted to approach each other, the less is the time available to respond to any emergency, and the fewer are the options available for resolving the emergency safely (Hopkin, 1995).

An apparent alternative option for handling more traffic would seem to be to employ more controllers and further partition the current region of airspace for which each controller or small team of controllers is responsible. Unfortunately, in the regions of densest traffic where the problems of handling more traffic are most acute, this process of partitioning has usually already reached its beneficial limits. Further partitioning may become self-defeating and counterproductive, wherever the consequent reductions in the controller's workload are outweighed by the extra work generated by the partitioning itself, in forms such as additional coordination, liaison, communications, handovers of responsibility, and shortened experience of each flight and its history. Further partitioning would also be unwelcome in cockpits, where it would lead to extra work through additional reselections of communications frequencies. Although some restructuring of sector boundaries may aid traffic flow locally, further partitioning of the airspace is generally not a practical option, nor is the loading of more traffic onto controllers while retaining the present control methods, because dealing with current heavy traffic keeps the controllers continuously busy and they cannot handle much more.

Therefore, the essence of the problem is that each controller must become responsible for handling more air traffic, but without any diminution of the existing high standards of safety and efficiency in the air-traffic control service, and preferably with some enhancement of them. Similarly, increased delays to traffic caused by air-traffic control are unacceptable. As current traffic levels keep the controller fully occupied, the implication is that somehow each controller must spend less time in dealing with each aircraft, without any impairment of standards (International Civil Aviation Organization, 1993). To achieve this, the controller needs help, much of which must come from automation and computer assistance (Hopkin, 1994a). As a first step, the broad human-factors implications of foreseeable technical developments have to be deduced (Hopkin, 1997). There is a growing awareness that some solutions to problems may not have universal validity, because cultural differences intervene (Mouloua & Koonce, 1997).

Human-factors aspects of air-traffic control automation have been considered in a series of texts, which vary according to their envisaged readerships, their objectives, and the circumstances that gave rise to them (Cardosi & Murphy, 1995; Hopkin, 1995; International Civil Aviation Organization, 1993; Isaac & Ruitenberg, 1999; Smolensky & Stein, 1997; Wickens, Mavor, Parasuraman, & McGee, 1998; Wise et al., 1991).

22.2 Automation and Computer Assistance

This chapter covers both automation and computer assistance in relation to the air-traffic controller, but does not treat these concepts as synonymous. It seems prudent to specify some distinguishing characteristics of each in relation to human factors (Hopkin, 1995).

Automation is used to refer to functions that do not require, and often do not permit, any direct human intervention or participation in them. The human controller generally remains unaware of the actual processes of automation, but may be aware only of its products, unless special provision has been made to notify the controller of its occurrence or progress. The products of automation may be applied by the controller, who is normally unable to intervene in the processes that lead to these products. Previous applications of automation to air-traffic control have mostly been with respect to very simple functions that are routine, continuous, or frequently repeated. Such automated functions include data gathering and storage, data compilation and correlation, the computation and presentation of summaries of data, the retrieval and updating of data, and data synthesis. A specific example is the provision of an aircraft's altitude within its label on the radar display. Most applications of automation are universal and unselective. Some limited selective automation to accommodate different information requirements of different tasks has been achieved, and is expected to become more common. When the selectivity permits human intervention, or is adaptive in accordance with the needs of individual controllers, it then constitutes computer assistance.

In computer assistance, the human tasks, roles, and functions are central in that they are the hub or focus of activities and are supported by the computer. The human controller must retain some means to guide and participate in the processes of computer assistance wherever the controller carries the legal responsibility for air-traffic control events. The concept of human-centered automation (Billings, 1991) represents a reaction against forms of computer assistance that have owed more to their technical feasibility than to user requirements, but human-centered automation has itself been criticized as narrow, because it fails to adequately cover organizational factors, job satisfaction, and motivation—the characteristics of humans which make them people (Brooker, 2005). In air-traffic control, computer assistance of cognitively complex human functions has always been preferred to their full automation. The latter has sometimes been proposed, as in the prevention of conflicts between flights by automated flight-profile adjustments without referral to either the controller or the pilot, but a combination of formidable technical difficulties, lack of user acceptability, and problems of legal responsibility has so far prevented its adoption. A defining characteristic of computer assistance is that some human participation is essential as a process or function cannot be completed without it, although the actual human role may be minimal, for example, to sanction an automated function.

Therefore, air-traffic control is now, and will remain, computer-assisted rather than automated in relation to those aspects of it which involve active participation by the human controller. Without the presence of the human controller there would be no air-traffic control in any current system or in any currently planned future system. All the evolutionary plans envisage the continuing involvement of the human controller (Federal Aviation Administration, 1995; Wickens et al., 1998). In this important respect, air-traffic control differs from some other large human–machine systems that can function automatically (Wise, Hopkin, & Stager, 1993a). The actual pace at which automation and computer assistance are introduced into air-traffic control has been slower than what might be expected. Time after time, practical realities of many kinds have forced the scaling down or abandonment of initially ambitious plans. Among these realities are the following: safety concerns, technical feasibility, escalating costs, difficult legal issues of responsibility, problems of human–machine matching, underestimates of the complexity of the required software, severe slippage in timescales during system procurement, training problems, the impossibility of predicting and allowing for every possible contingency and combination of events, the feasibility of manual reversion in the event of major system failure, and insufficient proof that real benefits in safety, efficiency, capacity, or costs will accrue from the planned changes (Hopkin, 1998).

The concepts of computer assistance have been considered for application in complex cognitive functions such as decision making, problem solving, predictions, planning, scheduling, and the allocation and management of resources (e.g., Vortac, Edwards, Fuller, & Manning, 1993). Some computer assistance can be selective, not responding identically to every individual and in every circumstance, but differently aiding particular jobs, tasks, functions, and human roles. The controller may sometimes retain the option of dispensing with it altogether. A characteristic of computer assistance in air-traffic control, though not an inevitable property of it, is that it is intended to aid the individual controller. Few current or pending forms of computer assistance are intended to aid controllers functioning as teams or their supervisors.

In some texts, the concept of automation embraces computer assistance, and different degrees of feasible human intervention are referred to as different levels of automation (Wickens, Mavor, & McGee, 1997). A further practical categorization of air-traffic control functions that are not wholly manual includes semimanual, semiautomated, automated, and fully automated functions (Cardosi & Murphy, 1995).

22.3 Technological Advances with Human-Factors Implications

22.3.1 Communications

At one time, spoken messages were the main means of communication between the controller and the pilot (Moser, 1959). These are still being studied (Prinzo & McClellan, 2005; Prinzo & Morrow, 2002), but they are being replaced by data transponded automatically or on request. Thus, both the need for human speech and the reliance on it are consequently reduced. In principle, time is saved, though not always in practice (Cardosi, 1993). Other kinds of information contained in speech may be lost, for example, those used to judge the competence or confidence of the speaker.

22.3.2 Radar

Radar provides a plan view of the traffic, and thus, shows the lateral and longitudinal separations between the aircraft in flight. The permitted separation minima between aircraft within radar coverage have usually been much less than those for aircraft beyond radar coverage, for example, in transoceanic flight. Modern secondary radar supplements the plan view with a label attached to each aircraft's position on the radar screen, showing its identity, destination, and aspects of its current state, such as its altitude (flight level), speed, and whether it is climbing, descending, or in level flight. The changing information on the label is updated frequently and automatically, with some significant changes being signaled to the controller.

22.3.3 Navigation Aids

Ground-based aids that can be sensed or interrogated from aircraft mark standard routes, airways, or corridors that often extend between major centers of population. Navigation aids that show the direction from the aircraft of sensors at known positions on the ground permit computations about the current track and heading of the aircraft. Other navigation aids use this information to make comparisons between aircraft, and thus, help the controller to maintain safe separations between them. The information available to the controller depends considerably upon the navigation aids in use.

22.3.4 Satellites

Data derived from satellites about the location of aircraft represent a technical advance that can transform the accuracy and coverage of the information available about air traffic. To accommodate and benefit from this increased accuracy, human tasks and functions must adapt to it not only in terms of task performance, but also of appropriately revised assessments of its trustworthiness and reliability. Some of the greatest benefits of satellite data for air-traffic control are where there has been no radar, for example, over oceans.

22.3.5 Automatically Transponded Data

These data are not obtained on request or as a machine response to a controller's actions, but are independent of controller activities. They can replace routine human actions and chores, but may also remove some adaptability and responsiveness to controller needs. The controller cannot know or access such data unless special provision for this has been made.

22.3.6 Datalinks

These send data continuously or very frequently between the aircraft and the air-traffic control system, independently of the pilot and the controller, who may be able to tap into them for information or be presented automatically with it. Thus, the associated human-factors problems are centered on what information derivable from datalinks is needed by which controllers under what circumstances, and on its forms of presentation and level of detail.

22.3.7 Information Displays

All the information needed by the controller about the aircraft cannot be presented within the labels on a radar display, without this information becoming too large and cluttered and generating visual problems of label overlap. Furthermore, much of the information is not readily adaptable to such forms of presentation. There have always been further kinds of information display in air-traffic control, such as maps and tabular displays. In the latter, aircraft can be listed or categorized according to a dimension, such as flight level, direction, route, destination, or other criterion appropriate for the controller's tasks. Tabular displays of automatically compiled and updated information can be suitable for presentation as windows in other air-traffic control displays.

22.3.8 Electronic Flight Progress Strips

These are a particular kind of tabular display, intended to replace paper flight progress strips. On the latter, the details of each aircraft for which the controller had responsibility appeared on a paper strip in a holder on a strip board, and were amended by hand. On the other hand, electronic strips can be generated and amended automatically, but the controller must use input devices instead of handwritten annotations to amend them. The full usage of paper flight strips has to be understood prior to the automation of strips (Durso, Batsakes, Crutchfield, Braden, & Manning, 2004). Electronic strips have posed some human-factors problems of design by revealing difficulties in electronically capturing the full functionality of paper flight strips, which are more complex than they seem (Hughes, Randall, & Shapiro, 1993). Also, the greater flexibility of electronic formats calls for some reappraisal of the desirability of providing different flight strip formats for different controller needs.

22.3.9 Data Input Devices

Through these devices, the controller enters information into the computer and the system, and initiates events. The type, sensitivity, and positioning of the chosen input devices must be appropriate for the tasks, displays, communications, and forms of feedback; the latter being essential for learning, gaining experience, and acquiring skills. Decisions about the input devices partly predetermine the kinds of human error that are possible and will occur in their use. Technical advances may extend the range of input devices available, for example, by introducing touch-sensitive surfaces or automated speech recognition. They raise the human-factors issues about the respective merits and disadvantages of alternative input devices, and their mutual compatibility within a single workspace (Hopkin, 1995).

22.4 Computations with Human-Factors Implications

22.4.1 Alerting

Various visual or auditory alerting signals can be provided as automated aids. They may serve as memory aids, prompts, or instructions to the controller, or may signify a state of progress or a change of state. They seek to draw the controller's attention to particular information or require the controller to respond. They are normally triggered whenever a predefined set of circumstances actually arises, and can be distracting if employed inappropriately or excessively.

22.4.2 Track Deviation

To save searching, automatic computations can compare the intended and actual track of an aircraft and signal to the controller whenever an aircraft deviates by more than a predetermined permissible margin. The controller is then expected to contact the pilot to ascertain the reasons for the deviation and correct it wherever appropriate. The significance and degree of urgency of a track deviation depend on the phase of the flight. It can become very urgent if it occurs during the final approach to landing.

22.4.3 Conflict Detection

This is also an aid to searching, which can facilitate judgement. Comparisons between aircraft are made frequently and automatically, and the controller's attention is drawn by changing the visual codings of any displayed aircraft that are predicted to infringe the separation standards between them within a given time or distance. Depending on the quality of the data about the aircraft, a balance is struck to give as much forewarning as possible without incurring too many false alarms. The practical value of the aid depends on correct computation and on getting this balance right. Sometimes, the position or time of occurrence of the anticipated conflict is depicted, but the aid may provide no further information.

22.4.4 Conflict Resolution

This aid takes conflict detection a stage further. The data used to specify a conflict can be applied to the data on other aircraft traffic to compute and present automatically one or more solutions to the conflict that meet all predefined criteria and rules. If more than one solution is offered to the controller, the order of computer preference usually follows the same rules. Nominally, the controller can still choose to devise and implement another solution, but controllers are trained and expected to accept the preferred computer solution in normal circumstances. It can be difficult for the controller who imposes a human solution to ascertain all the factors included in the automated one. However, this becomes necessary either if the automation has taken account of information unknown to the controller or if the controller possesses information that is unavailable to the computer, but invalidates its solution. One type of conflict detection aid warns the controller of a very imminent conflict and issues a single instruction to resolve it, which the controller is expected to implement at once.

22.4.5 Computer-Assisted Approach Sequencing

This form of assistance applies to flows of air-traffic approaching an airfield from diverse directions, and amalgamating into a single flow approaching one runway or into parallel flows approaching two or more parallel runways. By showing the predicted flight paths of arriving aircraft on the display, this aid either shows directly or permits extrapolation of their expected order of arrival at the amalgamation position and the gaps between the consecutive aircraft when they arrive there. The controller can request minor flight path or speed changes to adjust and smooth gap sizes, and ensure that the minimum vortex separation standards applicable to weight categories of aircraft during final approach are met.

22.4.6 Flows and Slots

Various schemes have evolved that treat aircraft as items in traffic flows, within which control is exercised by assigning slots and slot times to each aircraft. Separations can then be dealt with by reference to the slots. The maximum traffic-handling capacities of flows can be utilized, and tactical adjustments can be minimized by allowing the intersection or amalgamation of traffic flows as a part of the initial slot allocation.

22.4.7 Traffic-Flow Management

Although flow management as a concept may refer to a system that includes flows and slots, it is usually applied to larger traffic regions. It refers to the broad procedures that prevent excessive general congestion by limiting the total amount of traffic, diverting flows, or imposing quotas on departures, rather than by tactically maneuvering single aircraft. Traffic-flow management is normally more strategic than air-traffic control, and precedes it. It imposes its own training problems (Wise, Hopkin, & Garland, 1998). However, the role of traffic-flow management is tending to increase (Duytschaever, 1993; Harwood, 1993).

22.4.8 Free Flight

The principle of free flight relies heavily on automation, because most current systems do not give the controller access to all the data needed to confirm the computer calculations. Variants of free-flight principles are sometimes called random routing or direct routing. The intentions are for the pilot to specify airports, times of arrival and departure, and preferred route and flight profile, and for the computer to check and confirm that the proposed flight will not incur conflicts with other known flights and will not encounter major delays at the arrival airport. If all is well, the flight would be sanctioned, perhaps automatically. The flight is independent of the air-traffic control route structure, and if weather-permitting, would normally follow the shortest direct route, a segment of a great circle. The controllers would verify its continued safety, deal with discrepancies between its actual and planned track, and perhaps, introduce minor modifications to enhance system capacity. Free flight could be more efficient in saving time and fuel and by allowing the optimum flight profile to be flown if known. It is becoming more widespread. It represents a reversion to more tactical procedures dealing with single aircraft, at a time when other policies favor more strategic air-traffic control procedures dealing with flows of traffic. It requires accurate and timely information, and poses numerous human-factors issues, not only of practicality and roles, but of information access and legal responsibility. Much planning and research effort is being applied to free flight and its implications, and more is needed. For example, while Remington, Johnston, Ruthruff, Gold, and Romera (2000) found that conflict detection was not necessarily poorer with free flight and was sometimes actually better in free flight without fixed routes, Metzger and Parasuraman (2001) reported that conflict detection performance was degraded by passive control and high traffic density. The forms of free flight will not be identical everywhere, but depend on other factors such as route structures, traffic mixes and densities, and typical flight durations.

22.4.9 Associated Legal and International Requirements

Extensive human-factors knowledge of various kinds can be applied to air-traffic control (International Civil Aviation Organization, 1993), and many technological advances can be matched successfully with human capabilities and limitations to further the performance of air-traffic control tasks (Wise et al., 1991). Nevertheless, some practical constraints must be applied to meet the legal requirements of air-traffic control or agreed international practices, procedures, and rules, including the format and language of spoken communications.

22.4.10 Consequences for Responsibilities

When computer assistance to aid the controller is introduced, it is vital that all the responsibilities of the controller are met fully using the facilities provided. This is a practical constraint not only on the forms of computer assistance supplied, but also on their acceptability to controllers (Hopkin, 1995).

22.5 Options for Helping the Controller

The primary objective of all forms of automation and computer assistance provided for the air-traffic controller is to aid the controller's safe task performance (Hopkin, 1994a). The most favored forms aim to promote this objective by enabling the controller to handle more traffic. The controller can be assisted in the following main broad ways.

22.5.1 Full Automation of Functions

In this option, some functions and tasks are removed altogether from the controller. It applies especially to the frequent and routine gathering, storage, transfer, manipulation, and presentation of data. All these functions have often been automated extensively, so that in most modern air-traffic control systems, the controllers no longer spend much time on such tasks. An example is the provision of the identity of each aircraft within its label on the radar display.

22.5.2 Improvements to the Quality of the Data

These can be achieved in several ways. For example, the data can become more frequently updated, accurate, reliable, consistent, precise, valid, trustworthy, or acceptable. It is also necessary for the controller to know how much better the data are and what level of trust they should be accorded (Lee & See, 2004). Such knowledge can arise through training, learning from experience, a displayed indication, or understanding about the nature of the data. A main purpose is to render the controller's behavior appropriate for the actual quality of the data presented.

22.5.3 Reductions in the Time Needed

This option reduces the time required by the controller to perform particular functions or tasks. Several means to achieve this objective are available. The required time can be shortened by performing specific tasks or functions in less detail, less often, in condensed form, in simplified form, with less information, fewer actions, some parts omitted, or data compiled more succinctly. All these means are applied in air-traffic control, with the choice depending on the forms of computer assistance that can be provided, and on the ways in which they can be matched appropriately with human capabilities and limitations for the effective performance of the various tasks and functions.

22.5.4 Treating Aircraft as Traffic Flows

To control air traffic as flows rather than as single aircraft, a change from tactical to strategic air-traffic control is normally involved. Among the consequences seem to be fewer human interventions, less criticality in their timing because they may often be brought forward at the behest of the controller, and more emphasis on the prevention rather than the solution of problems. Although some believe that air-traffic control must evolve in this way, most current and pending forms of computer assistance are primarily tactical and applicable to single aircraft. Most aids for the control of flows are still in quite early stages of development, lacking such fundamental human-factors contributions as satisfactory codings to differentiate between the main defining parameters of traffic flows.

22.5.5 Sharing Human Functions with Machines

In this option, machines fulfill some aspects of functions or they help, prompt, guide, or direct the human. In the most popular initial variant of this option, which seems attractive but actually does not work very well, the machine does much of the work and the human monitors the machine's performance. Unfortunately, the human finds it difficult to maintain concentration indefinitely in a passive role with nothing active to do. This same passivity can incur some loss of information processing and understanding, which may be tolerable in normal circumstances, but becomes a liability in nonstandard ones. The introduction of any major form of computer assistance that affects the controller's tasks will change the controller's situational awareness (Garland & Hopkin, 1994) and require rematching of the human and machine databases. The human controller relies greatly on a detailed mental picture of the air traffic, which active task performance and manipulation of data help to sustain. Any forms of computer assistance that interfere with these processes may result in a reported loss of clarity or detail in the controller's mental picture of the traffic. An underlying premise, which dates back over 50 years (Fitts, 1951), whereby functions are shared by a static process of allocating them to human or machine, is being replaced by adaptive automation in which the allocation of functions to human or machine is flexible and not fixed (Wickens et al., 1998).

22.5.6 Expanding the Range of Machine Support

One approach is to employ machines to support human activities and to expand the forms of machine support offered, with increasing degrees of complexity. Monitoring, for which machines can be well suited, becomes a machine rather than a human role. The machine that gathers, stores, and compiles information automatically, also collates, summarizes, selects, and presents it automatically with timing and ordering appropriate for the tasks. Thus, it functions as a memory aid and prompt, and guides attention. Given the high-quality data, machines can often make better predictions than humans, so that controllers can use them to discover the consequences of proposed actions before implementing them. The controller, before accepting or rejecting such assistance, needs to know or be able to discover what information the computer has or has not taken into account. The machine can offer solutions to problems and can aid in decision making and planning. An example is a planning aid that utilizes improved perceptual information integration (Moertl et al., 2002), and provides automated sequencing for the virtual tokens that may supersede paper flight progress strips (Gronlund et al., 2002). Another example is the development of computer programs to establish baseline measures that can be applied to ascertain the effects of air-traffic control changes (Mills, Pfleiderer, & Manning, 2002). The machine can apply far more information far more quickly than the human. If a machine generates a preferred solution to a problem, it may seem to be a small technical step for it to recommend that solution for human acceptance, another small technical step for it to implement that solution automatically unless the notified controller chooses to intervene, a further small technical step for it to implement the solution automatically and then notify the controller afterwards, and a final small technical step not to notify the controller at all. However, in human-factors terms and in legal terms, these are all big steps, the last being full automation.

22.6 A Classification of Human–Machine Relationships

The relationships that are feasible or will become feasible in air-traffic control are listed as follows, to assist the recognition and categorization of those that actually occur and their comparison with alternative relationships that could apply or result from proposed changes (Hopkin, 1995). The relationships are listed in the approximate order in which they became or will become technically feasible. The list of relationships is expanding because technological innovations introduce new options from time-to-time without invalidating any of the existing options. The main possible human–machine relationships include the following:

1. The human adapts to the machine.
2. The human and the machine compete for functions.
3. The human is partly replaced by the machine.
4. The human complements the machine.
5. The human supports a failed machine.
6. The machine adapts to the human.
7. The human and the machine function as hybrids.
8. The human and the machine function symbiotically.
9. The human and the machine duplicate functions in parallel.
10. The human and the machine are mutually adaptive to each other.
11. The human and the machine are functionally interchangeable.
12. The human and the machine have flexible and not fixed relationships.
13. The human and the machine form a virtual air-traffic control world.
14. The machine takes over in the event of human incapacitation.

These are relationships. They exclude the extremes, where there is no machine but only the human and where there is no human but only the machine, neither of which seems likely in air-traffic control for the foreseeable future. The corollaries of the choice of human–machine relationships, particularly in terms of options excluded and associated with decisions taken, are not always recognized at the time (Hopkin, 1988a).

22.7 Relevant Human Attributes

Some human characteristics with no machine equivalent must be emphasized with regard to computer assistance. Otherwise, they are likely to be ignored in human–machine comparisons, considerations, and allocations of functions, if their relevance remains unacknowledged, irrespective of how important they may actually be. Many of these characteristics are becoming more widely recognized, although their influence is often still insufficient.

22.7.1 Common Human Attributes Related to the Workspace

Some attributes have been widely studied. Although human workload can be difficult to measure, there are always some limitations on human workload capacity that computers do not share (Costa, 1993). Human errors can be classified into distinctive categories that may be differentiated from typical machine errors and may be partly predictable when their origins can be traced to design decisions about the tasks and equipment (Reason, 1993; Reason & Zapf, 1994). Procedures for investigating human error in air-traffic control and other systems of comparable complexity have been described (Strauch, 2002). Humans become tired, experience fatigue (Mital & Kumar, 1994), and need breaks, rosters, and adjustments of circadian rhythms (Costa, 1991) in ways in which machines do not. Humans have social, recreational, and sleep needs, which rostering and work–rest cycles must accommodate (Hopkin, 1982). Excessive demands may induce stress in humans, although stress and workload are complex concepts, both in their causality and measurement (Hopkin, 1980a; Tattersall, Farmer, & Belyavin, 1991). Insufficient demands can induce boredom, the causes and consequences of which have been neglected in most air-traffic control research, although most commonsense assumptions about boredom appear to be unsupported and its effects on safety are particularly obscure (Hopkin, 1980b). The optimum physical environmental conditions for the controller, in terms of heating, lighting, décor, acoustics, airflow, temperature, humidity, radiation, and appearance may not accord with those for the computer, but they must be met (Hopkin, 1995). The machine must adapt to human anthropometric characteristics that determine recommended reach and viewing distances, and the requirements for human comfort and health (Pheasant & Haslegrave, 2005). If computer assistance is cost-effective, manual reversion in the event of its failure must usually entail some loss of efficiency, but safety must never be compromised. The feasibility of manual reversion in the event of machine failure has too often been studied as a one-way process instead of a two-way process. The real difficulty, that after manual reversion everyone may become too busy running the system in manual mode to spare any time to reload the machine with up-to-date data after it has been repaired, prior to switching back to it, has been neglected.

22.7.2 The Context of Work

For the human, work has rewarding properties in its own right. Decisions about the nature and conditions of work affect job satisfaction, which is influenced by the level of autonomy delegated to the individual controller, the opportunities to develop and apply particular skills, and the responsibilities and the means of exercising them, all of which can be changed greatly by computer assistance, sometimes inadvertently. Team roles and structures may also change, often because the computer assistance aids the individual controller but is introduced into contexts where much of the work has been done by teams, in which each controller's actions have been readily observable by others in the team. This observability is crucial for the successful development of professional norms and standards, which are strong motivating forces in air-traffic control, contributing to its professional ethos, morale, and camaraderie. These forces also imply the continued presence of some individual differences between controllers, so that controllers can, to some extent, develop individual styles that are used by colleagues and supervisors to judge how good they are as controllers. These judgements in turn influence decisions on training, promotion, and career development. The exploration and understanding of the full consequences of computer assistance in air-traffic control in terms of observability and the effects of its loss have been important for some time (Hopkin, 1995) and have now become urgent.

22.7.3 Attitudes

To the controller, computer assistance must not only be effective, supportive, and safe; it must be acceptable. The effects of the introduction of computer assistance on attitudes have been comparatively neglected in air-traffic control. Attitude formation covers how the attitudes are formed, how they are influenced, and how far they can be predicted and controlled (Rajecki, 1990). Much is known, for example, from advertising and marketing studies, about how to influence and manipulate attitudes, but this knowledge has rarely been applied to air-traffic control, and there is an ethical issue of whether it should be. However, selective technical means to apply it have been established (Crawley, Spurgeon, & Whitfield, 1980). What would be entailed is the deliberate application of current evidence about the characteristics of equipment items and forms of computer assistance in air-traffic control, which improves their user acceptability, including how they function, how they feel, and how they look. Attitudes toward any change, whether favorable or not, are formed quickly, and the methods employed in the initial introduction of any change and in the earliest stages of training can therefore be crucial. Attitudes, once formed, become resistant to contrary evidence. If controllers' initial attitudes toward computer assistance are favorable, then they will strive to make it work and to get the best from it, but if their initial attitudes toward the assistance are unfavorable, then they may become adept at demonstrating how unhelpful the computer assistance can be. The kinds of factor that can influence the acceptability of computer assistance are now becoming clearer. Among them are its effects on responsibility and autonomy, development and applicability of skills, job satisfaction, and the challenge and interest of the work. Controllers generally like being controllers, and their attitudes toward the work itself are often more favorable than their attitudes toward their conditions of work. Perhaps, some of the research effort devoted to optimizing the ergonomic aspects of the system might profitably be redeployed on studies of attitude formation, because the evidence available from existing sources should usually be sufficient to prevent serious ergonomic errors; however, without positive attitudes, the computer assistance could be ergonomically optimum, yet still be unacceptable to its users.

22.7.4 Degree of Caution

If controllers like a form of computer assistance, then they can become too enthusiastic about it and too reliant on it. A positive aspect of overenthusiasm can be a dedicated effort to make the computer assistance function as effectively as possible, although controllers may try to extend its usage to assist tasks that lie beyond the original design intentions and for which its usage has not been legally sanctioned. Controllers may welcome computer assistance, because it is superior to poor equipment that it replaces, but if the computer assistance is not in fact very good, then their positive attitudes toward the change may disguise and discourage the need for further practical improvements. Favorable attitudes may accrue for reasons that can include enhanced status or increased attractiveness, as in the case of color coding, and they can induce strong beliefs in the benefits of the system even when none can be demonstrated by objective measures of performance or safety. More sensitive and appropriate measures of the benefits of positive attitudes may relate to fewer complaints, improved collaboration, increased motivation and job satisfaction, and lower absenteeism and job attrition rates. Controllers are renowned for some skepticism toward novel forms of computer assistance. Many recall earlier forms of computer assistance, claimed to be helpful in reducing workload, which in practice had limited value and some of which actually added to the work. Controllers have learned to call for more tangible evidence of promised benefits, in forms such as prior demonstration or proof.

22.7.5 Disagreements between Measured Human Attributes

One justification for the employment of several measures to test the efficacy of computer assistance is the potential lack of agreement between them. This does not necessarily imply that one or other of the measures must be wrong, but the different kinds of evidence on which they draw may be contradictory. A common example in air-traffic control and elsewhere concerns the replacement of monochrome coding by the color coding of displayed information. Objective measures of task performance often reveal far fewer beneficial effects of color than the glowing subjective measures would lead one to expect. Furthermore, the tendency to dismiss one kind of measure as spurious must be resisted, as neither kind of measure is ever complete and fully comprehensive. More probable explanations are that the performance measures fail to cover some aspects of the tasks most affected subjectively by color coding, and that the subjective measures tap genuine benefits that are not measured objectively. Examples of the latter could include color coding as an aid to memory, as an entrenchment of understanding, and as a means to structure and declutter visual information—none of which might influence the chosen set of objective measures directly (Hopkin, 1994b; Reynolds, 1994).

22.7.6 Function Descriptions

Functions that seem identical or very similar when expressed in system concepts can be quite different when expressed in human terms. Attempts to introduce computer assistance for human air-traffic control tasks often reveal that their degree of complexity has been underestimated. The human controller who takes a decision without assistance chooses what evidence to seek, gathers it, applies rules and experience to it, reaches a decision, implements the decision, and fulfills all these functions actively. Therefore, the controller is well-placed to judge whether any given change in the evidence warrants reexamination of the decision. When the computer presents a decision for the controller to accept or reject, this may seem similar functionally and when described in system concepts, but it is not. The controller may not know, and often cannot readily discover, what evidence has been taken into account in the computer decisions, whether it is correct, and what new information would invalidate it. As the controller needs to process far less information to accept a computer decision than to reach the same decision without computer assistance, the assisted controller tends to understand and recall less about the decision and its circumstances (Narborough-Hall, 1987). In human terms, the processes of human and computer-assisted decision-making often cannot be equated.

22.8 Human-Factors Implications of Automation and Computer Assistance

22.8.1 Interface Designs

Many of the commonest human-factors problems that result from automation and computer assistance in air-traffic control occur under definable sets of circumstances. Making these explicit clarifies the origins of the problems, reveals the prevailing constraints, and suggests practical solutions. One of the most familiar human-factors problems arises when a function must be performed somehow, but no machine can be devised to perform it. Tasks may be assigned to humans not because they do them well, but because no machine can do them safely or at all. As technology advances, this problem recedes, but in modern systems the controller can do only what the computer allows the controller to do. If the specification of the human–machine interface makes no provision for an action, it may be impossible for the controller to implement it, no matter how correct it may be. Moreover, the controller's attempts to implement it are liable to be ruled by the computer as invalid actions. The effectiveness of computer assistance is critically dependent on the human–machine interface designs that enable human roles to be performed. For example, the human controller cannot be flexible unless the human–machine interface permits human flexibility (Hopkin, 1991a).

22.8.2 Attributes of Speech

An artifact of computer assistance is that many dialogs formerly conducted by the controller with pilots or other controllers have to be conducted through the human–machine interface, so that the information in the system is updated and can be applied to automated computations. Much of the further information incidentally contained in speech, on the basis of which pilots and controllers make judgements about each other, becomes no longer available (Hopkin, 1982). Few formal studies have ever been carried out to quantify and describe the nature and extent of the influence of this further information on the actions of the controller and on the conduct of air-traffic control. Its influence was never negligible and could become substantial, and the human-factors implications of its absence could be very significant. The reduction in spoken human communications in air-traffic control implies some loss of information gleaned from attributes of speech, such as accents, pace, pauses, hesitancies, repetitions, acknowledgments, misunderstandings, degree of formality, standardization, courtesies, choice of vocabulary, message formats, and the sequencing of items within a message. The basis for judgements of the speaker's confidence, competence, professionalism, and familiarity with the region is curtailed, yet these judgements may be important, particularly in emergencies.

On the other hand, it is conceivable that these judgements are so often wrong that the system is safer without them. All the categories of potentially useful information lost when speech is removed should be identified, not to perpetuate speech but to determine whether surrogates are needed and the forms that they should take.

22.8.3 Computer-Generated Workload

The basic objective of enabling the controller to deal with more aircraft implies that a criterion for the adoption of any form of computer assistance must be that it results in less work for the controller. Some previous forms of computer assistance have violated this precept, especially when messages that are spoken have also had to be entered as data into the system, or when quite simple and standard tasks have required cumbersome keying procedures. Such forms of computer assistance not only negate their main purpose, but are unpopular and can lead to counterproductive attitudes towards computer assistance in general. Meanwhile, efforts continue to assess the subjective workload objectively, for example, by measuring air-traffic control communications (Manning, Mills, Fox, Pfleiderer, & Mogilka, 2002), and by employing physiological measures to discriminate between acceptable mental workload and overloading (Wilson & Russell, 2003).

22.8.4 Cognitive Consequences

In retrospect, some of the main adverse cognitive consequences of various initial forms of computer assistance in air-traffic control were insufficiently recognized, although they are more familiar now. They can be allowed for either by designing the other human and machine functions so that certain penalties are acceptable, or by redesigning the computer-assisted tasks to avoid such effects, perhaps, by keeping the controller more closely involved in the control loops than the computer assistance strictly requires. The crucial influence of human cognitive functioning and information processing on the successful matching of human and machine has now received the attention that it deserves (Cardosi & Murphy, 1995; Wickens et al., 1997). As the application of computer assistance has consistently revealed that many of the human functions that it is intended to replace or supplement are much more complex than they seem superficially to be, it has proved difficult to capture their full functionality in many forms of computer assistance in their stead. Techniques such as cognitive task analysis can be applied (Hoffman & Woods, 2000), and the integration of disparate branches of psychology can be helpful (Hodgkinson, 2003). An example of more complex cognitive effects than those initially anticipated concerns the replacement of paper flight progress strips with electronic strips (Hopkin, 1991b; Vortac, Edwards, Jones, Manning, & Rotter, 1993). Most aspects of task performance with paper strips, their manipulation, and their updating are relatively easy to capture electronically, but a strip is a talisman, an emblem, a history, a record, and a separate object. Active writing on strips, annotation of them, offsetting them sideways, and initial placement of them in relation to other strips on the board, all help in understanding, memory, and the building of the controller's picture. Strips collectively denote current activities and future workload, and are observable and accessible to colleagues and supervisors. These and further aspects have proved more difficult to represent electronically. A recurring human-factors issue is to identify which functions of paper flight strips can and should be retained electronically, which can be discarded altogether, and which cannot be perpetuated in electronic form yet must still be retained in an alternative form.

22.8.5 Rules and Objectives

The preceding point about the complexity of some air-traffic control functions also applies to some rules, which can seem quite simple until they have to be written as software, at which point they begin to look complex. There may be many exceptions about them, and considerable complexity concerning the circumstances under which one rule overrides another. The objectives of air-traffic control are multiple. Not only must it be safe, orderly, and expeditious, but also, cost-effective, noise abating, fuel conserving, and job satisfying, responsive to the needs of its customers while never harming the wellbeing of those employed in it. With so many objectives, there is much scope for their mutual incompatibility, which the rules and relative weightings of rules attempt to resolve at the cost of some complexity.

22.8.6 Observability

Most forms of computer assistance have the incidental but unplanned consequence of rendering the work of the controller much less observable by the others, including immediate colleagues and supervisors. Air-traffic control as a team activity relies heavily on tacit understanding among the controllers. Each member of the team builds expectations about the activities of colleagues and learns to rely on them. Where the activities of colleagues can no longer be observed in detail, such reliance and trust become initially more difficult to develop and ultimately impossible to build in the same way. For example, a colleague may have difficulty in detecting whether a controller has accepted or rejected a computer solution of a problem, because acceptance and rejection may both involve similar key pressings. General key-pressing activity may remain observable, but not the particular keys that have been pressed. Loss of observability can make it more difficult for the controllers to appreciate the skills of colleagues, acquire new skills by observation, and demonstrate their own accomplishments to others. A complicating factor can be reduced flexibility in nonstandard circumstances, because the options provided within the menus and dialogs available through the human–machine interface are preset.

22.8.7 Concealment of Human Inadequacy

Many of the forms of support that computer assistance can provide have the inherent capability of compensating for human weaknesses, to the extent that they can disguise human incompetence and conceal human inadequacy. This can become very serious if it is compounded by controllers' inability to observe closely what their colleagues are doing. If a controller always accepts computer solutions to problems, this may indeed utilize the computer assistance most beneficially, but it is impossible for others to tell from that controller's activities whether or not the controller has fully understood the solutions that have been accepted. In a more manual system with less computer assistance and more observability, it is not possible for a controller to disguise such lack of knowledge from colleagues indefinitely. This is not an imputation on the professionalism of the controllers, of which they are rightly proud, nor is it a claim that this problem could become rife. However, it is a statement that important safeguards that are present now could be taken too much for granted, and could be undermined inadvertently by future changes made for other reasons.

22.8.8 Stress

Although the problems may have been exaggerated, human stress has been claimed for a long time to be associated with air-traffic control that has acquired a spurious reputation as a particularly stressful occupation (Melton, 1982). This does not mean that there is no stress in air-traffic control, for indeed there is, but its levels are not beyond those in many other walks of life. Initially, stress was usually attributed to time pressures and excessive workload, coupled with responsibilities without proper means to exercise them. Computer assistance can introduce its own forms of human stress, if the controller must rely on machine assistance that is not fully trusted, must use forms of assistance that function too complexly to be verified, or must fulfill functions that are incompletely understood but that the controller has no power to change.

22.8.9 Team Roles

Computer assistance in air-traffic control often changes many of the traditional roles and functions of the team, some of which may disappear altogether. This can be acceptable, provided that the full functionality of teams has been defined beforehand, so that team roles are not removed inadvertently by other events and their diminution does not arrive as an unwelcome surprise. The neglect of teams has been twofold. Most current and planned forms of computer assistance in air-traffic control are not designed for teams, and most forms of computer assistance designed for teams in other work contexts are not being proposed for air-traffic control application. Teams have many functions. They include the building and maintenance of tacit understandings, adaptability to colleagues, and local agreed air-traffic control practices. Through team mechanisms, controllers gain and retain the trust and respect of their peers, which depend on the need for practical interactions between the team members and on sufficient mutual observability of activities within the team. There is a need to develop tools to measure categories of communications between the controllers (Peterson, Bailey, & Willems, 2001), and the effects of automated decision support tools on controllers' communications with each other (Bailey, Willems, & Peterson, 2001). Computer assistance may render some of the traditional roles of the air-traffic control supervisor impractical. The future roles of supervision need planning according to known policies, and should not be changed by default.

22.8.10 Coping with Machine Failure

The functioning of many forms of computer assistance is not transparent to the controllers who use them. In particular, it is not apparent how they could fail, what they would look like if they did, or how it would be possible for the controller to discover which functions were still usable, because they remained unaffected by the failure. This is a crucial aspect of successful human–machine matching within the system. For many kinds of failure of computer assistance, no provision has been made to inform the user that the system is not functioning normally. The controller is not concerned with the minutiae of the reasons for failure, because it is not the controller's job to remedy it, but the controller does need to know the existence and ramifications of any failure and how far it extends.

22.8.11 Controller Assessment

In manual air-traffic control systems, the concept of the "good controller" is a familiar one. The criteria for this judgement have proved elusive, but there is usually quite high consensus among colleagues familiar with their work about who the best controllers are. Computer assistance has implications for the development of this concept of the good controller, because it may restrict the judgements on which it can be based. Lack of observability can also make decisions about careers, promotions, and retraining seem more arbitrary. Who would be the best controller using computer assistance—one who always accepts it, one who overrides it in ways that are predictable because they are rigid, one who overrides it frequently, or one who overrides it selectively but to the occasional discomfiture of colleagues? What criteria for promotion would be acceptable as fair when controllers are generally expected to adopt the forms of computer assistance provided?

22.8.12 Other Air-Traffic Control Personnel

For a long time, there has been an imbalance within air-traffic control concerning the impact of automation and computer assistance on those who work within it. Almost all of the limited human-factors resources have been concentrated on the air-traffic controller as an individual, neglecting supervisors, assistants, teams and their roles and functions, and technical and maintenance staffs. This imbalance is beginning to be redressed in some respects, but must not mean neglecting the controller.

22.9 Implications for Selection and Training

As the introduction of computer assistance progresses, questions about the continuing validity of the selection procedures for controllers are bound to arise (Della Rocco, Manning, & Wing, 1991). The simplest issue is whether an ability to work well with the new forms of computer assistance should become an additional measured requirement in the selection procedure. A more complex question is whether some of the abilities for which controllers have been selected in the past no longer remain sufficiently relevant to current or future air-traffic control jobs to justify their retention in selection. Issues concerning the circumstances under which it becomes necessary to adapt selection procedures are quite difficult to resolve, and the criteria for deciding when, how, and at what level intervention in these procedures becomes essential are poorly defined. Many current controllers were not selected for jobs like those now envisaged for them. It is not clear how far training could compensate for this discrepancy, or how far selecting different kinds of people must constitute the ultimate remedy. Attempts to employ validated personality measures in the selection of air-traffic controllers have a long history but are still being made (King, Retzlaff, Detwiler, Schroeder, & Broach, 2003). Perhaps, modifications of the chosen forms of computer assistance or greater flexibility in them could obviate many of the adjustments in the selection procedures that might otherwise be needed. Another application of automation to air-traffic control selection is the computerization of the selection test battery, coupled with demonstrations of its validity (Ramos, Heil, & Manning, 2001). However, the use of automation has created in new form of traditional worries about the effects of coaching and practice on the validity of test scores (Heil et al., 2002).

Both automation and computer assistance entail some changes in what the controller needs to know. The controller's professional knowledge, much of which is gained initially through training, must match the facilities provided. Therefore, the controller's training in relation to computer assistance has to cover how it is designed to be used, how the human and machine are intended to match each other, what the controller is expected to do, and what the controller needs to understand about the functioning of the computer assistance to work in harmony with it. Effective computer assistance also entails considerable practical training in how to access data, interrogate the computer, manipulate information, use menus and conduct dialogs, and learn all the options available through the human–machine interface. The taught procedures and instructions may have to be revised to realign the controller's actions with the computer assistance. The controller may need some human-factors knowledge (Hunt, 1997), and the distinction between what the controller is taught and what the controller learns may warrant reappraisal, the former referring to training content and the latter to on-the-job experience (Hopkin, 1994c).

Any changes in the machine database that affect the computer assistance of the controller always require some restoration of the optimum match between human and machine, in the form of corresponding changes in the human database that consists of the controller's knowledge, skills, experience, and professionalism. Changes may be needed to rematch the controller's situational awareness and mental picture of the traffic with the system (Mogford, 1994), taking account of the revised imagery that may have become more appropriate for the controller in the computer-assisted system (Isaac, 1994). These rematching processes begin with retraining, which obviously must accomplish the new learning required, but less obviously may require the discarding of old knowledge and skills, now rendered inapplicable but still thoroughly familiar through years of experience and practical application. Much less is known about how to train controllers to forget the old and irrelevant than about how to train them to learn the new; however, a potential hazard, particularly under stress or high workload, is the reversion to familiar former habits and practices that do not apply any more. If this can happen, it must not be dangerous. Although most emphasis is on new learning, some of the most urgent practical problems are concerned with how to make forgetting safe (Hopkin, 1988b).

Much effort is expended to ensure that all forms of computer assistance in air-traffic control are safe, efficient, successful, and acceptable, but they must also be teachable. Practical and cost-effective means must be devised to teach the new form of computer assistance to the whole workforce for whom it is intended. Learning to use it should not be laborious for that would prejudice its acceptability and raise training costs, and any difficulties in understanding its functioning will lead to its misinterpretation and misuse if they are not resolved. Training with computer assistance should always include appropriate forms of team training, work scheduling, and resource management, so that the performance of tasks with computer assistance fits snugly within all other facets of the job.

Training relies extensively on real-time simulation, which is also employed for most human-factors research in air-traffic control. Although real-time simulation is an essential tool, it is not a sufficient one for every purpose. A comprehensive list has been compiled comprising actual human-factors applications in operational evaluations, many of which also cover real-time simulations. The listing distinguishes between applications that are valid, applications where simulation may be helpful if supported by external confirmatory evidence, and applications for which simulation is inherently inappropriate as a technique and should not be used (Hopkin, 1990). This listing is subjected to modification in the light of subsequent experience, and shares its uncertain validity with many other human-factors activities concerning air-traffic control. Originally, validation of findings was considered essential, and it is still common in some activities such as selection procedures, but texts on the validation of human-factors recommendations for air-traffic control systems have revealed the extent of the uncertainties (Wise, Hopkin, & Stager, 1993a, 1993b), and point to the increasing difficulty of deriving independent validation criteria for human-factors recommendations as systems increase in their complexity and integrality. Possible approaches include the integration of validation techniques into design processes and the adaptation of certification procedures as validation tools (Wise, Hopkin, & Garland, 1994). Furthermore, methods for introducing more human-factors contributions into certification processes have been examined (Wise & Hopkin, 2000).

22.10 The Future

Automation and computer assistance for much current air-traffic control are still confined to quite routine human functions, but their envisaged future forms will affect many cognitive functions of the controller and could change the controller's job greatly. This means that air-traffic control is well placed to profit from the experience of others in contexts where computer assistance has already been applied more extensively. However, with regard to computer assistance, there is a prevailing impression of responding to external events as further technical innovations become practicable. It would be better to strive for the implementation of broad and principled policies about what the forms of computer assistance and the resultant human roles ought to be in air-traffic control.

Many attributes traditionally treated as exclusively human are also becoming machine attributes. These include intelligence, adaptability, flexibility, and a capacity to innovate. The rules, in so far as they exist, about the optimum matching of the human and the machine when both possess these attributes are not yet firm enough to be applied now uncritically to air-traffic control. However, some of the issues, such as the roles of adaptive machines, have been addressed (Mouloua & Koonce, 1997). If computer assistance reduces workload as it is intended to do, the controller will be driven less by immediate task demands and gain more control over workload and its scheduling. Excessive controller workload would then occur in the future only if it was self-inflicted, because excessively high workload could always be prevented if the controller employed the computer assistance in accordance with the designer's intentions. Thus, very high workload could signify that the controller needs further training. It would also be expected that more of the workload would become strategic rather than tactical, unless free flight became widespread.

It will become more important to understand the reasons for the controller's acceptance of computer assistance and satisfaction with it. An incidental consequence of more widespread computer assistance could be to make air-traffic control more similar to many other jobs, if the primary knowledge and skills required relate more to the manipulation of a human–machine interface than to its particular applications in air-traffic control. Currently, most knowledge, experience, and skill as an air-traffic controller do not transfer to other jobs. This may not remain true. Those employers who provide the best conditions of employment, the greatest satisfaction of human needs and aspirations in the workspace, and the forms of computer assistance that match human needs and responsibilities best, may attract the best applicants for the jobs, have the lowest job attrition rates, incur the lowest selection and training costs, and employ a workforce that is justifiably proud of its achievements and that others want to join. Such a development would extend further the nature of the human-factors objectives and contributions to air-traffic control.

References

Bailey, L. L., Willems, B. F., & Peterson, L. M. (2001). *The effects of workload and decision support automation on en-route R-side and D-side communication exchanges* (Rep. No. DOT/FAA/AM-01/20). Washington, DC: Federal Aviation Administration.

Billings, C. E. (1991). *Human-centered aircraft automation: a concept and guidelines* (Report No. NASA TM 10385). Moffett Field, CA: NASA Ames Research Center.

Brooker, P. (2005). Air traffic control automation: For humans or people. *Human Factors and Aerospace Safety, 5*(1), 23–41.

Cardosi, K. M. (1993). Time required for transmission of time-critical ATC messages in an en-route environment. *International Journal of Aviation Psychology, 3*(4), 303–313.

Cardosi, K. M., & Murphy, E. D. (Eds.). (1995). *Human factors in the design and evaluation of air traffic control systems* (Rep. No. DOT/FAA/RD-95/3). Washington, DC: Federal Aviation Administration, Office of Aviation Research.

Costa, G. (1991). Shiftwork and circadian variations of vigilance and performance. In J. A. Wise, V. D. Hopkin, & M. L. Smith (Eds.), *Automation and systems issues in air traffic control* (pp. 267–280). Berlin: Springer-Verlag, NATO ASI Series Vol. F73.

Costa, G. (1993). Evaluation of workload in air traffic controllers. *Ergonomics, 36*(9), 1111–1120.

Crawley, R., Spurgeon, P., & Whitfield, D. (1980). *Air traffic controller reactions to computer assistance: A methodology for investigating controllers' motivations and satisfactions in the present system as a basis for system design* (Vols. 1–3, AP Report 94). Birmingham, U.K.: University of Aston, Applied Psychology Department().

Della Rocco, P., Manning, C. A., & Wing, H. (1991). Selection of air traffic controllers for automated systems: Applications from today's research. In J. A. Wise, V. D. Hopkin, & M. L. Smith (Eds.), *Automation and systems issues in air traffic control* (pp. 429–451). Berlin: Springer-Verlag, NATO ASI Series Vol. F73.

Durso, F. T., Batsakes, P. J., Crutchfield, J. M., Braden, J. B., & Manning, C. A. (2004). The use of flight progress strips while working live traffic: Frequencies, importance, and perceived benefits. *Human Factors, 46*(1), 32–49.

Duytschaever, D. (1993). The development and implementation of the EUROCONTROL central air traffic control management unit. *Journal of Navigation, 46*(3), 343–352.

Federal Aviation Administration. (1995). *National plan for civil aviation human factors: An initiative for research and application.* Washington, DC: Author.

Fitts, P. M. (Ed.). (1951). *Human engineering for an effective air navigation and traffic control system.* Washington, DC: National Research Council, Committee on Aviation Psychology.

Garland, D. J., & Hopkin, V. D. (1994). Controlling automation in future air traffic control: The impact on situational awareness. In R. D. Gilson, D. J. Garland, & J. M. Koonce (Eds.), *Situational awareness in complex systems* (pp. 179–197). Daytona Beach, FL: Embry-Riddle Aeronautical University Press.

Gronlund, S. D., Canning, J. M., Moertl, P. M., Johansson, J., Dougherty, M. R. P., & Mills, S. H. (2002). An information organization tool for planning in air traffic control. *The International Journal of Aviation Psychology, 12*(4), 377–390.

Harwood, K. (1993). Defining human-centered system issues for verifying and validating air traffic control systems. In J. A. Wise, V. D. Hopkin, & P. Stager (Eds.). *Verification and validation of complex systems: Human factors issues* (pp. 115–129). Berlin: Springer-Verlag, NATO ASI Series Vol. F110.

Heil, M. C., Detwiler, C. A., Agen, R., Williams, C. A., Agnew, B. O., & King, R. E. (2002). *The effects of practice and coaching on the air traffic control selection and training test battery* (Rep. No. DOT/FAA/AM-02/24). Washington, DC: Federal Aviation Administration.

Hodgkinson, G. P. (2003). The interface of cognitive and industrial, work and organizational psychology. *Journal of Occupational and Organizational Psychology, 76*(1), 1–25.

Hoffman, R. R., & Woods, D. D. (Eds.). (2000). Studying cognitive systems in context: Preface to the special edition. *Human Factors, 42*(1), 1–7.

Hopkin, V. D. (1980a). The measurement of the air traffic controller. *Human Factors, 22*(5), 547–560.

Hopkin, V. D. (1980b). Boredom. *The Controller, 19*(1), 6–10.

Hopkin, V. D. (1982). *Human factors in air traffic control.* Paris: NATO AGARDograph No. 275.

Hopkin, V. D. (1988a). Air traffic control. In E. L. Wiener, & D. C. Nagel (Eds.), *Human factors in aviation* (pp. 639–663). San Diego, CA: Academic Press.

Hopkin, V. D. (1988b). Training implications of technological advances in air traffic control. In *Proceedings of Symposium on Air Traffic Control Training for Tomorrow's Technology* (pp. 6–26). Oklahoma City, OK: Federal Aviation Administration.

Hopkin, V. D. (1989). Implications of automation on air traffic control. In R. S. Jensen (Ed.), *Aviation psychology* (pp. 96–108). Aldershot, Hants: Gower Technical.

Hopkin, V. D. (1990). Operational evaluation. In M. A. Life, C. S. Narborough-Hall, & I. Hamilton (Eds.), *Simulation and the user interface* (pp. 73–83). London: Taylor & Francis.

Hopkin, V. D. (1991a). The impact of automation on air traffic control systems. In J. A. Wise, V. D. Hopkin, & M. L. Smith (Eds.), *Automation and systems issues in air traffic control* (pp. 3–19). Berlin: Springer-Verlag, NATO ASI Series Vol. F73.

Hopkin, V. D. (1991b). Automated flight strip usage: Lessons from the functions of paper strips. In *Proceedings of AIAA/NASA/FAA/HFS Symposium on Challenges in Aviation Human Factors: The National Plan* (pp. 62–64). Vienna, VA: American Institute of Aeronautics and Astronautics.

Hopkin, V. D. (1994a). Human performance implications of air traffic control automation. In M. Mouloua, & R. Parasuraman, (Eds.), *Human performance in automated systems: Current research and trends* (pp. 314–319). Hillsdale, NJ: Lawrence Erlbaum Associates.

Hopkin, V. D. (1994b). Color on air traffic control displays. *Information Display, 10*(1), 14–18.

Hopkin, V. D. (1994c). Organizational and team aspects of air traffic control training. In G. E. Bradley, & H. W. Hendrick (Eds.), *Human factors in organizational design and management* (Vol. 4, pp. 309–314). Amsterdam, the Netherlands: North Holland.

Hopkin, V. D. (1995). *Human factors in air traffic control.* London: Taylor & Francis.

Hopkin, V. D. (1997). Automation in air traffic control: Recent advances and major issues. In M. Mouloua, & J. M. Koonce (Eds.), *Human-automation interaction: Current research and practice* (pp. 250–257). Mahwah, NJ: Lawrence Erlbaum Associates.

Hopkin, V. D. (1998). Human factors in air traffic control. In R. Baldwin (Ed.), *Developing the future aviation system* (pp. 85–109). Aldershot, Hants: Ashgate.

Hughes, J. A., Randall, D., & Shapiro, D. (1993). Faltering from ethnography to design. In J. A. Wise, V. D. Hopkin, & P. Stager (Eds.), *Verification and validation of complex systems: Additional human factors issues* (pp. 77–90). Daytona Beach, FL: Embry-Riddle Aeronautical University Press.

Hunt, G. J. F. (Ed.). (1997). *Designing instruction for human factors training in aviation.* Aldershot, Hants: Avebury Aviation.

International Civil Aviation Organization. (1993). *Human Factors Digest No. 8: Human factors in air traffic control* (Circular 241-AN/145). Montreal, Canada: Author.

Isaac, A. R. (1994). Imagery ability and air traffic personnel. *Aviation, Space, and Environmental Medicine, 65*(2), 95–99.

Isaac, A. R., & Ruitenberg, B. (1999). *Air traffic control: Human performance factors.* Aldershot, Hants: Ashgate.

King, R. E., Retzlaff, P. D., Detwiler, C. A., Schroeder, D. J., & Broach, D. (2003). *Use of personality assessment measures in the selection of air traffic control specialists* (Rep. No. DOT/FAA/AM-03/20). Washington, DC: Federal Aviation Administration.

Lee, J. D., & See, K. A. (2004). Trust in automation: Designing for appropriate reliance. *Human Factors, 46*(1), 50–80.

Manning, C. A., Mills, S. H., Fox, C. M., Pfleiderer, E. M., & Mogilka, H. J. (2002). *Using air traffic control taskload measures and communication events to predict subjective workload* (Rep. No. DOT/FAA/AM-02/4). Washington, DC: Federal Aviation Administration.

Melton, C. E. (1982). *Physiological stress in air traffic controllers: A review* (Rep. No. DOT/FAA/AM-82/17). Washington, DC: Federal Aviation Administration.

Metzger, U., & Parasuraman, R. (2001). The role of the air traffic controller in future air traffic management: An empirical study of active control versus passive monitoring. *Human Factors, 43*(4), 519–528.

Mills, S. H., Pfleiderer, E. M., & Manning, C. A. (2002). *POWER: Objective activity and taskload assessment in en route air traffic control* (Rep. No. DOT/FAA/AM-02/2). Washington, DC: Federal Aviation Administration.

Mital, A., & Kumar, S. (Eds.). (1994). Fatigue. *Human Factors, 36*(2), 195–349.

Moertl, P. M., Canning, J. M., Gronlund, S. D., Dougherty, M. R. P., Johansson, J., & Mills, S. H. (2002). Aiding planning in air traffic control: An experimental investigation of the effects of perceptual information integration. *Human Factors, 44*(3), 404–412.

Mogford, R. (1994). Mental models and situation awareness in air traffic control. In R. D. Gilson, D. J. Garland, & J. M. Koonce (Eds.), *Situational awareness in complex systems* (pp. 199–207). Daytona Beach, FL: Embry-Riddle Aeronautical University Press.

Moser, H. M. (1959). *The evolution and rationale of the ICAO word spelling alphabet* (Report No. AFCRC-TN-59-54). Bedford, MA: USAF Research and Development Command.

Mouloua, M. & Koonce, J. M. (Eds.). (1997). *Human-automation interaction: Research and practice.* Mahwah, NJ: Lawrence Erlbaum Associates.

Narborough-Hall, C. S. (1987). Automation implications for knowledge retention as a function of operator control responsibility. In D. Diaper, & R. Winder (Eds.), *People and computers II* (pp. 269–282). Cambridge, Cambridge University Press.

Peterson, L. M., Bailey, L. L., & Willems, B. F. (2001). *Controller-to-controller communication and coordination taxonomy* (Rep. No. DOT/FAA/AM-01/19). Washington, DC: Federal Aviation Administration.

Pheasant, S., & Haslegrave, C. M. (2005). *Bodyspace: Anthropometry, ergonomics and the design of work.* London: CRC Press.

Prinzo, O. V., & McClellan, M. (2005). *Terminal radar approach control: Measures of voice communications* (Rep. No. DOT/FAA/AM-05/19). Washington, DC: Federal Aviation Administration.

Prinzo, O. V., & Morrow, D. G. (2002). Improving pilot/air traffic control voice communication in general aviation. *The International Journal of Aviation Psychology, 12*(4), 341–357.

Rajecki, D. W. (1990). *Attitudes.* Oxford: W. H. Freeman.

Ramos, R. A., Heil, M. C., & Manning, C. A. (2001). *Documentation of validity for the AT-SAT computerized test battery (2 Volumes)* (Rep. Nos. DOT/FAA/AM-01/5 & 6). Washington, DC: Federal Aviation Administration.

Reason, J. T. (1993). The identification of latent organizational failures in complex systems. In J. A. Wise, V. D. Hopkin, & P. Stager (Eds.), *Verification and validation of complex systems: Human factors issues* (pp. 223–237). Berlin: Springer-Verlag, NATO ASI Series Vol. F110.

Reason, J. T., & Zapf, D. (Eds.). (1994). Errors, error detection and error recovery. *Applied Psychology: An International Review, 43*(4), 427–584.

Remington, R. W., Johnston, J. C., Ruthruff, E., Gold, M., & Romera, M. (2000). Visual search in complex displays: Factors affecting conflict detection by air traffic controllers. *Human Factors, 42*(3), 349–366.

Reynolds, L. (1994). Colour for air traffic control displays. *Displays, 15*(4), 215–225.

Smolensky, S. W., & Stein, E. S. (Eds.). (1997). *Human factors in air traffic control.* San Diego, CA: Academic Press.

Strauch, B. (2002). *Investigating human error: Incidents, accidents, and complex systems.* Aldershot, Hants: Ashgate.

Tattersall, A. J., Farmer, E. W., & Belyavin, A. J. (1991). Stress and workload management in air traffic control. In J. A. Wise, V. D. Hopkin, & M. L. Smith (Eds.), *Automation and systems issues in air traffic control* (pp. 255–266). Berlin: Springer-Verlag, NATO ASI Series Vol. F73.

Vortac, O. U., Edwards, M. B., Fuller, D. K., & Manning, C. A. (1993). Automation and cognition in air traffic control: An empirical investigation. *Applied Cognitive Psychology, 7*, 631–651.

Vortac, O. U., Edwards, M. B., Jones, J. P., Manning, C. A., & Rotter, A. J. (1993). En-route air traffic controllers' use of flight progress strips: A graph theoretic analysis. *International Journal of Aviation Psychology, 3*(4), 327–343.

Wickens, C. D., Mavor, A. S., & McGee, J. P. (1997*). Flight to the future: Human factors in air traffic control.* Washington, DC: National Research Council.

Wickens, C. D., Mavor, A. S., Parasuraman, R., & McGee, J. P. (1998). *The future of air traffic control: Human operators and automation.* Washington, DC: National Research Council.

Wilson, G. F., & Russell, C. A. (2003). Operator functional state classification using multiple psychophysiological features in an air traffic control task. *Human Factors, 45*(3), 381–389.

Wise, J. A., & Hopkin, V. D., (Eds.). (2000). *Human factors in certification.* Mahwah, NJ: Lawrence Erlbaum Associates.

Wise, J. A., Hopkin, V. D., & Garland, D. J. (Eds.). (1994). *Human factors certification of advanced aviation technologies.* Daytona Beach, FL: Embry-Riddle Aeronautical University Press.

Wise, J. A., Hopkin, V. D., & Garland, D. J. (1998). Training issues in air traffic flow management. In R. Baldwin (Ed.), *Developing the future aviation system* (pp. 110–132). Aldershot, Hants: Ashgate.

Wise, J. A., Hopkin, V. D., & Smith, M. L. (Eds.). (1991). *Automation and systems issues in air traffic control.* Berlin: Springer-Verlag, NATO ASI Series Vol. F73.

Wise, J. A., Hopkin, V. D., & Stager, P. (Eds.). (1993a). *Verification and validation of complex systems: Human factors issues.* Berlin: Springer-Verlag, NATO ASI Series Vol. F110.

Wise, J. A., Hopkin, V. D., & Stager, P. (Eds.). (1993b). *Verification and validation of complex systems: Additional human factors issues.* Daytona Beach, FL: Embry-Riddle Aeronautical University Press.

V

Aviation Operations and Design

23

Air-Traffic Control/ Flight Deck Integration

Karol Kerns
MITRE

23.1 Introduction

In the commerce of air transportation, no single development has had wider significance than the introduction of two-way voice communications in the mid-1930s. This innovation made it possible for controllers and pilots to coordinate their activities day or night regardless of the weather. It also launched an acculturation process and lexicon that became the foundation for modern air-traffic control (ATC) operations. Now, every day in the National Airspace System (NAS), millions of radio transmissions are made between controllers and pilots to coordinate flight clearances, weather reports, and information on every conceivable equipment, traffic, or environmental factor that may affect operations. The end result of this communication is a shared understanding of the situation and mutual comprehension of one another's actions and intentions.

To an overwhelming degree, interpersonal communication forms the matrix of teamwork between controllers and pilots. But on the scale of today's operations, the overhead involved in orchestrating a team effort is rapidly exhausting the resources of the current system. Increasingly, frequency congestion and the collateral chore of keeping aircraft and ground computers up to date, limits the level of collaboration possible between controllers and pilots. For many years, visionaries have looked toward digital communications for a breakthrough in system productivity. By deploying a digital communications system, the same infrastructure can be used to connect people and computers in aircraft and on the ground seamlessly. Although voice transmissions remain paramount, a digital system would expand the bandwidth and modes available for communication, enabling transmission of data or images, as well as applications that help integrate information with automated functions.

In a process begun more than half a century ago, research and development (R&D) projects have matured into technological implementations and operational innovations, which alter what it means to be controller and pilot. This chapter examines the developing state of ATC/flight deck integration, a subject with broad reach. In a sense, ATC/flight deck integration is about the transition from interpersonal communication to seamless interconnection of machines and people. But in a larger sense, ATC/flight deck integration is more about fundamental changes in the operating process expected to follow from this transition. From either perspective, controllers and pilots are crucial constituents. By looking at ATC/flight deck integration in terms of the partnership between controller and pilot, I hope to capture and explore a representative cross section of the progress and latest thinking in this area.

When I first wrote about this subject, the material available consisted almost entirely of research findings—field studies of operational communications and simulation studies of future concepts and capabilities for data link. A decade has passed since then. New developments have crept into the operational system, changing the way controllers and pilots work together and enabling them to do things they have never done before. More than anything else, the oceanic data-link system, which provides controller–pilot data-link communications (CPDLC) and automatic dependent surveillance, laid the foundation for reshaping the ordinary conduct of ATC in remote and oceanic areas of airspace throughout the world. In oceanic airspace controlled by the U.S. Oakland and Anchorage centers, e.g., until the mid-1990s, controllers and pilots depended on intermediaries to relay crucial surveillance reports and clearance messages between them; a situation that left working methods virtually unchanged since the earliest days of ATC. Now direct communication and its interlinks with aircraft and ground computing technology are rapidly extending the tools of collaboration and the efficiency of modern air-traffic service into this formerly isolated environment.

Meanwhile, the recent appearance of advanced area navigation (RNAV) procedures in the terminal airspace illustrates how the process can also flow the other way. Technology may produce better integrated ATC/flight deck performance, depending on whether conditions in the operating environment favor it, but procedures can certainly establish new rules of engagement causing the contributions of the ATC and flight deck elements to work in greater concert. In contrast to the ocean, working methods in the terminal area represent the summit of tightly coordinated controller and pilot operations. Paradoxically, this degree of coordination is itself part of the problem because established ways of working deplete the controller's resources and the communication bandwidth too rapidly to keep up with the growing numbers of flights at major airports. In 2001, the Federal Aviation Administration (FAA) and industry joined forces to confront the problem by designing more complex but efficient arrival and departure paths, which would also spread the traffic over more of airspace. The premise was to redistribute the workload, reducing the controller's role in navigation assistance and enlarging the role of the pilot with support from advanced navigation systems onboard many aircraft. Operational experience with these procedures is beginning to shed interesting light on the integration problem. Among other things, this work illuminates the potential for even modest shifts in controller and pilot roles to produce diffuse effects, which could scarcely be imagined beforehand.

Finally, the research arena is opening new vistas on a future where collaborative ATC-planning and decision making is commonplace, separation of controller and pilot roles blurs, and role taking becomes more fluid. The fulcrum on which this turns is ATC/flight deck integration; the intellectual progenitor is the Free Flight Concept (RTCA, 1995). More than a decade has passed since the visionaries who conceived of Free Flight presented the concept to the aviation community, spawning an ambitious program of research. This work appears to be nearing an important threshold. One of the most advanced lines of R&D is designing a system that gives pilots a picture of the operational air situation. Such a system has a spectrum of useful applications from awareness of the relative positions of nearby traffic to extending pilots' ability to separate themselves. Applications of this cockpit display system have proceeded through various stages of operational trials and evaluation. In restricted environments (Alaska), relatively sophisticated suites of applications are already in use, enabling pilots to fly in remote airspaces with greater safety. And basic applications for improving pilots' situation awareness seem ready for use on a large scale.

The rest of this chapter examines the working partnership between controller and pilot as it relates to ATC/flight deck integration—the historical events and forces that formed controller and pilot inter-role coordination, the constraints and opportunities of new means available for coordination and collaboration, and the impact of expanded channels of communication on system operations.

23.2 Perspective on ATC/Flight Deck Integration

Historically, the need for ATC originated around airport areas and extended out into the approach and en route airspace. As the system expanded, operating environments differentiated, and roles and procedures moved toward greater specialization. For most of this history, the FAA exercised considerable autonomy over the strategy pursued to increase the capacity of the system while industry determined how to increase the business of air transportation. Now, however, the prospects for unilateral initiatives on either side to yield a significant payoff in capacity have dwindled or appear prohibitively expensive. Many in the community believe that the only way out of this bind is to mobilize all of the available resources, a change that is likely to alter the traditional model of traffic management.

What we refer to ATC/flight deck integration can be viewed as the next structural development in the partnership between controller and pilot. It hinges on an efficient means of connecting air and ground resources, and people and machines. With resources fully networked, the community expects to launch a spectrum of useful applications and reclaim lost productivity. An account of these applications, all of which depend on digital communications, is given in the next section. But first it is informative to examine research on operational issues related to voice communications as a rationale for improvements being sought in future system applications and capability. These include characteristic vulnerabilities to errors, limited availability of and access to service and information, onerous workload to manage information transfer, and a rigid allocation of functions that tends to overload the controller (FAA, 1990).

23.2.1 Communications Function and Dysfunction

Some operational environments are more prone to communications failures than others. What accounts for the difference is a high demand for transmissions relative to the opportunities available. The highest concentrations of transmissions occur in major terminal radar approach control (TRACON) and airport control tower environments (Cardosi & Di Fiori, 2004), during peak periods of demand, which occur two or three times a day. In en route airspace, only 7% of the sectors experience peaks in demand so intense that flight movement becomes impaired, according to a national survey. But as a result of this frequency congestion, approximately 10% of the flights in the NAS are delayed each day (Data Link Benefits Study Team, 1995; Massimini, Dieudonne, Monticione, Lamiano, & Brestle, 2000).

Of all the controller's duties, communications is the most prevalent. It is estimated that controllers spend as much as half of their time communicating (Mead, 2000). Analysis of recorded communications shows that during peak demand, controllers working in control towers can make as many as 12 transmissions per minute with an average of 8 transmissions per minute (Burki-Cohen, 1994). For pilots, communications ordinarily take up less (about 15%) of their time (Uckerman & Radke, 1983). But when unpredictable and inevitable complicating circumstances arise, such as bad weather, pilots are extremely sensitive to congestion. At these times, with everyone trying to talk at once, contention for the channel can make it virtually impossible to get in touch with the controller.

Errors in communication generally fall into one of two cases: (1) cases in which a pilot reads back a different clearance than the one that was issued and (2) cases in which the pilot asks the controller to repeat a previous transmission. Measured as a proportion of total transmissions, communications error rates tend to be consistently low, less than 1%, but they vary considerably across operating environments. The hourly error rate is a commonly applied measure of communications performance and a more telling indicator of how performance differs across operating environments. On this comparison, the hourly error rates observed in en route, TRACON, and airport data ranged from a high of 2 errors per hour for the TRACON down to 0.4 errors per hour for the airport (Cardosi, 1994, 1996). By this index, the magnitude of the performance discrepancy between environments appears striking. This estimated range of values would mean that the incidence of errors can increase by a factor of 5 between environments.

Why should the effectiveness of communications vary to such a degree? Congestion creates the conditions in which the underlying mechanisms are easiest to see. As demand approaches capacity, the voice system adapts by increasing the amount of information contained in each message and the rate of talking. Studies (Cardosi, Falzarano, & Han, 1999; Grayson & Billings, 1981; Monan, 1986, 1983) indicate that while errors tend to increase as clearances become more complex, the threshold at which this occurs depends significantly on the pilot's familiarity with the message. Performance is naturally better for routine messages, even lengthy ones, because the pilot is prepared; the pilot is actively listening and mentally attuned to expect a message with a specific content. In terms of speech rates, the findings show that when controllers were busiest, and presumably speaking most rapidly, they missed almost three times as many read-back errors as they did when least busy (Burki-Cohen, 1995). Although pilots complain about the difficulty of comprehending instructions when controllers speak too rapidly, there is nothing in the research to support a connection between speech rate and errors in a pilot's read back (Cardosi, 1994; Morrow, 1993).

23.2.2 Congestion Dynamics

As the frequency becomes saturated, the likelihood of transmission overlap increases to a point and then escalates sharply, resulting in blocked or partially blocked transmissions. It is easy to appreciate the dynamics of communication congestion, if we picture a positive feedback loop. Each incidence of blocked transmissions begets a proportionate number of retransmissions. This results in an exponential increase in blocked transmissions (Cardosi, 2003; Nadler, DiSario, Mengert, & Sussman, 1990; Prinzo & McClellan, 2005). A full accounting of the incidence of blocked transmissions in the NAS has not been made. The best evidence now available comes from recent analyses of actual controller–pilot transmissions recorded in TRACON facilities that put the incidence—average number of occurrences as a fraction of the total hours of arrival and departure communications—around four times per hour during high demand (DiFiore & Cardosi, 2002; Prinzo & McClellan, 2005).

With today's voice communications system, blocked transmissions are unavoidable. Pilots and controllers cannot predict when a channel is about to become occupied. Even when they follow procedures correctly and wait for an explicit cue indicating that a transmission is over before transmitting, it is impossible to avoid overlaps completely. The characteristic transmission delay of the ground radio is longer than that of the aircraft radio. This means that a pilot can transmit before it is humanly possible to realize that the controller is also transmitting. When both transmitters operate at the same time, the two users competing for the channel are unable to detect it. Other listeners on the channel might hear a squeal or garbled speech, or if relative signal strengths of the competing transmissions are different, the weaker transmission may be swamped. In addition to the overlapping transmissions, the voice channel can also be blocked unintentionally, if a radio transceiver gets stuck in the transmit mode.

At the center of the problem of blocked transmissions is one particular case in which circumstances create a serious threat to flight safety. Blocked transmissions that result in safety consequences have a characteristic "signature," according to incident reports collected in the FAA Aviation Safety Reporting System and the United Kingdom Mandatory Occurrence Reporting System. Almost exclusively, the type of transmission block from which a threat to flight safety develops is one in which two aircraft have responded to an ATC clearance intended for only one with the erroneous response masked to the controller by the correct response (Harris, 2000). One of the worst aviation disasters happened on the runway at Tenerife airport in March 1977, when two transmissions, warning one aircraft to wait for takeoff, coincided and were lost. A collision on the runway resulted in 583 fatalities.

For almost a decade, prevention of blocked transmissions has been recognized as an operational imperative according to controller and pilot professional organizations. But since then precious little progress has been made toward this goal. But so far, only one airport in the United Kingdom and a few European airlines are using a communications system with antiblocking technology (Cardosi & Di Fiore, 2004). And in the United States, the FAA has successfully demonstrated a communications system that provides antiblocking along with other advanced features, such as call waiting and caller ID; but at the moment implementation plans have stalled as the agency grapples with a tight budget.

23.2.3 Miscommunications

Although the connection between the complexity of a message and miscommunication is well established, a number of variables mediate this relationship (Cardosi & Fiori, 2004; Prinzo, 1996; Morrow, Lee, & Rodvold, 1993; Williams & Green, 1991). One is the type of message. More errors occur in messages conveying radio frequencies than in any other type of message, at least in TRACON and en route environments (Cardosi, 1993, 1996; Cardosi & Di Fiore, 2004). Messages containing voice frequencies represent 41% of the errors measured in en route and 26% of the errors measured in TRACON transmissions. This translates into a probability of an error once in every 100 voice frequency assignments. Chronic and frequent problems with certain types of messages are caused to a large degree by the pilots' unfamiliarity or inability to infer the content. This is borne out by research in the airport environment. Here, taxi instructions, the details of which are difficult to predict, are the most misunderstood type of messages (Burki-Cohen, 1995).

Auditory processing of communications appears to place a limit on the complexity of information that can be transmitted reliably and with accuracy. In TRACON and en route environments, there is a sharp divide in error rates between transmissions containing less than four elements and those with four or more (Cardosi & Fiori, 2004). Moreover, the effect is so strong that transmissions above the limit are from two to seven times as likely to produce an error as those below it. In the airport environment, however, the upward transition in error rate does not appear until messages contain seven or more elements, and the relationship between complexity and errors is not nearly as clear (Burki-Cohen, 1995; Cardosi, 1994).

A study of message complexity and phrasing alternatives (Burki-Cohen, 1994), which controlled for the effect of operating context, shows how the linguistic structure of the message affects processing. In this study, pilots listened to prerecorded clearance messages whose complexity varied between three and six pieces of information. Three phrasings of the numerical information were compared: one enunciated each digit in the message sequentially, "one eight thousand"; the second enunciated numbers in a grouped format, "eighteen thousand"; and the third combined both phrasings, stating messages twice using the sequential and grouped phrasing. In general, communication deteriorated as clearances got more complex but the restated phrasing protected even the most complex messages from errors. Meaningful pieces of information (altitude) were heard more accurately than less important pieces (frequencies and altimeters), regardless of complexity or phrasing. This illustrates one of the significant phenomena of voice communication, the selectivity of the listener. With voice, message processing is both sequential and hierarchical. As the message is spoken, the listener seeks out certain references that fit a repertoire of ready-made mental labels for essential information. Since the span of memory is limited, more meaningful information may supplant less meaningful information regardless of where it occurs in the syntax.

In most cases, miscommunications are merely a nuisance, but they appear with disturbing frequency as coconspirators in reports of safety-related incidents. Studies of operational errors, incidents where aircraft get closer to each than legally allowed, indicate that communication errors are implicated with exceptional frequency in reports documenting operational errors. Communication errors were cited in over 40% of operational errors in the tower environment (Cardosi & Yost, 2001), and in over a third (36%) of the operational errors in the en route environment (Rodgers & Nye, 1993). By contrasting the minor and more serious operational errors in the en route environment, researchers found that the exceptional frequency of communication errors was even more striking in more serious operational errors. Less than a third of all operational errors are classified as moderate or major safety threats; of those, over 40% mention communication errors as a factor.

23.2.4 Controller and Pilot Roles

The present system of ATC is based on a centralized, hierarchical structure of roles with the controller having a pivotal role in planning the movement of air-traffic and transmitting instructions to carry out the plan. With respect to ATC, the pilot's role, once a flight plan has been coordinated and cleared, is one of processing advisory information, accepting instructions, and acting upon them (Billings & Cheaney, 1981). Under this arrangement, the controller sets the pace at which flight movements occur and can often be overworked given the scale of modern air-traffic operations. Among the most arduous are controller's procedures for handcrafting the flow of arrivals and departures that can entail multiple coordinating cycles between controller and pilot to establish a flight on its final approach or departure route. For its time, this was a highly efficient method of moving traffic. As long as the controller has the best understanding of the traffic situation overall, the traditional roles generally offer the most efficient compromise among the conflicting demands of individual flights. Up to a point, the "Big Picture" perspective and pivotal role of the controller lends order and efficiency to the application of common operational procedures. But it is also true that the centralized structure becomes a bottleneck during periods of peak demand.

Two principles have defined the relationship between controller and pilot—reduce uncertainty by ritualizing role coordination and authority by information primacy. The evolution of standard operating procedures helped relieve the controller of the continuing press of certain types of decisions. It also affords similar relief to pilots by presenting them with predictable behavior to which they can adjust. According to Degani and Weiner (1994), however, the cumulative effect of this evolution, after more than half a century, is a simultaneous over- and under-proceduralization of operations. In time, many procedures were embedded in the culture and artifacts of the NAS, hardening into fixed constraints. When procedures, e.g., ATC-preferred routings, traffic flow, and airspace restrictions, become too rigid and are applied without question, coordination is precluded. Degani and Weiner (1994) refer to this as over-proceduralization. On the other hand, where there are huge variations among users and ATC facilities in the way things are done, this is called under-proceduralization. The lack of a common structure makes coordination difficult. Studies indicate that the ATC and flight deck operations are most poorly integrated during the ground movement of aircraft (Adam, Kelley, & Steinbacher, 1994). In the face of growing traffic, surface operations have come under increasing pressure. Taxi instructions tend to be delivered at inopportune times and are complicated. To an unusual degree airport layouts and taxi routes vary as do operator policies and crew procedures during ground operations. With this, there is a high probability that controller and pilot interactions will interfere with or preempt concurrent ATC or flight deck activities.

It is easy to see how controller and pilot roles are deeply entrenched in the distinct frameworks of knowledge and information surrounding each side's understanding of events. Air-traffic controllers and pilots mention at times something they call the Picture—a working [mental] model of the overall traffic situation. This Picture is the aperture that allows controllers and pilots to perceive a situation and take actions quickly and expertly. For several reasons, shared understanding of the situation, of the task, and of each other's background knowledge and expectations is an important key to collaboration (Farley, Hansman, Endsley, Amonlirdviman, & Vigeant-Langlois, 1988).

While years of refining the language of ATS communications has produced a high standard of accuracy and integrity, spoken language also draws a boundary around the realm of possible expression, beyond which it is difficult to negotiate. Voice communication permits a group of several pilots and a controller using a common frequency, to develop a limited degree of reciprocal awareness. Listeners, monitoring the channel, can piece together aspects of the air situation and anticipate instructions they will be given with adequate preparation time. But the process is not necessarily accurate (Midkiff & Hansman, 1993; Pritchett & Hansman, 1994).

Since communication is enormously facilitated by a shared workspace (Krauss & Fussell, 1990), development of applications that use cockpit displays of traffic information (CDTI) to enhance shared situation awareness has been avidly pursued for decades. Kreifeldt (1980) summarized a program of research on CDTI applications that explored alternative allocations of functions between ATC and flight deck in the terminal environment. On balance, the best overall allocation gave the job of setting up an arrival sequence, and transmitting the assigned positions and the flight they were to follow. Pilots were to use the CDTI to identify the flights ahead of them and maintain their place in the sequence. This operation, which was equivalent to the traditional method in terms of spacing aircraft efficiently and with less variability, also reduced the controller's workload. Although pilots reported a higher workload with CDTI, they much preferred the new distributed approach. But it can do more than that. A CDTI affords the system a capability to adjust and respond to sudden failures and contingencies more quickly than would be possible with the inherently long delay times of an air–ground loop.

The CDTI work is one of the three broad lines of R&D exploring new avenues of ATC/flight deck integration seeking to advance the broad goal of collaborative decision making. Another line is focusing on collaboration and integration of strategic decision making at regional or NAS levels, addressing the roles of the air carriers' operational control and FAA traffic flow management units. This work may someday affect the role sets of controllers and pilots. For this discussion, it is sufficient to note that such strategic initiatives ultimately flow down into coordination at the controller and pilot level.

A third line of R&D is helping to lay the foundation for integrating ATC computer-to-avionics computer exchanges with controller–pilot collaboration. Initial implementations of advanced RNAV procedures can be viewed as an early form of ATC/flight deck integration, although without a direct exchange of the flight path data between computers. Operational experience with the new roles shows promise of improvements in system efficiency. It is also instructive in exposing areas of the integration problem that loom large as challenges to be overcome.

23.3 Digital Communications

A comprehensive analysis of ATS communications distinguished four classes of services: air-traffic management, flight information, navigation, and surveillance (FAA Operational Requirements Team, 1994). Transactions grouped together in a class are similar in many ways. An essential similarity among those in the air-traffic management class, which accounts for the vast majority of communications involving controller and pilot, is a characteristic collaborative process.

This general process is significant because it captures the detailed interaction pattern by which communicators establish that a message has become part of their common ground. Controller–pilot data-link communications in this class should embody the same process. By doing so, controllers and pilots can apply what they already know making training and transition a bit easier. Throughout this section, this process emerges as a major theme in examining factors that account for progress and problems related to operational use of digital communications (Kerns, 1991).

23.3.1 Communications Capacity and Efficiency

In the early years of digital communications R&D, most of the attention was given to how it could be used to alleviate frequency congestion (Lee, 1989b). Researchers studying data link estimated its impact on the problem of frequency congestion. Talotta, Shingledecker, and Reynolds (1990) looked at how much time controllers spent on the voice channel under three levels of data-link equipage. In this study, en route controllers used data link to issue radio frequencies and altitude assignments; other communications were conducted via voice, regardless of aircraft equipage. Relative to a voice communications baseline, this study found a 28% reduction in controller time spent on the voice channel when 20% of the aircraft under control were data-link-equipped, and a 45% reduction when 70% of the aircraft were data-link-equipped. Comparable reductions in radio frequency utilization were also reported for terminal controllers (Blassic & Kerns, 1990; Talotta & Shingledecker, 1992a, 1992b).

Research shows that reductions in voice transmissions were not simply a function of the level of data-link equipage simulated in the air-traffic scenario. Using equivalent scenarios, results indicate that a dual voice and data-link communication system requires fewer total transmissions than an all-voice system. Studies consistently report that the dual system appears to reduce the occurrence of repeated messages and missed calls (Blassic & Kerns, 1990; Hinton & Lohr, 1988; Knox & Scanlon, 1991; Lozito, McGann, & Corker, 1993; Talotta et al., 1990).

From the flight deck perspective, studies show that data link had slight advantage over voice in terms of its efficiency for pilot communications tasks (Talotta et al., 1990; Uckerman & Radke, 1983). If the data-link system is interfaced to flight management functions of the aircraft, the advantage becomes conspicuous (Knox & Scanlon, 1991; Waller, 1992).

23.3.1.1 Access and Availability

The necessity of a voice link for ATS communications is indisputable. But the current analog system is extremely susceptible to curtailed capacity and message degradation when congested. Blocked transmissions, a phenomenon of congestion, can be virtually eliminated with a digital voice system evaluated by the FAA. This system uses digital signaling to mediate access to the channel, limiting access to one user at a time.

An evaluation of the digital voice system has already revealed some surprising insights into how antiblocking improves the performance of the voice service. By comparing it to the analog system, researchers found that despite its slightly higher throughput delays, the digital system allowed more successful transmissions without any increase in the number of blocked transmissions (Sollenburger, McAnulty, & Kerns, 2002; Zingale, McAnulty, & Kerns, 2003). Of all the study findings, one of the most significant was how much pilots favored the digital system over the analog system. What made the difference in favor of the digital voice system is the additional feedback it provides. Pilots could tell with certainty whether a transmission was possible by listening for a busy signal. When two aircraft transmit at the same time using the current voice communications system, it is much harder to tell that it happened and decide what, if anything, needs to be done. The busy tone provided by the digital system appears to be easier for pilots and controllers to understand. "Communication will be clearer with this design," one of the pilot participants explained. "It was almost like getting a receipt when your transmission went through. You know the message was sent."

Two conversations take place between controller and pilot each time the communications connection is transferred from one controller to the next. This procedure has remained unchanged since the earliest days of ATC. As an aircraft leaves one jurisdiction, the controller tells the pilot who to contact and what channel to tune next. After the new channel is tuned in the radio, the pilot calls the new controller and confirms the current and expected status of relevant tasks and instructions. In some areas, pilots do this as often as once every 3–4 min (RTCA, 1994, p. 18). And it is estimated that in the domestic NAS, transfer-of-communication (TOC) messages constitute between 10% and 25% of all message traffic (Talotta & Zurinskas, 1990). The simple fact of making so many transmissions in order to keep the link open seriously depletes the capacity available for other messages.

Data link can be used to transfer communications. Early operational experience shows that the data-link service is essentially effective with a few exceptions. When controllers began using the CPDLC TOCs, FAA evaluators noticed that sometimes the frequency assignments being uplinked to flights were incorrect (FAA, 2003). The data-link TOC is designed to work from a computerized address book stored in the ground system that contains the voice frequencies currently assigned to each operational position. For the controller to switch a connection, it is necessary to request a TOC by identifying the flight to the computer. The actual TOC message is prepared automatically behind the scenes by the software. And although it is possible to check and see what frequency was actually sent, there is no compelling reason for a controller to do this. While on the flight deck, the pilot's message to "Monitor" (the next frequency) omits a routine check of the new voice connection. If a bad frequency happens to get sent, it could take a while for either controller or pilot to discover it. In hindsight, it appears that some aspects of the collaborative process have been bypassed in this mode.

As a strategy for development, the designers of data link nearly always assumed that increases in human workload were a small front-end price for eventual, large workload reductions. In designing a data-link initial check-in (IC), for instance, they gave pilots a new task to perform. A time is probably coming when the aircraft automation will be deemed competent to correctly infer what altitude belongs in the IC report, but for several reasons that day has not yet arrived. As the only authoritative source of the assigned altitude data, the pilot composes the IC report in one of two ways: by entering the altitude data or by reviewing a default value preselected by the system. Pilots have reacted negatively to this because the data-link IC is far more likely to contend with some other task.

23.3.1.2 The Gist

From the beginning, FAA and industry experts were quick to recognize the enormous potential of using data link as a means of coordinating clearance information. As R&D matured, the focus of attention is shifting toward comprehensive solutions that help manage the transfer of information to its ultimate application in the operational process. Encoded for data-link transmission communications, data can be processed more readily by computers. For this reason, virtually, all of NAS operational evolution presupposes data link as the device for air–ground information exchange. The progression of services inspired by data link would bring about astounding changes in the economy of information transfers while increasing the consistency and precision of flight path definitions.

It is because of the uncompromising limits (four to six elements) imposed by voice that data link is seen as a better medium for communicating complex messages. The experience with data link shows that the visual medium successfully preserves more information; however, intelligibility can still be impaired by the display features such as organization and format. One of the earliest examples of this is the display used in the flight deck presentations of PDC messages. These were reproductions of the controller's display format, using the vernacular of the ATC computer system, which is extremely terse and idiomatic. Pilots needed a glossary to interpret the message.

Operational experience with the FANS system uncovered other features of the written language and display layout that resulted in comprehension problems for pilots. In the case of conditional clearances, the message began with a qualifying phrase, "at this location," followed by an instruction, "climb to this level." Pilots tended to misread these messages, overlooking the restriction and taking action immediately. Another intelligibility phenomenon observed in the FANS displays was the channeling influence of typography and layout. Evaluators discovered that when color or a font size was applied to emphasize specific display elements, there were mixed results. Pilots paid attention to the coded information, but the coding also tended to de-emphasize surrounding content without coding. Text formats that segmented messages unintelligently or arbitrarily, e.g., inserting extra space between content labels (flight level) and their associated data (350), also hindered comprehension (Lozito, Verma, Martin, Dunbar, & McGann, 2003).

Studies investigating graphical display formats showed that graphical displays improved avoidance of wind shear by pilots when compared to a voice presentation (Lee, 1991; Wanke & Hansman, 1990). Lee also found that flight crews provided only with conventional ATC transmission of weather information had difficulty discriminating conditions conducive to microburst events from less hazardous wind shear events, and that real-time updates of the data-linked information contributed to improved situation awareness for microburst events.

Over the past decade or two, scientists and industry experts gathered extensive evidence on breakdowns in the transfer of information between pilots and aircraft automation systems (FAA Human Factors Team, 1996). To date, research on the interplay between data-link, pilots, and aircraft automation is extremely scarce. In his doctoral research, Olson (1999) studied pilots' ability to detect errors in ATS communications using two types of data-link systems: one system used a manual procedure to load data from the data-link system in the Flight Management System (FMS); the other system had a feature that allowed the pilot to move and load data from the data-link system in the FMS. Two types of errors were simulated in the data-link clearances. One type, called a goal conflict, was a clearance that conflicted with other pilot goals, e.g., a clearance to descend to an altitude above its current altitude. The other type, called an implementation conflict, was a clearance that was acceptable but had an unintended or undesirable result when loaded into the FMS, e.g., a change to the route also deleted the vertical profile. Results showed that pilots were not good at detecting either type of problem, but fewer implementation conflicts were detected. The worst performance resulted when pilots were asked to detect implementation conflicts while using an automated data-link system. A common thread from studies of data-link and flight-deck automation is the importance of framing the clearance in terms of the current and future situations (Hahn & Hansman, 1992; Knox & Scanlon, 1991; Lozito et al., 1993; Sarter & Woods, 1992, 1994). One way to set up this framing would be to provide a preview of the projected result for comparison with the current situation.

23.3.2 Ways and Means

A relatively large body of research supports the use of consistent procedures when conducting voice and data communications (Kerns, 1991). Designers of the FANS and CPDLC data-link systems took this into account, making data-link operational dialogs closely resemble their counterparts in the voice mode. Early operational experience with predeparture clearance (PDC) using a different process indicates some problems associated with it. Controllers issue PDCs digitally to participating airlines through the airline communications network. In turn, the airline dispatch office has the responsibility for actual delivery of PDCs to flight crews, typically using the company data link and a cockpit display or printer. While digital PDCs have received broad support from the participants (Moody, 1990), incident reports cite procedural deviations in which crews failed to obtain their PDC or received the wrong PDC (Drew, 1994). Under the current PDC delivery system, voice communications can be used to verify that the PDC has been issued correctly; however, this procedure is not standard across airports.

In the course of comparing various options for combining voice and data link, a key theme across the findings was a recommendation that data link be used to replace specific voice transactions. A few years ago, scientists at NASA (Lozito et al., 2003) were able to show how skillful handling of a dual mode communications system takes more than learning the mechanics of interacting with each medium. The researchers wanted to see what would happen in the dual mode environment if pilots were unable to predict whether voice or data link would be used. They discovered that when modes were chosen randomly, a dual mode system was generally detrimental to performance—controller pilot dialogs took longer and messages were more apt to be misunderstood. By comparing how professional pilots conducted communications using three different systems—voice, data link, and a dual mode—they were able to demonstrate that, without rules, the dual system actually disrupted turn-taking and information transfer.

These findings impressed not only the scientists but also the operational experts. Until they were released, the conventional wisdom held that the dual mode system would make communications more efficient overall (Kerns, 1991, 1994). And although a dual mode system can have this effect, as is shown in many previous studies, these latest findings indicate that it would be a mistake to overlook the primary requirement for use conventions. Controllers and pilots will need to acquire a substantial body of knowledge and practice in order to master the use of data link. But the real gain in communicating power will only come if they also share a plan for using each mode.

This conclusion is borne out in a recent study. After surveying oceanic controllers from Oakland and Anchorage centers, Prasse (in press) found a similar connection between a dual mode communications system and controllers' assessments of their experience with data link. Controllers from Oakland center, where data link became the primary means of communication, overwhelmingly reported that it had made their job easier. While Anchorage controllers, who have both data link and voice available in parts of the airspace, were split and generally more qualified in their assessments.

In terms of crew procedures, research has supported a data-link procedure that entails verbal coordination between pilots prior to sending a data-link response to the message (Hahn & Hansman, 1992; Lee, 1989a; Lozito et al., 1993; Waller & Lohr, 1989).

For a controller team, data-link implementation creates new options for expanding the role of a second controller in en route and terminal environments. A study in which data link was evaluated by controller teams rather than single controllers found that data link promoted team cooperation and reallocation of tasks (Data Link Benefits Study Team, 1995; Shingledecker & Darby, 1995). Controller teams working on en route sectors used voice and data link for communications. Not only did the dual mode system enable controllers to balance communications workload, a reduction in the relative proportion of voice for air–ground communication enabled more planning and communication within the controller team. Controllers considered team performance to be superior when voice communication was supplemented by data link. Comparable results were obtained with a dual communications system in the terminal environment (Data Link Benefits Study Team, 1996).

23.3.3 Operational Evolution

Until recently, before new equipment was turned over to the users, standard practice among those who developed and approved it typically culminated with operability testing, verifying performance with a range of cases in the actual environment. This practice turned a corner when the FANS developers attempted to produce a system of interoperable ground and aircraft equipment. It now appears that the notion of an abrupt transition between development and operational use is overly simplistic. Passing the key decision point of certifying the airborne element of the system for operation, the FANS developers disbanded their interoperability team (Brown, 2001). Having done that, they soon noticed a sharp falloff in the performance of the total air–ground system, and with it, the confidence of the users. Out of desperation, the stakeholders reconstituted a multidisciplinary interoperability team, drawing on expertise from the air-traffic service provider, data-link communications service provider, controllers, pilots, and aircraft and equipment manufacturers. The reconstituted team kept authority over reports of suspected flaws in the system, working cooperatively to figure out the best way to repair performance and make the "fixes" work. In doing so, the FANS stakeholders recognized the necessity of a steady mechanism to monitor natural adaptations, maintain the operation, and indeed to promote increasing use of system capabilities.

At an early stage of FANS implementation, the interoperability team observed that the usual strategies for preparing controllers and pilots were not working out as planned. The team had adopted the training strategies and mechanisms that were generally abroad in the industry; each of the participating groups produced and delivered a training program through their preexisting networks. This resulted in a long, drawn-out process of diffusing FANS knowledge throughout the pilot and controller community, during which clear indications of problems started to crop up (Lozito, 1999). The most striking evidence comes from two examples showing how widespread misconceptions held by one group of users precipitated serious difficulties for the other. In the first example, pilots, unsure of whether their system had sent a message, reacted by continuing to send the same message, creating duplicate messages for the controller. In the other example, controllers chose to bypass the formatted data-link messages and compose their own using free text. In almost all of these cases, the system provided an equivalent, preformatted message. Contained in the structure of these formatted messages were links to the system's basic error protection and value-added processing functions. Using free text left messages open to errors and made it impossible to move the data into other automation systems without retyping. Almost certainly, the prolonged transition period hindered the development of a shared model of operations among controllers and pilots. The initial training was invaluable but insufficient: Immediately after training controllers and pilots had scant opportunities for practice, and experience continued to be sketchy for quite some time.

As a strategy for coping with ATC/flight deck integration problems, the FANS integration team collected ideas, explanations, and solutions from a variety of perspectives. The attempt to turn those inputs into mitigation plans required extensive negotiation and collaboration between engineering and operational experts on both air and ground sides of the system. Sometimes they changed a ground application or procedure, sometimes an avionics application or procedure, and sometimes a bit of both.

Thanks to the experience with FANS data link, we now have a protocol for four-way collaboration that works in the confines of oceanic operations. Extending this process into more demanding contexts found in other parts of the NAS will require the right tools and the right intuition. Although the overwhelming majority of clearance messages in the domestic en route and terminal environments is relatively simple, there are also many examples today of clearances that are quite intricate. And the relative importance of communicating clearances in this form is expected to grow in the future.

Some of the most intricate clearances currently used refer to arrival and departure procedures, which map flight paths in three dimensions. Today, controllers and pilots rely on very simple and static media to store and retrieve the details of the charted path. In the rapid-fire speech of real-time operations, they will use a short, public name to refer to the procedure. Visionaries imagine a day when four-dimensional paths transitioning aircraft between the airport and the en route environments can be readily adapted based on demand patterns and transmitted on the fly. That represents quite a spectacular advance in the technology of procedure design as well as the efficiency of ATC/flight integration.

While the system benefits may be huge, it is not easy to handle coordination in the transition and terminal airspace. Time is short and the interactions between controller and pilot have to be not only quick but also precise. Yet, giving instructions and extracting information from the FMS is neither of these (Crane, Prevot, & Palmer, 1999; Olson; 1999). Although by now, most pilots have learned to accept and even appreciate the involvement of the FMS in flight operations, despite its onerous workload (Lee, Sanford, & Slattery, 1997).

Because pilots contend with some of the most rigorous parts of their job during arrival and departure phases of flight, standardized protocols for collaboration are especially important. Implementation of RNAV procedures broadened our perspective on the difficulties involved in reaching across the interpersonal and technological divide between ATC and the flight deck. Presently, the flight deck side embodies a range of distinctions in equipment, including RNAV and FMS processing and capabilities, and crew procedures. Developers have begun to recognize that during the operational transition, anticipating and managing such differences is critical. Like the FANS integration team, FAA/industry implementation teams kept track of problem reports once a new RNAV procedure became operational. They discovered that, without a direct connection to diagnose programming discrepancies machine-to-machine, the process was improved by controllers and pilots using voice to cross-check and verify key elements such as departing runway and the first waypoint of a procedure. It soon became apparent that a guiding principle of use should be to minimize real-time modifications. To the extent possible, advance coordination of flight path data, and controller and pilot expectations appears to be the most robust approach.

Another important inference derived from controller and pilot reports concerns the importance of a cross-training to increase mutual understanding between the two occupational specialties. In a substantial number of the problems reported, it was clear that, despite variations in the details, a common thread ran through them: Controllers and pilots were fundamentally ignorant of basic facts about each others' operating environments. Some telling examples came from cases in which a controller used radar vectors to merge and space aircraft cleared for an RNAV arrival procedure. Well placed, such maneuvers can be very efficient. But, in general, once cleared for the procedure, the controller should plan on keeping an aircraft on it and use speed adjustments to adjust spacing between aircraft. This strategy is most compatible with an FMS operation, and within limits, controllers can be trained to accommodate it (Jarvis, Smith, Tennille, & Wallace, 2003; Tennille, Wallace, & Smith, 2004).

The final strategy presented in this section considers ATC/flight deck integration in the context of advanced surveillance and monitoring capabilities (Callantine, Lee, Mercer, Prevot, & Palmer, 2005; Grimaud, Hoffman, Rognin, & Zeghal, 2005; McAnulty & Zingale, 2005). While the literature in this area is large and growing rapidly, the core operational concepts focus on taking ADS-B data and integrating it with CDTI applications, and aircraft and ground decision aids to improve arrival spacing and conformance monitoring. Reviews of this work suggest that it offers vast opportunities for developing new operating models. At the same time, a shared picture of the situation, and connections that allow access on both sides to self-check and cross-check plans and performance, is a fundamental step.

At the next level of integration, the evolution of two-way data links for clearances, especially complex ones, and a decline in the use of voice transmissions to manage the continuity of communications service is required. This transition is crucial because it reverses a long, unsustainable trend in which transmissions made for the purpose of switching frequencies are rapidly depleting the channel as they become increasingly unwieldy. Nowadays, the drudgery of this one, extremely repetitive communications chore, takes more time from highly trained, seasoned controllers and pilots than any other single type of communication, and all future plans to expand the voice spectrum will only make this worse—five-digit channel designators are expanded to six. Done properly, this one accomplishment could upgrade service reliability, free users from a tedious, uninteresting task, and sharply curtail the waste of productive capacity.

In the estimation of users and safety experts alike, the most significant function of the digital voice system is antiblocking. Voice is an extremely flexible and efficient means for the controller and pilot to coordinate tasks and intentions, and it becomes particularly important in situations where procedural rules are inadequate for the task at hand or have broken down. This role for voice communications will almost certainly increase in importance as future communications use data link for direct coordination between automation systems. In the era of four-party communication, a considerable part of the collaborative effort between controllers and pilots may involve learning how their respective automation systems "understand" a message and comparing the two interpretations. Working through any discrepancies will depend on the flexibility of voice communications and the resourcefulness of pilots and controllers. They will be the ones who compensate if the automation system fails to correctly implement the joint intention.

23.4 Conclusion

Aircraft and ground automation systems are slowly and irreversibly creeping into more aspects of ATC. Still missing from the picture, however, is the means to link these new functions and allow them to be molded directly by the forces of the operating environment. This situation has parallels that go back as far as the introduction of the FMS. At the conclusion of his pioneering study of the impact of advanced aircraft automation on pilots, Earl Weiner (1989) offered this cogent observation:

> It is regrettable that from the beginning aircraft and ground-based ATC systems were designed, developed, and manufactured almost as if they were unrelated and independent enterprises. Even the current developments in ATC and flight guidance systems reflect this proclivity. The proper utilization of aircraft and airspace will only be achieved when aircraft designers and those who design and operate ground-based ATC work in closer harmony. It seems strange that in 1989 it is still necessary to say that.

Visionaries today think no less of ATC/flight deck integration as a way to create a bridge to the future NAS. Modern means of communications appear poised to bring greater parity to ATC and flight deck capabilities for monitoring and communication, leading to heightened mutual understanding and higher quality decisions. The next frontier in ATC/flight deck integration will be aimed at discovering how to configure the applications on each side to the best operational advantage.

References

Adam, G. L., Kelley, D. R., & Steinbacher, J. G. (1994). *Reports by airline pilots on airport surface operations: Part 1. Identified problems and proposed solutions for surface navigation and communications* (MITRE Technical Report MTR 94W60). McLean, VA: The MITRE Corporation.

Billings, C. E., & Cheaney, E. S. (1981). *Information transfer problems in the aviation system* (NASA Tech. Paper 1875). Moffett Field, CA: NASA Ames Research Center.

Blassic, E. J., & Kerns, K. (1990). *Controller evaluation of terminal data link services: Study 1* (MITRE Tech. Report MTR 90W215). McLean, VA: The MITRE Corporation.

Brown, J. A. (2001). *Human factors issues in controller-to-pilot data link communications.* Seattle, WA: The Boeing Corporation.

Burki-Cohen, J. (1994). *An analysis of tower (ground) controller-pilot voice communications* (Report No. DOT/FAA/AR-96/19). Washington, DC: U.S. Department of Transportation, Federal Aviation Administration.

Burki-Cohen, J. (1995, September). Say again? How complexity and format of air traffic control instructions affect pilot recall. *40th Annual Air Traffic Control Association Conference Proceedings* (pp. 225–229). Las Vegas, NV.

Callentine, T. J., Lee, P. U., Mercer, J., Prevot, T., & Palmer, E. (2005). Air and ground simulation of terminal area-FMS arrivals with airborne spacing and merging. *6th USA/Europe Air Traffic Management R&D Seminar.* Baltimore, MD: http://www.eurocontrol.fr/atmsem/

Cardosi, K. (1993). *An analysis of en route controller-pilot voice communications* (Report No. DOT/FAA/RD-93/11). Washington, DC: Federal Aviation Administration.

Cardosi, K. (1994). *An analysis of tower (local) controller-pilot voice communications* (Report No. DOT/FAA/RD-94/15). Washington, DC: Federal Aviation Administration.

Cardosi, K. (1996). *An analysis of TRACON (terminal radar approach control) controller-pilot voice communications* (Report No. DOT/FAA/AR-96/66). Washington, DC: Federal Aviation Administration.

Cardosi, K. (2003). Human factors challenges in the Terminal Radar Approach Control (TRACON) environment. (Report No. DOT/FAA/AR-02/17).

Cardosi, K., & Di Fiori, A. (2004). Communications metrics. *ATC Quarterly, 12*(4), 297–314.

Cardosi, K., Falzarano, P., & Han, S. (1999). *Pilot-controller communication errors: An analysis of aviation safety reporting system (ASRS) reports* (Report No. DOT/FAA/AR-98/17). Washington, DC: Federal Aviation Administration.

Cardosi, K., & Yost, A. (2001). *Controller and pilot error in airport operations: A review of previous research and safety data* (Report No. DOT/FAA/AR-00-57). Washington, DC: Federal Aviation Administration.

Crane, B., Prevot, T., & Palmer, E. (1999). Flight crew factors for CTAS/FMS integration in the terminal. *Proceedings of the Tenth International Symposium on Aviation Psychology.* Columbus, OH.

Data Link Benefits Study Team. (1995). *User benefits of two-way data link ATC communications: Aircraft delay and flight efficiency in congested en route airspace* (Rep. No. /FAA/CT-95/4). Washington, DC: Federal Aviation Administration.

Data Link Benefits Study Team. (1996). *User benefits of two-way data link ATC communications in terminal airspace* (Rep. No. /FAA/CT-9613). Washington, DC: Federal Aviation Administration.

Degani, A., & Weiner, E. (1994). *On the design of flight deck procedures* (NASA Contractor Report 177642). Moffett Field, CA: NASA Ames Research Center.

DiFiori, A., & Cardosi, K. (2002). [Analysis of TRACON controller-pilot voice communications]. Unpublished report.

Drew, C. (1994, March). PDC's: The problems with predeparture clearances. *ASRS Directline, 5,* 2–6.

Farley, T. C., Hansman, R. J., Endsley, M. R., Amonlirdviman, K., & Vigeant-Langlois, L. (1988). The effect of shared information on pilot/controller situation awareness and re-route negotiation. *2nd U.S.A./Europe Air Traffic Management R&D Seminar.* Orlando, FL: http://www.eurocontrol.fr/atmsem/

FAA (1990, November). *The national plan for aviation human factors.* Washington, DC: Federal Aviation Administration.

FAA (2003). *Controller-pilot data link communication build 1: Independent operational test and evaluation early operational assessment report* (Report No. FAA-ATQ-CPDLC-RP01-1-F). Washington DC: Federal Aviation Administration.

FAA Operational Requirements Team. (1994). *The aeronautical data link system operational concept.* Washington, DC: Federal Aviation Administration.

FAA Human Factors Team. (1996). *The interfaces between flight crews and modern flight deck systems.* Washington, DC: Federal Aviation Administration.

Grayson, R. L., & Billings, C. E. (1981). Information transfer between air traffic control and aircraft: Communication problems in flight operations. In C. E. Billings, & E. S. Cheaney (Eds.), *Information transfer problems in the aviation system* (NASA Technical Paper 1875). Moffett Field, CA: NASA Ames Research Center.

Grimaud, I., Hoffman, E., Rognin, L., & Zeghal, K. (2005). Spacing instructions in approach: Benefits and limits from an air traffic controller perspective. *6th USA/Europe Air Traffic Management R&D Seminar*. Baltimore, MD: http://www.eurocontrol.fr/atmsem/

Hahn, E. C., & Hansman, Jr., R. J. (1992). *Experimental studies on the effect of automation on pilot situational awareness in the data link ATC environment* (SAE Tech. Paper 922022). Warrendale, PA: SAE International.

Harris, D. (2000, February). *The potential effectiveness of VHF radio anti-blocking devices.* Paper presented to RTCA Special Committee 162, Washington, DC.

Hinton, D. A., & Lohr, G. A. (1988). *Simulator investigation of digital data link ATC communications in single-pilot operations* (NASA Tech. Paper 2837). Hampton, VA: NASA Langley Research Center.

Jarvis, E. J., Smith, T. M., Tennille, G. F., & Wallace, K. G. (2003). *Development of advanced area navigation terminal procedures: Issues and implementation strategies* (MITRE Paper MP03W191). McLean VA: The MITRE Corporation.

Kerns, K. (1991). Data link communication between controllers and pilots: A review and synthesis of the simulation literature. *The International Journal of Aviation Psychology, 1*(3), 181–204.

Kerns, K. (1994). *Human factors in ATC/flight deck integration: Implications of data link simulation research* (MITRE Paper MP94W98). McLean, VA: The MITRE Corporation.

Knox, C. E., & Scanlon, C. H. (1991). *Flight tests with a data link used for air traffic control information exchange* (NASA Tech. Paper 3135). Hampton, VA: NASA Langley Research Center.

Krauss, R. M., & Fussell, S. R. (1990). Mutual knowledge and communicative effectiveness. In J. Galegher, R. E. Kraut, & C. Egido (Eds.), *Intellectual teamwork: Social and technical bases of collaborative work* (pp. 111–145). Hillsdale, NJ: Erlbaum.

Kreifeldt, J. G. (1980). Cockpit displayed traffic information and distributed management in air traffic control. *Human Factors, 22*(6), 671–691.

Lee, A. T. (1989a). *Display-based communications for advanced transportation aircraft* (NASA Technical Memorandum 102187). Moffett Field, CA: NASA Ames Research Center.

Lee, A. T. (1989b). Human factors and information transfer. *Proceedings of the Second Conference, Human Error Avoidance Techniques* (pp. 43–48). Warrendale, PA: SAE International.

Lee, A. T. (1991). Aircrew decision-making behavior in hazardous weather avoidance. *Aviation, Space, and Environmental Medicine, 62*, 158–161.

Lee, K. K., Sanford, B. D., & Slattery, R. A. (1997 August). The human factors of FMS usage in the terminal area. *American Institute of Aeronautics and Astronautics Modeling and Simulation Technologies Conference.* New Orleans, LA.

Lozito, S. (1999). *Oceanic Data Link Lessons Learned.* Unpublished manuscript.

Lozito, S., McGann, A., & Corker, K. (1993). Data link air traffic control and flight deck environments: Experiment in flight crew performance. *Proceedings of the Seventh International Symposium on Aviation Psychology* (pp. 1009–1015). Columbus: The Ohio State University.

Lozito, S., Verma, S., Martin, L., Dunbar, M., & McGann, A. (2003). The impact of data link and mixed air traffic control environments on flight deck procedures. *Proceedings of the 5th USA/Europe R & D Symposium.* Budapest, Hungary: http://www.eurocontrol.fr/atmsem/.

Massimini, P. A., Dieudonne, J. E., Monticione, L. C., Lamiano, D. F., & Brestle, E. A. (2000, September). Insertion of controller-pilot data link communications into the National Airspace System: Is it more efficient? *IEEE Aerospace and Electronic Systems Magazine, 15*(9), 25–29.

McAnulty, D. M., & Zingale, C. M. (2005). *Pilot-based spacing and separation on approach to landing: The effect on air traffic controller workload and performance* (Report No. DOT/FAA/CT-05/14). William J. Hughes Technical Center, Atlantic City, NJ: Federal Aviation Administration.

Mead, R. (2000, October). Preliminary Eurocontrol test of air/ground data link, Phase II: Operational validation & early implementation. *Proceedings RTCA Symposium 2000*. Vienna, VA: RTCA.

Midkiff, A. H., & Hansman, Jr., R. J. (1993). Identification of important "Party Line" information elements and implications for situational awareness in the data link environment. *Air Traffic Control Quarterly*, *1*(1), 5–30.

Monan, W. P. (1983). *Cleared for the visual approach: Human factors problems in air carrier operations* (NASA Contractor Report 166573). Moffett Field, CA: NASA Ames Research Center.

Monan, W. P. (1986). *Human factors in Aviation operations: The hearback problem* (NASA Contractor Report 177398). Moffett Field, CA: NASA Ames Research Center.

Moody, J. C. (1990). *Predeparture clearance via tower workstation: Operational evaluation at Dallas/Ft. Worth and Chicago O'Hare* (MITRE Technical Report MTR 90W108). McLean, VA: The MITRE Corporation.

Morrow, D., Lee, A., & Rodvold, M. (1993). Analysis of problems in routine controller-pilot communications. *The International Journal of Aviation Psychology*, *3*(4), 285–302.

Nadler, E. D., DiSario, R., Mengert, P., & Sussman, D. (1990). *A simulation study of the effects of communication delay on air traffic control* (Report No. DOT/FAA/CT-90/6). Washington DC: Federal Aviation Administration.

Olson, W. A. (1999). *Supporting coordination in widely distributed cognitive systems: The role of conflict type, time pressure, display design and trust*. Unpublished Doctoral Dissertation. University of Illinois, Urbana-Champaign, IL.

Prasse, L. C. (in press). The effects of automation and data link on Oakland and Anchorage oceanic controllers. *ATC Quarterly*.

Pritchett, A. R., & Hansman, Jr., R. J. (1994). *Variations in party line information requirements for flight crew situation awareness in the data link environment* (Report No. ASL-94-5). Cambridge: The Massachusetts Institute of Technology.

Prinzo, O. V. (1996). *An analysis of approach control/pilot voice communications* (Report No DOT/FAA/AM-96/26). Washington, DC: Federal Aviation Administration.

Prinzo, O. V., & McClelland, M. (2005). *Terminal radar approach control: Measures of voice communications system performance* (Report No. DOT/FAA/AM-05/19). Washington, DC: Federal Aviation Administration.

Rodgers, M. D., & Nye, L. G. (1993). Factors associated with the severity of operational errors in Air Route Traffic Control Centers. In M. D. Rodgers (Ed.) *An analysis of the operational error database for Air Route Traffic Control Centers* (DOT/FAA/AM-93-22). Oklahoma City, OK: Civil Aeromedical Institute Federal Aviation Administration.

RTCA. (1994). *VHF air-ground communications system improvements alternatives study and selection of proposals for future action* (Document Number (DO)-225). Washington DC: RTCA.

RTCA. (1995). *Final report of RTCA task force 3: Free-flight implementation*. Washington DC: RTCA.

Sarter, N., & Woods, D. D. (1992). Pilot interaction with cockpit automation: Operational experience with the flight management system. *International Journal of Aviation Psychology*, *2*, 303–321.

Sarter, N., & Woods, D. D. (1994). Pilot interaction with cockpit automation II: An experimental study of pilots' model and awareness of the flight management system. *International Journal of Aviation Psychology*, *4*, 1–28.

Shingledecker, C. A., & Darby, E. R. (1995). Effects of data link ATC communications on teamwork and sector productivity. *Air Traffic Control Quarterly*, *3*(2), 65–94.

Sollenberger, R., McAnulty, D. M., & Kerns, K. (2002). *The effect of voice communications latency in high density, communications-intensive airspace* (DOT/FAA/CT/TN03/04). Atlantic City, NJ: William J. Hughes Technical Center Federal Aviation Administration.

Talotta, N. J., Shingledecker, C. A., & Reynolds, M. (1990). *Operational evaluation of initial data link en route services, Volume I* (Report No. DOT/FAA/CT-90/1, I). Washington, DC: Federal Aviation Administration.

Talotta, N. J., & Shingledecker, C. A. (1992a). *Controller evaluation of initial data link terminal air traffic control services: Mini -study 2, Volume I* (Report No. DOT/FAA/CT-92/2, I). Washington, DC: Federal Aviation Administration.

Talotta, N. J., & Shingledecker, C. A. (1992b). *Controller evaluation of initial data link terminal air traffic control services: Ministudy 3, Volume I* (Report No. DOT/FAA/CT-92/18, I). Washington, DC: Federal Aviation Administration.

Talotta, N. J., & Zurinskas, T. E. (1990). *The impact of data link on ATC communications.* Unpublished manuscript.

Tennille, G. F., Wallace, K. G., & Smith, T. M. (2004). *Status of advanced area navigation (RNAV) terminal procedures: Progress of STAR and SID design and implementation* (MITRE Paper MP04W231). McLean VA: The MITRE Corporation.

Uckerman, R., & Radke, H. (1983). *Evaluation of an airborne terminal for a digital data link in aviation* (Deutsche Forschungs-und Versuchsanstalt fr Luft-und Raumfahrt-FB83-05). Braunschweig, Germany: The DFVLR Institute for Flight Guidance.

Waller, M. C. (1992). *Flight deck benefits of integrated data link communication* (NASA Tech. Paper 3219). Hampton, VA: NASA Langley Research Center.

Waller, M. C., & Lohr, G. W. (1989). *A piloted simulation of data link ATC message exchange* (NASA Tech. Paper 2859). Hampton, VA: NASA Langley Research Center.

Wanke, C. R., & Hansman, Jr., R. J. (1990, May). *Operational cockpit display of ground measured hazardous windshear information* (Report No. ASL-90-4). Cambridge, MA: The Massachusetts Institute of Technology.

Weiner, E. L. (1989). *Human factors in advanced technology ("Glass Cockpit") transport aircraft* (NASA Contractor Report 177528). Moffett Field, CA: NASA Ames Research Center.

Williams, D. H., & Green, S. (1991). *Airborne four dimensional flight management in a time-based air traffic control environment* (NASA TM-4249). Springfield, VA: National Technical Information Service.

Zingale, C. M., McAnulty, D. M., & Kerns, K. (2003). *The effect of voice communications latency in high density, communications-intensive airspace phase II: Pilot perspective and comparison of analog and digital systems* (DOT/FAA/CT-TN04/02). Atlantic City, NJ: William J. Hughes Technical Center Federal Aviation Administration.

24

Intelligent Interfaces

John M. Hammer
Applied Systems Intelligence, Inc.

The usability of systems is determined by interface functionality and presentation. Much research in human factors has concentrated primarily on presentation—the surface aspects of the interface. Today, interface presentation and interaction have greatly improved to the extent that navigating the interface is much easier. Understanding complex system functionality from a presentation is difficult, and functionality alone has a dominant impact on usability. This problem is termed the *human factors of functionality*. It complements the study of mental models—how users adapt to and understand system functionality (Kieras, 1990; Rouse, Salas, & Cannon-Bowers, 1992). This chapter describes a user-centered architecture that adapts to users rather than requiring users to adapt to it. Its structure and conceptual organization differ radically from traditional automation.

How does functionality influence usability? First, the functionality partly determines what tasks the user performs and how these tasks are performed. The functions require inputs, and there are a variety of ways that inputs can be organized conceptually. The concepts presented determine in part how the user thinks about the interface. For example, the degree of control automation determines whether the user is continuously or intermittently involved in the task. The sensors and sensor-data processing determine how much data interpretation the user must perform. On the other hand, if the system is organized around the user's conceptual model and understands its own functions, their applicability, and how to aid the user, there is a potentially tremendous improvement in system usability (Hammer & Small, 1995; Geddes & Shalin, 1997; Rouse, Geddes, & Curry, 1987; Rouse, Geddes, & Hammer, 1990).

This chapter argues that a system that understands itself and the user is the next revolutionary step in avionics architecture. It covers some human factors problems in avionics functionality and describes a revolutionary avionics architecture that we feel can address these problems.

24.1 Problems with Existing Systems

It is widely accepted that automation does not eliminate human–machine interaction problems (Wiener & Curry, 1980). Instead, these problems are displaced or transformed from one type of problem to another. Typically, physical interaction problems are transformed into cognitive problems that are concerned with understanding the system. The following describes some problems that are covered in more detail elsewhere in this volume.

24.1.1 Automation Modes Work at Cross-Purposes

The modes of various automated systems can be set to work at cross-purposes with each other or the crew. The many possibilities include modes that should not be combined and modes that inhibit other modes or aircraft capabilities. For example, in the crash of China Airlines Flight 140, the crew mistakenly activated an automatic go-around without realizing it (Mecham, 1994). The autopilot tried to increase power and gain altitude while the crew attempted to maintain power and reduce altitude. Because the crew controlled the elevators and automation controlled the horizontal stabilizer, the nature of the conflict was not apparent. The aircraft eventually attained an unrecoverable state and then crashed. Either a fully manual landing or a fully automatic go-around would have prevented this accident.

24.1.2 Completely Automatic or Completely Manual Subsystems

Automation is now sufficiently capable that some subsystems could be completely automated, except when there is a malfunction that requires manual operation. Depending on the design philosophy, there may not be any intermediate levels of automation. In some ways, this simplicity is attractive. Only two modes—on or off—reduce training requirements. On the other hand, manual operation could impose a significant workload on the pilot who may be unfamiliar with operations that are usually automated and require no intervention. Some subsystems on the aircraft have multiple levels of automation. One example is flight-path control, which can be done manually or automatically by relatively simple autopilots or by sophisticated flight-management systems (FMSs). The levels of automation available in flight control recapitulate their history of introduction.

24.1.3 Automation Cannot Be Understood from Display

Interpreting system functions from displays can be difficult. There are several causes for this. First, finding the relevant information is difficult because most displays include more information than is relevant at any one time. Second, the consequences of many courses of action are not displayed (often, the automation does not know the consequences). Frequently, automation is so complex that displaying more complete information about what it is doing would worsen the information overload. At the same time, displaying limited information makes it difficult to understand the automation.

Although display shortcomings can be considered at fault, it is also possible that the functions themselves are too complicated to display, given the current understanding of display design. In fact, that is probably the case today, as our ability to conceive and implement functions has far exceeded our ability to display their state or consequences. Significant display improvements may be impossible due to the intrinsic complexity of traditional automation.

24.1.4 Automation Compensates for Worsening Failure While Complicating Recovery

Automation can also compensate for failures, but this compensation may mask a problem until it becomes severe. An example is the China Air Flight 006 where the autopilot compensated for reduced thrust from one engine (Wiener, 1988). When the autopilot was disengaged, the crew was apparently unaware of the extent to which the autopilot had compensated for the engine failure. Disengaging the autopilot removed this compensation. The crew did not reapply it, and the aircraft rolled into a dive that lost 30,000 of its 40,000 ft of altitude. The automation did not understand its own limitations or how authority should be transferred to the crew.

24.1.5 Surprising Engagement or Disengagement of Automation

At times, automation fails to engage as expected. The pilot will configure the aircraft and attempt to engage automation, but it refuses. Frequently, the automation checks some condition that must be true before it may be engaged. Because the pilot does not have access to the precondition, it is sometimes impossible to understand why engagement fails. Conversely, automation may change modes or disengage due to similar tests, and for the same reasons, the automation is difficult to understand. This problem is not confined to on/off engagement. Sometimes, the automation will engage but do something entirely unexpected.

24.1.6 Unnatural Behavior

Automation sometimes operates the aircraft in ways that are unlike human pilots. For example, if a human pilot wanted to achieve a particular altitude and speed at a particular navigation fix, the aircraft would be flown to achieve those goals a few miles in advance of passing through the fix. An FMS, on the other hand, would attempt to achieve the goals exactly at the fix. The crew needs to learn two ways of doing something: the natural way and the automation way.

24.1.7 Conclusion

So much control has been delegated to automation that it would be reasonable to consider it a member of the crew, at least for the purposes of discussing its interface. As such, it could be evaluated with respect to its cockpit resource management (CRM) skills (Foushee & Helmreich, 1988). Although it is unusually precise, it cannot explain its actions, nor is it aware of the interaction problems it causes. A human with such poor skills would be sent back to CRM training to improve these skills. Automation granted near-human levels of authority should also have near-human interaction skills.

24.2 How Functions Are Currently Designed

There are at least two views on how avionics functionality is designed today. In commercial air transport, the new design is based on the most recent design with whatever minimal changes are necessary or desirable. Change is minimized for several reasons. First, it reduces design cost and pilot retraining cost when pilots change aircraft. Second, minimal change means minimal opportunity to introduce major problems in the interface. Third, it allows vendor components to be reused with little change, which reduces the time to market.

The disadvantage to minimal change is that it traps the design at a local maximum. In the case of both computer hardware and software, there will be enormous strides in technological capabilities and concepts. Minimizing the introduction of change reduces the ability to make improvements. The greatest improvements possible would probably come from a rethinking of the functionality.

The second way that functionality would be designed would be through task analysis and related system analysis. Task analysis starts with a mission and decomposes it into a sequence of user tasks and attributes that form a model of the user's overall activity over time. This fine-grained description of user activity should provide an insight to the designer as to the consequences of particular functionality choices.

24.2.1 Problems with Task Analysis

From the designer's perspective, the task analysis does not tell one how to design. Instead, it is intended to show the consequences of design. Furthermore, whatever design feedback is given is probably useful only for local optimization. This is not bad in and of itself, but it implies that the assumed functionality on which the task analysis is built is itself not necessarily questioned by the task analysis. There is the same possibility of being trapped at a local functionality maximum as in evolutionary modification.

The attributes associated with individual tasks are often meant to further describe human performance and the demands on human performance. The difficulty here is that it is often difficult to justify the particular values assigned to the attributes. The results of the analysis can be no better than the inputs, as represented in these attributes. This approach is becoming increasingly problematic as the attributes used shift from physically observable measurements to unobservable cognitive concepts. A related problem, exemplified by the issue of pilot workload, concerns the dependency of the attributes on precise characteristics of the mission situation. For example, in a tactical air-to-air engagement, we might expect that pilot workload might be influenced more by the caliber of the opponent than with anything else. In general, workload might depend more on the situation than the tasks being performed (Suchman, 1987). If this is the case, and it seems to be true in the case of pilot workload, the wisdom of basing a workload analysis on tasks seems questionable. One might well base workload measurement on factors that exert more influence. This example raises the question of whether tasks are a suitable basis for answering the questions that are supposed to be answered by a task analysis.

The conceptual distance between function design and task analysis may be too far for any connection to be made. The concepts manipulated in task analysis are rather distant conceptually from the design of functions. Because of this, conclusions drawn in task analysis may indicate problems in function design but not solutions, at least for macroscopic design problems.

Due to cost, task analysis in practice seems to be performed only on a relatively small number of missions. If the task analysis incorporates many mission specifics, the results of the task analysis are really influenced more by the mission. If so, the resulting conclusions may be mission artifacts more than anything else. This, compounded with the limited number of missions studied, could result in a design for a mission that is never actually flown in the fielded system.

24.2.2 Impact of Conceptual Organization on Interface Functions

The concepts used to organize an interface influence its usability. Even something as simple as a route may be thought of either as a sequence of points or as a sequence of segments. A point perspective is more appropriate for arrivals at a particular point at a particular time. A segment perspective is more appropriate for considerations such as fuel usage and speed. Of course, the system may offer both point and segment perspectives due to functional requirements. This third option, perhaps to be labeled creeping featurism, presents additional functional complexity to the user. The system becomes over constrained, in the sense that the user may express more constraints than are physically possible to be realized. For example, once a start time and speed are specified, the end time is determined for a given segment.

The choice of point, segment, or both perspectives influences how routes are viewed, what inputs and outputs (categorically) are available and required, and what functions are available to the user.

The functionality determines what an input is and what a consequence is. In a tightly coupled system such as an aircraft, there are many system variables that one can choose to control. However, only a subset of these can be controlled, whereas the others are determined by the values chosen by the controlled set. For example, to reach a particular point, a pilot may control fuel consumption by selecting the most efficient climb rate or may control trajectory by flying directly to that point.

24.2.3 Choosing among Concept Spaces

One important decision from a human factors standpoint for the avionics software designer is designing a conceptual organization for an interface. Conceptual organization is defined as the objects that are manipulated, the possible manipulations, and the behavior that is manifested. This design choice is not merely selection of compatible components but rather a choice among incompatibles. Consider the example of mechanically and electronically steered radar antennas. An electronically steered antenna can be pointed much more rapidly. The slow pointing of a mechanically steered antenna so constrains the feasible patterns of movement that the interface concepts are tied to the movement patterns. The interface to an electronically steered antenna is not so constrained. Traditional concepts from mechanically steered antennas could be used, or the problem could be viewed entirely differently as a resource allocation problem (e.g., keep track of targets with these properties, search this sector if there are any remaining resources). Even if a resource allocation approach is taken, there are a tremendous number of potential conceptual organizations to be considered by the designer. Since software is far less constrained by the laws of physics than mechanical systems, the conceptual organization is practically limitless.

In general, electronic systems entirely controlled by avionics computers have an enormous design flexibility in how the functionality of the user interface is organized. This problem may in fact be insurmountable in the following sense. The traditional approaches to effective user–system operation include selection, training, human factors design, and aiding. Design is the principal focus of this chapter so far, yet it is unclear how a suitable solution can be found in this enormous design space. Indeed, the designer would have to be wise to be able to predict how a novel system would be used and what the effects would be of various types of possible functionality on system performance. Possibly, there is no solution to these problems that can be applied during design, at least as design is practiced now.

To conclude, functional design is conceptually difficult and perhaps intractable given the large amount of functionality to be hosted in modern automation. There is already more capability than human operators typically utilize. The next section describes an alternative approach to perfecting functionality. With aiding and training, perhaps functionality does not need to be perfect; maybe it just needs to understand itself, the user, and the situation.

24.3 An Associate System as a Solution

For the last decade, my colleagues and I have been investigating a new concept termed the intelligent interface. The intelligent interface goes beyond traditional interfaces, whatever their surface form, in that it contains intelligent functions that are intended only to help the user. The specific functions are: managing displayed information, watching for hazards, adaptively executing tasks on behalf of the crew, assessing the situation, and recommending responses. This set of functions—collectively termed an *associate system*—employs models of the user, the user's intentions, and the situation. The design goal for the associate system is to provide decision aiding with the competence and interactive skill of a highly trained user (e.g., a copilot).

Will an intelligent interface help to remedy the problems described earlier? It could be argued that automation is the problem and that intelligent automation may compound the problem. One description of the intelligent interface is a system to help the pilot use the systems on the aircraft. Obviously, increasing the level of automation could worsen the automation-related problems. This chapter discusses how intelligent interfaces should be designed to be successful.

There are several differences between intelligent interfaces and traditional automation. An intelligent interface contains models that enable it to be aware of some of the conditions that lead to the automation defects described previously. A second difference is in the automation philosophy by which the intelligent interface is designed. Its sole purpose is to help the pilot fly the aircraft. Although this claim could be made of traditional automation, there is a significant difference. Traditional automation helps the pilot operate the aircraft by taking tasks away from the pilot. These tasks are automated, and the pilot monitors this automation and makes occasional commands to that automation. The intelligent interface keeps the pilot in charge of tasks while supporting pilot decision-making. If the pilot changes intentions, the intelligent interface follows the pilot's lead.

24.3.1 Introduction to Intelligent Interface Processing

The heart of the issue is the depth of processing of the intelligent interface that causes it to avoid problems of traditional automation. In terms of system design, it is important to distinguish the depth and functionality of an intelligent interface from a traditional interface. A traditional interface portrays automation to the user. Typically, there are no changes made between the inputs/outputs of the automation and those made by the user. If the automation needs a target airspeed, the user will enter a target airspeed into the interface by typing, adjusting a knob, adjusting a slider on a mouse-based graphical user interface (GUI), and so on. The format may vary, but the information content is not changed by the traditional interface.

In an intelligent interface, the conceptual distance from user input to avionics input is much larger, and there is a considerable functionality between the user and the traditional automation. To continue the example, further intelligent processing will be done once the speed has been entered into the interface. First, the speed will be examined for possible hazards. This will involve bounds checking on the speed itself as well as examination of the impact of speed changes on other hazards that are currently being monitored. The speed changed will be interpreted in terms of previously identified user plans, and if there is a significant change, the displays themselves might be configured. The intelligent interface attempts to determine the meaning and consequences of the speed change before passing the change to the traditional avionics system.

None of this description of additional processing describes the depth or intelligence of the intelligent interface. For example, in evaluating the speed for hazard, the monitoring might consider the aircraft configuration (flags, speed brakes and spoilers, and gear) to avoid damage to the aircraft. It might consider location to avoid speeds over 250 knots near terminal airspace. It might consider the weather so that the aircraft is not flown too fast in turbulence. It will consider the flight plan to determine what impact speed changes might have on it. Even these various checks on airspeed could be themselves fairly elaborate. For example, flying too fast with the gear down might cause further consideration of whether the gear can in fact be retracted (or has it been damaged or perhaps must be left down). The diagnosis of the problem may be either "slow down" or "raise gear," as appropriate.

24.3.1.1 Interactions between Intelligent Interface, the Display System, and Traditional Avionics

Architecturally, the intelligent interface occupies a significant place between the traditional system and the user (Figure 24.1). There are two general types of inputs to the intelligent interface. First, as discussed in Figure 24.1, are the inputs that are ultimately bound for the traditional avionics system. The second general type of input is that needed explicitly by the intelligent interface.

To understand this second category, consider the role of the intelligent interface. The intelligent interface serves as an aid or assistant to the user. In this role, the expected form of interaction is approximately

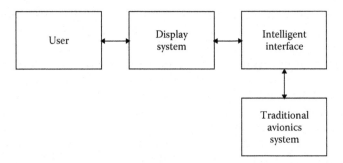

FIGURE 24.1   Intelligent interface stands between the user and the traditional system.

that of what would be expected of two humans on a mouse-based GUI, and so on. The format may vary, but the information content is not changed by the traditional interface.

The remainder of this chapter discusses the components and structure of an intelligent interface. The intelligent interface has models of crew intentions, crew information needs, and aircraft hazards. These models are intended to be intelligent enough to avoid many of the problems associated with traditional avionics.

24.3.2 Intentionality

One pointed criticism of modern avionics is its minimal or nonexistent understanding of what is happening overall. In other words, the avionics has no model that is useful in understanding what the flight crew is attempting to do, what responses by the crew are appropriate, or even the situation in which the crew finds itself. Virtually, all communication between humans takes advantage of or even depends on a contextual model. Because the avionics lacks such a model, communication with it is difficult. For example, extra communicative acts are required because there is no contextual model to fill in the gaps.

The reason for the minimal contextual model within the avionics is that the data representations within the avionics are intended primarily to support the avionics itself in its automatic control of the aircraft. Despite claims that another automation philosophy drives design, a detailed audit of the purposes for which each datum is represented would show a predominant bias toward supporting automation rather than the user. The reason for this bias is primarily that of the organizational forces in which the avionics software designer practices. Unless there are strong forces to the contrary, the avionics software design and representations will support the needs of the avionics software itself rather than those of the user.

As envisioned earlier, current automation uses control limits as an approximation to authority limits. These limits are typically quite modest in the coverage, at least with respect to robustness. Thus, one finds that the limits err by being too aggressive or too conservative, with the aforementioned China Air as an example of aggressiveness (in that the engine balance was never announced) and its failure to engage as an example of conservativeness.

24.3.2.1 Content of a Model of Intentions

The representations of a model of intentions depend on what uses are made of it to make contextual decisions about information, authority, hazards, and so forth. For example, to make decisions about information needs, the model must be able to recognize situations when a particular piece of information is and is not needed. Naturally, to accomplish this, the model depends on the structure of the domain, particularly the situations that occur, and the user actions that can be taken as well as the structure of the information—its meaning, breadth, and resolution.

The model of intentions is typically based on a structure that is similar to a task analysis that describes the missions that might be expected to occur. This generality is important because the intentional model should cover—to the extent possible—all possible situations so that there is no gap in the functional coverage of the intelligent interface. There are a number of differences between the structures used for manual task analysis, as practiced during design, and intentional models that are used for real-time decision aiding. First, a manual task analysis uses tasks that are identified by the designer during analysis. A model of intentions must have a structure that permits online task recognition by a computer. The feasibility of recognition is a primary concern of the designer of the intentional model. For example, distinguishing between a touch-and-go practice landing and a real landing in advance is impossible. Only after the fact can the two be distinguished, and after-the-fact intentional structures are somewhat less useful. The model is more useful if the recognized intentions are temporally leading or at least concurrent indications of activities, rather than trailing indications (Geddes, 1989).

The intentional model should be made an active component of the intelligent interface. It should react to the situation and activate or deactivate elements of the model structure to keep the model as an accurate description of what is happening. In recognizing transitions in the situation, it is advisable not to rely primarily on the passage of time, unlike a task analysis conducted for design purposes. In other words, the task should be recognized based on what is happening rather than on what happened previously. Those who have organized models temporally, although the models seem attractive, have found that models often get stuck in particular states when an out-of-the-ordinary turn of events occurs. Time should be used only when it truly is the mechanism that activates and deactivates elements. Both temporal processing and situational processing have found places in current models, although situational elements tend to be generally more descriptive than temporal elements.

A hierarchical intentional model is employed for several reasons. First, a hierarchical model can represent the situation at several different conceptual levels. As such, it can describe multiple intentions for a low-level action. For example, a military aircraft could jettison fuel either to reduce weight for landing or to send a visual signal when the jettisoned fuel is ignited with afterburners. Second, a hierarchical model may be able to represent a situation at a high level when model limitations prevent a low-level description. For example, determining that the pilot is attempting to land may be possible, but determining the runway may not.

The intentional model can be both descriptive and prescriptive. The descriptive model represents what the pilot is attempting to do. The prescriptive model represents a recommendation from decision aids to the pilot about what should be done. Although these two models are processed separately, they share a common representation to facilitate communication in the intelligent interface. The descriptive and prescriptive models can synergistically aid the pilot. When the pilot begins something new, the descriptive model can recognize it. This description is then specialized by the prescriptive model and displayed to the pilot. The result is a system that anticipates the information the pilot needs.

Our experience has been that the intentional model, shared as described earlier, has had a profound influence on the architecture of the intelligent interface. Designers, once exposed to the power of the model, tend to make heavy use of the model in functional processing. To those without this experience, it is difficult to appreciate how significant such a model can be in an intelligent interface. Because we as humans use an intentional model, it is difficult for us to appreciate the significant impact such a model makes on software without seeing it firsthand.

24.3.3 Information Models

Although modern avionics possesses considerable data, it has little idea of which data are actually information (data of value) to the pilot. As a result, the avionics is limited in its ability to change the information on the displays. Intelligent information management automatically selects the displayed information, its level of emphasis, and its format. There are several reasons for the importance of intelligent information management in a modern crew station. First, there is a tremendous volume of data available.

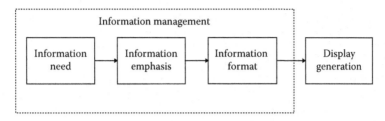

FIGURE 24.2 Information management processing sequence.

Current trends in data communication and data storage aboard aircraft promise to increase this volume. Second, most of the data is without value at any one particular time, although presumably all of it can be of value under some circumstances. As a result, the user can spend a considerable amount of time and effort selecting the appropriate data for display (Small & Howard, 1991).

24.3.3.1 Representational Needs in Information Modeling

The three representational needs in information modeling are, in order of importance, information need, emphasis, and format (Figure 24.2). Information need is modeling the information that is relevant in current and near-term situations. Information emphasis is determining which selected information should receive increased display emphasis. Information format modeling is determining how information that is relevant and possibly emphasized should be displayed. The range of display choices includes such dimensions as location, size, color, shape, symbology, and modality.

24.3.3.2 Information Need

The foremost modeling question is, what gives rise to information requirements? The most obvious requirement for information is task execution, which is described in the intention model. Indeed, one of the traditional uses of task analysis models was to determine information requirements. Both task analysis and intentional models are top-down methods of determining information requirements. Given such a model, information requirements are associated with tasks or intentions.

The second source of information requirements is significant situational changes, or events. The assessment module finds the few significant events among the many insignificant changes. Events are bottom-up sources of information because they are unanticipated within the intentional structure and are the result of noticing changes to low-level data. Information requirements arise from two source events and intentions. Combining these two sources is integral to selecting the right information to display.

There are dozens of dimensions that can be used to describe information. Some examples include the priority, the use (warning, control, and checking), and the type (navigation, weather, and system). Starting with these dimensions makes the problem of automatic information selection and formatting seem extraordinarily difficult. Our experience has been that most of these dimensions are not useful in information management. What has been more practical is to work backward from the decisions to the inputs necessary for those decisions (i.e., need, emphasis, and format).

Information arising from intentions and events can be in conflict, in that there may be more information required than fits within the available display area. Further, information requirements arising from a single type of source, such as intentions, can be in conflict with each other. Fundamentally, information selection is a resource allocation problem, and resource allocation usually means that there is competition for resources. Concepts such as priority and importance are essential to resolving these conflicts optimally.

Event-based information may or may not be of interest to the pilot at the moment the event is detected. A way to determine whether an event is of interest is to map it onto changes in capabilities of the aircraft: thrust, sensing, navigation, fuel, and so forth. In the intention model, interest in changes in the capabilities can be expressed. Determining whether there is any interest is simply a matter of looking at the capability concerns of the active tasks.

24.3.3.3 Information Emphasis

The emphasis function causes certain display elements to be given perceptual attributes that cause their salience to be higher. The exact attributes changed are determined in the format decision, which is discussed later. The emphasis decision merely decides what information should be emphasized, but not how it should be done. The remainder of this discussion on emphasis concentrates on how this decision can be made.

There are numerous reasons why information should be emphasized. One reason is the extreme consequences of not taking into account the information content. An example is failure to complete some item on a checklist, such as setting flaps for takeoff and landing. Another reason is doctrine. For example, current commercial air transport practice is to call out the altitude at 100 ft intervals during landings (in the United States). This information is already available on the altimeter; it is emphasized by a voice callout. In addition to emphasis, this procedure also presumably increases altitude awareness for one crew member beyond what would otherwise be the case. Emphasis is also required by unusual or urgent events to alert the crew to the unusual conditions and to secure a prompt response to the urgent event.

Correspondingly, representation of the need for emphasis can be included in several places within the models discussed thus far. The most frequent source of emphasis is in significant events. Typically, if an event's information is important enough to display (i.e., to change the displays in favor of this information), then emphasis is also required. Intentions can also serve as a convenient structure on which to associate the need to emphasize, particularly with regard to information that is emphasized due to policy. A third source, not unlike the first, is the error monitor (discussed later), which monitors for hazardous situations and produces events to notify the pilot.

24.3.3.4 Information Format

The final decision to be made about information is the display format or modality. This includes selection of modality (aural, visual, or both) and the display element to use (bar chart, digital tape, etc.). The motivation behind these decisions is to configure the most directly visible perceptual aspects of the displays to convey information to the user in a way that it is most suitable for its intended use. This process is most akin to the traditional human factors display design process. In fact, it could be considered an online version of the same. Any of the criteria used in conventional display design are potential candidates for the online version. Examples include

- The accuracy with which the information must be perceived
- Whether rates of change of the displayed information are needed
- Whether the information is to be used for checking or control

Research by Shalin and Geddes (1994) has shown considerable performance improvements by adapting the information format to the task.

The final shape of the modality selection depends a great deal on the display flexibility available to it. For example, if there are few display capabilities for varying displays of altitude, there is little need to consider it during design. A highly capable information manager can place heavy demands on display generation. In practice, the display programming effort has been at least as large as the information management effort. Display flexibility is less of a restriction in selection and emphasis of information because virtually all display generators have some way to perform both of these functions.

24.3.3.5 Conclusion

It is worth making a few points about visibility and appreciation for various types of functionality. Selection of information is a highly visible function. Its changes are immediately apparent on the display system, and if correct, they immediately give a positive impression. Emphasis, which consists of highlighting visual displays, is less apparent than selection, and formatting is the least apparent. The practical implication of this difference in functionality, or rather this perception of functionality, is that some consideration needs to be given during requirements generation to the perceived value of the various functions.

One criticism of traditional automation is that it takes too much authority and awareness away from the pilot. At first glance, the same claim could be leveled at information management because it controls displays automatically. There are several reasons why this claim does not hold up under scrutiny. The first reason is that the information manager is intended to improve the pilot's situation awareness by showing information that the pilot would have selected anyway. The pilot does miss out on the reasoning that went into display selection, but the displays selected should make this reasoning evident.

The second reason is that the pilot always has the authority to override the information manager. When this happens, the displays are under manual exclusive pilot control until certain conditions are met. Conditions might be that a certain amount of time has elapsed or the situation has changed significantly; however, the best condition to use is still an active research topic.

A third reason is that the behavior of the information manager can be adjusted to the domain in a way that does not diminish the pilot's authority. For example, display selection may remind the pilot by placing an icon at the edge of a screen. Automatically replacing one display with another would be reserved for the most immediate and serious problems. One approach to this essential problem is to develop a model of how crew members share information and would perform this task for each other (Zenyuh, Small, Hammer, & Greenberg, 1994).

24.3.4 Error Monitoring

Aviation has adopted several approaches to the problem of human error in the cockpit: selection, training, human factors design, and accident investigation. Selection tests candidates before and during training to select those most likely to succeed. Training attempts to reduce error by instilling correct practices and knowledge in pilots, and by practicing unusual or dangerous situations in simulators or with instructors so that pilots will be prepared for them should they really occur. Human factors design attempts to design crew stations and tasks to eliminate error-prone characteristics. Accident investigation is a feedback loop that investigates accidents and incidents to identify defects in all of the approaches.

These various approaches combine as a layered defense against human error. Selection, training, and human factors design operate before flight, and investigation after an accident or incident. One layer that has received less than its due is concurrent error monitoring that takes place during flight, especially as implemented in an intelligent interface.

Currently, concurrent detection of errors is implemented by redundant human operators and the somewhat limited practices implemented in today's avionics. One justification for multiple crewmembers, human air traffic controllers, and shared radio channels is as a check on human errors. To some extent, traditional avionics has limited checks for errors.

In traditional software, including avionics software, the perspective on error detection is virtually always that of the software itself. Software detects those errors that affect its processing and functionality, not necessarily those that represent errors of the pilot. The reason for this is that what error processing is present is embedded within the functional context of the avionics. In other words, the purpose of the avionics is to accomplish some control function, and error processing is possible only to the extent that it fits within the functional context. From an organizational context, the budget for error processing is controlled by those who seek increased functionality.

For example, consider the error of commanding the aircraft to land at the wrong airport. The avionics could check this command for plausibility before passing it to a lower-level control loop. It is only within the context of automation that checks are performed on commands. If the automation is turned off or the pilot flies manually, the checks in automation are not made. If the pilot enters a code for which there is no airport, the automation will reject the input because it is impossible for the automation to fly to a nonexistent destination. On the other hand, if the pilot enters a code for an unsuitable airport, there is no guarantee that the automation will detect such a problem because its purpose is to navigate, not detect errors. In traditional avionics design, error monitoring is an incidental, opportunistic function.

Some avionics, such as the ground proximity warning system (GPWS), have a functional purpose that is solely oriented toward error detection. There are several problems with this type of functionality in today's traditional avionics. First, there is too little of it. Second, it is not very intelligent. Third, it lacks independence from the functional aspects of traditional avionics. Finally, it does not consider consequences.

24.3.4.1 Comprehensive Coverage

A number of authorities have advocated an electronic cocoon around the aircraft. As long as the aircraft was operated within the safety of this cocoon, the crew would be free to do whatever it wanted (Small, 1995; Wiener & Curry, 1980). However, the crew would be alerted as the aircraft drew near the edge of the cocoon and prevented from unintentionally leaving the cocoon. We are still far from achieving this goal.

The goal of a complete, airtight safety cocoon seems to be theoretically unachievable in the sense that one could convincingly demonstrate that an aircraft could never be operated unsafely. There are simply many ways to fail. A more practical approach is to enumerate a large number of unsafe situations and actions and then prepare software to detect each one of them.

24.3.4.2 Intelligent Monitoring

Traditional GPWSs, which have yielded a reduction in accidents, have often been criticized for an excessive false alarm rate. Consider the information available to a GPWS unit. A radar altimeter measures the altitude directly underneath the aircraft. The aircraft's position can be measured by GPS/GLONASS and INS systems, and it would not be difficult to install ground altitude data on CDROM for the area over which the aircraft is expected to operate. Using these data recently made available, a true cocoon could be established, at least with respect to the problem that GPWS is intended to prevent.

The point is that to make an intelligent decision about the need for an alarm requires access to many sources of information, not just one. It is easy to build an alarm system that provides many false alarms, and then rely on the pilot to sort out the true alarms. The deleterious effects of excessive false alarms on human performance have been known for some time. Our contention is that more sophisticated processing of more inputs should reduce the false alarm rate and thus improve response to warnings.

24.3.4.3 Independence from Traditional Avionics Functionality

To be successful, error detection should be functionally independent of the traditional avionics. There are several reasons for this. First, the purposes of traditional avionics and error monitoring are dissimilar. To embed error monitoring within traditional avionics is to limit monitoring to those situations that are recognizable from within the traditional avionics perspective (i.e., the data it stores). Second, an error monitor must have data structures and models that meet its primary purpose of detecting and thus avoiding errors. From an object-oriented perspective, the separation of error monitoring from traditional avionics would be to give first-class status (i.e., object status) to errors. Finally, error monitoring should not depend on whether functions are enabled in traditional avionics.

24.3.4.4 Error Monitoring Must Be Done in a Module Dedicated to That Purpose

To consider GPWS again, its processing has no concept of consequences. Of course, the designers knew that flight below a certain altitude could have most severe consequences. However, none of that consequential reasoning is present in the GPWS unit itself. It merely compares the radar altitude to the threshold and sets off an alarm if the threshold is transgressed. As a result, GPWS can be considered to cause many false alarms, at least when evaluating the true state of the aircraft with respect to the distance to the ground. In other words, if the aircraft continues on its current trajectory, how far is it from the edge of the cocoon? The GPWS has no representation about cocoon borders.

The point is that a situation is a hazard only if the potential consequences are severe. Evaluating errors requires a structural orientation toward consequences within the monitor. Other approaches that have been tried include omission of prescribed actions and human error theory. Experience with the omission of actions is that the severity of the error is usually unknown without other information about consequences. Human error theory can suggest what might be done about repairing the error (e.g., omission or repetition errors are somewhat self-diagnosing) or explain why it happened. Understanding the cause for an error may be useful for the designer or the pilot (in a debrief), but it serves little purpose in alerting the pilot to a serious error (Greenberg, Small, Zenyuh, & Skidmore, 1995).

24.4 Summary of Associate Architecture

A high-level architecture for an intelligent interface has been described. The description represents a family of solutions, not an individual solution. The model structures described provide a sufficient framework for dealing with the problems of automation. One key property of the intelligent interface is that it increases the level of interactive intelligence in the avionics to correspond more nearly with the authority already granted. Historically, the intelligent interface represents the next generation of automation that is built on the current layers of FMSs and autopilots. The purpose of the intelligent interface is to support the pilot's decision-making. This differs from the purpose of traditional automation, which is to automate tasks for the pilot.

System engineering becomes an essential effort for any system constructed with an intelligent interface. To build an intelligent interface component requires a thorough understanding of the purpose, benefits, and employment of each subsystem component to be installed on the aircraft. This understanding is a necessary part of the system engineering because knowledge engineering about the subsystem is necessary. The questions asked include

- What are the effects of using the subsystem in each of its modes on the aircraft and environment? This is aimed at producing a device-level model of the subsystem.
- When is it appropriate to use the subsystem?
- How does the subsystem interact with other subsystems on the aircraft, especially with regard to the previous questions?
- When would using the subsystem be inappropriate or erroneous (as opposed to ineffectual or irrelevant)?

24.4.1 The Intelligent Interface as Run-Time System Engineering

It is widely suspected that those who construct new systems do not fully understand all the ramifications and implications of what they are designing. Answering these questions will challenge the designers of traditional avionics.

Those who have participated in the design of an intelligent interface have found that the scrutiny given the traditional avionics design can produce a more purposeful product. During design, a number of intelligent interface models are constructed as to how the entire system will be used from the pilot's perspective. This model building can yield benefits by improving the design as well as incorporating the intelligent interface functionality. For example, I was once preparing the knowledge base for an information manager that was to select from one of several available displays. It used information requirements that were associated with intention structures, and it picked the best display by matching its information display capabilities to the information requirements that had been accumulated from all active intentions.

While I was debugging the knowledge base, I noticed that some displays were never chosen and that other displays were frequently chosen. Naturally, this was assumed to be a fault of the knowledge base, as it was under development. After close observation of the display-selection algorithm, I came to the conclusion that the algorithm and knowledge base were correct. The problem was in the displays themselves. Some displays lacked elements that were always demanded. Other displays seemed to support situations that would never occur. To fix the problem, new display designs were prepared. The point of this example is that evaluation of the information content of displays was made possible only by computing a match of displays to situations. Although it would certainly be possible to prepare a written argument that the displays are well designed, computation was a more compelling proof.

The strength of this approach lies in the executable nature of the knowledge. It is not merely that the knowledge can then be applied via execution to produce simulations of the effects of the subsystems along with the associated knowledge. As such, it represents a powerful system-engineering capability that is especially useful to those who are responsible for the overall technical project administration. To succeed, those developing this type of system require the support of management to get answers to knowledge-engineering questions. These answers are not always simple to obtain but can benefit both the design and the operation of complex systems.

24.5 Implementations and Evaluations of Associate Systems

The second version of this chapter reviews various large-scale implementations of the concepts described previously. To conserve space, only key projects have been reviewed, with the criteria being implementation and evaluation of at least one of the key concepts in an aviation or similar domain.

24.5.1 Pilot's Associate

The Pilot's Associate was the earliest attempt to build the functions described above. The system was evaluated in a medium fidelity simulation with a limited visual field, which was of minimal importance due to the high altitude, beyond visual range mission. The evaluation pilots, who were supplied by the military customer, were initially skeptical of this new type of automation. They initially turned off all support from the associate system. Since the mission was quite difficult, they failed. Gradually, they enabled more and more of the support until they were able to succeed at the mission. By the end of the experiment, they concluded that the system was highly beneficial. They also felt that they had learned something from the recommendations produced by the automation (Smith & Geddes, 2002).

24.5.2 Hazard Monitor

One commercialization attempt following the original Pilot's Associate program was the Hazard Monitor (Bass, Ernst-Fortin, Small, & Hogans, 2004; Greenberg et al., 1995). The goal of this project was to develop an error monitor for military and commercial air transport aircraft. Key components included an assessor, an expectation monitor, and a notification arbitrator (Figure 24.3).

The purpose of the situation-assessment component was the same as described above: create higher-level, interpreted state from sensed data. The expectation monitor, which is unique to error monitoring, interprets the expectation network. This network is a finite-state machine in which each state has a set of expectations consisting of

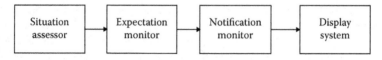

FIGURE 24.3   Hazard monitor's key components.

- A condition comparing expected aircraft state to actual state
- A severity level
- A message template for display to the pilot

An evaluation of the hazard monitor was conducted with commercial-airline-transport-rated pilots. Errors using the FMS were the focus. The pilots were initially skeptical of the concept (having been exposed to earlier, low-technology warning systems). They were, however, favorably disposed towards this more sophisticated system.

24.5.3 Operator Function Model

Mitchell and her colleagues have developed and refined the operator function model (OFM) since the late 1980s. The purpose of an OFM is intent interpretation, although its detailed structure differs considerably from the plan-goal graph technique described by Geddes (1989). Both OFM the and plan-goal graph model perform with accuracy above 90% in identifying operator intentions. OFMs have been used as a core model to study a variety of problems in complex systems: operator-decision support, intelligent tutoring systems, system engineering, and display design at the information requirements level.

A significant implication of the diversity of applications is that models of operator intention have great power that is not appreciated by those outside the field. First, these models are a core technology in an intelligent interface to a complex system. This is evidenced by the central role such models play in a variety of functions. Second, these models provide strong support for system engineering. A large, complex system that costs a billion dollars or more to develop can be portrayed from the operator's perspective as a graph with a few hundred nodes and links. To be able to see every concern of the operator at every level of abstraction on a single, large diagram is incredibly insightful. The system-engineering possibilities of these diagrams have not yet been fully explored.

24.5.4 Rotorcraft Pilot's Associate

Miller and Hannen (1999) describe the cockpit information manager (CIM) of the Rotorcraft Pilot's Associate. Because the CIM was based on the design of the information manager of the Pilot's Associate, remarks here will be limited to the simulation evaluation.* The experiment was conducted with a highly realistic simulation including a large visual field, realistic threats, several wingmen, realistic communication, and unexpected mission changes.

The controlled experiment had a single experimental factor—the availability of the CIM and related decision aids—and a blocking factor—whether the subject was first exposed to the treatment (CIM) or the control (baseline automation). The performance measures were based on the crew's subjective interpretation. The crews found the CIM to frequently, but not always, provide the right information at the right time. CIM behavior was frequently found to be predictable. TLX[†] workload measures were significantly better with CIM decision aiding. Since pilots are typically quite leery of new automation, these results should be seen as confirming the value of information management.

24.5.5 Crew Assistant Military Aircraft

Onken and his colleagues have developed an intelligent system for aiding pilots that has incorporated virtually all of the concepts described in this chapter. The intelligent interpreter function is based on a normative model implemented with Petri nets. The use of normative or prescriptive models is rather philosophically different from the descriptive approaches of plan-goal graphs (Geddes, 1989) or OFMs

---

* There was also a flight test of the RPA, but no released information is available.
† Task Load Index (TLX) is a subjective workload assessment tool developed by NASA.

(Mitchell, 1999). Implementation using Petri nets emphasizes that no consensus exists on the best technical approach to intent interpretation.

The error-monitoring function (Strohal & Onken, 1998) was closely integrated with an intent interpretation function, which is consistent with other work in this area. Activities that could not be recognized by the intent interpreter were analyzed by the error monitor. Activities are classified as errors using fuzzy, rule-based logic.

The information management function performs a single function: prioritizing and then displaying messages from the decision support system (Walsdorf & Onken, 1997). Map rendering, speech output, and speech input are also provided, but our classification of these functions leaves them outside the information manager and inside a display-rendering system.

Simulation-based evaluation of crew assistant military aircraft (CAMA) was performed recently after roughly a decade of development (Frey, Lenz, Putzer, Walsdorf, & Onken, 2001). Military-qualified transport and test pilots flew a low-level drop mission. Post-experiment debriefing found that the subjects had highly favorable opinions of CAMA's ability to detect pilot error and improve flight safety. They also found CAMA to be easy to understand.

24.6 Conclusion

Several recurring results were observed across the research reviewed here. Although some scientific communities would regard intent inference or automatic information management as impossible, a number of different projects have all been able to successfully create these functions. Generally, the functions work as claimed—as judged by a most critical audience—pilots. The functions work with a high degree of accuracy and effectiveness, but not perfectly. These functions are found to be useful by this same critical audience. Pilots generally report that the behavior of this type of human-centered automation is understandable. This result stands in strong contrast to the understandability of conventional automation.

No engineering consensus exists on the best technical approach to implement these functions. No head-to-head experimental evaluation of approaches has ever been performed. A relatively small number of organizations have demonstrated capability to build the type of intelligent interface described here. Building this type of interface requires a background in advanced information-processing techniques that are typically considered part of artificial intelligence. Nontheoretical approaches to autonomous systems, however, have been observed to be unsuccessful (Hammer & Singletary, 2004).

Given these observations, moving an intelligent interface into next-generation systems would seem to be a logical development. In actuality, this movement is slowly occurring in piecemeal fashion. Factors that slow its adoption include the relatively complicated aspects of the technology, the effort required to develop the systems, and competition from more traditional human factors approaches. As systems become more complex, risks become larger, and information overload predominates, the intelligent interface becomes an increasingly viable solution.

References

Bass, E. J., Ernst-Fortin, S. T., Small, R. L., & Hogans, J. (2004). Architecture and development environment of a knowledge-based monitor that facilitated incremental knowledge-based development. *IEEE Transactions on Systems, Man, and Cybernetics, 34*(4), 441–449.

Foushee, H. C., & Helmreich, R. L. (1988). Group interaction and flight crew performance. In E. L. Wiener, & D. C. Nagel (Eds.), *Human factors in aviation* (pp. 189–227). San Diego, CA: Academic Press.

Frey, A., Lenz, A., Putzer, H., Walsdorf, A., & Onken, R. (2001). In-flight evaluation of CAMA – The crew assistant military aircraft. *Deutscher Luft-und Raumfahrtkongress.* Hamburg, Germany.

Geddes, N. D. (1989). *Understanding human operator's intentions in complex systems.* Unpublished doctoral thesis, Georgia Institute of Technology, Atlanta.

Geddes, N. D., & Shalin, V. L. (1997). *Intelligent decision aiding for aviation.* The Linköping Institute of Technology, Linköping, Sweden.

Greenberg, A. D., Small, R. L., Zenyuh, J. P., & Skidmore, M. D. (1995). Monitoring for hazard in flight management systems. *European Journal of Operations Research, 84,* 5–24.

Hammer, J. M., & Singletary, B. (2004). Common shortcomings in software for autonomous systems. *American Institute of Aeronautics and Astronautics 1st Intelligent Systems Technical Conference.* Chicago, IL.

Hammer, J. M., & Small, R. L. (1995). An intelligent interface in an associate system. In W. B. Rouse (Ed.), *Human/technology interaction in complex systems* (Vol. 7, pp. 1–44). Greenwich, CT: JAI Press.

Kieras, D. E. (1990). The role of cognitive simulation on the development of advanced training and testing systems. In N. Fredrickson, R. Glaser, A. Lesgold, & M. G. Shafto (Eds.), *Diagnostic monitoring of skill and knowledge acquisition* (pp. 51–73). Hillsdale, NJ: Lawrence Erlbaum Associates.

Mecham, M. (1994, May 9). Autopilot go-around key to CAL crash. *Aviation Week & Space Technology,* pp. 31–32.

Miller, C. A., & Hannen, M. D. (1999). User acceptance of an intelligent user interface: A rotorcraft pilot's associate example. *Proceedings of the International Conference on Intelligent User Interfaces* (pp. 109–116). Redondo Beach, CA.

Mitchell, C. M. (1999). Model-based design of human interaction with complex systems. In Sage, A. P., & Rouse, W. B. (Eds.), *Handbook of system engineering and management.* New York: Wiley.

Rouse, W. B., Geddes, N. D., & Curry, R. E. (1987). An architecture for intelligent interfaces: Outline of an approach to supporting operators of complex systems. *Human-Computer Interaction, 3*(2), 87–122.

Rouse, W. B., Geddes, N. D., & Hammer, J. M. (1990). Computer-aided fighter pilots. *IEEE Spectrum, 27*(3), 38–41.

Rouse, W. B., Salas, E., & Cannon-Bowers, J. (1992). The role of mental models in team performance in complex systems. *IEEE Transactions on Systems, Man and Cybernetics, 22*(6), 1296–1308.

Shalin, V. L., & Geddes, N. D. (1994). Task dependent information management in a dynamic environment: Concept and measurement issues. *Proceedings of the IEEE International Conference on Systems, Man and Cybernetics* (pp. 2102–2107). San Antonio, TX.

Small, R. L., & Howard, C. W. (1991). A real-time approach to information management in a pilot's associate. *Proceedings of the Tenth Digital Avionics Systems Conference* (pp. 440–445). Los Angeles, CA.

Small, R. L. (1995). Developing the electronic cocoon. In Jensen, R. S., & Rokovan, L. A. (Eds.), *Proceedings of the Eighth International Symposium on Aviation Psychology* (pp. 310–314). Columbus, OH.

Smith, P. J., & Geddes, N. D. (2002). A cognitive systems engineering approach to the design of decision support systems. In Jacko, J. A., & Sears, A. (Eds.), *The human-computer interaction handbook: Fundamentals, evolving technologies and emerging applications* (pp. 656–676). Mahwah, NJ: Lawrence Erlbaum.

Strohal, M., & Onken, R. (1998). Intent and error recognition as a part of a knowledge-based cockpit assistant. *Proceedings of SPIE.* Orlando, FL.

Suchman, L. A. (1987). *Plans and situated actions.* Cambridge, NY: Cambridge University Press.

Walsdorf, A., & Onken, R. (1997). The crew assistance military aircraft (CAMA). *Proceedings of the Fourth Joint GAF/RAF/USAF Workshop on the Human-Electronic Crew.* Kreuth, Germany.

Wiener, E. L. (1988). Cockpit automation. In E. L. Wiener, & D. C. Nagel (Eds.), *Human factors in aviation* (pp. 433–461). San Diego, CA: Academic Press.

Wiener, E. L., & Curry, R. E. (1980). Flight-deck automation: Promises and problems. *Ergonomics, 23,* 955–1011.

Zenyuh, J. P., Small, R. L., Hammer, J. M., & Greenberg, A. D. (1994). Principles of interaction for intelligent systems. The human-computer crew: Can we trust the team? *Proceedings of the Third International Workshop on Human-Computer Teamwork.* Cambridge, U.K.

25

Weather Information Presentation

Tenny A. Lindholm
The National Center for Atmospheric Research

Before we can fully relate the aviation user's weather information needs to the function or task at hand, both now and in the future, we must comprehend and contrast the differences between the current aviation environment and whatever is envisioned for weather information available to the users, and then provide a vision of the future air-traffic control (ATC) system and associated weather information. The human factors and display allow user needs to evolve in the proper context. Thus, each element and user within the National Airspace System (NAS) will be considered, as well as the implications of their interactions within the system. The goal is to develop a true system-level understanding of weather to support decisions, which are very much varied. The approach taken here ensures that the functional interactions of the ATC system and its allocation of weather display capabilities is well understood as the system is modernized.

25.1 Aviation Weather Dissemination—Case Studies

On the afternoon of August 2, 1985, a wide-bodied jetliner crashed short of Runway 17L at Dallas-Fort Worth Airport, with considerable loss of human life. The only indication of a hazard to the flight crew was a moderate-to-severe rain shower just to the right of the approach course. The National Transportation Safety Board (NTSB) listed the occurrence of a small, short-lived but severe downburst now widely known as a microburst, as a probable cause (Fugita, 1986).

A number of years ago, a South American jetliner crashed while on approach to New York's John F. Kennedy Airport during marginal ceiling and visibility conditions. The aircraft arrived in the terminal area with just enough fuel and reserves to complete a normal sequence to landing. After unplanned and lengthy holding delays, the aircraft crashed about 10 miles short of the runway from fuel starvation. The NTSB identified the flight crew's lack of awareness of the evolving weather impact on normal sequencing as a contributing factor.

In 1988, a jetliner crashed on departure from Detroit with, again, considerable loss of life. The NTSB identified the flight crew's failure to properly configure the aircraft flaps and leading edge devices as the probable cause of this accident. However, the cockpit voice transcripts clearly indicate confusion by both the pilots as they tried to convert the encoded alphanumeric weather data to a graphic portrayal on a map to make the information more usable on departure. These actions could have contributed to flight-crew distraction while completing checklist actions. In fact, the cockpit voice recorder revealed that the captain of this flight remarked, "Not now, I'm weathered out," in response to updated alphanumerics just prior to departure (Sumwalt, 1992).

Finally, a medium-sized jetliner crashed in 1991 just south of the Colorado Springs Airport after encountering a severe roll to the right and immediate dive to the ground. The NTSB was initially unable to agree on a probable cause for this accident. However, severe horizontal vortices near the mountains coupled with aircraft rudder responses appear to be the primary causes of the accident.

Among the major air carriers, the NTSB reported that 35.6% of all the accidents between 1991 and 2000 were weather-related (NTSB, 2004a). With regard to general aviation, 26.1% of all the accidents, and 38.1% of all the fatal accidents, were weather-related (NTSB, 2004b). Apart from the obvious economic and societal costs associated with these numbers, improved weather information can potentially save NAS operators literally hundreds of millions of dollars annually through the elimination of needless ground holds for weather, unnecessary diversion of aircraft and associated system-wide disruption, more efficient routing, and better planning for ground operations and terminal sequencing. The Federal Aviation Administration (FAA) stated that 80% of all the delays of more than 15 min are caused by weather, resulting in an "economic loss" of $1 billion per year (FAA, 1992b). Airspace planning can and should become more strategic. However, to realize these benefits, airspace system designers must address the following top-level user needs relative to aviation weather:

The pilot (the ultimate end user) needs accurate, timely, and appropriate information.

The flight crew in our first case study required precise information regarding the occurrence, position, and intensity of a weather phenomenon that is very localized and short-lived. A weather detection and dissemination system for this type of hazard should meet these needs and nothing else, for this phenomenon requires immediate evasive action by the pilot, and there is no time for interpreting a complex display.

The system or structure that supports the ultimate end user—operations, meteorology, air-traffic control—needs accurate, timely, and appropriate information. In our second case study, the NTSB verified the existence of an information void confronted by this crew. The crew was unaware of the developing weather situation to the point that proper fuel management was related to a very low priority. The information needs for those supporting airspace operations are quite different, as discussed later, from those of the end user.

The weather information must be presented in an optimal form that makes it quickly and unambiguously used as a decision aid. This requires the system developer to understand the extrinsic as well as the cognitive aspects of each user's task. In our third case study, experienced pilots were bewildered by the long, complex alphanumeric teletype message that was trying to describe a complex, three-dimensional graphic. Pilots think in four dimensions (the fourth being time); hence, decision-support information, in most cases, should be presented similarly.

A mechanism should be in place to develop new weather "products" or refine the current ones to address new aviation hazards as they are identified. Our final case study leaves a field of questions unanswered regarding the possible existence of essentially clear air, terrain-induced, extremely severe wind phenomena. History has shown that the scientific community and aircraft operators can work extremely well together to precisely define new hazards, determine how to detect or predict them, and get appropriate information to the end user on time to initiate a proper response (Mahapatra & Zrnic, 1991).

In summary, to best serve the aviation weather user community, weather observations and forecasts must improve, aviation weather information dissemination must improve, and users of aviation weather must be trained in its proper use, given their functional role. These goals translate into human-factors issues that will challenge human-factors researchers and practitioners alike.

25.2 Human-Factors Challenges

The use of weather information by the pilot and other users is regulated heavily, and in some cases, is mandated in terms of source, format, validity, and geographic orientation. Unfortunately, weather forecasting is an inexact science, and aviation weather "products" in the past have been lacking in credibility. The combination of regulated use and lack of credible weather products have created a situation where different classes of users have different expectations, and even individuals within a particular class may indicate differing needs. This creates a human-factors challenge in that, in this inexact environment, the system developer must probe deeply into the cognitive use of such information from varied sources. That is, individual users will, through personal experiences, establish perceptual biases on how weather information currently affects their behavior to one that is rule based for time-critical weather encounters.

This general observation translates into a host of other issues faced by the human-factors community, which include the following:

This is a crucial first step: Identify who the users are and what function (this is the complete task) they perform. End users typically use weather information as a decision aid and are generally not meteorologists. To the system developer, this means that information presentation must be tailored to the exact needs and criticality. Users who are not end users fulfill a number of roles, such as in traffic management, being weather information provider, and in air-traffic control. We have precisely defined the different classes of users in the later sections.

Identify how weather information is to be used, both in today's system and that of the future. The system developer must establish a realistic vision about how the ATC system will permit operations in the future automated environment, and how classes of the users will be permitted to interact with the environment. Obviously, we can only predict the future to the best level we can, but the user cannot be expected to elicit his or her needs within an unknown system.

How should the weather information be displayed to the various classes of users? Too much information or improper display of the needed information is dangerous. The example case study where the flight crew was required to decode the alphanumerics to obtain the needed information is an excellent instance of improper information transfer, and other examples are abounding.

Carefully identify exactly what information is needed. We need to think of weather end users as process controllers at a top level (Wickens, 1984), who need information to manage and control the flight process. Approaching information needs through a functional breakout using systems engineering-analysis techniques objectively identifies information needs in a top-down fashion, separating subtasks such as course tracking and controlling the fight process. Conceptually, the aviation weather users can be placed along a continuum that indicates relative closeness to the actual aircraft operation. For example, a meteorologist at a regional National Weather Service (NWS) forecast center would be placed on one end of the continuum, and a pilot would be placed on the opposite end. In general, users fall onto the continuum according to how close they are to the actual aviation operations (National Research Council [NRC], 1994). The closer the user is to the actual operations, the lesser is the amount of analytical detail that the user needs to aid the decision process. In other words, the operators require decision aids, and the weather information providers require the analytical detail. Above all, the user should be given no more than what is needed.

How doe the system developer integrate new weather technologies into current airspace system functions? Two issues are crucial in terms of new technology insertion. First, the airspace management and control system is slowly evolving, whereas weather information technology is in the midst of a revolution. Second, users, for the most part, have never had the kind of weather products that are about to be introduced—extremely high resolution (spatially and temporally) and accurate—and they do not know how to use them to the best benefit.

Finally, how can a sound human systems engineering approach integrate with the needed scientific research to produce more advanced aviation weather products to handle the not-yet-defined atmospheric hazards? We have seen revolutionary success with the concepts of concurrent engineering in large system development. A similar hand-in-hand approach to air-traffic control and aviation weather-product development would help to ensure that the user needs are addressed throughout the development process.

When we talk about integrating user needs with the development process for an air-traffic control function, which includes advanced weather information, we are trying to capture the concept of shared situational awareness (SA) among all the national airspace users. If we can cause our transfer of weather information to represent a common mental model of the weather situation across the spectrum of aviation users, we can enhance decision making in a cooperative way. That is, decisions can be mutually arrived at based on the same perception of system state. This is a top-level goal; we need to address how this goal filters down into lower functional levels within the system, and how do functions, information needs, and goals interrelate at the lower levels. These broad questions suggest the need for a top-down systems engineering approach to user needs, as they relate to function. This concept will permeate the chapter.

We have approached these questions in greater detail as this chapter unfolds.

25.3 Transformation of Aviation Weather

It has been stated earlier that the aviation weather system is in the midst of a revolution, whereas the NAS is slowly evolving to high levels of automation. In a contradictory sense, the aviation weather revolution will also evolve to supply information when and wherever appropriate to support the NAS evolution. The revolutionary aspect of change relates to the precision, timeliness, system integration, display capabilities, and above all, accuracy of weather information provided to NAS users, and an implementation of these changes will begin in the very near term (McCarthy & Serafin, 1990, p. 4). It is important to summarize these phased changes to weather information dissemination so that we can properly address their implications on the various classes of users.

25.3.1 Gridded Data

The heart of the emerging aviation weather system is known as the Aviation Gridded Forecast System (AGFS), and is being validated and fielded by the NWS. The AGFS is a national, four-dimensional database consisting of atmospheric variables of most interest to aviation—wind speed and direction, clouds and other impediments to visibility (from temperature and relative humidity), turbulence, and icing. In addition, the gridded radar mosaics from a national Doppler weather radar network provide the three-dimensional views of convective systems, and the fourth dimension provides the forecast information.

The greatest impact on the aviation user is the availability of the data to support the weather products in all the three spatial dimensions. That is, the user can now "slice and dice" the atmosphere anywhere over the United States and view the graphic depictions of the aviation impact variables (AIVs) that are route- and altitude-specific to the user's particular needs. The concept of a national database also lends itself well to frequent updates from varied sources, greatly impacting the accuracy of weather information and forecasts.

25.3.2 Observations—Density, Timeliness

In an effort to increase the user credibility in aviation weather forecasting, a major revolution is taking place in the sensing of hazardous and operationally significant weather. In 1990, the spacing of the most basic observation (balloon radiosonde measurements) was about 250 miles. However, a number of new sensing mechanisms are planned to increase the atmospheric sampling by several orders of magnitude. Some of these include inflight sampling and data link, automated surface observing systems, wind profiler networks, and the Doppler weather radar network. These data will be used to provide users with better current information, and will also be used to increase the accuracy of weather forecasts in the three spatial dimensions (McCarthy, 1991).

25.3.3 Temporal and Spatial Resolutions

The present observing system was designed to provide information about large-scale weather-system phenomena that shape the evolving nature of the weather conditions. However, the weather events of most interest to aviation are of much smaller scales—tens of kilometers and less than 1 h in duration. With greater computing capability, increased observation density, and AGFS, spatial and temporal resolutions can be increased to better meet the needs of aviation users. For example, the mechanisms that create and propagate turbulence are very small, in the order of a few kilometers or less. Because of increased temporal resolution, the NAS user can expect more frequent forecasts based on updated data, and potentially better decision aids for preflight and enroute operations.

25.3.4 Forecasting Capabilities

With much higher-resolution input data on the state of the atmosphere, it does not necessarily follow that forecasting capabilities will improve. Forecasters are faced with the same dilemma, that is, data overload that we want to avoid with other users within the NAS. As part of the aviation weather revolution, the scientific community (FAA, 1992a; McCarthy, 1993) is concurrently developing the algorithms and automated processes to transform the huge amount of incoming raw data into the aviation impact variables, contained in the national four-dimensional AGFS, which could support the graphics portrayal to the user. An integral part of this effort is an ongoing verification program that documents an accuracy of the resulting information and recommends improvements to the algorithms to enhance accuracy.

The impact on the user will be almost immediate in terms of better forecasts and increased resolution. Furthermore, the accuracy and resolution improvements will continue for many years as driven by the needs of the user.

25.3.5 New Advanced Weather Products

The concept of "weather product" has emerged, and will continue as the user needs evolve. To illustrate, let us explore an example. A significant weather advisory describing the potential for moderate icing (very similar to the alphanumeric advisory presented to the pilots in our third case study) might be issued to a pilot in the following encoded format:

WA OR CA AND CSTL WTRS FROM YXC TO REO TO 140 SW UKI TO 120W FOT TO 120W TOU TO YXW LGT OCNL MDT RIME ICGICIP FRZLVL TO 180. CONDS SPRDG EWD AND CONT BYD 08Z.

First, this advisory is very difficult to read and understand. It requires the pilot to plot the corners of the affected area on a flight chart, and then observe if his route of flight passes through the volume. Second, once the pilot carries out the plotting exercise, he observes whether the affected volume encompasses a three-state area of up to 18,000 ft. Finally, when compared with the actual icing encounters, he might find the area of the actual icing conditions to be only 25 miles square. Hence, when we consider a weather product, we must think of the meteorological information tailored to route and altitude, that is spatially and temporally accurate, and is presented in a display concept that is appropriate to the user.

25.4 Advanced Aviation Weather and the NAS

Now, we need to address the evolving NAS by describing a vision of the future at a very top functional level. In what sort of system will the end user control his or her process, and what weather information needs will confront the non–end user?

25.4.1 NAS Evolution

A quick observation is essential before we could focus on the future NAS. For sometime to come, weather information will have to support the current, largely human-directed and workload-intensive NAS structure. A tremendous human-factors challenge exists with this task, because we can expect the growth of air traffic to continue, with little assistance to address the problem for the human in the loop. Weather information to the aviation user will have to overcome years of incredulity and perceptions, as well as be presented to the user such that it helps with task accomplishment, before we can expect any derived benefits to accrue.

Today's aviation weather system provides imprecise information, covers huge geographic areas, and often over predicts or entirely misses the adverse weather conditions. When weather conditions are marginal or rapidly changing, the safety, the efficiency, and the capacity of aviation operations are compromised (McCarthy, 1993, p. 1).

Today's aviation weather information is basically data rich—it requires some understanding of meteorology, is difficult to interpret relative to a particular situation, is not very timely, and generally is not route- or altitude-specific. We find the primary end user—the pilot—faced with a number of choices as to where to obtain weather information. The information is given in a fairly standard format, usually alphanumeric and compatible with typical data manipulation and transmission schemes such as a teletype. It often provides a textual description of weather hazards, clouds, winds, and other information of interest to the pilot over a much larger geographical area than needed, along with terminal weather conditions for departure and arrival. Though some hard copy or computer graphics are available, generally, these products require the pilot to take extraordinary efforts to obtain them. Little information, other than verbal updates, is available to the pilot in flight. This situation is true for the commercial as well as the general-aviation pilot. For the ground user involved with the air-traffic control and management, we find a better situation because computer graphics are more prevalent. However, they are on separate displays from the primary workstation, and require mental integration with the four-dimensional process being controlled. A considerable amount of data is available on paper, which at many times, has to be transferred manually to a graphics display prior to use. Information is routinely updated on the order of every 12 h, except for hazardous weather conditions that can be updated as often as necessary. This description is necessarily brief, but paints a picture of the weather information system that will be replaced gradually over the next decade.

The current weather system essentially feeds a manually operated NAS. Pilots are responsible for avoiding hazardous weather conditions appropriate for their operation and type of aircraft, and they do so in a tactical way, using a see-and-avoid concept or airborne weather radar. Automatic weather updates are not routinely provided to the pilot. Controllers maintain an overall awareness of hazardous weather conditions that might impact their area of responsibility, but are not required to separate the aircraft from the weather. The strategic planning of routes owing to weather does occur, but often, it is based on an incomplete picture of the weather state. As a result, traffic is usually organized in an "in-trail" structure, requiring aircraft to fly at less-than-optimal altitudes, speeds, and routings.

NAS modernization may result in the introduction of automated functions that can transition the controller from tactical control to strategic traffic management. The pilot can be transformed into a systems monitor who will no longer personally interact with the air-traffic control. These things will, of course, occur in carefully orchestrated stages over many years, and the aviation weather system must match the needs of each stage. A vision of the NAS of the future shows the aircraft separation being maintained by satellite communications; computer routing of aircraft to permit direct and optimal routings; extensive use of satellite and data-link communications for flight management and weather information; highly interactive and graphic displays for process management; and, overall, strategic planning being the rule instead of the exception. Furthermore, with the evolution of NAS, the preciseness and informational content of the aviation weather supporting it should also evolve.

The Aviation Digital Data Service (ADDS), now approved for operational use by the FAA and NWS, is a means to take advantage of the temporal and spatial improvements and the gridded four-dimensional nature of today's weather data. The ADDS is a highly user-interactive graphical tool available to all classes of users (described later), allowing the visualization of flight profile-specific aviation weather hazards—turbulence, icing, convection—as well as winds and temperature. The user can also overlay or request "legacy" NWS products including the graphical depictions of hazard information and textual products. This service is available 24 h a day, 7 days a week, at http://adds.aviationweather.gov.

Soon to come are the probabilistic forecasts of turbulence, convection, and inflight icing that will be integrated into the future decision-support tools for air-traffic management. Figure 25.1 is an example of an inflight icing depiction extracted from the ADDS Web site that shows the hazard levels up to 18,000 ft, similar to the textual description presented earlier, only on a national scale and much easier to understand. The user can "zoom" as needed and create cross sections along a planned route of flight at a particular altitude.

25.4.2 Users of Aviation Weather

The aviation weather users of today will change as the NAS changes, but generally, the functions performed will remain intact, possibly performed by computer or defined by other roles in the system. Therefore, it is very important to assume a task or functional orientation when allocating weather information needs. The following discussion of aviation weather users is by no means exhaustive, but conceptually illustrates the broad spectrum of user needs that are being addressed.

Airline and military users cover the type-of-user spectrum, from meteorologist to end user or pilot. Similarly, weather information needs span the spectrum from raw data to decision aids, and a careful consideration of these needs can literally make the difference between high payoff or miserable failure of the informational impact on operations.

Functions within the broad area of air-traffic control also require diverse approaches to defining the needs. Floor controllers within ARTCC sectors and Terminal/Radar Approach Control (TRACON), for example, are concerned with tactical control and separation of traffic in smaller geographical areas. Relative to weather, this function is perhaps limited by today's incomplete and imprecise weather information picture—that is, better information and presentation might expand the strategic planning role of these users. The ARTCC Traffic Management Unit (TMU) and Air-Traffic Control System Command Center (ATCSCC), on the other hand, are primarily concerned with strategic planning and traffic flow management at a national level. Weather information needs tend to be at a top-level for the most part, but also needs to include detailed and precise descriptions of weather conditions at key or pacing

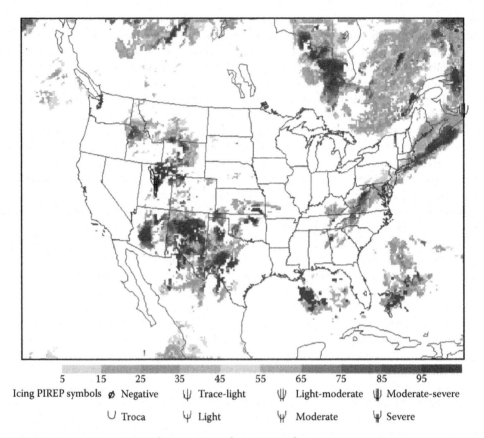

Icing PIREP symbols ø Negative   ⱷ Trace-light   ⱷ Light-moderate   ⱷ Moderate-severe

∪ Troca   ⱷ Light   ⱷ Moderate   ⱷ Severe

FIGURE 25.1   Current icing potential, composite surface to 18,000 ft.

airports across the country. The air-traffic control tower (ATCT) is much like the pilot—aids to support quick and accurate decision making, and not data, are clearly needed.

Conceptually, a completely different set of needs is represented by the many information producers and providers of today's aviation weather information, and these functions will continue to exist for the foreseeable future. These users need varying shades of data, because they are predominantly highly trained in meteorology and their primary role is to transform the atmospheric data into sometimes rigidly defined aviation weather products, and there is a continuum here also. For example, meteorologists in the NWS and Aviation Weather Center (AWC) rarely interact with aviation operations and/or end users. They generate the current and forecast weather products for passive transmission to other users and providers. On the other hand, Automated Flight Service Station (AFSS) specialists are primarily concerned with communicating weather information via briefings to pilots, both on the ground and airborne, in response to specific requests. Additionally, there are numerous commercial weather hardcopy graphics, text, verbal briefings, and computer graphics.

Perhaps, the most important point that can be made relative to a human-systems approach to aviation weather systems development is the following: in each step of the way, we can never be certain about how this new capability will impact a person's job performance. We have mentioned briefly that more precise, complete weather information might create a more efficient method of planning air traffic—that is, make planning more strategic—but we will never really know until a near full-capability prototype is evaluated operationally. As we set the stage for this important process, it is instructive to again consider the concept of user classifications.

25.5 Classes of Users

Recall the concept introduced earlier that places users along a continuum based on their relative closeness to actual aviation operations. This is how the system designer should look at classifying users according to the information needs. This is, of course, a good starting point for the entire process to be introduced shortly.

We have been using the term *end user* throughout to refer loosely to the pilot operating within the NAS. The end user represents one end of the continuum, and the function within the NAS. The function represented here is a seeker and consumer of information to support strategic and tactic decision making. This implies, from a workload management viewpoint, that this information should be immediately useful with little or no mental capacity required to make it match with the situation or transform it from data to useful information. It should be presented in such a way that it already matches the temporal and spatial situations driving its need. It is a decision aid. Users in this class are, of course, the flight crew, and some air-traffic control users such as ATCT specialists and traffic-management personnel in the TMUs and at the ATCSCC.

Meteorologists who transform the huge amount of atmospheric data into usable aviation products fall on the other end of the continuum. They use the finite, absolute detail to form the mental and physical models of what is happening, to generate specific weather events, and form products to convey information in specific formats to other users. They need structure and organization to the large amount of data they must assimilate. For this reason, data formatting must fit the assimilation process that they mentally use to forecast the weather. For example, as atmospheric trends are very important to forecasting, meteorologists will frequently ask for weather data and graphics to be presented in a movie loop or time-sequenced images. This aids the meteorologist in visualizing the developing weather pattern.

The users who fit between the extremes on the continuum are the information conduits to other users. These users include meteorologists (when filling their briefing role), AFSS specialists, ARTCC sector supervisors and controllers, airline dispatchers, and any other users who, as part of their function, are required to convey an understanding of the weather state to another consumer of information. Based on a particular function, a single user's needs can vary considerably in the course of performing a task.

25.6 Human-Factors Systems Design Issues

We have mentioned a number of theoretical human-factors principles and constructs, and identified, at a top level, some of the human-factors challenges faced by the system designer of aviation weather products and displays. Now, we need to address them further within a formal human-factors framework, but still in the context of aviation weather, while leading to a discussion of the process needed to address them properly.

25.6.1 Paradigm Fixation

Users are not always right. They cannot anticipate with complete accuracy on how they will use the new technology. This phenomenon can be called as *paradigm fixation*, and it occurs whenever new technology introduces a new informational environment for the user. As a designer, one must build a system that the users will want when it gets here, not build the system they want as they see things today (Lewis & Rieman, 1993). In practice, lengthy operational exposure during development is the only way to fully understand how new information will be assimilated into the current task structure. Even then, there may be a nagging doubt that operational usage did not expose one or more crucial information needs, or exercise every critical decision path.

We mentioned the "evolution" of automation within the NAS versus the "revolution" in aviation weather sensing, forecasting, and display. The natural tendency, for sometime to come, will be for the users to continue their business as usual. This presents a real difficulty to the system designer, who must somehow elicit the future weather information needs from a user who is working within the constraints of an NAS that is evolving in small but sure steps. Even in a rapid prototyping environment, where a near-fully capable system is exercised in an operational role, lack of confidence along with a lack of time to be creative with a new source of information and the comfort of the current task structure, will probably result in invalid user feedback.

25.6.2 Validation, Verification, Evaluation

Closely related to the preceding discussion, and absolutely crucial to the solution of any human engineering problem, is the issue of validation/verification. Validation is the process of assessing the degree to which a test or other instrument of measurement does indeed measure what it is supposed to measure (Hopkin, 1993). Woods and Sarter (1993) went further stating that validation, as an iterative evaluation, should be an integral part of system design rather than something tracked on at the end. It should help the designer to improve the system and not simply justify the resulting design. Validation as a process should provide "converging evidence on system performance." Verification, on the other hand, is the process of determining the truth or correctness of a hypothesis (Reber, 1985), or in this context, should explore how far major system elements, such as software, hardware, and interfaces, possess the properties of theories, or confirm the appropriateness by gathering environmental information informally (Hopkin, 1993).

These two general concepts can be further placed into the context of aviation weather products. There is an operational *validation* that must occur along with a continuous meteorological *verification* to measure and document how accurately algorithms describe the meteorological phenomena. Obviously, these two tasks go hand-in-hand. Furthermore, it is suggested that another concept is needed to complete the triad—*evaluation*. Evaluation is a means to determine how well the initial goals have been achieved (Hopkin, 1993). However, evaluation may also reach conclusions about feasibility, practicality, and user acceptance. Each—validation, verification, evaluation— has been mentioned separately only to relate the task at hand to a formal process, when indeed, the elements of all the three should be integrated into the design process, which may probably have to occur iteratively.

As always, certain "social issues" need to be considered in any val/ver/eval (validation/verification/evaluation) process. For aviation weather or any verifiable information that can directly have an impact on aviation safety, one must address the type of evaluation that is acceptable to the users and public, how much time and money should be used to test, and relatedly, when is it good enough to place in the public's hands verses the that are current, not-as-reliable information. Finally, what level of security and reliability will the public demand from the system? These questions will have an impact on how extensive an operational evaluation is permitted prior to "complete" verification (Wise & Wise, 1993), and will certainly have an effect on the perceived value of an evaluation.

A final issue—technical usability—is absolutely crucial and needs to be an integral part of evaluation. Basically, technical usability refers to traditional human-factors issues—display design, domain suitability, human–machine interface match to cognitive problem solution. If the system is difficult for the user in any way, then evaluation results will be confounded and difficult to parse. The val/ver/eval process should try to eliminate user annoyances or system-usability issues as soon as possible to keep the user focused on true operational utility.

To paraphrase past experience: "An ATC system can only be validated in operation" (Smoker, 1993, p. 524).

25.7 The "Criterion Problem"

The "criterion problem" directly relates to some of the issues already identified in conjunction with evaluating the utility of a weather information system against a set of goals. The definition of the criterion problem, as given by Fitts, is the problem of validating and verifying procedures and equipment against a goal, purpose, or set of aims (Fitts, 1951). Three related problems arise in the context of NAS modernization (Harwood, 1993).

The NAS goal is to provide safe, expeditious, and orderly flow of traffic. The goal of NAS modernization (and aviation weather improvement) is to enhance NAS safety, capacity, and efficiency. The problem here is to establish objective, concise measures of success that represents consensus of the user community.

There is a lack of knowledge of task structure of individual and controller teams in current and future ATC environments. Hence, the system developer must consider the resulting ATC environment after each incremental change to the NAS on the way to full modernization, and the resulting user needs.

There is a requirement for sensitive criterion measures when transitioning from old to new systems, to maintain continuity and safety. The system developer must be sensitive to consequences of the new system for controller task performance. Thus, the question becomes, "when is it good enough for testing and implementation?"

There are no answers to the questions raised by the "criterion problem." However, a systematic approach to the evaluation phase of development, to include extensive user and customer involvement and agree-upon criteria for success that are goal-oriented, will help. The following discussion on task structure should provide some guidance on relating task performance to user needs and evaluation of system utility.

25.8 Task Structure

This section summarizes some of the literature-related task to validation and evaluation, and places this knowledge in the context of aviation weather information. It is very important for the system developer to understand the physical and cognitive processes involved with user-task accomplishment. During system evaluation, the system developer should be as familiar with the task as the user, so that meaningful observations of how the system is being used can be made. Kantowitz (1992) suggested that external validation has three components or characteristics: A validation process must be representative of subjects, variables, and setting. This means that evaluation should occur on the job with real users. To extend this basic rule of thumb further, the development and evaluation of complex human–machine systems require not only an adequate representation (prototype) of the user interface, but also an appropriate understanding and representation of the task environment.

Going further, validation sometimes identifies the unanticipated interactions between the user, work environment, system, and outside environments, creating a need for redesign, or resulting in suboptimal system performance. Extending Kantowitz' suggestion, the "TEST" model identifies the variables and interactions that have to be addressed or controlled in design and validation. "TEST" is an acronym for *task, environment, subject* (i.e., do not use highly skilled or test subject; use normal people in normal work environment that includes normal lighting, fatigue, and stress), and *training* (i.e., you must train, but the user is not fully trained in the use of the system for a task, and system performance will improve on a learning curve). Measures of effectiveness are system performance, operator performance, workload reduction, skill acquisition, and development of individual differences and strategies (Jorna, 1993).

A definition of a *task* is the act of pursuing a desired goal through a set of operations by utilizing the potentials of the available system (Sanders & Roelofsma, 1993). This definition can be supplemented by the suggestion that there is a hierarchy of "subtasks" that are somehow dependent on each other for accomplishment. These dependencies might be predicated on simple subtask accomplishment, or possibly, on information about the state of the environment, and are the source of true information needs of the user.

Closely related to the task structure is the concept of a mental model, which very much guides task accomplishment by virtue of its regulative function on activity and a reflection of the world on which the subject acts (Dubois & Gaussin, 1993). Most definitions of a mental model include words, such as symbolic; relation–structure of the system it imitates (Waern, 1989); parallel to reality; and knowledge of the potential characteristics of a part of the external world (Payne, 1991). In general, the system developer should strive to understand the user's mental model of either the real world that the user is attempting to describe, or the prescribed task structure (as the user perceives it), and match the user interface and system software structure to the model. This has many implications on future use, including training and eventual user acceptance. A mismatch would explain why certain users, when given a new tool to incorporate into job accomplishment, initially have difficulty in assimilating it into their task, and why a suitably long evaluation period is necessary. A simple example can further illustrate this point. The AFSS Preflight position is responsible for generating and giving preflight weather briefings to pilots that are route- and altitude-specific. The structure and content of this briefing is regulated and rigid. The AFSS specialists have a very specific mental model of this structure, which guides the seeking of information to maintain awareness and develop specific briefings. A weather display system must match this structure in terms of how the user interfaces with the system and the type of weather information presented. If it does not, then severe training and acceptance penalties will surely result.

Often, a revolutionary decision aid or information source, such as aviation weather, is introduced operationally, which fundamentally changes the current task accomplishment or structure, or even eliminates the need to perform a particular task. In the interest of improving the overall system performance, the mental model justifiably should be adapted to accommodate the new capability. That is, the system developer should accept the fact that a fundamental change in doing business is necessary, and hence, should be willing to accept (perhaps) a significant training burden. The following statement by Stager (1993) should be kept in mind: "design requirement that is often overlooked is that the information provided at the interface must support the acquisition, maintenance, and updating of a *valid* mental model for the user" (p. 105). It must be ensured that the process defines a valid model of system operation for the user.

With the modernization of NAS, we can expect increasing levels of task automation, including the processing, display, and impact of weather information on system decision making. This situation introduces the question of how tasks are allocated between human and machine, or task sharing (Hancock, 1993; Kantowitz & Sorkin, 1987). There seems to be a universal agreement on the fact that functions or tasks should be performed by human and machine together. This implies that automation should provide an effective decision aid to the human, and not always make crucial decisions for the operator. By taking this approach, the operator is physically and cognitively entrained in the system operation, enhancing the overall situation awareness. However, this may not be a principle that can be applied to every case; but, it is especially applicable to aviation weather-information use by NAS users. In either case, if the concept of task allocation is used, a top-down understanding of the task is essential.

To summarize, in the same sense of a knowledge-based system development, the designer must not only observe, but really understand each action of the user, and what cognitive process and mental model that the user exercises as he or she invokes each task. This understanding also has direct application to the graphical user interface, display, and software structure design.

25.9 Display and Information Transfer Issues

There is a huge body of literature on computer–human interface and display design. Here, only some top-level principles that have been validated through research or operation experience and that have relevance to weather information presentation, have been covered.

A very important principle that follows from the previous discussion on task structure is: Ensure compatibility of analog display with the orientation of human mental representation. It must be remembered that most aviation users are spatially oriented. For example, digitized aircraft altitude requires significant mental processing to transform it into the analog conceptual representation, which incorporates error (Grether, 1949). The same applies to display movement—make route and height depictions congruent with the real world (e.g., east-to-west is shown right-to-left).

The concept of perceptual schema—the form of knowledge or mental representation that people use to assign stimuli to ill-defined categories, a general body of knowledge about a perceptual category, developing from perceptual experience with examples rather than a strict listing of features (Wickens, 1984)—is important when applied to information supporting effective ATC. Posner and Keele (1968) suggested that there are two components—a general representation of the mean, and some abstract representation of the variability. We find that experienced controllers and pilots have developed schema relative to system states that potentially have significant impact on system operation, such as weather hazards. Posner and Keele's research suggests that variability must be addressed directly in training and display design, and not just the prototypical case. For example, the system developer should fully understand the system "outlier" states that are important to the user, so that training and display design properly highlight them.

There is some advantage to top-down, context-driven processing when the user seeks information. That is, have the user work from top-level displays down to the level of detail needed, rather than just flash a "chunk" of information on the screen in response to a request. This is because there is strong research evidence that says that human processing generally follows this model (Wickens, 1984). Closely related is the concept of holistic processing (Wickens, 1984), which "describes a mode of information processing in which the whole is perceived directly rather than a consequence of this separate analysis of its constituent elements" (p. 164). According to Navon (1977), this does not mean that perceptual analysis of the whole precedes the analysis of the elements, but rather suggests that the conscious perceptual awareness is initially of the whole and that perception of the elements must follow from a more detailed analysis. Generally, aviation users seek to process stimuli in a holistic way. The reason for this is to relieve demands on short-term memory (the "whole" sticks in short-term memory, better than an enumeration of its parts). If a detail regarding the whole is important, it must be highlighted in some way.

Why do weather graphics seem so important to aviation users? Pure human-factors research suggests that providing information along one dimension—text, color, quality—and then expecting the human to make an absolute judgment about the stimulus is very difficult. When more dimensions are added, research suggests that less information is transmitted along each dimension, but more overall information is conveyed. This lessens the demands on the human; a graphic decision aid uses many dimensions for this reason. It also matches the structure of the mental model used by the user (Wickens, 1984). With graphics, it is important that the image given to the user or pilot be accurate at the first time. Research suggests that subjects place an undue amount of diagnostic weight to the early stimulus, called *anchoring*. Subsequent sources of evidence are not given the same amount of weight, but are used only to shift the anchor slightly in one direction or another (Wickens, 1984). Also, research suggests that the number of cues has a negative impact on response accuracy. This means cues must be informative and not so salient that it overrides the information content, reliable and limited to those that are truly diagnostic of the situation that one wants to convey. Once again, the user must be provided with no more information than what is needed (Wickens, 1984).

The implication is that information and displays must be unambiguous, context driven, and require little mental interpretation, such that the structure of the graphical user interface and display must match the user's model and task structure, and that the information transmitted must be accurate at the first time. A goal, of course, is to enhance user awareness of the weather state without negatively impacting the workload.

25.10 Workload and Time Sharing

Users within the NAS have learnt to be efficient at time sharing and allocating limited attentional resources. The current ATC system and weather displays have had an important role in this learning process, because they almost require the user to develop work-around strategies to be effective. Generally, research on workload suggests that there is a trade-off between maintaining SA and minimizing workload through automation. The trick is to optimally allocate the workload between human and computer-aiding, without removing the human from the cognitive control loop. The user's primary task is flying or controlling air traffic. Information and decision aids are provided to make this task easier, and they must be provided so that they can be used without the user having to devote excessive attentional resources and mental processing. In fact, all the preceding principles and concepts relate to this one, top-level goal.

We must think of users within the NAS as process controllers at varying degrees. This means that they spend a lot of time attempting to predict the future environmental states to support decision making. Computers and automation can have tremendous impact on the current NAS via computer-aiding (Wickens, 1984) and anticipating future goals and system responses based on those states. The cognitive element of task performance can perhaps best be left to the user.

25.11 The Process—Addressing the Issues

25.11.1 Model of System Development

Here, a model of system development is introduced, which addresses the issues that we have previously identified and discussed relative to aviation weather presentation. It is fairly generic, not very profound, but carries with it a practical human-factors focus that includes many of the necessary elements that we have been addressing (see Figure 25.2).

As can be seen, the model describes a circular, iterative process that includes numerous opportunities for the user to actively influence the system design. In the upper-left corner, a concept enters the process and exists as a specification at various stages in its life, while the process continues to operate on future iterations of the concept. In other words, this is a living process that really never ends. For the weather products, we have included scientific assessment and research to address the meteorological verification and systematic product improvement. The user needs are addressed through assessment, validation of straw man capability and rapid prototype capability, and feedback into science and engineering development. The engineering function performs system integration, software development, human-factors assessment, and display design. This model represents the true concurrent engineering with built-in preplanned product improvement.

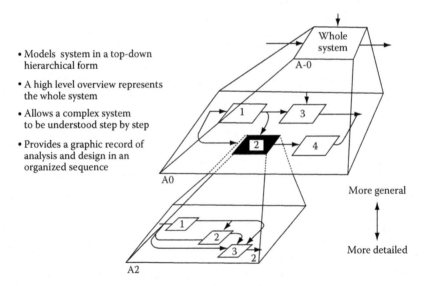

FIGURE 25.2 Top-down, systems engineering approach to weather system development.

Obviously, this concept of system development is an ideal that could be applied to just about anything. It particularly works well with aviation weather-product development, because we are essentially building a national capability from the foundation. The many opportunities for user involvement in the process are not just an ideal situation, but rather an absolute necessity, as we are unsure about the true economic benefit that the weather information and capability can exactly demonstrate. We have also addressed the difficulty in supporting the NAS in evolution, because the process is circular, and we have overcome the user's tendency to fixate on current task structures with lengthy operational evaluations and iteration.

Perhaps, one difficulty that we still need to address is how the system developer gets an objective handle on user tasks, functions, and the information needs to support their efficient accomplishment. The user's help is again invaluable, along with analytical tools to help to structure the investigation.

25.11.2 Systems Engineering Model

There are number of tools available that can be used to model the functions of a system, but actually, what is important is the process or way of thinking, and not the idea that one must use as a formal tool. The example in Figure 25.3 (Bachert, Lindholm, & Lytle, 1990) is from the integrated computer-aided manufacturing definition (IDEF) modeling language, which has the same structure and basis as the SAINT or MicroSAINT modeling packages. What is important is that the analysis focuses on the functional hierarchy of the system and that it proceeds in a top-down fashion to ensure a one-to-one mapping from top-level to lower-level functions. As one identifies each activity and breaks it down into lower-level activities, interrelationships begin to emerge between the activities, and the resources and information needed to support each activity become more specific. The method establishes a source document for tracking information requirements. It becomes an easy task, for example, to identify the resource-need differences as the functions begin to evolve with the NAS. Also, with new and most-changed systems, a task analysis using IDEF and tested with SAINT or MicroSAINT can be used to uncover shortcomings in the old system that needs changing. Thus, the interrelationships between the tasks and information resource needs will clearly emerge, from top to bottom, to support decision making (Sanders & Roelofsma, 1993).

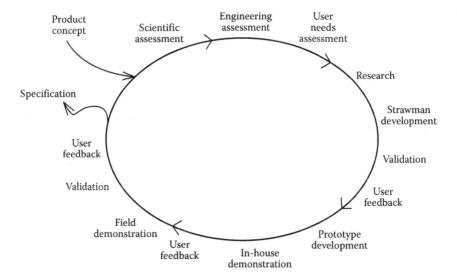

FIGURE 25.3 Iterative system development process.

25.11.3 Rapid Prototyping

Rapid prototyping has no value unless it is rapid. Incremental capability improvements need to be in front of the user as soon as possible. The development process, really, shows two phases of rapid proto-typing. First, an early capability is demonstrated in-house to gather early user feedback, which might actually occur several times. Second, a near-full-capability prototype is evaluated in an operational field demonstration over a suitably long period of time.

The idea of quick exposure is important, but an apparent paradox associated with rapid prototyping must be kept in mind. One must field the system to evaluate its effect on the user and his or her function. To do this, the system must be "realized" at various levels, such as the operational and display concept, the software, and the human interface. Once the designer is committed to a design represented by a particular realization, an acceptance of change becomes more difficult. This means evaluation questions asked late in the design process are very narrow, and tend to be focused on how to show that the system could work as opposed to finding out what the contextual limits of the system are and identifying the impact of the new device on the joint cognitive system and operational processes (Woods & Sarter, 1993). To know how to decouple the effects of different levels of realization from actual system effects during the evaluation is difficult, as mentioned previously in the context of the user interface. For example, things like system crashes or improper contrast and colors will elicit responses about the system that are not of interest or unfairly harsh, and entirely miss the issues that the evaluator is trying to resolve (Woods & Sarter, 1993).

The idea of early and continuous user involvement throughout the weather-product development cycle must be emphasized. User exposure and familiarity with various new weather capabilities are really the only way to overcome fixation with current paradigms and identify benefits derived from previously unthought-of ways of doing business. Users may be creative with new capabilities, and can provide the system designer with an insight that will never emerge from an engineering design team. On the negative side, attempts to incorporate user input late in the development process are much more difficult and expensive than building the user needs from the beginning. The $4 billion aviation weather modernization program is full of examples of creative user input, derived from actual operational experience, which collectively demonstrate a huge potential impact on airspace safety and efficiency.

By emphasizing on the early and often operational demonstrations using prototypes, the system developer maintains a task (function) orientation throughout development. The process of development using a task orientation can be carried out as follows: First, the descriptions of all the tasks are written and circulated to the users for comment (understand their cognitive process and resource or information needs); second, an interface design is roughed out and a scenario for each of the tasks is produced; third, the scenarios are represented using storyboards, without taking the process out of the context of the task. The "cognitive walkthrough" is sometimes helpful in understanding the mental processes being used in accomplishing a particular task (Lewis & Rieman, 1993). The task orientation is essential, however, there are some precautions to be followed:

As a developer, one cannot cover every task that the user is expected to perform, and without a top-down functional approach, one may probably miss the cross-task interactions (Lewis & Reiman, 1993). These functional and informational interactions are very important, because these will tend to identify areas where the duplication of tasks will occur, and efficiencies will arise out of better planning.

One should not fragment the tasks when evaluating new information capabilities with users. Field tests using operational tasks are better than laboratory tests using fragmented tasks. One needs the complete task (Lewis & Rieman, 1993).

It is also necessary to use process data as opposed to bottom-line data. Process data are observations of what the tests users are doing and thinking as they work through the tasks. Bottom-line data give a summary of what happened: How long did users take, were they successful, were there any errors? When evaluating in the field, process data validate the task process and even identify better ways of doing the task (Lewis & Rieman, 1993).

The bottom line is: Do not shortcut the rapid prototyping process—to do so will introduce rapid prototyping risks associated with too little training and lack of familiarity with the new weather capability, thereby inhibiting the user opportunity to develop new strategies that integrate performance evaluations and workload management (Jorna, 1993).

25.11.4 User Needs

At this point, we can begin to address definitively the aviation weather needs in today's ATC environment. In actual practice, the process described in Figure 25.2 has been very helpful in merging scientific discovery with what current and future aviation systems need in terms of weather information. We now bring all we have discussed in this chapter together to identify a starting point for pinpointing top-level user needs. Operational evaluations using prototypes will then have a departure point from which detail can be identified for a particular class of user.

In general, critical issues in designing any user interface are as follows: Sufficient information is available for each step in the decision sequence and the information is accessible within the context of the overall task structure (Clare, 1993). These are good rules to follow when developing a weather information and display system for the end user who requires a decision aid. To generalize for all classes of user, a weather information dissemination system must (Tonner & Kalmback, 1993):

Make the job easier
Be easy to work with
Be compatible with neighboring systems
Not lull controllers (users) into false sense of security
Keep the user in the decision loop, in charge, and well informed
Be reliable

25.11.5 Situation Awareness

Weather impacts ATC and operations more than any other state variable. Although reference to a weather information system will be accomplished by the user to support decisions, a display system must double as a continuous provider of information about the weather state, both now and in the future. Further, proper information about the weather must be provided across the spectrum of users to avoid the perceptions of contradictory state information and promote cooperative decision making. When we want to simply give all the users the same information in the proper context, with the goal of all users perceiving the same situational state, we are enhancing their shared SA.

The concept of shared SA is fairly simple—obviously, controllers and pilots cannot be expected to arrive at the same decision about, for example, routing, if both are depending on conflicting information on the location and severity of enroute thunderstorms. Endsley (1988) defined SA as the perception of the elements in the environment within a volume of time and space, and the comprehension of their meaning and projection of their status in the near future. Pew (1994) offered a concise definition of a situation: A situation is a set of environmental conditions and system states with which the participant is interacting, which can be characterized uniquely by its priority goals and response options. Keywords for the system developer are perception, time and space, projection into the near future, system state, priority goals, and response options. We must be sure of the user's perceptions of the information and display that we provide, so that it supports the proper decision. The information must be location-specific, and must provide some predictive element of the system state. Finally, the system developer must be absolutely sure of the user's goals and about the options that are driven by perceptions. These points are absolutely essential to maintain and enhance SA, and the weather information system should be expected to do just that.

Another point relative to SA is that weather information can be provided in three basic formats: graphics, usually derived from gridded weather aviation impact variables; icons showing location and spatial extent; and text. In general, the best format for the end user is defined by the amount of interaction with the information that the user can be expected to accomplished. For example, with three-dimensional graphics, the user can interact with the graphic using "slice-and-dice" as desired. With text and most icons, the user receives the information and acts with no interaction or further inquiry. Clearly, more user interaction is good for SA, but only to the extent that workload is impacted or there is a diminished value in terms of task accomplishment. However, the icon is a good indication of a state variable that needs immediate attention, and text is a provider of background information that requires no action. Thus, again we are forced to determine how the information relates to the task.

25.11.6 Needs by Class of User

Based on extensive operational-prototype experience with advanced weather products and display concepts, some verified top-level user needs are presented as follows.

Given NAS users' previous experience with an outdated, sometimes inaccurate, aviation weather system, validated data and information perhaps take on the highest priority to spur the user confidence in the consistent good quality of information. Before the users actually make strategic traffic-flow decisions based on advanced weather products and real benefits are derived, the information and decision aids will have to prove their utility. In general, we can also state that users who are not meteorologists (or whose tasks do not require them to interpret raw atmospheric data) need information and display concepts that require little or no interpretation. This means that most weather information will be presented in the form of decision aids to these users, and that some data will always be required to support certain functional needs.

In general, our end user (the pilot) needs highly focused information rather than data, and he or she needs decision aids relevant to the immediate situation rather than general advisories. Weather observations should be transformed into visualizations and decision aids that facilitate direct inferences and immediate action by the pilot. By distinguishing between weather information and decision aids, the concept of a hierarchy or pyramid of data and information is suggested, stratified by increasing direct relevance to aeronautic operations (NRC, 1994). The temporal and spatial needs are defined by the fact that the systems and phenomena of most interest to aviation and many other activities are of small-scale—tens of miles in size and often less than an hour in duration. Weather products include variations in wind speed and direction, clouds and other impediments to visibility, turbulence, icing, and convective systems such as thunderstorms (NRC, 1994).

User information and presentation concepts are always tailored to the task. For this reason, the meteorologist (or, the weather product generator function) requires considerable atmospheric data that is properly formatted to enhance and remain in the same format as the user's mental model of the forecasting process. For example, graphic looping is very useful for visualizing atmospheric processes in change and in large measure, aiding the forecasting task. Decision aids in the form of simple graphics or icons are probably not very useful in this environment.

Display concepts must meet the following general needs:

Aviation impact variables or state-of the-atmosphere variables (SAVs) for meteorologists and decision aids must be presented in easily comprehended formats, with visualizations.

The presentations must be available for flight planning through remote computer access, and must be an integral part of the information for controllers.

The system must facilitate user interaction with three-dimensional, route-specific, vertical cross-sections, so that pilots can easily study alternatives and make informed decisions. The same weather information, perhaps formatted differently, must be given to the distributors of weather information, such as the AFSS.

The information must be provided to the cockpit while airborne. Much is yet to be learnt on the exact product and format to provide, and the supporting communications or data-link infrastructure; however, we certainly know the top-level needs well enough, to begin a user needs investigation through our rapid prototyping process.

25.12 Perspectives on the Process

The process described in Figure 25.2 works very well in the development of any user-intensive system, and it really works for defining advanced weather products. Since the mid-1980s, we have collected a considerable amount of experience bringing together the needs from all classes of users within the NAS, and this work will continue for time. The most important ingredient for success is extensive user involvement from the beginning. As always, there are potential pitfalls, and hence, it is fitting that this chapter is concluded with some of the more crucial lessons learnt from exercising this process.

Advanced weather products represent a new information technology. With the introduction of any new process, source of information, or task structure, the system developer should temper user-stated needs with observations from actual operational experience. The user will initially state his or her needs from the perspective of how his or her task was performed in the past. By all means, all user input must be noted and considered, but with extra attention to contradictions between observed needs and user-stated needs. Always one should be aware that users initially have difficulty in using new information sources that are revolutionary in terms of content. Mistrust, difficulty in fitting into the current task structure, and the inherent delay involved in formulating new structures that use the new information are all valid reasons for this.

User feedback that comes from a laboratory experiment or an environment that is different from an operational setting, must be used with caution, where the user has had the opportunity to use the products for a significant period of time. The displays and advance weather technology should be exercised with the complete task and under identical physical conditions as those in the operational setting: that is, lighting, stress, interaction with other functions, workload surges, decision making, and planning expectation (Smoker, 1993, p. 524).

As most weather products will be presented as decision aids for the users, one must be aware of some difficulties with this form of information, as identified by Wickens (1984). The complexity of the aid can make it difficult to use, because many times, it depends on the ability of the user to "divide and conquer," or divide the overall decision into its constituent components for analysis in the context of the given aid. This method of processing can alter the nature of the problem to the point that it is difficult to evaluate the success of decision aid—that is, would the decision have been better with or without it?

Relative to the display of complex weather graphics, there are two very interrelated aspects of display design that must be evaluated with users in an integrated way—display functionality and display concept. The complex weather graphics that will be presented require some means for the user to manipulate, interact with, "slice and dice," zoom and pan, very easily and intuitively. The display concept refers to the structure built into the software that defines how the user interacts with the entire display. The concept or structure should strike a match with the user's mental model or defined the task structure as a starting point.

One should not allow basic human-factors issues, such as color/contrast, usability, display clarity, and functionality, to confound the results of attempting to evaluate the weather product for utility. These difficulties must be worked out carefully in-house prior to introducing the system to operational use.

Similarly, one should carefully verify the accuracy of meteorological informational in parallel with one's validation effort. The introduction of inaccurate and invalid weather products during rapid prototyping will very quickly destroy the credibility of the entire demonstration.

And finally, as a broad issue to guide implementation, airspace users and the service providers should agree to implement future airborne and ground systems as well as improved standards and procedures simultaneously, to ensure incremental benefits throughout the transition period (ATA, 1994). Advanced weather products are in a sense perishable; that is, if benefits are not shown quickly, support for better aviation weather will be lost. And without an NAS that is structured to use better weather information, benefits will be difficult, if not impossible to show.

References

Air Transport Association (ATA). (1994, April 29). *Air traffic management in the future air navigation system*. (White paper).

Bachert, R. F., Lindholm, T. A., & Lytle, D. D. (1990, October 1–4). *The training enterprise: A view from the top* (SAE Technical Paper Series 901943). Presented at the Aerospace Technology Conference and Exposition, Long Beach, CA.

Clare, J. (1993). Requirements analysis for human system information exchange. In J. A. Wise, V. D. Hopkin, & P. Stager (Eds.), *Verification and validation of complex systems: Human factors issues* (p. 333). Berlin: Springer-Verlag.

Dubois, M., & Gaussin, J. (1993). How to fit the man-machine interface and mental models of the operators. In J. A. Wise, V. D. Hopkin, & P. Stager (Eds.), *Verification and validation of complex systems: Human factors issues* (p. 385). Berlin: Springer-Verlag.

Endsley, M. R. (1988). Design and evaluation for situation awareness enhancement. In *Proceedings of the 32nd Annual Meeting of the Human Factors Society* (Vol. 1, pp. 97). Anaheim, CA: Human Factors Society.

Federal Aviation Administration (FAA). (1992a, April). *A weather vision to support improved capacity, efficiency and safety of the air space system in the twenty-first center* (FAA Code ASD-1). Washington, DC: Author.

Fitts, P. M. (Ed.). (1951). *Human engineering for an effective air-navigation and traffic control system*. Washington, DC: National Research Council.

Fugita, T. T. (1986). *DFW microburst*. Chicago: Satellite and Mesometeorology Research Project (SMRT).

Grether, W. F. (1949). Instrument reading I: The design of long-scale indicators for speed and accuracy of quantitative readings. *Journal of Applied Psychology, 33*, 363–372.

Hancock, P. A. (1993). On the future of hybrid human-machine systems. In J. A. Wise, V. D. Hopkin, & P. Stager (Eds.), *Verification and validation of complex systems: Human factors issues* (p. 73). Berlin: Springer-Verlag.

Harwood, K. (1993). Defining human-centered system issues for verifying and validating air traffic control systems. In J. A. Wise, V. D. Hopkin, & P. Stager (Eds.), *Verification and validation of complex systems: Human factors issues* (pp. 115–129). Berlin: Springer-Verlag.

Hopkin, V. D. (1993). Verification and validation: Concepts, issues, and applications. In J. A. Wise, V. D. Hopkin, & P. Stager (Eds.), *Verification and validation of complex systems: Human factors issues* (pp. 9–33). Berlin: Springer-Verlag.

Jorna, P. G. A. M. (1993). The human component of system validation. In J. A. Wise, V. D. Hopkin, & P. Stager (Eds.), *Verification and validation of complex systems: Human factors issues* (pp. 295–298). Berlin: Springer-Verlag.

Kantowitz, B. H. (1992). *Selecting measures for human factors research. Human factors psychology*. Amsterdam, the Netherlands: North-Holland.

Kantowitz, B. H., & Sorkin, R. D. (1987). Allocation of functions. In G. Salvendy (Ed.), *Handbook of human factors*. New York: Wiley.

Lewis, C., & Rieman, J. (1993). *Task-centered user interface design*. Boulder, CO: Textbook published as shareware.

Mahapatra, P. R., & Zrnic, D. S. (1991). Sensors and systems to enhance aviation Against weather hazards. *Proceedings of the IEEE, 79*, 1234–1267.

McCarthy, J. (1991, June 4–26). The aviation weather products generator. In *American Meteorological Society, 4th International Conference on Aviation Weather Systems*. Paris, France.

McCarthy, J. (1993, March 2–4). *A vision of aviation weather system to support air traffic management in the twenty-first century.* Presented to Flight Safety Foundation 5th Annual European Corporate and Regional Aircraft Seminar, Amsterdam, the Netherlands.

McCarthy, J., & Serafin, R. J. (1990, November 19–22). An advanced aviation weather system Based on new weather sensing technologies. In *Proceedings, 43rd International Air Safety Seminar*, Rome, Italy. Arlington, VA: Flight Safety Foundation.

National Research Council (NRC). (1994, March). *Weather for those who fly.* Prepared by the National Weather Service Modernization Committee, Commission on Engineering and Technical Systems, National Research Council. Washington, DC: National Academy Press.

National Transportation Safety Board. (2004a). *Annual review of aircraft accident data, U.S. air carrier operations. Calendar year 2000.* Washington, DC: Author.

National Transportation Safety Board. (2004b). *Annual review of aircraft accident data, U.S. general aviation. Calendar year 2000.* Washington, DC: Author.

Navon, D. (1977). Forest before trees: The presence of global features in visual perception. *Cognitive Psychology, 9*, 353–383.

Payne, S. J. (1991). A descriptive study of mental models. *Behaviour and Information Technology, 10*, 3–21.

Pew, R. W. (1994). Situation awareness: The buzzword of the '90s. *CSERIAC Gateway 5*(1), 2.

Posner, M. I., & Keele, S. W. (1968). On the genesis of abstract ideas. *Journal of Experimental Psychology, 77*, 353–363.

Reber, A. S. (1985). *The Penguin dictionary of psychology.* Harmondsworth, England: Penguin Books.

Sanders, A. F., & Roelofsma, P. H. M. P. (1993). Performance evaluation of human-machineSystems. In J. A. Wise, V. D. Hopkin, & P. Stager (Eds.), *Verification and validation of complex systems: Human factors issues* (p. 316). Berlin: Springer-Verlag.

Smoker, A. (1993). Simulating and evaluating the future—Pitfalls or success. In J. A. Wise, V. D. Hopkin, & P. Stager (Eds.), *Verification and validation of complex systems: Human factors issues* (p. 524). Berlin: Springer-Verlag.

Stager, P. (1993). Validation in complex systems: Behavioral issues. In J. A. Wise, V. D. Hopkin, & P. Stager (Eds.), *Verification and validation of complex systems: Human factors issues* (pp. 99–114). Berlin: Springer-Verlag.

Sumwalt, (Captain) R. L. III. (1992, January). Weather or not to go. *Professional Pilot, 26*(1), 84–89.

Tonner, J. M., & Kalmback, K. (1993). Contemporary issues in ATC system development. In J. A. Wise, V. D. Hopkin, & P. Stager (Eds.), *Verification and validation of complex systems: Human factors issues* (p. 492). Berlin: Springer-Verlag.

Waern, Y. (1989). *Cognitive aspects of computer supported tasks.* Chichester, England: John Wiley & Sons.

Wickens, C. D. (1984). *Engineering psychology and human performance.* Columbus, OH: Charles E. Merrill.

Wise, J. A., & Wise, M. A. (1993). Basic considerations in validation and verification. In J. A. Wise, V. D. Hopkin, & P. Stager (Eds.), *Verification and validation of complex systems: Human factors issues* (p. 88). Berlin: Springer-Verlag.

Woods, D. D., & Sarter, N. B. (1993). Evaluating the impact of new technology on human-machine cooperation. In J. A. Wise, V. D. Hopkin, & P. Stager (Eds.), *Verification and validation of complex systems: Human factors issues* (pp. 133–158). Berlin: Springer-Verlag.

26

Aviation Maintenance

Colin G. Drury
State University of
New York at Buffalo

The past decade has seen large changes in civil aviation, partly owing to external political and economic events, but also partly owing to new ideas being applied. The latter is the case with the application of human-factors engineering to the inspection and maintenance activities upon which air travel depends. Recognition of the importance of human factors in maintenance and inspection has lagged its application to the flight deck and air-traffic control, but the fact that 15% of civilian hull-loss accidents have "maintenance" as a contributing factor (Rankin, Hibit, Allen, & Sargent, 2000) has brought public attention. The precipitating event was the "Aloha incident" in 1988, where undetected multiple fuselage cracks allowed the upper skin of an airliner to peel open, when it was pressurized in flight (see description in Taylor, 2000).

Government and industry response can be observed in the development of the *National Plan for Aviation Human Factors* (FAA, 1993), the Gore Commission report on aviation safety (Gore, 1997), International Civil Aviation Organization (ICAO) promulgation of maintenance human factors (e.g., Hobbs & Williamson, 2002; ICAO, 1998), and including a current requirement for all countries to have human-factors training for maintenance personnel. In terms of literature, there have been a series of conferences devoted to human factors in aviation maintenance since 1990 (see http://hfskyway. com), books on the subject (Reason & Hobbs, 2003; Taylor & Christenson, 1998), and a special issue of *International Journal of Industrial Engineering* (Gramopadhye & Drury, 2000). Readers are referred to these sources for more detailed discussions.

26.1 The Maintenance and Inspection System

Before human-factors techniques can be applied appropriately in any system, the system itself must be well understood by the human-factors engineers. The following description of aviation maintenance and inspection emphasizes the philosophy behind the system design and the points where there is potential for operator error.

An aircraft structure is designed to be used indefinitely, provided that any defects arising over time are repaired correctly. Most structural components do not have a design life, but rely on periodic inspection and repair for their integrity. There are standard systems for ensuring structural safety (e.g., Goranson & Miller, 1989), but the one that most concerns us is that which uses engineering knowledge of defect types and their time histories to specify appropriate inspection intervals. The primary defects are cracks and corrosion (which can interact destructively at times), arising from repeated stretching of the structure from air or pressure loads and from weathering or harmful chemicals, respectively. Known growth rates of both the defect types allow the analyst to choose intervals for inspection at which the defects will be both visible and safe. Typically, more than one such inspection is called for between the visibility level and the safety level, to ensure some redundancy in the inspection process. As the inspection system is a human/machine system, continuing airworthiness has thus been redefined by the design process from a mechanical-engineering problem to a human-factors one. Inspection, like maintenance in general, is regulated by the Federal Aviation Administration (FAA) in the United States, the Civil Aviation Authority (CAA) in the United Kingdom, and equivalent bodies in other countries. However, enforcement can only be with regard to following the procedures (e.g., hours of training and record-keeping to show that tasks have been completed), and not regarding the effectiveness of each inspector. Inspection is also a complex socio-technical system (Taylor, 1990), and as such, can be expected to exert stresses on the inspectors and on other organizational players (Drury, 1985).

Maintenance and inspection are scheduled on a regular basis for each aircraft, with the schedule eventually being translated into a set of workcards for the aircraft when it arrives at the maintenance site. Equipment that impedes access is removed (e.g., seats, galleys), the aircraft is cleaned, and access hatches are opened. Subsequently, a relatively heavy inspection load to determine any problems (cracks, corrosion, loose parts) that will need repair is carried out. During inspection, each of these inspection findings is written up as a nonroutine repair (NRR) item. After some NRRs are repaired, an inspector must approve or "buyback" these repairs. Thus, the workload of inspectors is very high when an aircraft arrives, often necessitating overtime working, decreases when the initial inspection is complete, and slowly increases toward the end of the service owing to buybacks. Much of the inspection is carried out in the night shift, including routine inspections of aircraft between the last flight of the day and the subsequent first flight on the flightline.

Maintenance can be performed either in parallel with the inspection or following the raising of an NRR. Much maintenance is known to be required prior to inspection and can thus be scheduled before the aircraft arrives. In contrast to this scheduled maintenance, response to an NRR is considered unscheduled. At present, unscheduled maintenance represents a large and increasing fraction of the total repair activity, primarily owing to the aging of the civil fleet. In 1990, the average age of jet transport aircraft in the United States was 12.7 years, with over a quarter of the aircraft more than 20 years old (Bobo, Puckett, & Broz, 1996). From 1980 to 1988, as the aircraft fleet increased by 36%, the maintenance costs increased by 96%.

26.2 Human-Factors Analysis of the Maintenance and Inspection System

One early and thorough analysis of the inspection function (Lock & Strutt, 1985) used logical models of the process and field observations to understand the potential errors within the system. It is still the case that inspection and maintenance tasks need to be analyzed in more detail than the preceding systems description, if human-factors techniques are to be used in a logical fashion. At the level of function description, Tables 26.1 and 26.2 give a generic function listing for the activities in inspection and maintenance. It can be noted that not all "inspection" activities are performed by a person with the title of "inspector." Examples are transit checks, "A" checks, and avionics diagnostics, which are often performed by an aviation maintenance technician (AMT), also known as a mechanic. Each of the functions listed has different human-factors considerations as critical elements. Some, such as search, in Inspection, depend critically on vision and visual perception. Others, such as site access, in Repair, are motor responses where human motion and motor output are critical.

TABLE 26.1   Generic Task Description of Inspection

Function	Visual Inspection Example
Initiate	Read and understand workcard
	Select equipment
	Calibrate equipment
Access	Locate area on aircraft
	Move to worksite
	Position self and equipment
Search	Move eyes (or probe) across area to be searched
	Stop if any indication
Decision	Re-examine area of indication
	Evaluate indication against standards
	Decide whether indication is defect
Respond	Mark defect indication
	Write up NRR
	Return to search
Buyback	Examine repair against standards
	Sign-off if repair meets standards

TABLE 26.2   Generic Functions in Aircraft Repair

Function	Tasks
Initiate	Read and understand workcard
	Prepare tools, equipment
	Collect parts, supplies
	Inspect parts, supplies
Site access	Move to worksite, with tools, equipment, parts, supplies
Part access	Remove items to access parts
	Inspect/store removed items
Diagnosis	Follow diagnostic procedures
	Determine parts to replace/repair
	Collect and inspect more parts and supplies if required
Replace/repair	Remove parts to be replaced/repaired
	Repair parts if needed
	Replace parts
Reset systems	Add fluids supplies
	Adjust systems to specification
	Inspect adjustments
	Buyback, if needed
Close access	Refit items removed for access
	Adjust items refitted
	Remove tools, equipment, parts, unused supplies
Respond	Document repair

In principle, it is possible to proceed through each function and task, listing the major human subsystems involved, the error potential of each, and the design requirements for reducing these errors. Indeed, the first part of this exercise has been performed for inspection by a team working for the FAA's Office of Aviation Medicine, on the basis of field observations of many different inspection tasks (Drury, Prabhu, & Gramopadhye, 1990). The error mechanisms of interest in these systems were enumerated and studied by Latorella and Prabhu (2000). Drury (1991) provided an overview of these early error studies and included error breakdowns of the inspection function originally developed for the *National Plan for Aviation Human Factors* (FAA, 1993). As an example of the listing of possible errors, Table 26.3 shows those for the *initiate* function of Inspection. This function listing has been used directly as the basis for training programs for general aviation inspection (Jacob, Raina, Regunath, Subramanian, & Gramopadhye, 2004), as well as for the *Best Practices Guides* for a number of nondestructive inspection (NDI) systems (e.g., Drury & Watson, 2001).

Mere listing of possible errors is often less useful than classifying errors into the behavioral category or stage of human information processing involved. Examples of error classification schemes are abounding, such as Reason (1990), Hollnagel (1997), and Senders and Moray (1991), depending on the use of the data. More specifically, with regard to aviation maintenance, Reason and Hobbs (2003) listed the following:

- Recognition failures
- Memory lapses
- Slips of action
- Errors of habit
- Mistaken assumptions
- Knowledge-based errors
- Violations

As a technique for structuring the systematic application of human factors to aircraft inspection and maintenance, the error approach suffers from a fundamental flaw: In such a complex system, the number of possible errors is very large and effectively innumerable. In human-factors methodology, it is usual to make use of existing error data, if the system has been in operation long enough, to prioritize the errors. However, for aviation maintenance, the error-data collection systems have not been particularly useful in the past.

TABLE 26.3 Sample of Aircraft Maintenance and Inspection Errors by Task Step for the Initiate Task

Task	Error(s)
1.1 Correct instructions written	1.1.1 Incorrect instructions
	1.1.2 Incomplete instructions
	1.1.3 No instructions available
1.2 Correct equipment procured	1.2.1 Incorrect equipment
	1.2.2 Equipment not procured
1.3 Inspector gets instructions	1.3.1 Fails to get instructions
1.4 Inspector reads instructions	1.4.1 Fails to read instructions
	1.4.2 Partially reads instructions
1.5 Inspector understands instructions	1.5.1 Fails to understand instructions
	1.5.2 Misinterprets instructions
	1.5.3 Does not act on instructions
1.6 Correct equipment available	1.6.1 Correct equipment not available
	1.6.2 Equipment is incomplete
	1.6.3 Equipment is not working
1.7 Inspector gets equipment	1.7.1 Gets wrong equipment
	1.7.2 Gets incomplete equipment
	1.7.3 Gets nonworking equipment
1.8 Inspector checks/calibrates equipment	1.8.1 Fails to check/calibrate
	1.8.2 Checks/calibrate incorrectly

Currently, error reports are primarily used for documenting error situations for administrative purposes by internal or external regulatory agencies. All these reporting systems suffer from a number of problems with regard to feedback or corrective mechanisms at the systems level. First, they are driven by the external event of a problem being detected: If the problem is not detected, the error is not captured. In flight operations, in contrast, there are self-reporting mechanisms that capture a broader range of error events. These, such as ASRS, are now being used by maintenance and inspection personnel.

Second, the feedback of the digested error data to the users is not well human-factored. Often, the data are merely compiled rather than abridged, and hence, mechanics or inspectors must search large amounts of data with little reward. Typically, each incident is investigated, dealt with in isolation, and the compiled data is analyzed one-dimensionally, for example, by aircraft type, station, or time period. Such analyses cannot directly guide the interventions. Wenner and Drury (2000) were able to reanalyze the data on ground damage incidents from a major airline to provide more usable interventions. They cross-tabulated the incident type with the contributing factors, and used Chi-square tests to find the factors that were highly associated with particular types of incidents. This helped the management to focus on the intervention resources, where they have the highest probability of success.

Third, error reports in maintenance and inspection produced for administrative purposes are typically concerned with establishing accountability for an error and its consequences, rather than understanding the causal factors and situational context of the error. This type of information is not appropriate for use as performance feedback to inspectors or maintenance personnel, nor is it a helpful information for error-tolerant system design. Error-reporting schemes are developed within an organization and therefore, vary greatly among organizations. The framework of these error-reporting schemes is event-driven and developed iteratively; thus, additions are made only with the occurrence of a new error situation. To a large extent, the information recorded about a situation is constrained by the format of the error-reporting scheme. An error-reporting scheme should ideally be developed from a general theory of the task and the factors that shape how the task is performed. Principally, the behavioral characteristics of the operator, but ideally, organizational environment, job definition, workspace design, and the operator's physical, intellectual, and affective characteristics should also be considered. Much better error-analysis systems have now been developed to guide human-factors interventions. Allen and Marx (1994) proposed the maintenance error decision aid (MEDA) system, in which aircraft maintenance and inspection personnel self-report errors in a format compatible with human-factors analysis methods. This tool provides the bridge between systems interpretation in terms of error taxonomies (e.g., Latorella & Drury, 1992) and practical interventions across the whole maintenance and inspection system.

The success of the MEDA system has been recorded and evaluated (Rankin et al., 2000). In a series of later studies on how investigators investigate incidents (Drury & Ma, 2004), it was found that the use of either MEDA or another system tailored to aviation maintenance produced greater depth of investigation and more complete incident reports. MEDA is in use in over 60 airlines and maintenance organizations worldwide.

It can be noted that both the approaches of analyzing the tasks (e.g., Tables 26.1 and 26.2) for potential errors and analyzing errors/accidents to learn about the system and its failure modes, aid in understanding a complex system. As demonstrated by Hollnagel (1997), they are the two sides of the same coin, and the payout on that coin is the intervention to improve an already safe system.

26.3 A Classification of Human-Factors Interventions in Maintenance and Inspection

If the aim of applying human factors to aircraft inspection and maintenance is to improve both human performance and human well-being, then any interventions should address human/system mismatches, either potential or actual. Direct interventions can be logically only of two types: changing the operators to better fit the system, and changing the system to better fit the operators. The former are personnel subsystem interventions, whereas, the latter are hardware/software interventions (in terms of the SHELL model of ICAO, 1989, these would be classified as liveware and hardware/software/environment, respectively). In addition to such direct interventions, there are examples of system-level actions designed to enable system participants to understand, evaluate, and facilitate change within the system.

Since the increase in public concern for maintenance and inspection of human factors after the Aloha Airlines incident in 1988, there have been ongoing programs to identify and tackle human-factors issues in this field, led initially by the FAA and later by other organizations around the world. The function breakdown of the necessary activities (Tables 26.1 and 26.2) and the classification into systems-level, personnel/hardware, and software interventions, forms a convenient framework for the presentation of the literature describing these efforts. It also helps to point out where the programs exist, and hence, helps to guide future research and application.

The rows of Table 26.4 presents a merging of the function descriptions from Tables 26.1 and 26.2, in the order expected when an inspection activity discovers a defect that must be repaired. Scheduled maintenance activities would generally start at the Initiate maintenance function and omit the Inspection Buyback function. The columns in Table 26.4 represent the two alternative interventions, while the entries provide the framework for presentation of current interventions. In parallel to this effort have been research efforts, for example, aimed at understanding error mechanisms (Latorella & Drury, 1992; Prabhu & Drury, 1992) and speed/accuracy trade-off (Drury & Gramopadhye, 1992, in Section 5.3.4; Galaxy Scientific Corporation, 1993) in inspection.

TABLE 26.4   Classification of Interventions for Human Factors in Maintenance and Inspection

System Level Actions
1. Socio-technical systems analysis
2. MRM training for maintenance and inspection
3. Hangar-floor ergonomics programs
4. Development of human-factors audit programs
5. Characterization of visual inspection and NDI
6. Error analysis and reporting systems
7. Computer-based regulatory audits
8. Human-factors guide

Function-Specific Interventions

Function	Personnel Subsystem	Hardware/Software Subsystem
Initiate inspection		9. Workcard redesign
Inspection access		10. Restricted space changes
Search	11. Visual-search training	12. Task lighting design
Decision	13. Feedback for decision training	
	14. Individual differences in NDI	
Inspection response		15. Computer-based workcards
Initiate maintenance		
Maintenance site access		
Diagnosis	16. Diagnostic training	17. ITS computer-based job aid
Maintenance part access		
Replace/repair	18. International differences	
Reset system		
Inspection buyback	18. International differences	
Close access		
Maintenance response		

Note: ITS, intelligent tutoring system.

26.4 Human-Factors Actions and Interventions

This section provides additional detail on the entries in Table 26.4, showing human-factors considerations in each project. System-level actions are treated first to provide additional system overview information. For more details, see the review by Latorella & Prabhu (2000).

1. *Socio-technical systems (STS) analysis.* Within a complex system that is highly technical, labor-intensive, and highly regulated, there is still considerable room for alternative organizational designs, and Taylor's work in analysis of socio-technical systems in aviation maintenance (e.g., Taylor, 1990) has been the foundation of organizational changes, as well as the maintenance resource management (MRM) initiatives (see the following #2). Although individuals are usually highly motivated and conscientious in their work, communication patterns between groups and between shifts are often in need of improvement. The benefits of organizational changes that move decision-making closer to the work point have already been demonstrated in improved aircraft availability and fleet performance in a military context (Rogers, 1991).

2. *Maintenance resource management training.* The preceding STS analysis suggested the need for improved communication procedures. Hence, an early project was undertaken to provide crew resource management (CRM) training within the maintenance and inspection function on one airline and measure its results (Taylor, 1993). CRM had already been applied successfully to reduce crew coordination errors in flight crews (Heimreick, Foushee, Benson, & Russini, 1986). This work extended into a whole series of MRM studies with regard to different airlines (e.g., Taylor, 2000). The studies showed the importance of interpersonal communication at all levels, particularly by AMTs, who can be notoriously uncommunicative in their work. A complete book on the importance of communication in aviation maintenance (Taylor & Christenson, 1998) has been influential in the maintenance community, which detail many MRM programs and measure their success in teaching and fostering effective communication. Similar programs have been developed in Canada, for example, Dupont (1996) devised one based on prototypical error-prone situations, and named it as Dupont "The Dirty Dozen."

   The MRM interventions noted earlier have all involved hangar-floor programs. These have been used to train mechanics and other personnel to be aware of accident-prone situations, and give them the communications skills (e.g., assertiveness) necessary to remedy adverse situations. They have also been used to foster a "just culture" where root causes of incidents are the norm, rather than laying blame on individuals (e.g., Reason & Hobbs, 2003). Other training programs for maintenance have been devised by Walter (2000), based on task analysis, and by Endsley and Robertson (2000) using situation-awareness concepts, particularly related to team functioning.

3. *Hangar-floor human-factors programs.* The change process in ergonomics typically involves small groups of users and human-factors specialists performing analysis, redesign, and implementation on the users' own workplaces. At one airline partner, implementation was performed using the analyses already carried out as part of the restrictive space project (see Access section), which obtained good results. An existing methodology (Reynolds, Drury, & Broderick, 1994) was adapted for use at that partner airline to provide a more systematic model using the audit program (described earlier) for analysis, rather than the particular measures relevant to restrictive spaces.

4. *Development of human-factors audit programs.* The need for an ergonomics/human-factors evaluation system has been apparent for some time, and audit programs have been developed (e.g., Drury, 2001) to provide a rapid overview of the factors that are likely to impact human/system mismatches at each workplace. In the aircraft inspection and maintenance context, there is no fixed workplace, so that any audit program has to start with the workcard as the basic unit rather than the workplace. Such a system was produced in conjunction with two airline partners (Chervak & Drury, 1995; Lofgren & Drury, 1994) and tested for both large airliners and helicopters. The system was tested for reliability and modified wherever needed, before being validated against human-factors experts' judgments; and significant agreement was found. The system can

be used from either a paper data-collection form (with later data entry) or directly from a portable computer. The computer was used to compare the data collected against the appropriate standards, and to print out a report suitable for use in an existing airline audit environment. The report allowed the airline to focus the available change resources on major human/system mismatches. The completed ERNAP system has been thoroughly evaluated (Koli, Chervak, & Drury, 1998) and is available for download at http://hfskyway.faa.gov. Furthermore, other quite different systems exist, such as Reason's managing engineering safety health (MESH) system (Reason, 1997) that use ratings of environmental and organizational factors in the hangar.

5. *Characterization of visual inspection and NDI.* The process of inspection is, like other human activities, error-prone. Ultimately, inspectors can make two errors (Drury, 1991):

Type 1: Reporting an indication that is not a defect (false alarm)
Type 2: Not reporting an indication that is a defect (miss)

However, all of the processes within inspection (Table 26.1) can contribute to these errors, and hence, a detailed error analysis is required. Over the years, there have been attempts to quantify inspection reliability, so that models of crack growth can be combined with detection probabilities to optimize inspection intervals. Two recent studies on human and equipment performance in eddy-current inspection for cracks have been undertaken. The first study by Spencer and Schurman (1994) evaluated inspectors at nine facilities, and established the probability of detection (POD) curves against crack size for each. There were significant differences among the facilities, much of which were accounted for by the differences in the calibration and probing techniques. The second study was carried out by Murgatroyd, Worrall, and Waites (1994), who used computer-simulated signals in a laboratory setting, and observed no effects of a degraded inspection environment, but again found large individual differences among the inspectors. Such individual differences were also studied in laboratory experiments, reported in later discussions. More recent work has evaluated visual inspection under field conditions (Wenner, Wenner, Drury, & Spencer, 1997). Much of the human-factors knowledge of inspection has been incorporated into a series of *Best Practices Guides* for individual NDI techniques, as noted earlier.

6. *Error analysis and reporting systems.* The error-characterization work in inspection has continued in the broader context of maintenance (Allen & Marx, 1994). In one airline, maintenance, towing, pushback, and servicing errors accounted for over $16 million over a 3 year period, with the majority of errors being procedural. The most common errors were fitting of wrong parts and incorrect installation, along with failure to secure the aircraft after repair. As noted earlier, this led to development of the MEDA system, currently the most frequently encountered error analysis system in aviation maintenance (Rankin et al., 2000). Furthermore, other systems have been developed, such as the Aurora system (Marx, 1998) and the Five Rules of Causation (Marx & Watson, 2001).

7. *Audit system for regulators.* In addition to the ergonomics audit (described earlier), the concept of auditing has a long history in the regulatory environment, which provides an additional source of feedback to the maintenance and inspection system. Layton and Johnson (1993) reported on a job aid for these FAA inspectors, based on a pen computer. This system, Performance Enhancement System or PENS, contains most of the relevant federal aviation regulations in its database, as well as the details of aircraft operators and their aircraft. Thus, the FAA inspectors can rapidly enter heading data into a report, and can both rate and comment on the performance of the person being observed. A more recent system for regulators is the OASIS computer system (Hastings, Merriken, & Johnson, 2000) that allows FAA inspectors to access the current information on aircraft fleets and Federal Air regulations. This system was designed based on task analysis of Aviation Safety Inspectors, tested for usability, and evaluated to save almost 20% of the time for this overworked branch of the government.

8. *Human-factors guide for aviation maintenance.* With so much research and development activities on human factors in maintenance and inspection, there is an obvious need to get usable information for the nonspecialists within the system. Since 1992, a guide has been under development to codify the human-factors principles, techniques, and findings for the system participants, such as managers and supervisors of maintenance and inspection. This guide was produced in CDROM and hard copy forms in the mid-1990s (Maddox, 1995), and has formed the basis of training and interventions in the industry.

9. *Workcard redesign.* As existing workcards were often found to be unsatisfactory from a human-factors viewpoint, a project was undertaken to show how they could be improved. The first phase of this project (Patel, Drury, & Prabhu, 1993) used the human-factors literature to the determine principles of information design applicable to workcards, and to design new workcards embodying these principles. These new workcards were developed as job aids for two distinct types of inspection. For a C-check, which is a heavy inspection conducted infrequently, inspectors need detailed guidance on what defects to expect and which areas to search. For the more frequent A-checks, the inspection is typically the same every day (or more accurately, every night), and hence, a layered information system is needed. Here, a checklist provides procedural and sequence information to prevent procedural errors, and more detailed information is available behind the checklist for reference as needed. Evaluation of the C-check prototype showed highly significant improvements when the inspectors rated the workcard design (Patel, Drury, & Lofgren, 1994).

Since that study, there has been much interest in using AECMA Simplified English for workcards. Chervak, Drury, and Ouellette (1996) showed that Simplified English did reduce comprehension errors. Later, Chervak and Drury (2003) demonstrated that maintenance errors were also reduced in a simple maintenance task. A design tool for workcards (documentation design aid [DDA]) was developed by Drury and Sarac (1997), and has been used extensively (available for download at http://hfskyway.faa.gov). A more recent study of the DDA in outsourced aviation maintenance (Drury, Wenner, & Kritkausky, 1999) showed that it could reduce half of the comprehension errors. Currently, Simplified English is one of the several techniques being tested in overseas repair stations, as a defense against language errors in maintenance (Drury & Ma, 2004).

10. *Restrictive space changes.* Many aircraft inspection tasks must be performed in restrictive spaces owing to airframe structural constraints. A study at an airline partner measured the effect of restrictions on postural accommodations (e.g., movements), perceived discomfort, and perceived workload (TLX). It was found that it is possible to differentiate between good and poor workspaces using these measures, and to use the findings to initiate countermeasures in the form of improved access equipment (Reynolds, Drury, & Eberhardt, 1994). A classification scheme for restricted spaces was developed to assist this work, and was tested using laboratory simulations of inspection tasks (Reynolds, Drury, Sharit, & Cerny, 1994).

11. *Visual-search training.* A comprehensive series of projects used a workstation-based visual inspection simulator (Latorella et al., 1992) to test the various hypotheses about improvement of inspection training. For visual-search training, both improvements in defect conspicuity and improvements in search strategy were sought (Drury & Gramopadhye, 1992). Current inspection-training procedures are largely either classroom-based, covering theory and regulation, or on-the-job practice. Neither technique is most appropriate to the skills required in inspection, particularly, the search skills. One experiment tested a technique of practice on a visual-lobe testing task and showed that this practice transferred to search the performance for both similar and perceptually similar defects. The second experiment evaluated both performance feedback and cognitive feedback as techniques for improving search strategy and performance. It was found (Drury & Gramopadhye, 1992) that the two types of feedback have different effects, and hence, both may be needed to obtain the best results.

12. *Task lighting design.* To perform the inspection task effectively, the inspector must be able to detect the indication (e.g., crack or corrosion), which is often a difficult visual task. As search performance depends on detection off the optic axis, good lighting is extremely important to enhance the conspicuity of indications. Lighting can range from ambient, through portable, to personal (e.g., flashlights), but together, these must provide illumination of the structure with sufficient quantity and quality to give a high probability of detection. Using the existing hangar of an airline partner, detailed lighting surveys were carried out, and the results were used to determine the need for improvement. A multifactor evaluation of alternative light sources was performed, and a methodology was developed to allow airlines to specify particular devices that will supply adequate lighting and meet other safety and portability criteria (Reynolds, Gramopadhye, & Drury, 1992).

13. *Feedback training for decision.* Using the same eddy-current simulator as described by Latorella et al. (1992), Drury and Gramopadhye (1992) compared the different techniques available to help in training the inspectors to make complex, multifactorial judgments. In decision training, the experiments showed that an active training program significantly improved the number of correct decisions made on multiattribute indications, irrespective of whether the inspector was given specific standards in training or had to develop a template during training (Gramopadhye, Drury, & Sharit, 1993). Thus, it is more advantageous to train inspectors to make complex judgments about indications with many attributes (e.g., for corrosion, these could be area, depth, severity), if the inspector is actively involved in each decision, rather than passively watching another inspector making the decision.

14. *Individual differences in inspection.* As noted in the crack-detection studies discussed earlier, there are large differences in performance among the inspectors, and this has been known for many years in the industrial inspection literature (e.g., Gallwey, 1981; Drury & Wang, 1986). Owing to the possibility of selection tests for inspectors, Thackray (1995) ran a series of experiments to find correlates of performance on a simulated NDI task. The task chosen was the NDI task detailed by Latorella et al. (1992), which simulated eddy-current inspection of lap splice joints on an aircraft fuselage. Thackray found significant correlations between different aspects of performance and a number of pretest measures, of which the best predictor was the mechanical aptitude. However, the full implications of these findings are yet to be integrated into either aircraft inspection practice or the industrial inspection literature.

15. *Computer-based workcards.* Drury, Patel, and Prabhu (2000) described an implementation of improved workcards (discussed earlier) as a Hypertext program on a portable computer. The relevance to the response function of inspection was the automatic generation of much of the information needed on the NRR forms. Computer-based delivery of workcard information to the mechanic has been tested in a military context (Johnson, 1990), but the hypercard system developed here used the human-factors guidelines for information design derived earlier. There are obvious advantages from having an easily updated electronic delivery system, but it must also meet the inspectors' needs. In a direct evaluation against both original and improved paper-based workcards for part of an A-check, Drury et al. (2000) found an overwhelming support for the computer-based system over the original. However, it should be noted that about 80% of the improvement was also observed with regard to the improved paper-based workcards. Clearly, it is a good strategy to implement changes to the existing system without waiting for the benefits of electronic delivery.

16, 17. *Diagnostic improvements: Intelligent tutoring systems for training and job aiding.* The costs of incorrect diagnosis in aircraft systems are high. If the wrong unit is removed, then there is a cost of the test process for a good unit, as well as the cost of a delay, until the malfunctioning unit is found. Thus, training in fault diagnosis is a critical skill in ensuring both effectiveness and efficiency of the maintenance system. Johnson, Norton, and Utsman (1992) showed how computer-based training has evolved into Intelligent Tutoring Systems (ITS), in which models of the instructor and trainee are included in the software. Thus, in addition to system logic and data, the program for instruction contains person-models that allow more appropriate feedback and branching. An ITS was developed for the environmental control system of a Boeing-767-300 (Johnson et al., 1992), usable both as a job aid and a training device. An evaluation of this system, using 20 AMTs, compared the ITS with the instructor-led instruction, by comparing the performance on a posttraining examination (Johnson, 1990). No significant performance differences were found, showing that the system was at least as effective as the much more expensive instructor-led training.

With the evolution of technology allowing the use of portable computer systems at the work point, the basic logic and interface of such an ITS can become a useful job aid. Particularly, when interfaced with the central maintenance computer of a modern aircraft, it can support improved diagnosis techniques. Indeed, in the military, Johnson (1990) showed that a diagnosis task is dramatically improved in speed and accuracy with the use of a portable-computer-based job aid. Aircraft are now designed with on-board maintenance computer systems, so that the hardware support for such tasks is in place. However, human factors in design of the interface and logic are still required to ensure usability.

An additional project (Jones & Jackson, 1992) applied many of the intelligent tutoring systems developed for airline maintenance to an airways facilities environment. This advanced technology training system used the MITT Tutor (from Galaxy Scientific Corporation) to develop a troubleshooting training program for the air-traffic control beacon interrogator (ATCBI-4). The trainee was able to interact with a model of the ABI-4 and solve problems using various diagnostic procedures. The program allowed access to flow diagrams and oscilloscope traces, while monitoring trainee progress and errors.

18. *International differences in inspection.* The organization of the inspection/repair/ buyback process is different in the United States and the United Kingdom. A study of these differences (Drury & Lock, 1992) showed that integration between inspection and repair was emphasized in the United Kingdom, while organizational separation of these functions was considered desirable in the United States. Recent work (parallel to the preceding program) at an airline (Scoble, 1994) showed that it is possible to better integrate the repair and buyback functions with the inspection process within the existing United States context.

26.5 Future Challenges for Human Factors in Maintenance and Inspection

The function- and task-based approach detailed in this chapter was introduced to put human actions, and particularly human error, into a systems context of ensuring continuing airworthiness. In this way, the potential for human-factors interventions can be seen, alongside those of the physicists and engineers who specify the inspection intervals and who design the equipment for defect detection and repairs. The need for human-factors effort is clear, as it continues to be in flight operations. Maintenance and inspection error shows itself in spectacular system failures with depressing regularity.

As will be clear from the review of both system-level studies and function-specific interventions in the previous section, many valid studies have been carried out to bring human-factors techniques into a domain neglected for far too long. These are not the only efforts, but just those for which specific references can be cited. In a number of airlines, human factors has been introduced: Error-reporting, human-factors audits, new forms of work organization, and particularly, MRM training in almost all. In addition, aviation regulatory authorities, beyond the FAA and CAA already mentioned, are analyzing maintenance human factors in aircraft accidents. ICAO's concept of a human-factors model (the SHELL model) has moved from the cockpit into the hangar. ICAO has already mandated human-factors training for all maintenance personnel.

Thus, we conclude that we have applied human factors in a comprehensive manner; Table 26.4 shows just how spotty is our coverage of the essential functions. We can list the referenced interventions in only about a third of the cells of this table, and only a single intervention in most cells. However, when compared with the literature on human factors in flight operations, we have barely begun. Some of the cells of Table 26.4 can be covered with small extensions from other cells. Thus, the redesigned workcards for inspection should be applicable almost into the Initiate maintenance function. Similarly, the restricted space studies and improved task lighting go beyond the inspection.

However, 20 years after the study by Lock and Strutt (1985), and 15 years after the FAA's prominent involvement, we still require more studies at the systems level and demonstration projects at the function level. We also need to move from retrofitting the existing systems to designing out some of the error-prone situations in the new systems. Already new aircraft can be designed with anthropometric models in the CAD (computer-assisted design) system, much publicized for the Boeing 777 (Proctor, 1993). Such an intervention should prevent the creation of future restricted space problems. However, we also need to design human interfaces for new NDI systems, using task analytic techniques for new methods of repairing composites, applying STS design ideas to new contract repair stations, and helping design new computer systems for job control and workcard presentation.

The aviation industry has made itself into an extremely safe transportation system, but more is always demanded. As long as there are people in the system, there will be the potential for errors that arise from human–system interaction. Furthermore, human factors has a long way to go to ensure that the next level of system safety is reached.

Acknowledgments

This work was carried out under the contract from the Federal Aviation Administration, Flight Standards Service, under the direction of Jean Watson.

References

Allen, J., & Marx, D. (1994). Maintenance error decision aid project. In *Proceedings of the 8th FAA/OAM Meeting on Human Factors in Aviation Maintenance and Inspection: Trends and Advances in Aviation Maintenance Operations*, November 16–17, 1993 (pp. 101–116). Alexandria, VA.

Bobo, S. N., Puckett, C. H., & Broz, A. L. (1996). *Nondestructive inspection for aircraft*. Washington, DC: Federal Aviation Administration, DOT, Flight Standards Service.

Chervak, S., & Drury, C. G. (1995). Human factors audit program for maintenance. In *Human factors in aviation maintenance, phase 5: Progress report* (DOT/FAA/95/xx). Springfield, VA: National Technical Information Service.

Chervak, S. C., & Drury, C. G. (2003). Effects of job instruction on maintenance task performance. *Occupational Ergonomics, 3*(2), 121–132.

Chervak, S., Drury, C. G., & Ouellette, J. L. (1996). Simplified English for aircraft workcards. In *Proceedings of the Human Factors and Ergonomic Society 39th Annual Meeting 1996* (pp. 303–307). Nashville, TN.

Drury, C. G. (1985). Stress and quality control inspection. In C. L. Cooper, & M. J. Smith (Eds.), *Job stress and blue collar work* (Vol. 7, pp. 113–129). Chichester, U.K.: John Wiley.

Drury, C. G. (1991). Errors in aviation maintenance: taxonomy and control. In *Proceedings of the Human Factors Society 35th Annual Meeting*. (Vol. 1, pp. 42–46). San Francisco, CA.

Drury, C. G. (2001). Human factors audit. In G. Salvendy (Ed.), *Handbook of industrial engineering* (Chapter 42, pp. 1131–1155). New York: John Wiley & Sons.

Drury, C. G., & Gramopadhye, A. K. (1992). Training for visual inspection: Controlled studies and field implications. In *Proceedings of the Seventh FAA Meeting on Human Factors Issues in Aircraft Maintenance and Inspection* (pp. 135–146). Atlanta, GA.

Drury, C. G., & Lock, M. W. B. (1992). Ergonomics in civil aircraft inspection. In *Contemporary ergonomics 1992* (pp. 116–123). London: Taylor & Francis.

Drury, C. G., & Ma, J. (2004). Experiments on language errors in aviation maintenance. In *Proceedings of the Annual Meeting of the Human Factors and Ergonomics Society*. New Orleans, LA.

Drury, C. G., & Sarac, A. (1997). A design aid for improved documentation in aircraft maintenance. In *Proceedings of the 41st Annual Human Factors and Ergonomics Society Meeting* (pp. 1158–1162). Albuquerque, NM.

Drury, C. G., & Wang, M. J. (1986). Are research results in inspection tasks specific? In *Proceedings of the Human Factors Society 30th Annual Meeting* (Vol. 1, pp. 393–397). Dayton, OH.

Drury, C. G., & Watson, J. (2001). *Human factors good practices in borescope inspection*. Washington, DC: Office of Aviation Medicine, FAA.

Drury, C. G., Patel, S., & Prabhu, P. V. (2000). Relative advantage of portable computer-based workcards for aircraft inspection. *International Journal of Industrial Ergonomics, 26*, 163–176.

Drury, C. G., Prabhu, P. V., & Gramopadhye, A. K. (1990). Task analysis of aircraft inspection activities: methods and findings. In *Proceedings of the Human Factors Society 34th Annual Meeting* (pp. 1181–1185). Santa Monica, CA.

Drury, C. G., Wenner, C., & Kritkausky, K. (1999). Information design issues in repair stations. In *Proceedings of the Tenth International Symposium on Aviation Psychology*, May 3–6, 1999. Columbus, OH.

Dupont, G. (1996). First test flight of HPIM, Part 2. *Ground Effects, 1*(2), 3–9.

Endsley, M. R., & Robertson, M. M. (2000). Situation awareness in aircraft maintenance teams. *International Journal of Industrial Ergonomics, 26*, 301–325.

Federal Aviation Administration (FAA). (1993). *National plan for aviation human factors*. Washington, DC: U.S. Department of Transportation.

Galaxy Scientific Corporation. (1993). *Human factors in aviation maintenance, phase 2: Progress report* (Report No. DOT/FAA/AM-93/5). Springfield, VA: National Technical Information Service.

Gallwey, T. J. (1981). Selection task for visual inspection on a multiple fault type task. *Ergonomics, 25*, 1077–1092.

Goranson, U. G., & Miller, M. (1989). Aging jet transport structural evaluation programs. In *Proceedings of the 15th ICAF Symposium: Aeronautical Fatigue in the Electronic Era* (pp. 319–353). Jerusalem, Israel.

Gore, A. (1997). White House Commission on Aviation Safety and Security. In *Final Report to President Clinton*, Washington, DC: The White House, http://www.fas.org/irp/threat.

Gramopadhye, A., Drury, C. G., & J. Sharit (1993). Training for decision making in aircraft inspection. In *Proceedings of the 37th Annual Human Factors and Ergonomics Society Meeting* (Vol. 1, pp. 1267–1271). Seattle, WA.

Hastings, P. A., Merriken, M., & Johnson, W. B. (2000). An analysis of the costs and benefits of a system for FAA safety inspections. *International Journal of Industrial Ergonomics 26*, 231–248.

Heimreick, R. I., Foushee, H. C., Benson, R., & Russini, R. (1986). Cockpit management attitudes: Exploring the attitude-performance linkage. *Aviation, Space and Environmental Medicine, 57*, 1198–1200.

Hobbs, A., & Williamson, A. (2002). Unsafe acts and unsafe outcomes in aircraft maintenance. *Ergonomics, 45*(12), 866–882.

Hollnagel, E. (Ed.). (1997). *CREAM-cognitive reliability and error analysis method*. New York: Elsevier Science.

ICAO. (1989). *Human factors digest no. 1, fundamental human factors concepts* (Circular No. 216-AN/131). Montreal, Quebec, Canada: International Civil Aviation Organization.

ICAO. (1998). *Annex 1, 4.2.1.2.e*. Montreal, Quebec, Canada: International Civil Aviation Organization (JAA, JAR 66 AMC 66.25 module 9. Hoofddorp, the Netherlands: European Joint Aviation Authorities).

Jacob, R. J., Raina, S., Regunath, S., Subramanian, R. C., & Gramopadhye, A. K. (2004). Improving inspector's performance and reducing errors—General Aviation Inspection Training Systems (GAITS). In *Proceedings of the Human Factors and Ergonomics Society 48th Annual Meeting—2004*, September 2004. New Orleans, LA.

Johnson, W. S. (1990). Advanced technology training for aviation maintenance. In *Final Report of the Third FAA Meeting on Human Factors Issues in Aircraft Maintenance and Inspection* (pp. 115–134). Atlantic City, NJ.

Johnson, W. B., Norton, J. E., & Utsman, L. G. (1992). Integrated information for maintenance training, aiding and on-line documentation. In *Proceedings of the 36th Annual Meeting of the Human Factors Society* (pp. 87–91). Atlanta, GA: The Human Factors Society.

Jones, J. A., & Jackson, J. (1992). Proficiency training systems for airway facilities technicians. In *Report of a Meeting on Human Factors Issues in Aircraft Maintenance and Inspection: Science, Technology, and Management: A Program Review*. Atlanta, GA.

Koli, S., Chervak, S., & Drury, C. G. (1998). Human factors audit programs for nonrepetitive tasks. *Human Factors and Ergonomics in Manufacturing, 8*(3), 215–231.

Latorella, K. A., & Drury, C. G. (1992). A framework for human reliability in aircraft inspection. In *Meeting Proceedings of the Seventh Federal Aviation Administration Meeting on Human Factors Issues in Aircraft Maintenance and Inspection* (pp. 71–82). Atlanta, GA.

Latorella, K. A., & Prabhu, P. V. (2000). A review of human error in aviation maintenance and inspection. *International Journal of Industrial Ergonomics, 26*, 133–161.

Latorella, K. A. et al. (1992, October). Computer-simulated aircraft inspection tasks for off-line experimentation. In *Proceedings of the Human Factors Society 36th Annual Meeting* (Vol. 1, pp. 92–96). Santa Monica, CA: Human Factors Society.

Layton, C. F., & Johnson, W. B. (1993). Job performance aids for the flight standards service. In *Proceedings of the 37th Annual Meeting of the Human Factors Society, Designing for Diversity* (Vol. 1, pp. 26–29), Seattle, WA. Santa Monica, CA: Human Factors Society.

Lock, M. W. B., & Strutt, J. E. (1985). *Reliability of In-Service Inspection of Transport Aircraft Structures. CAA Paper 85013.* London: Civil Aviation Authority.

Lofgren, J., & Drury, C. G. (1994). Human factors advances at continental airlines. In *Proceedings of the 8th FAA/OAM Meeting on Human Factors in Aviation Maintenance and Inspection: Trends and Advances in Aviation Maintenance Operations* (pp. 117–138). Alexandria, VA.

Marx, D. A. (1998). *Learning from our mistakes: A review of maintenance error investigation and analysis systems.* Washington, DC: FAA, FAA/OAM.

Marx, D., & Watson, J. (2001). *Maintenance error causation.* Washington, DC: FAA, Office of Aviation Medicine.

Maddox, M. (Ed.). (1995). *The human factors guide for aviation maintenance 3.0.* Washington, DC: FAA, Office of Aviation Medicine.

Murgatroyd, R. A., Worrall, G. M., & Waites, C. (1994). *A study of the human factors influencing the reliability of aircraft inspection* (Report No. AEA/TSD/0173). Warrington, U.K.: AEA Technology.

Patel, S., Drury, C. G., & Prabhu, P. (1993, October). Design and usability evaluation of work control documentation. In *Proceedings of the 37th Annual Human Factors and Ergonomics Society Meeting* (Vol. 1, pp. 1156–1160). Seattle, WA.

Patel, S., Drury, C. G., & Lofgren, J. (1994). Design of workcards for aircraft inspection. *Applied Ergonomics, 25*(5), 283–293.

Prabhu, P., & Drury, C. G. (1992). A framework for the design of the aircraft inspection information environment. In *Meeting Proceedings of the Seventh Federal Aviation Administration Meeting on Human Factors Issues in Aircraft Maintenance and Inspection* (pp. 83–92). Atlanta, GA.

Proctor, P. (1993, November 22). Boeing 777 design targets gate mechanic. *Aviation Week and Space Technology, 139*(21), 60.

Rankin, W., Hibit, R., Allen, J., & Sargent, R. (2000). Development and evaluation of the maintenance error decision aid (MEDA) process. *International Journal of Industrial Ergonomics, 26*, 261–276.

Reason, J. (1990). *Human error.* Cambridge, MA: Cambridge University Press.

Reason, J. (1997). *Managing the risks of organizational accidents.* Aldershot, England: Ashgate Press.

Reason, J., & Hobbs, A. (2003). *Managing maintenance error, a practical guide.* Burlington, VA: Ashgate Publishing.

Reynolds, J. L., Drury, C. G., & Broderick, R. L. (1994). A field methodology for the control of musculoskeletal injuries. *Applied Ergonomics 1994, 25*(1), 3–16.

Reynolds, J. L., Drury, C. G., & Eberhardt, S. (1994). Effect of working postures in confined spaces. In *Proceedings of the 8th FAA/OAM Meeting on Human Factors Issues in Aircraft Maintenance and Inspection: Trends and Advances in Aviation Maintenance Operations* (pp. 139–158). Alexandria, VA.

Reynolds, J. L., Drury, C. G., Sharit, J., & Cerny, F. (1994). The effects of different forms of space restriction on inspection performance. In *Proceedings of Human Factors and Ergonomics Society 38th Annual Meeting* (Vol. 1, pp. 631–635). Nashville, TN.

Reynolds, J. L., Gramopadhye, A., & Drury, C. G. (1992). Design of the aircraft inspection/maintenance visual environment. In *Meeting Proceedings of the Seventh Federal Aviation Administration Meeting on Human Factors Issues in Aircraft Maintenance and Inspection* (pp. 151–162). Atlanta, GA.

Rogers, A. (1991). Organizational factors in the enhancement of military aviation maintenance. In *Proceedings of the Fourth International Symposium on Aircraft Maintenance and Inspection* (pp. 43–63). Washington, DC: Federal Aviation Administration.

Scoble, R. (1994, November). Recent changes in aircraft maintenance worker relationships. In *Meeting Proceedings of the 8th FAA/OAM Meeting on Human Factors in Aviation Maintenance and Inspection, Trends and Advances in Aviation Maintenance Operations* (pp. 45–48). Alexandria, VA.

Senders, J. W., & Moray, N. P. (1991). *Human error: Cause, prediction and reduction.* Hillsdale, NJ: Lawrence Erlbaum Associates.

Spencer, F., & Schurman, D. (1994). Human factors effects in the FAA eddy current inspection reliability experiment. In *Meeting Proceedings of the 8th FAA/OAM Meeting on Human Factors in Aviation Maintenance and Inspection, Trends and Advances in Aviation Maintenance Operations* (pp. 63–74). Alexandria, VA.

Taylor, J. C. (1990). Organizational context for aircraft maintenance and inspection. In *Proceedings of the Human Factors Society 34th Annual Meeting* (Vol. 2, pp. 1176–1180). Santa Monica, CA: The Human Factors Society.

Taylor, J. C. (1993). The effects of crew resource management (CRM) training in maintenance: an early demonstration of training effects on attitudes and performance. In *Human factors in aviation maintenance—phase two progress report* (Report No. DOT/FAA/AM-93/5, 159–181), Chapter 7, Springfield, VA: National Technical Information Service.

Taylor, J. C. (2000). The evolution and effectiveness of maintenance resource management (MRM). *International Journal of Industrial Ergonomics, 26*, 201–215.

Taylor, J. C., & Christenson, T. D. (1998). *Airline maintenance resource management, improving communications.* Warrendale, PA: SAC.

Thackray, R. (1995). Correlates of individual differences in non-destructive inspection performance. In *Human factors in aviation maintenance—phase four, volume 1 program report* (Report No. DOT/FAA/AM-95/14, 117–133). Springfield, VA: National Technical Information.

Wenner, C. A., & Drury, C. G. (2000). Analyzing human error in aircraft ground damage incidents. *International Journal of Industrial Ergonomics, 26*, 177–199.

Walter, D. (2000). Competency-based on-the-job training for aviation maintenance and inspection—a human factors approach. *International Journal of Industrial Ergonomics, 26*, 249–259.

Wenner, C. L., Wenner, F., Drury, C. G., & Spencer, F. (1997). Beyond "Hits" and "Misses": Evaluating inspection performance of regional airline inspectors. In *Proceedings of the 41st Annual Human Factors and Ergonomics Society Meeting* (pp. 579–583). Albuquerque, NM.

27

Civil Aviation Security*

Gerald D. Gibb
SRA-Galaxy International Corporation

Ronald John Lofaro
Embry-Riddle Aeronautical University

* The views and opinions expressed in the chapter are solely those of the authors. They do not, necessarily, represent those of any governmental, public, or private organizations.

27.1 Introduction

In the first edition of this book, the opening sentence of this chapter announced that it may very well be different from others in the book. That caveat is even more true in this second edition. The reason is simple: between the editions, the American homeland had been successfully attacked using commercial aircraft—September 11, 2001 happened. This event not only changed our lives in ways we are both aware and not aware of, but it completely transformed the landscape of civil aviation security. Today, 4 years later, we have a cabinet-level Department of Homeland Security (DHS), a newly created Transportation Security Administration (TSA), new public laws, as well as new federal aviation regulations* (FARs*), and a greatly changed role for the Federal Aviation Administration (FAA) in civil aviation security. In short, the awareness, mindset, federal budget, and civil aviation security structures and processes are completely new, in conjunction with the quantum leaps on both the technological and human factors aspects of security.

The authors of this chapter share a long history in the field of aviation security human factors. Dr. Gibb worked with Dr. Lofaro, who was with the FAA's aviation security R&D program, on airport security research and development in the early–mid 1990s, and has continued in that field for the past 15 years. He has become a leading expert in aviation security selection, training, and performance enhancement and evaluation. Both authors have more than 20 years of R&D backgrounds in the discipline, and have a long-standing coauthoring and presentation relationship. During that time, they have worked together, and published on, several major efforts. Thus, this chapter formally reunited them and allowed them to bring all their security expertise, training, and experiences to bear on an arena that sorely needed human factors intervention, an arena that impacts each of us who is a member of the American flying public, and those on the ground over which planes fly, an arena that has already suffered one 9/11 and must not suffer any repetition of that day.

Airport security, from events that began with the downing of Pan Am flight 103 over Lockerbie, Scotland, and continued for well over a decade, has brought important issues to the forefront. This chapter examines the contributions of human factors over the past decade and beyond. Since aviation security has changed dramatically since the first printing of this book, what follows initially is a brief, highly edited version of those portions of the first edition that either still have relevance or provide a historical context. The majority of the chapter is dedicated to the progress and advances in the past few years and the events that led to those achievements.

While every attempt is made to provide a comprehensive treatment of the area, the reader is well advised to understand that we are unable to provide depth and detail on what is considered sensitive information.

27.2 Terrorism

Terrorism is seen to encompass the use of physical force or psychological intimidation against innocent targets. Such methods are used to achieve the social, economic, political, and strategic goals of the terrorist or terrorist organizations. The millennium has seen two major organizations come to the fore in the terrorist efforts against the United States: the Taliban and the Al Qaeda. The Taliban, based in Afghanistan,

---

* The U.S. regulations covering aviation are all found in the Combined Federal Regulations (CFR), Title 14, Aeronautics and Space—these are commonly referred to as the FARs.

have now been fairly well neutralized; the Al Qaeda remains committed, organized, and aggressive, with a seemingly worldwide set of terrorist cells bent on terrorist activities designed to overthrow or cripple all governments that are not Islamic fundamentalists in nature, and a desire to inflict as much damage on the U.S. homeland as possible.

The U.S. Department of State defines terrorism as "premeditated, politically motivated violence perpetrated against noncombative targets by subnational or clandestine agents, usually intended to influence an audience." (Title 22; U.S. Code 2656(d)). This 1980 definition only lacks the element of the religious motivation of Al Qaeda. According to the U.S. Department of State view, "the term noncombatant target is interpreted to include, in addition to civilians, military personnel who, at the time of the incident, are unarmed and/or not on duty." The Department of State also considers as "acts of terrorism", attacks on military installations, on armed military personnel when a state of military hostilities does not exist at the site, such as bombings against United States bases in Europe, the Philippines, and elsewhere. To this, one may add the homicide bombings of the U.S. military personnel and innocent civilians by the media-styled "insurgents" in Iraq.

Terrorism has proven, and remains, a cost-effective tool that violates all law. Beginning in the late 1960s, terrorism has become a common occurrence on the international scene. The statistics on both domestic and international incidents were startling as the U.S. Congress Office of Technology Assessment (OTA, 1991) documented when the first edition of this book was written. During the 1970s, the total number of incidents worldwide included 8,114 people killed. The 1980s were even bloodier with over 70,000 killed and 48,849 injured worldwide. The 1980s ended with the destruction of Pan Am 103 over Lockerbie, Scotland. The 1990s saw the first attack on the World Trade Center (WTC), attacks on the USS Cole, Mogidishu, Ruwanda, the Khobar Towers, and more.

It became obvious that while terrorists continued to operate worldwide, the United States was becoming a favorite target, at home and abroad. The new millennium brought us September 11, with the destruction of both towers at the WTC by two hijacked U.S. flag-carrying airliners, the crashing of a third airliner into the Pentagon, and the fourth airliner being crashed by the actions of passengers into a field in Pennsylvania rather than into the White House or Senate building—potential intended targets. The result? Almost 3,000 Americans died, within the United States. The new millennium has continued apace worldwide, but the United States has not suffered another attack, as yet. It is hoped that the new laws, agencies, and procedures resulting from September 11 will keep that true.

27.3 Civil Aviation Security

27.3.1 History

The major threat to civil aviation security has changed dramatically in the past 20 years. The previous danger, circa 1970, was that of hijacking the plane and passengers for some kind of "ransom," usually the release of one or more captured terrorists. However, it also could be for the escape of the hijacker terrorists and could include the killing of some/all of the passengers as an object lesson. The Anti-Hijacking Act of 1974, Public Law 93-366, was designed to counter hijacking and the FAA was given the primary responsibility in this area. In fact, it was the carryover of what flight crews were then taught that was a factor in September 11, 2001: crews were instructed not to fight with the terrorists, to simply fly the plane to wherever the hijackers instructed, and let the FAA, FBI, and law enforcement handle the rest. However, TWA 847 in 1985 and Pan Am 103 in 1988 shifted the focus to that of the sabotage, bombing, and armed terrorist attacks against aircraft with the purpose of using the plane itself as a weapon of terror. It also became clear that the U.S. and its airline industry were prime targets of terrorists. One result was Public Law 101-604 passed in 1990: The Aviation Security Improvement Act.

In response to the Congressional mandate of PL 101-604, the FAA rapidly expanded its aviation security R&D service, located at the FAA Technical Center in New Jersey. As a part of this expansion, the Aviation Security Human Factors Program (ASHFP) was established in 1991. The ASHFP sought to develop guidelines, specifications, and certification criteria for human performance in aviation security systems. Major goals included improving the man–machine interface, and human input to decision making, and improving and assessing human and system performance and operational effectiveness. The ASHFP was the government's mainstay in aviation security HF until September 11. Further on in this chapter, we will discuss security programs that were begun by the ASHFP and continue today, albeit not under the aegis of the ASHFP.

In the millennium, we have come full circle in that hijacking has again taken center stage; only this time, the plane itself is the weapon, a flying bomb. There is no thought of prisoner exchange, ransom, or the like. The plane and all on board, to include the homicide bombers, are to be sacrificed to instill fear. Added to that is the threat of improvised explosive devices (IEDs) concealed and undetected in baggage and the landscape has changed again, as have the procedures and training on the flight side. We now have a federal air marshall program that put armed federal agents onboard selected aircraft; we have a program to strengthen flight deck doors, giving the flight crew time to try and land the plane. The Arming Pilots against Terrorism Act (APATA) amended subchapter 1 of Chapter 449 of Title 49, United States Code. It was passed as part of Homeland Security Act of 2002. APATA established the Federal flight deck officer (FFDO) program. This program trains and arms flight crew in case the doors are breached. No longer will the flight crew passively submit to onboard terrorist demands to take over the plane. Of course, these are last resort measures. The object of civil aviation security must be to prevent armed terrorists' boarding a flight or placing explosive devices on board in baggage. And, this is the responsibility of the baggage screeners.

27.3.2 Major Changes in Civil Aviation Security after September 11

To lay some groundwork for the remainder of this chapter, the following are the major changes that the events of that single moment in time engendered in the civil aviation security arena. The first is that the FAA no longer has the lead role and the responsibility for civil aviation security, as well as the R&D for civil aviation security. Reassigning the lead role and responsibility for civil aviation security was part of the initial actions in transforming the civil aviation security landscape. This began with the formation of the TSA within the department of transportation (DOT) via PL 107-71: The Aviation and Transportation Security Act (ATSA) of November 2001. This law established TSA within DOT. Later acts and laws set up the DHS (Department of Homeland Security Act of 2002), whose head, The U.S. Secretary of Homeland Security, is a presidential cabinet officer. There was the Intelligence Reform and Terrorism Prevention Act of 2004. These acts, laws, and organizations were in response to September 11 and perceived failures and needs in civil aviation security. It must be noted that this is an evolving process, and as late as November 2004 and possibly April 2005, some significant changes again occurred or were proposed. As one example, at the time of this writing, the third TSA administrator in as many years has stepped down amid intense pressure from the Congress to restructure the agency.

Before we take a brief look at the TSA and the DHS, a major change in security was instituted as part of the TSA, a change that is now reverting to its previous structure. Prior to September 11, the baggage screeners at most commercial airports (some 450) were employees of various private, for-profit security companies. The companies ranged from very large complex organizations with training staffs to very small, geographically local firms, dependent primarily on the size of the airport they served. Some passenger security checkpoints were staffed by commuter air carrier personnel. The security companies were contracted by the air carriers, putting a buffer layer between the air carrier and the FAA as to security breaches, violations, and enforcement.

This changed with the formation of the TSA. Congress enacted legislation to federalize the baggage screener workforce. Further, it was decided that this new baggage screener workforce, some 48,000–60,000 (initially), must be in place within a year. This led, understandably, but not effectively, to large-scale recruiting, processing, and testing programs to staff the workforce. Predictably, the former private company screeners were advantaged in passing the selection tests, as prior training and experience were beneficial factors in job sample, skill-based tests. Former screeners under the FAA/air carrier systems had monetary and benefits incentives to take these tests and remain as baggage screeners as they could become federal employees at greatly enhanced salaries.

Under the pre–September 11 system, airport security screeners were characteristically paid at or slightly above the prevailing state and/or federal minimum wage standards with minimal, if any, health, vacation, or related benefits. The private companies were looking for entry-level or retired people who were often scheduled to work 30–39 h per week (i.e., not enough to be full-time and entitled to benefits). An undesirable impact of this approach was a job turnover rate that was extraordinarily high, exceeding 70% per year at some companies. There was only a small nucleus of long-time experienced employees at these private companies around which to build a skilled workforce.

Within the year congressionally mandated to federalize the screener workforce, the TSA indeed produced a "new" workforce—all government employees and all a product of a testing and training program rushed from conception to implementation in only a few weeks. At the time of the writing of this chapter (mid-2005), the media focused on two new reports, one from the Government Accounting Office (GAO—the auditing arm of the Congress, 2005), and one from the Inspector General (IG) of DHS (DHS, March 2005), that dealt with the effectiveness of the TSA baggage screeners at civil aviation airports. The overwhelming conclusions reached the state that government screeners perform no better than those screeners in place on September 11, and that significant vulnerability still exists (an earlier GAO report is considered in further detail later). Those reports should be examined, and one should draw one's own independent conclusions. However, any review of the documents should focus on raising valid questions regarding what performance is assessed, what metrics were in place, and the nature of the data. In sum, these reports should be examined with a critical human factors perspective.

The DHS is composed of multiple agencies and organizations. The DHS attempts to leverage resources and have a coordinative, collaborative, and connective function, across federal, state, and local-level organizations, which have security responsibilities. The DHS has, as component agencies, the TSA, the U.S. Coast Guard, U.S. Citizenship and Immigration services (USCIS), the U.S. Secret Service, and the Federal emergency management agency (FEMA). The DHS also has internal organizations, which work on the science and technology needed in security in conjunction with national laboratories; the information analysis and infrastructure protection needed to deter or prevent acts of terrorism by assessing vulnerabilities in relation to threat capabilities and U.S. border protection. Obviously, the DHS has links with the FBI and the FAA. Since the DHS is a new organization, changes in its structure and function may occur by the time this chapter gets printed.

27.3.3 The U.S. General Accounting Office: Governmental Evaluation of the Progress Made

The U.S. GAO is the federal government's auditing, evaluation, and investigative arm, and conducts investigations at the request of the Congress. Although there are many reports regarding aviation security, this agency has published in recent years, the September 24, 2003 report to the Chairman, Subcommittee on Aviation, Committee on Transportation and Infrastructure, House of Representatives was a key document reviewing airport security in the post–September 11 era (GAO, 2003). A brief review of its preliminary findings is worthwhile, as it sets out the challenges ahead. Since the report was preliminary, the GAO did not make any recommendations.

The GAO was tasked to conduct an ongoing evaluation of the TSA's progress in ensuring that passenger screeners were effectively trained and supervised, that their performance in detecting threats was measured, and that their performance was compared with that in the contract screening pilot program.* While the GAO acknowledges that the TSA had met many challenges (i.e., hiring, training, and deploying more than 50,000 federal airport security screeners in less than 1 year), a number of shortcomings were highlighted. In particular

- The TSA had not fully developed or delivered recurrency training to maintain or improve the skills of the screener workforce.
- No supervisory training program was developed and deployed.
- The agency collected little data on the threat detection performance of the screener workforce. The audit revealed that the majority of that data was from covert tests conducted by the TSA's Office of Internal Affairs and Program Review.
- TSA was not using the Threat Image Projection system (TIP: described in detail later in the chapter) developed by the FAA to assess screener performance.
- No annual certification program was implemented at the time of the report.
- The TSA developed a performance management information system (PMIS) to maintain a database on screening operations, but the PMIS contained little data on screener performance.
- Performance at the contract screening pilot airports was not yet initiated, nor was a plan in place in how to evaluate and measure performance of screeners at those sites.

While these are not encouraging results, more than 2 years since the events of September 11, the challenges identified by the GAO indicate a number of areas where human factors contributions can be made. Throughout this chapter, how these weaknesses are addressed and the efforts invested to correct these deficiencies are detailed.

27.3.4 The Future?

As of 2005, some large-scale changes have occurred or are on the horizon. As an example, the new screener partnership program (SPP) may have profound effects. This will allow, as of November 2004, airports to opt out of having federal TSA screeners and to replace them once again with private sector companies' screeners. One of the rationales for the SPP is that testing found little difference in the TSA screeners' performance and that of screeners from a private screening company. And, in mid-2005, several news services have reported that the TSA may be either dissolved or restructured into a very narrow role as a part of a massive restructuring of the DHS. However, regardless of the employing organization, the human factors issues of selecting, training, performance assessment, retention, man–machine interface, technology insertion, and so forth will continue to stay front and center in the goal to improve airport security screening.

As another example of the ever-changing security landscape, let us again briefly look at the TSA. Before the DHA was formed and subsumed the TSA, the TSA itself subsumed many organizations and agencies, such as the U.S. Coast Guard, and several of the national laboratories such as Lawrence Livermore. The TSA was designed to be the flagship organization in the $20 billion effort to protect air travelers. It still is the lead horse, so to speak, in this effort. However, in the 2006 (proposed) federal budget, the TSA will lose many of its signature programs to the DHS. Added to that is the fact that, of this writing, the current TSA director has been asked to step down, and the signs seem to point to the TSA being eliminated as a distinct entity or being placed in a more narrow role in the foreseeable future.

As could be expected with the magnitude and urgency of the responsibilities thrust on the TSA, there have been missteps. Yet, to date, the TSA has played a significant role in enhancing the nation's

---

* The contract screening pilot program includes five airports, each representing a specific category of airport (e.g., large international hub, large domestic hub, etc.), that maintain private screening contractors as opposed to federally hired screeners. These airports were set aside as a comparative testbed. Although the screener workforce at these locations are employed by a private commercial entity; the training, standards, testing, certification, equipment, and so forth is the same as for federally hired screeners. The largest airport in the group is San Francisco International Airport.

transportation and aviation security. It is not our role in this chapter to play seers or predictors of political events. Therefore, we will end this section here.

27.4 Human Performance and the Human–Machine Interface in X-Ray Screening

Here men from the Planet Earth first set foot upon the moon. July 20 1969.

Neil A. Armstrong, Michael Collins, Edwin E. Buzz Aldrin

27.4.1 Introduction

Space travel to the distant lunar surface and airport security screening—what common bond do these two seemingly disparate technologies share? Aviation, they are both inextricably tied to the field of aviation. And they both emerged at relatively the same point in history. Within a few scant years of man setting foot on the surface of the moon, x-ray devices became commonplace in commercial airports throughout the United States. In 1973, the inception of the passenger and carry-on baggage screening program became mandatory as a result of legislation aimed at preventing the rash of hijackings.

In the three decades since those milestones, space travel has evolved to heights few could dream about when those immortal words "the Eagle has landed" were first spoken. Today, the international community is building a habitable, permanent outpost in space! The accomplishments and achievements in airport security screening pale in comparison. The pace of technological advancement in these two domains is unquestionably dissimilar. Although the civil aviation security community is powerless to alter past history, human factors can advance the pace of development moving forward.

In this section, we will look at the role human factors has had in changing one of the most enigmatic areas of aviation security—x-ray threat detection, an area that is highly reliant on the skill of the operator. Evaluating human performance in this task has proven to be one of the most difficult challenges faced by experts.

27.4.2 Humans, Machines, and the Interface

The effectiveness of x-ray screening in maintaining international aviation security can be attributed to two main elements—personnel and equipment. Although it can be argued that policy, procedures, and governmental regulations play some role, it is the interaction between the operator and the machine that bears the burden of detecting and preventing threats from entering the aircraft. The human–machine interface is the domain of human factors engineering. Machine refers to hardware and software, encompassing the operational x-ray devices, simulation and computer-based instruction delivery systems and mechanisms used to provide on-the-job training and performance monitoring. And although some strides can be made in selecting the appropriate individuals as airport screeners (Fobes et al., 1995; Schwaninger, 2003a; Lofaro et al., 1994a, 1994b; Lofaro, Gibb, & Garland, 1994; Rubinstein, 2001; Neiderman, 1997), most resources regarding personnel were placed on training and evaluation. The question then becomes: Have we optimized the relationship and capabilities between the human and the machine? Sadly, the answer to that straightforward inquiry is an emphatic "no."

27.4.3 Equipment

On the hardware side, airport x-ray devices have not improved measurably since their introduction as black and white shadowgraph images 30 years ago. Despite the advantages brought about by color images that distinguish between organic, inorganic, and high-density metals, improvements in image resolution and the addition of sophisticated image analysis functions and explosive detection software in some models (Fobes & Lofaro, 1994), x-ray remains a two-dimensional shadowgraph depiction of the contents of baggage. Bag factors such as clutter, presence of electronic devices, complexity, and infinite orientations of objects hinder the ability of airport screeners to detect potential threats—particularly IEDs. And the development of three-dimensional x-ray images is still in its infancy. And while computed tomography (CTX) provides "slices" of images, such displays are still two-dimensional. Both conventional and CTX technologies often result in the use of manual or trace detection methods to resolve the state of questionable items in baggage.

27.4.4 The Human Element

On the personnel side, there are several reasons for why identifying threats from x-ray images are so difficult, and most are considered human factors issues. In general, most causes can be summed simply as the difficulties encountered in discerning the relatively rare occurrences of a threat against the background of inert objects. In layman's terms, it is finding the proverbial needle in a haystack.

The image analyses features (e.g., inverse imaging, organic stripping, magnification) have been engineered to better discriminate among the contents within a bag. The addition of explosive detection systems (EDS) software to x-ray machines is intended to provide decision aids for operators. The introduction and use of the TIP was developed to maintain screener vigilance, provide increased exposure to threats, offer feedback, and increase expectations of finding threats—albeit if most are fictitious items presented on the x-ray monitors. (We will return to an in-depth look at the TIP later on in this chapter.) These tools ostensibly promote better performance among airport screeners by addressing specific human factors concerns (Barrientos et al., 2002).

27.5 X-Ray Screening: Toward Functional Assessment of Performance and Training Systems

27.5.1 Introduction

One of the most challenging and important tasks of the human factors professional is to assess human performance. This task is especially difficult when it concerns measuring the x-ray threat detection performance of airport security screeners. It is a critical responsibility to measure such performance, as it drives personnel actions, such as the need for recurrent/remedial training, identification of systemic and individual performance weaknesses, and both initial and annual certification processes. Ostensibly, performance can be defined as the ability of the screener to detect and prevent conventional weapons, explosives, hazardous materials, and other potentially dangerous goods from passing the security checkpoints and onto aircraft. Measuring that ability is another matter.

27.5.2 Performance Assessment

Performance is multidimensional in aviation security. From a strict pragmatic point of view, a threat is either missed or detected—a dichotomous outcome. However, as is common in all inspection tasks (Drury & Sinclair, 1983), each trial is also characterized by the response made in the absence of a threat. That is, when a bag contains no threat, does the screener correctly clear the bag as safe or falsely declare it contains a threat. Consequently, four response possibilities exist: (a) hit (threat present, screener detects it), (b) miss (threat present, screener fails to detect), (c) correct rejection (no threat present, screener clears the bag), and (d) false alarm (no threat present, screener declares a threat exists). These response categorizations represent the theoretical model of signal detection theory (SDT; Green & Swets, 1988).

The SDT further offers a number of other classical performance metrics that are derived from the four basic response choices including probability of detection (Pd, or the number of hits as a ratio to the number of threats presented), probability of a false alarm (Pfa, or the number of false alarms to the number of events [bags] less the number of threats), d' (operator sensitivity), and β (beta or operator bias). The former two are most commonly used and form the basic performance criteria established by the TSA. These measures, however, speak only of the accuracy of the screening decision. In an operational world, throughput rate or the number of bags per hour plays some role, as expediency in moving passengers and their belongings to their aircraft is also important.

27.5.3 Initial Efforts at Performance Testing

Before the turn of the millennium, screeners were characteristically "tested" by the use of test objects that were hidden in bags and placed within the stream of passenger bags. The FAA had at one time used a series of eight test objects (e.g., encapsulated handgun and grenade), and in the latter part of the 1990s included the use of the modular bomb set.* Airline ground security coordinators, and the FAA civil aviation security office field agents, trainers, and security vendor supervisors also carried out these test scenarios. There were problems as to the accuracy of the results of such tests. Often, the only item in the test bag (e.g., briefcase) was one of the test items. Screeners, rather than recognizing the test item, often realized that a briefcase with but one item was probably a test. The limited nature of the contents of the bag, therefore, served as a cue.

There were numerous other problems with such an approach:

- Screeners often recognized the personnel who conducted the evaluations, and therefore were cued to the impending test.
- The pool of test objects was extremely limited and could easily be memorized.
- Testing could only be conducted on a limited basis.
- The integrity of the test was highly reliant on the skill and expertise of those conducting the evaluations, with standardization difficult if not impossible.
- A minimum number of evaluations were required monthly and often occurred predictably at the end of the month.
- Only a single measure of performance (MOP) was obtained for but one individual during the test process.

The extent of these problems became apparent when the FAA personnel did what was characterized as black tests, in which airline passengers were used to carry test bags. There was no cue of a test situation and test items were placed in baggage that was cluttered with other, nonthreat objects. The results of such tests are classified, but it can be safely stated that the detection rates were considerably below the desired levels. Although some off-line evaluation tools were becoming available (e.g., SPEARS [Screener Proficiency Evaluation and Reporting System]), the task of assessing operational performance was still out of reach.

An effective system of performance evaluation was only one component of a multitude of requirements. The literature (e.g., Drury & Sinclair, 1983; Davies & Parasuraman, 1982) is well endowed with empirical studies that demonstrate poor performance and a rapid decline of vigilance in low signal-to-noise environments—as is the case with a comparatively low frequency of potential threats to the vast number of passenger bags that are screened. A significant challenge was to improve both performance and vigilance in a monitoring and inspection task, an area well documented at which humans perform poorly. Concomitantly, there was an intense need to train screeners in threat recognition beyond what was represented by test articles.

---

* MBS: a kit containing an array of common bomb components such as power sources, timing devices, simulant masses, and initiators that could be assembled to mimic improvised explosive devices.

27.5.4 Human Factors Goals

Human factors elements are more complex and intricate to address. How should performance be assessed and what metrics are applicable? What projection rates are optimal to serve a number of diverse goals? How do we integrate both evaluation and training concepts into a single system? Is it possible to develop a national performance standard for x-ray threat detection?

It has now been over a decade since the first prototypical TIP system was tested at Chicago O'Hare's International Concourse, and although the technology for the TIP has advanced tremendously in that time frame, many human performance questions are still under study.

From a human factors perspective, what if it was possible to

- Eliminate cueing problems caused by test personnel introducing physical threat targets during the testing process
- Provide screeners with a technology that, by introducing threats on a more frequent and unpredictable basis, would increase vigilance
- Launch an evaluation program that was objective, fair, and could be used with screener personnel regardless of the size of the airport, the geographic location, or the employer
- Deploy a system that could actually train screeners while doing their operational job
- Expose screeners to a vast array of threats and weapons that would not be possible in a conventional training environment
- Develop a vehicle that, as new threats evolved, could quickly be introduced to an entire workforce without deploying a national trainer contingent

These goals became the underlying technical specifications to build a national performance evaluation system for x-ray threat detection. And although the primary functions of the system were to provide continuous training, maintain screener vigilance, improve threat detection skills, and evaluate performance, the TIP had far reaching implications for other human factors efforts. We shall explore those implications in the next section.

27.6 Threat Image Projection: System, Challenges, and Value

27.6.1 Overview

Although not a panacea, the TIP program held open great promise for addressing many human factors concerns. Conceptually, the TIP is not difficult to describe. At a mechanical level, it involves capturing x-ray images of threats, storing them in a "library," and projecting those threat images into the images of actual baggage as they are presented on x-ray monitors. While the threat images are fictitious (not physically present when projected), they are nevertheless "real" since they are genuine images of actual threats. The threat images (known as TIPs) are inserted into the bag images before they are observed by the screener on the monitor and for all intents and purposes appear integrated in the bags.

The SDT, although a convenient model to apply for the development of a performance evaluation system for x-ray screening, is not without pitfalls. As already indicated, the throughput rate must have some account in the process. In the human performance measurement arena, the trade-off between speed and accuracy of performance must be addressed. Accordingly, since the decision process to either clear or suspect a bag must occur within some time parameters, the TIP was designed such that the belt must be stopped within a relatively short time (set as the amount of time it takes, an image to travel across the monitor and exit the x-ray chamber), and the decision must be made within a specified time interval thereafter, or the event is considered a miss (if a TIP was present).

Nevertheless, the SDT model is still not a perfect fit for an evaluation protocol within the TIP. Screeners have other valid procedural options with each bag when a decision cannot be made as to clearing or suspecting a bag. For any number of reasons, they may subject the bag to manual physical inspection, change the orientation, and resubmit the bag to an additional x-ray scan, or request a secondary search using other technologies. There is no adaptation of the SDT model that takes into account a secondary level of decision making. As sophisticated as the TIP software is, it can only project threat images—it does not have the capability to determine what the screener is examining. Therefore, the only possible options to "code" screener performance are derived from the basic 2 × 2 SDT model (threat presence or absence × screener response or no response). This forms the foundation of the performance assessment system within the TIP.

27.6.2 Can Training and Evaluation Occur Simultaneously within TIP?

Another significant challenge that faced human factors engineers was the development of a system that could serve both an evaluation and a training need. Irrespective of the problems with a direct application of the SDT model to assessing screener performance, the TIP system represented tremendous strides forward in quantifying screener x-ray threat detection performance. However, could the system design also encompass a training function? Indeed yes!

Guns, knives, martial arts implements, tools, and so forth have distinctive and readily recognizable shapes—in spite of the diversity within each of these threat subcategories. Threats of this nature are amenable to classroom training without requiring the need to have a sample of each item. We expect stimulus generalization, in the classical conditioning sense of the term, to occur in training process. As an illustration, scissors have a fairly recognizable shape and one need not show thousands of scissors to the screener personnel to develop the skills to detect such threats. Albeit, there are some differences in the saliency of an object, even if of the same threat subcategory, because of the size, types of materials used in construction, angle, and so forth.

27.6.3 From Conventional Weapons to Bombs

An overarching concern, however, is the detection of so-called IEDs (improvised explosive devices). This threat can take on any number of configurations and uses multiple components. For example, a mechanism to trigger such an implement can range anywhere from a mechanical, spring-loaded device to a barometric sensor to a timer no more complex than a microchip. There are infinite number of permutations possible, and it is therefore highly impractical, if not impossible, to demonstrate all possibilities in a standard training environment. However, it is possible to integrate training of such devices into the operational environment.

27.6.4 The TIP Solution

The challenge is to (1) provide sufficient training to understand and recognize IEDs/bomb components in a general sense, (2) provide exposure to such threats on a frequent enough basis to develop detection skills, and (3) conduct this training without the need for trainer oversight and guidance. We have successfully achieved this process with the TIP. Through careful specification of functional requirements to equipment vendors regarding the use of textual feedback messages, highlighting projected TIP images, and incorporating a performance-monitoring system that uses the SDT as the foundation, an effective online training system resulted (Neiderman et al., 1997).

Feedback concepts within a TIP are straightforward. Specific input keys are available on the operator control panel of all TRX systems. When depressed, it indicates to the system that the operator believes that there is a suspect item present. If a TIP (not a physical threat) is present, a positive textual feedback message is displayed, indicating the subcategory of the threat (e.g., gun, knife, and bomb). If not present, the feedback indicates "this is not a test" and directs the screener to follow normal procedures. If a TIP is missed, indicated by the screener failing to stop the conveyer belt within the allotted time or failing to indicate a threat is present after the belt is stopped within a set time parameter, a negative message is shown. The screener cancels out these messages by depressing the appropriate key.

The true training value is not in the textual message, but in the feedback that appears after the message is canceled out. Since the operator's task is a visual inspection task, the use of text information alone provides minimal feedback. Therefore, whenever there is a TIP project, whether missed or not, the threat is highlighted with a rectangular box outlining the area of the image where the TIP has been inserted. This serves two purposes: (1) it provides an opportunity to see the threat against the background noise of the bag contents, and (2) it reaffirms the validity of the textual feedback.

The question may come to mind as to why the TIP image is highlighted when the operator correctly identified the threat and received a positive textual feedback message. Quite simply, the software is only capable of comparing the operator's response to whether a TIP was projected or not. It is conceivable that the operator responded correctly in the presence of a TIP image, but actually did not see the image. In other words, the operator indicated that he or she suspected the bag, but the response was based on another stimulus in the bag. The visual feedback assures that the actual threat image is identified regardless of response. Yes, from a performance assessment perspective, the screener would be incorrectly "credited" with a "hit"—but the training value outweighs the slight imperfections within the system. The TIP system may not recognize this mischaracterization of response, but the operator does and learns from the event.

27.6.5 Enhancing Detection Rates via TIP

Human factors issues extend far beyond that of performance measurement and training with a TIP. Although extensive laboratory research investigating factors impacting inspection performance have been well documented, there is a marked paucity of studies examining how such variables affect x-ray threat detection performance. Recent studies at the Transportation Security Laboratory in Atlantic City, NJ (Barrientos, Dixon, & Fobes, 2000; Barrientos, 2002), have focused on the impact of different threat to bag ratios.

The preliminary study (Barrientos et al., 2000) attempted to set threat to bag ratios at one threat in every 25–300 bag images in increments of 25 (in all, 12 different ratios) using 200 unique threat images. A number of technical problems beset this initial effort, principally in achieving the desired presentation ratios. The findings nevertheless indicated that there was a reasonable possibility that the TIP projection frequency may impact performance (i.e., Pd, Pfa, and d'*). A 1:150 ratio produces the highest d' value. And although lower ratios (i.e., 1:25, 1:50, etc.) resulted in nearly as strong Pd values, they also generated the highest false alarm rates (Pfa).

In 2002, the Barrientos team conducted a more comprehensive study using 55 trained screeners who were "screened" using both visual acuity and color discrimination tests. The later study employed the same type of x-ray equipment and considered only five different ratios (25, 50, 100, 150, and 300 bag images to each TIP) and a larger threat image set. These were fixed ratios and the capability to establish a range around those settings, or inserting randomly dispersed TIPs, was not used. Barrientos and his team were able to set the ratios remotely to avoid cueing the participants to changes in the ratio. Each ratio ran for one to two weeks.

Once again, difficulties were encountered in attempting to achieve the higher ratio settings of 150 and 300. At the higher settings, the drift between the set ratio and the actual projection ratio increases dramatically. Consequently, Barrientos collapsed the data for the higher settings in one "extreme"

---

* Pd is probability of detection or number of successfully detected TIPs as a ratio to the number projected. Pfa is the probability of a false alarm computed as the number of screener indications a TIP is projected (when one has not) in relationship to the number of bags screened less the number of TIPs projected. The *d'* value is a complex statistical derivation that examines Pd in relationship to Pfa.

condition. The lowest false alarm rates were observed at the highest threat-to-bag ratios, and the converse was true for the lowest ratios—generating the highest Pfa values. A marked decrease in Pfa was demonstrated between the 1:25 and 1:50 ratios. The reverse was true for Pd, as the highest Pd values generally were found with the lowest ratios. When both Pd and Pfa are taken together, as with the d' metric, the most optimal performance was seen at the 1:100 ratio.

It is possible that TIP events may well be more predictable at lower threat-to-bag ratios, contributing to higher detection rates—in conjunction with higher false alarm rates that will artificially increase hit rates by detecting more threats by chance. It follows directly from the SDT that as the expectancy of an event increases, as it would at lower ratios, higher false alarms can be expected. The reader is reminded, however, that the SDT, and the computation of the d' metric, places equal weights on Pd and Pfa components. In an aviation security environment, the cost of missing a potential threat far outweighs a false alarm, which has as its primary consequence a slight increase in the inspection time of a bag image.

27.6.6 TIP and Today's Screeners: The HF Issues

Much of the human factors work today focuses on developing national performance criteria for the TIP, and in improving threat detection performance. There are a host of issues involved in developing performance criteria that must apply to a workforce in excess of 40,000 TSA screeners distributed across the nation's 459 commercial airports. The most salient challenges include

- Addressing potential performance differences that result from the deployment of four different x-ray platforms, which have different input controls and displays
- Determining the effect of experience on performance and concomitantly the minimum amount of time on job or number of TIPs presented before performance is evaluated
- Understanding variations caused by airports and traffic loads on performance since bag volume drives TIP projections
- Deriving a true false alarm rate as many "false alarms" result in bag checks, and re-examination by x-ray or other secondary screening methods generated by physical characteristics of the bag image
- Appreciating the impact of local policy, procedures, workarounds, or checkpoint configurations on performance
- Identifying optimal TIP to bag image ratios
- Providing suitable off-line training and simulation exercises to improve skill levels

Efforts dedicated to developing performance criteria are not independent of human factors work that seeks to improve x-ray threat detection skills. Currently, an initiative is underway to use the TIP performance to identify screeners who have consistently demonstrated performance that is well above the norm. These operators may well hold the key to identifying successful strategies, techniques, approaches, cues, and cognitive processes that result in exceptional threat detection performance. A number of techniques from the cognitive task analysis domain and simulation are being used to extract critical information that will ultimately be transformed into advanced training programs.

In the final analysis, the TIP remains a landmark program and a human factors success story. Many challenges and issues remain on the road ahead, but it unequivocally stands as a solid system for objectively and fairly measuring screener performance. On any given day, the nation's screener workforce is simultaneously evaluated, trained, and kept vigilant.

27.6.7 TIP: Utility and Implications for Other Human Factors Efforts

The TIP system is truly one of the largest, most comprehensive, relatively unbiased performance assessment programs implemented. On any given month, an excess of 1.5 million TIP events are projected in the United States alone. The software captures a multitude of performance measures ranging from classic

SDT metrics to throughput, and processing speed to detailed information about the threat images missed or successfully detected. As such, a TIP represents an extraordinary wealth of performance criteria. The criteria have implications far beyond the intended use of the system.

The TIP performance metrics have utility for

- Validating personnel selection tests and optimizing applicant selection processes
- Evaluating various training approaches, systems, and methodologies, including simulation and part-task trainers
- Determining proficiency, certification, and other qualifications
- Diagnostic purposes, such as identification of workforce deficiencies, training gaps, and needs to reallocate training resources
- Potential use in advancement/promotion/employment termination decisions
- Indicator for remedial or refresher training
- Possible application in awarding performance incentives
- Assessing the impact of new displays, controls, image enhancement tools, and emerging technologies
- Quantifying the effects of fatigue, shift length, environmental factors, transfer of training, and other performance influences
- Establishing an empirical basis for the development of threat detection training programs

In short, TIP metrics serve a wide range of criteria needs for human factors research and development work. Two such areas, personnel selection test development and x-ray threat detection training program development, are worth discussing further to illustrate.

X-ray screening of bag images has long been acknowledged as the most difficult task for airport security screeners (Kaempf, Klinger, & Hutton, 1996; Kaempf, Klinger, & Wolf, 1994; McClumpha, James, Hellier, & Hawkins, 1994). And early efforts to develop and validate selection instruments (Fobes et al., 1995), or to determine the cognitive processes used in x-ray image interpretation (Lofaro et al., 1994a, 1994b; Lofaro, Gibb, & Garland, 1994; Kaempf et al., 1994), were often hampered by the lack of objective MOPs. The Kaempf team, for example, evaluated expertise using a combination of experience and peer recommendations. Although the cognitive processes that may contribute to success in x-ray threat detection were fairly well understood through job and task analyses (Fobes et al., 1995; Kaempf et al., 1994; Donald, 2004), the shortcomings of validating instruments developed to assess those qualities could not be overcome without a reliable criterion.

Personnel selection instruments, although sometimes considering personality and motivational traits (Donald, 2004), characteristically have emphasized cognitive abilities required in x-ray image analysis (Rubinstein, 2001; Fobes et al., 1995; Donald, 2004). Cognitive assessment instruments considered thus far include visual analysis, recognition of anomalies, sustained vigilance (Donald, 2004), flexibility of closure, perceptual speed (Lofaro et al., 1994a, 1994b; Lofaro, Gibb & Garland, 1994), field dependence–independence, pattern recognition (Fobes et al., 1995), perceptual rotation, cognitive extrapolation of object identification from "slices," cognitive dissection, and cognitive integration (Smolensky, Gibb, Banarjee, & Rajendram, 1996). Consequently, the vast majority of effort and interest has been in the development of selection tools that have predictive validity with x-ray threat detection performance.

The TIP system serves two basic functions in airport security screener selection: (1) the database can be queried to generate job performance measures for comparison against selection test performance metrics, and (2) performance to specific TIP images can be compared with specific cognitive traits (i.e., spatial rotation—as the TIP system encompasses thousands of threat images from multiple visual perspectives ranging from the canonical to various rotations about three axes). Such research efforts were not possible until recently, as we had no suitable reliable criteria for x-ray threat detection performance.

Perhaps, the most innovative application of the TIP criteria, however, is using the data to empirically establish a foundation in constructing an x-ray threat detection training program. Because of the comprehensive content within the database structure, human factors engineers have a wealth of information available to analyze performance along multiple dimensions. On the one hand, there is objective, quantitative data to stratify or rank order screeners based on their performance. Performance data are available not just for individuals, but for threat images as well. When an image is projected, on what machine, when, to whom, and the result (e.g., hit or miss) is captured and stored. This allows the human factors professional to conduct many of the analyses traditionally associated with test construction (i.e., item analyses, discriminant analysis, validity testing, and factor analytical approaches; Anastasi, 1954). These same statistical approaches and processes have other applications as well.

Moreover, these data are well suited to building difficulty indices for various individual threat images and threat subcategories. Since the number of projections, hits, and misses is maintained for each threat image, the relative difficulty of a threat image can be estimated in much the same way as Pd is computed for screener performance—the number of hits to number of projections. While the threat image will always appear in a different bag for each projection, the factors not associated with the threat image that affect performance (e.g., bag clutter, density, bag size, and type of content) can be assumed as randomly distributed.

With thousands of projections of each threat image across all airports, in a diversity of bags, to thousands of screeners, the threat image Pd becomes an excellent measure of image difficulty level. Not surprisingly, analytical studies (limited circulation security sensitive documents) have demonstrated that the individual threat image Pds are remarkably consistent month over month. Correlation coefficients for threat image Pds have consistently exceeded $r = +0.95$ when examined across monthly reporting periods.

Together, these two important facets of performance—that of the individual screener and that of threat images—offer the possibility to isolate individuals who perform consistently above the norms while also identifying threat images that are the most difficult to detect. These data permit the human factors professional to link human performance to threat image performance, and in doing so, allow a direct comparison of how well the strongest performers detect the most difficult threats. In many respects, this is similar to identifying what items provide the best discriminability in psychological or achievement test construction. Threat images that have stable but low Pds, but that superior performers detect at high levels, are indicative of threat images that discriminate performance well. However, the goal is not to develop an x-ray threat detection performance instrument with excellent psychometric properties, but to develop a training program that builds or improves these skills.

Through the use of x-ray simulators (to display threat/bag image combinations) and cognitive task analyses techniques, the researcher has the capability to methodically capture discrete cues, strategies, approaches, techniques, and image analysis processes that highly skilled security screeners use to identify complex threats. An analysis of human performance data using a prescribed set of parameters identifies consistently outstanding performing personnel, while an analysis of threat image data allows the selection of difficult-to-detect threats. The research question is no longer whether or not screeners can detect such threat images, but *how* they detect such items. Analytical approaches have already confirmed that they are exceptional at threat detection, particularly at the most difficult of threat images. Therefore, their methods and cognitive processes become the data of interest that form the basis of an effective threat detection training program.

Although transformation of research data into training programs is probably best left to instructional systems designers, the role of human factors in such an endeavor is critical in all the phases leading up to the actual development of the training. Nevertheless, without a performance assessment system implemented, such as the TIP, both personnel selection test validation and training program development efforts as described here are not possible. Indeed, in this short section, we have demonstrated how the human factors engineer uses tools and unique methodologies to design a training program. But in reality, the training was in effect developed by those with the greatest expertise and skill—the exceptional aviation security screeners.

X-ray screening is but one technology used to protect aviation assets and the traveling public. The value of a strong transportation security system lies in the use of overlapping measures. And as with the x-ray technology, the role of human factors has been no less important.

27.7 Other Threat Detection Methods: Handheld Metal Detectors, Walk-Through Metal Detectors, and Physical and Explosive Trace Detection Screenings

27.7.1 Introduction

X-ray equipment is only one layer or process used in airport security for screening. Most travelers are well acquainted with the other mechanisms that are in place. Ever forget to remove your cell phone or keys after placing your belongings on the x-ray conveyor belt? Were you distracted or thinking about racing to the departure gate to get on the standby list for a first-class upgrade? Chances are if you travel often enough you heard that distinctive alarm on the walk-through metal detector (WTMD) and got the opportunity to spend a bit more time with screener personnel than you planned. If you were fortunate enough, or unfortunate enough depending on your perspective, to observe the other screening operations you might be asking yourself "how could human factors play a role in what appears as a procedural and highly routine task"?

Good question. First, let us expand the scope of this perspective.

27.7.2 Human Factors in Screener Performance

Human performance assessment has numerous functions (evaluating training, retention and promotion decisions, examining the efficacy of the human–machine interface, etc.); however, the focus in aviation security human factors is on the quality of performance of the individual. It is not necessarily for the purpose of rank ordering individuals, allocating work assignments, or career advancement decisions—but on initial and annual certification, and on improving performance. The certification process is to assure that the individual can perform to a standard without oversight, while improving performance has far graver consequences. Stated simply, human error in aviation security can result in the failure to detect a single threat that may in turn have immense consequences. It is, therefore, not only the totality of performance that is always of concern, but rather each individual discrete action of the process. In a sense, each tree is as important as the forest.

An illustration is in order here. Take the example of an individual who passes through the WTMD and alarms. Procedurally, the individual may be provided two opportunities to pass through the WTMD, before a handheld metal detector (HHMD; referred commonly as a wand) screening and a physical search (pat down) are required to clear the alarm. If the individual was wearing a hat, and the HHMD, when passed over the head, indicated an alarm (presence of metal); procedurally, the correct process is to identify the cause of the alarm. If an assumption is made that a metal band on the outer rim of the hat caused the alarm, that is an error. The performance error was not visually inspecting the headgear. This individual may have concealed a threat underneath the headgear that was missed. Every other aspect of the screening process could have been performed to precise specifications, and overall performance would be high, but it is that single error that could result in a security breach.

This scenario applies uniformly to all aspects of the process, whether it involves the screening of passengers or baggage. The challenge for human factors engineers is, therefore, to isolate all aspects of each process and develop appropriate instrumentation for evaluating each process—HHMD, WTMD, physical search, or ETD. Yet, the evaluation process must be standardized and suitable for performing evaluations in the operational environment.

In 1994, an Embry-Riddle Aeronautical University team (Gibb, Wabiszewski, Kelly, & Simon, 1996), using task analysis procedures and adapting an evaluation model in common use for certifying pilots, developed a simple but effective tool to assess the HHMD performance. Interestingly, the primary purpose of their work was not to develop an assessment instrument, but to evaluate the impact of a training intervention on screener performance. The assessment instrument was a tool developed to provide quantifiable data to determine the effectiveness of the training.

The assessment instrument identified all the procedural and decision elements of the HHMD process. Parameters were developed such that each element could objectively be rated as either correct or an error. The simplicity of the design used was in that no judgment was required on the part of an evaluator to "score" each element of the process—it was simply a matter of whether the action was completed or not. The evaluation could be completed in real time as the screening occurred since the sequence of the process was also standardized. The instrument could identify specific elements of human error that occurred with high frequency, or, alternatively, could be used to generate a more global MOP by comparing the number of elements performed correctly to the total number of opportunities for error with each screening process.

27.7.3 The Canadian Experience

Although the HHMD evaluation tool was used for field research in a number of studies, it was not until the Canadian Air Transport Security Authority (CATSA) developed its Point Leader course (frontline supervisory personnel) shortly after the tragic events of September 11 that such assessment tools saw operational application. Human factors, education, and instructional systems design professionals teamed to develop the CATSA Point Leader training program. Because of the role of human factors training specialists, significant emphasis was placed on training supervisors to evaluate performance, identify deficiencies, and improve performance.

This innovative program was witness to substantial strides forward in two areas: human factors training and performance monitoring by supervisory personnel. The four-day training program included topics in coaching and correcting, leadership implemented, enhancing skills, perception and communication, and performance evaluation. A number of performance assessment instruments for WTMD, HHMD, and ETD were developed to support the capability of supervisors (point leaders) to reduce human error and improve the skills of their teams. Considerable emphasis was directed toward in-class practicum, role playing, and scenarios to reinforce these newly acquired skills.

This brief section demonstrates how human factors played a significant role in both performance assessment and training development, across other areas of aviation security. It is through the assessment process, and the training of frontline personnel, that human error can be reduced. And as indicated earlier, human error equates to threat penetration in this environment with disastrous consequences.

Many of the recently deployed technologies, training programs, and practices in use today reflect major contributions from human factors engineering. Let us examine some of those influences in the next few sections.

27.7.4 Design, Test and Evaluation, Human–Machine Interface: The Legacy of Human Factors in Improving Airport Security

Perhaps, a spin-off on a recent television commercial sums it up best: We (human factors engineers) do not make the equipment for aviation security—we just make it better. Human factors engineers at the Transportation Security Laboratory in Atlantic City, NJ, and a host of colleagues in academia and private industry, take part in virtually every aspect of equipment and software deployed to airports from cradle to grave. The process often begins with the development of specifications destined to vendors, participation in the design and engineering phases, development of the training, laboratory, and operational field testing, and often culminating in performance or system measurement and later refinement of the technologies.

Modern U.S. airports deploy many sophisticated technologies to thwart terrorist activities. One such area that has exploded (no pun intended) in the past decade has been in explosives detection. Whether involving checked baggage and cargo, or passenger carry-on articles, technologies to identify dangerous explosive substances have been widely deployed at even the nation's smallest airports. The success of those systems is partly due to the human factors efforts behind them. One of the greatest success stories can be found in the recent advent of explosive trace detection devices, commonly called ETDs.

ETDs made their first appearance in the aviation security arena in the mid-1990s. Although the technology itself was not cutting edge, as gas chromatography was an established process for identification of trace amounts of substances, deployment outside relatively sterile laboratory environments was a challenge. Airport environments are laden with dust and moisture, security screeners are not trained laboratory technicians, and the test process had to be rapid and repeated countless times each day. The task was daunting—identify devices that would stand up to the rigors of public areas, could effectively be used by the workforce (including calibrating and performing routine maintenance), and yet would be effective tools in the identification of explosive substances while not impeding the flow of commerce (you and your bags). A human factors perspective translates those goals into (a) develop a methodology that empirically evaluates the devices in an operational environment, (b) determine and resolve human–machine interface issues, and (c) evaluate operational effectiveness.

27.7.5 A Final Word: Passenger Checkpoint and Checked Baggage Configurations

The design and equipment configurations of both checkpoint and checked baggage screening areas can have considerable effects on system effectiveness. Inadequate communication systems, inability to maintain chain-of-custody of passengers or their belongings, habitual patterns of "randomly" selecting items for additional inspections, failure to maintain situational awareness, and other human performance issues can provide avenues for circumventing security systems. Such impediments can often occur by how the screening areas are designed or where the equipment is located. Several excellent resources are available that provide sound methodologies for proper equipment and system design (e.g., Bailey, 1982; Van Cott & Kinkade, 1972). A host of additional environmental factors (glare, ventilation, ambient noise levels, exhaust fumes, lighting, temperature, etc.) can profoundly affect human performance as well, particularly for vigilance tasks.

One of the most dramatic illustrations of environmental effects is found in the early deployment of the CTX technology (computed tomography x-ray). Most air travelers are familiar with seeing these mini-van-sized machines in airport check-in lobbies recently. However, when the first machines were deployed, they were installed on the in-line checked baggage systems. These systems were located below passenger areas in the bowels of airports. Unfortunately, the size, the weight, and the infrastructure requirements of those machines, including the need to integrate them into the checked baggage systems relegated the equipment to areas not conducive to a visual monitoring task. In one installation, operators initially endured severe environmental conditions—heavy fumes from passing air carrier Tugs™, continuous intense noise from luggage conveyer systems, dramatic seasonal temperature changes, poor lighting, and high dust levels. It was not unusual to observe an operator wearing gloves with a space heater nearby in winter months.

The human factors engineer has several responsibilities in this area. Because of the specialized training, and an understanding of human performance, we are well suited to assist with the design phases of equipment and workspace layouts, conduct human factors audits to redress problematic areas, and to mitigate environmental factors when possible. The rapid explosion of evolving technologies used in the development of aviation security equipment will have little effect on enhancing security if there is an inadequate interface between the equipment and the operator, or if deployed in environments that are not suitable for the tasks assigned to the human operator.

27.8 A Human Factors Test and Evaluation Paradigm

27.8.1 Introduction

Clearly, the evaluation of the technology involves areas beyond the immediate scope of human factors (e.g., chemistry and process engineering); however, the fundamental principles of the test and evaluation process serve as the underlying foundation. Human factors engineers are well equipped to provide the expertise needed to move a technology through the design, evaluation, acquisition, and deployment stages.

Most efforts of this magnitude and complexity (e.g., Fox et al., 1997) begin with the development of a test and evaluation plan (TEP). The TEP details the scientific process that will be used to objectively evaluate a system and includes at a minimum—the critical operational issues and concerns (COICs) that are addressed, the MOPs, the precise methodology of the study, instrumentation, and the statistical analysis that is applied. An understanding of human factors principles is critical to thoroughly identify all the COICs and in comprehensively laying out the MOPs. Knowledge of human–machine interface issues, human error, and training, transfer of training, and so forth helps to establish a test plan that maintains the human operator as a system component. After all, nearly all systems rely on the successful interaction between the operator and the remainder of the system. The TEP has many similarities with a thesis or a dissertation proposal. It is a written record that documents our first goal—developing a methodology to evaluate a technology.

However, what distinguishes a TEP from most other research proposals is the depth of detail, detail that is the domain of human factors. To illustrate, what type of training and standardization should be afforded to the data collection team? What types of data shall be collected and how can error variance be mitigated? How will instrumentation be developed or used that minimizes subjectivity, and produces high inter-rater reliability coefficients? How is human error and efficiency defined? These are the types of questions that are addressed in any thorough TEP.

27.8.2 The Paradigm in Process

We go back to our example of evaluating ETD technologies. A number of variables must be assessed. For example, we may want to know what the potential for human error is on each device, ability to respond and interpret warnings, ease of calibration and testing, and so forth. Of course, this is an operational evaluation that will ultimately result in a procurement decision, so one must further consider cost factors, RMA elements (reliability, maintainability, availability), stakeholder considerations (length of time for processing samples from passenger bags), and technical issues (calibration drift and sensitivity). Each of the domains most likely has numerous aspects, all of which must be assessed and evaluated.

A human factors team carefully sets out all the variables that will be addressed and creatively designs each of the MOPs. In many cases, an MOP may have multiple levels. For example, is human error addressed on such a system? The question is better phrased "what can the operator do that decreases the efficacy of the system?" ETDs are extraordinary sensitive devices that require strict adherence to specific procedures to reach the maximum value. Some measures of human error could be stated as

- Is the equipment calibrated and properly tested?
- Was the sample taken properly?
- Was the sample inserted into the device properly?
- Did the screener maintain custody of the passenger?
- Did the passenger have access to the bag or the article during the testing process?
- Did the screener interpret the aural and/or visual machine information correctly?
- Was the table swabbed with alcohol after an alarm (to prevent cross-contamination)?

These illustrations are stated in a broad sense—defining the scope of human performance and error. And they are by no means comprehensive. Each area is further detailed into its specific elements. Defining "was the sample taken properly" can be described into its constituent elements, the subtasks. Often task analyses techniques are an effective method and can be accomplished using a number of sources for the data—direct observation, manufacturer specification, policy and procedure documents, etc. In our example, some possible subtasks would include sampling all required areas of a bag, providing correct information to the passenger, sampling in only one direction (vice scrubbing motion), and so forth.

Once all elements of each MOP are specified (and remember, this is for all the variables from cost factors to human performance to stakeholder considerations), it is then possible to construct the instrumentation—the tools that will be used to obtain the evaluation data. This is done in the context of maintaining standardization across the field data collection team, and simplifying the process as much as possible. Therefore, regardless of the MOP, it becomes a task in the domain of human factors engineering.

Let us illustrate. In defining our goals, we stated that operational effectiveness included not impeding the flow of commerce. Obviously, air carriers (and passengers) would have a tremendous stake in the process, and unduly long screening processes would delay passengers, and ultimately flights. So, operational effectiveness related to passenger/bag flow is a COIC that must be assessed and have associated MOPs. Some COICs could be stated as

- Does the use of ETD devices increase or decrease the flow rate of individuals clearing the screening checkpoint?
- Are there differences between ETD devices with regard to processing time completing the trace detection process?

In practice, we actually had four COICs to fully address that domain, each with one or two associated MOPs (Fox et al., 1997). The question then becomes "how does one address those questions (COICs) while maintaining standardization across a field data collection team and yet generating quantitative data that is defensible?" Ultimately, the answer must result in data collection protocols and instrumentation!

These rather basic COICs were chosen to illustrate the process. The first examines whether this type of technology has an impact, whereas the second literally pits devices of a similar technology against one another. Whether or not the deployment of ETDs has an impact on passenger flow rate is easily measured, as there are several approaches to define this impact. Some metrics could include the average time required for a passenger to transit the security checkpoint, the processing time for target inspection items (e.g., electronics), the length of the passenger queue before entering the checkpoints, and so forth. Each of the metrics can be assessed empirically in quantitative terms. In addressing the first COIC, the data are obtained from airport security checkpoints lanes with ETDs deployed adjacent to those using conventional legacy systems in place. The second COIC is simply a comparison of checkpoint lanes using different ETDs. It goes without saying that a number of intervening variables must be controlled for or counterbalanced (e.g., passenger traffic flow adjusted for heavy and light periods, types of bags, and staffing levels).

Thus far, we have demonstrated how defensible quantitative data may be obtained, but how do we achieve standardization in the data collection process? The time it takes a passenger to transit the security checkpoint appears straightforward, arm data collection teams with stopwatches and clipboard to record the data. However, although the timepieces may assure standardization for the underlying scale of time (e.g., seconds), this does not assure that there is standardization across each measurement taken. When is the data for this metric initiated? When is it terminated? The parameters of the measurement must be specified and easily identified. In this case, the initiation was set as the moment the passenger stepped over a temporary colored tape placed on the floor for the purposes of the study. The termination of the transit time was identified as when a screener released the passenger from the screening process (identified by a specific verbal instruction to the passenger). A similar protocol was designed for the time required to inspect a bag using both conventional and ETD search techniques (e.g., the bag check call from the x-ray position to release of the bag to the passenger).

We have demonstrated, using a rather elementary pair of COICs, how to (a) define appropriate MOPs, (b) implement appropriate metrics, and (c) define the parameters and protocols. This cycle is repeated until the full gamut of all COICs has been addressed. This provides the guidance needed to develop appropriate data collection forms (or software on Palm Pilots to build the database simultaneously with the data collection process). With these elements in place, training the field data collection teams becomes straightforward. The goal is to design metrics and protocols, as we illustrated here, such that the process in the operational field test environment is one of recording data, rather than interpreting events and interjecting subjectivity into the process. The findings of particular field evaluation and procurement efforts can be found elsewhere (Dickey et al., 1998a, 1998b).

The evaluation process highlighted numerous human error and human–machine interface issues quite inadvertently, and consequently without an initial plan to obtain such data. Although the human factors professional can anticipate human error in any new system or component, it is not always possible to foresee the nature of those problems until prototypes are first deployed and used by operators in the operational environment. So goes the story with the evaluation and the acquisition of ETDs into airport security. A policy decision was made, in consultation with human factors specialists, engineers, and technicians, to begin the process with a beta test. Vendors submitted devices that met minimum requirements, provided some initial training to operators, and installed the equipment at a handful of test airport checkpoints. The beta tests were the first introduction of the ETD technology into airports. Rather than commence an evaluation effort that results directly in a procurement decision, beta tests allow for identifying system integration issues, addressing those challenges, improving the devices, and then moving forward with a more complete test and evaluation process.

27.8.3 ETD: Findings

Although this chapter is not the appropriate mechanism to discuss the full scope of all ETD human–machine interface issues in detail, it is useful to highlight some of the more profound findings. These issues were common to most of the initial seven devices included in the evaluation process:

- Lack of audible or visual signals to indicate the device alarmed on a designated substance—particularly against the ambient airport noise
- Insertion of the sample into the device orifice required unnatural hand movements, generating high proportions of invalid samples
- Devices could be operated out of calibration without any warning
- Inability to respond appropriately to alarms because of display issues
- Procedural complexities in using the devices
- Overly complex and poor procedures for calibration
- Insufficient training in operation, maintenance, and inspection
- Results that were difficult to interpret

As a result of such beta tests, and close cooperation between field test personnel, program managers, and vendor engineering staffs, many of the human–machine interface, training, and design challenges were satisfactorily resolved. This quickly led to the redevelopment and redeployment of systems that were far better suited to the intended operator group.

The successful integration of ETDs into the airport security environment was a remarkable demonstration of how human factors contributed to the design, test and evaluation, and improvement of the human–machine interface. These devices can be found at nearly every U.S. airport and are used in both carry-on and checked baggage screening. Their success in the airport security environment later transcended aviation and has found applications in other arenas. The ETDs are often found used by customs agents, courthouses, and the military.

27.8.4 Thoughts, Questions, and Issues

We have yet to come full circle with our application in this section of the chapter. At the beginning of this section, we discussed three goals: develop a methodology that empirically evaluates the devices in an operational environment, determine and resolve human–machine interface issues, and evaluate operational effectiveness. We examined in tandem structuring a methodology and implementing our protocols in the operational environment to obtain the empirical data, and examined at least some aspects of "effectiveness," but we have yet to discuss the role of human factors in the human–machine interface issues. After all, doesn't our role extend beyond the evaluation process into the domain of improving a technology? Could our evaluation process highlight interface issues specific to the technology? Could we determine if the workforce successfully uses the technology? Are there differences in the potential for human error between various ETD devices? By raising these issues, it is then possible to formulate specific MOPs, develop protocols and instruments to obtain these data, and once again provide a basis for evaluation.

27.9 Aviation Security Training: Contributions of Human Factors

27.9.1 Overview

Until now, training has only appeared briefly in the discussions of aviation security human factors. What role has human factors engineering played in the training of airport security personnel? Generally, when human factors engineers have been associated with training issues, the focus is more often than not on applied research, development of training or simulation systems, evaluation, or performing the related task and/or job analyses. In the next section, we will explore our role in the aviation security training arena.

27.9.2 SPEARS and SET: From 1990 until Today

Much attention was devoted to the TIP and in many respects this system is a powerful training tool. But contributions toward aviation security training extend far beyond the TIP system. In the past decade, human factors specialists have been intimately involved in building simulation systems (Schwaninger, 2003b), creating supervisor courses (SET; Cammaroto et al., 1996), developing screener, supervisor, and management training (CATSA; Gibb, Gibb, Owens, Poryzees, & Miller, 2003; Gibb & Lofaro, 1994), and establishing continuous performance improvement programs (SPEARS; Fobes & Lofaro, 1995). Numerous empirical, analytical, field, and task analysis studies supported many of these efforts.

In the mid-1990s, several research and development efforts were underway targeting the improvement of screener performance and training. Before continuing, the reader is reminded that before 2002, private security contractors predominantly staffed the screener workforce, numbering around 19,000 personnel in the United States. Employee turnover was extraordinarily high and presented a challenging hurdle to overcome. After all, what value do training and performance enhancements have in a revolving door personnel system when more experienced screeners attrite and are replaced by a constant flow of novices? The use of a federal employee workforce has aided significantly in stemming the problem of screener retention. And those early efforts a decade earlier formed were not in vain, as they provided the foundation for implementations that are successful today with a more stable workforce.

Two of those efforts from the 1990s are worth discussing—SPEARS, or the Screener Proficiency Evaluation and Reporting System, and SET, an acronym for Supervisor Effectiveness Training (Fobes et al., 1995). SPEARS evolved as a mandate from the U.S. Congress to the FAA to enhance airport security by improving human performance. The Congress had specifically identified and directed the use of human factors engineering solutions to achieve that goal. SET, however, evolved out of a human factors research effort to identify and quantify the causes of airport security employee turnover (Lofaro & Gibb, 1994). Through the use of scaling methodologies and a modified Delphi technique (Lofaro, 1992), poor supervisor skills and relationships were identified as major contributors to screener turnover.

SPEARS was instrumental in laying the groundwork for much of the improvements that followed in the next 10 years. Under the SPEARS banner, several key tasks were accomplished including (a) detailing a function flow analysis of the checkpoint processes, (b) preparing a task analysis for the x-ray screening function, (c) compiling all available aviation security literature, and (d) obtaining user perspectives through an extensive series of interviews. The methodologies, tools, and information developed under this program are evident in many of the training programs implemented in the United States and Canada today.

SET marked a dramatic change in how security supervisory personnel were trained. SET was an 8-h training program that was developed to provide practical training in communication, leadership, and supervisor skills (Fobes et al., 1995). Characteristically, supervisors were selected (promoted) on the basis of longevity vice any specific training or experience in a managerial role. Consequently, they often lacked any formal supervisory training. A nationwide survey of airport security supervisors indicated that very few had previous positions or experience in such a role (Gibb et al., 1995). The program was developed by human factors engineers and had emphasized conflict resolution, principles of effective communication, goal setting, leadership models, and improving subordinate performance (Cammaroto et al., 1996). Although the program was implemented and evaluated, demonstrating its effectiveness, national implementation was never accomplished. The workforce was very much fragmented into many private security vendors, and implementation of national-level programs was not easily accomplished. Prior to September 11, there was no uniform training force as exists today.

SET, like SPEARS, were programs in their infancy and provided the foundation for things yet to come. SPEARS concepts were later integrated into systems such as TIP and off-line computer-based training programs while SET became the underlying foundation of many of the CATSA training programs for screeners, point leaders (supervisors), and managers (CATSA, 2003a, 2003b, 2003c). Both prongs of training approaches are worth a short review.

Shortly after the events of September 11, the Canadian government shifted the responsibility and oversight for airport security from Transport Canada to a newly formed Crown corporation known as the CATSA.* The CATSA is the Canadian counterpart of the U.S. DHS's TSA, although there are significant political, structural, and legal differences between the two entities. Probably, the most substantial difference is private vendors employ the Canadian airport security screener workforce, whereas the United States (with the exception of five test protocol airports) is a federalized system.

27.9.3 Again, the Canadian Experience

The CATSA was faced with enormous tasks—upgrade the training of the existing workforce to a new set of standards, and produce a new training system and courses for three different levels of screeners, point leaders, and security managers. While presenting challenges that were unprecedented, these tasks also offered opportunities for the human factors community that is rare. It afforded the freedom to design, develop, and implement training programs that could capitalize on nearly 20 years of research in aviation security human factors.

Security screener courses, in addition to providing critical skills and knowledge in screening procedures, emphasized a team approach to improving effectiveness and efficiency. Considerable courseware was dedicated toward conflict management, listening skills, teamwork, situational awareness, and mentoring. In essence, there was a powerful element of interaction skills to supplement technical training.

---

* A Crown corporation is an entity owned by the government but is self-supporting. The U.S. Postal Service is a similar entity in the United States.

Each of the three levels of screener training moved the trainee into higher levels of acting in the role of a team leader. Combined, the three levels of training constituted 8 days of formal training that were distributed over a several-month period with intervening on-the-job training and performance assessment.

However, the most significant introduction of human factors principles was witnessed at the point leader and management levels of training. Point leaders, charged with the direct supervision of screeners, were provided with extensive training in perception, leadership, coaching and correcting, performance evaluation, and enhancing the skills of subordinates. This was a landmark program in that these individuals learned and practiced skills that were specific to identifying performance deficiencies and correcting them. The evaluation tools that were provided, actually evolved from assessment tools developed for research purposes!

Management courses included in-depth training in performance management, team building, team performance assessment, continuous improvement processes, goal setting and planning, and leadership. But perhaps, the most innovative aspect of the 6 day program was the implementation of crew resource management (CRM) for security checkpoints. Adapting principles and concepts developed for both air carrier and U.S. Navy programs, the CRM training sought to incorporate strong teamwork elements and contribute to overall security effectiveness.

27.10 The Role of Simulation in Security Training

While these developments were taking place in the more traditional forms of training, efforts were underway in Zurich, Switzerland, in the simulation arena. Development and enhancement of x-ray threat detection skills have long been known as two of the most difficult proficiencies for screeners to master. And while experience on the job plays an important role in attaining these skills, off-line approaches were required to both develop the initial skills and provide supplemental experience.

Researchers at the University of Zurich (Schwaninger, 2002, 2003a, 2003b), building on laboratory work in visual cognition research, were instrumental in developing a simulator platform that enveloped a number of interesting characteristics. Foremost, x-ray images were presented and categorized along three primary dimensions: (a) rotation (viewpoint dependency), (b) superposition or the degree of occlusion of one object by others, and (c) complexity. These dimensions in part determine the difficulty in identifying threats by manipulating the saliency of a threat against the background of the bag contents (in the SDT framework [Green & Swets, 1988], the saliency of the signal against the noise). Second, the simulation software employed adaptive learning algorithms such that the difficulty levels would advance in response to the success of the individual trainee. Although not evaluated on a wide scale, improvements in threat detection at the Zurich International Airport have been achieved (Schwaninger, 2004). Several other simulation software platforms have been deployed over the past few years (e.g., Safe Passage, and Smart Approach Smart Screen) or are currently in development (TRX Simulator), but independent, empirical evaluations are not readily available.

Recent work at the University of Illinois (McCarley, Kramer, Wickens, Vidoni, & Booth, 2004), examining visual skill acquisition and scanning patterns of subjects trained to screen x-ray images of bags for threats, concluded that there was little evidence that practice improves the effectiveness of visual scanning. Although the simulated task was only tangentially related to the actual job (the threat set only included knives in cluttered bags), sensitivity improved and response times decreased reliably over the test sessions. Test sessions were held over a several-day period, as each session included 300 trials of 240 clear bags (no threats) and 60 threat bags each. (An average experienced screener examines approximately 175 bag images in a 30 min shift.) Their results further indicated that object recognition performance might be stimulus specific, as detection performance significantly degraded with the introduction of new threat objects. They recommended that screening training for x-ray threat detection be more appropriately focused on developing recognition skills that can be generalized across a greater spectrum of threats. Currently, the TSA has initiated new research programs into identifying the strategies, techniques, approaches, cues, image analysis feature uses, and perceptual-cognitive processes used by highly effective screeners. The availability of TIP performance data was found extremely beneficial in identifying individuals who perform consistently above the norms. Perhaps, the research exemplified in these studies in time may ultimately move screener training and simulation to a new level.

It has not been possible within the confines of a chapter section to describe in depth all the contributions to aviation security training that are directly attributable to human factors engineering research and development efforts. We, however, hope that the scope and complexity of the work discussed here has given the reader an appreciation for what has been done, and how the profession will advance transportation security in the future. One must remember that a formal human factors program in aviation security is scarcely more than 15 years old.

27.11 Domestic Passenger Profiling

27.11.1 Introduction

This chapter would be remiss and less than complete if we did not discuss passenger profiling. In the aftermath of September 11, there has been a constant stream of "experts," "analysts," "Monday morning quarterbacks," "talking heads," etc, who have produced reams of words, which purport to explain why September 11 happened, and, secondly, assure how it will not happen again. Sadly, the explanations or assurances have been less than satisfactory. The September 11 Commission, a partisan, "bi-partisan" group, which included a person as a commissioner who probably should have been a witness, produced a report, which was a best seller. In time, it may join the Warren Report as a document that did little to provide resolution. In this chapter, we do not pretend to have all the answers. Having said all this, a major point must be made. Both authors have signed confidentiality oaths as to not revealing classified data they worked with or had access to. Therefore, at times, the level of detail presented may not be completely satisfying to the reader. Be assured it is not always satisfying to the authors to provide less than a comprehensive treatment of any area.

27.11.2 History

An OTA Report (1992) discussed in some detail the role of humans in passenger profiling. The OTA Report said that there were two general approaches to passenger profiling: one comparing passenger demographic, and the other the background data (age, sex, nationality, travel itinerary, etc.) to historic or recent, intelligence-driven "threat profiles." The other is based on the examiner's psychological assessment of the passenger, taking into account nervousness, hostility, or other suspicious characteristics. It is instructive to note how this OTA Report was both responded to, and then ignored.

In response to this Report and the recommendation that R&D in profiling should be done by the FAA, the FAA's ASHFP was involved, 1993–1994, in an effort with a major air carrier to develop a two-pronged passenger profiling system, to be called the CAPPS (computer-assisted passenger profiling system). One prong was termed the manual domestic passenger profiling system (MDPPS). On the one hand, passengers who were in frequent flyer programs and about whom the carrier had some (long-term) background information, were identified by a computer program and given a "pass" to bypass additional screening before boarding. This further screening would have involved such things as the HHMD screening, physical searches, open baggage inspections, etc. On the other hand, the DMPPS was designed to develop and use an initial profile of passengers who required additional screening. The concept was application of a quick checklist to each passenger to see if the need existed for what was called "special handling" or additional security.

Before we go further: a passenger profile, containing elements that over-zealous civil rights activists seem to object to, can have many forms, but only one goal—safety. Many profiles are a combination of behavioral descriptors (avoids eye contact and prowls the boarding area) and objective data, such as buying a one-way ticket or paying cash. Some profiles have a rank-order approach with a red-flag total, e.g., any passenger who has two or three of the first five behavioral or objective data descriptors on the profile must be given special handling. Some profiles simply put numerical weights on the descriptors and assign a numerical total above which the potential passenger is to have special handling. In all cases, however, any objective data must include race, gender, and age. Why? As the OTA Report in 1992 said: Because these are known terrorist traits; they come from past experiences and historical incidents and have not changed. For 30 years, the profile of a terrorist was—and still is—a young male of Middle-Eastern descent, usually under 40.

To return: The FAA/air carrier profiling effort of the early 1990s was done in conjunction with FBI and INS agents who provided the descriptors and provided the rank ordering. The profiling system was to be a combination of the passive (frequent flyer type data) and the proactive (a checklist of descriptors with a red-flag total) approach. This system was developed in a rough form and a feasibility test was conducted. At that point, to go further required high-level FAA and Secretary of Transportation approval. The project was not continued, as it was deemed discriminatory. Thus, the OTA Report was now to be ignored.

The FAA/air carrier profile developed in the early 1990s, had the potential to become a valuable tool in detecting 19 of the 20 September 11 terrorists, as it would have identified them as requiring special handling, especially given that there were four or five such individuals on each of the three ill-fated flights. We live in a free society, one which places a considerable value on personal freedoms. However, it may also defeat some of the tools and techniques put in place to protect citizens. We are glad to report that the CAPPS seems to have been recently revivified and is termed CAPPS II. Nevertheless, no matter how good our screeners become at finding weapons and IEDs, the last resort is the ability to stop the persons who are terrorists bent on destroying life from boarding a plane. To deny placing profiling as a weapon in the safety arsenal, seems to undermine a significant part of civil aviation security.

27.12 General Aviation: The New Threat in the Millennium?

27.12.1 Overview

The first point to be made is what the FAA classifies as General Aviation (GA). It is a definition by exclusion: the GA is everything that is nonmilitary and everything where there are no revenue-paying passengers (civil aviation). This means that the GA includes everything from balloons and blimps to Cessna 172's to business jets (bizjets) and corporate jets. And, there is the rub. A twofold threat exists as bizjets and corporate jets are in the GA category. The first part of the threat revolves around the size and the speed of such jets. They are fairly heavy, fast, and can carry a large amount of fuel, meaning they can do considerable damage if used as a flying bomb. They can also carry a significant amount of cargo. If that "cargo" is explosives, the damage possibilities increase geometrically. A typical mid-size bizjet is the Lear 145; it can take off weighing 20,000 lb and fly at nearly 500 mph. The Sabreliner variant used by the military can take off weighing 18,500 lb and fly at nearly 450 mph. It should be apparent that the damage that one of these bizjets, with or without explosives, can do to buildings, bridges, tunnels, as well as the occupants and users of such structures is significant. While not in the weight/size class of a B-767, a Lear or Saberliner bizjet flying at 450 mph can be a much more lethal terrorist weapon than what the public considers a typical GA plane: a slow (say 120 mph), lightweight, piston-engine, high-wing monoplane such as a Cessna 172 or a Piper Super Club.

At this point, the reader may say, but, there are security measures in place at airports to stop this use of GA aircraft by terrorists. Let us answer that, for it is the second part of the threat posed by GA aircraft. Using the FAA Administrator's Fact Book, we see that, while there are 539 "certificated" civil aviation airports with full-up security measures (ranging from an O'Hare to a Norfolk), there are also 4,000 public-use airports with minimal if any security and there are 14,000 private-use airports with no security to speak of at all. In the private-/public-use airport mix, some 4,500 have paved runways, which may make it possible for them to be used by a bizjet. These airports have nothing, security-wise, resembling a certificated airport. Aside, it is possible to use some of the "larger" twin-engine, nonjet GA craft as weapons.

While they lack the size and the speed of a bizjet, they can be deadly if loaded with explosives. However, it is the size and the speed of a bizjet that cause concern.

To return to the public-/private-use airports: While corporate and bizjets often use one of the 539 certificated airports, most frequently their operations are in a separate/separated part of the field (sometimes called a GA annex) and are not covered by the security systems (portals et al.) in place in the civil aviation portion of the airport. The fact is that the security level at GA operations ranges from nonexistent to low. If terrorists could access a bizjet near an urban center, its speed would almost surely preclude it being intercepted before it wreaked havoc on some part of that urban center.

27.12.2 Current Status

In 2002, the FAA used a consortium of researchers at universities from Alaska to Florida, to try and get a handle on the actual security/threat issues posed by the GA, look at possible solutions, and make recommendations. The Report that came out of this effort looked at the GA gamut…from what was the definition of the GA, to the numbers of aircraft in the GA, to security measures in place at GA airports. An input was solicited for organizations that had a stake in this effort (RAA, AOPA, GAMA et al.). The Report was turned over to the TSA. It can be safely said that the state of the GA security was low; the interested organizations were aware of that and were not "happy" with it…they hoped the FAA and/or the TSA would fund security equipment/personnel, etc., to close the GA security gaps. At this time and to our knowledge, there has been no action taken as to the GA security.

The May 12 incursion into one of the most closely guarded and restricted airspaces, the U.S. Capital airspace, by a (GA) Cessna 150 is a telling commentary. We were all witness to scenes of hordes of panicked people rushing through the streets of Washington, DC. We saw, on TV, two F-16s poised to shoot down the Cessna. Yet, it had penetrated the restricted airspace already and, were it faster (say a Lear or a Saberliner typejet), and in the hands of terrorists, it would have been able to wreak destruction.

27.13 Conclusions and Recommendations

We begin with the caveat that this chapter was written at a time when changes in civil aviation security systems are proceeding apace. Further, it is certain that terrorists want to and will strike again. Therefore, what we recommend may have already occurred by the time this chapter gets printed or, due to advances and changes, may be, as they say, "overcome by events." The conclusions seem easy…the aviation security system is highly dependent on the skills and the motivation of the screeners. Reports from two previous presidential commissions have reiterated these points forcefully (1990, 1997). The GAO (2003) has indicated that these are key areas of concern in their recent audit of the civil aviation security system. That takes us squarely in the realm of human factors.

Finally, we do not have the luxury of a long time-frame in which to accomplish this. Again, as they say, it needs to be done yesterday. The alternative is not only to suffer another September 11, but to suffer many of them. The resultant negative impact on the flying public's confidence, indeed, the confidence and morale of the American people, is incalculable. As with the events of September 2001, the air carriers will suffer economically and the entire economy will again go into a tailspin. Having painted that gloomy but accurate prognosis—what does human factors have to offer to preclude such events?

The chapter terminates not with a conclusion, for there is none, but with a challenge. The history of aviation security human factors is short, emerging from circumstances of worldwide tensions, and the challenge extended is to contribute the skills, the talents, and the innovations to address the complexities of building an effective civil aviation security system. There is much work that remains ahead. At this point, the selection, training, and performance assessments of these screeners require a careful study, enhancement, and a renewed emphasis. In a system where most of the current tasks are constrained by human limitations, the development of pioneering threat detection technologies less reliant on visual search and monitoring is mandatory. This implies a continual relationship with manufacturer engineering staffs from the conceptual through the deployment phases. The requirements to design, deliver, evaluate, and refine superior training programs for all levels of system users have never been greater. The impact of environment factors on performance (e.g., glare, vibration, ambient noise, and so forth) is not well understood. Nor do we appreciate, or have we quantified, the effects of shift length, work cycles, fatigue, currency, or other job elements on threat detection performance. The consequences of a possible reprivatization on workforce turnover are unknown. The TIP and other simulation systems are barely beyond their infancy, and considerable work is needed to produce more effective second and subsequent generation systems. Such are the challenges that demand the contributions of the field of human factors engineering.

References

Anastasi, A. (1954). *Psychological testing.* New York: Macmillan Publishing Company.

Bailey, R. (1982). *Human performance engineering: A guide for system designers.* Englewood Cliffs, NJ: Prentice-Hall Publishing.

Barrientos, J. (2002). The effect of threat image projection event rates on airport security screener performance. Master's thesis, Embry-Riddle Aeronautical University, Daytona Beach, FL.

Barrientos, J. et al. (2002). *Functional requirements for threat image projection systems on x-ray machines* (Technical Report No. DOT/FAA/AR-01/109). U.S. Department of Transportation, Atlantic City International Airport, NJ: DOT/William J. Hughes Technical Center.

Barrientos, J., Dixon, M., & Fobes, J. (2000). *Test and evaluation report for the optimal threat image projection ratio study* (Technical Report No. DOT/FAA/AR-01/109). U.S. Department of Transportation, Atlantic City International Airport, NJ: DOT/William J. Hughes Technical Center.

Cammaroto, R. et al. (1996). *The supervisor effectiveness and training program: Description and curriculum.* Washington, DC: Office of Aviation Research.

Davies, D. R., & Parasuraman, R. (1982). *The psychology of vigilance.* San Diego, CA: Academic Press.

Dickey, R. et al. (1998a). *Test and evaluation report on the comparison of explosive detection devices at multiple airport locations.* Washington, DC: U.S. Department of Transportation, Federal Aviation Administration.

Dickey, R. et al. (1998b). *Conduct of a study to evaluate traced detection devices at multiple airport locations.* Washington, DC: U.S. Department of Transportation, Federal Aviation Administration.

Donald, C. (2004, July/August). Enhancing aviation X-ray screening: Through human factors applications. *Intersec: The Journal of International Security.*

Drury, C. G., & Sinclair, M. A. (1983). Human and machine performance in an inspection task. *Human Factors, 25,* 391–399.

Fobes, J. et al. (1995). *Airport security screener and checkpoint security supervisor training workshops* (Technical Report No. DOT/FAA/AR-95/35). U.S. Department of Transportation, Atlantic City International Airport, NJ: DOT/William J. Hughes Technical Center.

Fobes, J. et al. (1995). *Initial development of selection instruments for evaluating applicants as potential airline passenger baggage screeners for conventional X-ray technology* (Technical Report No. DOT/FAA/CT-95/41). U.S. Department of Transportation, Atlantic City International Airport, NJ: DOT/William J. Hughes Technical Center.

Fobes, J. L., Lofaro, R. J., Berkowitz, N., Dolan, J., & Fischer, D. S. (1994). *Test and evaluation plan for the manual domestic passive profiling system (MDPPS)* (Technical Report No. DOT/FAA/CT-94/22). U.S. Department of Transportation, Federal Aviation Administration, Atlantic City International Airport, NJ: DOT/William J. Hughes Technical Center.

Fobes, J. L., Lofaro, R. J., Berkowitz, N., Dolan, J., & Fischer, D. S. (1994). *Test and evaluation report for the manual domestic passive profiling system (MDPPS)* (Technical Report No. DOT/FAA/CT-94/86). U.S. Department of Transportation, Federal Aviation Administration, Atlantic City International Airport, NJ: DOT/William J. Hughes Technical Center.

Fobes, J., & Lofaro, R. (1994). *Test and evaluation report for improvised explosive detection systems* (DOT/FAA/CT-94/112). U.S. Department of Transportation, Federal Aviation Administration, Atlantic City International Airport, NJ: DOT/William J. Hughes Technical Center.

Fobes, J., & Lofaro, R. (1995). Screener proficiency evaluation and reporting system (SPEARS): Human factors functional requirements and procedures. Human Factors Program Aviation Security Research and Development Service and the Lawrence Livermore National Laboratory of the University of California.

Fox, F. et al. (1997). Test and evaluation plan for the explosive trace detection equipment study. Washington, DC: U.S. Department of Transportation, Federal Aviation Administration.

Gibb, G. et al. (1995). *Job satisfiers and dissatisfiers of airport passenger baggage screeners: Identifying causes of employee turnover.* Daytona Beach, FL: Embry-Riddle Aeronautical University.

Gibb, G., & Lofaro, R. (1994). *Airport security screener and checkpoint security supervisor training workshop* (DOT/FAA/CT-94/109). U.S. Department of Transportation, Federal Aviation Administration, Atlantic City International Airport, NJ: DOT/William J. Hughes Technical Center.

Gibb, G., Wabiszewski, C., Kelly, S., & Simon, X. (1996). *An evaluation of the supervisor effectiveness training program.* U.S. Department of Transportation, Federal Aviation Administration, Atlantic City International Airport, NJ: DOT/William J. Hughes Technical Center.

Gibb, G., Gibb, K., Owens, T., Poryzees, C., & Miller, J. (2003b). *CATSA National Training Program: Pre-Board Screening Manager Curriculum.* Ottawa, Ontario, Canada: Canadian Air Transport Security Authority.

Gibb, G., Gibb, K., Owens, T., Poryzees, C., & Miller, J. (2003c). *CATSA National Training Program: Screener L1–L3 Curriculums.* Ottawa, Ontario, Canada: Canadian Air Transport Security Authority.

Government Accounting Office. (2003). *Airport passenger screening: Preliminary observations on progress made and challenges remaining* (GAO-03-1173). Washington, DC: GAO.

Government Accounting Office. (2005). *Aviation security: Screener training and performance measurement strengthened, but more work remains* (GAO-05-457). Washington, DC: GAO.

Green, D., & Swets, J. (1988). *Signal detection theory and psychophysics* (2nd ed.). Los Altos, CA: Peninsula Publishing.

Kaempf, G., Klinger, D., & Hutton, R. (1996). Performance measurement for airport security personnel. Klein Associates Inc, Fairborn, OH. Final report prepared for Embry-Riddle Aeronautical University, Center for Aviation/Aerospace Research, Daytona Beach, FL.

Kaempf, G., Klinger, D., & Wolf, S. (1994). Development of decision-centered interventions for airport security checkpoints. Final report prepared under contract no DTRS-57-93-C-00129 for the Federal Aviation Administration. Klein Associates Inc., Fairborn, OH.

Lofaro, R. (1992). A small group Delphi paradigm. *Human Factors Society Bulletin, 35,* 2.

Lofaro, R. (1999). Human factors in civil aviation security. In D. J. Garland, J. Wise, & V. D. Hopkin (Eds.), *Handbook of aviation human factors* (1st ed.). Mahwah, NJ: Lawrence Erlbaum Publishers.

Lofaro, R. et al. (1994a). *A protocol for selecting airline passenger baggage screeners* (Technical Report No. DOT/FAA/CT-94/107). U.S. Department of Transportation, Atlantic City International Airport, NJ: DOT/William J. Hughes Technical Center.

Lofaro, R. et al. (1994b). *Review of the literature related to screening airline passenger baggage* (Technical Report No. DOT/FAA/CT-94/108). U.S. Department of Transportation, Atlantic City International Airport, NJ: DOT/William J. Hughes Technical Center.

Lofaro, R., & Gibb, G. (1994). *Job tenure factors for airline passenger screener personnel* (Technical Report No. DOT/FAA/CT-94/109). U.S. Department of Transportation, Atlantic City International Airport, NJ: DOT/William J. Hughes Technical Center.

Lofaro, R., Gibb, G., & Garland, D. (1994c). *A protocol for selecting airline passenger baggage screeners* (DOT/FAA/CT-94/110). U.S. Department of Transportation, Federal Aviation Administration, Atlantic City International Airport, NJ: DOT/William J. Hughes Technical Center.

McCarley, J., Kramer, A., Wickens, C., Vidoni, E., & Booth, W. (2004). Visual skills in airport security screening, *Psychological Science, 15*, 302–306.

McClumpha, A., James, M., Hellier, E., & Hawkins, R. (1994). *Human factors in X-ray baggage screening* (IAM Report 720). Farnborough, Hampshire, U.K.: Defence Research Agency Center for Human Sciences.

Neiderman, E. (1997). *Test and evaluation plan for airport demonstration of selection tests for x-ray operators* (Technical Report No. DOT/FAA/AR-97/29). U.S. Department of Transportation. Atlantic City International Airport, NJ: DOT/William J. Hughes Technical Center.

Neiderman, E., Fobes, J., Barrientos, J., & Klock, B. (1997). *Functional requirements for threat image projection systems on x-ray machines* (Technical Report No. DOT/FAA/AR-97/67). U.S. Department of Transportation, Atlantic City International Airport, NJ: DOT/William J. Hughes Technical Center.

Office of Inspector General, Department of Homeland Security. (2005, March). *Follow-up audit of passenger and baggage screening procedures at domestic airports* (Report No. OIG-05-16). Washington, DC: U.S. Department of Homeland Security.

Report of the President's Commission on Aviation Safety and Security, Government Printing Office, February 1997.

Report of the President's Commission on Aviation Security and Terrorism, Government Printing Office, May, 1990.

Rubinstein, J. (2001). *Test and evaluation plan: X-ray image screener selection test* (Technical Report No. DOT/FAA/AR-01/47). U.S. Department of Transportation, Atlantic City International Airport, NJ: DOT/William J. Hughes Technical Center.

Schwaninger, A. (2002). Visual cognition. *Airport, 3*.

Schwaninger, A. (2003a). Evaluation and selection of airport security screeners. *Airport, 2*, 14–15.

Schwaninger, A. (2003b). Training of airport security screeners. *Airport, 5*, 11–13.

Schwaninger, A. (2004). The enhancement of human factors. *Airport*, 30–36.

Smolensky, M., Gibb, G., Banarjee, A., & Rajendram, P. (1996). Initial development of selection instruments for evaluating applicants as potential airline passenger baggage screeners using CT-Scan technology. Report prepared for Aviation Security Human Factors Research and Development Division, Federal Aviation Administration, Atlantic City, NJ.

U.S. Congress, Office of Technology Assessment. (1991, July). *Technology against terrorism: the federal effort* (Report No. OTA-ISC-511). Washington, DC: U.S. Government Printing Office.

U.S. Congress, Office of Technology Assessment. (1992, January). *Technology against terrorism: Structuring security* (Report No. OTA-ISC-481). Washington, DC: U.S. Government Printing Office.

United States General Accounting Office. (2003). Airport passenger screening: Preliminary observations on progress made and challenges remaining. Report to the Chairman, Subcommittee on Aviation, Committee on Transportation and Infrastructure, U.S. House of Representatives, Washington, DC.

Van Cott, H., & Kinkade, R. (1972). *Human engineering guide to equipment design*. Washington, DC: McGraw-Hill Company.

28

Incident and Accident Investigation

Sue Baker
U.K. Civil Aviation Authority

Any discussion on aviation-related incident and accident investigation invariably prompts a number of questions, many of which raise fairly fundamental issues about the nature and purpose of the investigation process. For example, should time and resources be expended on the investigations of incidents rather than focusing all the effort on the major aviation accidents? What is the underlying purpose of investigations and who should conduct them, and, if a full-scale field investigation is conducted, what benefits can be gained from this as against a more limited and less resource intensive "desk-top" enquiry? One of the aims of this chapter will be an attempt to answer these questions and to consider, in some detail, the practice and process of investigation in the aviation sphere. The information on which this chapter is based is drawn from first-hand experience of the investigation of air-traffic control (ATC)-related incidents and accidents in the United Kingdom, but it seems reasonable to assume that the points raised have a general application extending beyond the ATC area or any one particular state.

To convey an insight into what incident investigation is and what it does, it may be helpful to consider what incident investigation is not. First and foremost, it should not be an exercise in the apportioning of blame. The individual does not work in a vacuum. Mistakes are made in the context of the system, and, unless the system itself is considered during an investigation, the whole process is likely to be of dubious value. Blaming and/or punishing an individual serves no valuable function for the person concerned. All these may only maintain the status quo and thus, the circumstances under which further errors may occur, doing little or nothing to rectify the shortcomings or prevent future occurrences. A report published by the U.K. Air Accident Investigation Branch in 1990 illustrates the point. In the accident in question, a BAC 1-11 had been inadvertently fitted with the wrong-sized windscreen retaining bolts during maintenance. At around 17,000 ft, the affected windscreen separated from the aircraft, and, in the ensuing de-pressurization, the pilot was partially sucked through the gap. The accident gave rise to a number of human-factors concerns regarding the maintenance procedures that are outside the scope of this chapter. However, what is of direct interest here is the manner in which this admittedly highly unusual situation was handled by one of the air-traffic controllers involved. Specifically, doubts were cast on the quality of the training received by the controller in incident and accident handling. The recommendations of the subsequent report, backed up by the data from other, less serious occurrences, led to a reappraisal and reconfiguration of emergency training for controllers within the United Kingdom. The point is that data from this accident, together with that gathered from other occurrences, did not point to a negligent or blameworthy controller, but rather to a system deficiency that needed to be rectified if further similar events were to be avoided. However, to lay the responsibility for every error at the door of the system is as ill-advised in terms of an investigation as to blame everything on the individual. Generally, to ignore the system-related factors is to ignore the opportunity to make the system-based improvements with respect to the whole organization and its staff, and not just to provide a stop gap, quick fix, on an individual level. Undoubtedly, problems will be discovered and individual errors can be found, but these need to be viewed as chances for improvement, and not as opportunities for punishment. Perhaps, somewhat paradoxically, incident and accident investigation need not only discover information on deficiencies in the system, and the lessons learned from the successful handling of an incident or accident are equally valuable. A recent investigation involving an aircraft with engine problems seeking a diversion for a speedy landing was skillfully and expeditiously handled by a trainee controller, who had recently undergone a period of training in the handling of emergencies in accordance with the recommendation made in the BAC 1-11 accident report. It is important that successful performance be given as much "publicity" as inadequate performance, not only for its motivating effect, but also because it illustrates that improvements can be made to existing systems.

Thus, incident investigation is not a justification for punishment. Equally, it is not, or at least should not be, simply an academic data-gathering exercise. The collection and analysis of data, together with the knowledge gained and conclusions drawn about individual and system performance and problems, should be undertaken with the aim of improving flight safety. Therefore, the provision of accurate and adequate feedback on the lessons learned from the investigations is vitally important.

It has now become a truism that incidents and accidents tend to be the result of a chain of causal events and/or contributory events. To look at these chains is to describe the system, and not the individual. Human-factors input is of value, because it can be one of the ways in which the scope of error causation is extended from the so-called person at the sharp end, to a consideration of the wider aspects underlying the organization and its structure and function.

It has been suggested (ICAO Circular 247-AN/148) that this extension of emphasis shifts the "blame" for incident causation from the individual who perpetrated the visible error to the decisions made at management level. The logical extension of this, it is suggested, is a failure to recognize or accept individual culpability or responsibility, as the onus for all incidents and accidents could be firmly laid at the door of management. However, this line of argument misses the point.

Individuals do make errors, sometimes without any evident predisposing factors in the system. There is also little doubt that on some, fortunately rare, occasions, individuals or groups will deliberately violate rules and procedures (Reason, 1989). Such a situation, once discovered, obviously requires remedial action. However, remediation at the individual level is only ever going to prove of limited value. At best, it may help the individual to mend his ways, but it is likely to do little in terms of future prevention in more general terms. The individual exists in the system, and overlooking the possibility of system-based antecedents in error occurrence is to overlook the opportunity to take more far-reaching preventative measures. The major point has to be, however, that the investigator should not approach the investigation with preconceived ideas regarding the causal factors nor attempt to validate some existing, perhaps prematurely formed hypothesis or pet theory. The presence of more than one human-factors specialist in the team may also help to ensure that the conclusions reached are not a function of one individual's perspective. The opportunity to discuss incidents and accident data with peers and to "bounce" ideas off colleagues goes some way toward preventing an idiosyncratic approach.

28.1 Incidents vs. Accidents

The decision to investigate incidents as well as accidents should not be taken lightly. The investigation of incidents and accidents is a specialized, resource-intensive activity. It is reasonable to ask whether the end justifies the means in terms of any benefits gained when seen against the outlay of resources. It has been asserted that "an incident investigation can often produce better accident prevention results than can an accident investigation" (ICAO Circular 240-AN/144). If this assertion is true and many investigators believe that it is, then the cost and effort involved in the investigation of incidents must still be justified.

The first and most obvious reason for investigating incidents as well as accidents is that there are more of them. This allows the investigators to build up a wider picture of the problems encountered and also to gain an understanding of any trends. A database developed in this way gives the investigator a baseline to assess whether subsequent occurrences are a further indication of a known problem or are an unfortunate "one-off." The more data available, the firmer is the basis on which conclusions and decisions can be made.

From the human-factors perspective, the behavior manifested by the individuals or groups involved in incidents may not differ greatly from that observed in accident scenarios. This has certainly been the case in the United Kingdom. Taken for granted, the gravity of an accident will add another dimension to the situation in which the controller or pilot finds him or herself, but, generally, the cognitive failures, problems in decision making, communications breakdown, distraction, and all the other factors which contribute to the sum total of behavior in an accident, will also be present in the incidents. As the major reason for investigation is the promotion of lessons learned to prevent future similar occurrences, knowledge gathered before an accident occurs can be seen to justify the effort and resources expended.

It could possibly be argued that a thorough investigation of a small number of accidents would yield data of such "quality" that decisions could be made on the basis of this small, but detailed data set. It is certainly true that generalizable lessons can be learned from accident investigations, but it is also true that the focusing of attention on such a limited set of occurrences may overlook the opportunities offered by the incident investigation to prevent such accidents in the first place. In addition, homing in on a limited number of instances does not provide the type of overall picture of system health which can be gained through more numerous, but still rigorous, incident investigations.

28.2 Data Quality

Although the need to investigate the human-factors aspects of incidents and accidents is gaining wider acceptance, there is still a degree of apprehension in some quarters resulting from the perception of human-factors findings as "speculative" and the assessment of human factors data as being of a lower order of credibility than more "physical" data, such as instrument readings, cockpit voice recordings (CVR), engine damage, or even body parts. While human-factors data are viewed in this light, reports are likely to present an incomplete account of the antecedents of the incidents and accidents. What is worse is that, human-factors issues that are left uninvestigated and unaddressed can form no part of the lessons learned for the future. While it is true that the evidence associated with human-factors findings may not be as tangible in some respects, as illustrated by the data described earlier, investigation of human-factors issues is invaluable in shedding light, not only on what occurred, but also why it occurred, especially when the event involved human error and not just mechanical failure.

A full-scale investigation, looking at all the aspects, including human factors, can provide the optimum opportunity for the collection of good quality data. In most aviation incidents, a wide range of information sources is available to the investigators, for example, radiotelephony (RTF) recordings and transcripts, video/CD recordings of radar displays, and controller and pilot reports. When coupled with visits to the units concerned and face-to-face interviews with the personnel involved, a picture of the whole context in which an incident or accident actually occurred can be obtained. This broader picture is essential to a system-based approach, allowing for a consideration of each of the factors that contributed to the occurrence and, equally important, the interaction among them.

28.3 Data Collection

Incident and accident investigation is, by definition, post hoc, involving a reconstruction of the events, action, and decisions which took place at that time. The accuracy of the reconstruction will depend to a great extent on the quality of the data gathered and the relevance of the questions asked. It is particularly difficult to conduct an analysis later from data which have not been specifically gathered for the purpose. There are a number of aspects to data collection. Some of these are considered in the following section.

28.3.1 Who Collects the Data?

So far in this chapter, the assumption has been made that human-factors data will be collected by a human-factors specialist, though this may not necessarily be the case. Indeed, it has been argued that "most accidents and incidents are investigated by investigators who are trained as 'generalists'" and that the human-factors investigators need not be "physicians, psychologists, sociologists, or ergonomists" (ICAO Circular 240-AN/144). This attitude is particularly unfortunate in a climate in which greater efforts are made to look closely at each aspect of the system. It is highly unlikely that anyone would suggest that the engineering or avionics side of an investigation could be conducted by a generalist. The generalist approach is certainly not the case in the United Kingdom, where a human-factors specialist is an integral part of the investigation team, at least where ATC-related events are concerned.

To accept the principle that anyone with training can conduct human-factors investigations, is to denigrate the role of human factors in the investigations and is also likely to lead to the collection of data of a lower quality than the one that might otherwise have been achieved. Many of the issues arising from the investigation of incidents and accidents are essentially in the realm of psychology, and one can include questions of decision making, problem solving, perception, attention, and so on. Furthermore, one can also add equipment design and ergonomic aspects to this realm. These are specialist areas whose understanding is not easily acquired without an appropriate educational background. However, there are other areas of expertise that possibly need to be developed on the job. These would relate to the specific details of the investigative process, for example, the role of the various interested parties, or the legal and licensing aspects, coupled with at least a broad familiarization with aviation and ATC. However, whether the skill in investigation is an art or a science and whether some individuals have a particular facility in this area is open to debate. The ideal situation would be for a potential human-factors investigator to take up the task with a prior background and experience in the human-factors field as a basic requirement. To this, the job training in those aspects of the task not already acquired can be added. It would seem logical to develop a multidisciplinary team of investigators, each with his or her own area of specialization which can be enhanced by familiarization training in the tasks performed by their colleagues. This cross fertilization could facilitate the working of the team and the data-gathering process.

28.3.2 What Data Are Collected?

Reference has already been made to the data sources available to aviation incident and accident investigators. The data of most interest to each member of the investigation team will depend to some extent on the area of specialization of the particular team members. From the human-factors point of view, transcripts and recordings of RTF communication will be equally essential as the written reports from the perspective of the individual controllers and/or pilots concerned. This allows the investigators to appreciate the background of an incident or accident, and prepares the way for the later stages of the investigation, i.e., unit visits and face-to-face interviews. From the point of view of the human-factors investigator, this background information is invaluable, as it allows an opportunity to draw on the expertise of the team colleagues who will probably be more familiar with the specific aviation-related aspects of the event. As a result of this preparation, the human-factors investigator may be in a better position to frame human-factors questions relevant to the context in which the incident or accident occurred.

28.4 Other Methods of Data Collection

The notion presented in this chapter is that the optimal method of conducting the human-factors side of incident and accident investigation is for the human-factors specialist to be present as an integral part of a team of experts each with possibly different, but complementary, areas of expertise. However, there are other means of gathering data and some of these are discussed as follows.

28.4.1 Checklists

It can be possible to provide a nonspecialist with a checklist, against which human-factors data could be gathered. However, the data collected could, in all probability, be a function of the nature of the comprehensiveness of the checklist items, than any real, in-depth understanding of the occurrence in question. The checklist approach has a number of disadvantages:

(a) The data are likely to be rather "coarse grained," in that they would not reflect the contributory factors in any great detail.

(b) The data would be limited to the contents of the checklist, rather than reflecting the nature of the specific incident or accident.

(c) While the checklist may be useful for noting the more tangible items of record, such as hours worked, weather conditions, and so on, the approach would not lend itself so readily to an understanding of the less evident data that are vital to an investigation. In this category, one might include the more cognitive aspects of the performance displayed by the individuals concerned in the event which are, arguably, best investigated by the human-factors specialist.

(d) A standardized checklist approach is also less likely to pick up on the more generic issues involved, which may not be immediately apparent.

(e) If the data initially gathered are prone to the shortcomings already mentioned, this will have serious implications on any subsequent uses of those data. If the right questions are not asked at the outset, it could prove difficult if not impossible, to retrieve the necessary information at a later stage. Attempts have been made in the past to conduct human-factors analyses of the incidents and accidents from the occurrence reports. Many of these attempts have been flawed by virtue of the fact that the source material has not been collected from a human-factors perspective.

28.4.2 Self-Reporting Schemes

A further means of gathering data, without the necessity for a full-scale field investigation, is to enlarge the scope of the self-reports completed by personnel involved in the occurrence. Currently, in the United Kingdom, ATC personnel involved in incidents and accidents are subjected to a mandatory reporting program and should complete a report covering the incident as they saw it, including aspects such as time on shift, shift start, equipment serviceability, etc. This could be extended to include additional subjective data, such as perceived workload, distractions, and more detailed data on the nature of the incident itself. However, asking the individuals concerned to effectively conduct a human-factors analysis of their own behavior is fraught with problems. First, there is the question of reporter bias that always needs to be taken into account when individuals report on their own behavior. The incident may naturally be described from one viewpoint that may or may not accord with the events that actually occurred. In addition to the more obvious memory problems, individuals involved in incidents or accidents are likely to try to make sense of what may otherwise appear to be an illogical situation. If we accept the premise that, generally, individuals will not deliberately make errors, then a situation in which an error occurs may be, almost by definition, a meaningless situation. Consequently, individuals placed in the situation of having to report on their own errors may attempt to present the event in a much more rational light. This is not to suggest that they are necessarily lying, but rather that they are attempting to understand a situation or an error for which they have no rational explanation.

28.4.3 Confidential Reporting Schemes

In addition to more formal incident-reporting programs, a number of states also run confidential reporting schemes that allow individuals to air their concerns on aviation safety in a confidential manner. Such schemes, for example, CHIRP in the United Kingdom, CAIR in Australia, and the ASRS scheme in the United States, are valuable in drawing attention to human-factors issues, often before the problems reported have manifested themselves in incidents or accidents. However, it has to be borne in mind that data gathered via these schemes are not of the same type as that gathered during an incident investigation. Reporters are very much a self-selected group, motivated by their view of events to report in a more public forum but, for whatever reason, unable or unwilling to utilize the more formal reporting channels. These reports are, therefore, likely to be even more prone to reporter bias and the problems mentioned earlier than the other methods already described, although the very act of reporting can serve as a cathartic function for the reporters. There is also the question of how far such reports can be progressed through the system, as there could well be a conflict between verifying the veracity of the reports and adhering to the stated pledge to maintain confidentiality. However, these caveats do not denigrate the value of these schemes in providing an educational function for other pilots or controllers, through which they can learn from the mistakes of others. They also serve as a useful "barometer" of the current state of aviation safety as perceived by those actually doing the job.

28.4.4 International Aspects

In the United Kingdom, in 2003, approximately 3000 ATS-related incidents and accidents were reported. Of these, around 500 were found to have an ATC *causal* element. All of these were investigated to determine the causal and contributory factors and to put in place the appropriate remedial measures. Among the total, around 60 were the subject of a full-field investigation involving a close examination of all the available data, together with site visits to the ATC facilities involved and interviews with relevant members of the staff. Resource allocation necessarily means that decisions have to be taken regarding the selection of occurrences to be investigated at this level, and priority is normally given to those assessed as involving the most risk. From a human-factors perspective, this may not necessarily be the best criterion, but an examination of the events investigated over a 10 year period might suggest that the existing human-factors database is fairly representative of the problems inherent in the U.K. ATC. The scope of the field investigation is such that a comprehensive picture is obtained of the event and its causation and mechanisms are in place to allow the feedback of the lessons learned. However, even 60 events per annum represent a relatively small data set to draw conclusions and make recommendations. Therefore, the availability of other confirmatory data is highly desirable. Communication between investigation agencies from different states is a valuable source of information. The exchange of information and ideas can only serve to strengthen the quality of the investigative process in general. Attempts are in hand to achieve some form of commonality in databases to facilitate the transfer of information, and with the current state of technology, this is an achievable aim. However, what is likely to prove more difficult is achieving some commonality in the method and process of incident investigation, including the human-factors side. Different states vary in the methods adopted and the number and scope of the events covered, with attention often being focused on accidents at the expense of the seemingly less serious, but nevertheless important, incidents. However, if common terminology and taxonomies can be agreed for the storage of investigation data, this would go some way toward overcoming the disadvantages of differences in the method. It has already been suggested in this chapter that data can be viewed as varying in quality depending on the manner in which they are collected and by whom. With the improved liaison between the investigation bodies, resulting in easier and frequent data transfer among them, the fact that not all data are regarded as "equal" and that, ideally, the sources of data are specified when data transfer occurs, will be important.

28.5 Open Reporting

The *raison d'etre* of any investigation should be the better understanding of causality with a view to future improvements and the prevention of similar occurrences. Fundamental to this is the existence of a fair and open reporting culture; i.e., a system in which those involved in aviation can report issues of concern without the fear or risk of being punished for admitting genuine errors. Such a system is not easy to achieve, but once in place, is all too easy to destroy. The current trend toward "criminalizing" the investigative process is guaranteed to destroy any faith that individuals may have that they will be fairly treated. A recent occurrence in Japan is a case in point (Takeda, 2004). On January 31, 2001 there was a near mid-air collision in Japanese airspace between a JAL B747 and DC10. The investigation was conducted by the Aircraft and Railway Accident Investigation Commission who published their report on July 12, 2002. By May 7, 2003, the Tokyo Metropolitan Police Dept. had opened a criminal case and the occurrence report was passed to the prosecutors, contrary to ICAO stipulations (Annex 13). Despite the fact that there were a number of contributory factors in the occurrence, including the pilot of one of the aircraft failing to follow a TCAS Resolution Advisory (RA), on March 30, 2004 the two controllers on duty at the time of the incident (one a trainee) were charged with professional negligence.

This is not an isolated occurrence. Similar criminalization has occurred in Italy with respect to the Linate accident, and also in France and Holland. The aftermath of a serious incident or accident is often accompanied by a wish to find someone to blame, as clearly demonstrated by the tragic death of the controller involved in the Ueberlingen mid-air; however, while criminal proceedings may satisfy the desire of an organization or the public to attribute blame, they do nothing to further the cause of air safety and, in fact, do a great deal of harm when they threaten or, indeed, eradicate open reporting.

28.6 Investigation Framework

It is important during an investigation that care is taken to ensure that no relevant information is overlooked in the data-gathering process. For this reason, it is sometimes proposed that investigators adopt a particular model as an investigative framework. Many of these are not models in the accepted sense of the term, i.e., they have little or no predictive capability and are, at best, a set of guidelines which can be used to inform the investigative process. In fact, they tend to represent explicit statements of the good practice that any investigator worth the name should be utilizing. Models may serve as a structure in which the nonspecialist can collect data. However, it could be argued that they have only limited utility in the most important aspects of investigation—namely the evaluation, prioritization, and interpretation of the data. It is in these areas where the specialist investigator comes into his or her own.

The problem of bias has already been mentioned from the perspective of the reporter. However, investigators can have biases too and it is essential that he or she is aware of the danger of bringing a biased approach to a particular investigation or set of circumstances, as well as forming hypotheses before the relevant data have been sifted and analyzed. The decisions as to what data are relevant in the causal chain and what can be safely left out of the equation are an exercise of judgment that forms one of the most important aspects of the investigation process. Any specialist may tend to see things in terms of his or her own field of specialization and interpret the data accordingly.

The point is that there would be a number of facets to the overall picture of the incident. Any facet ignored, or allowed to dominate to the detriment of the rest, may produce an outcome in terms of the feedback from the accident which may be biased and essentially flawed. The construction of multidisciplinary teams of specialists working together helps to militate against bias and may also prevent the formation of premature hypotheses on the part of any one investigator.

28.7 Feedback

The investigation of incidents and accidents has a number of phases—background preparatory work, initial data gathering, evaluation, prioritization, and interpretation. These are all essential elements in the investigative process, but are only subgoals in fulfilling the overall aim of the investigation, that is, the prevention of future incidents and accidents. The fulfillment of this aim demands the provision of clear, logical and, above all, practicable feedback. An incident or accident from which no lessons are learned is an occurrence which has been wasted in air-safety terms.

The point made in relation to the provision of feedback is that "it is probably worth making a distinction between safety regulation (i.e., the establishment of goals and standards) and safety management (which relates to the day to day application of those goals and standards in the operational environment)" (Baker, 1992). There are a number of target areas to which feedback from incidents and accidents can, and should, be addressed. First and foremost, this involves the providers of the services in question, in this case, ATC. The individuals and their management involved in the occurrence have a need and a right to be informed about the findings of any investigation. Those responsible for safety regulation and the setting of safety standards also need first-hand information on the state of the health of the systems that they are regulating. Incident investigation is a reactive process which indicates, *post facto*, that whatever safeguards were in place have failed in some way, since an incident or accident has occurred. However, as already stressed, the reactive nature of the incident investigation does not preclude its additional proactive role. One of the major ways in which this proactive role can be realized is in the provision of input to research activities. Investigation of incidents and accidents can, and should, provide quite detailed information on each facet of the system under investigation. The expertise that a human-factors specialist brings to the investigation of behavior and performance can, for example, be invaluable in informing the development of interfaces and decision-support aids for both the pilot and the controller. The very fact that the investigation process concentrates on the events in which the system has broken down serves to illustrate those areas demanding most attention, and helps to focus on those aspects of the task where research could most usefully be targeted. It is essential that lessons learned are effectively passed forward to those in a position to make decisions regarding the development and procurement of future systems. A knowledge of the past problems and inadequacies should help in the development of the sort of informed opinion that can ask the appropriate, and often awkward, questions regarding the system designers and equipment retailers to ensure that problems identified with past systems are not perpetuated in the future.

28.8 System Development and Evaluation

It has been argued that the development and evaluation of new systems should necessarily involve a comparative study of the performance of the old system by comparison with the new (Baker & Marshall, 1988), so that the advantages of the new system can be more readily and clearly demonstrated. This comparative approach, though desirable, is time-consuming and expensive. Expediency and cost often require that only the new developmental system is tested, and the adequate evaluation of the pros and cons of the new system vis-à-vis, the old, is frequently omitted. However, during the investigation process, much can be learned about the relative merits and demerits of the existing systems as they are perceived by the end user. As a result, it should be possible to indicate the strengths and weaknesses of current systems to system designers, and indicate those areas which need improvement and those which function well. However, the success of this process does require a symbiotic relationship between the human-factors investigator and the designers and evaluators of the systems. The problems inherent in this approach have been pointed out elsewhere; for example, Baker and Marshall made the point that "However desirable a co-operation between designers and human factors experts might be, human factors specialists are still not sufficiently involved in the design phase with the result that, often, the anticipated benefits from the system in question are not spelled out in any clearly testable way. Typically, psychologists are simply requested to validate or demonstrate the advantages of a new system." In the United Kingdom, at least, very few ATC-related incidents can be traced directly to problems related to inadequate or less than optimal equipment, and rather, poor equipment and facilities tend to be implicated as contributory, not causative factors. Nevertheless, investigations do reveal areas in which equipment development is needed. A good deal of attention has been focused, for example, on alerting systems that inform the pilot or controller of an impending collision. This is well and good, and very much necessary if airborne collisions are to be avoided. However, historically, relatively less attention has been focused on the development of systems that aid the planning and decision-making aspects of the ATC tasks, that is, to prevent the development of situations in which conflicts arise in the first place, although this is changing. The investigation of the human-factors aspects of incidents and accidents can be helpful here in highlighting those aspects of the planning and decision-making process most in need of support.

However, feedback is not restricted to ergonomic- and equipment-related issues. The adoption of the system approach discussed earlier, facilitates the gathering of information on all the aspects of ATC functioning. Human-factors recommendations that ensue from investigations can range from fairly basic "quick fixes" to more far-reaching issues involving, for example, aspects such as training or the role of management. In Reason's terms (Reason, 1989), both the "active" failures and "latent" problems in the system need to be addressed, and careful sifting and analysis of the information gathered from the investigations can reveal not only those areas in which failures have already occurred and errors been made, but also those aspects of the system which, if left unaddressed could well lead to problems in the future. The existence of these generic problems that may not have manifested themselves directly in an incident or whose connection to an occurrence may seem somewhat tenuous is often difficult to demonstrate. This is one area where the advantages of incident as well as accident investigation is most evident. It may be difficult to demonstrate, for example, on the basis of one accident, that a particular problem exists. However, if it can be shown that similar conclusions have been reached as a result of the more numerous incident investigations, then the case for a closer examination of the problem and perhaps, the initiation of research will be greatly strengthened.

28.9 Conclusion

The role of human factors in incident and accident investigation has received increased attention in recent years. Even then, the extent to which human-factors considerations are taken into account during the investigation process varies from state to state. This chapter has focused on the investigation of civil ATC-related incidents, and accidents in the United Kingdom, where a human-factors specialist is routinely included as part of a multidisciplinary investigation team.

The motivation for conducting investigations extends beyond discovering the cause of any one incident or accident. The main focus has to be on the lessons learned with a view to the prevention of similar incidents or accidents in the future. The greater the volume of the information which can be gathered, the more complete would be the picture which can be gained and the firmer would be the basis for any recommendations for future improvements. The additional knowledge gained from investigating incidents, in addition to less-frequent accidents, is invaluable in compiling the overall picture. However, there is a worrying trend appearing in a number of states, in that the investigation process is becoming increasingly criminalized. This does nothing to further the cause of air safety and, in fact, does a great deal of harm when it threatens or, indeed, eradicates open reporting. For open reporting to become a reality, those involved need to have faith that genuine mistakes will not be punished. If that faith is destroyed along with the open-reporting culture, then a significant means of improving aviation safety will be lost.

The collation and analysis of data, together with the compilation of reports and recommendations arising from a specific incident or accident is not the end of the story. An incident or accident from which no lessons are learned is a wasted event. There has to be practicable and accurate feedback that has to be acted upon. It is therefore essential that efficient mechanisms exist, not only to disseminate information to those individuals and/or organizations where it can do most good in terms of prevention, but also to monitor that the feedback has been utilized.

A successful investigation demands a balanced approach to the problem. Each of the team of experts involved in the investigation will have his or her own area of expertise, none of which should be allowed to assume undue priority and importance in the investigative process. However, the underlying causal factors in the incident and accident occurrence can vary. Accidents involving engine failure as the root causal factor, for example, will give rise to a different findings with different emphases, than those in which training or ground equipment are primarily implicated.

The inclusion of human factors as a potential issue in incident and accident occurrence has come fairly late on the investigative scene. However, to ignore the human-factors aspects of these events will, almost inevitably, lead to an unbalanced and incomplete picture in attempting to determine, not only what happened, but why it happened.

References

Baker, S. (1992). The role of incident investigation in system validation. In J. A. Wise, V. D. Hopkin, & P. Stager (Eds.), *Verification and validation of complex systems: Human factors issues. NATO ASI series*. Berlin: Springer-Verlag.

Baker, S., & Marshall, E. (1988). Evaluating the man-machine interface-the search for data. In J. Patrick, & K. D. Duncan (Eds.), *Training, human decision making and control*. Amsterdam, the Netherlands: North-Holland.

ICAO Annex 13 (5.12-Non-disclosure of records). (2001, July). *Aircraft accident and incident investigation* (9th ed.). Montreal, Canada: ICAO.

ICAO Circular 240-AN/144 Human Factors Digest No.7. (1993). *Investigation of human factors incidents and accidents*. Montreal, Quebec, Canada: ICAO.

ICAO Circular 247-AN/148 Human Factors Digest No.10. (1993). *Human factors, management and organization*. Montreal, Quebec, Canada: ICAO.

Reason, J. T. (1989). The contribution of latent human failures to the breakdown of complex systems. *Philosophical Transactions of the Royal Society (London), B, 327*, 475–484.

Air Accidents Investigation Branch. Report on the accident to BAC One-Eleven, G-BJRT over Didcot, Oxfordshire on 10 June 1990. Air Accidents Investigation Branch, Department of Transport, HMSO.

Takeda, O. (2004). A review of JAL B747/DC10 near mid-air collision. *5th International Air Traffic Control and Flight Operations Symposium*, June 2–4, Keelung, Taiwan.

29

Forensic Aviation Human Factors: Accident/Incident Analyses for Legal Proceedings

Richard D. Gilson
University of Central Florida

Eugenio L. Facci
University of Central Florida

In the early morning of October 8, 2001, the local air-traffic controllers at Milan Linate airport in Italy were operating at capacity for quite some time when one of the worst aviation disasters to ever take place in Europe was about to hit. A dense layer of fog had formed overnight and was stationed over the busy airport, adding workload for the controllers as they had to cope with an inoperative radar for control of ground operations. At about 6:10 am D-IEVX, a Cessna Citation was taxiing out for an intended demonstration flight to Paris. The pilot misinterpreted his position and accidentally entered runway 36, around the time when Scandinavian Airlines (SAS) flight 686 was being cleared for takeoff. During the take-off roll, the Scandinavian MD-87 seemed to gain visual contact with the Citation about 1 second prior to impact, as data recorders indicated an abrupt pitch up pressure on the control column combined with an unintelligible exclamation. In the subsequent runway collision, occurring at about 150 kts for the Scandinavian airliner, the Citation was chopped into three parts, while the SAS MD87 lost the right engine and the right main landing gear. The Scandinavian flight crew managed to get the plane airborne and keep marginal control of the aircraft for a few seconds, before they eventually slammed into a baggage-handling building located after the departure end of the runway. The impact forces and subsequent fires caused the death of 110 people onboard SAS 686, and 4 onboard the Citation, in addition to 4 airport employees on the ground.

Two and a half years later, in May 2004, a criminal court judge in Milan sentenced the controller handling D-IEVX at the time of the accident to serve an eight-year prison term and to pay substantial monetary damages to the parties that suffered losses. Personnel from the airport authority were also sentenced to serve similar terms. Overall, the judge found the management company culpable of sloppy safety practices, noncompliance with existing safety procedures, and poor maintenance of required taxi signs. Specific details were not given about how the judge came to subpartition responsibility among the personnel who were eventually found guilty. The air-traffic controller was convicted on accounts that he failed to read back a taxi clearance from the flight crew of D-IEVX.

When an aircraft crashes, two main needs arise. First, someone should determine whether the cause of the accident might trigger more accidents in the future. Second, others may have to determine if some party was at fault in the crash and whether compensation should be awarded to the parties that suffered losses. Forensic aviation human factors aim to serve both the purposes,* and in this chapter, we introduce readers to this discipline that encompasses a wide array of knowledge and practices from the fields of law, psychology, and aviation operations.

The chapter's layout is as follows. We start by providing some introductory concepts and describing, in Section 29.2, what forensic aviation human-factors experts do, who they are, and how accident investigations in general are conducted. Section 29.3 examines some actual accidents and their legal proceedings. Finally, Section 29.4 covers the discussion topics that are relevant to the determination of responsibility. We have also examined what criteria are used, what issues are currently being debated in the legal arena, and thereby provide a background for the understanding of forensic aviation human factors.

29.1 Introduction

29.1.1 Accident Investigations: Goals and Players

When someone has an accident[†] with an aircraft, automobile, or with any other system or device, people want to know why. The primary motivation[‡] for this is the prevention of their and others' involvement in similar mishaps. Humans do make mistakes with machine systems, and are listed as the probable cause of accidents more often than not.[§] However, who actually is to be blamed, why, and how to avert a reoccurrence may not be obvious or easy to uncover. A closer look is often needed through forensic human factors, particularly in complex systems such as aviation.

Courts of law serve as one of the several means for determination of cause and blame, and may spawn potential remedies. As those involved in the judicial process typically have no training in human factors or in aviation, forensic human-factors experts specializing in aviation are needed to analyze, explain, and give probabilistic opinions, so that judges and juries can make informed decisions. Although litigation is considered by many as a negative, a positive outcome may be to help in preventing further occurrences.

---

* As we will describe in more detail later, some suggest that now forensic aviation human factors serve more to determine the responsibilities than to prevent future accidents.
† An accident or mishap refers to an undesirable, unintentional, and unexpected event. However, used herein, an accident (or mishap) is *not* an unforeseeable random event, totally without cause, as an act of God. Forensics presumes and seeks out underlying causes or contributing factors and assumes that future care could make such accidents preventable.
‡ Certainly, there are other less altruistic motivations. The public has almost a morbid interest in aviation accidents, some people involved have mercenary pursuits, and others may even seek vengeance or absolution.
§ Human error as the primary cause of aviation accidents is frequently put at anywhere from 60% to 80%, depending on the source.

It is important to understand that all those who play a role in accident investigations and their aftermath have different roles and varying degrees of power or influence. Government agencies have the weight of law behind mandated changes, but must wait to amass statistical evidence and to make a persuasive case concerning the relatedness of what appear to be similar crashes. The process of proposed rule-making has to overcome many checks and balances designed to guard against unnecessary fixes or unwanted regulation that might arise from a truly isolated mishap, that is, a sampling error. Civil lawsuits, on the other hand, evaluate single cases, but carry no regulatory authority. Indeed, the fact that many final judgments in civil lawsuits remain limited to the single case suggests governmental dismissal of the cause as an act of God or as a unique event generally unlikely to reoccur (e.g., preventable carelessness). It is also possible that there were flaws in the final judgment because of the lack of knowledge or distortion of the meaning of facts about aviation or human factors. The extent to which the judgment was valid, yet not made regulatory, may serve as warning to designers and users alike that certain actions or inaction may carry the added risk of penalties. Human-factors experts in aviation can assist in all these processes, from design concepts to postcrash analysis as well as through recommended remedies. However, as it is the civil legal system that most often (vs. government agencies) retains the human-factors experts for aviation forensic services, this chapter focuses primarily on civil proceedings in an attempt to provide the readers with an understanding of the issues involved.

29.1.2 Forensics

The terminology of forensics deserves some explanation. First, the word *forensics* stems from forum, the marketplace, or assembly place of an ancient Roman city, which formed the center of judicial or public business. As a noun, it implies oratory, rhetoric, and argumentative discourse, such as what might be used in a debate. This use implies an advocacy position.

On the other hand, as an adjective, forensics is usually associated with judicial inquiries or with judicial evidence given in courts of law. For example, forensics medicine is considered as a science that deals with the relation and application of medical facts to legal problems, as in the role of "Quincy," the fictitious forensics pathologist in the long-running television series. In it, as in real life, an expert's opinion becomes legally admissible evidence, given under oath. The implications of the testimony are argued by the attorneys and ultimately are used in legal judgments.

These differences in meaning, we believe, sparked a controversial position statement and subsequent brouhaha over the value of forensics human factors in *Ergonomics in Design*. In an editorial, Daryle Jean Gardner-Bonneau (1994a) insinuated that forensics in human factors was less than a worthwhile scientific pursuit, by taking the unusual action of deliberately removing "Formative Forensics" from the masthead of *Ergonomics in Design*. Gardner-Bonneau's complaint appears to be that forensics human factors has "little science" and that human factors/ergonomics professionals, as expert witnesses, are "often called on to consider isolated aspects of a case and render judgments based on limited information... [thus] the waters of forensics practice, from [the] editor's perspective, are simply too murky to play in" (p. 3). Howell (1994), in a published article within that same issue entitled "Provocations: The human factors expert mercenary," also supported this view by suggesting that "the credibility of the human factors discipline [was being questioned]... as a result of our rapidly growing presence in the forensics area" (p. 6). It seems that both writers consider forensics as a noun, suggesting a narrow, almost quarreling/advocacy view for the benefit of one side—hardly the basis of an unbiased scientific inquiry and analysis.

A fusillade of letters offered rebuttal in the subsequent issue of *Ergonomics in Design*. In one letter by Richard Hornick (a former president of the Human Factors Society, and of the Human Factors and Ergonomics Society, Forensics Professional Group), the editorial position was severely criticized as potentially damaging to that society as well as to individual forensics practitioners (Hornick, 1994). Hornick argued that judicial scrutiny and challenges to an expert's opinion(s) "far exceed those [scientific peer reviews] that occur in academia and industry... [and that the] legal arena provides a powerful tool to correct flaws [in product/workplace design and as a defense against wrongful suits or unfair claims]" (p. 4). Other letters in the issue claimed that forensics work "mediate[s] scientific findings and practical requirements..., for real-world problems" (Beaton, 1994, p. 5) and that "forensics work drives you back to the basics of the subject" (Corlett, 1994, p. 5). Editor Gardner-Bonneau responded in her own letter (Gardner-Bonneau, 1994b) that "that human factors analysis *can* [her emphasis] have [value] in forensics work... [and that she] encourage[s] submissions that emphasize the application of human factors methodologies and techniques to the analysis of forensics cases" (p. 35). Finally, Deborah A. Boehm-Davis, then President of the Human Factors and Ergonomics Society, also took the time to write (Boehm-Davis, 1994) to "make it clear that the statements made in the editorial *do not* [her emphasis] represent the Human Factors and Ergonomics Society's position on the practice of forensics. The Society encourages the practice of sound forensics and sponsors a technical group that supports human factors ergonomics professionals who work in this field" (p. 35).

The intent here is to use forensics as a subspecialty of human factors relating scientific principles/facts to the analysis of a specific accident or mishap. At issue is the determination of what happened, why it happened, and how to help prevent reoccurrence of the underlying problem (if indeed there is one, beyond an "act of God"). This does not merely mean rendering professional views and expert opinions based on the theory and knowledge. As Senders (1994) advocated, it should involve field studies and empirical data, wherever possible. This approach is no different from using the procedures, investigative techniques, and findings/knowledge of other disciplines and subspecialties to forensics, such as forensics ballistics, forensics chemistry, forensics psychiatry, or the like.

29.2 Forensic Aviation Human-Factors Practice

A typical case for a forensic human-factors practitioner starts with a phone call from an attorney with a brief description of the facts (and perhaps the contentions) of the case. The request is normally to review the facts and circumstances of a specific mishap to determine the human-factors contributions/causation focusing on system use, interface design, procedures, training, human error(s), and so forth. Sometimes the request is to analyze a specific design or an associated warning as to the likelihood of human errors (misunderstanding and misuse) or designed-induced error(s). Occasionally, there is a request to specifically comment on the opinions offered by an opposing expert(s).

Without commitment on either side, follow-up material is usually sent for an initial review. Most of the time, the minimum information needed is the National Transportation Safety Board (NTSB) factual or at least the preliminary report (Notably, litigation usually follows the lengthy investigative process by the NTSB that results in the final NTSB Factual Report. Interestingly, the supporting documentation produced by the NTSB investigators is destroyed).

The next contact is usually by telephone from the expert, with his or her verbal assessment of preliminary opinions. Based on that conversation, if there is an agreement by the attorney that the expertise may help in the case and if the expert has a reliable scientific basis to take on the case, then there is some potential for a commitment. There are many reasons at this point for a refusal, including the expert's opinion being adverse to the attorney's interests, various conflicts that might create potential bias, incompatible scheduling demands, and so on. Before an agreement, it is appropriate to indicate what the general nature of testimony might entail as such in the area of human-factors design (e.g., display and information processing), or in human factors of flight operations (e.g., flight procedures, pilot behavior, pilot expectancies from air-traffic control [ATC], etc.). It is also appropriate to indicate areas outside the expertise, such as in crashworthiness or in accident reconstruction (bent metal analysis). Most often, testimony is in the form of expectancies, likelihoods, or probability that something, more likely than not, did occur or that should have been reasonably foreseeable for the manufacturer or user.

The level of commitment of an expert might be as a background consultant (available for questions from the hiring attorney) or as a declared expert available for deposition and trial. With such an agreement, a full analysis gets underway with materials sent by the attorney. Information may also be requested by the expert, including other information that is available, or information that needs to be obtained, such as depositions, proposed tests, experiments, and so forth. The following are some typical materials that may be available through the discovery process (each party in the lawsuit must inform the other about the documents that they have, in a process known as discovery).

29.2.1 Factual Data

NTSB factual report
Federal Aviation Administration (FAA) ATC plots of the flight path and ATC communications
Cockpit voice recorder (CVR) communications
Photographs of the accident site and the wreckage
Visual flight rules (VFR) charts or instrument flight rules (IFR) charts (as appropriate)
Aircraft information
Pilot Operating Handbook (POH)
Weight and balance information
Airframe, engine, propeller (or other component) logbooks, pilot/crew information
FAA pilot certificates and medical records
Pilot logbooks and training records
Weather information
Weather reports and forecasts
Testimony
Statements of eye or ear witnesses
Opposing expert reports
Depositions taken by attorneys on all sides

This mass of material arrives at various times, in sundry order, and at times by the box load. How is it organized? How is it verified, given the natural errors in assembling and recording so much information, the different perceptions by people seeing or hearing the same thing, and even outright fabrication of the truth by people with real self-interests? (Presumably, misrepresentations are an uncommon experience for most experts coming from a scientific setting.)

The organizational headings listed earlier work well for referencing material in most cases, but verification of conflicting data is more difficult. In general, the pattern of evidence can be viewed in a statistical context with occasional "outliers." The preponderance of the evidence centers on the median of the data and should anchor an opinion, unless there are extraordinary and explainable reasons for doing otherwise. One unsubstantiated conflict should raise a question and deserves focused consideration, but without corroboration, it should not unduly sway an expert's view. Most often, crash-witness statements are in conflict and can be checked against one another using something akin to law enforcement techniques. Law enforcement officers often interview people separately to highlight discrepancies and to look for converging evidence. This is not to say that an apparent discrepancy should be dismissed as untrue or just wrong. However, in real life, most events have some basis, right or wrong, and even the smallest troublesome detail may lead to a breakthrough of understanding. Nevertheless, there should be caution for relying too much on any one piece of evidence; sometimes, the underlying basis will never be known.

After a preliminary review, if there are gaps in existing evidence that need to be filled before a suitable human-factors analysis can be completed, then it is appropriate, if not expected, for an expert to actively seek additional facts (in coordination with the attorney). Information might be available from data searches, additional interrogatories or depositions, through videotaped observations, or even from modest experiments. Experts who should seek additional data that might be available, but do not do so, are open to spirited questioning by the other attorneys. They will be suspicion that such passive behavior can be manipulated and therefore, is not appropriate for an expert, or that this is really indicative of the expert "not asking the question, because of fear of getting the answer (an unwanted one)." It is true that regardless of the findings, all such pieces of evidence are subjected to discovery rules, as they should be in the search for truth. Generally, just like scientific research, if the theory is correct, then new data will fit. However, the converse is also true, signaling the need for a new approach and serious discussions with the attorney. It is safe to say that there are no attorneys who want to hear something negative about their side of the case, but they probably would all agree it is best to know the downsides of a case before a deposition is taken (questioning by the other party's attorney) and certainly, before testimony at trial.

29.2.2 The Impact of Forensics Human Factors

Human factors itself are relatively new as an applied science, and originally, often aimed at problems or hazards in aviation systems. Forensic human factors is newer in courts of law and apply to a number of other domains in addition to aviation. Typically, interest is in the discovery of potential misuse or danger of products from the legal standpoint of liability,* for example, in the context of human factors, design-induced errors, procedural flaws, or inadequate training or warnings. Behavior with products may range from what is normally prudent, to what should be reasonably expected (even if wrong) and what may be even deliberate risk-taking activities.

As the law has always focused on people with problems, the emergence of human factors in the legal system was probably inevitable and should provide a valuable service. Ideally, beyond the determination of fault, the legal system is designed to be one of a number of means to change future behavior (whether that of the manufacturer, seller, or user). However, it must be noted that to change behavior (by design, prevention, procedures, warnings, training, etc.), the problem must be well understood. Evidence regarding aviation operations and the even lesser-known field of human factors is often unfamiliar to the courts. Evaluation of what went wrong in these contexts often must be explained by expert testimony as part of the evidence.

Judges and juries (as well as attorneys and litigants) expect more than a reconstruction of the events and an analysis of the crash dynamics. They want a reliable determination of the *why* (of behavior), along with the *what* (happened). Details about how the design was intended to be used and how it actually was used provide salient evidence that often tilts the balance in decisions. Were errors the result of the individual or because of the design itself? Were there attention lapses, slowed reactions, inaccurate perceptions, wrong expectancies, hazardous activities stemming from deliberate behavior by the individual involved? Or were the problems induced by defective design, inadequate protections/warnings, deficient instructions/training, or even improper integration with other systems?

The determination of the problem and restitution for it occurs at many stages in legal proceedings, such as most potential lawsuits that do not come to full fruition. Even before an accident ever occurs, the manufacturer may reevaluate and modify a design for better usability and safety using an external human-factors expert. After an accident, but before a lawsuit, a manufacturer or insurance company may question a human-factors expert about its culpability or its relative exposure. During a lawsuit, settlement questions arise requiring expert opinions.

As a goal, specific human-factors input into the original design stages should help to insure that a product or system is easy and safe to use. This is done in automobile manufacturing, as it fits changes into the competitive marketplace, creating maximum user preference and minimal user problems. Actually, automobile manufacturing changes are constantly being rethought—responding to evolving market forces. With mega-volume sales and yearly changes in design, some automobile manufacturers actually have groups of full-time human-factors practitioners as their staff. These efforts have paid off, with the successful selling of safety, once thought impossible, as well as convenience, economy, and comfort. Similarly, with high per-unit cost of a commercial or military aircraft, large companies such

---

* Liability in this context generally suggests failure to exercise ordinary or reasonable care in the manufacture or use of a product.

as Boeing and others also use staffs of human-factors experts throughout the design and modification stages. Also, in commercial aviation and beyond, there is the pull from sales potentials and the incentives of profit. The consumer drives the marketplace and spawns product developments with purchasing power. New developments are proudly announced and advertised with gusto.

In contrast, general aviation manufacturers, with annual sales in the hundreds and dwindling, are forced to continue to use specialized designs often 40 years old (before human factors became separate from engineering design). Employment of full-time human-factors professionals is not a practical consideration, although their engineering staffs may have been updated with human-factors courses. Fortunately, with so few changes being made, there is plenty of experience with what works and what modifications are needed. Moreover, there are a number of mechanisms in place to provide feedback regarding problems with existing designs. These include FAA Airworthiness Directives, Service Difficulties Reports, Manufacturer's Service Bulletins, NASA's Aviation Safety Reporting System (ASRS), various NTSB recommendations, exchanges in technical and trade literature, insurance company publications, and a variety of magazines for pilots, aircraft owners, mechanics, fixed-base operators, and so on. These are evidence that today's general aviation suffers from remedies that come by way of push, not pull. The push comes primarily from government agencies with new regulations and the threat of penalties, acting as an unnatural force for change. Fixes, if deemed appropriate, are disseminated in drab technical literature, published as advisories or mandated by regulation. Thus, without strong market forces and with few accidents, changes in general aviation are traditionally slow to emerge, and these, for the most part, are evolutionary or regulatory refinements, and not revolutionary leaps ahead.

Despite all this, accidents do occur, ratcheting up demands for narrow analyses and remedies. News media reports, followed by NTSB Accidents Reports (or Military Safety Boards), manufacturers' mishap reports, and legal proceedings, typically help to illuminate what happened, but not necessarily why. Characteristically, they do not incorporate specialized human-factors considerations. Unfortunately, almost any aircraft accident creates intense human interest (when compared with the automobile accident) and often a media call for hurried fixes.

Detractors of this media hysteria are abounding. Some denounce it as creating a "tombstone mentality," delaying the evaluation of the efficacy/soundness of a design until damage, injury, or death has occurred. Others maintain that an atmosphere of fear of litigation has paralyzed the industry, causing the demise of general aviation. Their argument is that advances in technology become prime facie evidence of: design shortcomings, halting even the discussion of proposed modifications by a "circle the wagons" mentality. Even others condemn the process as relinquishing important complex-design decisions to lay juries and inconsistent tort law mandates, while dividing any joint efforts by those most directly involved, namely, the manufacturers and the users.

Rightly or wrongly, this means that human-factors experts currently have (with a few exceptions) their greatest input into aviation via postaccident analyses in response to the questions raised by the legal system. In addition, with aviation-product liability reform that limits the liability to 18 years after manufacture, there has been somewhat a shift from blaming the older aircraft involved in accidents to blaming component manufacturers, maintenance facilities, and service organizations (private and public). In any event, human-factors analyses will still be required.

For now, forensics aviation human-factors experts, working within the system, can bring about thoughtful application of scientific knowledge and analysis to specific problems. Their purpose should be to educate juries who are called upon to make legal judgments. Despite the fact that plaintiff and defense experts may disagree, the didactic process is helpful in revealing most sides of what are often complex issues. Imperfect as it may be, this process provides checks and balances of both sides, ironically in much the same way as the open literature provides checks and balances for scientific disputes.

Even without the prospect of civil exposure or penalties, progressive manufacturers will continue to improve their product line, including enhancements in safety. Companies seek human-factors experts for design assistance, and for positive marketing advantages, such as better panel layouts, more efficient controls, improved seating comfort, reduced cabin noise, and so on. However, at the same time, such experts always should be aware of the goal of designing a reasonably safe product, including consideration of foreseeable hazards, alternative design features against those hazards, and evaluation of such alternatives (Sanders & McCormick, 1993). Sanders and McCormick (1993) also pointed out that human-factors design and evaluation often are not straightforward, with a note of caution that alternatives may introduce other hazards. For example, if warnings are chosen as a means of highlighting potential hazards, then there may be a problem of warning overload, diluting the importance of vital information.

Of course, the best solution is to create enough sales in general aviation to enlarge the feedback loop so that market forces dominate again. That would create the incentive for people to buy airplanes again in large enough quantities, so that human-factors research and input during design and manufacture could make aviation systems, easier, safer, and more reliable to use. Better designs should beget more sales, and more sales will inevitably bring about better designs, spiraling the market outward and not inward as it has been for far more than a decade. Moreover, with more aviation activity in the mainstream of public life, distorted views should also diminish. For example, public interest in discovering the reason for almost any one of the nearly 40,000 lives lost each year through traffic accidents is disproportionate to the scrutiny given to almost any aviation mishap (admittedly, there are a number of reasons, rational, and otherwise). However, certainly, a life lost on the highway is not worth less than a life lost in the air. The goal obviously should be to responsibly reduce all risk as much as possible. However, the law recognizes, and the public should too, that there is a trade-off between risk and benefit, and that no system or product is absolutely safe or reliable. Human factors can and should help to reduce risk. For now, this is realistically only possible within the systems in place, regardless of their imperfections.

As specifically applied to aviation, it seems appropriate to consider forensic human factors as a broad view of inquiry beyond (i.e., not limited to) the courts of law and including occurrences that may surface in a variety of ways.* Regardless of how problems are known, the external analyses of human behavior with aviation systems, whether involving accidents, mishaps, incidents, or errors, are often extensive. They utilize various forms of investigative techniques, and now include human-factors analyses as a part of the whole process.

Unfortunately, aviation is nearly unique in modern societies—that is, outside military mission environments. It demands high levels of performance with the penalty for error being extensive damage, injury, or death. As the old saying goes, mistakes in aviation are often unforgiving. Yet, mistakes with actual equipment/systems, whether highlighted by incidents, mishaps, or accidents, may provide the truest forum for evaluation of human-factors designs, procedures, and training methodology. As such, forensics provides invaluable feedback to correct problems.†

Accordingly, forensic inquiries in various disciplines are indispensable ingredients in aviation-accident investigation. For the most part, aviation accidents occur in a three-dimensional space. However, inevitably, our investigations take place in the two-dimensional confines of the ground. Despite the added details of the events in the so-called black boxes, human-factors analyses are often inductive in nature and are hampered by loss of evidence, whether by death of those involved, by memory losses of those injured, by deliberately ambiguous statements, or even by rationalizing or intentional misstatements.

---

* For example, NTSB accident reports; National Aeronautics and Space Administration/Aviation Safety Reporting System (NASA/ASRS) incident reports; Federal Aviation Administration (FAA) Service Difficulty Reports (SDRs), FAA Condition Difficulty Reports (CDRs), and FAA enforcement actions; incidents made known in the technical literature and commercial magazines, books, or videos; through manufacturer's field reports; insurance company's databases; pilot organization publications, such as the Aircraft Owners and Pilot Association (AOPA Pilot), the AOPA Air Safety Foundation (AOPA/ASF), the Experimental Aircraft Association (EAA), the Flight Safety Foundation (FSF), etc.; professional organizations such as the *Forum* of the International Society of Air Safety Investigators (ISAST); military investigatory agencies and publications; and the list goes on and on.
† If forensics is the sole technique to initiating fixes, it properly deserves the criticism derisively insinuated by the moniker "tombstone mentality."

Nonetheless, forensic human factors in aviation legitimately takes its place as a method for integrating remaining evidence, specific facts, and circumstances to determine what, and (more important!) why human factors are more often than not the primary cause of an accident. Not surprisingly, as a result of forensic human-factors work in aviation, design or training remedies are often recommended.

According to many, forensic human factors in aviation started with aviation's first fatality, Lt. Selfridge, a U.S. Army pilot killed after crashing in a Wright Flyer in the early days of aviation. That accident analysis led to the search for preventative solutions, such as the use of helmets in the early days of flying, eventually developing into the overall consideration of methods for survivability and crashworthiness. Similarly, many non-combat-related accidents during World War II brought human factors to the forefront with design changes and classic texts.

What has grown from these seeds is the stimulus for applied experimental research and true evaluation of products in environment. What remains is the need to evaluate risk before the accident, although headway in this area has been made through a large-scale adaptation of the "critical incident" in databases, such as the NASA ASRS, and the FAA SDR.

For decades the statistical trends have illustrated continuing improvements in aircraft accident rates, primarily owing to system/equipment reliability. Human fallibilities have also lessened, but at a slower rate. Thus, the contribution of the human factor in accidents has grown proportionately when compared with the aircraft faults, underscoring the importance of understanding why. The following illustrates the growing importance of human factors, even a decade ago.

According to a January 3, 1994, editorial in *Aviation Week and Space Technology*, Human-factors engineering is required if the air transport industry is to achieve an acceptable level of safety in the next century. Human error is the cause of the vast majority of civil aircraft accidents.

Perhaps, the complication of sophisticated systems mitigates the benefits of improved designs and training. When infrequent and unanticipated faults do occur, they become blurry enigmas [engendering vague responses] for human supervisors. For an infamous, nonaviation human-factors example, one only needs to look at what occurred during the "Three Mile Island" nuclear incident. Forensic human-factors experts can reveal underlying areas of inadequacies in perception, attention, memory, situation awareness, and so forth, adding evidence to an accident investigator's reconstruction of the physical evidence and background data.

29.2.3 Accident Investigation Practice

When an accident occurs, a number of agencies and institutions are usually interested in conducting an investigation, and several investigations may be launched at the same time. In the United States alone, manufacturers and the aircraft operator will probably be interested in doing an investigation, along with unions, sometimes with military corps, and government agencies. However, in any case of a general aviation or commercial aircraft accident, the NTSB has by statute, the exclusive investigative authority.* The NTSB—also known as the Board—is a landmark in the field of accident investigation, and forensic human-factors practitioners may benefit from a basic understanding of the way it operates as an agency, as that affects the way forensic human-factors inquiries are carried out.

The NTSB conducts very extensive investigations—as a matter of fact, on every aircraft accident occurring in the United States—that rely on a clear distinction between the collection of factual information and their analysis.† During the collection of factual information, which starts with the on-scene investigation, an effort is made toward gathering as much data as possible to describe "what" happened.

---

* The NTSB can delegate that authority to the FAA. Various parties, usually including manufacturers, unions, and relevant experts are also often invited to participate in the investigation.
† In simple terms, factual information describes "what" happened. An example of factual information is, "The captain elected to land on runway 27." Analysis, instead, provides a discussion of "why" something happened. An example is "...the captain's decision to land on runway 27 may have been affected by fatigue."

Afterward, during the analysis part of the investigation, the Board aims to explain "why" the accident happened. That distinction, which is also mirrored in the way accident reports are written, bears on the work of forensic human-factors practitioners, as only factual information can be brought to court in a legal case. The NTSB analysis cannot be used for such a purpose. This is one of the reasons why forensic human-factors experts come into play, to help the court determine whether a certain behavior is culpable or not.

29.3 Exemplars of Postaccident Investigations

Many postaccident investigations now include aviation human-factors consultants and experts, in addition to traditional accident reconstruction experts.* Aviation human-factors forensics experts can serve to explain the human behavior (right or wrong) with the equipment that people use and depend upon. In most cases, the question is, why were the errors made? Could the errors have been avoided? Was the behavior typical—to be expected within the specific set of circumstances—or was it improper? Were there contributory factors involved, such as fatigue, stress, intentional misuse, or even alcohol or drugs? Could the design be more error tolerant, and if so what is the trade-off?

Attorneys may want an analysis of human- and design-induced errors, both for their own understanding and to provide a credible approach to initiate or defend a lawsuit. Were errors the result of the individual or because of the design itself? Human errors might include attention lapses, slowed reactions, inaccurate perceptions, risk-taking activities, and wrong expectancies stemming from inadequate situation awareness. Design-induced errors might include problems induced by defective design, hidden hazards, inadequate protection or warnings, deficient instructions or training, or even improper integration with other systems.

Juries may expect more than a reconstruction of the events and an analysis of the crash dynamics. They may want a reliable determination of the *why* (of behavior) behind the *what* (happened). Details about how the design was intended to be used and how it actually was used contribute salient evidence that often tilts the balance in their decisions.

29.3.1 Postaccident Investigative Procedures

Scientists serving as forensics experts initially spend much of their time reviewing a mountain of diverse information from various sources, the aftermath of most aircraft accident investigations involving fatalities. The task subsequently shifts to selecting and piecing together what appear to be causal factors into a cohesive theory of what happened and why, together with supporting evidence. Often this is followed by an active search for other sources of information to fill in inevitable gaps. This search can comprise flight demonstrations or even specific experiments carried out to verify or test a theory.

The following are the three descriptions of accidents involving aviation human-factors forensics analyses. As you will see, although the analyses were successful in identifying human errors, design defects, and operational deficiencies, the outcomes may be far from definitive to some readers. Most accidents do not cleanly fall into any particular category; indeed, most often there are multiple causal and contributing factors leading to a fatal crash. Beyond the presentation of evidence, the dynamics of legal proceedings bring into play the personal, social, and economic factors of those directly involved. This is a mix whose outcome may be baffling to some, yet it is a true reflection of real life. In all fairness, the process seems to hit the mark most of the time.

Human Error Exemplar. A 71-year-old pilot was the sole occupant of a high-performance single-engine airplane proceeding into night IMC (instrument meteorological conditions) from Raleigh, NC,

---

* The International Society of Air Safety Investigators (ISASI) has a Human Factors Working Group to address "Issues arising from examining human factors in accident investigations"; see McIntyre (1994).

to his home base located at a South Carolina Airport. This was the fifth flight leg of a 12 h business day. Although the evidence indicated that he had flown at least one coupled ILS (instrument landing system with the autopilot engaged) approach on that day, other information indicated that he had little overall night IFR flight experience. He commenced a Nondirectional Beacon (NDB) 21 instrument approach with the autopilot heading and pitch command engaged for track and descent. The pilot stopped the descent using the altitude hold function at 1200–1300 ft, well above the minimum descent altitude (MDA), after ATC issued a low-altitude alert (the airplane was off course, low for that position, and in the clouds). The pilot then started a climb for a missed approach by pulling back against the altitude hold. The autopilot resisted the altitude change by counter trimming, eventually reaching full nose-down trim. After several minutes of holding 40–50 lb of control yoke back force (according to subsequent test flights), the autopilot was disconnected electrically by pulling the trim and autopilot circuit breakers, and even momentarily turning off the master switch, but the pilot never retrimmed the aircraft manually. The down force remained unabated, as evidenced by panting sounds transcribed from the ATC audiotape. Confusion of both the pilot and air-traffic controllers as to the source of problem led to the interpretation that the autopilot was stuck "on" and pitching down because of a runaway "hard over" condition.

The radio transmissions are dramatic. "I'm in trouble I was using my autopilot ... and I can't get it off... ah autopilot is ah all is hung the trim...I... I'm fighting it like a bastard trying to keep the thing up... (panting sounds)" "... I pulled every circuit breaker I can find ... Negative [answering to ATC suggestion to turn off/on the master switch] I moved the master switch (unintelligible) turned it on." (There was no indication of an attempt to manually retrim, or to use the autopilot malfunction checklist that calls for retrim.)

For the next 32 min, the airplane was vectored and flown erratically while various "solutions" were tried, including resetting the autopilot circuit-breaker pitch command, while leaving the trim circuit breaker still pulled. This configuration pitted the autopilot clutch force (about 20–30 lb) against the airplane's maximum down trim, leaving about 20 lb of force on the control yoke. Eventually, at between 700 and 1200 ft, just below the clouds (according to weather reports, radar data, and intermittent communication), the pilot maneuvered to within sight of the Chapel Hill Airport on a base leg approach. In preparation for landing, he apparently disengaged the reset autopilot circuit breaker and the powerful out-of-trim condition reappeared in full force. With obvious control difficulty and distraction, he overshot the airport and reengaged the autopilot pitch control to abate the out-of-trim forces. Efforts to maneuver for another visual approach were convoluted, but finally led back to the airport. However, when the pilot apparently again turned off the autopilot for landing, this time clearly exhausted, he lost control and crashed to his death. There was no factual evidence of any actual airplane, autopilot, or system interface malfunction or failure prior to ground impact.

This pilot's confusion and "mental set" with the autopilot were consistent with his past behavior during other episodes. Previously, he had tried to troubleshoot the autopilot in the air, on one occasion by climbing above the clouds to VFR conditions "on top" and resetting the autopilot "to see if it would do it again." On another occasion, he reportedly "played with the switches" and presumably, the circuit breakers. On even another occasion, while using the autopilot for a coupled ILS approach, he apparently lost control to the point where the passenger (a VFR pilot) briefly took over control. Yet, there were "no mechanical problems" as demonstrated by the fact that the second coupled ILS approach was completed without difficulty. In each of these incidents, there was no mention by anyone onboard that the flight manual (the FAA-approved Airplane Flight Manual Supplement [AFMS]) was ever referred to and no evidence that postflight maintenance or instruction was ever sought. Apparently, this pilot had become highly dependent on the autopilot for flying. At the same time, he clearly misunderstood how the autopilot system worked, how to test it, or how to disengage it, and he made no efforts to learn more.

This pilot exacerbated his problem by his failure to use basic operational procedures, that is, aviate first, so as to concentrate all his mental and physical resources on landing, and then, troubleshoot only when safely on the ground. By dividing his attention and not concentrating fully on flying the aircraft, as evidenced by his haphazard ground track, he undoubtedly prolonged the flight, leading to the eventual crash. By attempting to troubleshoot while flying, he added to his mental workload and created more control problems; for example, by turning off the master switch, he also turned off the cabin lights, making controlled flight at night all the more difficult. He was distracted from compliance with ATC clearances (aimless headings, unauthorized altitude changes, disregarded low-altitude alerts), which more than likely eliminated his opportunity to land safely with a surveillance approach at the Raleigh-Durham airport. The transcript showed that controllers remarked among themselves on their reservations about the pilot's ability to handle his aircraft safely.

In his fatigued state, this pilot forgot the basics of flying manually. If he had concentrated on flying, he might have simply tried manual trim with the trim wheel, thereby removing the pitch down force and the problem. On the contrary, he flew desperately for more than 30 min simply because he did not neutralize the out-of-trim forces. Tragically, with just a few hand movements, he could have re-trimmed the airplane at any time and landed normally.

The complaint alleged that the autopilot malfunctioned and that the disconnects were defectively designed. Further, it was alleged that the aircraft manufacturer was negligent for choosing this autopilot for this airplane. This case was settled before it went to trial, even before the expense of depositions. Attorneys on all sides were finally convinced that the primary cause was the pilot's misuse of the autopilot system and his failure to comply with the FAA-approved AFMS.

Design Error Exemplar. A medium-size twin-engine helicopter departed from an oil platform in the North China Sea, with five people onboard including two pilots, one American, one Chinese. The American pilot was flying on the right side (the PIC position in this helicopter). Shortly after liftoff at about 200 ft above the water, one engine experienced what is described as a turbine burst. The engine blades and housing came apart with a loud bang, a fire flash, black smoke, and debris that fell from the right side of the aircraft, all heard or seen by witnesses on the oil platform. The CVR indicated the Chinese word "fire" and the master fire warning light presumably illuminated along with a 250 Hz continuous tone. The red fire light was located in a master warning panel (a "four-pack" design capable of showing fire, caution, and left-or right-engine failure). The crew responded according to the prescribed procedures first by pushing the master fire light/switch to silence the alarm sound (the master fire light remains "on"). The fire suppression procedures then called for the pilot to reach the overhead to find the lighted "T" handle signifying the affected engine. The CVR indicates that the American pilot did this, pulling back the lighted "T" handle, which in turn retarded the engine lever, shut off the fuel supply, and discharged the contents of the fire extinguisher bottle into the engine. All this took about 15.2 s, perhaps long, but it included the element of surprise and perhaps confusion caused by a language barrier between the two pilots.

Just before (1.6 s) the statement by the American pilot "pulling number two," an alternating 550/700-Hz tone signaled an engine out. The sound itself does not convey the engine number. This alarm sound is designed to trigger when N1 (rpm) drops below 59%, accompanied by the illumination of a warning light indicating the affected engine. The actual affected engine was number 1 (on the left side). Yet, the pilot was responding as if the fire was in the number 2 engine (on the right side). By the time the engine-out alarm sounded, the pilot may have been looking overhead, or his response may have been already mediated and in motion. Therefore, even if he did see the number 1 engine-out warning light, he may have merely silenced the sound by pushing the light/switch and did not mentally process the number at that point in time. In any event, 8.5 s after the irreversible action of "pulling number two" (and probably after realizing that the number 1 engine-out light had come "on"), the last words on the CVR, an expletive phrase, "f-ing number two was the wrong engine," came. With the loss of all power and low altitude, a power-off autorotation was not possible and the helicopter crashed into the sea 5–7 s later, killing all the five people onboard.

What at first looked like a case of pilot human error is actually more complicated. The postcrash investigation revealed that the turbine burst had sent shrapnel through the firewall between the two engines, creating holes large enough for light to go through. The infrared-light fire detectors in both the engine compartments were activated, in turn signaling both "T" handle alarms. The right-side American pilot apparently saw the closest number 2 "T" handle lighted (which visually overlaps the number 1 "T" handle from the pilot's vantage point) and presumably, thought it was the engine signaled by the master fire alarm. Evidently, the pilot did not look further because he found what he was directed to look for by the procedures, a lighted "T" handle.

If the pilot had been alerted to validate the affected engine by means other than the "T" handles, or if the pilot had been directed to look at both "T" handles before responding, he could have done so, as time was available. In retrospect, without any response, power for continued single-engine flight was not only possible but would have been routine (both engines power the main and tail rotors through the same shafts). Moreover, the engine compartment fire in all likelihood could have been contained well beyond the time needed for validation of the affected engine (via other instruments) and a shutdown. Other precautionary actions were available as well, for example, a return to the platform or deployment of floats and a water landing.

Documents in the aircraft manufacturer's possession, obtained by subpoena, indicated that other turbine bursts like this (but not of this consequence) were known to have occurred before, in this helicopter model. However, there was no indication in the operational procedures or in other information available to pilots (pilot publication's, alerts or advisories, service letters or bulletins, airworthiness directives, etc.) that both "T" handles could be lighted, one as a false alarm, which if responded to would lead to a total power loss. The manufacturer had designed and installed a "five-pack" master warning panel in the subsequent model helicopter, showing both the left and right fire alarms (note that "five-packs" were state-of-the-art at the time for comparable helicopters from other manufacturers). If a turbine burst sets off both the master fire alarms (side by side in the pilot's direct view), then both would be acknowledged by the pilot before shutting down an engine, thereby alerting the pilot to check before further action. This design was never offered as a retrofit to the prior model involved in this crash.

The complaint alleged that the engine was flawed and that the helicopter's warning system design misled the pilot to shut down the wrong engine, resulting in the death of the pilots and the passengers. Further, it was alleged that this event was foreseeable based on past similar incidents. The jury decided that neither the pilot crew nor the helicopter operator was in any way at fault, because its pilot had been misled to shut down the wrong engine.

System Error Exemplar. An ILS approach by the pilot of a high-performance single-engine airplane to the North Bend Airport, Oregon, resulted in a crash nearly 3 miles beyond the threshold of the runway. The final approach to Runway 4 was initiated from an assigned holding pattern at 4000 ft, about 3000 ft AGL. The final approach, as recorded on radar, was well-aligned along the localizer (LOG) course up to the last radar point near the landing threshold for Runway 4. However, the final descent path was always extremely high above the glideslope (GS) and continued beyond the missed approach point (MAP) in a descent. The crash site was along the extended LOG course at an elevation just above the decision height (DH), in an area enshrouded in clouds according to a nearby witness at the time of the occurrence.

Why did this acknowledged careful pilot overfly the entire airport into higher terrain beyond? The evidence suggests that he may have been confused by the information on the approach chart and by the procedures required after the approach clearance. With respect to the approach chart, he may have mistaken the distance from the final approach fix to the North Bend Vortac (6.3 nautical miles away) as the distance to the North Bend Airport (only 2.5 nm away). Both have the same three-letter identifier, OIK, which would be entered as such into the airplane's LORAN (long-range navigation) receiver. With a 3900-ft altitude leaving the holding pattern over the final approach fix and only 2.5 nm straight ahead to the runway, even a rapid descent would place this airplane high over the touchdown zone, as it did. With the actual position unknown to the pilot, he continued ahead in an unknowingly futile pursuit of the narrow (1.4 degree) GS beam down to just above DH at the crash site. Oddly, given the particular offset (to the side) location of the Vortac in relation to the airport, the pilot never read less than 2.5 nm even near the crash site. Under the high workload and stress of a single-pilot IFR (instrument flight rules) operation, this pilot might have misinterpreted his location as always being outside the outer marker (OM) (2.5 nm from the airport).

Why was not this pilot prompted by cockpit indications regarding his true position and his misinterpretation? First, flying the ILS does not depend on identifying geographical locations, such as the OM or the middle marker (MM). The ILS simply depends on intercepting and following the LOC (left-right) course and the GS (up-down) course down to the DH (262 ft). At this point, a decision to land or execute a missed approach is made dependent on whether the runway environment is in sight. Notably, this pilot did identify the OM as the holding fix during probably five or six times around the racetrack pattern, while waiting for the weather to improve. However, when cleared for the ILS 4 approach, the OM was no longer relevant and he apparently focused his attention on the descent needed to acquire the GS, straight ahead. Second, because the airplane was in fact well above the GS at the point of crossing, what should have been the actual geographic location of the MAP cockpit identification of it would not be possible (although the location was clear on ATC radar). Cockpit indications of the ILS MAP depend on being "on" GS and at DH altitude. Neither had been reached because of nonstandard procedures during the approach. Thus, the pilot, unaware, continued to descend beyond. Finally, timing the approach from the OM (and using speed for distance), although not required for an ILS, could have provided a clue about the location, but not if the pilot thought he had 6.3 nm to go (vs. the actual 2.5 nm).

Undoubtedly, both the parties (pilot and ATC) were confused by the approach procedure. The holding pattern (and altitude) had been verbally assigned by ATC (it was not depicted on the approach chart), and the inbound portion of the hold was aligned precisely along the final approach course toward the airport. Normally, except for the actually depicted holding patterns with altitude profiles printed on the approach chart, the procedure turn or radar vectoring is required to provide the maneuvering room for a descent to the GS intercept altitude (here at 1300 ft). When the approach clearance was originally issued, the pilot was turning directly inbound, making the procedure turn appear superfluous, except for the excessive altitude. ATC communication transcripts did not reveal any discussion or concern by the two controllers on duty about the unusual radar path/altitude, nor did they offer any radar assistance. On radar, it was apparent that the point-by-point position/altitude was very high along the approach path and unlikely to result in a successful landing on Runway 4. There were a total of 12 radar "hits" covering about 4 miles up to the runway threshold. Two of the specific "hits" pinpointed the airplane almost exactly over the actual MAP and over the runway touchdown zone, both extremely high. Strangely, even the appearance of the airplane itself on the radar could have signified that something was wrong. The final approaches to Runway 4 in this area were not even seen by ATC, because they are usually well below the radar coverage.

The ATC system, one that was intentionally designed to be redundant by depending on the vigilance of both controllers and pilots, is precisely the reason for the frequent occurrence of this type of ILS accident. The Airman's Information Manual* (AIM) states that "The responsibilities of the pilot and the controller intentionally overlap in many areas providing a degree of redundancy. Should one or the other fail in any manner, this overlapping responsibility is expected to compensate, in many cases, for failures that may affect safety." In this case, a breakdown of redundancy did occur, with several opportunities to avert this crash lost, some available only to the controllers. Notably, this was the second ILS 4 approach to the North Bend Airport for this aircraft. A similar sounding aircraft was heard by an ear witness about 30 min earlier, very close to the crash site, just about the time of this airplane's first missed approach.

The complaint alleged that despite the obvious pilot errors, the ATC failed to warn of the danger. This litigation entailed extensive discovery through various investigations, depositions, and expert's reports, which are now a requirement in some federal cases. A settlement took place just before trial.

---

* Airman's Information Manual (AIM), Pilot/Controller Roles and Responsibilities, paragraph 400.

29.4 Determining Responsibilities

The practice of forensic aviation human factors can present, at times, daunting challenges. Experts have to consider the facts of a crash, set out factors suspected to have played a role, examine the mutual and system-wide relations, determine the overall impact of every factor on the accident sequence and their contribution to the eventual outcome—the accident. Finally, they may be asked to suggest how responsibilities should be split among all the actors/factors in a manner that is consistent with the contextual legal system. In this final section, we have provided a reasoning framework to help understand the tasks and challenges of a forensic human-factors inquiry.

29.4.1 Criteria

The main goal of forensic aviation human factors is to assist juries in the determination of whether or not someone is responsible for an accident. While this discipline encompasses a broad array of knowledge spanning from law to psychology to aviation operations, it eventually comes down to a very simple question for the fact finders—culpable or not?

To answer that question, there are many issues that may be taken into account and that may help us draw a line. The typical questions that the forensic practitioners face are something along the lines of: was the observed behavior likely or not? Was there known risk? Was it reasonable? Were there external factors that influenced the behavior? How much control did the pilot/operator had on those external conditions?

29.4.1.1 Plaintiff-Defense Approaches

Although this chapter addresses forensics, direct reference to legal principles has been avoided up till now, as the primary focus of this book is on aviation and human factors. However, some discussions on legal issues are necessary to explain the legal context of forensics, but are given here in a compilation of words of others, because the authors are not attorneys.

Products liability is the legal term used to describe a type of civil lawsuit. In court, usually before a lay jury, an injured party (the plaintiff) seeks to recover damages for personal injury or loss of property from a manufacturer or seller (the defendant), as the plaintiff believes that the injuries or damages resulted from a defective product. Products liability falls under case law where each new court decision adds, changes, clarifies, or sometimes obscures the prior legal precedents. Civil actions are unlike criminal proceedings, where there is a presumption of innocence with proof of guilt required beyond reasonable doubt, well beyond the 50% mark. In civil actions, judgments can be made on the preponderance of the evidence, that is, on a 50% tilt or on an apportionment of blame among parties, by percentages. Just as there are few absolutes in life, there are few absolutes in the law; therefore, what is "reasonable" and "what is likely" are often heard.

Kantowitz and Sorkin (1983, pp. 629–630, 633) stated emphatically that

> There is no such thing as a perfectly safe product… Instead, there must be a balance between the potential harm a product may cause and the benefits to society of a plentiful supply of products…. As laws and judicial interpretations evolve, the definition of a reasonably safe product changes…. In a product liability lawsuit an injured party—the plaintiff—brings suit against a manufacturer or seller—the defendant—who has provided the allegedly defective product…. There is no absolute standard for unreasonable danger…. Expert witness testimony is often used to establish the degree of danger associated with a product. But even experts disagree. It is not unusual to find the human factors specialist testifying for the plaintiff at odds with the specialist for the defendant.

In the prior accident exemplars, the legal complaints ask who has responsibility, usually a euphemism for monetary damages. In the case of autopilot, it was alleged that the autopilot manufacturer allowed a defective design and that the aircraft manufacturer was negligent for choosing this autopilot for this aircraft. Surosky (1993, p. 29) stated:

> A product defect is one that makes it unreasonably dangerous for the user, or dangerous to an extent beyond what an ordinary user might contemplate. The basis for product liability is usually a defective design, defective manufacturing, or failure to warn of the hazards of an inherently dangerous product.... Contributory negligence is the failure on the part of the injured party to exercise ordinary care in self-protection, where such carelessness (along with any negligence by the defendant) is a direct cause of injury.

The settlement included some payment made by the autopilot manufacturer, but there was no payment made by the airplane manufacturer.

In the case of the helicopter turbine burst, it was alleged that the engine had a known manufacturing defect and that the helicopter manufacturer knew about this defect, but failed to design a proper warning for the failure. The complaint was essentially one of product liability. The engine was flawed, and the helicopter design misled the pilot to shut down the wrong engine, resulting in the death of the pilots and passengers. Further, it was alleged that this event was foreseeable based on past similar events. Kantowitz and Sorkin (1983, pp. 632–633) stated:

> The first step in establishing product liability in cases where no express warranty or misrepresentation is involved is to prove that the product was defective.... Product defects arise from two sources. First, a flaw in the manufacturing process may cause a defect. This results in a defective product that does not meet the manufacturer's own quality-control standards.... Second a product may have a design defect.... The manufacturer is liable for any defects in the product, including those that were unknown at the time of manufacture (but defects that were reasonably foreseeable [e.g., probable errors by the operator] in normal use or even from misuse).

The jury held the engine manufacturer responsible for the crash and awarded damages.

In the case of the instrument approach crash, it was alleged that the air-traffic controllers had unique knowledge about the ensuing danger (via radar), but failed to warn the pilot of the danger. Further, even though it was admitted that the pilot made errors, it was contended that the pilot was, in part, misled into those errors by the placement of government-installed navigational aids, ambiguous charts, and procedures. In effect, the allegations were of negligence on the part of the government. Kantowitz and Sorkin (1983, pp. 630–631) stated:

> To establish negligence, the plaintiff must prove that the conduct of the defendant involved an unreasonably great risk of causing damage. The plaintiff must prove that the defendant failed to exercise ordinary or reasonable care. A manufacturer has a legal obligation to use new developments that will decrease the level of risk associated with his or her product. Failure to keep abreast of new technology can be grounds for negligence.... There is no absolute standard for "unreasonably great risk." It is quite possible that the same defendant' would be judged negligent in one court and innocent in another court....

According to Joseph Nail, a lawyer member of the NTSB, as quoted by Barlay (1990, p. 125):

> In many instances, the public can sue the government because of the Federal Tort Claims Act, but there is an exemption called Discretionary Function to contend with. A government agency cannot be sued for making the wrong judgment. But if a government employee is negligent, the government is liable.

Clearly, there was not a single cause here, but errors made by both the pilot and the air-traffic controllers representing a system breakdown. This case resulted in a substantial monetary out-of-court settlement.

29.4.1.1.1 *The Debate about Strict Liability*

Many suggest that aviation product liability law is in disarray. Perhaps, this is an overstatement, but there are conflicting messages that legal circles send to the human-factors community. One of the biggest "bones of contention" is the legal concept of strict liability, which disconnects the issue of blame, risk, and error. Aviation human-factors experts usually help to explain why certain behavior occurred, versus what was expected, so that juries can judge what is reasonable. For some, strict liability judgments appear to bypass reason.

With regard to blame, strict liability,* often called liability without fault, allows for damage recovery without the need to show negligence. It is applicable if a product is found defective and unreasonably dangerous regardless of the care in design, manufacturing, or preparation of instructions, given that the user had not mismaintained, misused, or abused the product. In essence, manufacturers are held responsible to analyze their products for any inherent danger, and are in the best position to foresee risks that reasonably might arise with their normal use.

This (strict liability) theory of recovery says that if a product is defective and the defect makes it unreasonably dangerous, and that defect causes an injury or damage to someone who has not misused or abused the product, and the product is in substantially the same condition as it was when it left the manufacturer's hands, then regardless of how much care the manufacturer used in designing and building the product, the manufacturer is still liable for money damages to the injured or killed person or his family (Wolk, 1984, p. 166).

Of course, it is to be decided on what constitutes the misuse. With regard to risk, Holahan (1986) indicated:

> The roots of the product liability dilemma do not lie in aviation, but in the social-political attitude that has been growing in America for the last 25 years. Americans want guarantees of a riskless society. They demand that products be absolutely safe, even when they are misused. And when they have accidents, some one has to pay for them.... Our obsession for a riskless society has permeated the judicial system to the point where it ha s been instrumental in overturning our former negligence-based tort law, where once cases were judged on the basis of the comparative negligence of the parties involved. Replacing the negligence standard has been the doctrine of *Strict Liability*, which holds the manufacturer liable for whatever he did even though no negligence was involved.... Strict liability says "honest mistakes don't count any more" and the doctrine gives rise to the allowance of evidence (in most states) which judges yesterday's designs by today's technical know-how (p. 78).

29.4.1.2 Intentional Misuse

The concept of strict liability seems to be applicable, but only to a vast variety of accident circumstances. There is a specific situation that may be considered to limit the reach and applicability of strict liability, that is, when accidents are caused by an intentional and willful misuse of equipment, as manufacturers have virtually no capability to limit the hazards associated with such a misuse. Intentional misuse, as a legal concept, contrasts with that of inadvertent misuse occurring when the equipment is misused in a nonwillful manner. While there may be real-life cases where it is difficult to distinguish the two situations, the legal establishment may want to rely on intentionality as a criterion to determine whether or not liability claims can be brought against manufacturers.† The following is a real case of what may constitute an example of intentional misuse.

---

* In effect, negligence tests the conduct of the defendant; strict liability tests the quality of the product; and implied/ expressed *warranty* tests the performance of the product against representations (made by the seller).
† Readers should not be misled to believe that intentional misuse refers to an intention to cause an accident. Instead, we refer to an intention to misuse equipment—by willfully deviating from prescribed procedures—which in turn may eventually lead to an accident. To that regard, intentional misuse is a concept that relates to the process—using some equipment—rather than the outcome—causing an accident.

Early in 2000, a six-seat twin-engine airplane crashed moments after takeoff from a major airport in Texas. The plane was observed lifting off the ground, pitching up to an approximately 70 degree nose-up attitude, and entering a brief series of pitch up and down oscillations before eventually impacting the ground in a nose low, left wing low attitude. All the occupants onboard the aircraft perished in the crash. Postaccident examinations revealed that the "home-made" control lock systems had not been removed prior to takeoff. Specifically, an unapproved control column pin was found installed in the cockpit's flight control column and bent aft. Witnesses also reported that the plane took off without conducting a run-up check. The pilot himself had informed ATC that he would not need a run up before the takeoff. Although the control lock was supposed to be removed as the first step in the pre-flight checks, this pilot was known to taxi the airplane with the control lock being engaged.

Thus, with regard to the concept of strict liability, it may be considered that the legal system is intended to serve a necessary dual purpose. On one side, it has to protect people at a personal level, so that the rights of individuals, considered as single persons, are assured. However, on a broader level, it must also ensure that economic activities can be run in a fruitful, smooth, and efficient manner. Intensive application of strict liability law may threaten to place an unbearable burden on entire industries and professional figures that may not be able to withstand the impact of systematic awards of compensation and the rising insurance costs that stem from such. For instance, many in aviation blame the legal system for the near demise of general aviation, whose sales have dropped by more than 90%, owing to Draconian since the heydays around 1980. Arguments range from single-event law suits imposing judgments against manufacturers to liability insurance multiplying new airplane prices to beyond the reach of nearly everyone. Defense advocates state that this stifles innovation (where changes are a tacit admission of a design inadequacy) and sales. Others argue that a steady improvement in the accident rates is the evidence that legal process is enhancing safety. Plaintiffs point to cases forcing changes in design that might not have been addressed otherwise, such as improved seat and restraint designs, better component reliability, detection of structural defects, and refinements for control stability.

29.4.1.3 Reform

Regardless of one's position, legislative reform is in effect. The result of the General Aviation Revitalization Act has led some manufacturers to resume single-engine airplane production. This act is essentially one of repose. After 18 years, the airplane and its components have proven themselves by law, barring lawsuits (immunity from product liability actions) against the manufacturer.

However, questions are abounding. Undoubtedly, part replacement, overhaul, or modification starts a new 18-year time-clock, but does this also restart the calendar for the subsystem that part goes into, such as the engine or a flight control system? Will the design of such a subsystem be removed from immunity in a design-defect lawsuit, because the replaced part depends on and effectively revitalizes scrutiny of the original design? Will interacting subsystems also be affected? What happens if the pilot's operating handbook (POH) is updated or replaced with revised information (e.g., performance figures, checklists, warnings, etc.); does it affect liability for the manual or even for the aircraft it represents, or both? Will legal sights be refocused on defendants not covered by this federal statute, such as parts and component manufacturers, maintenance facilities, the FAA/ATC, fixed-base operators (FBOs), fuel distributors, training organizations, and flight instructors or designated pilot examiners?

The act also does not bar claims, for injury or damage to persons or property on the ground, only for those who were onboard (presumably who voluntarily have accepted a degree of risk). Further, it does not protect a manufacturer who knowingly misrepresents to the FAA or who conceals a defect that could cause harm. Therefore, even beyond the 18-year time frame, accident victims or their families still will have the rights and legal recourse. However, the argument may shift to what is a hidden defect (and a failure to warn), what is a misrepresentation, and of course, when is a defect harmful or "unreasonably dangerous." Obviously, these are future points of contention.

The fact that the General Aviation Revitalization Act legislates only for general aviation aircraft with fewer than 20 seats and is not used in scheduled passenger-carrying operations raises even other legal questions. It has been predicted that there will be challenges to this law on many grounds. One compelling argument is that the law seems to favor a small subset of the consumer product industry. Indeed, if the concept of repose is correct, then some contend that it should be applied to all product areas including lawn mowers, automobiles, and household appliances. Finally, there may be other higher issues to be resolved, namely, complaints that "tort law," under the jurisdiction of state systems and courts, supersedes and makes irrelevant federal authority (FAA) to set standards for safety. Such legal precedents, in effect, set different standards that vary from state to state, governing aircraft that cross state boundaries. Moreover, judgments are made by lay juries, who are mostly unknowledgeable with regard to aviation, rather than by federal agencies set up to represent the aviation industry. Others argue that these federal agencies are inadequate and that it is improper to use as a defense the fact that the product design was FAA certified.

It is unlikely that these mixed messages will disentangle in the near future. However, many advocate that placing aviation-accident cases under federal jurisdiction, along with all the other federal regulations that govern aviation, is the correct action. Thus, one suggestion is to place aviation under federal tort law. Clearly, the involvement of lawyers in aviation is not to be ignored. Therefore, it seems reasonable that aviation experts, including those in human factors (the highest risk area of aviation) remain involved in the legal process and outcomes. One way to accomplish this is through forensic aviation human factors.

There are issues that the human-factors experts need to explain in the context of aviation accidents that have happened and will happen. Humans (both designers and product users) have not yet discovered all the ways to make mistakes. Therefore, what happened and why will continue to direct inquiries and future research. Human error comes from various sources and in various forms, spanning the range from accident-prone products (low degree of avoidability) to accident-prone people (low degree of alertness). Whether a product (aviation or otherwise) induces an error or fails to prevent one, or whether error is self-induced by commission or omission, mistakes and blame for them will continue to be argued in the court.

With regard to the prevention of error, "Murphy's law" and its ubiquitous corollaries are always written in the negative: "If it can go wrong, it will go wrong." Murphy's laws never suggest that it could not go wrong. It is inevitable that humans will err and that machines will break (Sanders & McCormick, 1987). Nickerson (1992, p. 241) made the point that "There is no such thing as a risk-free existence, and the attempt to eliminate risk entirely can produce only disappointment and frustration." Nickerson (1992, p. 333) also stated that "No one with a rudimentary understanding of probability believes that zero-risk policies are attainable."

With the understanding that zero errors are impossible, people in human factors have developed specific strategies to reduce the likelihood or consequence of human errors (Sanders & McCormick, 1987). Designs can be fail-safe (or error tolerant) to reduce the consequence of error without necessarily reducing the likelihood of errors. There are even exclusion designs that eliminate certain errors or that minimize the possibility of most errors. Both the approaches make the reality of errors more acceptable, although still unwanted.

29.4.2 Considerations for Responsibility Determination

Since its early days, aviation has changed constantly and dramatically. Some of these changes have had, and will continue to have in the future, a tangible impact on the practice of forensic aviation human factors. There are several issues that seem to deserve increasing attention in this regard, and we will briefly discuss them here in an attempt to integrate increasing complexity, accident investigation, and forensics, with respect to a transnational legal context.

29.4.2.1 Emerging Complexity

Airplanes become more and more complex, fly in an increasingly complex environment and rely on more and more complex systems and procedures. Thus, the likelihood that an accident will be caused by a combination of factors rather than a single cause is constantly increasing. When we did not have sophisticated investigation techniques to identify all the factors that played a role in a crash, we were obviously not prompted to investigate how responsibilities should be shared among those unknown factors.* Yet, today we can rely on very powerful techniques that enable us to identify a large number of causal and contributing factors and analyze their mutual and system-wide influences. This makes it easier to serve one of the purposes of forensic human factors—making sure that similar types of accidents will not occur again in the future. However, it is also inevitably harder today to determine how responsibilities should be split among all the agents. We have more factors involved, which means that we will have more choices to make, and we have more complex systems, signifying that more complex analyses must be conducted. These decisions need to be made within a coherent legal system that has evolved consistently with the circumstances and scenarios that the human-factors experts are called to analyze.

29.4.2.2 Causation and Conviction in Complex Systems

This brings us to a highly debated and interesting issue, especially in the context of European laws—how can we determine causality and what conditions need to exist to find someone responsible? Ancient Roman law resolved the issue by determining that a demonstrated link between an action and a resulting harm was a foundation for culpability. This principle is still a cornerstone of legal systems around the world and continues to exert significant influence on how responsibility is weighted today in a number of countries. The problem is how to approach this principle in the contemporary context of complex systems, where we have multiple convoluted layers of agents and factors, all influencing each other. When there is no direct link and no single factor that by itself caused the crash, then the determination of how much each of them contributed to the crash is left to a rough estimation at the best. It appears that the causal link proposed by the Roman law may need to be reevaluated here, to be properly applied to this new, more complex context.

With this same line of reasoning, some point out that the principle of causation and conviction may at times not be fully aligned with the way the aviation safety system operates today, as the system is designed to rely extensively on principles such as redundancy, reliability and robustness. In other words, while the aviation safety system is built to absorb the errors that single individuals inevitably make, so that a consistently safe outcome can be delivered,† those same errors seem not to be allowed in the legal arena. An example of this disagreement was highlighted as part of the Milan Linate runway collision investigation, the one that we described in the chapter's opening paragraph. In that occasion, the air-traffic controller was convicted on the basis that he made one mistake,‡ an event that is preemptively accounted for by the aviation safety system, but not by the legal system. Given the surrounding legal context, the judge may have actually been required to issue a conviction sentence for that case, once he had determined that the controller's mistake directly contributed to the accident. While this may seem like a theoretical consideration, in actual fact, this issue may deserve some attention in the coming years if we do not want operators to be faced with the conflicting goals of being operationally safe on one side and legally safe on the other. It may be observed that, after all, the safety system allows a small amount of

---

* Some may argue that a few decades ago, the dynamics of accidents were usually simpler anyway, usually not requiring or prompting the development of advanced investigative techniques.

† It may be useful to recall the "Swiss cheese" model of aviation safety (Reason, 1990), which compares the aviation safety system with a pile of Swiss cheese slices, where the holes in the slices represent individual errors. Only when, under extremely rare circumstances, all the holes happen to be aligned—i.e., all the potential errors occur at the same time—we are able to see through the entire pile, which in the model, is the manifestation of an accident.

‡ He failed to require read back of a taxi clearance.

errors by individuals exactly because errors are inevitable in a fast-paced environment such as aviation. If we require operators to make no errors at all, as legal cases seem to suggest, then we may also need to lower the operational demands that we place on those operators. In the case of the Milan runway collision, this may have implied a reduction in the amount of traffic handled by controllers in a given time.

29.4.2.3 The Legal Context

What must be considered in the practice of forensics is the specific legal context within which we are operating. This consideration is particularly relevant when considering forensic practice in a non-U.S. environment or transnational investigations. Some have observed, for instance, that in a Roman-law setting, there is an extensive use of a framework that can fruitfully be applied to accident investigations, a framework that prompts us to categorize crashes into three main different types.* The first type of accident occurs when we have individuals not respecting prescribed procedures and therefore causing an accident, as in intentional misconduct with the control lock. In this case, the person who failed to adhere to prescribed procedures is the possible focus of culpability assessment. The second scenario is that of a whole company or organization not adhering to federally mandated procedures or establishing inadequate procedures or operational practices, as in the case of the Milan accident. It is likely that the company as a whole will here be examined to assess responsibility, with the potential for senior officers to be considered guilty if they made decisions that could be strongly linked to the inadequacy of the company's status quo. A third scenario could occur when investigators find that government-mandated and industry-wide procedures were inadequate. In this case, the governing agency, the government itself, or nobody at all† may be called to pay for the damages.

However, dividing accidents into three broad categories through this kind of analysis is a strategy that may not hold up as nicely in the context of civil law as it does in that of Roman law. As a matter of fact, in a civil-law trial, standing behind the meeting of government standards in and of itself, is usually not a successful defense, because such standards are considered to represent what is minimally acceptable as safe. Most jurors at least want an explanation of why the company chose only to meet those standards. They then carefully judge the supporting evidence for those positions, such as behavioral expectancies, prior incident or mishap surveys, design recommendations based on human-factors research findings, or even specific experiments/demonstrations definitively showing proof. Interestingly, the dual goals of human factors and ease of use and safety are generally disassociated in legal arguments. In most cases, the safety of a product often has little to do with how easy it is to use, although it seems that good designs are usually simple ones.

Thus, the bottom line of a forensic human-factors inquiry and thus, the crux of a lawsuit, is the judgment of "what is reasonable." A legal action usually will either prevail or not, depending on whether the design in question is needed, reasonable to use, and prudently safe in the eyes of an average juror. For example, a product may be potentially dangerous but may be needed and reasonable to use, such as an airplane or even a knife, if its benefits outweigh the risk of use. On the other hand, it may not be defendable if at the time of manufacture, safer alternatives were feasible and were not substantially outweighed by other factors such as utility and cost. Problems/errors that might be encountered in normal intended use or even foreseeable misuse must be considered. The latest developments, particularly with respect to safety, should be incorporated into new designs and, if possible, made available as

---

* The reality of facts makes accidents look much more blurry. Having a reasoning framework, though, helps the human-factors practitioner to venture into a legal case much like a flashlight that helps the explorer to walk through the darkness of a jungle.

† The judge may consider that the circumstances that led to the accident were unforeseeable, therefore, not prompting the development of adequate countermeasure procedures, and thus, resulting in none being blamed. Also, if investigators find that industry practices and procedures were inadequate, then it may be very difficult to pinpoint, within such large systems, the specific players who contributed with their action to the creation of the inadequacy.

modifications for prior designs. For example, a potential danger that is not obvious and that cannot be prevented needs a warning that is obvious, easily understood, and heeded.(there is no warning about the possibility of a cut from a sharp knife, because it is patently an obvious danger).

29.5 Conclusive Remarks

In this chapter, we have discussed a variety of topics, some of them in a broad manner and some of them in a more technical and detail-oriented fashion, in an attempt to provide the readers with a comprehensive understanding of forensic aviation human factors and its practices, players, and implications. It may be clear at this point that forensic aviation human factors is a discipline with a specific focus—helping judges determine culpability—but a broad base, one that requires experts in this field to be knowledgeable in law, psychology, aviation operations, and sometimes many other areas. Most importantly, forensic human factors is a multifaceted and fluid discipline which is influenced by the events, dynamics, and ideas that are generated by our society. Making sure that the discipline evolves in a harmonious manner with respect to the world that forensic experts are called to analyze is one of the greatest challenges ahead. Addressing this challenge in a comprehensive and thoughtful manner would allow us to treat legal issues fairly, while using what has been learnt to improve safety and create the conditions for the entire industry to prosper.

References

Barlay, S. (1990). *The final call: Why airline disasters continue to happen* (p. 125). New York: Pantheon Books.

Beaton, R. J. (1994, July). Letters from our readers. *Ergonomics in Design*, 4–5.

Boehm-Davis, D. A. (1994, July). HFES responds. *Ergonomics in Design*, 35.

Corlett, E. N. (1994, July). Letters from our readers. *Ergonomics in Design*, 5.

Corriere della Sera (2004). *Disastro di Linate, pene severe per tutti gli imputati*. April 17, 2004, online edition. (Not cited in the text).

Editorial (1994, January 3). *Aviation Week and Space Technology*, p. 66.

Gardner-Bonneau, D. J. (1994a, April). Comment from the editor. *Ergonomics in Design*, p. 3.

Gardner-Bonneau, D. J. (1994b, July). The editor responds. *Ergonomics in Design*, p. 35.

Holahan, J. (1986, January 1). Product liability. *Aviation International News*, p. 78.

Hornick, R. J. (1994, July). Letters from our readers. *Ergonomics in Design*, p. 4.

Howell, W. C. (1994, April). Provocations: The human factors expert mercenary. *Ergonomics in Design*, p. 6.

Kantowitz, B. H., & Sorkin, R. D. (1983). *Human factors: Understanding people-system relationships*. New York: Wiley.

McIntyre, J. A. (1994, September). Perspectives on human factors. The ISASI perspective. *Forum*, 27(3), 18.

Nickerson, R. S. (1992). *Looking ahead: Human factors challenges in a changing world*. Hillsdale, NJ: Lawrence Erlbaum Associates.

Reason, J. (1990). *Human Error*. New York: Cambridge University Press.

Sanders, M. S., & McCormick, E. J. (1987). *Human factors in engineering and design* (6th ed.). New York: McGraw-Hill.

Sanders, M. S., & McCormick, E. J. (1993). *Human factors in engineering and design* (7th ed.). New York: McGraw-Hill.

Senders, J. W. (1994, April). Warning assessment from the scientist's view. *Ergonomics in Design*, 6–7.

Surosky, A. E. (1993). *The expert witness guide for scientists and engineers*. Malabar, FL: Krieger.

Wolk, A. A. (1984, October). Points of law: Product liability—Aviation's nemesis of conscience (a personal opinion). *Business and Commercial Aviation*, p. 166.

Index